T0263666

THE
ENCYCLOPEDIA
▼▼▼ OF ▼▼▼
INTEGER
SEQUENCES

THE
ENCYCLOPEDIA
▼▼▼ OF ▼▼▼
INTEGER
SEQUENCES

N. J. A. Sloane
Mathematical Sciences Research Center
AT&T Bell Laboratories
Murray Hill, New Jersey

Simon Plouffe
Départment de Mathématiques et d'Informatique
Université du Québec à Montréal
Montréal, Québec

ACADEMIC PRESS
San Diego   New York   Boston   London   Sydney   Tokyo   Toronto

The cartoon in Fig. M0126 is copyright 1968 by United Feature Syndicate, Inc. (reproduced with permission). The illustration in Fig. M3987 appeared as the front cover of issue 7 (October 1973) of the periodical *Popular Computing*, published from 1973 to 1977 in Calabasas, California, by Fred Gruenberger. The illustration in Fig. M4482 is reproduced with permission from C. J. Bouwkamp and A. J. W. Duijvestijn, *Catalogue of Simple Perfect Squared Squares of Orders 21 Through 25*, Eindhoven University Technology, Department of Mathematics, Report 92-WSK-03, November 1992. All the other figures, including the cover image, are the work of the first author.

A computer readable index to the sequences in this volume has been prepared. [ISBN: 0-12-558632-9 (MS-DOS format); 0-12-558631-0 (Macintosh format).] For price and availability contact Academic Press at 1-800-321-5068.

Academic Press
A Division of Harcourt Brace & Company
525 B Street, Suite 1900, San Diego, California 92101-4495

United Kingdom Edition published by
Academic Press Limited
24-28 Oval Road, London NW1 7DX

Library of Congress Cataloging-in-Publication Data

Sloane, N. J. A. (Neil James Alexander), 1939-
        Encyclopedia of integer sequences / N.J.A. Sloane, Simon Plouffe.
        p.   cm.
    Includes index.
    ISBN 0-12-558630-2
    1. Sequences (Mathematics)   2. Numbers, Natural.   I. Plouffe,
Simon.   II. Title.
QA246.5.S66   1995
512'.72-dc20   94-42041
                                               CIP

Transferred to digital printing 2005

Contents

Preface

In spite of the large number of published mathematical tables, until the appearance of *A Handbook of Integer Sequences* (HIS) in 1973 there was no table of sequences of integers. Thus someone coming across the sequence 1, 1, 2, 5, 15, 52, 203, 877, 4140, . . . , for example, would have had difficulty in finding out that these are the Bell numbers, that they have been extensively studied, and that they can be generated by expanding $e^{e^{x}-1}$ in powers of x. The 1973 book remedied this situation to a certain extent, and the *Encyclopedia of Integer Sequences* is a greatly expanded version of that book. The main table now contains 5488 sequences of integers (compared with 2372 in the first book), collected from all branches of mathematics and science. The sequences are arranged in numerical order, and for each one a brief description and a reference are given. Figures interspersed throughout the table illustrate the most important sequences. The first part of the book describes how to use the table and gives methods for analyzing unknown sequences.

Who will use this book? Anyone who has ever been confronted with a strange sequence, whether in an intelligence test in high school, e.g.,

$$1, 11, 21, 1211, 111221, 312211, 13112221, \ldots$$

(guess![1]), or in solving a mathematical problem, e.g.,

$$1, 1, 2, 5, 14, 42, 132, 429, 1430, 4862, \ldots$$

(the Catalan numbers), or from a counting problem, e.g.,

$$1, 1, 2, 4, 9, 20, 48, 115, 286, 719, \ldots$$

(the number of rooted trees with n nodes), or in computer science, e.g.,

$$0, 1, 3, 5, 9, 11, 14, 17, 25, 27, \ldots$$

(the number of comparisons needed to sort n elements by list merging), or in physics, e.g.,

$$1, 6, 30, 138, 606, 2586, \ldots$$

---

[1] For many more terms and the explanation, see the main table.

(susceptibility coefficients for the planar hexagonal lattice[2]), or in chemistry, e.g.,

$$1, 1, 4, 8, 22, 51, 136, 335, 871, 2217, \ldots$$

(the number of alkyl derivatives of benzene with $n = 6, 7, \ldots$ carbon atoms), or in electrical engineering, e.g.,

$$3, 7, 46, 4436, 134281216, \ldots$$

(the number of Boolean functions of n variables), will find this encyclopedia useful.

If you encounter an integer sequence at work or at play and you want to find out if anyone has ever come across it before and, if so, how it is generated, then this is the book you need!

In addition to identifying integer sequences, the *Encyclopedia* will serve as an index to the literature for locating references on a particular problem and for quickly finding numbers like the number of partitions of 30, the 18th Catalan number, the expansion of π to 60 decimal places, or the number of possible chess games after 8 moves. It might also be useful to have around when the first signals arrive from Betelgeuse (sequence M5318, for example, would be a friendly beginning).

Some quotations from letters will show the diversity and enthusiasm of readers of the 1973 book. We expect the new book will find even wider applications, and look forward to hearing from readers who have used it successfully.

"I recently had the occasion to look for a sequence in your book. It wasn't there. I tried the sequence of first differences. It *was* there and pointed me in the direction of the literature. Enchanting" (Herbert S. Wilf, University of Pennsylvania).

"I also found N. J. A. Sloane's *A Handbook of Integer Sequences* to be an invaluable tool. I shall say no more about this marvelous reference except that every recreational mathematician should buy a copy forthwith" (Martin Gardner, *Scientific American*, July 1974).

---

[2] Also called the triangular lattice.

"Incomparable, eccentric, yet very useful. Contains thousands of 'well-defined and interesting' infinite sequences together with references for each. Sequences are arranged lexicographically and (to minimize errors) typeset from computer tape. If you ever wondered what comes after 1, 2, 4, 8, 17, 35, 71, ..., this is the place to look it up" (Lynn A. Steen, Telegraphic Review, *American Mathematical Monthly*, April 1974).

Nontechnical readers wrote to bless us, to speak of reading the book "cover to cover," or to remark that it was getting a great deal of use to the detriment of household chores and so on. Specialists in various fields had other tales to tell.

Anthony G. Shannon, an Australian combinatorial mathematician, wrote: "I must say how impressed I am with the book and already I am insisting that my students know their way around it just as with classics such as Abramowitz and Stegun."

Researchers wrote: "Our process of discovery consisted of generating these sequences and then identifying them with the aid of Sloane's *Handbook of Integer Sequences*" (J. M. Borwein, P. B. Borwein, and K. Dilcher, *American Mathematical Monthly*, October 1989).

Allen J. Schwenk, a graph theorist in Maryland wrote: "I thought I had something new until your book sent me to the Riordan reference, where I found 80% of my results and so I abandoned the problem."

We received letters describing the usefulness of the *Handbook* from such diverse readers as: a German geophysicist, a West Virginian astronomer, various graduate students, physicists, and even an epistemologist.

Finally, Harvey J. Hindin, writing from New York concluded a letter by saying:

"There's the Old Testament, the New Testament, and the *Handbook of Integer Sequences*."

Abbreviations

Abbreviations for the references are listed in the bibliography. References to journals give volume, page number, year.

$a(n)$	nth term of sequence
$A(x)$	generating function for sequence, usually the ordinary generating function $A(x) = \sum a_n x^n$, occasionally the exponential generating function $A_E(x) = \sum a_n x^n / n!$
AND	logical "AND", sometimes applied to binary representations of numbers
B_n	Bernoulli number (see Fig. M4189)
b.c.c .	body-centered cubic lattice (see [SPLAG 116])
binomial transform	of sequence $a_0, a_1, \ldots$ is sequence $b_0, b_1, \ldots$ where $$b_n = \sum_{k=0}^{n} \binom{n}{k} a_k$$
$C(n)$ or C_n	nth Catalan number (see Fig. M1459)
$C(n,k)$ or $\binom{n}{k}$	binomial coefficient (see Fig. M1645)
E.g.f.	exponential generating function $A_E(x) = \sum a_n x^n / n!$
Euler transform	of sequence $a_0, a_1, \ldots$ is sequence $b_1, b_2, \ldots$ where $$\sum_{n=0}^{\infty} a_n x^n = \prod_{n=1}^{\infty} \frac{1}{(1 - x^n)^{b_n}}$$
$\exp x$	e^x
$F(n)$ or F_n	nth Fibonacci number (see Fig. M0692)

$p(n)$ or p_n	usually nth prime, occasionally nth partition number, but in latter case always identified as such
q	a prime or prime power
Ref	reference(s)
Rev.e.g.f.	reversion of exponential generating function
Rev.o.g.f.	reversion of ordinary generating function
w.r.t.	with respect to
XOR	logical "EXCLUSIVE OR", usually applied to binary representations of numbers
Λ_n	n-dimensional laminated lattice (see [SPLAG, Chap. 6])
$\mu(n)$	Möbius function (see M0011)
π	ratio of circumference of circle to diameter (see Fig. M2218)
$\prod$	a product, usually from 1 (or 0) to infinity, unless indicated otherwise
$\sigma(n)$	sum of divisors of n (see M2329)
$\sum$	a sum, usually from 0 (or 1) to infinity, unless indicated otherwise
τ	the golden ratio $(1 + \sqrt{5})/2$ (see M4046)
$\phi(n)$	Euler totient function (see Fig. M0500)
$\uparrow$	exponentiation
$!$	factorial symbol: $0! = 1$, $n! = 1.2.3. \cdots .n$, $n \geq 1$ (see Fig. M4730)
#	number
$[x]$	largest integer not exceeding x
$\lceil x \rceil$	smallest integer not less than x

$_mF_n$	hypergeometric series (see [Slat66]):

$$_mF_n([r_1, r_2, \ldots, r_m];\ [s_1, s_2, \ldots, s_n]; x)$$

$$= \sum_{k=0}^{\infty} \frac{(r_1)_k (r_2)_k \cdots (r_m)_k}{(s_1)_k \cdots (s_n)_k}\ \frac{x^k}{k!}\ ,$$

where $(r)_0 = 1$, $(r)_k = r(r+1) \cdots (r+k-1)$, for $k = 1, 2, \ldots$

f.c.c.	face-centered cubic lattice (see [SPLAG 112])
g.c.d.	greatest common divisor
G.f.	generating function, usually the ordinary generating function $A(x)$
h.c.p.	hexagonal close packing (see [SPLAG 113])
l.c.m.	least common multiple
Lgd.e.g.f.	logarithmic derivative of exponential generating function
Lgd.o.g.f.	logarithmic derivative of ordinary generating function
Möbius transformation	of sequence $a_1, a_2, \ldots$ is sequence $b_1, b_2, \ldots$, where

$$b_n = \sum_{d|n} \mu\left(\frac{n}{d}\right) a_d\ ,$$

and $\mu(n)$ is the Möbius function M0011

multiplicative encoding	of a triangular array $\{t(n, k) \geq 0;\ n = 0, 1, \ldots$ and $0 \leq k \leq n\}$ is the sequence whose nth term is

$$\prod_{k=0}^{n} p_{k+1}^{t(n,k)}\ ,$$

whose $p_1 = 2$, $p_2 = 3, \ldots$ are the primes

n	either a typical subscript , as in M0705: "$a(n) = a(n-1) + 2a(n-3)$", or a typical term in the sequence, as in M0641: "$6n-1$, $6n+1$ are twin primes"
O.g.f.	ordinary generating function $A(x)$
OR	logical "OR", usually applied to binary representations of numbers
p	a prime

Chapter 1

Description of the Book

It is the fate of those who toil at the lower employments of life, to be driven rather by the fear of evil, than attracted by the prospect of good; to be exposed to censure, without hope of praise; to be disgraced by miscarriage, or punished for neglect, where success would have been without applause, and diligence without reward.

Among these unhappy mortals is the writer of dictionaries; whom mankind have considered, not as the pupil, but the slave of science, the pioneer of literature, doomed only to remove rubbish and clear obstructions from the paths of Learning and Genius, who press forward to conquest and glory, without bestowing a smile on the humble drudge that facilitates their progress. Every other authour may aspire to praise; the lexicographer can only hope to escape reproach, and even this negative recompense has yet been granted to very few.

Samuel Johnson, Preface to the "Dictionary," 1755

This epigraph, copied from the 1973 book, still applies!

1.1 Description of a Typical Entry

The main table is a list of about 5350 sequences of integers. A typical entry is:

M1484 $1, 1, 2, 5, 15, 52, 203, 877, 4140, 21147, 115975, 678570,$ $4213597, 27644437, 190899322, 1382958545, 10480142147,$ $82864869804, 682076806159, 5832742205057$
Bell or exponential numbers: $a(n + 1) = \Sigma a(k)C(n, k)$. See Fig M4981. Ref MOC 16 418 62. AMM 71 498 64. PSPM 19 172 71. GO71. [0,3; A0110, N0585]

$$\text{E.g.f.: } \exp(e^x - 1)$$

and consists of the following items:

1

M1484	The sequence identification number in this book
$1, 1, 2, 5, 15, 52, \ldots$	The sequence itself
Bell or exponential numbers	Name or descriptive phrase
$a(n+1)$ $= \sum a(k)C(n,k)$	A recurrence: $a(n)$ is the nth term, $C(n,k)$ is a binomial coefficient, and the sum is over the natural range of the dummy variable, in this case over $k = 0, 1, \ldots, n$
See Fig M4981	Further information will be found in the figure accompanying sequence M4981
Ref	References
MOC 16 418 62	*Mathematics of Computation*, vol. 16, p. 418, 1962
AMM 71 498 64	*American Mathematical Monthly*, vol. 71, p. 498, 1964
	For other references, see the bibliography
0,3	The offset [inside square brackets]: the first number, 0, indicates that the first term given is $a(0)$, and the second number, 3, that the third term of the sequence is the first that exceeds 1 (the latter is used to determine the position of the sequence in the lexicographic order in the table)
A0110	Absolute identification number for the sequence
N0585	(If present) the identification number of the sequence in the 1973 book [HIS]
E.g.f.	Further information about the sequence (typically a generating function or recurrence) may be displayed following the sequence; in this case "E.g.f." indicates an exponential generating function — see Section 2.4.

We have attempted to give the simplest possible descriptions. In the descriptions, phrases such as "The number of" or "The number of distinct" have usually been omitted. Since there are often several ways to interpret "distinct", there may be more than one sequence with the same name. The principal sequences are described in detail, while less information is given about subsiduary ones. The indices are usually $0, 1, 2, \ldots$ or $1, 2, 3, \ldots$, or sometimes the primes $2, 3, 5, \ldots$. The first number in square brackets at the end of the description gives the initial index.

1.2 Arrangement of Table

The entries are arranged in lexicographic order, so that sequences beginning 2, 3, ... come before those beginning 2, 4, ..., etc. Any initial 0's and 1's are ignored when doing this.

1.3 Number of Terms Given

Whenever possible enough terms are given to fill two lines. If fewer terms are given it is because either no one knows the next term (as in sequences M0219, M0223, M0233, M0240, M0582, M5482, for example), or because although it would be straightforward to calculate the next term, no one has taken the trouble to do so (as in sequences M0115, M0163, M0406, M0686, M0704, M5485, etc.). We encourage every reader to pick a sequence, extend it, and send the results to the first author, whose address is given in Section 2.2. Of course some sequences are known to be hard to extend: see Fig. M2051. The current status of any sequence can be found via the email servers mentioned in Section 2.9.

1.4 References

To conserve space, journal references are extremely abbreviated. They usually give the exact page on which the sequence may be found, but neither the author nor the title of the article. To find out more the reader must go to a library; to get the most out of this book, it should be used in conjunction with a library.

Journal references usually give volume, page, and year, in that order. (See the example at beginning of this chapter.) Years after 1899 are abbreviated, by dropping the 19. Earlier years are not abbreviated. Sometimes to avoid ambiguity we use the more expanded form of: journal name (series number), volume number (issue number), page number, year.

References to books give volume (if any) and page. (See the example at the beginning of this chapter.)

The references do not attempt to give the discoverer of a sequence, but rather the most extensive table of the sequence that has been published.

In most cases the sequence will be found on the page cited. In some instances, however, for instance when we have not seen the article (if it is in an obscure conference proceedings, or more often because the sequence was taken from a pre-publication version) the reference is to the first page of the article. Our policy has been to include all interesting sequences, no matter how obscure the reference. In a few cases the reference does not describe the sequence itself but only a closely-related one.

1.5   What Sequences Are Included?

To be included, a sequence must satisfy the following rules (although exceptions have been made to each of them).

Rule 1.   The sequence must consist of nonnegative integers.

Sequences with varying signs have been replaced by their absolute values.

Interesting sequences of fractions have been entered by numerators and denominators separately.

Arrays have been entered by rows, columns or diagonals, as appropriate, and in some cases by the multiplicative representation described in Fig. M1722.

Some sequences of real numbers have been replaced by their integer parts, others by the nearest integers.

The only genuine exceptions to Rule 1 are sequences such as M0728, M1551, which are integral for a considerable number of terms although eventually becoming nonintegral.

Rule 2.   The sequence must be infinite.

Exceptions have been made to this rule for certain important number-theoretic sequences, such as Euler's ideoneal (or suitable) numbers, M0476. Many sequences, such as the Mersemne primes, M0672, which are not yet known to be infinite, have been given the benefit of the doubt.

Rule 3.   The first nontrivial term in the sequence, i.e. the first that exceeds 1, must be between 2 and 999.

The position of the sequence in the lexicographic order in the table is determined by the terms of the sequence beginning at this point.

The artificial sequences M0004 and M5487 mark the boundaries of the table.

Rule 3 implicitly excludes sequences consisting of only 0's and 1's. However, for technical reasons related to the sequence transformations discussed in Sect. 2.7, a few 0, 1 sequences have been included. They appear at the beginning of the main table.

Rule 4. The sequence must be well-defined and interesting. Ideally it should have appeared somewhere in the scientific literature, although there are many exceptions to this. Enough terms must be known to distinguish the sequence from its neighbors in the table, although one or two exceptions to this have been made for especially important sequences.

The selection has inevitably been subjective, but the goal has been to include a broad variety of sequences and as many as possible.

1.6 The Figures

The figures interspersed through the table give further information about certain sequences. Our aim, not fully achieved, was that taken together the figures and the table entries would give at least a brief description of the properties of the most important sequences. By combining the entry for the *subfactorial* or *rencontres* numbers, M1937, for instance, with the information from Fig. M1937, one can obtain a definition, exact formula, generating function and a recurrence for these numbers.

The figures serve two other purposes. One is to provide a short discussion of certain especially interesting families of sequences (such as "self-generating" sequences, Fig. M0436; famous hard sequences, Fig. M2051; or our favorite sequences, Fig. M2629).

The other is to display the most important *arrays* of numbers and the sequences to which they give rise — see Fig. M1645, for example, which describes some of the many sequences connected with the diagonals and even the rows of Pascal's triangle. These figures compensate to a certain extent for the fact that the book does not catalogue arrays of numbers.

Chapter 2

How to Handle a Strange Sequence

We begin with tests that can be done "by hand", then give tests needing a computer, and end by describing two on-line versions of the Encyclopedia that can be accessed via electronic mail.

2.1 How to See If a Sequence Is in the Table

Obtain as many terms of the sequence as possible. To look it up in the table, first omit all minus signs. Then find the first nontrivial term in the sequence, i.e. the first that exceeds 1. The terms beginning at this point determine where the sequence is placed in the lexicographic order in the table.

For example, to locate 1, 1, 1, 1, 1, 2, 1, 2, 3, 2, 3, . . ., the underlined number is the first nontrivial term, so this sequence should be looked up in the table at 2, 1, 2, 3, 2, 3, . . . (it is M0112).

For handling arrays, rationals or real numbers, see Section 1.5.

2.2 If the Sequence Is Not in the Table

- Try examining the differences between terms, as discussed in Section 2.5, and look for a pattern.

- Try transforming the sequence in some of the ways described in Section 2.7, and see if the transformed sequence is in the table.

- Try the further methods of attack that are mentioned in Sections 2.6 and 2.8.

- Send it by electronic mail to superseeker@research.att.com, as described in Section 2.9. This program automatically applies many of the

7

tests described in this chapter.

- If all these methods fail, and it seems that the sequence is neither in the Encyclopedia nor has a simple explanation, please send the sequence and anything that is known about it, including appropriate references, to the first author[1] for possible inclusion in the table.

2.3 Finding the Next Term

Suppose we are given the first few terms

$$a_0 \quad a_1 \quad a_2 \quad a_3 \quad a_4 \quad a_5 \quad a_6$$

of a sequence, and would like to find a rule or explanation for it. If nothing is known about the history or provenance of the sequence, nothing can be said, and any continuation is possible. (Any $n + 1$ points can be fitted by an nth degree polynomial.)

But the sequences normally encountered, and those in this book, are distinguished in that they have been produced in some intelligent and systematic way. Occasionally such sequences have a simple explanation, and if so, the methods discussed in this chapter may help to find it. These methods can be divided roughly into two classes: those which look for a systematic way of generating the nth term a_n from the terms $a_0, \ldots, a_{n-1}$ before it, for instance by a recurrence such as $a_n = a_{n-1} + a_{n-2}$, i.e. methods which seek an *internal* explanation; and those which look for a systematic way of going from n to a_n, e.g. a_n is the number of divisors of n, or the number of trees with n nodes, or the nth prime number, i.e. methods which seek an *external* explanation. The methods in Sect. 2.5 and some of those in Sect. 2.6 are useful for attempting to discover internal explanations. External explanations are harder to find, although the transformations in Sect. 2.7 are of some help, in that they may reveal that the unknown sequence is a transformation of a sequence that has already been studied in some other context.

In spite of the warning given at the beginning of this section, in practice it is usually clear when the correct explanation for a sequence has been found. "Oh yes, of course!", one says.

There is an extensive literature dealing with the mathematical problems of defining the complexity of sequences. We will not discuss this subject here, but simply refer the reader to the literature: see for example Feder et al. [PGIT 38 1258 92],

---

[1]Address: N.J.A. Sloane, Room 2C-376, Mathematical Sciences Research Center, AT&T Bell Labs, 600 Mountain Avenue, Murray Hill, NJ 07974 USA; electronic mail: njas@research.att.com; fax: 908 582-3340.

Fine [IC 16 331 70], [FI1], Martin-Lof [IC 9 602 66], Ziv [Capo90 366], Lempel and Ziv [PGIT 22 75 76], as well as a number of other papers by Ziv and his collaborators that have appeared in [PGIT].

2.4 Recurrences and Generating Functions

Let the sequence be $a_0, a_1, a_2, a_3, \ldots$. Is there a systematic way of getting the nth term a_n from the preceding terms $a_{n-1}, a_{n-2}, \ldots$? A rule for doing this, such as $a_n = a_{n-1}^2 - a_{n-2}$, is called a *recurrence*, and of course provides a method for getting as many terms of the sequence as desired.

When studying sequences and recurrences it is often convenient to represent the sequence by a power series such as

$$A(x) = a_0 + a_1 x + a_2 x^2 + a_3 x^3 + \cdots ,$$

which is called its (*ordinary*) *generating function* (o.g.f. or simply g.f.), or

$$E(x) = a_0 + a_1 \frac{x}{1!} + a_2 \frac{x^2}{2!} + a_3 \frac{x^3}{3!} + \cdots ,$$

its *exponential generating function* (or e.g.f.). (These are formal power series having the sequence as coefficients; questions of convergence will not concern us.)

For example, the sequence M2535: 1, 3, 6, 10, 15, ... of triangular numbers has

$$A(x) = \frac{1}{(1-x)^3} ,$$

$$E(x) = \left(1 + 2x + \frac{x^2}{2}\right) e^x .$$

Generating functions provide a very efficient way to represent sequences.

A great deal has been written about how generating functions can be used in mathematics: see for example Bender & Goldman [IUMJ 20 753 71], Bergeron, Labelle and Leroux [BLL94], Cameron [DM 75 89 89], Doubilet, Rota and Stanley [Rota75 83], Graham, Knuth and Patashnik [GKP], Harary and Palmer [HP73], Leroux and Miloudi [LeMi91], Riordan [R1], [RCI], Stanley [Stan86], Wilf [Wilf90]. (See also the very interesting work of Viennot [Vien83].)

Once a recurrence has been found for a sequence, techniques for solving it will be found in Batchelder [Batc27], Greene and Knuth [GK90], Levy and Lessman [LeLe59], Riordan [R1], and Wimp [Wimp84].

For example, consider M0692, the Fibonacci numbers: 1, 1, 2, 3, 5, 8, 13, 21, 34, ... These are generated by the recurrence $a_n = a_{n-1} + a_{n-2}$, and from this it is not difficult to obtain the generating function

$$1 + x + 2x^2 + 3x^3 + 5x^4 + \cdots = \frac{1}{1 - x - x^2},$$

and the explicit formula for the nth term:

$$a_n = \frac{1}{\sqrt{5}} \left[\left(\frac{1 + \sqrt{5}}{2} \right)^{n+1} - \left(\frac{1 - \sqrt{5}}{2} \right)^{n+1} \right].$$

2.5   Analysis of Differences

This is the best method for analyzing a sequence "by hand". In favorable cases it will find a recurrence or an explicit formula for the nth term of a sequence, or at least it may suggest how to continue the sequence. It succeeds if the nth term is a polynomial in n, as well as in many other cases.

If the sequence is

$$a_0, a_1, a_2, a_3, a_4, \dots ,$$

then its first differences are the numbers

$$\Delta a_0 = a_1 - a_0, \quad \Delta a_1 = a_2 - a_1, \quad \Delta a_2 = a_3 - a_2, \quad \dots ,$$

its second differences are

$$\Delta^2 a_0 = \Delta a_1 - \Delta a_0, \quad \Delta^2 a_1 = \Delta a_2 - \Delta a_1, \quad \Delta^2 a_2 = \Delta a_3 - \Delta a_2, \quad \dots ,$$

and so on. The 0th differences are the original sequence: $\Delta^0 a_0 = a_0$, $\Delta^0 a_1 = a_1$, $\Delta^0 a_2 = a_2, \dots$; and the kth differences are

$$\Delta^k a_n = \Delta^{k-1} a_{n+1} - \Delta^{k-1} a_n$$

or, in terms of the original sequence,

$$\Delta^k a_n = \sum_{i=0}^{k} (-1)^i \binom{k}{i} a_{n+k-i} . \tag{2.1}$$

Therefore if the differences of some order can be identified, Eq. (2.1) gives a recurrence for the sequence. Furthermore, if the differences a_m, Δa_m, $\Delta^2 a_m$, $\Delta^3 a_m, \dots$ are known for some fixed value of m, then a formula for the nth term is given by

$$a_{n+m} = \sum_{k=0}^{n} \binom{n}{k} \Delta^k a_m . \tag{2.2}$$

The array of numbers

a_0		a_1		a_2		a_3		a_4		$\cdots$
	Δa_0		Δa_1		Δa_2		Δa_3		$\cdots$	
		$\Delta^2 a_0$		$\Delta^2 a_1$		$\Delta^2 a_2$		$\cdots$		
			$\Delta^3 a_0$		$\Delta^3 a_1$		$\cdots$			
				$\cdots$						

is called the *difference table of depth 1* for the sequence.

Example (i). M3818, the pentagonal numbers:

n	1	2	3	4	5	6	7	8
a_n	1	5	12	22	35	51	70	92
Δa_n	4	7	10	13	16	19	22	
$\Delta^2 a_n$	3	3	3	3	3	3		
$\Delta^3 a_n$	0	0	0	0	0			

Since $\Delta^2 a_n = 3$, $\Delta a_{n+1} - \Delta a_n = 3$, i.e. $a_{n+2} - 2a_{n+1} + a_n = 3$, which is a recurrence for the sequence. An explicit formula is obtained from Eq. (2.2) with $m = 1$:

$$a_{n+1} = 1 + 4\binom{n}{1} + 3\binom{n}{2} = 1 + 4n + 3\frac{n(n-1)}{2} = \frac{1}{2}(n+1)(3n+2) .$$

In general, if the rth differences are zero, a_n is a polynomial in n of degree $r - 1$.

Example (ii). M3416, Eulerian numbers:

n	0	1	2	3	4	5	6	7
a_n	0	1	4	11	26	57	120	247
Δa_n	1	3	7	15	31	63	127	
$\Delta^2 a_n$	2	4	8	16	32	64		

Here $\Delta^2 a_n = 2^{n+1}$, $\Delta a_n = 2^{n+1} - 1$, and $a_n = 2^{n+1} - n - 2$. Equation (2.2) gives the same answer.

Example (iii). M1413, the Pell numbers:

n	1	2	3	4	5	6	7
a_n	1	2	5	12	29	70	169
Δa_n		1	3	7	17	41	99
$\Delta^2 a_n$			2	4	10	24	58
$\frac{1}{2}\Delta^2 a_n$			1	2	5	12	29

Since $\frac{1}{2}\Delta^2 a_n = a_n$, Eq. (2.1) gives the recurrence $a_{n+2} - 2a_{n+1} - a_n = 0$. Calculating further differences shows that $\Delta^k a_1 = 2^{\lfloor k/2 \rfloor}$ and so Eq. (2.2) gives the formula

$$a_{n+1} = \sum_{k=0}^{n} \binom{n}{k} 2^{\lfloor k/2 \rfloor} .$$

If no pattern is visible in the difference table of depth 1, we may take the leading diagonal of that table to be the top row of a new difference table, the *difference table of depth* 2, and so on. For example, the difference table of depth 1 for

$$0, 2, 9, 31, 97, 291, 857$$

is

0		2		9		31		97		291		857
	2		7		22		66		194		566	
		5		15		44		128		372		
			10		29		84		244			
				19		55		160				
					36		105					
						69						

No pattern is visible, so we compute the difference table of depth 2:

0		2		5		10		19		36		69
	2		3		5		9		17		33	
		1		2		4		8		16		

Success! If we denote the sequence $0, 2, 5, \ldots$ by $b_0, b_1, b_2, \ldots$, then we see that $\Delta^2 b_n = 2^n$, $b_n = 2^n + n - 1$, and the original sequence is

$$a_n = \sum_{k=0}^{n} \binom{n}{k} (2^k + k - 1) .$$

In general, the relationship between the top row of a difference table

$$
\begin{array}{cccccc}
b_0 = a_0 & a_1 & a_2 & a_3 & a_4 & a_5 \\
b_1 & & \cdot & \cdot & \cdot & \cdot \\
& b_2 & & \cdot & \cdot & \cdot \\
& & b_3 & & \cdot & \cdot \\
& & & b_4 & & \cdot \\
\end{array}
$$

and the leading diagonal is given by

$$
a_n = \sum_{k=0}^{n} \binom{n}{k} b_k , \quad b_n = \sum_{k=0}^{n} (-1)^{n-k} \binom{n}{k} a_k . \tag{2.3}
$$

2.6 Other Methods for Hand Analysis

• Try transforming the sequence in various ways — see Sect. 2.7. • Is the sequence close to a known sequence, such as the powers of 2? If so, try subtracting off the known sequence. For example, M3416 (again): 0, 1, 4, 11, 26, 57, 120, 247, 502, 1013, 2036, 4083, ... The last four numbers are close to powers of 2: 512, 1024, 2048, 4096; and then it is easy to find $a_n = 2^n - n - 1$.

• Is a simple recurrence such as $a_n = \alpha a_{n-1} + \beta a_{n-2}$ (where α, β are integers) likely? For this to happen, the ratio $\rho_n = a_{n+1}/a_n$ of successive terms must approach a constant as n increases. Use the first few values to determine α and β and then check if the remaining terms are generated correctly.

• If the ratio ρ_n has first differences which are approximately constant, this suggests a recurrence of the type $a_n = \alpha n a_{n-1} \cdots$. For example, M1783: 0, 1, 2, 7, 30, 157, 972, 6961, 56660, 516901, . . . has successive ratios 2, 3.5, 4.29, 5.23, 6.19, 7.16, 8.14, 9.12, . . . with differences approaching 1, suggesting $a_n = n a_{n-1} + ?$. Subtracting $n a_{n-1}$ from a_n, we obtain the original sequence 0, 1, 2, 7, 30, 157, 972, . . . again, so $a_n = n a_{n-1} + a_{n-2}$.

This example illustrates the principle that whenever $\rho_n = a_{n+1}/a_n$ seems to be close to a recognizable sequence r_n, one should try to analyze the sequence $b_n = a_{n+1} - r_n a_n$.

• A recurrence of the form $a_n = n a_{n-1} +$ (small term) can be identified by the fact that the 10th term is approximately 10 times the 9th. For example, M1937: 0, 1, 2, 9, 44, 265, 1854, 14833, <u>133496</u>, <u>1334961</u>, . . ., $a_n = n a_{n-1} + (-1)^n$.

• The recurrence $a_n = a_{n-1}^2 + \cdots$ is characterized by the fact that each term is about twice as long as the one before. For example, M0865: 2, 3, 7, 43, 1807, 3263443, 10650056950807, . . ., and $a_n = a_{n-1}^2 - a_{n-1} + 1$.

• Does the sequence, or one obtained from it by some simple operation, have many factors? Consider the sequence 1, 5, 23, 119, 719, 5039, 40319, ... As it stands, the sequence cannot be factored, since 719 is prime, but the addition of 1 to all the terms gives the highly composite sequence $2, 6 = 2 \cdot 3, 24 = 2 \cdot 3 \cdot 4$, $120 = 2 \cdot 3 \cdot 4 \cdot 5, \ldots$, which are the factorial numbers, M1675.

• The presence of only small primes may also suggest binomial coefficients. For example, M1459, the Catalan numbers: $1, 1, 2, 5, 14 = 2 \cdot 7, 42 = 2 \cdot 3 \cdot 7$, $132 = 4 \cdot 3 \cdot 11, 429 = 3 \cdot 11 \cdot 13, 1430 = 2 \cdot 5 \cdot 11 \cdot 13, 4862 = 2 \cdot 11 \cdot 13 \cdot 17, \ldots$ and

$$a_n = \frac{1}{n+1} \binom{2n}{n}$$

(see Fig. M1459).

• Is there a pattern to the exponents in the prime factorization of the terms? E. g. M2050: $2 = 2^1$, $12 = 2^2 3^1$, $360 = 2^3 3^2 5^1$, etc.

• Sequences arising in number theory are sometimes *multiplicative*, i.e. have the property that $a_{mn} = a_m a_n$ whenever m and n have no common factor. For example, M0246: 1, 2, 2, 3, 2, 4, 2, 4, ..., the number of divisors of n.

• If the sequence is *two-valued*, i.e. takes on only two values X and Y (say), check if any of the six *characteristic sequences* can be recognized. The characteristic sequences, all essentially equivalent to the original sequence, are:

1. Replace X's and Y's by 1's and 2's

2. Replace X's and Y's by 2's and 1's

3. The sequence giving the positions of the X's

4. The sequence giving the positions of the Y's

5. The sequence of run lengths

6. The derivative sequence, i.e. the positions where the sequence changes

For example, the sequence

$$2, 2, 3, 3, 3, 2, 2, 2, 2, 2, 3, 3, 3, 3, 3, 3, 3, 2, 2, 2, 2, 2, 2, 2, 2, 2, 2, 2, 3, 3, 3, 3, \ldots$$

contains runs of lengths

$$2, 3, 5, 7, 11, \ldots$$

which suggests the prime numbers as a possible explanation.

• Write the terms of the sequence in base 2, or base 3, ..., or base 8, and see if any pattern is visible. E.g. M2403: 0, 1, 3, 5, 7, 9, 15, 17, 21, ..., the binary expansion is a palindrome.

• If the terms in the sequence are all single digits, is it the decimal expansion of a recognizable constant? See Fig. M2218. If only digits in the range 0 to $b - 1$ occur, is it the expansion of some constant in base b?

• Can anything be learned by considering the English words for the terms of the sequence? M1030 and M4780 are typical examples of sequences that can be explained in this way.

• There are a number of techniques for attempting to find a recurrence or generating function for a sequence. Most of these are best carried out by computer: see Sect. 2.8.

• **The quotient-difference algorithm.** One such method, however, can be carried out by hand. This procedure will succeed if the sequence satisfies a recurrence of the form

$$a_n = \sum_{i-1}^{r} c_i a_{n-i} , \qquad (2.4)$$

where r and $c_1, \ldots, c_r$ are constants. The following description is due to Lunnon [Lunn74], who calls it the *quotient-difference algorithm*, since it is similar to a standard method in numerical analysis (cf. Gragg [SIAR 14 1 72], Henrici [Henr67], Jones and Thron [JoTh80]). The algorithm is also described by Conway and Guy [CoGu95]. Given a sequence $a_0, a_1, \ldots$, we form an array $\{S_{m,n}\}$ with $S_{0,n} = 1$ for all n, $S_{1.n} = a_n$, and in general

$$S_{mn} = \det \begin{bmatrix} a_n & a_{n+1} & \cdots & a_{n+m-1} \\ a_{n-1} & a_n & \cdots & a_{n+m-2} \\ . & . & \cdots & . \\ a_{n-m+1} & & \cdots & a_n \end{bmatrix} . \qquad (2.5)$$

Any entry X in the array is related to its four neighbors

$$\begin{array}{ccc} & N & \\ W & X & E \\ & S & \end{array}$$

by the rule

$$X^2 = NS + EW , \qquad (2.6)$$

and this can be used to build up much of the array, falling back on (2.5) when (2.6) is indeterminate. A recurrence of the form (2.4) holds if the $(r + 1)$th row $S_{r+1,n}$ is identically zero.

For example, M2454: 1, 1, 1, 3, 5, 9, ... gives rise to the array

$$
\begin{array}{ccccccccc}
1 & 1 & 1 & 1 & 1 & 1 & 1 & 1 & 1 \\
1 & 1 & 1 & 3 & 5 & 9 & 17 & 31 & 57 \\
  & 0 & -2 & 4 & -2 & -4 & 10 & -8 & \\
  &   & 4 & 4 & 4 & 4 & 4 & & \\
  &   &   & 0 & 0 & 0 & & &
\end{array}
$$

Row 4 is identically zero, and indeed

$$a_n = a_{n-1} + a_{n-2} + a_{n-3} .$$

Zeros cause a problem in building the table, since then both sides of (2.6) vanish. Lunnon shows that the zeros always form square " windows", as illustrated in the following array for the sequence of Fibonacci numbers minus one (cf. M1056):

$$
\begin{array}{cccccccccccccc}
1 & 1 & 1 & 1 & 1 & 1 & 1 & 1 & 1 & 1 & 1 & 1 & 1 & 1 \\
4 & -4 & 1 & -2 & 0 & -1 & 0 & 0 & 1 & 2 & 4 & 7 & 12 & 20 \\
  & 12 & -7 & 4 & -2 & 1 & 0 & 0 & 1 & 0 & 2 & 1 & 4 & \\
  &   & 1 & -1 & 1 & -1 & 1 & -1 & 1 & -1 & 1 & -1 & & \\
  &   &   & 0 & 0 & 0 & 0 & 0 & 0 & 0 & 0 & & &
\end{array}
$$

There are simple rules for working past a window of zeros, found by J. H. Conway, and included here at his suggestion (see also [CoGu95]). To work past an isolated zero

$$
\begin{array}{ccccc}
 & & N' & & \\
 & & N & & \\
W' & W & 0 & E & E' \\
 & & S & & \\
 & & S' & &
\end{array}
$$

we use the rule that $N^2 S' + N' S^2 = W^2 E' + W' E^2$. To work around a larger window such as

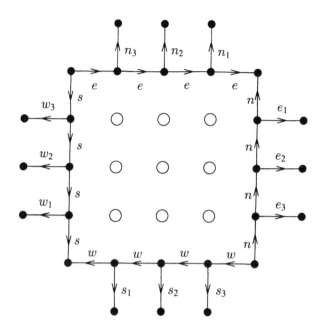

we let $n, s, e, w, n_1, s_1, \ldots$ denote the ratios of the entries at the head and tail of the appropriate arrow. Then the rules are that

$$ ns = \pm ew $$

($+$ for even-sized windows, $-$ for odd-sized), and

$$ \frac{s_1}{s} = \frac{n_1}{n} - \frac{w_1}{w} + \frac{e_1}{e}, $$
$$ \frac{s_2}{s} = \frac{n_2}{n} + \frac{w_2}{w} - \frac{e_2}{e}, $$
$$ \frac{s_3}{s} = \frac{n_3}{n} - \frac{w_3}{w} + \frac{e_3}{e}, $$

etc.

However, if a computer is available, it is generally easier to use the gfun package (Sect. 2.8) than the quotient difference algorithm.

Getu et al. [SIAD 5 497 92] show that in some cases one can learn more by decomposing the matrix on the right-hand side of (2.5) into a product of lower triangular, diagonal, and upper triangular matrices.

• Is there any other way in which the nth term of the sequence could be produced from the preceding terms? Does the sequence fall into the class of what are loosely called *self-generating* sequences? A typical example is M0257: 1, 2,

2, 3, 3, 4, 4, 4, 5, . . ., in which a_n is the number of times n appears in the sequence. See Figs. M0436, M0557 for further examples.

• Is this a Beatty sequence? If α and β are positive irrational numbers with $1/\alpha + 1/\beta = 1$, then the *Beatty sequences*

$$[\alpha], [2\alpha], [3\alpha], \ldots \quad \text{and} \quad [\beta], [2\beta], [3\beta], \ldots$$

together contain all the positive integers without repetition (see Fig. M1332). The following test for Beatty sequences is due to R. L. Graham. If $a_1, a_2, \ldots$ is a Beatty sequence, then the values of $a_1, \ldots, a_{n-1}$ determine a_n to within 1. Look at the sums $a_1 + a_{n-1}, a_2 + a_{n-2}, \ldots, a_{n-1} + a_1$. If all these sums have the same value, V say, then a_n must equal V or $V + 1$; but if they take on the two values V and $V + 1$, and no others, then a_n must equal $V + 1$. If anything else happens, it is not a Beatty sequence. For example, in the Beatty sequence M2322: 1, 3, 4, 6, 8, 9, . . ., we have $a_1 + a_1 = 2$ so a_2 must be 2 or 3 (it is 3); $a_1 + a_2 = 4$ so a_3 must be 4 or 5 (it is 4); $a_1 + a_3 = 5$ and $a_2 + a_2 = 6$, so a_4 must be 6 (it is); and so on.

2.7   Transformations of Sequences

One of the most powerful techniques for investigating a strange sequence is to transform it in some way and see if the resulting sequence is either in the table or can be otherwise identified. (A more elaborate procedure, at present prohibitively expensive, would apply these transformations both to the unknown sequence and to all the sequences in the table, and then look for a match between the two lists.)

For example, the sequence 1, 4, 5, 11, 10, 20, 14, 27, 24, 34, . . . (of no special interest, invented simply to illustrate this point), is not in the table. But the Möbius transform of it (defined below) is 1, 3, 4, 7, 9, 12, 13, 16, 19, 21, . . . which is M2336, the sequence of numbers that are of the form $x^2 + xy + y^2$.

This section describes some of the principal transformations that can be applied. Although any single transformation can be performed by hand, a thorough investigation using these methods is best carried out by computer. The program superseeker described in Sect. 2.9 tries many such transformation.

Our notation is that $a_0, a_1, a_2, \ldots$ is the unknown sequence, and $A(x)$ and $A_E(x)$ are its ordinary and exponential generating functions; $b_0, b_1, b_2, \ldots$ is the transformed sequence with o.g.f. $B(x)$ and e.g.f. $B_E(x)$.

We begin with some elementary transformations. The reader will easily invent many others of a similar nature. (Superseeker actually tries over 100 such transformations.)

- Translations: $b_n = a_n + c$; $b_n = a_n + n + c$; $b_n = a_n - n + c$; where c is $-3, -2, -1, 0, 1, 2,$ or 3.

- Rescaling: $b_n = 2a_n$; $b_n = 3a_n$; $b_n = a_n$ divided by the g.c.d. of all the a_i's; the same after deleting a_0; the same after deleting a_0 and a_1; $b_n = a_n/n!$ (if integral). If all a_n are odd, set $b_n = (a_n - 1)/2$.

- Differences: $b_n = \Delta a_n$; $b_n = \Delta^2 a_n$; etc. If a_n divides a_{n+1} for all n, set $b_n = a_{n+1}/a_n$.

- Sums of adjacent terms: $b_n = a_n + a_{n-1}$; $b_n = a_n + a_{n-2}$.

- Bisections: $b_n = a_{2n}$; $b_n = a_{2n+1}$; trisections: $b_n = a_{3n}$; $b_n = a_{3n+1}$, $b_n = a_{3n+2}$, etc.

- Reciprocal of generating function: $B(x) = 1/A(x)$. For the combinatorial interpretation of b_n in this case see Cameron [DM 75 91 89].

- Other operations on $A(x)$: $B(x) = A(x)^2$; $1/A(x)^2$; $A(x)/(1-x)$ [so that $b_n = \sum_{k \leq n} a_k$]; $A(x)/(1-x)^2$; etc.

- Similar operations on $A_E(x)$: $B_E(x) = A_E(x)^2$; $1/A_E(x)$; etc.

- Complementary sequences. Those numbers not in the original sequence. Also $b_n = n - a_n$; $b_n = \binom{n}{2} - a_n$.

The following transformations are rather more interesting.

- **Exponential and logarithmic transforms.** Several versions are possible, but the usual one transforms $a_1, a_2, a_3, \ldots$ into $b_1, b_2, b_3, \ldots$ via

$$1 + \sum_{n=1}^{\infty} \frac{b_n x^n}{n!} = \exp\left(\sum_{n=1}^{\infty} \frac{a_n x^n}{n!}\right), \tag{2.7}$$

i.e.

$$1 + B_E(x) = \exp A_E(x). \tag{2.8}$$

There is a combinatorial interpretation. For example, if a_n is the number of connected labeled graphs on n nodes, M3671, then $b_n = 2^{\binom{n}{2}}$, M1897, is the total number of connected or disconnected labeled graphs on n nodes. More generally, if a_n is the number of connected labeled graphs with a certain property, then b_n is the total number of labeled graphs with that property. Eq. (2.7) is Riddell's formula for labeled graphs (Harary and Palmer [HP73 8]).

Of course the inverse transformation is

$$\sum_{n=1}^{\infty} \frac{a_n x^n}{n!} = \log\left(1 + \sum_{n=1}^{\infty} \frac{b_n x^n}{n!}\right) . \tag{2.9}$$

In this situation we say that $b_1, b_2, \ldots$ is the *exponential transform* of $a_1, a_2, \ldots$, and that $a_1, a_2, \ldots$ is the *logarithmic transform* of $b_1, b_2, \ldots$.

● **The Euler transform.** For unlabeled graphs a different pair of transformations applies. If two sequences $a_1, a_2, a_3, \ldots$ and $b_1, b_2, b_3, \ldots$ are related by

$$1 + \sum_{n=1}^{\infty} b_n x^n = \prod_{i=1}^{\infty} \frac{1}{(1 - x^i)^{a_i}} , \tag{2.10}$$

or equivalently

$$1 + B(x) = \exp\left(\sum_{k=1}^{\infty} \frac{A(x^k)}{k}\right) , \tag{2.11}$$

then we say that $\{b_n\}$ is the *Euler* transform of $\{a_n\}$, and that $\{a_n\}$ is the *inverse Euler* transform of $\{b_n\}$.

Calculations are facilitated by introducing an intermediate sequence $c_1, c_2, \ldots$ defined by

$$c_n = \sum_{d|n} d a_d , \tag{2.12}$$

or

$$c_n = n b_n - \sum_{k=1}^{n-1} c_k b_{n-k} , \tag{2.13}$$

with

$$a_n = \frac{1}{n} \sum_{d|n} \mu\left(\frac{n}{d}\right) c_d , \tag{2.14}$$

where μ is the Möbius function (see M0011 and Fig. M0500). Using these formula $\{b_n\}$ can be obtained from $\{a_n\}$, or vice versa. The c_n have generating function

$$\log(1 + B(x)) = \sum_{n=1}^{\infty} c_n \frac{x^n}{n} . \tag{2.15}$$

There are many applications of this pair of transforms. In graph theory, if a_n is the number of connected, unlabeled graphs with some property, then b_n is the total number of graphs (connected or not) with the same property. In this context (2.11) is sometimes called Riddell's formula for unlabeled graphs (cf. Cadogan [JCT B11 193 71], Harary and Palmer [HP73 90]).

For example, if a_n ($n \geq 1$) is the number of connected unlabeled graphs with n nodes, M1657: 1, 1, 2, 6, 21, ..., then b_n ($n \geq 1$) is the total number of unlabeled graphs with n nodes, M1253: 1, 2, 4, 11, ... The intermediate sequence c_n is M2691: 1, 3, 7, 27, ...

There are also number-theoretic applications: b_n is the number of partitions of n into integer parts of which there are a_1 different types of parts of size 1, a_2 of size 2, and so on. E.g. if all $a_n = 1$, then b_n is simply the number of partitions of n into integer parts (M0663). If $a_n = 1$ when n is a prime and 0 when n is composite, b_n is the number of partitions of n into prime parts (M0265). An important example of the $\{b_n\}$ sequence is M0266, which arises in connection with the Rogers-Ramanujan identities — see Andrews [Andr85], Andrews and Baxter [AMM 96 403 89]. Andrews [Andr85] discusses a number of other number-theoretic applications, and Cameron [DM 75 89 89] gives further applications in other parts of mathematics.

• **The Möbius transform.** If sequences $a_1, a_2, a_3, \ldots$ and $b_1, b_2, b_3, \ldots$ are related by

$$b_n = \sum_{d|n} \mu\left(\frac{n}{d}\right) a_d , \tag{2.16}$$

$$a_n = \sum_{d|n} b_d , \tag{2.17}$$

where the summations are taken over all positive integers d that divide n, we say that $\{b_n\}$ is the *Möbius* transform of $\{a_n\}$, and that $\{a_n\}$ is the *inverse Möbius* (or *sum-of-divisors*) transform of $\{b_n\}$. Equations (2.16), (2.17) are called the *Möbius inversion formulae*. (The sequences in (2.12) and (2.14) are related in this way.) Two equivalent formulations are

$$\sum_{n=1}^{\infty} a_n x^n = \sum_{n=1}^{\infty} b_n \frac{x^n}{1-x^n} , \tag{2.18}$$

$$\sum_{n=1}^{\infty} \frac{a_n}{n^s} = \zeta(s) \sum_{n=1}^{\infty} \frac{b_n}{n^s} , \tag{2.19}$$

where

$$\zeta(s) = \sum_{n=1}^{\infty} \frac{1}{n^s} = \prod_{p \text{ prime}} \frac{1}{1 - \frac{1}{p^s}} \tag{2.20}$$

is the Riemann zeta function.

Again there are many applications. For combinatorial applications see Rota [ZFW 2 340 64] (as well as several other papers reprinted in [GeRo87]), Bender and Goldman [AMM 82 789 75], and Stanley [Stan86]. For number-theoretic applications see for example Hardy and Wright [HWI §17.10] — the right-hand side of (2.18) is called a *Lambert series*.

Examples. (i) If $b_n = 1, 1, 1, \ldots$, a_n = number of divisors of n (M0246). (ii) If $b_n = 1, 0, 0, \ldots$, a_n = Möbius function (M0011). (iii) If $b_n = n$, a_n = Euler totient function (M0299). (iv) If $b_{2n} = 0$, $b_{2n+1} = (-1)^n 4$, then a_n = number of ways of writing n as a sum of two squares (M3218).

• **The binomial transform.** If $a_0, a_1, a_2, \ldots$ and $b_0, b_1, b_2, \ldots$ are related as in Eq. (2.3), we say that $\{a_n\}$ is the *binomial* transform of $\{b_n\}$, and that $\{b_n\}$ is the *inverse binomial* transform of $\{a_n\}$. Equivalently, the exponential generating functions are related by

$$A_E(x) = e^x B_E(x) . \tag{2.21}$$

As we saw in Sect. 2.5, these transformations arise in studying the differences of a sequence. The leading diagonal of the difference table of a sequence is the inverse binomial transform of the sequence.

Examples. If $a_n = 3^n$, $b_n = 2^n$, and more generally, if $a_n = k^n$, $b_n = (k-1)^n$.

The Bell numbers 1, 1, 2, 5, 15, 52, ... (M1484) are distinguished by the property that they are shifted one place by the binomial transform: $a_n = b_{n+1}$ [BeS194].

• **Reversion of series.** Given a sequence $a_1, a_2, a_3, \ldots$ we can form a generating function

$$y = x(1 + a_1 x + a_2 x^2 + \cdots) , \tag{2.22}$$

and by expressing x in terms of y obtain a new sequence $b_1, b_2, b_3, \ldots$ by writing

$$x = y(1 - b_1 y - b_2 y^2 - \cdots) . \tag{2.23}$$

This process is called *reversion of series*, and explicit formulae expressing b_n in terms of $a_1, \ldots, a_n$ can be found for example in [AS1 16], [RCI 149], [TMJ 2 73 92]. This transformation is its own inverse. For example, if the a_n are the Fibonacci numbers 1, 2, 3, 5, 8, ... (M0692), the b_n are 1, 2, 5, 15, 51, 188, ... (M1480). It is amusing that the latter sequence is also the binomial transform of the Catalan numbers (M1459). An alternative version of this transformation is: given $a_0 = 1$, $a_1, \ldots$ we set $y = \sum_{i=0}^{\infty} a_i x^{i+1}$, whose reversion is $x = \sum_{i=0}^{\infty} b_i x^{i+1}$, producing the transformed sequence $b_0 = 1, b_1, \ldots$.

• **Other transforms.** A pair of transforms of the form

$$a_n = \sum C_{n,k} b_k, \quad b_n = \sum D_{n,k} a_k$$

can be defined whenever we find integer arrays $\{C_{n,k}\}$ and $\{D_{n,k}\}$ satisfying the orthogonality relation

$$\sum C_{m,k} D_{k,n} = \begin{cases} 1 & m = n, \\ 0 & m \neq n. \end{cases}$$

Riordan's book [RCI] gives many such examples, including transforms that are based on Chebyshev and Legendre polynomials.

We conclude by mentioning that the pair of transforms based on Stirling numbers seems to be worth investigating further, particularly in the context of enumerating permutations. In this case we have

$$a_n = \sum_{k=0}^{n} s(n,k) b_k, \quad b_n = \sum_{k=0}^{n} S(n,k) a_k, \tag{2.24}$$

where the coefficients are Stirling numbers of the first and second kinds, respectively (see Figs. M4730, M4981; also [R1 48], [RCI 90], [GKP 252], [BeSl94]).

2.8 Methods for Computer Investigation of Sequences

As we have already mentioned, a thorough investigation of the transformations of a sequence described in the previous section is best done by computer.

• `Gfun`. At the heart of the following techniques is an algorithm of Cabay and Choi [SIAC 15 243 86] that uses Padé approximations to take a truncated power series

$$c_0 + c_1 x + c_2 x^2 + \cdots + c_{n-1} x^{n-1} \tag{2.25}$$

with rational coefficients, and determines a rational function $p(x)/q(x)$, where $p(x)$ and $q(x)$ are polynomials with rational coefficients, whose Taylor series expansion agrees with (2.25) and in which $\deg p + \deg q$ is minimized. If $\deg p + \deg q < n - 2$, we say this is a "good" representation of (2.25) (for then $p(x)/q(x)$ contains fewer constants than the original series).

The Cabay-Choi algorithm is incorporated in the Maple `convert/ratpoly` procedure. Bergeron and Plouffe [EXPM 1 307 92] observed that this provides an efficient way to search for a wide class of generating functions for sequences. Given a sequence $a_0, a_1, \ldots, a_{n-1}$, one can form the o.g.f. $A(x)$ and e.g.f. $A_E(x)$, and see if either have a "good" rational representation. If not, one can try again with the *logarithmic derivates* $A'(x)/A(x)$ and $A'_E(x)/A_E(x)$, and with many other transformed generating functions. In this way Bergeron and S.P. were able to find generating functions such as $1/(2 - e^x)$ for M2952.

This work was carried much further by S.P. in his thesis [Plou92], which gives over 1000 generating functions, recurrences and formulae for the 4500 sequences in a 1991 version of the present table. Some of these are immediate, others can be proved with difficulty, but a considerable number are still only conjectural. The simplest of these (but not the conjectural ones) have now been incorporated in the table. To have included the rest, which are usually quite complicated, would have greatly increased the length of this book.

The gfun Maple package of Salvy and Zimmermann [SaZi94] incorporates and greatly extends the ideas of Bergeron and S.P. With gfun, one can (among many other things) check very easily:

(a) whether there is a "good" rational function representation for the o.g.f. or e.g.f. of a sequence, or for their logarithmic derivatives, or their reversions;

(b) whether the generating function $y(x)$ of any of these types satisfies a polynomial equation or a linear differential equation with polynomial coefficients;

(c) whether the coefficients of any of these generating functions satisfy a linear recurrence with polynomial coefficients;

and many other things. The package contains a number of commands that make it easy to manipulate sequences and power series and to convert between different types. The superseeker program described in Section 2.9 makes good use of gfun.

• **Look for sequences in the table that are close to the unknown sequence.** There are a number of ways to do this. Let $a = a_0, a_1, \ldots, a_{n-1}$ be the unknown sequence, and $b = b_0, b_1, \ldots, b_{m-1}$ a typical sequence in the table. We truncate the longer sequence so they both contain the same number of terms, n. Then we may ask:

(a) Which sequences in the table are closest in L_1 norm, i.e. minimize

$$\sum_{i=0}^{n-1} |a_i - b_i| \ ?$$

(b) Is there a sequence in the table such that

$$|a_i - b_i| \leq 1 \quad \text{for all} \quad i \ ?$$

Or for which $|a_i - b_i|$ is a constant sequence?

(c) Which sequences in the table are closest in Hamming distance? (Write a and b as strings of decimal digits and spaces, and count the places where they differ.)

(d) Which sequences in the table are most closely *correlated* with the unknown sequence? I.e., which maximize the squared correlation coefficient

$$r^2 = \frac{1}{(n-1)^2 s_a^2 s_b^2} \left(\sum_{i=0}^{n-1} (a_i - \bar{a})(b_i - \bar{b}) \right)^2 ,$$

where

$$\bar{a} = \frac{1}{n} \sum_{i=0}^{n-1} a_i, \quad s_a^2 = \frac{1}{n-1} \sum_{i=0}^{n-1} (a_i - \bar{a})^2$$

are the mean and variance of a, with similar definitions for $\bar{b}$ and s_b^2.

Notes: Among other things, (a) will detect small errors in calculation; (b) will detect sequences whose definition differs by a constant from one in the table; (c) will detect typing errors; (d) is the most time-consuming of these tests, and will detect a sequence of the form $a = pb + q$, where b is in the table and p and q are constants.

Another possible test of this type is to see if a is a subsequence of some sequence in the table, but we have not found this useful.

The remaining tests in this section are more speculative. However, once in a while they may find an explanation for a sequence that has not succumbed to any other test.

• **Apply the Berlekamp-Massey or Reed-Sloane algorithms.** Suppose the sequence takes on only a small number of different values, e.g. $\{0, 1, 2, 3\}$. By regarding the values as the elements of a finite field (the Galois field $GF(4)$ would be appropriate in this case) we may think of the sequence as a sequence from this field. The Berlekamp-Massey algorithm is an efficient procedure for finding the shortest linear recurrence with coefficients from the field that will generate the sequence — see Berlekamp [Be68 Chap. 7] and Massey [PGIT 15 122 69].

(Other references that discuss this extremely useful algorithm are Dickinson et al. [PGAC 19 31 74], Berlekamp et al. [UM 5 305 74], Mills [MOC 29 173 75], Gustavson [IBMJ 20 204 76], McEliece [McEl77], MacWilliams and Sloane [MS78 Chap. 9], and Brent et al. [JAlgo 1 259 80].) This algorithm would discover for example that the sequence

$$0, 1, 2, 1, 3, 0, 3, 0, 1, 3, 3, 2, 3, 3, 3, 1, 2, 0, 1, 1, 0, 0, \ldots$$

is generated by the linear recurrence

$$a_n = \omega(a_{n-1} + a_{n-2} + a_{n-3})$$

over $GF(4)$, where we take $GF(4)$ to consist of the elements $\{0, 1, \omega, \omega^2\}$, with $\omega^2 = \omega + 1$, and write 2 for ω, 3 for ω^2. The Reed-Sloane algorithm [SIAC 14 505 85] is an extension of this algorithm which applies when the terms of the sequences are integers modulo m, for some given modulus m. For example, this algorithm would discover that the sequence

$$1, 2, 4, 3, 1, 3, 6, 7, 4, 4, 1, 5, 3, 0, 5, 6, \ldots$$

is produced by the recurrence

$$a_n = a_{n-1} + 2a_{n-2} + 3a_{n-3} \pmod{8}.$$

• **Apply a data compression algorithm.** Feed the sequence to a data compression algorithm, such as the Ziv-Lempel algorithm as implemented in the Unix commands `compress` or `gzip`.

If the sequence is compressed to a much greater extent than a comparable random sequence of the same length would be, there is some structure present that can be recovered by examining the compression algorithm (see for example [BCW90]).

For example, `gzip` compresses M0001 from 150 characters to 36 characters, whereas a random binary sequence of the same length typically is compressed only to 60 bits. So if a 150-character binary sequence is compressed to (say) 45 bits or less, one can be sure it has some concealed structure.

It would be worth running this test on any stubborn sequence which contains only a limited set of symbols. By experimenting with random sequences of the same length and containing the same symbols, one can determine their average compressibility. If the stubborn sequence is compressed to a greater degree than this then it has some hidden structure.

• **Compute the Fourier transform of the sequence.** An article by Loxton [Loxt89] demonstrates that the Fourier transform of a sequence can reveal much about how it is generated. This is a topic that deserves further investigation.

2.9   The On-Line Versions of the Encyclopedia

There are two on-line versions of the Encyclopedia that can be accessed via electronic mail. The first is a simple look-up service, while the second tries very hard to find an explanation for a sequence. Both make use of the latest and most up-to-date version of the main table.

To use the simple look-up service, send email to

> sequences@research.att.com

containing lines of the form

> lookup 5 14 42 132 429

There may be up to five such lines in a message. The program will automatically inform you of the first seven sequences in the table that match each line. If there are no "lookup" lines, you will be sent an instruction file.

Notes. When submitting a sequence, separate the terms by spaces (not commas). It may be advisable to omit the initial term, since there are often different opinions about how a sequence should begin. (Does one start counting graphs, say, at 0 nodes or at 1 node? Do the Lucas numbers begin 1, 3, 4, 7, 11, ... or 2, 1, 3, 4, 7, 11, ...?) Omit all minus signs, since they have been omitted from the table. If you receive seven matches to a sequence, try again giving more terms. For more details, see [Sloa94].

The second server not only looks up the sequence in the table, it also tries hard to find an explanation for it, using many of the tricks described in this chapter (and possibly others — at the time of writing the program is still being expanded). To use this more powerful program, send email to

> superseeker@research.att.com

containing a line of the form

> lookup 1 2 4 6 10 14 20 26 36 46 60 74 94 114 140 166

The program will apply many tests, and report any potentially useful information it discovers.

Notes. The word "lookup" should appear only once in the message. The terms of the sequence should be separated by spaces (not commas). For this program the sequence should be given from the beginning. Minus signs *should be included*, since most of the programs will make use of them. If possible, give from 10 to 20 terms. If you receive seven matches from the table, try again giving more terms.

2.10   The Floppy Disk

A floppy disk containing every sequence in the table (although not their descriptions) is available from the publisher. Please contact Academic Press at

1–800–321–5068 for information regarding the floppy disk to accompany *The Encyclopedia of Integer Sequences*. Please indicate desired format by referring to the ISBN for Macintosh (0-12-558631-0) or for IBM/MSDOS (0-12-558632-9).

The disk contains a line such as

```
M[1916] := [A6226, 1, 2, 9, 18, 118]:
```

for each sequence. The first number gives the sequence number in this book, the second gives the absolute identification number for the sequence, and the remaining numbers are the sequence itself.

This disk will enable readers to study the sequences in their own computers. Of course the book will still be needed for the descriptions of the sequences and the references.

Chapter 3

Further Topics

3.1 Applications

We begin by describing some typical ways in which the 1973 book [HIS] has been used, as well as some applications of the sequence servers mentioned in Section 2.9. (Even though at the time of writing the latter have been in existence for only a few months, there have already been some interesting applications). It is to be expected that the present book will find similar applications.

The most important way the table is used is in discovering whether someone has already worked on your problem. Discrete mathematics has grown exponentially over the last thirty years, and so there is a good chance that someone has already looked at the same problem, or an equivalent one. In this respect the book serves as an index, or field guide, to a broad spectrum of mathematics. If the answers to the first few special cases of a problem can be described by integers, and someone has considered the problem worth studying, there is a good chance you will find the sequence of numbers in this book. Of course if not, and if superseeker can't do anything with it, you should send in the sequence so that it can be added to the table — see Sect. 2.2 for instructions. Apart from anything else, this stakes out your claim to the problem! But, more important, you will be performing a service to the scientific community.

As with any dictionary (and as predicted by the epigraph to Chapter 1), most such successful uses go unrecorded. The reader simply stops working on the problem, as soon as he or she has been pointed to the appropriate place in the literature.

In many cases the book has led to mathematical discoveries. The following stories are typical.

• R. L. Graham and D. H. Lehmer were investigating the permanent P_n of Schur's matrix, the $n \times n$ matrix (α^{jk}), $0 \leq j, k \leq n - 1$, where $\alpha = e^{2\pi i/n}$, and found that the initial values $P_1, P_3, P_5, \ldots$ were

$$1, -3, -5, -105, 81, \ldots$$

(P_n is 0 if n is even). As it happened, this sequence (M2509) was in the Supplement [Supp74] to the 1973 book, and N. J. A. S. was able to refer Graham and Lehmer to an earlier paper by D. H. Lehmer, where the same sequence had arisen! This provided an unexpected connection with circulant matrices [JAuMS A21 496 76].

• Extract from a letter about the 1973 book: "After reading about your book in *Scientific American*, I ordered a copy. Several of my friends looked at the book and stated they thought it was interesting but doubted its usefulness. A few days later I was attempting to determine the number of spanning trees on an n by m lattice. In working out the 2 by m case, I determined the first numbers in the sequence to be 1, 4, 15, 56. Noticing that both sequences No. 1420 and 1421 started this way, I worked out another term, 209; thus sequence No. 1420 seemed to fit. After much thought I was able to establish a complicated recursion relationship which I was later able to show was equivalent to the recursion you gave for No. 1420. ... In closing I would like to say that your book has already proved to be worthwhile to me since it provided guidelines for organizing my thoughts on this problem and suggested a hypothesis for the next term of the sequence. I'm sold!" (Alamogordo, New Mexico).

• While investigating a problem arising from cellular radio, Mira Bernstein, Paul Wright and N. J. A. S. were led to consider the number of sublattices of index n of the planar hexagonal lattice. For $n = 1, 2, 3, \ldots$ they calculated that these numbers were $1, 1, 2, 3, 2, 3, 3, 5, \ldots$. To their surprise, the table revealed that this sequence, M0420, had arisen in 1973 in an apparently totally different context, that of enumerating maps on a torus (Altshuler [DM 4 201 73]), and supplied a recurrence that they had overlooked. (However, it is only fair to add that the earlier paper did not find the elegant exact formula for the nth term that is given in [BSW94]. There is also an error in the values given in the earlier paper: $\chi(16)$ should be 9, not 16.)

• C. L. Mallows was interested in determining the number of statistical models with n factors, in particular linear hierarchical models that allow 2-way interactions. For $n = 1, 2, \ldots$ he found the numbers of such models to be

$$2, 4, 8, 19, 53, 209 \, .$$

This sequence was not at that time in the table, but `superseeker` (see Sect. 2.9) pointed out that these numbers agreed with the partial sums of M1253, the number of graphs on n nodes. With this hint, Mallows was instantly able to show that this explained his sequence (which is now M1153).

• R. K. Guy and W. O. J. Moser [GuMo94] report a successful application of `superseeker` in finding a recurrence for the number of subsequences of $[1, 2, \ldots, n]$ in which every odd number has at least one even neighbor. The first try with the program was unsuccessful, because of an error in one of their terms, but when the corrected sequence

$$1, 1, 3, 5, 11, 17, 39, 61, 139, \ldots$$

(now M2480) was submitted, superseeker used gfun to find the elegant generating function

$$\frac{1 + x + 2x^3}{1 - 3x^2 - 2x^4}.$$

• Inspection of the log file for the sequence servers on March 28, 1994 shows that at least one high-school student used the program to identify a sequence (M2638) for her homework.

Another important application of the book is to suggest possible connections between sequences arising in different areas, as in the Mallows story above. Here is a typical (although ultimately unsuccessful) example.

• The dimensions of the spaces of primitive Vassiliev knot invariants of orders $1, \ldots, 9$ form the sequence

$$1, 1, 1, 2, 3, 5, 8, 12, 18$$

the next term being presently unknown (see Birman [BAMS 28 281 93], Bar-Natan [BarN94]). This sequence coincides with the beginning of M0687, which gives the number of ways of arranging n pennies in rows of contiguous pennies, each touching two in the row below. Alas, further investigation by D. Bar-Natan has shown that next term in the former sequence is at least 27, and so these sequences are in fact *not* the same.

• As already mentioned in Sect. 2.8, S.P.'s thesis [Plou92] contains many conjectures about possible generating functions. For example, M2401, the size of the smallest square into which one can pack squares of sizes $1, 2, \ldots, n$, appeared to have generating function

$$(1 - z)^{-3}(1 - z^2) \prod_{m=4}^{\infty} (1 - z^{2m+1})(1 - z^{2m})^{-1},$$

which agreed with the 17 values known at the time [UPG D5]. This prompted R. K. Guy [rkg] to calculate some further terms, and to show that in fact this generating function is not correct. At present no general formula is known for this sequence.

For an example of a conjectured generating function (for M2306) that turned out to be correct, see Allouche et al. [AABB].

3.2 History

I[1] started collecting sequences in 1965 when I was a graduate student at Cornell University. I had run across several sequences whose asymptotic behavior I needed to determine, so I was hoping to find recurrences for them. Although John Riordan's book [R1] was full of sequences, the ones I was interested in did not seem to be there. Or were they? It was hard to tell, certainly some very similar sequences were mentioned. So I started collecting sequences on punched cards. Almost thirty years later, the collection is still growing (although it is no longer on punched cards.)

Over the course of several years I systematically searched through all the books and journals in the Cornell mathematics library, and then the Bell Labs library, when I joined the Labs in 1969. A visit to Brown University, with its marvelous collection of older mathematics books and journals, filled in many gaps. I never did find the sequences I was originally looking for, although of course they are now in the table (M4558 was the one I was most interested in: $0, 1, 8, 78, 944, 13800, \ldots$ a very familiar sequence! It essentially gives the average height of a rooted labeled tree.)

The first book [HIS] was finally published by Academic Press in 1973, and a supplement [Supp74] was issued a year later. Over the next fifteen years new material poured in, and by 1990 over a cubic meter of letters, articles, preprints, postcards, etc., had accumulated in my office. I made one attempt to revise the book in 1980, with the help of two summer students, Bob Hinman and Tray Peck, and managed to transfer the 1973 table from punched cards to magnetic disk, and started processing the new material. But at the end of that summer other projects intervened (cf. [MS78], [SPLAG]). Ten years later the amount of material waiting to be processed was overwhelming.

Fortunately S.P. wrote to me in 1991, offering to help with a new edition, and this provided the stimulus that, four years later, has produced the new book. It very nearly never happened!

3.3 Differences from the 1973 Book

- Size: There are now 5488 sequences, compared with 2372 in [HIS].

- Format: In [HIS], every sequence was normalized so as to begin $1, n$, with $2 \leq n \leq 999$, an initial 1 being added as a marker if necessary. Now the sequence can begin in any way, subject only to Rule 3 of Sect. 1.5.

---

[1]The first person seems appropriate here (N.J.A.S.).

• The descriptions are much more informative. Many generating functions have been included. One of the benefits of the transition from punched cards to magnetic disk has been an enlarged character set. Before, only upper case letters could be used; now, all standard mathematical symbols are available.

• All known errors in [HIS] have been corrected. In almost every case these were errors in the source material, not in transcription. Some erroneous or worthless sequences have been omitted.

• There is also a technical change. In the older mathematical literature 1 was regarded as a prime number, whereas today it is not. This has necessitated changes to a few sequences. M3352 for example now begins $4, 9, 11, \ldots$ rather than $2, 4, 9, 11, \ldots$ as in [HIS].

3.4 Future Plans

• The table should be modified so as to include minus signs. Unfortunately to do this thoroughly would require re-examining thousands of sequences, and this book has already been delayed long enough.

• It would be nice to have a series of essays, one for each family of sequences (Boolean functions, partitions, graphs, lattices, etc.), showing how the sequences are related to each other and which are fundamental. This would clarify the sequences that one should concentrate on when looking for generating functions, finding more terms, and so on. The late Victor Meally spent a great deal of time on such a project, and every square centimeter of his copy of [HIS], now in the Strens collection of the University of Calgary library, is annotated with cross-references between sequences, tables, diagrams, and so on — in other words a greatly expanded version of the Figures in the present book. It would be worthwhile doing this in a systematic way. Such commentaries could easily fill a companion volume.

• It would also be useful to classify the sequences into various categories, a multiple classification that would indicate:

- subject (graphs, partitions, etc.),
- type (enumerative, number-theoretic, dependent on base 10 representation, frivolous, etc.), and
- method of generation (ranging from "explicit formula", "recurrence", etc., to "the next term not known").

It is surprisingly difficult to give precise definitions for some of these classes —

there are explicit formulae for the nth prime, for instance, and the most intractable enumeration problem can be encoded into a recurrence if one defines enough variables (see for example [JCT 5 135 68]).

There are however a number of mathematically well-defined classes of sequences, for instance generalized periodic sequences (MacGregor [AMM 87 90 80]), k-automatic sequences (Cobham [MST 3 186 69; 6 164 72]), k-regular sequences (Allouche and Shallit [TCS 98 163 92]), differentiably finite sequences (Stanley [EJC 1 175 80]), constructibly differentiably finite sequences (Bergeron and Reutenauer [EJC 11 501 90]), etc., which could be used as a basis for a more rigorous classification. We should also mention the recent studies of integer sequences that have been made by Lisoněk [Liso93], Sattler [Satt94] and Théorêt [Theo94], [Theo95].

- There are many other features that could be added to the table, such as:

  - Maple, Macsyma, Mathematica, Pari, etc. procedures to generate as many terms of the sequence as desired (if available), or

  - a complete list of all known terms (if it is difficult to generate);

  - generating functions or recurrences in every case for which they are known;

  - a description of the asymptotic behavior of the sequence, and other interesting mathematical properties;

  - full details of the source for each sequence (author, title, etc.), or even,

  - the full text of the article or an extract from the book where the sequence appeared.

Finally, what about a table of arrays? Much remains to be done!

3.5 Acknowledgments

We thank the more than 400 correspondents who have contributed sequences to this book over the past twenty-five years. It would not be appropriate to list all their names here, but without their help, the book would not be as complete as it is.

We are especially grateful to Mira Bernstein, John Conway, Susanna Cuyler, Martin Gardner, Richard Guy (for a correspondence that spans more than 25 years), Colin Mallows, Robert Robinson, Jeffrey Shallit, and Robert Wilson (who has been our most prolific contributor of new sequences) for their assistance, as well as the late Victor Meally, John Riordan and Herman Robinson. Friends in the Unix room at Bell Labs, especially Andrew Hume and Brian Kernighan, have helped in innumerable ways. The gfun package of Bruno Salvy and Paul Zimmermann (see Section 2.87) has been of great help. Sue Pope typed the introductory chapters and produced LATEXversions of many of the figures. In the summer of 1980 Bob Hinman and Theodore Peck helped in converting the sequences from the punched card format of the 1973 book.

The staff of the Bell Labs library, especially Dick Matula, have been very helpful. This great library is one of the few in the world where one can find comprehensive collections in mathematics, engineering, physics and chemistry under one roof.

Above all, N. J. A. S. wishes to thank the Mathematical Sciences Research Center at AT&T Bell Labs for its continuing support over the years. Thank you, Robert Calderbank, Mike Garey, Ron Graham, Andrew Odlyzko and Henry Pollak.

M0005   0, 0, 0, 2, 0, 0, 0, 0, 0, 0, ...

THE TABLE OF SEQUENCES

SEQUENCES OF 0's AND 1's

M0000   0, 0, 0, 0, 0, 0, 0, 0, 0, 0, 0, 0, 0, 0, 0, 0, 0, 0, 0, 0, 0, 0, 0, 0, 0, 0, 0, 0, 0, 0, 0, 0, 0,
0, 0, 0, 0, 0, 0, 0, 0, 0, 0, 0, 0, 0, 0, 0, 0, 0, 0, 0, 0, 0, 0, 0, 0, 0, 0, 0, 0, 0, 0, 0, 0, 0
The zero sequence. [0,1; A0004]

M0001   0, 1, 0, 1, 0, 1, 0, 1, 0, 1, 0, 1, 0, 1, 0, 1, 0, 1, 0, 1, 0, 1, 0, 1, 0, 1, 0, 1, 0, 1, 0, 1, 0,
1, 0, 1, 0, 1, 0, 1, 0, 1, 0, 1, 0, 1, 0, 1, 0, 1, 0, 1, 0, 1, 0, 1, 0, 1, 0, 1, 0, 1, 0, 1, 0, 1, 0
A simple periodic sequence. [0,1; A0035]

M0002   1, 0, 0, 0, 0, 0, 0, 0, 0, 0, 0, 0, 0, 0, 0, 0, 0, 0, 0, 0, 0, 0, 0, 0, 0, 0, 0, 0, 0, 0, 0, 0, 0,
0, 0, 0, 0, 0, 0, 0, 0, 0, 0, 0, 0, 0, 0, 0, 0, 0, 0, 0, 0, 0, 0, 0, 0, 0, 0, 0, 0, 0, 0, 0, 0, 0, 0
The characteristic function of 0: $a(n) = 0^n$. [0,1; A0007]

M0003   1, 1, 1, 1, 1, 1, 1, 1, 1, 1, 1, 1, 1, 1, 1, 1, 1, 1, 1, 1, 1, 1, 1, 1, 1, 1, 1, 1, 1, 1, 1, 1, 1,
1, 1, 1, 1, 1, 1, 1, 1, 1, 1, 1, 1, 1, 1, 1, 1, 1, 1, 1, 1, 1, 1, 1, 1, 1, 1, 1, 1, 1, 1, 1, 1, 1, 1
The simplest sequence of positive numbers: the all 1's sequence. [0,1; A0012]

SEQUENCES BEGINNING . . ., 2, 0, . . .

M0004   2, 0, 0, 0, 0, 0, 0, 0, 0, 0, 0, 0, 0, 0, 0, 0, 0, 0, 0, 0, 0, 0, 0, 0, 0, 0, 0, 0, 0, 0, 0, 0, 0,
0, 0, 0, 0, 0, 0, 0, 0, 0, 0, 0, 0, 0, 0, 0, 0, 0, 0, 0, 0, 0, 0, 0, 0, 0, 0, 0, 0, 0, 0, 0, 0, 0, 0
The first sequence in the main table. [0,1; A0038]

M0005   0, 0, 0, 2, 0, 0, 0, 0, 0, 0, 0, 0, 0, 0, 0, 0, 0, 0, 0, 6, 0, 0, 0, 0, 0, 0, 0, 0, 0, 0, 0, 0, 0,
0, 0, 12, 0, 0, 0, 0, 0, 0, 0, 0, 0, 0, 0, 0, 0, 0, 12, 0, 0, 0, 0, 0, 0, 0, 0, 0, 0, 0, 0, 0, 0, 0, 6
Theta series of diamond lattice with respect to mid-point of edge. Ref JMP 28 1653 87.
[0,4; A5926]

M0006   0, 2, 0, 0, 0, 0, 0, 4, 0, 0, 0, 0, 0, 4, 0, 0, 0, 0, 0, 4, 0, 0, 0, 0, 0, 2, 0, 0, 0, 0, 0, 4, 0,
0, 0, 0, 0, 4, 0, 0, 0, 0, 0, 4, 0, 0, 0, 0, 0, 6, 0, 0, 0, 0, 0, 0, 0, 0, 0, 0, 0, 4, 0, 0, 0, 0, 0, 4, 0
Theta series of hexagonal net with respect to mid-point of edge. Ref JMP 28 1654 87. [0,2;
A5929]

M0007   0, 0, 0, 2, 0, 0, 0, 0, 0, 4, 0, 0, 0, 0, 0, 4, 0, 0, 0, 0, 0, 8, 0, 0, 0, 0, 0, 6, 0, 0, 0, 0, 0,
2, 0, 4, 0, 0, 0, 8, 0, 4, 0, 0, 0, 4, 0, 0, 0, 0, 0, 4, 0, 8, 0, 0, 0, 10, 0, 4, 0, 0, 0, 4, 0, 0, 0, 0, 0
Theta series of hexagonal close-packing with respect to edge within layer. Ref JCP 83 6528
85. [0,4; A5871]

M0008   0, 0, 0, 2, 0, 0, 0, 0, 0, 4, 0, 0, 0, 0, 0, 4, 0, 0, 0, 2, 0, 4, 0, 0, 0, 4, 0, 2, 0, 0, 0, 4, 0,
4, 0, 0, 0, 4, 0, 4, 0, 0, 0, 2, 0, 4, 0, 0, 0, 4, 0, 8, 0, 0, 0, 4, 0, 4, 0, 0, 0, 4, 0, 8, 0, 0, 0, 6, 0
Theta series of hexagonal close-packing with respect to edge between layers. Ref JCP 83
6529 85. [0,4; A5888]

M0009   2, 0, 0, 0, 0, 4, 8, 0, 4, 0, 82, 0, 0, 0, 112, 0, 132, 0
Number of *n*-node vertex-transitive graphs which are not Cayley graphs. Ref ARS 30 175
90. JAuMS A56 53 94. [10,1; A6792]

M0010   1, 0, 0, 0, 0, 0, 2, 0, 0, 0, 0, 24, 26, 0, 0, 48, 252, 720, 438, 192, 984, 1008, 12924,
19536, 3062, 8280, 26694, 153536, 507948, 406056, 79532, 729912, 631608, 9279376
Magnetization for face-centered cubic lattice. Ref JMP 6 297 65. JPA 6 1510 73. DG74
420. [0,7; A3196]

M0011   2, 0, 0, 1, 0, 2, 0, 1, 1, 2, 0, 1, 0, 2, 2, 1, 0, 1, 0, 1, 2, 2, 0, 1, 1, 2, 1, 1, 0, 0, 0, 1, 2,
2, 2, 1, 0, 2, 2, 1, 0, 0, 0, 1, 1, 2, 0, 1, 1, 1, 2, 1, 0, 1, 2, 1, 2, 2, 0, 1, 0, 2, 1, 1, 2, 0, 0, 1, 2
1 + μ(*n*), where μ is the Moebius function. Ref HW1 234. IrRo82 19. [1,1; A7423]

M0012   1, 1, 0, 1, 2, 0, 0, 1, 1, 2, 0, 0, 2, 0, 0, 1, 2, 1, 0, 2, 0, 0, 0, 0, 3, 2, 0, 0, 2, 0, 0, 1, 0,
2, 0, 1, 2, 0, 0, 2, 2, 0, 0, 0, 2, 0, 0, 0, 1, 3, 0, 2, 2, 0, 0, 0, 0, 2, 0, 0, 2, 0, 0, 1, 4, 0, 0, 2, 0
Number of ways of writing *n* as a sum of 2 squares. Ref QJMA 20 164 1884. GR85 15.
[1,5; A2654, N0001]

M0013   2, 0, 0, 1, 1, 2, 1, 2, 1, 1, 1, 1, 0, 1, 1, 0, 1, 1, 1, 1, 1, 1, 2, 1, 1, 1, 2, 1, 1, 1, 1, 1, 1,
1, 1, 0, 1, 1, 1, 1, 0, 1, 1, 1, 1, 1, 1, 1, 1, 1, 1, 1, 2, 1, 1, 1, 1, 1, 2, 1, 1, 1, 1, 1, 1, 1, 1, 1, 1, 1
1 + coefficient of x^n in $\Pi(1-x^k)$, $k=1..\infty$ (essentially the expansion of the Dedekind
function $\eta(x)$). Ref AS1 825. Scho74 70. [0,1; A7706]

M0014   1, 2, 0, 0, 10, 36, 0, 0, 720, 5600, 0, 0, 703760, 11220000, 0, 0
Number of self-complementary graphs with *n* nodes. Ref JLMS 38 103 63. [4,2; A0171,
N0780]

M0015   1, 0, 0, 0, 2, 0, 0, 16, 18, 0, 168, 384, 314, 1632, 6264, 9744, 10014, 86976,
205344, 80176, 1009338, 3579568, 4575296, 8301024, 54012882, 112640896, 5164464
Magnetization for body-centered cubic lattice. Ref JMP 6 297 65. JPA 6 1511 73. DG74
420. [0,5; A3193]

M0016 1, 0, 1, 1, 0, 0, 2, 0, 1, 0, 0, 1, 2, 0, 0, 1, 0, 0, 2, 0, 2, 0, 0, 0, 1, 0, 1, 2, 0, 0, 2, 0, 0,
0, 0, 1, 2, 0, 2, 0, 0, 0, 2, 0, 0, 0, 0, 1, 3, 0, 0, 2, 0, 0, 0, 0, 2, 0, 0, 0, 2, 0, 2, 1, 0, 0, 2, 0, 0
Number of divisors of $n \equiv 1$ mod 3 minus number of divisors of $n \equiv 2$ mod 3. Ref MES 31
67 01. L1 8. [1,7; A2324, N0002]

M0017 0, 1, 2, 0, 1, 2, 2, 1, 2, 4, 3, 4, 4, 5, 5, 4, 5, 8, 6, 8, 7, 8, 8, 9, 9, 10, 10, 15, 11, 12,
12, 11, 12, 16, 13, 16, 14, 15, 15, 17, 16, 17, 17, 19, 18, 19, 19, 20, 20, 21, 21, 23, 22, 23
2-part of number of graphs on n nodes. Ref CaRo91. [1,3; A7149]

M0018 0, 1, 2, 0, 1, 2, 4, 2, 0, 1, 2, 2, 5, 4, 2, 0, 1, 2, 6, 2, 6, 6, 4, 2, 0, 1, 2, 4, 5, 2, 8, 4, 4,
4, 2, 0, 1, 2, 2, 2, 3, 2, 10, 8, 6, 12, 4, 2, 0, 1, 2, 6, 5, 6, 4, 2, 6, 7, 6, 4, 11, 4, 2, 0, 1, 2, 10
Period of continued fraction for square root of n. Ref BR73 197. [1,3; A3285]

M0019 0, 0, 2, 0, 1, 2, 5, 0, 6, 1, 4, 2, 2, 5, 5, 0, 3, 6, 6, 1, 1, 4, 4, 2, 7, 2, 41, 5, 5, 5, 39, 0,
8, 3, 3, 6, 6, 6, 11, 1, 40, 1, 9, 4, 4, 4, 38, 2, 7, 7, 7, 2, 2, 41, 41, 5, 10, 5, 10, 5, 5, 39, 39, 0
Number of tripling steps to reach 1 in '$3x + 1$' problem. See Fig M2629. Ref UPNT E16.
rwg. [1,3; A6667]

M0020 1, 2, 0, 2, 0, 5, 6, 9, 0, 3, 1, 5, 9, 5, 9, 4, 2, 8, 5, 3, 9, 9, 7, 3, 8, 1, 6, 1, 5, 1, 1, 4, 4,
9, 9, 9, 0, 7, 6, 4, 9, 8, 6, 2, 9, 2, 3, 4, 0, 4, 9, 8, 8, 8, 1, 7, 9, 2, 2, 7, 1, 5, 5, 5, 5, 3, 4, 1, 8, 3
Decimal expansion of $\zeta(3) = \Sigma \ 1/n^3$. See Fig M2218. Ref FMR 1 84. [1,2; A2117]

M0021 1, 2, 0, 2, 1, 0, 0, 0, 0, 2, 1, 0, 1, 2, 0, 0, 0, 0, 0, 0, 0, 0, 0, 0, 0, 0, 0, 2, 1, 0, 1, 2, 0,
0, 0, 0, 1, 2, 0, 2, 1, 0, 0, 0, 0, 0, 0, 0, 0, 0, 0, 0, 0, 0, 0, 0, 0, 0, 0, 0, 0, 0, 0, 0, 0, 0, 0, 0
$C(2n,n)$ mod 3. [0,2; A6996]

M0022 1, 0, 1, 1, 0, 2, 0, 2, 1, 1, 2, 1, 2, 2, 2, 2, 3, 2, 4, 3, 4, 4, 4, 5, 5, 5, 6, 5, 6, 7, 6, 9, 7,
9, 9, 9, 11, 11, 11, 13, 12, 14, 15, 15, 17, 16, 18, 19, 20, 21, 23, 22, 25, 26, 27, 30, 29, 32
Partitions of n into distinct primes. Ref PNISI 21 186 55. PURB 107 285 57. [0,6; A0586,
N0004]

M0023 1, 0, 1, 1, 1, 1, 0, 2, 0, 2, 2, 4, 3, 2, 8, 1, 8, 8, 12, 11, 4, 25, 4, 24, 21, 40, 31, 16, 82,
14
From symmetric functions. Ref PLMS 23 309 23. [0,8; A2121, N0005]

M0024 1, 0, 0, 2, 0, 2, 3, 2, 6, 4, 9, 14, 11, 26, 29, 34, 62, 68, 99, 140, 169, 252, 322, 430,
607, 764, 1059, 1424, 1845, 2546
From symmetric functions. Ref PLMS 23 315 23. [0,4; A2125, N0006]

M0025 0, 1, 1, 2, 0, 3, 1, 1, 0, 3, 3, 2, 2, 4, 0, 5, 2, 2, 3, 3, 0, 1, 1, 3, 0, 2, 1, 1, 0, 4, 5, 2, 7,
4, 0, 1, 1, 2, 0, 3, 1, 1, 0, 3, 3, 2, 2, 4, 4, 5, 5, 2, 3, 3, 0, 1, 1, 3, 0, 2, 1, 1, 0, 4, 5, 3, 7, 4, 8
Values of Dawson's chess. Ref PCPS 52 517 56. WW 89. [0,4; A2187, N0007]

M0026 1, 1, 0, 0, 1, 0, 2, 0, 3, 1, 5, 2, 10, 5, 16, 13, 28, 24, 50, 46, 84, 87, 141, 153, 241,
266, 396, 459, 653, 766, 1070, 1267, 1725, 2075, 2762, 3342, 4397, 5330, 6918, 8432
Representation degeneracies for boson strings. Ref NUPH B274 544 86. [0,7; A5290]

M0027 1, 1, 0, 1, 0, 2, 0, 3, 1, 6, 2, 4, 9, 18, 8, 30, 16, 56, 32, 101, 64, 191, 128, 351, 256, 668, 512, 1257, 1026, 2402, 2056
Generalized Fibonacci numbers. Ref FQ 27 120 89. [1,6; A6209]

M0028 1, 1, 0, 0, 1, 0, 1, 0, 2, 0, 3, 1, 6, 2, 9, 6, 16
Bosonic string states. Ref CU86. [0,9; A5307]

M0029 2, 0, 3, 2, 6, 7, 14, 20, 35, 54, 90, 143, 234, 376, 611, 986, 1598, 2583, 4182, 6764, 10947, 17710, 28658, 46367, 75026, 121392, 196419, 317810, 514230, 832039, 1346270
Fibonacci numbers $\pm$ 1. Ref HO85a 129. [0,1; A7492]

M0030 0, 0, 1, 2, 0, 4, 0, 16, 16, 32, 64, 64, 256, 0, 768, 512, 2048, 3072, 4096, 12288, 4096, 40960, 16384, 114688, 131072, 262144, 589824, 393216, 2097152, 262144
Berstel sequence: $a(n+1)=2a(n)-4a(n-1)+4a(n-2)$. Ref Robe92 193. [0,4; A7420]

M0031 1, 1, 0, 1, 2, 0, 4, 1, 3, 2, 8, 0, 10, 4, 0, 2, 14, 3, 16, 2, 0, 8, 20, 0, 13, 10, 8, 4, 26, 0, 28, 4, 0, 14, 8, 3, 34, 16, 0, 2, 38, 0, 40, 8, 6, 20, 44, 0, 31, 13, 0, 10, 50, 8, 16, 4, 0, 26, 56
Moebius transform applied thrice to natural numbers. Ref EIS § 2.7. [1,5; A7432]

M0032 1, 1, 0, 1, 0, 0, 1, 2, 0, 4, 7, 0, 12, 8, 0, 80, 84, 0, 820
Number of cyclic Steiner triple systems of order $2n+1$. Ref GU70 504. [0,8; A2885, N0393]

M0033 0, 1, 2, 0, 4, 9, 18, 17, 0, 24, 35, 36, 12, 40, 11, 0, 13, 56, 30, 79, 45, 39, 67, 100, 0, 133, 83, 48, 53, 104, 138, 7, 163, 100, 26, 0, 28, 116, 217, 9, 248, 104, 17, 80, 79, 8, 139
The minimal sequence (from solving $n^3 - m^2 = a(n)$). Ref AB71 177. [1,3; A2938, N0008]

M0034 1, 0, 2, 0, 5, 9, 21, 42, 76, 174, 396, 888, 2023, 4345, 9921, 22566
Self-avoiding walks on square lattice. Ref JCT A13 181 72. [0,3; A2976]

M0035 2, 0, 8, 24, 72, 240, 896, 3640, 15688, 70512
Energy function for square lattice. Ref PHA 28 926 62. [0,1; A2909, N0009]

M0036 2, 0, 9, 4, 5, 5, 1, 4, 8, 1, 5, 4, 2, 3, 2, 6, 5, 9, 1, 4, 8, 2, 3, 8, 6, 5, 4, 0, 5, 7, 9, 3, 0, 2, 9, 6, 3, 8, 5, 7, 3, 0, 6, 1, 0, 5, 6, 2, 8, 2, 3, 9, 1, 8, 0, 3, 0, 4, 1, 2, 8, 5, 2, 9, 0, 4, 5, 3, 1
Decimal expansion of Wallis' number. Ref ScAm 250(4) 22 84. [1,1; A7493]

M0037 1, 1, 2, 0, 9, 35, 230, 1624, 13209, 120287, 1214674, 13469896, 162744945, 2128047987, 29943053062, 451123462672, 7245940789073, 123604151490591
Logarithmic numbers: expansion of $-\ln(1-x)e^{-x}$. Ref TMS 31 77 63. [1,3; A2741, N0010]

M0038 1, 0, 0, 2, 0, 12, 14, 90, 192, 792, 2148, 7716, 23262, 79512, 252054, 846628, 2753520, 9205800, 30371124, 101585544, 338095596
Magnetization for cubic lattice. Ref JMP 6 297 65. JPA 6 1511 73. DG74 420. [0,4; A2929, N0011]

M0039 1, 0, 0, 0, 2, 0, 28, 64, 39, 224, 884, 1368, 1350, 12272, 28752, 11944, 138873, 494184, 640856, 1111568, 7363194, 15488224, 1198848, 93506112
Low temperature antiferromagnetic susceptibility for b.c.c. lattice. Ref DG74 422. [0,5; A7218]

SEQUENCES BEGINNING . . ., 2, 1, . . .

M0040 1, 2, 1, 0, 1, 1, 1, 2, 2, 0, 2, 1, 2, 2, 1, 0, 0, 1, 0, 1, 1, 0, 2, 1, 1, 0, 1, 1, 1, 1, 1, 2, 1, 2, 0, 1, 2, 1, 1, 2, 2, 2, 0, 0, 2, 0, 1, 2, 0, 2, 2, 1, 0, 2, 1, 1, 2, 1, 2, 2, 2, 0, 2, 2, 1, 2, 2, 2, 1
Digits of integers written to base 3. [1,2; A3137]

M0041 0, 0, 1, 0, 1, 1, 0, 1, 1, 2, 1, 0, 1, 1, 1, 2, 3, 1, 0, 1, 1, 1, 2, 1, 3, 2, 3, 4, 1, 0, 1, 1, 1, 1, 2, 1, 3, 2, 3, 4, 5, 1, 0, 1, 1, 1, 1, 2, 1, 2, 3, 1, 4, 3, 2, 5, 3, 4, 5, 6, 1, 0, 1, 1, 1, 1, 1, 2, 1
Numerators of Farey series of orders 1, 2, ... Cf. M0081. Ref HW1 23. LE56 1 154. NZ66 141. [0,10; A6842]

M0042 0, 1, 0, 1, 0, 1, 0, 1, 2, 1, 0, 1, 2, 3, 2, 1, 0, 1, 2, 3, 4, 3, 2, 3, 2, 1, 0, 1, 2, 3, 4, 5, 6, 5, 4, 3, 2, 3, 2, 1, 0, 1, 2, 3, 4, 5, 4, 5, 6, 5, 6, 5, 6, 7, 6, 5, 4, 3, 2, 3, 2, 3, 2, 3, 2, 1, 2, 3, 4
Liouville's function $L(n)$. Ref PURB 3 48 50. [0,9; A2819, N0012]

M0043 1, 1, 2, 1, 0, 2, 0, 1, 3, 0, 2, 2, 0, 0, 0, 1, 2, 3, 2, 0, 0, 2, 0, 2, 1, 0, 4, 0, 0, 0, 0, 1, 4, 2, 0, 3, 0, 2, 0, 0, 2, 0, 2, 2, 0, 0, 0, 2, 1, 1, 4, 0, 0, 4, 0, 0, 4, 0, 2, 0, 0, 0, 0, 1, 0, 4, 2, 2, 0
Glaisher's J numbers. Ref MES 31 86 01. L1 9. [1,3; A2325, N0013]

G.f. of Moebius transform: $(1 + x^2) / (1 + x^4)$.

M0044 0, 0, 0, 1, 0, 2, 1, 0, 2, 1, 0, 2, 1, 3, 2, 1, 3, 2, 4, 3, 0, 4, 3, 0, 4, 3, 0, 4, 1, 2, 3, 1, 2, 4, 1, 2, 4, 1, 2, 4, 1, 5, 4, 1, 5, 4, 1, 5, 4, 1, 0, 2, 1, 0, 2, 1, 5, 2, 1, 3, 2, 1, 3, 2, 4, 3, 2, 4, 3
Values of Grundy's game. Ref PCPS 52 525 56. WW 97. [0,6; A2188, N0014]

M0045 1, 0, 1, 2, 1, 0, 2, 2, 0, 1, 3, 2, 0, 2, 3, 1, 0, 3, 3, 0, 2, 4, 2, 0, 3, 3, 0, 1, 4, 3, 0, 3, 5, 2, 0, 4, 4, 0, 2, 5, 3, 0, 3, 4, 1, 0, 4, 4, 0, 3, 6, 3, 0, 5, 5, 0, 2, 6, 4, 0, 4, 6, 2, 0, 5, 5, 0, 3, 6
Representations of n as a sum of Lucas numbers. Ref BR72 58. [1,4; A3263]

M0046 0, 2, 1, 0, 4, 2, 3, 1, 0, 6, 3, 2, 180, 4, 1, 0, 8, 4, 39, 2, 12, 42, 5, 1, 0, 10, 5, 24, 1820, 2, 273, 3, 4, 6, 1, 0, 12, 6, 4, 3, 320, 2, 531, 30, 24, 3588, 7, 1, 0, 14, 7, 90, 9100, 66
Solution to Pellian: smallest y such that $x^2 - ny^2 = 1$. Cf. M2240. Ref DE17. CAY 13 434. L1 55. [1,2; A2349, N0015]

M0047 1, 1, 1, 0, 0, 1, 1, 2, 1, 1, 0, 1, 0, 0, 1, 1, 2, 2, 2, 1, 0, 0, 1, 1, 2, 1, 2, 1, 2, 0, 0, 2, 1, 3, 2, 3, 1, 1, 1, 1, 3, 3, 3, 2, 2, 1, 0, 1, 2, 2, 3, 3, 3, 1, 1, 0, 0, 2, 2, 3, 2, 3, 2, 1, 1, 1, 2, 2, 3
Expansion of 1 / 1155th cyclotomic polynomial. [0,8; A7273]

M0048 0, 1, 1, 0, 1, 1, 0, 1, 1, 2, 1, 1, 0, 1, 1, 0, 1, 3, 2, 1, 1, 2, 1, 1, 0, 1, 1, 0, 1, 1, 0, 1, 1, 4, 3, 1, 2, 3, 1, 2, 1, 3, 2, 1, 1, 2, 1, 1, 0, 1, 1, 0, 1, 1, 0, 1, 1, 2, 1, 1, 0, 1, 1, 0, 1, 5, 4, 1, 3
$a(2n)=a(n)$, $a(2n+1)=a(n+1)-a(n)$. Ref AAMS 5 16 (809-10-185) 84. [0,10; A5590]

M0049   1, 1, 1, 1, 2, 1, 1, 1, 1, 1, 1, 1, 2, 1, 1, 1, 1, 2, 1, 1, 1, 1, 2, 1, 2, 1, 1, 1, 1, 2, 1, 1, 1,
1, 1, 1, 1, 2, 1, 1, 1, 1, 2, 1, 2, 1, 1, 1, 1, 1, 1, 1, 2, 1, 1, 1, 1, 2, 1, 2, 1, 1, 1, 1, 2, 1, 1, 1, 1
A continued fraction. Ref JNT 11 217 79. [0,5; A6466]

M0050   1, 1, 1, 1, 2, 1, 1, 1, 1, 1, 1, 2, 2, 1, 2, 2, 2, 1, 1, 2, 1, 2, 2, 3, 3, 2, 3, 2, 2, 2, 1, 2, 1,
3, 3, 3, 4, 3, 3, 2, 3, 3, 1, 2, 3, 4, 4, 3, 4, 3, 4, 3, 3, 3, 3, 3, 4, 5, 5, 3, 3, 4, 4, 3, 2, 4, 3, 4, 4
Partitions of n into not more than 5 pentagonal numbers. Ref LINR 1 14 46. MOC 2 301
47. [1,5; A2637, N0016]

M0051   1, 1, 1, 1, 1, 1, 1, 1, 1, 1, 1, 2, 1, 1, 1, 1, 1, 2, 1, 2, 1, 1, 1, 1, 2, 1, 1, 1, 1, 1, 2, 2,
1, 1, 2, 1, 1, 1, 2, 1, 2, 1, 4, 1, 1, 2, 1, 1, 2, 2, 1, 1, 1, 1, 1, 2, 1, 1, 1, 1, 2, 2, 1, 1, 1, 2, 2, 3
Class number of real quadratic field with discriminant n. Ref BU89 236. [2,12; A3652]

M0052   0, 1, 1, 1, 2, 1, 1, 1, 1, 2, 2, 1, 1, 1, 2, 1, 1, 1, 1, 2, 1, 2, 1, 1, 3, 1, 1, 1, 1, 2, 2, 1, 2,
1, 2, 1, 1, 1, 1, 2, 2, 1, 2, 2, 2, 1, 1, 1, 1, 3, 1, 1, 1, 1, 2, 1, 1, 1, 1, 2, 2, 2, 1, 1, 2, 2, 1, 1, 1
Leonardo logarithm of n. Ref HM68. MOC 23 460 69. ACA 16 109 69. [1,5; A1179,
N0017]

M0053   1, 1, 1, 2, 1, 1, 1, 1, 2, 2, 1, 2, 2, 1, 1, 2, 2, 3, 2, 2, 2, 2, 1, 1, 3, 3, 3, 3, 2, 2, 2, 1, 3,
4, 2, 4, 3, 3, 2, 2, 3, 4, 3, 2, 4, 2, 2, 4, 5, 3, 5, 3, 5, 3, 1, 4, 5, 3, 3, 4, 3, 4, 2, 4, 6, 4, 4, 4
Partitions of n into not more than 4 squares. Ref LINR 1 10 46. MOC 2 301 47. [1,4;
A2635, N0018]

M0054   1, 1, 1, 1, 1, 2, 1, 1, 1, 2, 1, 1, 1, 1, 1, 2, 1, 2, 1, 1, 2, 2, 1, 1, 2, 1, 2, 1, 1, 1, 2, 1, 2,
1, 2, 1, 1, 1, 2, 2, 1, 1, 2, 1, 1, 2, 1, 2, 3, 4, 1, 2, 1, 2, 1, 2, 1, 1, 2, 1, 1, 2, 1, 2, 2, 1, 1, 2, 2
Class number of $Q(\sqrt{n})$, n squarefree. Ref BU89 236. [2,6; A3649]

M0055   1, 1, 1, 1, 1, 1, 1, 1, 1, 1, 1, 1, 2, 1, 1, 1, 2, 1, 1, 1, 1, 2, 1, 1, 1, 1, 1, 4, 1, 1, 1, 2, 1,
1, 1, 2, 1, 1, 1, 2, 1, 1, 1, 2, 3, 1, 1, 1, 1, 1, 3, 2, 1, 2, 1, 1, 2, 1, 1, 2, 1, 1, 1, 3, 1, 1, 1, 2, 1
Class number of real quadratic field with discriminant $4n+1$. Cf. M0066. Ref BU89 236.
[1,12; A3650]

M0056   1, 1, 1, 1, 2, 1, 1, 1, 2, 1, 2, 1, 2, 2, 1, 1, 2, 1, 2, 2, 1, 2, 1, 2, 1, 2, 1, 3, 1, 1, 2,
2, 2, 2, 1, 2, 2, 2, 1, 3, 1, 2, 2, 2, 1, 2, 1, 2, 2, 2, 1, 2, 2, 2, 2, 2, 1, 3, 1, 2, 2, 1, 2, 3, 1, 2, 2
Number of distinct primes dividing n. Ref AS1 844. [1,6; A1221, N0019]

M0057   0, 1, 1, 1, 1, 1, 1, 1, 2, 1, 1, 1, 2, 1, 2, 2, 2, 2, 2, 2, 2, 2, 3, 4, 2, 3, 3, 3, 3, 3, 4, 3
Minimal norm of n-dimensional unimodular lattice. Ref BAMS 23 384 90. SPLAG xxiv.
[0,9; A5136]

M0058   2, 1, 1, 1, 2, 1, 2, 3, 3, 3, 1, 1, 3, 4, 2, 1, 3, 4, 1, 5, 3, 5, 5, 2, 4, 5, 3, 4, 2, 6, 1, 7, 7,
1, 3, 7, 5, 4, 5, 7, 8, 6, 8, 7, 7, 6, 3, 7, 9, 7, 9, 8, 1, 3, 9, 5, 6, 3, 7, 10, 1, 6, 4, 10, 7, 9, 5, 9
y such that $p = (x^2 + 27y^2)/4$. Cf. M3754. Ref CU04 1. L1 55. [3,1; A2339, N0043]

M0059   1, 2, 1, 1, 1, 2, 1, 2, 8, 1, 25, 1, 5, 1, 22, 1, 8, 1, 1, 9, 1, 1, 4, 1, 2, 1, 2, 1, 2, 2, 1, 1,
1, 1, 2, 1, 6, 2, 46, 1, 12, 1, 32, 1, 2, 3, 2, 3, 55, 1, 12, 3, 8, 1, 1, 11, 1, 4, 1, 1, 1, 2, 1, 1, 7
Continued fraction for fifth root of 5. [1,2; A2951]

M0060 1, 1, 1, 2, 1, 1, 1, 2, 2, 1, 1, 2, 1, 1, 1, 2, 1, 2, 1, 2, 1, 1, 1, 2, 2, 1, 2, 2, 1, 1, 1, 2, 1,
1, 1, 2, 1, 1, 1, 2, 1, 1, 1, 2, 2, 1, 1, 2, 2, 2, 1, 2, 1, 2, 1, 2, 1, 1, 1, 2, 1, 1, 1, 2, 2, 1, 1, 1, 2, 1
1 if n is squarefree, else 2. [1,4; A7424]

M0061 1, 1, 1, 2, 1, 1, 1, 2, 2, 1, 1, 2, 2, 1, 1, 1, 1, 1, 2, 2, 1, 1, 2, 2, 2, 1, 1, 1, 2, 1, 2,
2, 1, 1, 1, 2, 2, 1, 4, 1, 2, 1, 2, 2, 1, 1, 4, 2, 1, 2, 1, 4, 2, 1, 2, 1, 1, 2, 2, 2, 2, 2, 1, 1, 2, 2, 2
Number of genera of quadratic field with discriminant $-4n+1$. Cf. 3642. Ref BU89 224.
[1,4; A3641]

M0062 0, 1, 1, 2, 1, 1, 1, 2, 3, 3, 1, 6, 1, 4, 4, 5, 1, 3, 1, 6, 2, 5, 1, 4, 2, 6, 2, 1, 1, 14, 1, 2, 5,
7, 2, 3, 1, 6, 2, 3, 1, 13, 1, 4, 6, 7, 1, 5, 3, 2, 3, 8, 1, 12, 2, 4, 2, 3, 1, 10, 1, 8, 2, 3, 2, 11, 1
Length of aliquot sequence for n. See Fig M0062. Ref MA71 558. [1,4; A3023]

‖‖‖

Figure M0062. ALIQUOT SEQUENCES.

Let $s(n)$ denote the sum of the **aliquot divisors** of n, i.e. the divisors of n other than n itself.
A number is called **deficient** (M0514), **perfect** (M4186), or **abundant** (M4825) according as
$s(n)$ is less than, equal to, or greater than n. Before the advent of computers it was believed that
starting at any number n, the sequence $n, s(n), s(s(n)), s(s(s(n))),\ldots$ would eventually
cycle. Sequence M0062 gives the number of steps for the sequence to cycle, beginning at
$n = 1, 2, 3,\ldots$. For example, $n = 30$ takes 14 steps before it cycles:

$$30 \to 42 \to 54 \to 66 \to 78 \to 90 \to 144 \to 259 \to$$
$$45 \to 33 \to 15 \to 9 \to 4 \to 3 \to 1 \to 1 \to \ldots.$$

However, it is now conjectured that there are many numbers for which this sequence never
terminates, although so far no such number has been found! (See [UPNT B6].) So strictly
speaking M0062 may not be a proper sequence. **Amicable** pairs (M5414 & M5435), such as
220 and 284, belong to cycles of length 2: $s(220) = 284$, $s(284) = 220$.

‖‖‖

M0063 1, 1, 1, 2, 1, 1, 1, 3, 2, 1, 1, 2, 1, 1, 1, 4, 1, 2, 1, 2, 1, 1, 1, 3, 2, 1, 3, 2, 1, 1, 1, 5, 1,
1, 1, 5
Number of numbers $\le n$ with same prime factors as n: $a(\Pi p^r)=\Pi p^{r-1}$. Ref AMM 78
680 71. [1,4; A5361]

M0064 1, 1, 1, 2, 1, 1, 1, 3, 2, 1, 1, 2, 1, 1, 1, 5, 1, 2, 1, 2, 1, 1, 1, 3, 2, 1, 3, 2, 1, 1, 1, 7, 1,
1, 1, 4, 1, 1, 1, 3, 1, 1, 1, 2, 2, 1, 1, 5, 2, 2, 1, 2, 1, 3, 1, 3, 1, 1, 1, 2, 1, 1, 2, 11, 1, 1, 1, 2, 1
Number of Abelian groups of order n. Ref MZT 56 21 52. [1,4; A0688, N0020]

M0065 2, 1, 1, 1, 3, 9, 23, 53, 115, 237, 457, 801, 1213, 1389, 445, 3667, 15081, 41335,
95059, 195769, 370803, 652463, 1063359, 1570205, 1961755, 1560269, 1401991
Inverse binomial transform of primes. Ref EIS § 2.7. [1,1; A7442]

M0066 1, 1, 1, 1, 2, 1, 1, 2, 1, 1, 1, 2, 2, 1, 2, 2, 1, 2, 2, 1, 1, 1, 2, 2, 1, 1, 2, 1, 2, 1, 2, 2,
3, 4, 1, 1, 2, 2, 1, 2, 2, 1, 2, 1, 2, 2, 2, 2, 1, 2, 2, 2, 1, 4, 1, 1, 2, 1, 3, 2, 2, 1, 2, 2, 1, 2, 1, 1
Class number of real quadratic field with discriminant $4n$, $n\equiv 2,3 \pmod 4$. Cf. M0055. Ref
BU89 241. [2,5; A3651]

M0067 2, 1, 1, 2, 1, 2, 1, 1, 2, 2, 1, 1, 1, 2, 2, 2, 1, 1, 1, 1, 2, 2, 1, 2, 2, 2, 1, 1, 1, 1, 1, 1, 2,
2, 2, 2, 1, 2, 2, 2, 1, 1, 1, 1, 1, 2, 1, 1, 2, 1, 2, 1, 1, 2, 2, 2, 2, 2, 1, 2, 1, 2, 1, 2, 2, 1, 1, 1, 2
Liouville's function: parity of number of primes dividing n (with multplicity). Ref MOC 14
311 60. Robe92 279. [1,1; A7421]

M0068 2, 1, 1, 2, 1, 2, 1, 2, 1, 1, 2, 1, 2, 1, 1, 2, 1, 2, 1, 1, 2, 1, 2, 1, 2, 1, 1, 2, 1, 2, 1, 1, 2,
1, 2, 1, 2, 1, 1, 2, 1, 2, 1, 1, 2, 1, 2, 1, 2, 1, 1, 2, 1, 2, 1, 1, 2, 1, 2, 1, 1, 2, 1, 2, 1, 2, 1, 2, 1, 1
Fixed under $1 \to 21$, $2 \to 211$. See Fig M0436. Ref MMAG 36 180 63. [1,1; A1030,
N0021]

M0069 1, 1, 2, 1, 1, 2, 1, 2, 1, 6, 1, 2, 3, 2, 1, 1, 1, 1, 2, 8, 2, 1, 8, 2, 1, 2, 1, 3, 4, 18, 1, 2, 1,
1, 10, 3, 1, 2, 1, 1, 1, 2, 2, 1, 2, 1, 6, 1, 3, 8, 2, 10, 5, 16, 2, 1, 2, 3, 4, 3, 1, 3, 2, 2, 1, 11, 16
$(p-1)/x$, where $2^x \equiv 1 \mod p$. Ref KAB 35 666 13. Krai24 1 131. L1 10. [3,3; A1917,
N0022]

M0070 1, 2, 1, 1, 2, 1, 2, 2, 1, 2, 2, 1, 1, 1, 2, 1, 1, 2, 2, 1, 2, 1, 1, 2, 1, 2, 2, 1, 1, 2, 1, 1, 2, 1,
2, 2, 1, 2, 2, 1, 1, 1, 2, 1, 2, 2, 1, 2, 1, 1, 2, 1, 1, 2, 2, 1, 2, 2, 1, 1, 1, 2, 1, 2, 2, 1, 2, 2, 1, 1, 2, 1
At step n, if last digit is x, append $a(n)$ copies of $3-x$. [1,2; A6928]

M0071 1, 2, 1, 1, 2, 1, 2, 3, 2, 1, 3, 2, 2, 3, 2, 2, 4, 2, 3, 4, 2, 3, 5, 1, 4, 5, 2, 3, 5, 1, 3, 5, 3,
3, 5, 3, 5, 7, 3, 5, 7, 4, 4, 7, 3, 3, 7, 4, 3, 9, 5, 3, 7, 5, 3, 8, 5, 4, 8, 5, 3, 7, 5, 3, 9, 4, 3, 12, 6
Decompositions of $2n$ into sum of 2 lucky numbers. Ref MMAG 29 119 59. SS64. MOC
19 332 65. [1,2; A2850, N0023]

M0072 2, 1, 1, 2, 1, 4, 2, 0, 1, 0, 7, 3, 8, 30, 8, 23, 2, 14, 10, 28, 2, 18, 4, 24, 8, 12, 4, 0, 28,
33, 28, 44, 40, 2, 22, 16, 18, 25, 7, 22, 12, 12, 10, 30, 8, 42, 27, 34, 32, 47, 27, 39, 25
Coefficient of x^p in $\Pi(1-x^k)^2(1-x^{11k})^2$. Ref JRAM 221 218 66. [2,1; A2070, N0024]

M0073 0, 1, 2, 1, 1, 2, 2, 1, 1, 0, 3, 2, 1, 2, 1, 1, 2, 2, 2, 2, 2, 1, 3, 1, 0, 1, 3, 2, 2, 2, 1, 3, 2,
0, 2, 1, 1, 4, 2, 1, 3, 2, 2, 2, 2, 1, 4, 2, 1, 1, 3, 3, 3, 3, 4, 1, 5, 3, 2, 4, 3, 2, 4, 2, 1, 2, 4, 2, 3
Partitions of n into a prime and a square. Ref AMN 47 30 43. [1,3; A2471, N0025]

M0074 1, 2, 1, 1, 2, 2, 1, 2, 1, 1, 2, 1, 2, 2, 1, 1, 2, 1, 1, 1, 2, 2, 1, 2, 1, 1, 2, 2, 1, 1, 1, 2, 1,
1, 2, 2, 1, 2, 1, 1, 2, 1, 2, 2, 1, 1, 2, 1, 1, 1, 2, 2, 1, 2, 1, 1, 2, 2, 1, 2, 2, 2, 1, 1, 2, 1, 2, 2, 1
The Linus sequence: $a(n)$ avoids longest repetition. See Fig M0126. Ref nsh. [1,2; A6345]

M0075 1, 2, 1, 1, 2, 2, 1, 2, 1, 2, 2, 2, 1, 1, 1, 2, 1, 1, 1, 1, 2, 2, 2, 2, 1, 2, 2, 1, 1, 2, 1, 2, 1,
1, 2, 1, 1, 2, 2, 2, 1, 2, 1, 1, 1, 2, 2, 1, 1, 1, 1, 2, 1, 2, 2, 1, 2, 2, 2, 2, 2, 1, 1, 2, 2, 1, 1, 2
A maximally unpredictable sequence. Ref AMM 99 374 92. [0,2; A7061]

M0076 1, 1, 2, 1, 1, 2, 2, 1, 2, 2, 1, 3, 2, 1, 2, 3, 2, 2, 2, 1, 4, 3, 2, 2, 2, 2, 3, 3, 1, 4, 4, 2, 2,
3, 2, 3, 4, 2, 3, 3, 2, 4, 3, 2, 4, 4, 2, 4, 4, 1, 4, 5, 1, 2, 3, 4, 6, 4, 3, 2, 5, 2, 3, 3, 3, 6, 5, 2, 2
Partitions of n into not more than 3 triangular numbers. Ref LINR 1 12 46. MOC 2 301 47.
[1,3; A2636, N0027]

M0077 1, 1, 2, 1, 1, 2, 2, 1, 2, 1, 2, 1, 2, 1, 2, 2, 4, 1, 2, 2, 1, 2, 2, 2, 2, 1, 2, 2, 1, 1, 2, 4,
1, 1, 4, 2, 2, 2, 4, 1, 4, 2, 4, 1, 2, 4, 1, 2, 4, 4, 2, 1, 2, 1, 2, 2, 2, 1, 1, 2, 4, 4, 2, 2, 2, 4, 4, 3
Class number of quadratic forms with discriminant n. Ref BU89 236. [2,3; A3646]

M0078 1, 1, 1, 2, 1, 1, 2, 3, 2, 1, 2, 3, 1, 2, 3, 4, 2, 2, 2, 4, 2, 1, 4, 5, 2, 2, 3, 4, 2, 2, 4, 6, 2, 2, 4, 6, 1, 2, 5, 5, 3, 2, 2, 6, 3, 2, 6, 7, 3, 3, 4, 4, 2, 3, 5, 8, 2, 1, 5, 7, 2, 3, 7, 7, 3, 3, 2, 7, 3
Orbits of Lorentzian modular group. Ref FP91. [1,4; A5793]

M0079 1, 1, 1, 2, 1, 1, 2, 3, 2, 1, 3, 3, 1, 2, 4, 4, 2, 2, 3, 5, 2, 1, 6, 5, 2, 3, 4, 4, 3, 2, 6, 7, 2, 2, 6, 7, 1, 3, 8, 5
Orbits of Lorentzian modular group. Ref FP91. [1,4; A5794]

M0080 2, 1, 1, 3, 2, 1, 1, 2, 4, 2, 1, 2, 2, 1, 2, 5, 2, 2, 1, 3, 2, 1, 1, 3, 6, 2, 2, 2, 2, 2, 1, 3, 2, 2, 2, 7, 3, 1, 2, 4, 2, 2, 1, 2, 4, 1, 1, 4, 6, 3, 2, 3, 2, 2, 2, 3, 2, 2, 1, 4, 2, 1, 3, 8, 5, 2, 1, 3, 2
Reduced indefinite binary forms of determinant − n. Ref SPLAG 362. [1,1; A6375]

M0081 1, 1, 1, 1, 2, 1, 1, 3, 2, 3, 1, 1, 4, 3, 2, 3, 4, 1, 1, 5, 4, 3, 5, 2, 5, 3, 4, 5, 1, 1, 6, 5, 4, 3, 5, 2, 5, 3, 4, 5, 6, 1, 1, 7, 6, 5, 4, 7, 3, 5, 7, 2, 7, 5, 3, 7, 4, 5, 6, 7, 1, 1, 8, 7, 6, 5, 4, 7, 3
Denominators of Farey series of orders 1, 2, ... Cf. M0041. Ref HW1 23. LE56 1 154. NZ66 141. [0,5; A6843]

M0082 1, 1, 1, 1, 2, 1, 1, 3, 3, 1, 1, 4, 6, 4, 1, 1, 5, 10, 10, 5, 1, 1, 6, 15, 20, 15, 6, 1, 1, 7, 21, 35, 35, 21, 7, 1, 1, 8, 28, 56, 70, 56, 28, 8, 1, 1, 9, 36, 84, 126, 126, 84, 36, 9, 1, 1, 10
Pascal triangle read by rows. See Fig M1645. [0,5; A7318]

M0083 1, 1, 1, 1, 1, 1, 2, 1, 1, 8, 1, 1, 1, 21, 7, 1, 1
Translation planes of order q, q = prime power. Ref LNM 158. MaRo94. [2,7; A7375]

M0084 1, 1, 1, 2, 1, 2, 1, 1, 1, 2, 1, 2, 2, 1, 2, 2, 1, 2, 1, 1, 4, 1, 1, 2, 2, 1, 2, 4, 1, 1, 2, 4, 1, 2, 1, 2, 1, 1, 2, 4, 2, 2, 2, 4, 3, 1, 2, 1, 2, 2, 3, 2, 1, 4, 1, 1, 4, 1, 2, 4, 2, 1, 1, 6, 2, 1, 2, 4, 1
Class number of real quadratic forms with discriminant $4n + 1$. Cf. 3648. Ref BU89 236. [1,4; A3647]

M0085 1, 1, 1, 1, 1, 1, 1, 1, 1, 2, 1, 2, 1, 1, 1, 3, 3, 3, 1, 3, 1, 2, 1, 1, 3, 2, 4, 1, 1, 2, 5, 1, 3, 2, 1, 1, 3, 3, 4, 2, 2, 1, 2, 1, 7, 2, 5, 2, 1, 5, 2, 3, 1, 5, 3, 6, 1, 1, 4, 4, 2, 7, 1, 5, 1, 3, 3, 3, 3
Classes per genus in Q($\sqrt{-n}$), n squarefree. Ref BU89 224. [1,10; A3639]

M0086 1, 2, 1, 2, 1, 1, 2, 1, 2, 1, 1, 2, 1, 2, 1, 2, 1, 1, 2, 1, 2, 1, 1, 2, 1, 2, 1, 2, 1, 1, 2, 1, 2, 1, 1, 2, 1, 2, 1, 1, 2, 1, 2, 1, 1, 2, 1, 2, 1, 2, 1, 1, 2, 1, 2, 1, 1, 2, 1, 2, 1, 2, 1, 1, 2, 1, 2, 1, 1, 2, 1, 2, 1, 1
$[(n+1)\sqrt{2}] - [n\sqrt{2}]$. Ref drh. [1,2; A6337]

M0087 2, 1, 2, 1, 1, 2, 1, 2, 1, 2, 1, 1, 2, 1, 2, 1, 1, 2, 1, 2, 1, 2, 1, 1, 2, 1, 2, 1, 1, 2, 1, 2, 1, 1, 2, 1, 2, 1, 1, 2, 1, 2, 1, 2, 1, 1, 2, 1, 2, 1, 1, 2, 1, 2, 1, 2, 1, 1, 2, 1, 2, 1, 1, 2, 1, 2, 1
$[(n+1)\sqrt{2}+\frac{1}{2}] - [n\sqrt{2}+\frac{1}{2}]$. Ref drh. [1,1; A6338]

M0088 2, 1, 2, 1, 1, 4, 1, 1, 6, 1, 1, 8, 1, 1, 10, 1, 1, 12, 1, 1, 14, 1, 1, 16, 1, 1, 18, 1, 1, 20, 1, 1, 22, 1, 1, 24, 1, 1, 26, 1, 1, 28, 1, 1, 30, 1, 1, 32, 1, 1, 34, 1, 1, 36, 1, 1, 38, 1, 1, 40, 1
Continued fraction for e. Ref PE29 134. [1,1; A3417]

M0089 1, 2, 1
A simple periodic sequence. [0,2; A0034]

M0090   1, 1, 1, 1, 1, 2, 1, 2, 1, 2, 1, 2, 2, 2, 1, 1, 2, 2, 2, 2, 1, 1, 2, 1, 1, 1, 4, 2, 2, 2, 2, 1, 2,
    1, 2, 2, 2, 2, 4, 2, 1, 1, 4, 2, 1, 2, 2, 1, 2, 2, 2, 1, 2, 1, 4, 1, 2, 2, 2, 1, 4, 1, 2, 1, 2, 2, 2, 1, 1
Number of genera of quadratic field with discriminant $-n$. Ref BU89 224. [3,6; A3640]

M0091   1, 2, 1, 2, 1, 2, 2, 0, 2, 2, 1, 0, 0, 2, 3, 2, 2, 0, 0, 2, 2, 0, 0, 2, 1, 0, 2, 2, 2, 2, 1, 2, 0,
    2, 2, 2, 2, 0, 2, 0, 4, 0, 0, 0, 1, 2, 0, 0, 2, 0, 2, 2, 1, 2, 0, 2, 2, 0, 0, 2, 0, 2, 0, 2, 2, 0, 4, 0, 0
Expansion of $\Pi\ (1-x^k)^2$. Ref PLMS 21 190 1889. [0,2; A2107, N0028]

M0092   1, 2, 1, 2, 1, 2, 2, 0, 2, 2, 1, 2, 4, 4, 1, 4, 2, 4, 0, 2, 2, 2, 1, 0, 4, 8, 5, 4, 0, 2, 7, 8, 1,
    4, 2, 4, 3, 0, 4, 0, 8, 4, 6, 2, 2, 2, 8, 4, 3, 8, 2, 8, 6, 10, 1, 0, 0, 0, 5, 2, 12, 14, 4, 8, 4, 2, 7, 4
Expansion of $\Pi(1-x^k)^2(1-x^{11k})^2$, $k \geq 1$. Ref JRAM 221 218 66. [0,2; A6571]

M0093   1, 2, 1, 2, 1, 2, 2, 3, 4, 1, 5, 3, 6, 8, 3, 4, 8, 3, 0, 2, 8, 4, 4, 13, 9, 5, 18, 2, 2, 8, 3, 10,
    0, 4, 2, 19, 14, 7, 8, 0, 20, 4, 1, 8, 2, 15, 7, 8, 26, 10, 26, 18, 10, 2, 10, 28, 29
Related to partitions. Ref JRAM 179 128 38. [0,2; A2300, N0029]

M0094   0, 1, 1, 2, 1, 2, 1, 3, 2, 2, 1, 3, 1, 2, 2, 4, 1, 3, 1, 3, 2, 2, 1, 4, 2, 2, 3, 3, 1, 3, 1, 5, 2,
    2, 2, 4, 1, 2, 2, 4, 1, 3, 1, 3, 3, 2, 1, 5, 2, 3, 2, 3, 1, 4, 2, 4, 2, 2, 1, 4, 1, 2, 3, 6, 2, 3, 1, 3, 2
Number of primes dividing n. Ref AS1 844. [1,4; A1222, N0031]

M0095   1, 1, 1, 2, 1, 2, 1, 3, 2, 2, 1, 4, 1, 2, 2, 5, 1, 4, 1, 4, 2, 2, 1, 7, 2, 2, 3, 4, 1, 5, 1, 7, 2,
    2, 2, 9, 1, 2, 2, 7, 1, 5, 1, 4, 4, 2, 1, 12, 2, 4, 2, 4, 1, 7, 2, 7, 2, 2, 1, 11, 1, 2, 4, 11, 2, 5, 1, 4
Number of ways of factoring n. Ref AS1 844. [1,4; A1055, N0032]

M0096   1, 1, 2, 1, 2, 1, 4, 2, 5, 3, 5, 4, 1, 3, 7, 4, 7, 6, 2, 9, 7, 1, 2, 8, 4, 1, 10, 9, 5, 2, 12, 11,
    9, 5, 8, 7, 10, 6, 1, 3, 14, 12, 7, 4, 10, 5, 11, 10, 14, 13, 1, 8, 5, 17, 16, 4, 13, 6, 12, 1, 5, 15
y such that $p = x^2 + y^2$, $x \leq y$. Cf. M0462. Ref CU04 1. AMM 56 526 49. [2,3; A2331,
N0033]

M0097   1, 1, 2, 1, 2, 1, 4, 3, 13, 5, 1, 1, 8, 1, 2, 4, 1, 1, 40, 1, 11, 3, 7, 1, 7, 1, 1, 5, 1, 49, 4,
    1, 65, 1, 4, 7, 11, 1, 399, 2, 1, 3, 2, 1, 2, 1, 5, 3, 2, 1, 10, 1, 1, 1, 1, 2, 1, 1, 3, 1, 4, 1, 1, 2, 5
Continued fraction for Euler's constant γ. Ref LE59. MOC 25 387 71. [0,3; A2852, N0034]

M0098   1, 1, 1, 2, 1, 2, 1, 5, 2, 2, 1, 5, 1, 2, 1, 14, 1, 5, 1, 5, 2, 2, 1, 15, 2, 2, 5, 4, 1, 4, 1, 51,
    1, 2, 1, 14, 1, 2, 2, 14, 1, 6, 1, 4, 2, 2, 1, 52, 2, 5, 1, 5, 1, 15, 2, 13, 2, 2, 1, 13, 1, 2, 4, 267
Number of groups of order n. Ref AJM 52 617 30. HS64. CM84 134. [1,4; A0001, N0035]

M0099   2, 1, 2, 2, 1, 2, 1, 2, 2, 1, 2, 2, 1, 2, 1, 2, 2, 1, 2, 1, 2, 2, 1, 2, 1, 2, 2, 1, 2, 2, 1, 2, 2,
    1, 2, 1, 2, 2, 1, 2, 1, 2, 2, 1, 2, 2, 1, 2, 1, 2, 2, 1, 2, 1, 2, 2, 1, 2, 2, 1, 2, 1, 2, 2, 1, 2, 2, 1
$[(n+1)\tau] - [n\tau]$, $\tau = (1+\sqrt5)/2$. Ref AMM 64 198 57. drh. IJCM 26 36 88. [1,1;
A1468, N0036]

M0100   1, 2, 1, 2, 2, 1, 2, 2, 1, 2, 1, 2, 2, 1, 2, 2, 1, 2, 1, 2, 2, 1, 2, 1, 2, 2, 1, 2, 2, 1, 2, 1, 2,
    2, 1, 2, 1, 2, 2, 1, 2, 2, 1, 2, 1, 2, 2, 1, 2, 2, 1, 2, 1, 2, 2, 1, 2, 1, 2, 2, 1, 2, 2, 1, 2, 2, 1, 2, 2, 1
$[(n+1)\tau + \frac{1}{2}] - [n\tau + \frac{1}{2}]$, $\tau = (1+\sqrt5)/2$. Ref drh. [1,2; A6340]

M0101 1, 1, 1, 2, 1, 2, 2, 1, 3, 2, 2, 3, 1, 3, 3, 2, 4, 2, 3, 3, 1, 4, 3, 3, 5, 2, 4, 4, 2, 5, 3, 3, 4,
1, 4, 4, 3, 6, 3, 5, 5, 2, 6, 4, 4, 6, 2, 5, 5, 3, 6, 3, 4, 4, 1, 5, 4, 4, 7, 3, 6, 6, 3, 8, 5, 5, 7, 2, 6
Representations of n as a sum of distinct Fibonacci numbers. Ref FQ 4 305 66. BR72 54.
[0,4; A0119, N0037]

M0102 1, 0, 1, 1, 2, 1, 2, 2, 2, 1, 2, 2, 3, 2, 1, 1, 2, 2, 3, 3, 2, 1, 2, 2, 2, 1, 1, 1, 2, 3, 4, 4, 3,
2, 1, 1, 2, 1, 0, 0, 1, 2, 3, 3, 3, 2, 3, 3, 3, 3, 2, 2, 3, 3, 2, 2, 1, 0, 1, 1, 2, 1, 1, 1, 0, 1, 2, 2, 1
Merten's function: $\Sigma\mu(k)$, $k \le n$, μ = Moebius function. Ref WIEN 106(2A) 843 1897.
L1 7. [1,5; A2321, N0038]

M0103 1, 1, 2, 1, 2, 2, 2, 1, 2, 2, 3, 2, 3, 2, 2, 1, 2, 2, 3, 2, 3, 3, 3, 2, 3, 3, 3, 2, 3, 2, 2, 1, 2,
2, 3, 2, 3, 3, 3, 2, 3, 3, 4, 3, 4, 3, 3, 2, 3, 3, 4, 3, 4, 3, 3, 2, 3, 3, 3, 2, 3, 2, 2, 1, 2, 2, 3, 2, 3
Optimal cost function between two processors at distance n. Ref Algo93 158. [1,3; A7302]

M0104 0, 0, 1, 2, 1, 2, 2, 2, 2, 3, 3, 3, 2, 3, 2, 4, 4, 2, 3, 4, 3, 4, 5, 4, 3, 5, 3, 4, 6, 3, 5, 6, 2,
5, 6, 5, 5, 7, 4, 5, 8, 5, 4, 9, 4, 5, 7, 3, 6, 8, 5, 6, 8, 6, 7, 10, 6, 6, 12, 4, 5, 10, 3, 7, 9, 6, 5, 8
From Goldbach problem: decompositions of $2n$ into sum of two odd primes. Ref GR38 19.
L1 80. [1,4; A2375, N0040]

M0105 0, 1, 1, 2, 1, 2, 2, 3, 1, 2, 2, 3, 2, 3, 3, 4, 1, 2, 2, 3, 2, 3, 3, 4, 2, 3, 3, 4, 3, 4, 4, 5, 1,
2, 2, 3, 2, 3, 3, 4, 2, 3, 3, 4, 3, 4, 4, 5, 2, 3, 3, 4, 3, 4, 4, 5, 3, 4, 4, 5, 4, 5, 5, 6, 1, 2, 2, 3, 2
Number of 1's in binary expansion of n. Ref FQ 4 374 66. ANY 175 177 70. [0,4; A0120,
N0041]

M0106 0, 0, 0, 0, 1, 0, 2, 1, 2, 2, 4, 1, 5, 4, 4, 4, 7, 4, 8, 5, 7, 8, 10, 5, 10, 10, 10, 9, 13, 8,
14, 11, 13, 14, 14, 10, 17, 16, 16, 13, 19, 14, 20, 17, 17, 20, 22, 15, 22, 20, 22, 21, 25, 20
$[n/2] + 1$ − number of divisors of n. Ref NE72 186. [1,7; A3165]

M0107 2, 1, 2, 2, 6, 6, 18, 16, 48, 60, 176, 144, 630, 756, 1800, 2048, 7710, 7776, 27594,
24000, 84672
Primitive polynomials of degree n over GF(2). Ref BE65 296. BE68 84. [1,1; A0020,
N0132]

M0108 1, 2, 1, 2, 3, 1, 2, 1, 2, 3, 1, 2, 3, 4, 1, 2, 1, 2, 3, 1, 2, 1, 2, 3, 1, 2, 3, 4, 1, 2, 1, 2, 3,
1, 2, 3, 4, 1, 2, 3, 4, 5, 1, 2, 1, 2, 3, 1, 2, 1, 2, 3, 1, 2, 3, 4, 1, 2, 1, 2, 3, 1, 2, 1, 2, 3, 1, 2, 3
Generated by $1 \rightarrow 12, 2 \rightarrow 123, 3 \rightarrow 1234$, etc. starting from 1. Ref jpropp. [1,2; A7001]

M0109 1, 1, 1, 1, 2, 1, 2, 3, 1, 3, 1, 4, 1, 1, 2, 4, 5, 5, 1, 2, 3, 6, 3, 1, 5, 2, 4, 1, 7, 5, 3, 5, 7,
1, 5, 7, 3, 1, 4, 5, 6, 8, 1, 2, 7, 9, 4, 5, 3, 5, 2, 1, 9, 5, 6, 7, 10, 11, 3, 1, 4, 11, 6, 7, 8, 9, 7, 1
x such that $p = (x^2 - 5y^2)/4$. Cf. M3758. Ref CU04 1. L1 55. [5,5; A2343, N0042]

M0110 0, 1, 2, 1, 2, 3, 2, 1, 2, 3, 4, 3, 2, 3, 2, 1, 2, 3, 4, 3, 4, 5, 4, 3, 2, 3, 4, 3, 2, 3, 2, 1, 2,
3, 4, 3, 4, 5, 4, 3, 4, 5, 6, 5, 4, 5, 4, 3, 2, 3, 4, 3, 4, 5, 4, 3, 2, 3, 4, 3, 2, 3, 2, 1
Number of 1's in Gray code for n. Ref SIAC 9 144 80. [0,3; A5811]

$a(2n+1) = 2a(n) - a(2n) + 1$, $a(4n) = a(2n)$, $a(4n+2) = 1 + a(2n+1)$.

M0111 0, 1, 2, 1, 2, 3, 2, 3, 2, 1, 2, 3, 2, 3, 4, 3
Weight of balanced ternary representation of n. Ref SIAC 9 150 80. [0,3; A5812]

M0112 1, 1, 1, 1, 1, 2, 1, 2, 3, 2, 3, 2, 4, 2, 1, 5, 2, 2, 4, 4, 3, 1, 4, 7, 5, 3, 4, 6, 2, 2, 8, 5, 6, 3, 8, 2, 6, 10, 4, 2, 5, 5, 4, 4, 3, 10, 2, 7, 6, 4, 10, 1, 8, 11, 4, 5, 8, 4, 2, 13, 4, 9, 4, 3, 6, 14
Class number of forms with discriminant $-n$. Ref BU89 224. [3,6; A6641]

M0113 1, 1, 2, 1, 2, 3, 3, 1, 2, 3, 3, 4, 5, 5, 4, 1, 2, 3, 3, 4, 5, 5, 4, 5, 7, 8, 7, 7, 8, 7, 5, 1, 2, 3, 3, 4, 5, 5, 4, 5, 7, 8, 7, 7, 8, 7, 5, 6, 9, 11, 10, 11, 13, 12, 9, 9, 12, 13, 11, 10, 11, 9, 6, 1
Numerators of Farey tree fractions. Cf. M0437. Ref PSAM 46 42 92. [1,3; A7305]

M0114 2, 1, 2, 3, 4, 8, 8, 18, 18, 38, 28, 142, 72, 234, 360, 669, 520, 2606, 1608, 7293
Necklaces. Ref IJM 8 269 64. [1,1; A2730, N0044]

M0115 1, 2, 1, 2, 3, 6, 8, 16, 24, 42, 69, 124, 208, 378, 668, 1214, 2220, 4110, 7630, 14308, 26931
Necklaces with n beads. Ref IJM 5 663 61. [0,2; A1371, N0045]

M0116 1, 2, 1, 2, 3, 6, 9, 18, 30, 56, 99, 186, 335, 630, 1161, 2182, 4080, 7710, 14532, 27594, 52377, 99858, 190557, 364722, 698870, 1342176, 2580795, 4971008, 9586395
Degree n irreducible polynomials (n-bead necklaces). See Fig M3860. Cf. M0564. Ref IJM 5 663 61. JSIAM 12 288 64. [0,2; A1037, N0046]

M0117 1, 1, 2, 1, 2, 4, 0, 5, 2, 4, 0, 10, 0, 12, 4, 13, 6, 12, 0, 18, 12, 20, 20, 36, 20, 36, 16, 44, 32, 60, 40, 73, 50, 56, 40, 58, 44, 52, 60, 84, 36, 112, 88, 108, 136, 132, 152, 178, 136
Shifts left under AND-convolution with itself. Ref BeS194. [0,3; A7461]

M0118 2, 1, 2, 5, 1, 1, 2, 1, 1, 3, 10, 2, 1, 3, 2, 24, 1, 3, 2, 3, 1, 1, 1, 90, 2, 1, 12, 1, 1, 1, 1, 5, 2, 6, 1, 6, 3, 1, 1, 2, 5, 2, 1, 2, 1, 1, 4, 1, 2, 2, 3, 2, 1, 1, 4, 1, 1, 2, 5, 2, 1, 1, 3, 29, 8, 3, 1
Continued fraction for Khintchine's constant. Ref MOC 20 446 66. [1,1; A2211, N0047]

M0119 1, 1, 2, 1, 2, 5, 8, 2, 1, 3, 10, 7, 3, 15, 4, 1, 4, 17, 170, 4, 5, 197, 24, 5, 1, 5, 26, 16, 70, 11, 1520, 6, 23, 35, 6, 1, 6, 37, 25, 6, 32, 13, 3482, 20, 4, 24335, 6, 7, 1, 7, 50, 36, 182
Solution to Pellian: x such that $x^2 - ny^2 = \pm 1, \pm 4$. Cf. M0398. Ref DE17. CAY 13 434. L1 55. [1,3; A6704]

M0120 1, 1, 2, 1, 2, 5, 8, 3, 1, 3, 10, 7, 18, 15, 4, 1, 4, 17, 170, 9, 55, 197, 24, 5, 1, 5, 26, 127, 70, 11, 1520, 17, 23, 35, 6, 1, 6, 37, 25, 19, 32, 13, 3482, 199, 161, 24335, 6, 7, 1, 7
Solution to Pellian: x such that $x^2 - ny^2 = \pm 1$. Cf. M0399. Ref DE17. CAY 13 434. L1 55. [1,3; A6702]

M0121 2, 1, 2, 5, 17, 92, 994, 28262, 2700791
Threshold functions of n variables. Ref PGEC 19 821 70. MU71 38. [0,1; A0619, N0048]

M0122 2, 1, 2, 9, 96, 2690, 226360, 64646855, 68339572672
Threshold functions of n variables. Ref PGEC 19 821 70. MU71 38. [0,1; A2079, N0049]

M0123 1, 1, 2, 1, 2, 15, 1, 5, 7, 971, 20, 276
No-3-in-line problem for equilateral triangle array of side *n*. Ref rhb. [1,3; A7402]

M0124 1, 1, 1, 2, 1, 2, 1382, 4, 3617, 87734, 349222, 310732, 472728182, 2631724,
  13571120588, 13785346041608, 7709321041217, 303257395102
Numerators of multiples of Bernoulli numbers. Ref RO00 331. FMR 1 74. [1,4; A2431,
N0050]

M0125 1, 2, 1, 3, 1, 1, 1, 3, 3, 3, 1, 3, 1, 3, 5, 3, 1, 5, 1, 3, 7, 3, 1, 7, 1, 3, 9, 3, 1, 9, 1, 3, 11,
  3, 1, 11, 1, 3, 13, 3, 1, 13, 1, 3, 15, 3, 1, 15, 1, 3, 17, 3, 1, 17, 1, 3, 19, 3, 1, 19, 1, 3, 21, 3
Continued fraction for *e*/2. Ref KN1 2 601. [1,2; A6083]

M0126 0, 1, 1, 2, 1, 3, 1, 1, 3, 2, 1, 6, 3, 2, 1, 3, 1, 1, 6, 3, 2, 4, 1, 1, 3, 2, 1, 3, 1, 6, 4, 2, 1,
  2, 4, 3, 1, 8, 3, 2, 1, 6, 3, 2, 1, 3, 1, 1, 6, 3, 2, 4, 1, 1, 3, 2, 1, 3, 1, 30, 6, 3, 2, 4, 1, 1, 3, 2, 1
The Sally sequence: the length of repetition avoided in M0074. See Fig M0126. Ref nsh.
[1,4; A6346]

||||||||||||||||||||||||||||||||||||||||||||||||||||||||||||||||||||||||||||||||||||||||||||||||||||||||||||||||||||||||||||||||||||||||

Figure M0126. LINUS AND SALLY SEQUENCES.

  These sequences were invented by Nathaniel Hellerstein [nsh] after seeing this Linus cartoon. In M0074, $a(n + 1)$ is chosen so as to avoid repeating the longest possible substring of $a(1) \cdots a(n)$. (This differs slightly from Linus's sequence.) The Sally sequence, M0126, gives the length of the run that was avoided.

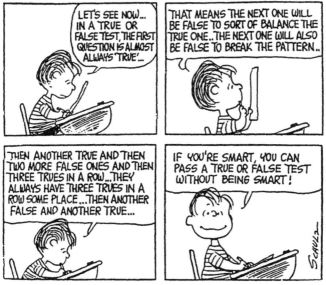

© 1968 United Feature Syndicate, Inc.

M0127 1, 2, 1, 3, 1, 2, 1, 4, 1, 2, 1, 3, 1, 2, 1, 5, 1, 2, 1, 3, 1, 2, 1, 4, 1, 2, 1, 3, 1, 2, 1, 6, 1,
   2, 1, 3, 1, 2, 1, 4, 1, 2, 1, 3, 1, 2, 1, 5, 1, 2, 1, 3, 1, 2, 1, 4, 1, 2, 1, 3, 1, 2, 1, 7, 1, 2, 1, 3, 1
Number of 2's dividing $2n$. Ref MMAG 40 164 67. TCS 9 109 79. TCS 98 187 92. [1,2;
A1511, N0051]

M0128 1, 1, 2, 1, 3, 1, 3, 2, 5, 1, 6, 3, 2, 2, 8, 3, 9, 2, 3, 5, 11, 1, 10, 6, 9, 3, 14, 2, 15, 4, 5,
   8, 6, 3, 18, 9, 6, 2, 20, 3, 21, 5, 6, 11, 23, 2, 21, 10, 8, 6, 26, 9, 10, 3, 9, 14, 29, 2, 30, 15, 3
Reduced totient function (divided by 2). Cf. M0298. Ref CAU (2) 12 43. L1 7. [3,3;
A2616, N0052]

M0129 1, 1, 1, 1, 2, 1, 3, 1, 3, 2, 9, 1, 10, 2, 4, 3, 19, 1, 20, 2, 6, 4, 32, 1, 21, 7, 16, 7, 84, 1,
   85, 9, 18, 11, 35, 3, 161, 15, 30, 6, 212, 2, 214, 15, 12, 19, 260, 3, 154, 11, 62, 31, 521, 5
Coprime chains with largest member n. Ref PAMS 16 809 65. [1,5; A3139]

M0130 1, 2, 1, 3, 1, 4, 1, 5, 2, 6, 4, 7, 7, 8, 11, 9, 16, 11, 22, 15, 29, 31, 37, 33, 46, 49, 57,
   71, 72, 100, 94, 137, 127, 183, 176, 240, 247, 312, 347, 406
7th order maximal independent sets in path graph. Ref YaBa94. [1,2; A7381]

M0131 1, 1, 1, 2, 1, 3, 1, 4, 2, 3, 1, 8, 1, 3, 3, 8, 1, 8, 1, 8, 3, 3, 1, 20, 2, 3, 4, 8, 1, 13, 1, 16,
   3, 3, 3, 26, 1, 3, 3, 20, 1, 13, 1, 8, 8, 3, 1, 48, 2, 8, 3, 8, 1, 20, 3, 20, 3, 3, 1, 44, 1, 3, 8, 32
Perfect partitions of n, or ordered factorizations of $n + 1$. Ref R1 124. HO85a 141. [0,4;
A2033, N0053]

M0132 1, 2, 1, 3, 1, 4, 2, 5, 4, 6, 7, 7, 11, 9, 16, 13, 22, 20, 29, 31, 38, 47, 51, 69, 71, 98,
   102, 136, 149, 187, 218, 258, 316, 360, 452, 509, 639, 727, 897, 1043
5th order maximal independent sets in path graph. Ref YaBa94. [1,2; A7380]

M0133 1, 1, 1, 2, 1, 3, 1, 12, 2, 3, 1, 20, 1, 3, 3, 54, 1, 34, 1, 44, 3, 3, 1, 764, 2, 3, 30, 140,
   1, 283, 1, 4470, 3, 3, 3, 10416, 1, 3, 3, 10820, 1, 2227, 1, 2060, 62, 3, 1, 958476, 2, 44, 3
Minimal plane trees with n terminal nodes. Ref MST 12 264 79. [1,4; A6241]

M0134 1, 2, 1, 3, 2, 1, 2, 3, 1, 2, 1, 3, 2, 3, 1, 3, 2, 1, 2, 3, 1, 2, 1, 3, 2, 1, 2, 3, 1, 3, 2, 3, 1,
   2, 1, 3, 2, 1, 2, 3, 1, 2, 1, 3
A square-free ternary sequence. Ref TH06 14. gs. [1,2; A5678]

M0135 1, 1, 2, 1, 3, 2, 1, 3, 4, 4, 2, 5, 5, 4, 2, 5, 3, 1, 5, 6, 7, 1, 4, 2, 8, 5, 7, 8, 1, 6, 7, 8, 9,
   4, 9, 5, 3, 10, 10, 7, 6, 10, 2, 5, 11, 10, 5, 7, 10, 12, 4, 12, 9, 8, 2, 11, 3, 6, 13, 13, 11, 1, 13
From quadratic partitions of primes. Ref KK71 243. [5,3; A2973]

M0136 2, 1, 3, 2, 1, 4, 1, 2, 5, 7, 4, 2, 6, 5, 1, 2, 3, 11, 6, 1, 5, 3, 10, 7, 12, 1, 2, 9, 4, 13, 7,
   6, 5, 9, 14, 16, 8, 11, 2, 7, 3, 4, 10, 1, 17, 19, 2, 8, 14, 16, 1, 13, 20, 9, 3, 8, 5, 6, 11, 14
y such that $p = x^2 - 5y^2$. Cf. M3739. Ref CU04 1. L1 55. [5,1; A2341, N0054]

M0137 1, 2, 1, 3, 2, 1, 4, 3, 2, 5, 1, 4, 3, 6, 2, 5, 1, 4, 7, 3, 6, 2, 5, 8, 1, 4, 7, 3, 6, 9, 2, 5, 8,
   1, 4, 7, 10, 3, 6, 9, 2, 5, 8, 1, 4, 7, 10
Signature sequence of $\sqrt{2}$. Ref Kimb94. [1,2; A7336]

M0138 1, 1, 1, 2, 1, 3, 2, 1, 4, 3, 2, 5, 1, 6, 4, 3, 7, 2, 8, 5, 1, 9, 6, 4, 10, 3, 11, 7, 2, 12, 8, 5, 13, 1, 14, 9, 6, 15, 4, 16, 10, 3, 17, 11, 7, 18, 2, 19, 12, 8, 20, 5, 21, 13, 1, 22, 14, 9, 23, 6
Fractal sequence obtained from Fibonacci numbers. Ref Kimb94a. [0,4; A3603]

M0139 1, 1, 2, 1, 3, 2, 1, 5, 2, 1, 4, 6, 3, 2, 7, 4, 3, 1, 7, 4, 9, 1, 8, 5, 10, 4, 7, 3, 2, 5, 8, 12, 2, 1, 9, 11, 8, 4, 7, 2, 1, 14, 6, 9, 5, 11, 13, 2, 14, 16, 4, 11, 8, 3, 2, 7, 10, 17, 12, 11, 1, 7
y such that $p = x^2 - 2y^2$. Cf. M0607. Ref CU04 1. L1 55. [2,3; A2335, N0055]

M0140 2, 1, 3, 2, 3, 1, 2, 1, 3, 1, 2, 3, 2, 1, 3, 2, 3, 1, 2, 3, 2, 1, 3, 1, 2, 1, 3, 2, 3, 1, 2, 1, 3, 1, 2, 3, 2, 1, 3, 1, 2, 1, 3, 2, 3, 1, 2, 3, 2, 1, 3, 2, 3, 1, 2, 1, 3, 1, 2, 3, 2, 1, 3, 2, 3, 1, 2, 3, 2
A square-free ternary sequence. Ref DUMJ 11 6 44. gs. SA81 10. [1,1; A5679]

M0141 1, 1, 2, 1, 3, 2, 3, 1, 4, 3, 5, 2, 5, 3, 4, 1, 5, 4, 7, 3, 8, 5, 7, 2, 7, 5, 8, 3, 7, 4, 5, 1, 6, 5, 9, 4, 11, 7, 10, 3, 11, 8, 13, 5, 12, 7, 9, 2, 9, 7, 12, 5, 13, 8, 11, 3, 10, 7, 11, 4, 9, 5, 6, 1
From Stern-Brocot tree: $a(2n+1) = a(n)$, $a(2n) = a(n) + a(n-1)$. Ref ELM 2 95 47. WW 114. TCS 98 187 92. [0,3; A2487, N0056]

M0142 1, 1, 2, 1, 3, 2, 3, 2, 1, 4, 5, 4, 1, 6, 3, 5, 7, 6, 7, 2, 8, 1, 7, 3, 6, 8, 5, 6, 3, 9, 8, 5, 4, 10, 11, 2, 11, 6, 4, 10, 12, 9, 12, 11, 1, 9, 13, 2, 7, 13, 4, 12, 13, 14, 11, 7, 9, 10, 4, 15, 14
y such that $p = x^2 + 3y^2$. Cf. M0166. Ref CU04 1. KNAW 54 14 51. [3,3; A1480, N0057]

M0143 1, 0, 1, 1, 1, 2, 1, 3, 2, 3, 3, 5, 4, 6, 5, 5, 8, 9, 10, 11, 11, 10, 14, 18, 19, 18, 20, 20, 25, 30, 35, 34, 32, 32, 43, 43, 57, 56, 51, 55, 67, 78, 87, 87, 80, 82, 97, 125, 128, 127, 128
Partitions of n into relatively prime parts ≥ 2. Ref mlb. [0,6; A7359]

M0144 1, 2, 1, 3, 2, 4, 1, 3, 5, 2, 4, 6, 1, 3, 5, 7, 2, 4, 6, 1, 8, 3, 5, 7, 2, 9, 4, 6, 1, 8, 3, 10, 5, 7, 2, 9, 4, 11, 6, 1, 8, 3, 10, 5, 12, 7, 2, 9, 4, 11, 6, 13
Signature sequence of $\sqrt{3}$. Ref Kimb94. [1,2; A7337]

M0145 1, 1, 2, 1, 3, 2, 4, 1, 5, 3, 6, 2, 7, 4, 8, 1, 9, 5, 10, 3, 11, 6, 12, 2, 13, 7, 14, 4, 15, 8, 16, 1, 17, 9, 18, 5, 19, 10, 20, 3, 21, 11, 22, 6, 23, 12, 24, 2, 25, 13, 26, 7, 27, 14, 28, 4, 29
Fractal sequence obtained from powers of 2. Ref Kimb94a. [0,3; A3602]

M0146 1, 0, 1, 1, 2, 1, 3, 2, 4, 3, 5, 4, 7, 5, 8, 7, 10, 8, 12, 10, 14, 12, 16, 14, 19, 16, 21, 19, 24, 21, 27, 24, 30, 27, 33, 30, 37, 33, 40, 37, 44, 40, 48, 44, 52, 48, 56, 52, 61, 56, 65, 61
Alcuin's sequence: expansion of $1/(1-x^2)(1-x^3)(1-x^4)$. Ref AMM 86 477 79; 86 687 79. Oliv93 58. [0,5; A5044]

M0147 0, 1, 0, 1, 0, 1, 1, 1, 2, 1, 3, 2, 4, 4, 5, 7, 7, 11, 11, 16, 18, 23, 29, 34, 45, 52, 68, 81, 102, 126, 154, 194, 235, 296, 361, 450, 555, 685, 851, 1046, 1301, 1601, 1986, 2452
$a(n) = a(n-2) + a(n-5)$. Ref MMAG 41 17 68. [0,9; A1687, N0059]

M0148 1, 1, 0, 1, 0, 2, 1, 3, 2, 6, 4, 9, 8, 18, 16, 32, 32, 61, 64, 115, 128, 224, 258, 431, 520, 850, 1050, 1673, 2128, 3328, 4320
Generalized Fibonacci numbers. Ref FQ 27 120 89. [1,6; A6208]

M0149 1, 2, 1, 3, 2, 7, 5, 18, 13, 47, 34, 123, 89, 322, 233, 843, 610, 2207, 1597, 5778, 4181, 15127, 10946, 39603, 28657, 103682, 75025, 271443, 196418, 710647, 514229
$a(n) = (1 + a(n-1)a(n-2))/a(n-3)$. Ref MAG 69 264 85. [0,2; A5247]

M0150 1, 1, 1, 1, 1, 2, 1, 3, 3, 2, 1, 2, 1, 4, 4, 4, 1, 4, 1, 4, 3, 2, 1, 4, 3, 5, 4, 2, 1, 3, 1, 3, 5, 2, 3, 3, 1, 4, 5, 2, 1, 3, 1, 5, 2, 4, 1, 2, 5, 3, 5, 2, 1, 2, 5, 2, 3, 2, 1, 3, 1, 6, 2, 3, 5, 2, 1, 4, 6
Iterates of a number-theoretic function. Ref MOC 23 181 69. [1,6; A2217, N0060]

M0151 1, 1, 2, 1, 3, 3, 2, 4, 1, 3, 5, 5, 5, 3, 6, 1, 7, 3, 6, 5, 3, 7, 6, 9, 9, 5, 8, 4, 10, 9, 7, 3, 11, 3, 9, 1, 11, 12, 8, 10, 12, 9, 11, 5, 9, 13, 3, 6, 1, 13, 3, 2, 10, 8, 15, 15, 7, 9, 13, 1, 15, 14
y such that $p = (x^2 + 11y^2)/4$. Cf. M2206. Ref CU04 1. L1 55. [3,3; A2347, N0061]

M0152 1, 1, 1, 2, 1, 3, 3, 2, 4, 1, 5, 2, 4, 3, 1, 7, 5, 3, 6, 2, 8, 5, 3, 8, 2, 10, 2, 5, 5, 3, 10, 7, 4, 10, 1, 11, 5, 8, 2, 13, 4, 9, 4, 3, 14, 4, 7, 5, 12, 2, 15, 6, 7, 12, 4, 13, 2, 11, 3, 14, 4, 8, 8
Class number of quadratic field with discriminant $-4n + 1$. Cf. M0225. Ref BU89 224. [1,4; A6642]

M0153 1, 0, 2, 1, 3, 3, 7, 6, 14, 15, 26, 31, 51, 60, 95, 116, 171, 215, 308, 385, 541, 683, 932, 1183, 1591, 2012, 2673, 3381, 4429, 5599, 7266
Representation degeneracies for boson strings. Ref NUPH B274 546 86. [2,3; A5292]

M0154 1, 0, 0, 1, 0, 1, 1, 1, 2, 1, 3, 4, 3, 7, 7, 8, 14, 15, 21, 28, 33, 47, 58, 76, 103, 125, 169, 220, 277, 373
From symmetric functions. Ref PLMS 23 314 23. [0,9; A2124, N0062]

M0155 2, 1, 3, 4, 7, 11, 18, 29, 47, 76, 123, 199, 322, 521, 843, 1364, 2207, 3571, 5778, 9349, 15127, 24476, 39603, 64079, 103682, 167761, 271443, 439204, 710647, 1149851
Lucas numbers (beginning at 2): $L(n) = L(n-1) + L(n-2)$. See Fig M0692. M2341. Ref HW1 148. HO69. C1 46. [0,1; A0032]

M0156 2, 1, 3, 4, 9, 7, 15, 12, 18, 17, 29, 20, 39, 25, 33, 34, 57, 30, 65, 38, 53, 47, 81, 40, 86, 59, 80, 60, 107, 41, 125, 78, 103, 79, 123, 66, 155, 95, 123, 90, 177, 75, 189, 110, 132
Moebius transform of primes. Ref EIS § 2.7. [1,1; A7444]

M0157 0, 1, 1, 1, 1, 2, 1, 3, 5, 7, 4, 23, 29, 59, 129, 314, 65, 1529, 3689, 8209, 16264, 83313, 113689, 620297, 2382785, 7869898, 7001471, 126742987, 398035821
$a(2n+1) = a(n+2)a(n)^3 - a(n-1)a(n+1)^3$,
$a(2n) = (a(n+2)a(n)a(n-1)^2 - a(n)a(n-2)a(n+1)^2)$. Ref FQ 17 17 79. [0,6; A6769]

M0158 1, 2, 1, 3, 6, 2, 16, 9, 23, 58, 6, 128, 109, 147, 512, 70, 954, 1233, 815, 4096, 1650, 6542, 13141, 3243, 32768, 23038, 42498, 131072, 3577, 258567, 272874, 251414
Parenthesized one way gives the powers of 2: (1), (2), (1+3), ..., another way the powers of 3 : (1), (2+1), (3+6), ... Ref kbrown. [0,2; A6895]

M0159 2, 1, 3, 12, 59, 354, 2535, 21190, 202731, 2183462, 26130441, 343956264, 4938891841, 76827253854, 1287026203647, 23100628140676, 442271719973507
Logarithm of e.g.f. for primes. [1,1; A7447]

M0160 2, 1, 3, 16, 380, 1227756, 400507805615570
Nondegenerate Boolean functions of n variables. Ref PGEC 12 464 63. MU71 38. [0,1; A0618, N0063]

M0161 1, 2, 1, 4, 1, 2, 1, 8, 1, 2, 1, 4, 1, 2, 1, 9, 1, 2, 1, 4, 1, 2, 1, 8, 1, 2, 1, 4, 1, 2, 1, 10, 1,
  2, 1, 4, 1, 2, 1, 8, 1, 2, 1, 4, 1, 2, 1, 9, 1, 2, 1, 4, 1, 2, 1, 8, 1, 2, 1, 4, 1, 2, 1, 12, 1, 2, 1, 4
Hurwitz-Radon function. Ref LA73a 131. [1,2; A3484]

If $n = 2^{4b+c} d$, d odd, then $a(n) = 8b + 2^c$.

M0162 1, 2, 1, 4, 1, 2, 1, 8, 1, 2, 1, 4, 1, 2, 1, 16, 1, 2, 1, 4, 1, 2, 1, 8, 1, 2, 1, 4, 1, 2, 1, 32,
  1, 2, 1, 4, 1, 2, 1, 8, 1, 2, 1, 4, 1, 2, 1, 16, 1, 2, 1, 4, 1, 2, 1, 8, 1, 2, 1, 4, 1, 2, 1, 64, 1, 2, 1
Highest power of 2 dividing n. [1,2; A6519]

M0163 1, 1, 1, 2, 1, 4, 1, 6, 3, 7, 6, 8, 5, 11, 3, 15, 8, 18, 13, 20, 9, 24, 8, 32, 17, 38, 23, 41,
  21, 50, 20, 62, 29, 71, 41, 81
Taylor series from Ramanujan's Lost Notebook. Ref LNM 899 44 81. [0,4; A6306]

M0164 1, 1, 1, 2, 1, 4, 2, 1, 3, 2, 2, 2, 1, 3, 2, 3, 2, 2, 2, 4, 1, 3, 3, 2, 2, 2, 2, 4, 2, 2, 2, 3, 3,
  1, 3, 3, 2, 4, 3, 2, 2, 4, 2, 2, 2, 2, 3, 2, 2, 3, 3, 3, 5, 1, 2, 3, 2, 3, 3
Min. of largest partial quotient of cont. fraction for k/n, $(k,n) = 1$. Ref MFM 101 309 86.
jos. [2,4; A6839]

M0165 0, 1, 2, 1, 4, 2, 2, 2, 2, 4, 1, 2, 4, 2, 2, 2, 4, 2, 1, 2, 2, 1, 2, 2, 4, 4, 2, 2, 1, 2, 1, 2, 2,
  4, 2, 2, 4, 1, 2, 2, 2, 2, 1, 2, 2, 2, 2, 4, 2, 2, 4, 2, 2, 2, 1, 1, 2, 4, 1, 2, 2, 4, 2, 2, 2, 2
Related to Fibonacci numbers. Ref HM68. MOC 23 459 69. ACA 16 109 69. [1,3; A1176, N0064]

M0166 0, 2, 1, 4, 2, 5, 4, 7, 8, 5, 2, 7, 10, 1, 10, 8, 2, 7, 4, 13, 1, 14, 8, 14, 11, 7, 14, 13, 16,
  8, 11, 16, 17, 7, 2, 19, 4, 17, 19, 11, 1, 14, 5, 10, 22, 16, 4, 23, 20, 8, 23, 13, 10, 5, 16, 22
x such that $p = x^2 + 3y^2$. Cf. M0142. Ref CU04 1. KNAW 54 14 51. [3,2; A1479, N0065]

M0167 1, 0, 1, 1, 2, 1, 4, 2, 6, 5, 9, 7, 16, 11, 22, 20, 33, 28, 51, 42, 71, 66, 100, 92, 147,
  131, 199, 193, 275, 263, 385, 364, 516, 511, 694, 686, 946, 925, 1246, 1260, 1650, 1663
Partitions of n in which no part occurs just once. Ref HO85a 242. [0,5; A7690]

M0168 1, 0, 1, 0, 2, 1, 4, 3, 8, 7, 15, 15, 28, 30, 51, 58, 92, 108, 163, 196, 285, 348, 490,
  605, 833, 1034, 1396, 1740, 2313, 2887, 3789, 4730
Representation degeneracies for boson strings. Ref NUPH B274 546 86. [1,5; A5291]

M0169 1, 2, 1, 4, 5, 46, 37, 176, 289, 1450, 641, 24652, 15061, 18734, 110125
$a(n) \equiv a(k) \bmod n - k$ for all k. Ref sab. [1,2; A2987]

M0170 1, 2, 1, 4, 6, 4, 17, 32, 44, 60, 70, 184, 476, 872, 1553, 2720, 4288, 6312
Shapes of height-balanced trees with n nodes. Ref IPL 17 18 83. [1,2; A6265]

M0171 2, 1, 4, 7, 24, 62, 216, 710, 2570, 9215, 34146, 126853, 477182
Chessboard polyominoes with n squares. Ref wfl. [1,1; A1933, N0066]

M0172 0, 0, 0, 0, 0, 1, 0, 1, 2, 1, 4, 9, 12, 27, 50, 99, 188, 386, 740, 1528, 3012, 6192,
   12376, 25594, 51628, 107135, 218100, 453895, 930812, 1943281, 4009512, 8394915
3-connected planar maps with n edges. Ref trsw. [1,9; A6445]

M0173 2, 1, 4, 10, 36, 108, 392, 1363, 5000, 18223, 67792, 252938, 952540
One-sided chessboard polyominoes with n cells. Ref wfl. [1,1; A1071, N0067]

M0174 2, 1, 4, 13, 44, 135, 472, 1492, 5324, 17405, 63944, 215096, 799416, 2752909,
   10310384, 36443256, 137263244, 489166324, 1860249448, 6739795717
Witt vector *2!. Ref SLC 16 106 88. [1,1; A6173]

M0175 2, 1, 5, 2, 1, 1, 1, 1, 2, 12, 8, 2, 1, 4, 1, 1, 2, 2, 9, 6, 2, 2, 1, 25, 3, 2, 1, 1, 3, 1, 17, 3,
   1, 2, 2, 2, 1, 4, 1, 1, 2, 1, 2, 2, 7, 1, 2, 1, 1, 34, 8, 5, 1, 1, 1, 54, 4, 10, 2, 2, 2, 2, 1, 4, 3, 1, 2
$(p-1)/x$, where $10^x \equiv 1 \mod p$. Ref Krai24 1 131. [3,1; A6556]

M0176 0, 1, 1, 2, 1, 5, 4, 29, 13, 854, 685, 730001, 260776, 532901720777,
   464897598601, 283984244007552571082330, 678544668225760539925129
$a(n) = a(n-2)^2 - a(n-1)$. Ref EUR 39 39 78. [0,4; A5605]

M0177 2, 1, 5, 7, 17, 25, 52, 77, 143, 218, 371, 564, 920, 1380, 2168
Representation degeneracies for Neveu-Schwarz strings. Ref NUPH B274 547 86. [1,1;
A5297]

M0178 1, 2, 1, 6, 3, 1, 5, 3, 2, 1, 2, 3, 4, 3, 1, 9, 3, 6, 7, 8, 1, 10, 3, 2, 3, 4, 5, 1, 4, 3, 8, 7, 5,
   9, 7, 1, 14, 3, 4, 7, 4, 2, 9, 4, 1, 2, 3, 4, 7, 8, 12, 16, 9, 3, 1, 12, 3, 14, 7, 6, 4, 8, 6, 3, 2, 1, 6
Remoteness numbers for Tribulations. Ref WW 502. [1,2; A6019]

M0179 1, 1, 0, 2, 1, 6, 4, 16, 15, 38, 46, 93, 118, 220, 294, 496, 687, 1101, 1533, 2371,
   3315, 4969, 6960, 10194, 14213, 20469, 28407, 40277, 55610, 77871, 106847, 148046
Representation degeneracies for Neveu-Schwarz strings. Ref NUPH B274 544 86. [0,4;
A5299]

M0180 1, 0, 0, 1, 2, 1, 6, 12, 46, 92, 341, 1787, 9233, 45752, 285053, 1846955
Ways of placing n nonattacking queens on $n \times n$ board. See Fig M0180. Ref SL26 47.
Well71 238. CACM 18 653 75. [1,5; A2562, N0068]

M0181 1, 1, 2, 1, 8, 7, 32, 31, 96, 97
Coefficients of order n in Baker-Campbell-Hausdorff expansion. Ref JMP 32 421 91. [1,3;
A5489]

M0182 1, 0, 0, 0, 1, 0, 2, 1, 8, 11, 56, 160, 777, 3259
4-connected polyhedral graphs with n faces. Ref Dil92. [4,7; A7026]

||||||||||||||||||||||||||||||||||||||||||||||||||||||||||||||||||||||||||||||||||||||||||||||||||||||||||||||||||||||||||||||||||||||||||||||||||||||||

Figure M0180. NON-ATTACKING QUEENS.

For $n \geq 4$, it is possible to place n queens on an $n \times n$ chessboard so that no queen is attacked by another. Sequence M1958 gives the total number of ways and M0180 the number of inequivalent ways in which this can be done. One of the 12 inequivalent solutions on an 8×8 board is shown here. M1761 gives the corresponding numbers for rooks.

||||||||||||||||||||||||||||||||||||||||||||||||||||||||||||||||||||||||||||||||||||||||||||||||||||||||||||||||||||||||||||||||||||||||||||||||||||||||

M0183   1, 1, 1, 2, 1, 12, 1, 2, 3, 10, 1, 12, 1, 28, 15, 2, 1, 12, 1, 10, 21, 22, 1, 12, 5, 26, 3,
  28, 1, 60, 1, 2, 33, 34, 35, 12, 1, 38, 39, 10, 1, 84, 1, 22, 15, 46, 1, 12, 7, 10, 51, 26, 1, 12
Related to nth powers of polynomials. Ref ACA 29 246 76. [1,4; A5730]

M0184   2, 1, 30, 3, 29, 2, 4, 28, 2, 1, 3, 17, 4, 2, 12, 8, 2, 4, 9, 3, 28, 2, 11, 4, 2, 3, 3, 2, 8, 2,
  10, 27, 2, 8, 10, 3, 3, 2, 4, 5, 2, 1, 3, 7, 4, 2, 9, 26, 2, 4, 7, 3, 9, 2, 7, 4, 2, 3, 3, 4, 97, 2, 8
Length of Beanstalk game. Ref MMAG 59 263 86. [1,1; A5693]

SEQUENCES BEGINNING . . ., 2, 2, . . .

M0185   0, 0, 1, 0, 0, 1, 1, 2, 2, 0, 0, 1, 0, 0, 1, 1, 2, 2, 1, 2, 2, 3, 3, 4, 3, 3, 4, 0, 0, 1, 0, 0, 1,
  1, 2, 2, 0, 0, 1, 0, 0, 1, 1, 2, 2, 1, 2, 2, 3, 3, 4, 3, 3, 4, 1, 2, 2, 3, 3, 4, 3, 3, 4, 4, 5, 5, 4, 5, 5
Partitioning integers to avoid arithmetic progressions of length 3. Ref PAMS 102 766 88.
[0,8; A6997]

$$a(3n+k) = [(3a(n)+k)/2], \ 0 \leq k \leq 2.$$

M0186   1, 0, 1, 2, 2, 0, 2, 4, 3, 0, 1, 6, 28, 72, 139, 242, 407, 722, 1215, 2348, 3753, 7186,
  9558, 21800, 16576, 61234, 7978, 226136, 446034, 1118180, 3180033, 6428640
Percolation series for directed square lattice. Ref JPA 21 3200 88. [0,4; A6462]

M0187 2, 2, 0, 4, 2, 0, 4, 0, 0, 4, 4, 0, 2, 2, 0, 4, 0, 0, 4, 4, 0, 4, 0, 0, 6, 0, 0, 0, 4, 0, 4, 4, 0, 4, 0, 0, 4, 2, 0, 4, 2, 0, 0, 0, 0, 8, 4, 0, 4, 0, 0, 4, 0, 0, 4, 4, 0, 0, 4, 0, 2, 0, 0, 4, 4, 0, 8, 0, 0
Theta series of planar hexagonal lattice w.r.t. edge. See Fig M2336. Ref JCP 83 6523 85. [0,1; A5881]

M0188 0, 0, 0, 0, 0, 0, 0, 0, 0, 0, 1, 0, 0, 1, 1, 0, 1, 0, 1, 1, 1, 2, 2, 1, 0, 2, 1, 2, 2, 3, 2, 1, 3, 3, 3, 1, 2, 4, 3, 3, 3, 5, 3, 4, 3, 5, 4, 3, 1, 2, 5, 5, 4, 4, 5, 5, 5, 6, 5, 7, 4, 7, 5, 3, 5, 9, 5, 7, 7
Genus of $\Gamma_0(n)$. Ref NBS B67 62 63. [1,22; A1617, N0069]

M0189 1, 2, 2, 1, 1, 1, 2, 2, 2, 2, 1, 2, 2, 1, 1, 1, 2, 1, 1, 1, 1, 2, 2, 1, 1, 1, 1, 2, 2, 2, 2, 2, 1, 1, 1, 2, 2, 2, 2, 2, 2, 2, 1, 1, 1, 1, 1, 1, 2, 2, 2, 2, 2, 2, 2, 2, 1, 1, 1, 1, 1, 1, 1, 1, 1, 2, 2, 2, 2, 2
Image of n after $2k$ iterates of '$(3x+1)/2$' map (k large). See Fig M2629. Ref UPNT E16. [1,2; A6513]

M0190 1, 2, 2, 1, 1, 2, 1, 2, 2, 1, 2, 2, 1, 1, 2, 1, 1, 2, 2, 1, 2, 1, 1, 2, 1, 2, 2, 1, 1, 2, 1, 1, 2, 1, 2, 2, 1, 2, 2, 1, 1, 2, 1, 1, 2, 1, 2, 2, 1, 2, 1, 1, 2, 1, 2, 2, 1, 1, 2, 1, 2, 2, 1, 2, 1, 1, 2
$a(n)$ is length of nth run in the sequence. See Fig M0436. Ref AMM 73 681 66. PSAM 46 66 92. VA91 233. [1,2; A0002, N0070]

M0191 1, 1, 1, 1, 1, 1, 2, 2, 1, 1, 3, 3, 1, 1, 2, 1, 3, 4, 1, 2, 3, 1, 1, 3, 4, 2, 2, 2, 1, 2, 5, 2, 1, 2, 1, 5, 6, 1, 4, 2, 7, 1, 1, 3, 3, 3, 4, 2, 3, 5, 3, 1, 1, 7, 4, 2, 2, 2, 8, 7, 2, 3, 4, 4, 1, 5, 3, 7
Classes per genus in quadratic field with discriminant $-4n$, $-n \equiv 2, 3 \pmod 4$. Cf. M2263. Ref BU89 231. [1,7; A3638]

M0192 0, 1, 1, 2, 1, 1, 4, 2, 1, 5, 2, 2, 3, 1, 6, 4, 5, 1, 4, 2, 3, 7, 2, 4, 7, 1, 4, 4, 1, 1, 12, 6, 1, 5, 2, 8, 7, 5, 2, 4, 1, 11, 4, 8, 9, 13, 4, 2, 7, 1, 2, 14, 1, 3, 4, 4, 5, 11, 8, 2, 7, 3, 18, 10, 1
Cycles in a certain permutation. Ref STNB 20 9 84. [0,4; A6694]

M0193 1, 2, 2, 1, 2, 1, 1, 2, 2, 1, 1, 2, 1, 2, 2, 1, 2, 1, 1, 2, 1, 2, 2, 1, 1, 2, 2, 1, 2, 1, 1, 2, 1, 2, 2, 1, 2, 1, 1, 2, 1, 2, 2, 1, 1, 2, 2, 1, 2, 1, 1, 2, 1, 2, 2, 1, 2, 1, 1, 2, 1, 2, 2, 1, 1, 2, 1, 2, 2, 1, 2, 1, 1, 2, 1
Thue-Morse sequence: follow $a(0), .., a(2^k - 1)$ by its complement. See Fig M0436. Ref TAMS 22 84 21. GH55 105. JCT A13 90 72. SA81 6. Loth83 23. [0,2; A1285, N0071]

M0194 1, 1, 1, 2, 2, 1, 2, 1, 2, 2, 2, 2, 1, 4, 2, 1, 2, 2, 4, 1, 4, 2, 2, 2, 2, 2, 4, 1, 2, 1, 2, 2, 2, 4, 2, 1, 2, 2, 4, 4, 1, 4, 4, 1, 2, 2, 4, 4, 1, 2, 1, 4, 2, 2, 2, 2, 4, 2, 2, 2, 2, 4, 1, 8, 2, 1, 2, 4
Number of genera of $Q(\sqrt{-n})$, n squarefree. Ref BU89 224. [1,4; A3643]

M0195 1, 1, 1, 2, 2, 1, 2, 1, 2, 4, 2, 4, 1, 4, 2, 3, 6, 6, 4, 3, 4, 4, 2, 2, 6, 4, 8, 4, 1, 4, 5, 2, 6, 4, 4, 2, 3, 6, 8, 8, 1, 8, 4, 7, 4, 10, 8, 4, 5, 4, 3, 4, 10, 6, 12, 2, 4, 8, 8, 4, 14, 4, 5, 8, 6, 3
Class number of $Q(\sqrt{-n})$, n squarefree. Ref BS66 425. BU89 224. [1,4; A0924, N0072]

M0196 1, 1, 1, 1, 2, 2, 1, 2, 2, 2, 3, 2, 2, 4, 2, 2, 4, 2, 3, 4, 4, 2, 3, 4, 2, 6, 3, 2, 6, 4, 3, 4, 4, 4, 6, 4, 2, 6, 4, 4, 8, 4, 3, 6, 4, 4, 5, 4, 4, 6, 6, 4, 6, 6, 4, 8, 4, 2, 9, 4, 6, 8, 4, 4, 8, 8, 3, 8, 8
Classes of primitive binary forms of discriminant $-4n$. Ref NBS B67 62 63. [1,5; A0003, N0073]

M0197 1, 1, 1, 2, 2, 1, 2, 3, 1, 3, 1, 2, 4, 3, 4, 2, 1, 3, 5, 5, 4, 2, 5, 4, 5, 1, 6, 4, 2, 1, 4, 7, 7,
6, 7, 2, 4, 6, 5, 7, 8, 8, 3, 7, 1, 5, 6, 3, 2, 8, 9, 7, 6, 8, 9, 4, 2, 5, 8, 1, 10, 10, 3, 7, 5, 2, 8, 10
y such that $p = x^2 + 7y^2$. Cf. M0197. Ref CU04 1. L1 55. [7,4; A2345, N0074]

M0198 1, 2, 2, 1, 2, 4, 2, 0, 1, 4, 2, 2, 2, 4, 4, 0, 2, 2, 2, 2, 4, 4, 2, 0, 1, 4, 0, 2, 2, 8, 2, 0, 4,
4, 4, 1, 2, 4, 4, 0, 2, 8, 2, 2, 2, 4, 2, 0, 1, 2, 4, 2, 2, 0, 4, 0, 4, 4, 2, 4, 2, 4, 2, 0, 4, 8, 2, 2, 4
Moebius transform applied twice to sequence 1,0,0,0,.... Ref EIS § 2.7. [1,2; A7427]

M0199 1, 2, 2, 1, 2, 7, 6, 1, 1, 0, 1, 2, 9, 40
Generalized Moore graphs with $2n$ nodes. Ref LNM 748 23 79. bdm. [2,2; A5007]

M0200 1, 1, 1, 2, 2, 1, 3, 2, 3, 1, 3, 1, 30, 1, 4, 1, 2, 9, 6, 4, 1, 1, 2, 7, 2, 3, 2, 1, 6, 1, 1, 1,
25, 1, 7, 7, 1, 1, 1, 1, 266, 1, 3, 2, 1, 3, 60, 1, 5, 1, 8, 5, 6, 1, 4, 20, 1, 4, 1, 1, 14, 1, 4, 4, 1
Continued fraction for cube root of 4. Ref JRAM 255 122 72. [1,4; A2947]

M0201 1, 2, 2, 1, 3, 7, 5, 6, 10, 27, 50, 42, 30, 41, 148, 241, 345, 303, 167, 275, 858, 1685,
2342, 2813, 2316, 536, 2914, 8228, 14531, 20955, 24370, 22393, 10265, 13839, 53386
Inverse Euler transform of primes. Ref EIS § 2.7. [0,2; A7441]

M0202 1, 0, 2, 2, 1, 4, 1, 4, 2, 2, 3, 2, 2, 4, 3, 2, 4, 2, 4, 4, 4, 2, 5, 2, 6, 2, 5, 0, 4, 2, 6, 4, 4,
2, 7, 0, 8, 2, 3, 2, 6, 2, 8, 4, 6, 2, 7, 2, 10, 2, 8, 0, 6, 2, 8, 2, 6, 0, 7, 2, 12, 4, 5, 2, 10, 0, 12
Decompositions of n into sum of 2 primes. Ref PLMS 23 315 23. [0,3; A2126, N0075]

M0203 0, 1, 2, 2, 1, 36, 1, 485, 27, 98, 23, 86, 13, 16, 577, 16, 13823, 14439374, 9, 1,
91167, 25899588, 95, 4443, 125, 1414178, 57, 330206, 74879, 1975104, 6913, 67, 33
From fundamental unit of $Z[(-n)^{1/4}]$. Ref MOC 48 49 87. [1,3; A6828]

M0204 1, 1, 0, 0, 2, 2, 1, 66, 3, 6, 1, 10, 2, 4, 51, 1, 11512, 4956000, 0, 8, 81146, 1629810,
3, 281, 13, 126466, 7, 39122, 2218, 217672, 663, 6, 4, 6768, 65, 6356, 54441051012990
From fundamental unit of $Z[(-n)^{1/4}]$. Ref MOC 48 49 87. [1,5; A6831]

M0205 0, 0, 0, 0, 0, 1, 0, 0, 0, 1, 0, 1, 0, 1, 1, 1, 0, 1, 0, 2, 2, 2, 0, 2, 1, 3, 2, 3, 1, 4, 2, 4, 3,
5, 4, 7, 3, 6, 5, 8, 6, 10, 6, 10, 9, 12, 9, 15, 11, 16, 14, 18, 14, 22, 19, 25, 22, 27, 23, 33, 29
Partitions of n into products of 2 primes. Ref JNSM 9 220 69. [1,20; A2100, N0076]

M0206 1, 1, 2, 2, 2, 1, 0, 1, 2, 0, 2, 0, 1, 2, 2, 1, 0, 2, 2, 2
Difference between two partition functions. Ref ADV 61 160 86. [0,3; A6140]

M0207 1, 1, 1, 1, 1, 2, 2, 2, 1, 2, 3, 2, 2, 2, 3, 3, 2, 3, 4, 2, 1, 4, 5, 4, 2, 2, 4, 4, 3, 4, 5, 4, 1,
4, 7, 3, 3, 4, 5, 6, 3
Reduced binary quadratic forms of discriminant $-n$. Ref DA82 144. [3,6; A6371]

M0208 2, 2, 2, 2, 2, 2, 2, 2, 2, 2, 2, 2, 2, 2, 2, 2, 2, 2, 2, 2, 2, 2, 2, 2, 2, 2, 2, 2, 2, 2, 2, 2, 2,
2, 2, 2, 2, 2, 2, 2, 2, 2, 2, 2, 2, 2, 2, 2, 2, 2, 2, 2, 2, 2, 2, 2, 2, 2, 2, 2, 2, 2, 2, 2, 2, 2, 2, 2, 2
The all 2's sequence. [1,1; A7395]

M0209 1, 1, 1, 1, 1, 1, 1, 1, 2, 2, 2, 2, 2, 2, 2, 2, 3, 3, 3, 3, 3, 3, 3, 3, 4, 4, 4, 5, 5, 5, 5, 5, 6,
6, 6, 7, 7, 7, 7, 7, 8, 8, 9, 9, 9, 9, 9, 10, 10, 10, 11, 11, 11, 12, 12, 13, 13, 13, 14, 14, 14
Partitions of n into cubes. [0,9; A3108]

M0210 1, 2, 2, 2, 2, 2, 2, 4, 2, 2, 2, 2, 4, 2, 4, 4, 2, 4, 4, 2, 2, 2, 4, 4, 2, 2, 2, 4, 2, 4, 2, 2, 4,
6, 4, 2, 2, 4, 4, 2, 4, 4, 2, 2, 2, 4, 4, 4, 4, 2, 4, 2, 4, 2, 4, 2, 2, 4, 2, 6, 4, 4, 2, 4, 4, 2, 4, 2, 2
Class number of quadratic forms with discriminant $4n$, $n \equiv 2,3 \pmod 4$. Cf. M0084. Ref
BU89 241. [2,2; A3648]

M0211 1, 1, 2, 2, 2, 2, 2, 2, 4, 2, 2, 2, 4, 4, 2, 2, 2, 2, 4, 2, 2, 4, 2, 2, 2, 4, 4, 4, 4, 2, 2, 4, 4,
2, 4, 2, 2, 4, 2, 2, 2, 4, 8, 2, 2, 4, 2, 4, 2, 2, 4, 4, 4, 2, 2, 4, 4, 2, 4, 2, 2, 4, 2, 2, 4, 8, 2, 4, 2
Number of genera of quadratic field with discriminant $-4n$, $-n \equiv 2,3 \pmod 4$. Cf. M0061.
Ref BU89 231. [1,3; A3642]

M0212 2, 2, 2, 2, 2, 2, 2, 2, 46, 46, 46, 46, 46, 46, 406, 718, 950, 950, 950, 1698, 1698,
1698, 1698, 1698, 3990, 3990, 3990, 53510, 77970, 89478, 89478, 89478, 89478, 89478
Class numbers of quadratic fields. Ref MOC 24 447 70. [3,3; A1991, N0863]

M0213 1, 2, 2, 2, 2, 2, 3, 2, 3, 3, 3, 4, 3, 2, 4, 3, 4, 4, 3, 3, 5, 4, 4, 6, 4, 3, 6, 3, 4, 7, 4, 5, 6,
3, 5, 7, 6, 5, 7, 5, 5, 9, 5, 4, 10, 4, 5, 7, 4, 6, 9, 6, 6, 9, 7, 7, 11, 6, 6, 12, 4, 5, 10, 4, 7, 10, 6
Decompositions of $2n$ into sum of 2 primes. Ref L1 79. SS64. MOC 19 332 65. [1,2;
A1031, N0077]

M0214 1, 1, 2, 2, 2, 2, 2, 3, 3, 2, 4, 4, 2, 4, 4, 4, 4, 3, 4, 6, 4, 2, 6, 6, 3, 6, 6, 4, 6, 4, 6, 7, 4,
4, 8, 8, 2, 6, 8, 6, 8, 4, 4, 10, 6, 4, 10, 8, 5, 7
Reduced binary forms of determinant n. Ref SPLAG 360. [1,3; A6374]

M0215 1, 1, 2, 2, 2, 2, 2, 3, 3, 3, 4, 5, 6, 6, 7
The coding-theoretic function $A(n,14,9)$. See Fig M0240. Ref PGIT 36 1337 90. [14,3;
A5861]

M0216 1, 1, 2, 2, 2, 2, 3, 3, 3, 3, 4, 4, 4, 5, 5, 6, 8
The coding-theoretic function $A(n,12,7)$. See Fig M0240. Ref PGIT 36 1337 90. [12,3;
A5857]

M0217 1, 1, 0, 1, 1, 1, 1, 1, 2, 2, 2, 2, 3, 3, 3, 4, 5, 5, 5, 6, 7, 8, 8, 9, 11, 12, 12, 14, 16, 17,
18, 20, 23, 25, 26, 29, 33, 35, 37, 41, 46, 49, 52, 57, 63, 68, 72, 78, 87, 93, 98, 107, 117
Partitions of n into distinct odd parts. Ref PLMS 42 553 36. CJM 4 383 52. [0,9; A0700,
N0078]

M0218 0, 1, 2, 2, 2, 2, 3, 3, 4, 5, 7, 8, 10, 13, 19, 24, 36, 53
n-dimensional determinant 3 lattices. Ref SPLAG 402. [0,3; A5139]

M0219 1, 1, 1, 1, 1, 1, 1, 1, 2, 2, 2, 2, 3, 3, 4, 5, 8, 9, 13, 16, 28, 40, 68, 117, 297, 665
n-dimensional determinant 1 lattices. Ref SPLAG 49. [0,9; A5134]

M0220 1, 1, 2, 2, 2, 2, 3, 3, 5, 5, 6, 6, 9, 10, 13, 15, 19
The coding-theoretic function $A(n,12,8)$. See Fig M0240. Ref PGIT 36 1337 90. [12,3; A5858]

M0221 1, 1, 1, 1, 2, 2, 2, 2, 3, 4, 4, 4, 5, 6, 6, 6, 8, 9, 10, 10, 12, 13, 14, 14, 16, 19, 20, 21, 23, 26, 27, 28, 31, 34, 37, 38, 43, 46, 49, 50, 55, 60, 63, 66, 71, 78, 81, 84, 90, 98, 104
Partitions of n into squares. Ref BIT 19 298 79. [0,5; A1156, N0079]

M0222 1, 1, 1, 2, 2, 2, 2, 3, 4, 4, 5, 6, 8, 9, 11
The coding-theoretic function $A(n,14,10)$. See Fig M0240. Ref PGIT 36 1337 90. [14,4; A5862]

M0223 1, 1, 1, 2, 2, 2, 2, 4, 2, 4, 5, 5, 6
Simplicial arrangements of n lines in the plane. Ref GR72 7. [3,4; A3036]

M0224 0, 1, 2, 2, 2, 2, 4, 4, 2, 2, 4, 4, 6, 4, 4, 6, 8, 6, 6, 6, 4, 8, 8, 8, 8, 8, 8, 6, 10, 8, 10, 10, 8, 12, 8, 10, 14, 12, 10, 12, 16, 10, 18, 14, 12, 14, 16, 14, 16, 14, 10, 16, 20, 14, 12, 16, 14
of (j,k): $j+k=n$, $(j,n)=(k,n)=1$, j,k squarefree. Ref AMM 99 573 92. [1,3; A7457]

M0225 1, 1, 2, 2, 2, 2, 4, 4, 4, 2, 6, 6, 4, 4, 4, 2, 6, 8, 4, 4, 6, 4, 2, 6, 8, 8, 8, 8, 4, 4, 10, 8, 4, 4, 4, 10, 12, 4, 8, 4, 14, 4, 8, 6, 6, 12, 8, 8, 6, 10, 12, 4, 4, 14, 8, 8, 8, 4, 8, 16, 14, 8, 6, 8
Class number of quadratic field with discriminant $-4n$, $n \equiv 1,2 \pmod 4$. Cf. M0152. Ref BU89 231. [1,3; A6643]

M0226 1, 1, 1, 1, 1, 1, 1, 2, 2, 2, 2, 4, 4, 8, 16, 32
The coding-theoretic function $A(n,8)$. See Fig M0240. Ref PGIT 36 1338 90. [1,8; A5866]

M0227 1, 2, 2, 2, 3, 2, 2, 2, 4, 2, 2, 2, 2, 4, 2, 2, 2, 2, 3, 4, 2, 2, 2, 2, 2, 2, 4, 2, 2, 2, 2, 2, 2, 4, 4, 2, 2, 2, 2, 2, 2, 2, 4, 2, 2, 2, 2, 2, 2, 2, 4, 4, 2, 2, 2, 2, 2, 2, 2, 2, 4, 2, 2, 2, 3
Occurrences of n as a binomial coefficient. Ref AMM 78 385 71. OG72 96. [2,2; A3016]

M0228 1, 1, 2, 2, 2, 3, 3, 3, 4, 4, 5, 7, 7, 8, 9, 10, 13, 14, 16
The coding-theoretic function $A(n,10,6)$. See Fig M0240. Ref PGIT 36 1336 90. [10,3; A5854]

M0229 1, 1, 2, 2, 2, 3, 3, 4, 3, 4, 3, 6, 4, 5, 6, 6, 4, 7, 5, 8, 8, 7, 5, 12, 6, 8, 7, 10, 6, 13, 7, 10, 10, 10, 10, 14, 8, 11, 12, 16, 8, 17, 9, 14, 14, 13, 9, 20, 11, 16, 14, 16, 10, 19, 14, 20
Primitive sublattices of index n in hexagonal lattice. Ref DM 4 216 73. BSW94. [1,3; A3050]

M0230 1, 1, 1, 1, 2, 2, 2, 3, 4, 4, 4, 4, 5, 6, 6, 7, 8, 8, 8, 8, 8, 9, 10, 10, 11, 12, 12, 12, 13, 14, 14, 15, 16, 16, 16, 16, 16, 16, 17, 18, 18, 19, 20, 20, 20, 21, 22, 22, 23, 24, 24, 24, 24
A well-behaved cousin of the Hofstadter sequence. Ref DM 105 227 92. TYCM 24 105 93. [0,5; A6949]

M0231 1, 1, 1, 2, 2, 2, 3, 4, 4, 6, 7, 10, 13, 21
The coding-theoretic function $A(n,14,11)$. See Fig M0240. Ref PGIT 36 1337 90. [14,4; A5863]

M0232 1, 1, 2, 2, 2, 3, 4, 5, 6, 8, 10, 13, 16
The coding-theoretic function $A(n,10,7)$. See Fig M0240. Ref PGIT 36 1336 90. [10,3; A5855]

M0233 0, 2, 2, 2, 4, 2, 4, 4, 4, 4, 4, 6, 8, 6, 6, 6, 8, 6, 8, 8, 8, 8, 8, 10, 12, 10, 10, 10, 12, 10, 12
Highest minimal distance of self-dual code of length $2n$. Ref PGIT 36 1319 90. SPLAG xxiv. [0,2; A5137]

M0234 1, 1, 2, 2, 2, 4, 4, 4, 6, 7, 7, 10, 11, 11, 15, 17, 17, 22, 24, 25, 32, 35, 36, 44, 48, 50, 60, 66, 68, 81, 89, 92, 107, 117, 121, 141, 153, 159, 181, 197, 205, 233, 252, 262, 295
Partitions of n into triangular numbers. [1,3; A7294]

M0235 1, 1, 1, 1, 1, 1, 1, 1, 2, 2, 2, 4, 4, 4, 7, 7, 8, 12, 12, 16, 21, 21, 31, 37, 38, 58, 65, 71, 106, 114, 135, 191, 201, 257, 341, 359, 485, 605, 652, 904, 1070, 1202, 1664, 1894, 2237
A generalized Fibonacci sequence. Ref FQ 4 244 66. [0,9; A1584, N0080]

M0236 1, 1, 1, 2, 2, 2, 4, 4, 5, 7, 8, 10, 16, 25
The coding-theoretic function $A(n,12,9)$. See Fig M0240. Ref PGIT 36 1337 90. [12,4; A5859]

M0237 1, 1, 2, 2, 2, 4, 5, 4, 8, 8, 7, 11, 8, 13, 4, 11, 12, 8, 12, 2, 13, 7, 22, 2, 8, 13, 26, 4, 26, 29, 17, 27, 26, 7, 33, 20, 16, 22, 29, 4, 13, 22, 25, 14, 22, 37, 18, 46, 42, 46, 9, 41, 12
Josephus problem: survivors. Ref JRM 10 124 77. [1,3; A7495]

M0238 1, 1, 1, 2, 2, 2, 4, 6, 6, 8, 11, 13, 17, 24, 28, 36
Rotatable partitions. Ref JLMS 43 504 68. [1,4; A2722, N0081]

M0239 2, 2, 2, 4, 6, 10, 18, 32, 56, 102, 186, 341, 630, 1170, 2184, 4096, 7710, 14563, 27594, 52428, 99864, 190650, 364722, 699050, 1342177, 2581110, 4971026, 9586980
$[2^n/n]$. [1,1; A0799, N0082]

M0240 1, 1, 1, 1, 1, 2, 2, 2, 4, 6, 12, 24, 32, 64, 128, 256
The coding-theoretic function $A(n,6)$. See Fig M0240. Ref PGIT 36 1338 90. [1,6; A5865]

M0241 2, 2, 2, 6, 12, 20, 30, 46, 74, 122, 200, 324, 522, 842, 1362
Cyclic n-bit strings with no alternating substring of length >2. Ref DM 70 295 88. [1,1; A7039]

M0242 1, 2, 2, 3, 2, 2, 3, 2, 5, 2, 3, 2, 6, 3, 5, 2, 2, 2, 2, 7, 5, 3, 2, 3, 5, 2, 5, 2, 6, 3, 3, 2, 3, 2, 2, 6, 5, 2, 5, 2, 2, 2, 19, 5, 2, 3, 2, 3, 2, 6, 3, 7, 7, 6, 3, 5, 2, 6, 5, 3, 3, 2, 5, 17, 10, 2, 3
Least positive primitive root of nth prime. Ref AS1 864. [1,2; A1918, N0083]

||||||||||||||||||||||||||||||||||||||||||||||||||||||||||||||||||||||||||||||||||||||||||||||||||||||||||||||||||||||||||||||||||

Figure M0240.   ERROR-CORRECTING CODES.

$A(n, d)$ denotes the maximal number of vectors of n 0's and 1's with the property that any two vectors differ in at least d places. $A(n, d, w)$ has a similar definition, but with the additional constraint that every vector must contain precisely w 1's. Such **codes** can correct $[(d-1)/2]$ errors. See [Be68], [MS78], [PGIT 36 1334 90] for further information. The table contains many such sequences. For example M0240 gives the known values of $A(n, 6)$. The twelfth term, $A(12, 6) = 24$, is realized by the following code:

```
000000000000   100100011101
011011100010   110010001110
001101110001   101001000111
010110111000   110100100011
001011011100   111010010001
000101101110   111101001000
000010110111   101110100100
010001011011   100111010010
011000101101   100011101001
011100010110   110001110100
001110001011   101000111010
010111000101   111111111111
```

The construction of the code corresponding to the last entry known in this sequence ($A(16, 6)$ = 256) has recently been considerably simplified — see [BAMS 29 218 93]. So has the last entry known in the $A(n, 4)$ sequence, M1111 — see [DCC 41 31 94].

||||||||||||||||||||||||||||||||||||||||||||||||||||||||||||||||||||||||||||||||||||||||||||||||||||||||||||||||||||||||||||||||||

M0243   1, 2, 2, 3, 2, 2, 3, 2, 5, 2, 3, 2, 7, 3, 5, 2, 2, 2, 2, 7, 5, 3, 2, 3, 5, 2, 5, 2, 11, 3, 3, 2, 3, 2, 2, 7, 5, 2, 5, 2, 2, 2, 19, 5, 2, 3, 2, 3, 2, 7, 3, 7, 7, 11, 3, 5, 2, 43, 5, 3, 3, 2, 5, 17, 17, 2
Least positive prime primitive root of nth prime. Ref RS9 2. AS1 864. [1,2; A2233, N0084]

M0244   1, 1, 2, 2, 3, 2, 3, 3, 3, 3, 4, 3, 4, 3, 4, 4, 5, 3, 4, 4, 4, 4, 5, 4, 5, 4, 4, 4, 5, 4, 5, 5, 5, 5, 5, 4, 5, 4, 5, 5, 6, 4, 5, 5, 5, 5, 6, 5, 5, 5, 5, 6, 5, 6, 4, 6, 5, 5, 5, 6, 5, 6, 5, 5, 6, 6, 5, 6, 6, 6
Related to $\phi(n)$. Ref BAMS 35 837 29. [1,3; A3434]

M0245   1, 1, 2, 2, 3, 2, 3, 4, 2, 2, 7, 2, 6, 9, 2, 2, 3, 2, 4, 2, 5, 2, 3, 3, 5, 2, 2, 3, 6, 3, 9, 3, 3, 4, 2, 5, 5, 4, 2, 2, 3, 2, 2, 5, 2, 2, 4, 9, 3, 6, 3, 2, 7, 3, 3, 2, 2, 2, 5, 3, 6, 2, 7, 2, 10, 2, 5, 10
Least negative primitive root of nth prime. Ref AS1 864. [1,3; A2199, N0085]

M0246   1, 2, 2, 3, 2, 4, 2, 4, 3, 4, 2, 6, 2, 4, 4, 5, 2, 6, 2, 6, 4, 4, 2, 8, 3, 4, 4, 6, 2, 8, 2, 6, 4, 4, 4, 9, 2, 4, 4, 8, 2, 8, 2, 6, 6, 4, 2, 10, 3, 6, 4, 6, 2, 8, 4, 8, 4, 4, 2, 12, 2, 4, 6, 7, 4, 8, 2, 6
$d(n)$, the number of divisors of n. Ref AS1 840. [1,2; A0005, N0086]

M0247 1, 2, 2, 3, 2, 4, 2, 4, 4, 4, 2, 6, 2, 4, 4, 6, 2, 8, 2, 6, 4, 4, 2, 8, 6, 4, 6, 6, 2, 8, 2, 8, 4,
4, 4, 12, 2, 4, 4, 8, 2, 8, 2, 6, 8, 4, 2, 12, 8, 12, 4, 6, 2, 12, 4, 8, 4, 4, 2, 12, 2, 4, 8, 12, 4, 8
Parabolic vertices of $\Gamma_0(n)$. Ref NBS B67 62 63. [1,2; A1616, N0087]

M0248 1, 0, 2, 2, 3, 2, 5, 2, 5, 4, 5, 2, 8, 2, 9, 8, 9, 2, 9, 2, 11, 8, 9, 2
$a(n)$ is number of k for which $C(n,k)$ is not divisible by n. Ref jhc. [0,3; A7012]

M0249 1, 2, 2, 3, 3, 3, 4, 3, 4, 5, 4, 5, 4, 4, 6, 5, 6, 6, 5, 6, 4, 5, 7, 6, 8, 7, 6, 8, 6, 7, 8, 6, 7,
5, 5, 8, 7, 9, 9, 8, 10, 7, 8, 10, 8, 10, 8, 7, 10, 8, 9, 9, 7, 8, 5, 6, 9, 8, 11, 10, 9, 12, 9, 11, 13
Representations of n as a sum of Fibonacci numbers. Ref FQ 4 304 66. [0,2; A0121, N0088]

M0250 1, 2, 2, 3, 3, 3, 4, 4, 4, 4, 5, 5, 5, 5, 5, 6, 6, 6, 6, 6, 6, 7, 7, 7, 7, 7, 7, 7, 8, 8, 8, 8, 8,
8, 8, 8, 9, 9, 9, 9, 9, 9, 9, 9, 10, 10, 10, 10, 10, 10, 10, 10, 10, 10, 11, 11, 11, 11, 11, 11
n appears n times: $a(n) = [\frac{1}{2} + \sqrt{(2n)}]$. See Fig M0436. Ref MMAG 38 186 65. KN1 1 43.
GKP 97. [1,2; A2024, N0089]

M0251 1, 1, 1, 1, 2, 2, 3, 3, 3, 4, 4, 4, 4, 5, 6, 6, 6, 6, 6, 7, 8, 8, 9, 9, 9, 9, 9, 9, 10, 11, 11,
12, 12, 12, 13, 13, 13, 13, 13, 13, 14, 15, 16, 16, 17, 17, 17, 18, 18, 18, 18, 19, 19, 19
$a(n) = a(a(n-1)) + a(n - a(n-1))$. Ref AMM 95 555 88. [1,5; A5707]

M0252 1, 1, 2, 2, 3, 3, 3, 4, 4, 4, 5, 5, 5, 5, 5, 5, 6, 6, 6, 6, 7
Chromatic number of path with n nodes. Ref ADM 41 21 89. [1,3; A6670]

M0253 1, 1, 1, 2, 2, 3, 3, 3, 4, 5, 5, 5, 5, 6, 7, 7, 8, 8, 8, 8, 9, 10, 11, 11, 12, 12, 12, 13,
13, 13, 13, 13, 13, 14, 15, 16, 16, 17, 18, 18, 19, 19, 19, 20, 20, 20, 20, 21, 21, 21, 21, 21
$a(n) = a(a(n-1)) + a(n - a(n-1))$. Ref MMAG 63 11 90. [1,4; A5350]

M0254 1, 1, 1, 1, 1, 2, 2, 3, 3, 3, 4, 5, 6, 7, 8, 9, 10, 12, 14, 16, 18, 20, 23, 26, 30, 34, 38,
42, 47, 53, 60, 67, 74, 82, 91, 102, 114, 126, 139, 153, 169, 187, 207, 228, 250, 274, 301
Partitions of n into parts $6n + 1$ or $6n - 1$. [0,6; A3105]

M0255 0, 1, 2, 2, 3, 3, 4, 3, 4, 4, 5, 4, 5, 5, 5, 4, 5, 5, 6, 5, 6, 6, 6, 5, 6, 6, 6, 6, 7, 6, 7, 5, 6,
6, 7, 6, 7, 7, 7, 6, 7, 7, 7, 7, 7, 7, 8, 6, 7, 7, 7, 7, 8, 7, 8, 7, 8, 8, 8, 7, 8, 8, 8, 6, 7, 7, 8, 7, 8
Minimum multiplications to compute nth power. Ref KN1 2 446. [1,3; A3313]

M0256 0, 1, 2, 2, 3, 3, 4, 4, 4, 4, 5, 5, 6, 6, 6, 6, 7, 7, 8, 8, 8, 8, 9, 9, 9, 9, 9, 9, 10, 10, 11,
11, 11, 11, 11, 11, 12, 12, 12, 12, 13, 13, 14, 14, 14, 14, 15, 15, 15, 15, 15, 15, 16, 16, 16
$\pi(n)$, the number of primes $\leq n$. Ref AS1 870. [1,3; A0720, N0090]

M0257 1, 2, 2, 3, 3, 4, 4, 4, 5, 5, 5, 6, 6, 6, 6, 7, 7, 7, 7, 8, 8, 8, 8, 9, 9, 9, 9, 9, 10, 10, 10,
10, 10, 11, 11, 11, 11, 11, 12, 12, 12, 12, 12, 12, 13, 13, 13, 13, 13, 13, 14, 14, 14, 14, 14
$a(n)$ is the number of times n occurs. See Fig M0436. Ref AMM 74 740 67. GKP 66.
UPNT E25. JNT 40 1 92. [1,2; A1462, N0091]

M0258 1, 1, 2, 2, 3, 3, 4, 4, 4, 5, 5, 5, 6, 6, 6, 7, 7, 7, 7, 8, 8, 8, 8, 9, 9, 9, 9, 10, 10, 10, 10,
  10, 11, 11, 11, 11, 11, 12, 12, 12, 12, 12, 13, 13, 13, 13, 13, 13, 14, 14, 14, 14, 14, 14, 15
A self-generating sequence. Ref MMAG 52 265 79. jos. [0,3; A5041]

M0259 1, 1, 2, 2, 3, 3, 4, 4, 4, 5, 5, 6, 6, 6, 6, 7, 7, 7, 8, 8, 8, 8, 9, 9, 9, 10, 10, 10, 10, 10,
  11, 11, 11, 11, 12, 12, 12, 12, 12, 13, 13, 13, 13, 13, 14, 14, 14, 14, 15, 15, 15, 15, 15, 15
Integer part of square root of *n*th prime. Ref AS1 2. [1,3; A0006, N0092]

M0260 1, 0, 0, 1, 1, 1, 1, 1, 2, 2, 3, 3, 4, 4, 5, 6, 7, 8, 10, 11, 13
Related to Rogers-Ramanujan identities. Ref AMM 96 403 89. [0,9; A6141]

M0261 1, 0, 1, 1, 1, 1, 2, 2, 3, 3, 4, 4, 6, 6, 8, 9, 11, 12, 15, 16, 20, 22, 26, 29, 35, 38, 45,
  50, 58, 64, 75, 82, 95, 105, 120, 133, 152, 167, 190, 210, 237, 261, 295, 324, 364, 401
Partitions of *n* into parts $5n+2$ or $5n+3$. Ref Andr76 238. AMM 95 711 88; 96 403 89.
[0,7; A3106]

M0262 0, 1, 1, 1, 1, 1, 1, 2, 2, 3, 3, 4, 4, 6, 7, 11, 14, 24, 30
n-dimensional determinant 2 lattices. Ref SPLAG 400. [0,8; A5138]

M0263 1, 1, 2, 2, 3, 3, 4, 5, 5, 6, 6, 7, 8, 8, 9, 9, 10, 11, 11, 12, 13, 13, 14, 14, 15, 16, 16,
  17, 17, 18, 19, 19, 20, 21, 21, 22, 22, 23, 24, 24, 25, 25, 26, 27, 27, 28, 29, 29, 30, 30, 31
The female of a pair of recurrences. See Fig M0436. Cf. M0278. Ref GEB 137. [0,3;
A5378]

M0264 1, 1, 2, 2, 3, 3, 4, 5, 5, 6, 8, 9, 10, 11, 10, 13, 17, 19, 21, 22, 21, 24, 32, 37, 37, 38,
  40, 45, 55, 65, 69, 66, 64, 75, 86, 100, 113, 107, 106, 122, 145, 165, 174, 167, 162, 179
Partitions of *n* into relatively prime parts (allowing a part = 1). Ref mlb. [0,3; A7360]

M0265 1, 0, 1, 1, 1, 2, 2, 3, 3, 4, 5, 6, 7, 9, 10, 12, 14, 17, 19, 23, 26, 30, 35, 40, 46, 52, 60,
  67, 77, 87, 98, 111, 124, 140, 157, 175, 197, 219, 244, 272, 302, 336, 372, 413, 456, 504
Partitions of *n* into prime parts. Ref PNISI 21 183 55. AMM 95 711 88. [0,6; A0607,
N0093]

M0266 1, 1, 1, 1, 2, 2, 3, 3, 4, 5, 6, 7, 9, 10, 12, 14, 17, 19, 23, 26, 31, 35, 41, 46, 54, 61,
  70, 79, 91, 102, 117, 131, 149, 167, 189, 211, 239, 266, 299, 333, 374, 415, 465, 515, 575
Partitions of *n* into parts $5n+1$ or $5n-1$. Ref Andr76 238. AMM 95 711 88; 96 403 89.
[0,5; A3114]

M0267 1, 2, 2, 3, 3, 4, 5, 6, 8, 9, 12, 14, 18, 22, 27, 34, 41, 52, 63, 79, 97, 120, 149, 183,
  228, 280, 348, 429, 531, 657, 811
Twopins positions. Ref GU81. [5,2; A5686]

M0268 1, 1, 2, 2, 3, 3, 4, 6, 6, 7, 9, 12, 16, 21, 21, 23, 24, 30, 30
The coding-theoretic function $A(n,8,5)$. See Fig M0240. Ref PGIT 36 1336 90. [8,3;
A5851]

M0269 0, 0, 1, 2, 2, 3, 3, 5, 4, 6, 5, 7, 6, 7, 7, 10, 8, 9, 9, 12, 10, 11, 11, 14, 12, 14, 13, 15, 14, 17, 15, 19, 16, 20, 17, 19, 18, 19, 19, 26, 20, 22, 21, 23, 22, 23, 23, 30, 24, 26, 25, 28
2-part of number of tournaments on n nodes. Ref CaRo91. [1,4; A7150]

M0270 1, 2, 2, 3, 3, 5, 5, 7, 8, 10, 11, 15, 16, 20, 23, 28, 31, 38, 42, 51, 57, 67, 75, 89, 99, 115, 129, 149, 166, 192, 213, 244, 272, 309, 344, 391, 433, 489, 543, 611, 676, 760, 839
Expansion of a permanent. Ref dhl. [1,2; A3113]

M0271 1, 1, 1, 1, 2, 2, 3, 3, 5, 6, 8, 8, 12, 13, 17, 19, 26, 28, 37, 40, 52, 58, 73, 79, 102, 113, 139, 154, 191, 210, 258, 284, 345, 384, 462, 509, 614, 679, 805, 893, 1060, 1171
Partitions of n into non-prime parts. Ref JNSM 9 91 69. [0,5; A2095, N0094]

M0272 1, 1, 1, 1, 2, 2, 3, 3, 5, 6, 8, 9, 11, 14, 19, 22
Rotatable partitions. Ref JLMS 43 504 68. [1,5; A2723, N0095]

M0273 1, 0, 2, 2, 3, 4, 1, 8, 1, 10, 9, 16, 18, 12, 42, 4, 58, 38, 82, 88, 54, 188, 18, 248, 151, 334, 338, 260, 760, 120
From symmetric functions. Ref PLMS 23 309 23. [0,3; A2122, N0096]

M0274 0, 1, 2, 2, 3, 4, 3, 4, 4, 3, 4, 4, 5, 5, 4, 6, 5, 6, 6, 6, 4, 6, 7, 6, 6, 5, 7, 6, 10, 4, 7, 8, 5, 5, 6, 7, 6, 6, 6, 8, 6, 6, 6, 5, 5, 6, 7, 7, 7, 6, 7, 6, 5, 7, 6, 7, 9, 7, 7, 7, 9, 5, 7, 10, 7, 7
Consecutive quadratic nonresidues mod p. Ref BAMS 32 284 26. [2,3; A2308, N0097]

M0275 1, 2, 2, 3, 4, 4, 4, 4, 5, 5, 6, 6, 7, 8, 8, 8, 8, 8, 8, 9, 9, 9, 9, 10, 10, 11, 11, 11, 11, 12, 12, 13, 13, 13, 13, 14, 14, 14, 14, 15, 16, 16, 16, 16, 16, 16, 16, 16, 17, 17, 17
3-free sequences. Ref MOC 26 768 72. [1,2; A3002]

M0276 1, 1, 2, 2, 3, 4, 4, 4, 5, 6, 7, 7, 8, 8, 8, 8, 9, 10, 11, 12, 12, 13, 14, 14, 15, 15, 15, 16, 16, 16, 16, 16, 17, 18, 19, 20, 21, 21, 22, 23, 24, 24, 25, 26, 26, 27, 27, 27, 28, 29, 29, 30
The Hofstadter-Conway $10,000 sequence: $a(n) = a(a(n-1)) + a(n - a(n-1))$. Ref drh. CO89. AMM 98 6 91. [1,3; A4001]

M0277 1, 1, 2, 2, 3, 4, 4, 4, 5, 6, 7, 8, 8, 8, 8, 9, 10, 11, 12, 13, 14, 15, 16, 16, 16, 16, 16, 16, 16, 16, 16, 17, 18, 19, 20, 21, 22, 23, 24, 25, 26, 27, 28, 29, 30, 31, 32, 32, 32, 32, 32
$a(2n+1) = a(n+1) + a(n)$, $a(2n) = 2a(n)$. Ref JRM 22 91 90. [1,3; A6165]

M0278 0, 0, 1, 2, 2, 3, 4, 4, 5, 6, 6, 7, 7, 8, 9, 9, 10, 11, 11, 12, 12, 13, 14, 14, 15, 16, 16, 17, 17, 18, 19, 19, 20, 20, 21, 22, 22, 23, 24, 24, 25, 25, 26, 27, 27, 28, 29, 29, 30, 30, 31
The male of a pair of recurrences. See Fig M0436. Cf. M0263. Ref GEB 137. [0,4; A5379]

M0279 1, 1, 2, 2, 3, 4, 5, 6, 7, 8, 10, 11, 13, 14, 16, 18, 20, 22, 24, 26, 29, 31, 34, 36, 39, 42, 45, 48, 51, 54, 58, 61, 65, 68, 72, 76, 80, 84, 88, 92, 97, 101, 106, 110, 115, 120, 125
Denumerants: expansion of $1/(1-x)(1-x^2)(1-x^5)$. Ref R1 152. [0,3; A0115, N0098]

M0280 1, 1, 2, 2, 3, 4, 5, 6, 7, 8, 11, 12, 15, 16, 19, 22, 25, 28, 31, 34, 40, 43, 49, 52, 58, 64, 70, 76, 82, 88, 98, 104, 114, 120, 130, 140, 150, 160, 170, 180, 195, 205, 220, 230
Denumerants. Ref R1 152. [0,3; A0008, N0099]

M0281 1, 1, 1, 2, 2, 3, 4, 5, 6, 8, 10, 12, 15, 18, 22, 27, 32, 38, 46, 54, 64, 76, 89, 104, 122, 142, 165, 192, 222, 256, 296, 340, 390, 448, 512, 585, 668, 760, 864, 982, 1113, 1260
Partitions of n into distinct parts. Ref AS1 836. [0,4; A0009, N0100]

$$\text{G.f.: } \prod_{k=1}^{\infty} (1 + x^k).$$

M0282 1, 1, 2, 2, 3, 4, 5, 7, 9, 11, 14, 18, 23, 29, 38, 47, 59, 76, 95, 120, 154, 191, 241, 310, 383, 483, 620, 767, 968, 1242, 1535, 1937, 2486, 3071, 3875, 4972, 6143, 7752
A nonlinear recurrence. Ref KN1 3 208. [0,3; A3073]

M0283 1, 1, 1, 1, 2, 2, 3, 4, 5, 7, 9, 11, 15, 18, 23, 30, 37, 47, 58, 71, 90, 110, 136, 164, 201, 248, 300, 364, 436, 525, 638, 764, 919, 1090, 1297, 1549, 1845, 2194, 2592, 3060
Maximum of a partition function. Ref JIMS 6 112 42. PSPM 19 172 71. [0,5; A2569, N0101]

M0284 1, 1, 1, 2, 2, 3, 4, 5, 7, 9, 12, 16, 21, 28, 37, 49, 65, 86, 114, 151, 200, 265, 351, 465, 616, 816, 1081, 1432, 1897, 2513, 3329, 4410, 5842, 7739, 10252, 13581, 17991
$a(n) = a(n-2) + a(n-3)$. Ref JA66 90. MMAG 41 17 68. [0,4; A0931, N0102]

M0285 1, 1, 1, 1, 2, 2, 3, 4, 6, 7, 11, 16, 24
The coding-theoretic function $A(n,12,10)$. See Fig M0240. Ref PGIT 36 1337 90. [12,5; A5860]

M0286 1, 1, 0, 1, 1, 2, 2, 3, 4, 6, 8, 11, 16, 23, 32, 46, 66, 94, 136, 195, 282, 408, 592, 856, 1248, 1814, 2646, 3858, 5644, 8246, 12088
Generalized Fibonacci numbers. Ref FQ 27 120 89. [1,6; A6207]

M0287 1, 1, 1, 2, 2, 3, 4, 6, 9, 12, 17, 21
The coding-theoretic function $A(n,10,8)$. See Fig M0240. Ref PGIT 36 1336 90. [10,4; A5856]

M0288 1, 2, 2, 3, 4, 6, 9, 14, 22, 35, 56, 90, 145, 234, 378, 611, 988, 1598, 2585, 4182, 6766, 10947, 17712, 28658, 46369, 75026, 121394, 196419, 317812, 514230, 832041
Fibonacci numbers + 1. Ref JA66 97. [0,2; A1611, N0103]

M0289 1, 1, 1, 2, 2, 3, 4, 7, 9, 16, 25, 55, 103, 261, 731
Number of self-dual binary codes of length $2n$. Ref PGIT 24 738 78. JCT A28 52 80. JCT A60 183 92. [0,4; A3179]

M0290 0, 0, 0, 1, 1, 1, 2, 2, 3, 5, 6, 7
4-in a row orchard problem with n trees. See Fig M0982. Ref GA88 116. [1,7; A6065]

M0291 1, 1, 2, 2, 3, 5, 6, 9, 13, 14, 15, 20, 20, 22, 25, 30, 31, 37, 40, 42, 50, 52, 54, 63, 65, 67, 76, 80, 82, 92, 96, 99, 111, 114, 117, 130, 133, 136, 149, 154, 157, 171, 176, 180, 196
The coding-theoretic function $A(n,6,4)$. See Fig M0240. Ref DM 20 1 77. JCT A26 278 79. PGIT 36 1335 90. [6,3; A4037, N0104]

M0292 1, 1, 1, 2, 2, 3, 5, 11, 32, 163, 1680
Linear geometries on n points with ≤ 3 points per line. See Fig M1197. Ref JCT A49 28 88. [0,4; A5426]

M0293 2, 2, 3, 6, 0, 6, 7, 9, 7, 7, 4, 9, 9, 7, 8, 9, 6, 9, 6, 4, 0, 9, 1, 7, 3, 6, 6, 8, 7, 3, 1, 2, 7, 6, 2, 3, 5, 4, 4, 1, 8, 3, 5, 9, 6, 1, 1, 5, 2, 5, 7, 2, 4, 2, 7, 0, 8, 9, 7, 2, 4, 5, 4, 1, 0, 5, 2, 0, 9
Decimal expansion of square root of 5. Ref RS8 XVIII. MOC 22 234 68. [1,1; A2163, N0105]

M0294 2, 2, 3, 7, 5, 11, 103, 71, 661, 269, 329891, 39916801, 2834329, 75024347, 3790360487, 46271341, 1059511, 1000357, 123610951, 1713311273363831
Largest factor of $n! + 1$. Ref SMA 14 25 48. MOC 26 569 72. [0,1; A2583, N0312]

M0295 0, 0, 1, 1, 2, 2, 3, 7, 15, 12, 30, 8, 32, 162, 21
Solutions of $x + y = z$ from $\{1, 2, ..., n\}$. Ref GU71. [1,5; A2848, N0106]

M0296 1, 2, 2, 4, 2, 4, 2, 4, 6, 2, 6, 4, 2, 4, 6, 6, 2, 6, 4, 2, 6, 4, 6, 8, 4, 2, 4, 2, 4, 14, 4, 6, 2, 10, 2, 6, 6, 4, 6, 6, 2, 10, 2, 4, 2, 12, 12, 4, 2, 4, 6, 2, 10, 6, 6, 6, 2, 6, 4, 2, 10, 14, 4, 2, 4
Differences between consecutive primes. Ref AS1 870. [1,2; A1223, N0108]

M0297 1, 2, 2, 4, 2, 4, 4, 8, 2, 4, 4, 8, 4, 8, 8, 16, 2, 4, 4, 8, 4, 8, 8, 16, 4, 8, 8, 16, 8, 16, 16, 32, 2, 4, 4, 8, 4, 8, 8, 16, 4, 8, 8, 16, 8, 16, 16, 32, 4, 8, 8, 16, 8, 16, 16, 32, 8, 16, 16, 32
Gould's sequence: $\Sigma\ (C(n,k) \mod 2)$. Ref GO61. TCS 98 188 92. [0,2; A1316, N0109]

M0298 1, 1, 2, 2, 4, 2, 6, 2, 6, 4, 10, 2, 12, 6, 4, 4, 16, 6, 18, 4, 6, 10, 22, 2, 20, 12, 18, 6, 28, 4, 30, 8, 10, 16, 12, 6, 36, 18, 12, 4, 40, 6, 42, 10, 12, 22, 46, 4, 42, 20, 16, 12, 52, 18
Reduced totient function $\psi(n)$: least k such that $x^k \equiv 1 \pmod{n}$ for all x prime to n. Ref CAU (2) 12 43. L1 7. [1,3; A2322, N0110]

M0299 1, 1, 2, 2, 4, 2, 6, 4, 6, 4, 10, 4, 12, 6, 8, 8, 16, 6, 18, 8, 12, 10, 22, 8, 20, 12, 18, 12, 28, 8, 30, 16, 20, 16, 24, 12, 36, 18, 24, 16, 40, 12, 42, 20, 24, 22, 46, 16, 42, 20, 32, 24
Euler totient function $\phi(n)$: count numbers $\leq n$ and prime to n. See Fig M0500. Ref AS1 840. MOC 23 682 69. [1,3; A0010, N0111]

M0300 1, 1, 2, 2, 4, 3, 3, 1, 5, 1, 1, 4, 10, 17, 1, 14, 1, 1, 3052, 1, 1, 1, 1, 1, 1, 2, 2, 1, 3, 2, 1, 13, 5, 1, 1, 1, 13, 2, 41, 1, 4, 12, 1, 5, 2, 7, 1, 1, 3, 33, 2, 1, 1, 1, 1, 1, 1, 3, 2, 2, 1, 15, 12
Continued fraction for cube root of 5. Ref JRAM 255 124 72. [1,3; A2948]

M0301 1, 1, 1, 2, 2, 4, 3, 7, 4, 11, 6, 15, 7, 24, 8, 29, 12, 40, 13, 51, 14, 68, 19, 76, 20, 107, 23, 116, 29, 147, 30, 175, 31, 215, 39, 229, 45, 297, 46, 312, 55, 387, 56, 435, 57, 513, 73
Shifts two places under inverse Moebius transformation. Ref EIS § 2.7. [1,4; A7439]

$$a(n+2) = \Sigma\ a(k), \quad k \mid n.$$

M0302 1, 2, 2, 4, 3, 8, 4, 14, 9, 22, 8, 74, 14, 56, 48, 286, 36, 380, 60, 1214, 240, 816, 188, 15506, 464, 4236, 1434
Transitive graphs with n nodes. Ref ARS 30 174 90. bdm. [1,2; A6799]

M0303 1, 2, 2, 4, 3, 8, 6, 22, 26, 176
Regular graphs with n nodes. Ref ST90. [1,2; A5176]

M0304 1, 2, 2, 4, 4, 4, 2, 1, 2, 1, 4, 1, 1, 4, 4, 2, 1, 4, 4, 2, 2, 1, 1, 2, 4, 2, 1, 1, 1, 1, 2, 4, 4, 2, 2, 1, 1, 1, 2, 4, 2, 4, 4, 2, 2, 2, 4, 4, 1, 1, 1, 4, 4, 2, 2, 2, 4, 2, 4, 2, 2, 4, 4, 1, 1, 1, 1, 4, 4
Image of n after $3k$ iterates of '$3x+1$' map (k large). See Fig M2629. Ref UPNT E16. [1,2; A6460]

M0305 1, 1, 1, 1, 1, 2, 2, 4, 4, 6, 6, 8, 10, 14, 18, 24, 29, 36, 44, 58, 72, 91, 113, 143, 179, 227, 287, 366, 460, 578, 732, 926, 1174, 1489, 1879, 2365, 2988, 3780, 4788, 6049, 7628
From the '$3x+1$' problem. See Fig M2629. Ref MAG 71 273 87. rkg. [0,6; A5186]

M0306 1, 0, 1, 1, 2, 2, 4, 4, 6, 7, 10, 11, 16, 17, 23, 26, 33, 37, 47, 52, 64, 72, 86, 96, 115, 127, 149, 166, 192, 212, 245, 269, 307, 338, 382, 419, 472, 515, 576, 629, 699, 760, 843
Expansion of a generating function. Ref CAY 10 415. [0,5; A1996, N0112]

M0307 1, 2, 2, 4, 4, 6, 7, 11, 12, 16, 18, 25, 28, 36, 41, 53, 59, 73, 82, 102, 115, 138, 155, 186, 209, 246, 275, 324, 363, 420, 468, 541, 605, 691, 768, 877, 976, 1103, 1222, 1380
Oscillates under partition transform. Cf. M0308. Ref BeSl94. EIS § 2.7. [1,2; A7212]

M0308 1, 2, 2, 4, 4, 7, 8, 12, 13, 18, 21, 29, 33, 43, 49, 63, 71, 91, 103, 128, 143, 176, 198, 241, 271, 324, 363, 431, 483, 569, 636, 743, 827, 960, 1068, 1236, 1371, 1573, 1742
Oscillates under partition transform. Cf. M0307. Ref BeSl94. EIS § 2.7. [1,2; A7213]

M0309 1, 0, 1, 1, 2, 2, 4, 4, 7, 8, 12, 14, 21, 24, 34, 41, 55, 66, 88, 105, 137, 165, 210, 253, 320, 383, 478, 574, 708, 847, 1039, 1238, 1507, 1794, 2167, 2573, 3094, 3660, 4378
Partitions of n with no part of size 1. Ref TAIT 1 334. AS1 836. [0,5; A2865, N0113]

M0310 1, 0, 0, 0, 1, 1, 2, 2, 4, 4, 7, 8, 14, 16, 25, 31
Bosonic string states. Ref CU86. [1,7; A5308]

$$\text{G.f.: } \prod (1 - x^k)^{-c(k)}, \; c(k) = 0,0,0,1,1,2,2,3,3,4,4,....$$

M0311 1, 1, 2, 2, 4, 4, 7, 10, 16, 17, 21, 28, 40, 56, 77
The coding-theoretic function $A(n,8,6)$. See Fig M0240. Ref PGIT 36 1336 90. [8,3; A5852]

M0312 1, 2, 2, 4, 4, 8, 9, 18, 23, 44, 63, 122, 190, 362, 612, 1162, 2056, 3914, 7155, 13648, 25482, 48734, 92205, 176906, 337594, 649532, 1246863, 2405236, 4636390
Necklaces with n beads. Ref IJM 5 662 61. JAuMS 33 12 82. [1,2; A0011, N0114]

M0313 1, 2, 2, 4, 4, 8, 10, 20, 30, 56, 94, 180, 316, 596, 1096, 2068, 3856, 7316, 13798, 26272, 49940, 95420, 182362, 349716, 671092, 1290872, 2485534, 4794088, 9256396
Necklaces with n beads. Ref IJM 5 662 61. [1,2; A0013, N0115]

M0314   1, 0, 2, 2, 4, 4, 10, 10, 19, 23, 38, 47, 75, 92, 140, 179, 257, 329, 466, 595, 821,
1055, 1426, 1828, 2442, 3117, 4112, 5244, 6836, 8685
Representation degeneracies for boson strings. Ref NUPH B274 546 86. [3,3; A5293]

M0315   1, 2, 2, 4, 5, 7, 9, 12, 16, 20, 25, 32, 39, 49, 58, 73, 86, 105, 123, 149, 175, 207,
241, 284, 331, 385, 444, 515, 592, 682, 777, 894, 1015, 1160, 1310, 1492, 1683, 1903
Number of partitions of n into parts of sizes $\{a(\ )\}$ is $a(n)$. Ref BeSl94. [1,2; A7209]

M0316   1, 1, 2, 2, 4, 5, 7, 9, 13, 16, 22, 27, 36, 44, 57, 70, 89, 108, 135, 163, 202, 243, 297,
355, 431, 513, 617, 731, 874, 1031, 1225, 1439, 1701, 1991, 2341, 2731, 3197, 3717
Partitions of n into parts prime to 3. Ref PSPM 8 145 65. [0,3; A0726, N0116]

M0317   1, 1, 1, 1, 2, 2, 4, 5, 8, 11, 18, 25, 40, 58, 90, 135, 210, 316, 492, 750, 1164, 1791,
2786, 4305, 6710, 10420, 16264, 25350, 39650, 61967, 97108
Generalized Fibonacci numbers. Ref FQ 27 120 89. [1,5; A6206]

M0318   1, 2, 2, 4, 5, 9, 12, 21, 30, 51, 76, 127, 195, 322, 504, 826, 1309, 2135, 3410, 5545,
8900, 14445, 23256, 37701, 60813, 98514, 159094, 257608, 416325, 673933, 1089648
Packing a box with n dominoes. Ref AMM 69 61 62. [1,2; A1224, N0117]

M0319   1, 1, 2, 2, 4, 5, 9, 12, 23, 34
Self-dual 2-colored necklaces with $2n$ beads. Ref PJM 110 210 84. [1,3; A7147]

M0320   1, 1, 1, 0, 1, 1, 2, 2, 4, 5, 10, 14, 26, 42, 78, 132, 249, 445, 842, 1561, 2988, 5671,
10981, 21209, 41472, 81181, 160176, 316749, 629933, 1256070, 2515169, 5049816
Series-reduced trees with n nodes. See Fig M0320. Ref AMA 101 150 59. HA69 232. dgc.
JAuMS A20 502 75. [0,7; A0014, N0118]

Figure M0320.   SERIES-REDUCED TREES.

M0320 gives the number of **series-reduced** (or **homeomorphically irreducible**) **trees**
with n nodes, i.e. trees in which no node has degree 2. (See also Fig. M0791.) A generating
function can be found in [HP78 62], [JAuMS A20 495 75]. M0768 gives the number of series-
reduced **planted** trees.

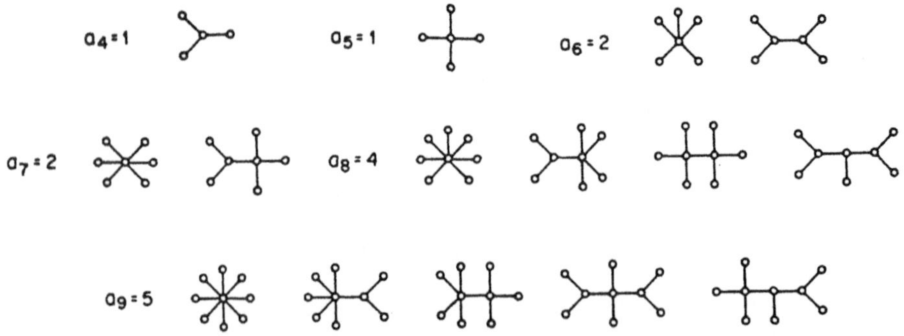

M0321 1, 2, 2, 4, 6, 6, 8, 10, 14, 20, 26, 34, 46, 62, 78, 102, 134, 176, 226, 302, 408, 528, 678, 904, 1182, 1540, 2012, 2606, 3410, 4462, 5808, 7586, 9898, 12884, 16774, 21890
Length of nth term in M4780 (a recurrence of order 71). Ref CoGo87 176. FPSAC 264. [1,2; A5341]

M0322 0, 1, 2, 2, 4, 6, 6, 11, 16, 20, 28, 41, 51, 70, 93, 122
Symmetrical planar partitions of n. Ref MA15 2 332. [1,3; A0784, N0119]

M0323 1, 2, 2, 4, 6, 8, 18, 20, 56, 48, 178, 132, 574, 348, 1870, 1008, 6144, 2812, 20314, 8420, 67534, 24396, 225472, 74756, 755672, 222556, 2540406, 693692, 8564622
Symmetric folds of a strip of n stamps. See Fig M4587. Cf. M1205. Ref JCT 5 151 68. [1,2; A1010, N0120]

M0324 1, 1, 2, 2, 4, 6, 10, 16, 30, 52, 94, 172, 316, 586, 1096, 2048, 3856, 7286, 13798, 26216, 49940, 95326, 182362, 349536, 671092, 1290556, 2485534, 4793492, 9256396
Shift registers. Ref GO67 172. BR80. [0,3; A0016, N0121]

M0325 1, 1, 1, 1, 2, 2, 4, 6, 10, 17, 29, 51, 89, 159, 284, 512, 930, 1692, 3101, 5698, 10515, 19464, 36143, 67296, 125622, 235050, 440756, 828142, 1558955, 2939761
Shifts 2 places left under weigh-transform. Ref BeSl94. EIS § 2.7. [1,5; A7560]

M0326 1, 1, 1, 2, 2, 4, 6, 11, 18, 37, 66, 135, 265, 552, 1132, 2410, 5098, 11020, 23846, 52233, 114796, 254371, 565734, 1265579, 2841632, 6408674, 14502229, 32935002
Free 3-trees with n nodes. Ref CAY 9 451. rcr. [1,4; A0672, N0122]

M0327 1, 1, 1, 2, 2, 4, 6, 12, 20, 39, 71, 137, 261, 511, 995, 1974, 3915, 7841, 15749, 31835, 64540, 131453, 268498, 550324, 1130899, 2330381, 4813031, 9963288
Series-reduced rooted trees with n nodes. Ref AMA 101 150 59. dgc. [0,4; A1679, N0123]

M0328 1, 2, 2, 4, 6, 12, 20, 40, 74, 148, 284, 568, 1116, 2232, 4424, 8848, 17622, 35244, 70340, 140680, 281076, 562152, 1123736, 2247472, 4493828, 8987656, 17973080
$a(2n+1)=2a(2n)$, $a(2n)=2a(2n-1)-a(n)$. [0,2; A3000]

M0329 1, 2, 2, 4, 7, 12, 16, 32
The coding-theoretic function $K(n,1)$. Ref JLMS 44 60 69. CRP 301 137 85. [1,2; A0983, N0124]

M0330 1, 0, 0, 2, 2, 4, 8, 4, 16, 12, 48, 80, 136, 420, 1240, 2872, 7652, 18104, 50184
Symmetric solutions to queens problem. Ref PSAM 10 93 60. [1,4; A0017, N0125]

M0331 1, 1, 2, 2, 4, 8, 13, 25, 44, 83, 152, 286, 538, 1020, 1942, 3725, 7145, 13781, 26627, 51572, 100099, 194633, 379037, 739250, 1443573, 2822186, 5522889
Integers $\leq 2^n$ of form x^2+16y^2. Ref MOC 20 567 66. [0,3; A0018, N0126]

M0332 1, 1, 1, 0, 2, 2, 4, 8, 18, 120
Sesquirotational Kotzig factorizations. Ref ADM 12 72 82. [1,5; A5702]

M0333 2, 2, 4, 9, 4, 4, 4, 10, 8, 4, 4
Number of genera of forms with | determinant | $= n$. Ref SPLAG 387. [1,1; A5141]

M0334 1, 1, 1, 2, 2, 4, 9, 22, 85
Deficiencies of partial Steiner triple systems of order n. Ref ARS 20 6 85. [3,4; A6182]

M0335 2, 2, 4, 10, 16, 28, 48, 76, 110, 144, 182, 222, 264, 310, 356, 408, 468, 536, 610,
684, 762, 842, 924, 1010, 1096, 1188, 1288, 1396, 1510, 1624, 1742, 1862, 1984, 2110
2nd differences are periodic. Ref TCPS 2 220 1827. [0,1; A2082, N0127]

M0336 2, 2, 4, 10, 18, 32, 58, 98, 164, 274, 442, 704, 1114, 1730, 2660, 4058, 6114, 9136,
13554, 19930
Representation degeneracies for Raymond strings. Ref NUPH B274 548 86. [1,1; A5304]

$$\text{G.f.: } \Pi (1 - x^k)^{-c(k)}, \ c(k) = 1,1,3,3,4,3,4,2,4,2,4,2,....$$

M0337 1, 2, 2, 4, 10, 28, 84, 264, 858, 2860, 9724, 33592, 117572, 416024, 1485800,
5348880, 19389690, 70715340, 259289580, 955277400, 3534526380, 13128240840
Expansion of $(1-4x)^{1/2}$. Ref TH09 164. FMR 1 55. [0,2; A2420, N0128]

M0338 1, 1, 1, 2, 2, 4, 10, 47, 472
Maximal partial Steiner triple systems of order n. Ref ARS 20 6 85. [3,4; A6181]

M0339 1, 0, 1, 2, 2, 4, 12, 22, 58, 158, 448, 1342, 4199, 13384, 43708, 144810, 485704,
1645576, 5623571, 19358410, 67078828, 233800162, 819267086, 2884908430
Polyhedral graphs with n edges. Ref MOC 37 524 81. trsw. Dil92. [6,4; A2840, N0129]

M0340 2, 2, 4, 12, 30, 60, 154, 404, 1046, 2540, 6720, 17484, 46522, 120300, 323800,
856032, 2315578, 6151080, 16745530, 44921984
Expansion of critical exponent for walks on tetrahedral lattice. Ref JPA 14 443 81. [1,1;
A7181]

M0341 1, 0, 2, 2, 4, 14, 52, 555, 1257
Strong starters in cyclic group of order $2n + 1$. Ref DM 56 59 85. [3,3; A6205]

M0342 1, 0, 2, 2, 4, 26, 2, 198, 12, 96, 14, 28, 2, 4, 204, 7, 58332, 11821890, 6, 36,
440140, 8909082, 46, 1405, 12, 220224, 2, 411912, 57396, 28184, 360, 74, 4, 77790, 390
From fundamental unit of $Z[(-n)^{1/4}]$. Ref MOC 48 49 87. [1,3; A6829]

M0343 2, 2, 5, 2, 5, 2, 17, 0, 3, 0, 5, 2, 29, 2, 3, 0, 3, 0, 11, 0, 7
Sum of n consecutive primes starting at $a(n)$ is prime (0 if impossible). Ref JRM 18 247
86. [1,1; A7610]

M0344 1, 2, 2, 5, 3, 8, 11, 22, 25, 53, 76, 151, 244, 435, 749, 1314, 2367, 4239, 7471,
13705
Goldbach partitions for powers of 2. Ref BIT 15 242 75. [3,2; A6307]

M0345 1, 1, 1, 2, 2, 5, 3, 10, 7, 18, 7, 64, 13, 51, 44, 272, 35, 365, 59, 1190, 235, 807, 187,
15422, 461, 4221, 1425
Connected transitive graphs with n nodes. Ref ARS 30 174 90. bdm. [1,4; A6800]

M0358 1, 0, 1, 1, 2, 2, 6, 9, 20, 37, ...

M0346 1, 1, 2, 2, 5, 4, 7, 7, 11, 9, 8, 6, 9, 4, 6, 22, 10, 4, 8, 4
Primitive groups of degree n. Ref LE70 178. [1,3; A0019, N0130]

M0347 1, 1, 1, 2, 2, 5, 4, 17, 22, 167
Connected regular graphs with n nodes. Ref ST90. [1,4; A5177]

M0348 1, 0, 2, 2, 5, 5, 11, 13, 24, 29, 48, 61, 96, 122, 182, 236, 339, 440, 617, 800, 1099, 1422, 1920, 2479, 3302, 4244, 5587, 7157, 9327
Representation degeneracies for boson strings. Ref NUPH B274 546 86. [4,3; A5294]

M0349 1, 1, 1, 2, 2, 5, 5, 12, 12, 27, 28, 64, 67, 147, 158, 348, 373, 799, 879, 1886, 2069, 4335, 4864
Symmetric filaments with n square cells. Ref PLC 1 337 70. [0,4; A2014, N0131]

M0350 1, 1, 1, 2, 2, 5, 9, 19, 38, 86, 188, 439, 1026, 2472, 5997, 14835, 36964, 93246, 236922, 607111, 1565478, 4062797, 10599853, 27797420, 73224806, 193709710
n-node trees with a forbidden limb. Ref HA73 297. [1,4; A2990]

M0351 0, 2, 2, 5, 9, 21, 43, 101, 226, 556, 1333, 3365, 8500, 22007, 57258, 151264, 401761, 1077063, 2902599, 7871250, 21440642, 58672581, 161155616, 444240599
Endpoints in trees with n nodes. Ref DM 12 364 75. [1,2; A3228]

M0352 1, 1, 2, 2, 5, 10, 45, 284, 3960, 110356, 6153615, 640014800, 120777999811, 41158185726269, 25486682538903526, 28943747337743989421
n-node graphs without points of degree 2. Ref rwr. [0,3; A5637]

M0353 1, 1, 2, 2, 6, 4, 18, 16, 48, 60
Primitive polynomials of degree n over $GF(2)$. Ref MA63 303. PW72 476. [1,3; A5992]

M0354 2, 2, 6, 6, 10, 20, 28, 46, 78, 122, 198, 324, 520, 842, 1366
Cyclic n-bit strings containing no runs of length > 2. Ref DM 70 295 88. [1,1; A7040]

M0355 0, 1, 0, 1, 1, 2, 2, 6, 8, 18, 30, 67, 127, 275, 551, 1192, 2507, 5475, 11820, 26007, 57077, 126686, 281625, 630660, 1416116, 3195784, 7232624, 16430563, 37429146
Bicentered boron trees with n nodes. Ref CAY 9 451. rcr. [1,6; A0673, N0133]

M0356 1, 0, 0, 1, 0, 1, 1, 2, 2, 6, 8, 26, 45, 148, 457
Indecomposable self-dual binary codes of length $2n$. Ref PGIT 24 738 78. JCT A28 52 80. JCT A60 183 92. [0,8; A3178]

M0357 1, 1, 2, 2, 6, 9, 17, 30, 54, 98, 183, 341, 645, 1220, 2327, 4451, 8555, 16489, 31859, 61717, 119779, 232919, 453584, 884544, 1727213, 3376505, 6607371
Integers $\leq 2^n$ of form $x^2 + 12y^2$. Ref MOC 20 567 66. [0,3; A0021, N0134]

M0358 1, 0, 1, 1, 2, 2, 6, 9, 20, 37, 86
Centered hydrocarbons with n atoms. Ref BS65 201. [1,5; A0022, N0135]

M0359 2, 2, 6, 14, 30, 62, 126, 246, 472
Fermionic string states. Ref CU86. [0,1; A5310]

M0360 2, 2, 6, 14, 34, 82, 198, 478, 1154, 2786, 6726, 16238, 39202, 94642, 228486,
    551614, 1331714, 3215042, 7761798, 18738638, 45239074, 109216786, 263672646
Companion Pell numbers: $a(n) = 2a(n-1) + a(n-2)$. Ref AJM 1 187 1878. FQ 4 373 66.
BPNR 43. [0,1; A2203, N0136]

M0361 1, 1, 0, 2, 2, 6, 16, 20, 132, 28, 1216, 936, 12440, 23672, 138048, 469456,
    1601264, 9112560, 18108928, 182135008, 161934624, 3804634784, 404007680
Expansion of $\exp(-x - \frac{1}{2}x^2)$. Ref CJM 7 168 55. [0,4; A1464, N0137]

M0362 1, 1, 1, 2, 2, 6, 17, 69
Hypertournaments on n elements under signed bijection. Ref KN91. [1,4; A6250]

M0363 1, 1, 1, 2, 2, 6, 17, 79
Hypertournaments on n elements under preisomorphism. Ref KN91. [1,4; A6249]

M0364 1, 2, 2, 6, 28, 160, 1036, 7294, 54548, 426960, 3463304, 28910816, 247104976,
    2154192248, 19097610480, 171769942086, 1564484503044, 14407366963440
Planar tree-rooted maps with n edges. Ref JCT B18 244 75. [0,2; A4304]

M0365 1, 2, 2, 6, 38, 390, 6062, 134526, 4172198, 178449270, 10508108222,
    853219050726, 95965963939958
Colored graphs. Ref CJM 22 596 70. rcr. [1,2; A2027, N0138]

M0366 1, 0, 2, 2, 7, 5, 26, 22, 91, 79, 326, 301, 1186, 1117, 4352, 4212, 16119, 15849,
    60174, 60089, 226146, 228426, 854803
Diagonally symmetric polyominoes with n cells. Ref DM 36 203 81. [3,3; A6748]

M0367 1, 0, 2, 2, 7, 8, 20, 27, 56, 79, 145, 212, 361, 530, 858, 1260
Representation degeneracies for Neveu-Schwarz strings. Ref NUPH B274 547 86. [0,3;
A5300]

M0368 1, 1, 2, 2, 7, 10, 20, 36, 65, 118, 221, 409, 776, 1463, 2788, 5328, 10222, 19714,
    38054, 73685, 142944, 277838, 540889, 1054535, 2058537, 4023278
Integers $\leq 2^n$ of form $x^2 + 10y^2$. Ref MOC 20 563 66. [0,3; A0024, N0139]

M0369 0, 0, 1, 0, 2, 2, 7, 10, 29, 52, 142, 294, 772, 1732, 4451, 10482, 26715, 64908,
    165194, 409720, 1044629
Dyck paths of knight moves. Ref DAM 24 218 89. [0,5; A5223]

M0370 2, 2, 7, 11, 25, 38, 78, 122, 219, 344, 579, 894, 1446, 2198
Representation degeneracies for Neveu-Schwarz strings. Ref NUPH B274 547 86. [2,1;
A5298]

M0371 1, 2, 2, 8, 4, 14, 21, 35, 49, 158, 191, 425, 828, 1864, 3659, 8324, 17344, 39601,
87407, 199984, 453361, 1053816, 2426228, 5672389, 13270695, 31277150, 73874375
Symmetries in unrooted 3-trees on $n + 1$ vertices. Ref GTA91 849. [1,2; A3612]

M0372 1, 1, 1, 2, 2, 8, 8, 32, 57, 185, 466, 1543, 4583, 15374, 50116, 171168
Trivalent 3-connected bipartite planar graphs with $4n$ nodes. Ref JCT B38 295 85. [2,4;
A7083]

M0373 1, 1, 2, 2, 8, 8, 112, 656, 5504, 49024, 491264, 5401856, 64826368, 842734592,
11798300672, 176974477312, 2831591702528, 48137058811904, 866467058876416
Expansion of $e^{-2x} / (1 - x)$. Ref R1 210. [0,3; A0023, N0140]

M0374 1, 1, 2, 2, 8, 12, 52, 86, 400, 710, 3404, 6316, 30888, 59204, 293192, 576018
Projective meanders. See Fig M4587. Ref SFCA91 295. [0,3; A6663]

M0375 2, 2, 8, 12, 88, 176, 2752, 8784
Self-complementary oriented graphs with n nodes. Ref KNAW 73 443 70. [3,1; A2785,
N0141]

M0376 1, 0, 2, 2, 8, 14, 36, 112, 216, 928, 1440, 8616, 11520, 87864, 100800
Bishops on an $n \times n$ board. Ref LNM 560 212 76. [2,3; A5633]

M0377 1, 2, 2, 8, 28, 20, 43, 143, 249, 546, 1223, 2703, 8107, 18085, 44013, 114919,
327712, 800937, 2146066, 5827711, 15923828, 43886143, 121888966, 340209504
Symmetries in unrooted 4-trees on $n + 1$ vertices. Ref GTA91 849. [1,2; A3616]

M0378 2, 2, 8, 68, 3904, 37329264
Nondegenerate Boolean functions of n variables. Ref MU71 38. [0,1; A3181]

M0379 2, 2, 8, 72, 1536, 86080, 14487040, 8274797440, 17494930604032
Threshold functions of n variables. Ref MOC 16 471 62. PGEC 19 821 70. MU71 38. [0,1;
A0615, N0142]

M0380 1, 1, 1, 2, 2, 9, 6, 118, 568, 4716, 38160, 358126, 3662088, 41073096, 500013528,
6573808200, 92840971200, 1402148010528
From nth derivative of x^x. Ref AMM 95 705 88. [1,4; A5168]

M0381 1, 0, 1, 1, 2, 2, 9, 11, 37, 79, 249, 671, 2182, 6692, 22131, 72405, 243806, 822788,
2815119, 9679205, 33527670
3-connected nets with n edges. Ref JCT 7 157 69. AMM 80 886 73. Dil92. [6,5; A2880,
N0143]

M0382 0, 0, 0, 0, 1, 0, 2, 2, 9, 17, 77, 261, 1265, 5852
Polyhedral graphs with n faces and minimal degree 4. Ref Dil92. [4,7; A7024]

M0383 1, 2, 2, 10, 14, 42, 90, 354, 758, 2290, 6002, 18410, 51310, 154106, 449322,
1384962, 4089174, 12475362, 37746786, 116037642, 355367310, 1097869386
Symmetries in planted (1,3) trees on $2n$ vertices. Ref GTA91 849. [1,2; A3609]

M0384 2, 2, 10, 28, 207, 1288, 10366, 91296
Hit polynomials. Ref RI63. [2,1; A1885, N0144]

M0385 2, 2, 10, 218, 64594, 4294642034, 18446744047940725978,
340282366920938463334247399005993378250
Nondegenerate Boolean functions of n variables (inverse binomial transform of M1297).
Ref HA65 170. MU71 38. [0,1; A0371, N0145]

M0386 2, 2, 10, 52246, 2631645209645100680142
Invertible Boolean functions. Ref PGEC 13 530 64. [1,1; A1038, N0146]

M0387 1, 1, 1, 2, 2, 12, 147
Species of Latin squares of order n. Ref HP73 231. [1,4; A3090]

M0388 1, 1, 1, 2, 2, 15, 39, 449, 2758
From analyzing an algorithm. Ref skb. [1,4; A6929]

M0389 1, 1, 1, 2, 2, 17, 1, 91
Queens problem. Ref SL26 49. [1,4; A2567, N0147]

M0390 0, 2, 2, 18, 66, 374, 1694, 9822, 51698
Nontrivial Baxter permutations of length $2n - 1$. Ref MAL 2 25 67. [1,2; A1183, N0148]

M0391 2, 2, 20, 38, 146, 368, 1070, 2824, 7680, 19996
Susceptibility for square lattice. Ref PHA 28 924 62. [0,1; A2907, N0149]

M0392 1, 1, 1, 1, 2, 2, 22, 563, 1676257
Types of Latin squares of order n. Ref R1 210. FY63 22. JCT 5 177 68. [0,5; A1012,
N0150]

M0393 0, 2, 2, 108, 2028, 32870, 1213110
Special permutations. Ref JNT 5 48 73. [0,2; A3110]

SEQUENCES BEGINNING . . ., 2, 3, . . .

M0394 2, 3, 0, 2, 5, 8, 5, 0, 9, 2, 9, 9, 4, 0, 4, 5, 6, 8, 4, 0, 1, 7, 9, 9, 1, 4, 5, 4, 6, 8, 4, 3, 6,
4, 2, 0, 7, 6, 0, 1, 1, 0, 1, 4, 8, 8, 6, 2, 8, 7, 7, 2, 9, 7, 6, 0, 3, 3, 3, 2, 7, 9, 0, 0, 9, 6, 7, 5, 7
Decimal expansion of natural logarithm of 10. Ref RS8 2. [1,1; A2392, N0151]

M0395 0, 0, 2, 3, 0, 11, 0, 17, 15, 14, 51
A partition function. Ref JNSM 9 103 69. [0,3; A2099, N0152]

M0396 1, 0, 1, 2, 3, 0, 12, 40, 100, 0, 1225, 6460, 28812, 0, 1037232, 9779616
Alternating sign matrices. Ref LNM 1234 292 86. [1,4; A5160]

M0397 2, 3, 0, 25, 152, 1350, 12644, 131391, 1489568, 18329481, 243365514,
3468969962, 52848096274, 857073295427, 14744289690560, 268202790690465
From discordant permutations. Ref KYU 10 13 56. [3,1; A2634, N0153]

M0398 0, 1, 1, 0, 1, 2, 3, 1, 0, 1, 3, 2, 1, 4, 1, 0, 1, 4, 39, 1, 1, 42, 5, 1, 0, 1, 5, 3, 13, 2, 273,
1, 4, 6, 1, 0, 1, 6, 4, 1, 5, 2, 531, 3, 1, 3588, 1, 1, 0, 1, 7, 5, 1, 66, 12, 2, 20, 13, 69, 1, 5, 8
Solution to Pellian: y such that $x^2 - n\,y^2 = \pm 1$, ± 4. Cf. M0119. Ref DE17. CAY 13 434.
L1 55. [1,6; A6705]

M0399 0, 1, 1, 0, 1, 2, 3, 1, 0, 1, 3, 2, 5, 4, 1, 0, 1, 4, 39, 2, 12, 42, 5, 1, 0, 1, 5, 24, 13, 2,
273, 3, 4, 6, 1, 0, 1, 6, 4, 3, 5, 2, 531, 30, 24, 3588, 1, 1, 0, 1, 7, 90, 25, 66, 12, 2, 20, 13
Solution to Pellian: y such that $x^2 - n\,y^2 = \pm 1$. Cf. M0120. Ref DE17. CAY 13 434. L1
55. [1,6; A6703]

M0400 1, 1, 0, 2, 3, 1, 0, 6, 3, 5, 0, 2, 11, 1, 8, 6, 5, 1, 8, 10, 5, 3, 10, 8, 3, 7, 14, 0, 1, 29, 0,
26, 7, 25, 4, 18, 11, 9, 8, 14, 11, 3, 18, 20, 11, 5, 20, 18, 9, 5, 28, 14, 17, 9, 12, 26, 7, 1, 44
$a(n) = \lfloor a(n-1) + 2a(n-2) - n \rfloor$. Ref PC 4 42-13 76. [1,4; A5210]

M0401 1, 1, 1, 2, 3, 1, 1, 4, 5, 1, 3, 1, 3, 1, 1, 8, 15, 3, 7, 4, 5, 2, 3, 3, 6, 2, 3, 2, 3, 1, 1, 16,
19, 7, 10, 5, 15, 4, 5, 7, 15, 3, 7, 4, 5, 2, 3, 5, 13, 3, 5, 4, 7, 1, 3, 3, 5, 2, 3, 1, 3, 1, 1, 32, 47
A problem in parity. Ref IJ1 11 163 69. [1,4; A2784, N0154]

M0402 0, 0, 0, 1, 0, 0, 1, 1, 0, 1, 0, 1, 1, 0, 0, 1, 1, 1, 2, 3, 1, 2, 1, 3, 2, 1, 2, 5, 1, 1, 3, 5, 2,
3, 1, 5, 5, 2, 2, 6, 2, 2, 5, 6, 3, 3, 3, 8, 4, 3
Indecomposable positive definite ternary forms of determinant n. Ref SPLAG 398. [1,19;
A6376]

M0403 1, 2, 3, 1, 2, 1, 3, 2, 3, 1, 3, 2, 1, 2, 3, 1, 2, 1, 3, 2, 1, 2, 3, 1, 3, 2, 3, 1, 2, 1, 3, 2, 3,
1, 3, 2, 1, 2, 3, 1, 3, 2, 3, 1, 2, 1, 3, 2, 1, 2, 3, 1, 2, 1, 3, 2, 3, 1, 3, 2, 1, 2, 3, 1, 2, 1, 3, 2, 1
A square-free ternary sequence. Ref gs. [1,2; A5680]

M0404 1, 2, 3, 1, 2, 3, 4, 2, 1, 2, 3, 3, 2, 3, 4, 1, 2, 2, 3, 2, 3, 3, 4, 3, 1, 2, 3, 4, 2, 3, 4, 2, 3,
2, 3, 1, 2, 3, 4, 2, 2, 3, 3, 3, 2, 3, 4, 3, 1, 2, 3, 2, 2, 3, 4, 3, 3, 2, 3, 4, 2, 3, 4, 1, 2, 3, 3, 2, 3
Least number of squares needed to represent n. [1,2; A2828, N0155]

M0405 1, 2, 3, 1, 2, 3, 4, 5, 1, 2, 3, 2, 3, 4, 5, 1, 2, 3, 4, 5, 3, 4, 5, 2, 1, 2, 3, 2, 3, 4, 2, 3, 2,
3, 4, 1, 2, 3, 4, 3, 4, 3, 2, 2, 3, 4, 3, 4, 1, 2, 3, 2, 2, 3, 4, 5, 6, 2, 3, 3, 3, 4, 5, 2, 1, 2, 3, 4, 2, 3
Representing n as sum of increasing powers. Ref BIT 12 342 72. [1,2; A3315]

M0406   1, 2, 3, 1, 3, 2, 1, 2, 3, 2, 1, 3, 1, 2, 3, 1, 3, 2, 1, 3, 1, 2, 3, 2, 1, 2, 3, 1, 3, 2
A nonrepetitive sequence. Ref Robe92 18. [1,2; A7413]

M0407   1, 2, 3, 1, 3, 2, 3, 1, 2, 3, 2, 1, 3, 1, 2, 1, 3, 2, 3, 1, 2, 3, 2, 1, 2, 3, 1, 2, 1, 3, 2, 3, 1,
   3, 2, 1, 3, 1, 2, 3, 2, 1, 2, 3, 1, 3, 2, 1, 3, 1, 2, 1, 3, 2, 3, 1, 2, 3, 2, 1, 2, 3, 1, 2, 1, 3, 2, 3, 1
A nonrepetitive sequence. Ref YAG 2 204. JCT A13 90 72. [1,2; A3270]

M0408   1, 2, 3, 1, 4, 1, 5, 1, 1, 6, 2, 5, 8, 3, 3, 4, 2, 6, 4, 4, 1, 3, 2, 3, 4, 1, 4, 9, 1, 8, 4, 3, 1,
   3, 2, 6, 1, 6, 1, 3, 1, 1, 1, 1, 12, 3, 1, 3, 1, 1, 4, 1, 6, 1, 5, 1, 2, 1, 3, 3, 11, 8, 1, 139, 8, 2, 8
Continued fraction for cube root of 3. Ref JRAM 255 120 72. [1,2; A2946]

M0409   0, 1, 2, 3, 1, 4, 3, 2, 0, 5, 2, 3, 1, 4, 3, 2, 0, 5, 2, 3, 1
The game of contours. Ref WW 553. [0,3; A6021]

M0410   0, 1, 2, 3, 1, 4, 3, 2, 1, 4, 2, 6, 4, 1, 2, 7, 1, 4, 3, 2, 1, 4, 6, 7, 4, 1, 2, 8, 5, 4, 7, 2, 1,
   8, 6, 7, 4, 1, 2, 3, 1, 4, 7, 2, 1, 8, 2, 7, 4, 1, 2, 8, 1, 4, 7, 2, 1, 4, 2, 7, 4, 1, 2, 8, 1, 4, 7, 2, 1
The game of Kayles. Ref PCPS 52 516 56. WW 91. [0,3; A2186, N0156]

M0411   1, 1, 1, 2, 3, 1, 5, 4, 3, 3, 9, 2, 11, 5
Number of pairs x,y such that $y-x=2$, $(x,n)=1$, $(y,n)=1$. Ref MTS 67 11 58. [1,4;
A2472, N0157]

M0412   0, 2, 3, 1, 8, 10, 11, 9, 12, 14, 15, 13, 4, 6, 7, 5, 32, 34, 35, 33, 40, 42, 43, 41, 44,
   46, 47, 45, 36, 38, 39, 37, 48, 50, 51, 49, 56, 58, 59, 57, 60, 62, 63, 61, 52, 54, 55, 53, 16
Nim product $2n$. Ref ONAG 52. [0,2; A6015]

M0413   1, 1, 0, 2, 3, 1, 11, 15, 13, 77, 86, 144, 595, 495, 1520, 4810, 2485, 15675, 39560,
   6290, 159105, 324805, 87075, 1592843, 2616757, 2136539, 15726114, 20247800
Reversion of g.f. for Fibonacci numbers 1,1,2,3,5,... Cf. M0692. [1,4; A7440]

$$(n + 3)\, a(n + 2) = -(2n + 3)\, a(n + 1) - 5n\, a(n), \quad a(1) = 1, \quad a(2) = -1.$$

M0414   0, 2, 3, 2, 0, 1, 7, 2, 6, 8, 22, 7, 0, 33, 3, 14, 51, 46, 19, 12, 94, 42, 23, 113, 150, 54,
   48, 345, 116, 109, 403, 498, 140, 219, 1057, 326, 259, 1271, 1641, 308, 656, 3396
From symmetric functions. Ref PLMS 23 297 23. [1,2; A2120, N0158]

M0415   1, 1, 1, 0, 1, 1, 1, 1, 0, 1, 2, 3, 2, 0, 2, 4, 4, 3, 1, 3, 6, 7, 5, 0, 5, 9, 10, 7, 1, 7, 14, 16,
   11, 1, 11, 20, 22, 16, 2, 15, 29, 33, 23, 2, 23, 41, 45, 32, 4, 30, 57, 64, 45, 4, 43, 78, 86, 60
Hauptmodul series for $\Gamma(5)$. Ref JPA 21 L984 88. [0,11; A7325]

$$\text{G.f.:} \quad \Pi\, (1-x^{5k-1})(1-x^{5k-4}) \,/\, (1-x^{5k-2})(1-x^{5k-3}).$$

M0416   1, 2, 3, 2, 1, 2, 2, 4, 2, 2, 1, 0, 4, 2, 3, 2, 2, 4, 0, 2, 2, 0, 4, 2, 3, 0, 2, 6, 2, 2, 1, 2, 0,
   2, 2, 2, 2, 4, 2, 0, 4, 4, 4, 0, 1, 2, 0, 4, 2, 0, 2, 2, 5, 2, 0, 2, 2, 4, 4, 2, 0, 2, 4, 2, 2, 0, 4, 0, 0
Excess divisor function for $12n+1$. Ref PLMS 21 190 1889. [0,2; A2175, N0159]

M0417 1, 2, 3, 2, 1, 2, 3, 4, 2, 1, 2, 3, 4, 3, 2, 3, 4, 5, 3, 2, 3, 4, 5, 4, 3, 4, 5, 6, 4, 3, 4, 5, 6,
5, 4, 5, 6, 7, 5, 2, 3, 4, 5, 4, 3, 4, 5, 6, 4, 1, 2, 3, 4, 3, 2, 3, 4, 5, 3, 2, 3, 4, 5, 4, 3, 4, 5, 6, 4
Letters in Roman numeral representation of n. [1,2; A6968]

M0418 1, 1, 1, 2, 3, 2, 2, 4, 4, 4, 4, 4, 3, 5, 4, 3, 5, 5, 6, 6, 4, 6, 7, 4, 4, 7, 7, 6, 5, 5, 7, 8, 6,
5, 4, 7, 6, 6, 6, 6, 6, 6, 6, 4, 7, 6, 7, 7, 7, 5, 6, 6, 6, 7, 7, 6, 7, 8, 7, 10, 7, 9, 9, 7, 10, 5, 5
Consecutive quadratic residues mod p. Ref BAMS 32 284 26. [2,4; A2307, N0160]

M0419 1, 2, 3, 2, 2, 6, 1, 0, 6, 4, 5, 6, 2, 2, 6
L-series for an elliptic curve. Ref LNM 1111 228 85. [1,2; A7653]

M0420 1, 1, 2, 3, 2, 3, 3, 5, 4, 4, 3, 8, 4, 5, 6, 9, 4, 8, 5, 10, 8, 7, 5, 15, 7, 8, 9, 13, 6, 13, 7,
15, 10, 10, 10, 20, 8, 11, 12, 20, 8, 17, 9, 17, 16, 13, 9, 28, 12, 17, 14, 20, 10, 22, 14, 25
Sublattices of index n in hexagonal lattice. Ref DM 4 216 73. BSW94. [1,3; A3051]

M0421 0, 0, 1, 2, 3, 2, 3, 4, 4, 4, 5, 6, 5, 4, 6, 4, 7, 8, 3, 6, 8, 6, 7, 10, 8, 6, 10, 6, 7, 12, 5,
10, 12, 4, 10, 12, 9, 10, 14, 8, 9, 16, 9, 8, 18, 8, 9, 14, 6, 12, 16, 10, 11, 16, 12, 14, 20, 12
Decompositions of $2n$ into sum of two odd primes. Ref FVS 4(4) 7 27. L1 80. [1,4; A2372, N0161]

M0422 1, 2, 3, 2, 4, 6, 3, 6, 9, 2, 4, 6, 4, 8, 12, 6, 12, 18, 3, 6, 9, 6, 12, 18, 9, 18, 27, 2, 4, 6,
4, 8, 12, 6, 12, 18, 4, 8, 12, 8, 16, 24, 12, 24, 36, 6, 12, 18, 12, 24, 36, 18, 36, 54, 3, 6, 9, 6
Entries in nth row of Pascal's triangle not divisible by 3. Ref TCS 98 188 92. [0,2; A6047]

M0423 2, 3, 2, 5, 2, 3, 7, 2, 11, 13, 2, 3, 5, 17, 19, 2, 23, 7, 29, 3, 31, 2, 37, 41, 43, 47, 5,
53, 59, 2, 11, 61, 3, 67, 71, 73, 79, 13, 83, 89, 2, 97, 101, 103, 107, 7, 109, 113, 17, 127
Related to highly composite numbers. Ref RAM1 115. [1,1; A0705, N0162]

M0424 0, 2, 3, 2, 5, 2, 7, 2, 9, 2, 11, 2, 13, 2, 15, 2, 17, 11, 19, 22, 21, 35, 23, 50, 25, 67,
36, 86, 58, 107, 93, 130, 143, 155, 210, 191, 296, 249, 403, 342, 533, 485, 688, 695, 879
7th order maximal independent sets in cycle graph. Ref YaBa94. [1,2; A7389]

M0425 0, 2, 3, 2, 5, 2, 7, 2, 9, 2, 11, 2, 13, 9, 15, 18, 17, 29, 19, 42, 28, 57, 46, 74, 75, 93,
117, 121, 174, 167, 248, 242, 242, 341, 359, 462, 533, 629, 781, 871, 1122, 1230, 1584
5th order maximal independent sets in cycle graph. Ref YaBa94. [1,2; A7388]

M0426 0, 2, 3, 2, 5, 2, 7, 2, 9, 7, 11, 14, 13, 23, 20, 34, 34, 47, 57, 67, 91, 101, 138, 158,
205, 249, 306, 387, 464, 592, 713, 898, 1100, 1362, 1692, 2075, 2590, 3175, 3952, 4867
3rd order maximal independent sets in cycle graph. Ref YaBa94. [1,2; A7387]

M0427 1, 1, 2, 3, 2, 5, 2, 21, 2, 3, 1, 55, 3, 13, 2, 21, 2, 85, 1, 57, 2, 1, 1, 8855, 2, 15, 2, 39,
1, 29, 10, 651, 2, 1, 2, 935, 1, 37, 2, 399, 1, 2665, 1, 129, 2, 1, 1, 416185, 2, 21, 2, 15, 1
Related to nth powers of polynomials. Ref ACA 29 246 76. [1,3; A5731]

M0428 2, 3, 2, 5, 3, 7, 2, 3, 5, 11, 3, 13, 7, 5, 2, 17, 3, 19, 5, 7, 11, 23, 3, 5, 13, 3, 7, 29, 5,
31, 2, 11, 17, 7, 3, 37, 19, 13, 5, 41, 7, 43, 11, 5, 23, 47, 3, 7, 5, 17, 13, 53, 3, 11, 7, 19, 29
Largest prime dividing n. Ref AS1 844. [2,1; A6530]

M0429 0, 2, 3, 2, 5, 5, 7, 10, 12, 17, 22, 29, 39, 51, 68, 90, 119, 158, 209, 277, 367, 486,
644, 853, 1130, 1497, 1983, 2627, 3480, 4610, 6107, 8090, 10717, 14197, 18807, 24914
Perrin sequence: $a(n)=a(n-2)+a(n-3)$. Ref AMM 15 209 08. JA66 90. FQ 6(3) 68 68.
MOC 39 255 82. [0,2; A1608, N0163]

M0430 1, 1, 1, 1, 2, 3, 2, 6, 6, 7, 14, 16, 20, 34, 42, 56, 84, 108, 152, 214, 295, 398, 569,
763, 1094, 1475, 2058, 2878, 3929, 5493
Numbers of complexity n. Ref FQ 27 16 89. [1,5; A5421]

M0431 1, 1, 1, 1, 1, 1, 1, 1, 1, 1, 1, 2, 3, 3, 1, 1, 1, 1, 1, 1, 1, 1, 1, 1, 1, 23
Number of laminated lattices of dimension n (next term probably exceeds 75000). Ref
SPLAG 159. [0,12; A5135]

M0432 2, 3, 3, 2, 3, 3, 3, 2, 3, 3, 3, 2, 3, 3, 2, 3, 3, 3, 2, 3, 3, 3, 2, 3, 3, 3, 2, 3, 3, 2, 3, 3, 3,
2, 3, 3, 3, 2, 3, 3, 3, 2, 3, 3, 2, 3, 3, 3, 2, 3, 3, 3, 2, 3, 3, 3, 2, 3, 3, 3, 2, 3, 3, 3, 2, 3, 3, 3, 2, 3
A self-generating sequence: there are $a(n)$ 3's between successive 2's. Ref MMAG 67 157
94. [1,1; A7538]

M0433 1, 1, 2, 3, 3, 3, 5, 7, 6, 6, 10, 12, 11, 13, 17, 20, 21, 21, 27, 34, 33, 36, 46, 51, 53,
58, 68, 78, 82, 89, 104, 118, 123, 131, 154, 171, 179, 197, 221, 245, 262, 279, 314, 349
Mock theta numbers. Ref TAMS 72 495 52. [0,3; A0025, N0164]

M0434 1, 1, 0, 1, 2, 3, 3, 4, 3, 3, 1, 2, 3, 2, 5, 8, 12, 11, 17, 16, 21, 25, 26, 25, 30, 32, 29,
32, 32, 31, 30, 29, 21, 23, 11, 17, 5, 4, 13, 15, 28, 29, 52, 53, 76, 78, 104, 105, 142, 139
Unique attractor for sliding Moebius transform. Ref BeSl94. EIS § 2.7. [1,5; A7554]

$$a(n+1) \ = \ \Sigma \ \mu(n/d) \ a(d) \, ; \ d \mid n.$$

M0435 1, 2, 3, 3, 4, 4, 5, 5, 6, 6, 7, 7, 8, 8, 9, 9, 9, 10
From interval orders. Ref SIAA 2 128 81. [1,2; A5410]

M0436 0, 1, 1, 2, 3, 3, 4, 4, 5, 6, 6, 7, 8, 8, 9, 9, 10, 11, 11, 12, 12, 13, 14, 14, 15, 16, 16,
17, 17, 18, 19, 19, 20, 21, 21, 22, 22, 23, 24, 24, 25, 25, 26, 27, 27, 28, 29, 29, 30, 30, 31
Hofstadter G-sequence: $a(n)=n-a(a(n-1))$. See Fig M0436. Ref FQ 15 317 77. GEB
137. JNT 30 238 88. IJCM 26 36 88. [0,4; A5206]

$$a(n) \ = \ [\, n \, \tau \,] \ - \ n \, , \ \text{where} \ \tau \ = \ (1 \, + \, \sqrt{5})/2.$$

M0437 2, 3, 3, 4, 5, 5, 4, 5, 7, 8, 7, 7, 8, 7, 5, 6, 9, 11, 10, 11, 13, 12, 9, 9, 12, 13, 11, 10,
11, 9, 6, 7, 11, 14, 13, 15, 18, 17, 13, 14, 19, 21, 18, 17, 19, 16, 11, 11, 16, 19, 17, 18, 21
Denominators of Farey tree fractions. Cf. M0113. Ref PSAM 46 42 92. [1,1; A7306]

||||||||||||||||||||||||||||||||||||||||||||||||||||||||||||||||||||||||||||||||||||||||||||||||||||||||||||||||||||||||||||||||||||||||||||||

Figure M0436.   SELF-GENERATING SEQUENCES.

These sequences are produced by simple yet unusual recurrence rules. They have been called (rather arbitrarily) **self-generating** sequences. The Ulam numbers (see Fig. M0557) could also have been included here.

(1) M0436, rediscovered several times, has the curious recurrence $a(0) = 0$, $a(n) = n - a(a(n-1))$. See the references cited for the strange properties of this sequence and its connection with the Fibonacci numbers. M0449, M0464, M0263, M0278 have similar rules.

(2) Let $A = \{a_0 = 1, a_1, a_2, \ldots\}$ be a sequence of 1's and 2's. If every 1 in A is replaced by 1, 2 and every 2 by 2, 1, a new sequence A' is obtained. Imposing the condition that $A' = A$ forces A to be the **Thue-Morse sequence** M0193. (This can also be constructed in many other ways.) M0068, M0190 have similar definitions.

(3) Let b_n be the number of times n occurs in A, for $n = 1, 2, \ldots$. If $b_n = n$ we obtain M0250, and if $b_n = a_{n-1}$ we get M0257. Seq. M2438 is related to the latter sequence.

(4) Other sequences of a similar nature are M2306, M2335, M0436, etc.

||||||||||||||||||||||||||||||||||||||||||||||||||||||||||||||||||||||||||||||||||||||||||||||||||||||||||||||||||||||||||||||||||||||||||||||

M0438   1, 1, 2, 3, 3, 4, 5, 5, 6, 6, 6, 8, 8, 8, 10, 9, 10, 11, 11, 12, 12, 12, 12, 16, 14, 14, 16,
16, 16, 16, 20, 17, 17, 20, 21, 19, 20, 22, 21, 22, 23, 23, 24, 24, 24, 24, 24, 32, 24, 25, 30
Hofstadter Q-sequence: $a(n) = a(n - a(n-1)) + a(n - a(n-2))$. Ref GEB 138. AMM 93 186 86. CO89. [1,3; A5185]

M0439   1, 2, 3, 3, 4, 5, 5, 6, 7, 8, 8, 8, 9, 9, 10, 10, 11, 11, 12, 12, 13, 13, 14, 14, 15, 15, 16,
17, 17, 18, 18, 18, 19, 20, 20, 20, 21, 21, 21, 22, 22, 22, 23, 23, 24, 24, 24, 25, 25, 26, 26
4-free sequences. Ref MOC 26 768 72. [1,2; A3003]

M0440   1, 1, 2, 3, 3, 4, 5, 6, 6, 6, 7, 8, 9, 10, 10, 11, 11, 12, 13, 14, 15, 16, 16, 16, 16, 17,
18, 18, 19, 20, 21, 22, 23, 24, 24, 25, 25, 26, 26, 26, 27, 28, 29, 29, 29, 30, 31, 32, 33, 34
$a(n) = a(a(n-1) - 1) + a(n + 1 - a(n-1))$. Ref JRM 22 90 90. [1,3; A6161]

M0441   1, 1, 2, 3, 3, 4, 5, 6, 6, 7, 7, 8, 9, 10, 10, 11, 12, 12, 13, 14, 15, 16, 16, 17, 17, 18,
19, 19, 20, 20, 21, 22, 23, 24, 24, 25, 26, 26, 27, 28, 29, 29, 30, 30, 30, 31, 32, 33, 34, 35
$a(n) = a(a(n-2)) + a(n - a(n-2))$. Ref AMM 98 19 91. [1,3; A5229]

M0442   1, 1, 2, 3, 3, 5, 9, 16, 28, 50, 89, 159, 285, 510, 914, 1639, 2938, 5269, 9451,
16952, 30410, 54555, 97871, 175588, 315016, 565168, 1013976, 1819198, 3263875
Binary codes with n letters. Ref PGIT 17 309 71. [1,3; A1180, N0165]

M0443   1, 2, 3, 4, 1, 4, 3, 2, 1, 2, 3, 2, 1, 4, 3, 4, 1, 2, 3, 4, 1, 4, 3, 4, 1, 2, 3, 2, 1, 4, 3, 2, 1,
2, 3, 4, 1, 4, 3, 2, 1, 2, 3, 2, 1, 4, 3, 2, 1, 2, 3, 4, 1, 4, 3, 4, 1, 2, 3, 2, 1, 4, 3, 4, 1, 2, 3, 4, 1
A nonrepetitive sequence. Ref AMM 72 383 65. JCT A13 90 72. [1,2; A3324]

M0444   1, 1, 1, 2, 3, 4, 3, 5, 3, 6, 1, 2, 6, 7, 4, 5, 8, 3, 9, 7, 6, 9, 1, 2, 6, 11, 4, 10, 9, 3, 12, 9,
12, 13, 8, 3, 14, 12, 13, 6, 1, 2, 12, 11, 5, 15, 16, 9, 3, 13, 8, 15, 12, 17, 16, 6, 14, 15, 10, 3
y such that $p = x^2 + 2y^2$. Cf. M2264. Ref CU04 1. L1 55. MOC 23 459 69. [2,4; A2333, N0166]

M0445   1, 2, 3, 4, 4, 4, 3, 2, 1, 2, 3, 2, 1, 2, 1, 1, 1, 1, 1, 1, 1, 1, 1, 1, 1, 1, 1, 1, 1, 1, 1, 1, 1, 1,
1, 1, 1, 1, 1, 1, 1, 1, 1, 1, 1, 1, 1, 1, 1, 1, 1, 1, 1, 1, 1, 1, 1, 1, 1, 1, 1, 1, 1, 1, 1, 1, 1, 1, 1
Minimal determinant of n-dimensional norm 2 lattice. Ref SPLAG 180. [0,2; A5102]

M0446   1, 1, 1, 2, 3, 4, 4, 4, 5, 5, 5, 6, 6, 6, 7, 8, 9, 9, 9, 10, 11, 12, 12, 12, 13, 14, 15, 15,
15, 16, 16, 16, 17, 17, 17, 18, 19, 20, 20, 20, 21, 21, 21, 22, 22, 22, 23, 24, 25, 25, 25, 26
A self-generating sequence. Ref FQ 10 507 72. [1,4; A3160]

M0447   1, 1, 1, 2, 3, 4, 4, 4, 5, 6, 6, 7, 8, 8, 8, 8, 9, 10, 10, 11, 12, 13, 13, 14, 15, 16, 16, 16,
16, 16, 16, 16, 17, 18, 18, 19, 20, 21, 21, 22, 23, 24, 25, 25, 26, 27, 28, 29, 29, 29, 29, 29
$a(n) = a(a(n-3)) + a(n-a(n-3))$. Ref CO89. JRM 22 89 90. [1,4; A6158]

M0448   0, 1, 2, 3, 4, 4, 5, 4, 4, 5, 6, 6, 7, 7, 7, 5, 6, 6, 7, 7, 8, 8, 9, 7, 6, 7, 5, 6, 7, 8, 9, 6, 7,
8, 9, 6, 7, 8, 9, 9, 10, 10, 11, 10, 9, 10, 11, 8, 7, 8, 9, 9, 10, 7, 8, 8, 9, 9, 10, 10, 11, 11, 10
Complexity of n. Ref BS71. [1,3; A5208]

M0449   0, 1, 1, 2, 3, 4, 4, 5, 5, 6, 7, 7, 8, 9, 10, 10, 11, 12, 13, 13, 14, 14, 15, 16, 17, 17, 18,
18, 19, 20, 20, 21, 22, 23, 23, 24, 24, 25, 26, 26, 27, 28, 29, 29, 30, 31, 32, 32, 33, 33, 34
Hofstadter H-sequence: $a(n) = n - a(a(a(n-1)))$. See Fig M0436. Ref GEB 137. [0,4; A5374]

M0450   1, 1, 2, 3, 4, 4, 5, 6, 7, 8, 8, 8, 9, 10, 11, 12, 13, 13, 14, 14, 15, 16, 17, 18, 19, 20,
20, 21, 22, 22, 22, 22, 23, 24, 25, 26, 27, 28, 29, 29, 30, 31, 32, 32, 33, 33, 34, 34, 35, 36
$a(n) = a(a(n-1)-2) + a(n+2-a(n-1))$. Ref JRM 22 90 90. [1,3; A6162]

M0451   1, 2, 3, 4, 4, 5, 6, 7, 8, 8, 9, 10, 11, 12, 12, 13, 14, 15, 16, 16, 16, 16, 16, 17, 18, 18,
19, 20, 21, 21, 22, 22, 23, 24, 24, 25, 26, 27, 28, 28, 29, 30, 31, 32, 32, 32, 32, 32, 33, 33
5-free sequences. Ref MOC 26 768 72. [1,2; A3004]

M0452   1, 2, 3, 4, 4, 7, 7, 6, 9, 13, 10, 13, 10, 7, 11, 15, 10, 15, 9, 12, 7, 17, 12, 18, 16, 14,
19, 20, 19, 12, 15, 20, 10, 20, 18, 22, 19, 13, 12, 13, 17, 29, 18, 33, 20
Unitary harmonic means. Ref PAMS 51 7 75. [1,2; A6087]

M0453   1, 2, 3, 4, 5, 3, 7, 4, 6, 5, 11, 4, 13, 7, 5, 6, 17, 6, 19, 5, 7, 11, 23, 4, 10, 13, 9, 7, 29,
5, 31, 8, 11, 17, 7, 6, 37, 19, 13, 5, 41, 7, 43, 11, 5, 23, 47, 6, 14, 10, 17, 13, 53, 9, 11, 7
Smarandache numbers: n divides $a(n)!$. Cf. M1669. Ref AMM 25 210 18. [1,2; A2034, N0167]

M0454   1, 1, 1, 1, 2, 3, 4, 5, 5, 5, 5, 6, 7, 7, 7, 8, 9, 9, 10, 11, 11, 12, 12, 12, 12, 13, 13, 14,
15, 16, 16, 16, 16, 17, 18, 19, 19, 20, 20, 21, 22, 23, 23, 23, 24, 24, 24, 24, 24, 25, 25, 26
$a(n) = a(a(n-4)) + a(n-a(n-4))$. Ref JRM 22 89 90. [1,5; A6159]

M0455   1, 0, 1, 1, 1, 1, 2, 3, 4, 5, 5, 5, 6, 7, 8, 9, 9, 9, 10, 11, 11, 11, 12, 13, 14, 15, 15, 15,
16, 17, 17, 17, 18, 19, 19, 19, 20, 21, 21, 21, 22, 23, 24, 25, 25, 25, 26, 27, 27, 27, 28, 29
$a(n+1) = a(n) + a(a(a(..(n-1)..)))$, depth $[n/2]$. Ref rgw. [0,7; A7599]

M0456 0, 2, 3, 4, 5, 5, 6, 6, 6, 7, 7, 7, 8, 8, 8, 8, 8, 8, 9, 9, 9, 9, 9, 9, 9, 9, 9, 10, 10, 10, 10,
10, 10, 10, 10, 10, 11, 11, 11, 11, 11, 11, 11, 11, 11, 11, 11, 11, 11, 11, 11, 11, 11, 11, 12
Minimal size of separating family for n-set. Ref HO85a 225. [1,2; A7600]

M0457 1, 2, 3, 4, 5, 5, 6, 6, 6, 7, 8, 7, 8, 8, 8, 8, 9, 8, 9, 9, 9, 10, 11, 9, 10, 10, 9, 10, 11, 10,
11, 10, 11, 11, 11, 10, 11, 11, 11, 11, 12, 11, 12, 12, 11, 12, 13, 11, 12, 12, 12, 12, 13, 11
Complexity of n. Ref AMM 93 189 86. [1,2; A5245]

M0458 0, 1, 1, 2, 3, 4, 5, 5, 6, 6, 7, 8, 8, 9, 10, 11, 11, 12, 13, 14, 15, 15, 16, 17, 18, 19, 19,
20, 20, 21, 22, 23, 24, 24, 25, 25, 26, 27, 27, 28, 29, 30, 31, 31, 32, 32, 33, 34, 34, 35, 36
$a(n) = n - a(a(a(a(n-1))))$. Ref GEB 137. [0,4; A5375]

M0459 1, 2, 3, 4, 5, 5, 6, 7, 8, 9, 9, 10, 11, 12, 13, 13, 14, 15, 16, 17, 17, 18, 19, 20, 21, 22,
22, 22, 23, 23, 23, 24, 25, 25, 26, 27, 28, 28, 29, 30, 31, 31, 31, 32, 33, 34, 34, 35, 36, 37
6-free sequences. Ref MOC 26 768 72. [1,2; A3005]

M0460 1, 1, 2, 3, 4, 5, 5, 6, 7, 8, 9, 10, 10, 10, 11, 12, 13, 14, 15, 16, 16, 17, 17, 18, 19, 20,
21, 22, 23, 24, 24, 25, 26, 26, 27, 28, 29, 30, 31, 32, 33, 34, 34, 34, 34, 35, 36, 37, 37, 38
$a(n) = a(a(n-1)-3) + a(n+3-a(n-1))$. Ref JRM 22 90 90. [1,3; A6163]

M0461 2, 3, 4, 5, 5, 7, 6, 6, 7, 11, 7, 13, 9, 8, 8, 17, 8, 19, 9, 10, 13, 23, 9, 10, 15, 9, 11, 29,
10, 31, 10, 14, 19, 12, 10, 37, 21, 16, 11, 41, 12, 43, 15, 11, 25, 47, 11, 14, 12, 20, 17, 53
Sum of primes dividing n (with repetition). Ref MOC 23 181 69. Robe92 89. [2,1; A1414,
N0168]

M0462 1, 2, 3, 4, 5, 6, 5, 7, 6, 8, 8, 9, 10, 10, 8, 11, 10, 11, 13, 10, 12, 14, 15, 13, 15, 16,
13, 14, 16, 17, 13, 14, 16, 18, 17, 18, 17, 19, 20, 20, 15, 17, 20, 21, 19, 22, 20, 21, 19, 20
x such that $p = x^2 + y^2$, $x \le y$. Cf. M0096. Ref CU04 1. AMM 56 526 49. [2,2; A2330,
N0169]

M0463 1, 1, 1, 1, 1, 2, 3, 4, 5, 6, 6, 6, 6, 6, 7, 8, 8, 8, 8, 9, 10, 10, 11, 12, 13, 14, 14, 14, 14,
14, 14, 14, 14, 15, 16, 16, 17, 18, 19, 20, 21, 21, 21, 21, 22, 23, 24, 24, 24, 24, 25, 26, 26
$a(n) = a(a(n-5)) + a(n-a(n-5))$. Ref JRM 22 89 90. [1,6; A6160]

M0464 0, 1, 1, 2, 3, 4, 5, 6, 6, 7, 7, 8, 9, 9, 10, 11, 12, 12, 13, 14, 15, 16, 16, 17, 18, 19, 20,
21, 21, 22, 23, 24, 25, 26, 26, 27, 27, 28, 29, 30, 31, 32, 32, 33, 33, 34, 35, 35, 36, 37, 38
$a(n) = n - a(a(a(a(a(n-1)))))$. See Fig M0436. Ref GEB 137. [0,4; A5376]

M0465 1, 1, 2, 3, 4, 5, 6, 6, 7, 8, 9, 10, 11, 12, 12, 12, 13, 14, 15, 16, 17, 18, 19, 19, 20, 20,
21, 22, 23, 24, 25, 26, 27, 28, 28, 29, 30, 30, 31, 32, 33, 34, 35, 36, 37, 38, 39, 39, 40, 41
$a(n) = a(a(n-1)-4) + a(n+4-a(n-1))$. Ref JRM 22 90 90. [1,3; A6164]

M0466 1, 2, 3, 4, 5, 6, 7, 1, 2, 3, 4, 5, 6, 7, 8, 2, 3, 4, 5, 6, 7, 8, 9, 3, 4, 5, 1, 2, 3, 4, 5, 4, 5,
6, 2, 3, 4, 5, 6, 5, 6, 7, 3, 4, 5, 6, 7, 6, 7, 8, 4, 5, 6, 2, 3, 4, 5, 6, 5, 6, 7, 3, 4, 1, 2, 3, 4, 5, 6
n is a sum of $a(n)$ cubes. Ref JRAM 14 279 1835. L1 81. [1,2; A2376, N0170]

M0467 1, 2, 3, 4, 5, 6, 7, 6, 6, 10, 11, 12, 13, 14, 15, 8, 17, 12, 19, 20, 21, 22, 23, 18, 10, 26, 9, 28, 29, 30, 31, 10, 33, 34, 35, 24, 37, 38, 39, 30, 41, 42, 43, 44, 30, 46, 47, 24, 14
Mosaic numbers. Ref BAMS 69 446 63. CJM 17 1010 65. [1,2; A0026, N0171]

M0468 1, 2, 3, 4, 5, 6, 7, 6, 7, 8, 9, 10, 11, 12, 11, 12, 13, 14, 15, 16, 17, 18, 19, 18, 19, 20, 21, 20, 19, 18, 19, 18, 19, 20, 21, 22, 23, 24, 25, 24, 25, 26, 27, 28, 29, 30, 29, 30, 31, 32
Summation related to binary digits. Ref INV 73 107 83. [1,2; A5599]

M0469 1, 2, 3, 4, 5, 6, 7, 8, 9, 1, 0, 1, 1, 1, 2, 1, 3, 1, 4, 1, 5, 1, 6, 1, 7, 1, 8, 1, 9, 2, 0, 2, 1, 2, 2, 2, 3, 2, 4, 2, 5, 2, 6, 2, 7, 2, 8, 2, 9, 3, 0, 3, 1, 3, 2, 3, 3, 3, 4, 3, 5, 3, 6, 3, 7, 3, 8, 3, 9
The almost-natural numbers. Ref Krai53 49. MMAG 61 131 88. [1,2; A7376]

M0470 1, 2, 3, 4, 5, 6, 7, 8, 9, 1, 1, 1, 1, 1, 1, 1, 1, 1, 1, 1, 2, 2, 2, 2, 2, 2, 2, 2, 2, 2, 3, 3, 3, 3, 3, 3, 3, 3, 3, 3, 4, 4, 4, 4, 4, 4, 4, 4, 4, 4, 5, 5, 5, 5, 5, 5, 5, 5, 5, 5, 6, 6, 6, 6, 6, 6, 6, 6, 6, 6
Initial digits of integers. Ref MST 6 167 72. [1,2; A0030]

M0471 1, 2, 3, 4, 5, 6, 7, 8, 9, 10, 11, 12, 13, 14, 15, 1, 2, 3, 4, 5, 6, 7, 8, 9, 10, 11, 12, 13, 14, 15, 16, 2, 3, 4, 5, 6, 7, 8, 9, 10, 11, 12, 13, 14, 15, 16, 17, 3, 4, 5, 6, 7, 8, 9, 10, 11, 12
Least number of 4th powers to represent n. Ref JRAM 46 3 1853. L1 82. [1,2; A2377, N0172]

M0472 1, 2, 3, 4, 5, 6, 7, 8, 9, 10, 11, 12, 13, 14, 15, 16, 17, 18, 19, 20, 21, 22, 23, 24, 25, 26, 27, 28, 29, 30, 31, 32, 33, 34, 35, 36, 37, 38, 39, 40, 41, 42, 43, 44, 45, 46, 47, 48, 49
The natural numbers. [1,2; A0027, N0173]

M0473 0, 1, 2, 3, 4, 5, 6, 7, 8, 9, 10, 11, 12, 13, 14, 15, 16, 17, 18, 19, 20, 21, 22, 23, 24, 25, 26, 27, 28, 29, 30, 31, 32, 33, 34, 35, 36, 37, 38, 39, 42, 43, 45, 46, 47, 48, 50, 51, 52
$n^2 + n + 41$ is prime. [0,3; A2837, N0174]

M0474 1, 2, 3, 4, 5, 6, 7, 8, 9, 10, 11, 12, 15, 20, 22, 24, 30, 33, 36, 40, 44, 48, 50, 55, 60, 66, 70, 77, 80, 88, 90, 99, 100, 101, 102, 104, 105, 110, 111, 112, 115, 120, 122, 124, 126
Divisible by each nonzero digit. Ref JRM 1 217 68. [1,2; A2796, N0175]

M0475 2, 3, 4, 5, 6, 7, 8, 9, 10, 12, 13, 14, 16, 18, 19, 20, 21, 22, 25, 27, 28, 30, 32, 34, 37, 38, 40, 42, 44, 45, 48, 50, 51, 54, 58, 61, 62, 64, 65, 67, 72, 74, 75, 75
k-arcs on elliptic curves over $GF(q)$. Ref HW84 51. [2,1; A5524]

M0476 1, 2, 3, 4, 5, 6, 7, 8, 9, 10, 12, 13, 15, 16, 18, 21, 22, 24, 25, 28, 30, 33, 37, 40, 42, 45, 48, 57, 58, 60, 70, 72, 78, 85, 88, 93, 102, 105, 112, 120, 130, 133, 165, 168, 177, 190
Euler's ideoneal or suitable numbers (a finite sequence): n such that p odd having unique representation as $x^2 + ny^2$, $(x,y) = 1$, implies p prime. Ref BS66 427. ELM 21 83 66. MINT 7 55 85. [1,2; A0926, N0176]

M0477 1, 2, 3, 4, 5, 6, 7, 8, 9, 10, 12, 14, 15, 16, 18, 20, 21, 24, 25, 27, 28, 30, 32, 35, 36, 40, 42, 45, 48, 49, 50, 54, 56, 60, 63, 64, 70, 72, 75, 80, 81, 84, 90, 96, 98, 100, 105, 108
Divisible by no prime greater than 7. [1,2; A2473, N0177]

M0478 0, 0, 0, 0, 0, 1, 2, 3, 4, 5, 6, 7, 8, 9, 10, 12, 14, 16, 18, 20, 22, 24, 26, 28, 30, 32, 34, 36, 38, 40
Low discrepancy sequences in base 5. Ref JNT 30 69 88. [1,7; A5358]

M0479 1, 1, 1, 1, 1, 1, 1, 1, 2, 3, 4, 5, 6, 7, 8, 9, 10, 12, 15, 19, 24, 30, 37, 45, 54, 64, 76, 91, 110, 134, 164, 201, 246, 300, 364, 440, 531, 641, 775, 939, 1140, 1386, 1686, 2050
$a(n) = a(n-1) + a(n-9)$. Ref AMM 95 555 88. [0,9; A5711]

M0480 1, 2, 3, 4, 5, 6, 7, 8, 9, 10, 12, 18, 20, 21, 23, 24, 27, 30, 36, 40, 42, 45, 48, 50, 54, 60, 63, 67, 70, 72, 80, 81, 84, 90, 100, 102, 104, 108
Power-sum numbers. Ref JRM 18 275 86. [1,2; A7603]

M0481 1, 2, 3, 4, 5, 6, 7, 8, 9, 10, 12, 18, 20, 21, 24, 27, 30, 36, 40, 42, 45, 48, 50, 54, 60, 63, 70, 72, 80, 81, 84, 90, 100, 102, 108, 110, 111, 112, 114, 117, 120, 126, 132, 133, 135
Niven (or Harshad) numbers: divisible by the sum of their digits. Ref Well86 171. MMAG 63 10 90. [1,2; A5349]

M0482 1, 2, 3, 4, 5, 6, 7, 8, 9, 11, 12, 15, 24, 36, 111, 112, 115, 128, 132, 135, 144, 175, 212, 216, 224, 312, 315, 384, 432, 612, 624, 672, 735, 816, 1111, 1112, 1113, 1115, 1116
Divisible by the product of its digits. Ref rgw. [1,2; A7602]

M0483 1, 1, 1, 1, 1, 1, 1, 1, 2, 3, 4, 5, 6, 7, 8, 9, 11, 14, 18, 23, 29, 36, 44, 53, 64, 78, 96, 119, 148, 184, 228, 281, 345, 423, 519, 638, 786, 970, 1198, 1479, 1824, 2247, 2766
$a(n) = a(n-1) + a(n-8)$. Ref AMM 95 555 88. [0,9; A5710]

M0484 0, 1, 2, 3, 4, 5, 6, 7, 8, 9, 11, 22, 33, 44, 55, 66, 77, 88, 99, 101, 111, 121, 131, 141, 151, 161, 171, 181, 191, 202, 212, 222, 232, 242, 252, 262, 272, 282, 292, 303, 313, 323
Palindromes. [0,3; A2113, N0178]

M0485 1, 2, 3, 4, 5, 6, 7, 8, 9, 13, 14, 15, 16, 18, 19, 24, 25, 27, 28, 31, 32, 33, 34, 35, 36, 37, 39, 49, 51, 67, 72, 76, 77, 81, 86
Decimal expansion of 2^n contains no 0 (probably 86 is last term). Ref Mada66 126. [1,2; A7377]

M0486 0, 1, 2, 3, 4, 5, 6, 7, 8, 9, 19, 18, 17, 16, 15, 14, 13, 12, 11, 10, 20, 21, 22, 23, 24, 25, 26, 27, 28, 29, 39, 38, 37, 36, 35, 34, 33, 32, 31, 30, 40, 41, 42, 43, 44, 45, 46, 47, 48
Decimal Gray code for n. Ref MAG 50 122 66. GA86 18. [0,3; A3100]

M0487 1, 2, 3, 4, 5, 6, 7, 8, 9, 24, 43, 63, 89, 132, 135, 153, 175, 209, 224, 226, 262, 264, 267, 283, 332, 333, 334, 357, 370, 371, 372, 373, 374, 375, 376, 377, 378, 379, 407, 445
Powerful numbers (2): a sum of positive powers of its digits. Ref rgw. [1,2; A7532]

M0488 1, 2, 3, 4, 5, 6, 7, 8, 9, 153, 370, 371, 407, 1634, 8208, 9474, 54748, 92727, 93084, 548834, 1741725, 4210818, 9800817, 9926315, 24678050, 24678051, 88593477
Armstrong numbers: equals sum of nth powers of its n digits. Ref LA81. JRM 14 87 81. rgw. [1,2; A5188]

M0489 1, 2, 3, 4, 5, 6, 7, 8, 9, 190, 209, 48, 247, 266, 195, 448, 476, 198, 874, 3980, 399,
   2398, 1679, 888, 4975, 1898, 999, 7588, 4988, 39990, 8959, 17888, 42999, 28798, 57995
Smallest multiple of n whose digits sum to n. [1,2; A2998]

M0490 1, 2, 3, 4, 5, 6, 7, 8, 10, 11, 12, 13, 14, 15, 16, 17, 18, 20, 21, 22, 23, 24, 25, 26, 27,
   28, 30, 31, 32, 33, 34, 35, 36, 37, 38, 40, 41, 42, 43, 44, 45, 46, 47, 48, 50, 51, 52, 53, 54
Natural numbers in base 9. [1,2; A7095]

M0491 1, 2, 3, 4, 5, 6, 7, 8, 10, 12, 14, 16, 19, 21, 23, 25, 30, 44, 46, 48, 50, 55, 65, 73, 74,
   77, 84, 86, 91, 95, 97, 114, 122, 123, 126
A self-generating sequence. Ref JCT A12 65 72. [1,2; A3045]

M0492 1, 1, 1, 1, 1, 1, 1, 2, 3, 4, 5, 6, 7, 8, 10, 13, 17, 22, 28, 35, 43, 53, 66, 83, 105, 133,
   168, 211, 264, 330, 413, 518, 651, 819, 1030, 1294, 1624, 2037, 2555, 3206, 4025, 5055
$a(n) = a(n-1) + a(n-7)$. Ref AMM 95 555 88. [0,8; A5709]

M0493 2, 3, 4, 5, 6, 7, 9, 10, 11, 12, 13, 14, 15, 16, 17, 18, 19, 20, 21, 22, 23, 24, 25, 26,
   28, 29, 30, 31, 32, 33, 34, 35, 36, 37, 38, 39, 40, 41, 42, 43, 44, 45, 46, 47, 48, 49, 50, 51
The non-cubes: $n + [(n + [n^{1/3}])^{1/3}]$. Ref MMAG 63 53 90. Robe92 11. [1,1; A7412]

M0494 1, 2, 3, 4, 5, 6, 7, 9, 10, 11, 12, 14, 16, 17, 20, 21, 22, 25, 27, 29, 31, 32, 36, 39, 40,
   42, 45, 46, 47, 49, 51, 54, 55, 56, 57, 60, 61, 65, 66, 67, 69, 71, 77, 84, 86, 87, 90, 94, 95
$n^2 - n - 1$ is prime. Ref PO20 249. L1 46. [1,2; A2328, N0179]

M0495 1, 1, 1, 2, 3, 4, 5, 6, 7, 9, 11, 15, 20, 27, 35, 44, 56, 73, 91, 115, 148, 186, 227, 283,
   358, 435, 538, 671, 813, 1001, 1233, 1492, 1815, 2223, 2673, 3247, 3933, 4713, 5683
Arrangements of pennies in rows. Ref PCPS 47 686 51. QJMO 23 153 72. rkg. [1,4;
A5577]

M0496 1, 1, 1, 1, 1, 1, 2, 3, 4, 5, 6, 7, 9, 12, 16, 21, 27, 34, 43, 55, 71, 92, 119, 153, 196,
   251, 322, 414, 533, 686, 882, 1133, 1455, 1869, 2402, 3088, 3970, 5103, 6558, 8427
$a(n) = a(n-1) + a(n-6)$. Ref AMM 95 555 88. [0,7; A5708]

M0497 1, 2, 3, 4, 5, 6, 7, 9, 18, 33
Decimal expansions of 2^n and 5^n contain no 0's (probably 33 is last term). Ref OgAn66
89. [1,2; A7496]

M0498 1, 2, 3, 4, 5, 6, 7, 10, 11, 12, 13, 14, 15, 16, 17, 20, 21, 22, 23, 24, 25, 26, 27, 30,
   31, 32, 33, 34, 35, 36, 37, 40, 41, 42, 43, 44, 45, 46, 47, 50, 51, 52, 53, 54, 55, 56, 57, 60
Natural numbers in base 8. [1,2; A7094]

M0499 0, 1, 1, 2, 3, 4, 5, 6, 7, 10, 13, 16, 22
n-dimensional determinant 4 lattices. Ref PRS A418 18 88. [0,4; A5140]

M0500 1, 2, 3, 4, 5, 6, 8, 9, 10, 11, 12, 14, 15, 16, 18, 20, 21, 22, 23, 24, 26, 27, 28, 29, 30, 32, 33, 35, 36, 39, 40, 41, 42, 44, 46, 48, 50, 51, 52, 53, 54, 55, 56, 58, 60, 63, 64, 65, 66
Values of Euler totient function (divided by 2). See Fig M0500. Cf. M0987. Ref BA8 64. AS1 840. [2,2; A2180, N0180]

|||||||||||||||||||||||||||||||||||||||||||||||||||||||||||||||||||||||||||||||||||||||||||||||||||||||||||||||||||||||||||||||||||||||||||||||||||||||

Figure M0500. EULER TOTIENT FUNCTION.

Euler's **totient** function $\phi(n)$ is the number of positive integers $\leq n$ that are relatively prime to n (M0299). Then $\phi(1) = 1$ and, for $n > 1$,

$$\phi(n) = n \prod_{p \mid n} \left(1 - \frac{1}{p} \right) ,$$

where the product is over all primes dividing n [NZ66 37]. The set of values $\{\phi(n)\}$ forms M0987. For $n \geq 3$, $\phi(n)$ is even, and the set of values $\{\frac{1}{2}\phi(n)\}$ gives M0500.

|||||||||||||||||||||||||||||||||||||||||||||||||||||||||||||||||||||||||||||||||||||||||||||||||||||||||||||||||||||||||||||||||||||||||||||||||||||||

M0501 2, 3, 4, 5, 6, 8, 9, 10, 12, 14, 17, 18, 20, 24, 26, 28, 30
Barriers for $\omega(n)$. Ref UPNT B8. [1,1; A5236]

M0502 1, 2, 3, 4, 5, 6, 8, 9, 14, 15, 16, 22, 28, 29, 36, 37, 54, 59, 85, 93, 117, 119, 161, 189, 193, 256, 308, 322, 327, 411, 466, 577, 591, 902, 928, 946, 1162, 1428, 1708, 1724
$45.2^n - 1$ is prime. Ref MOC 23 874 69. [1,2; A2242, N0181]

M0503 0, 0, 0, 0, 1, 1, 1, 2, 3, 4, 5, 6, 8, 10, 11, 13, 16, 18, 20, 23, 26, 29, 32, 35, 39, 43, 46, 50, 55, 59, 63, 68, 73, 78, 83, 88, 94, 100, 105, 111, 118, 124, 130, 137, 144, 151, 158
Genus of complete graph on n nodes: $\lceil (n-3)(n-4)/12 \rceil$. Ref PNAS 60 438 68. [1,8; A0933, N0182]

M0504 0, 0, 0, 0, 1, 2, 3, 4, 5, 6, 8, 10, 12, 14, 16, 18, 20, 22, 24, 26, 28, 30, 32, 34, 36, 38, 40, 42, 44, 46, 48, 50, 52, 54, 56, 58, 60, 62, 64, 66, 68, 70, 72, 74, 76, 78, 80, 82, 84, 86
Low discrepancy sequences in base 4. Ref JNT 30 69 88. [1,6; A5377]

$$\text{G.f.:} (x^4 + x^{10}) / (1 - 2x + x^2).$$

M0505 1, 2, 3, 4, 5, 6, 8, 10, 12, 15, 16, 17, 20, 24, 30, 32, 34, 40, 48, 51, 60, 64, 68, 80, 85, 96, 102, 120, 128, 136, 160, 170, 192, 204, 240, 255, 256, 257, 272, 320, 340, 384
Polygons constructible with ruler and compass. Ref GA01 460. VDW 1 187. B1 183. [1,2; A3401]

M0506 1, 2, 3, 4, 5, 6, 8, 10, 12, 15, 17, 19, 29, 31, 33, 43, 47, 51, 54, 58, 68, 69, 78, 79, 86, 95, 99, 110, 113, 117, 133
A self-generating sequence. Ref JCT A12 64 72. [1,2; A3044]

M0507 1, 1, 1, 1, 1, 2, 3, 4, 5, 6, 8, 11, 15, 20, 26, 34, 45, 60, 80, 106, 140, 185, 245, 325,
431, 571, 756, 1001, 1326, 1757, 2328, 3084, 4085, 5411, 7168, 9496, 12580, 16665
$a(n) = a(n-1) + a(n-5)$. Ref BR72 119. FQ 14 38 76. [0,6; A3520]

M0508 0, 1, 2, 3, 4, 5, 6, 8, 12, 10, 12, 13, 15, 18, 21, 24, 32, 22, 23
Edges in minimal broadcast graph with n nodes. Ref SIAD 1 532 88. [1,3; A7192]

M0509 1, 2, 3, 4, 5, 6, 9, 8, 7, 10, 15, 12, 25, 18, 27, 16, 11, 14, 21, 20, 35, 30, 45, 24, 49,
50, 75, 36, 125, 54, 81, 32, 13, 22, 55, 28
Write $n-1$ in binary; power of p_k in $a(n)$ is # of 1's that are followed by $k-1$ 0's. Ref jhc.
[1,2; A5940]

M0510 1, 2, 3, 4, 5, 6, 9, 8, 7, 10, 17, 12, 33, 18, 11, 16, 65, 14, 129, 20, 19, 34, 257, 24,
13, 66, 15, 36, 513, 22, 1025, 32, 35, 130, 21, 28
Inverse of M0509. Ref jhc. [1,2; A5941]

M0511 1, 2, 3, 4, 5, 6, 10, 11, 12, 13, 14, 15, 16, 20, 21, 22, 23, 24, 25, 26, 30, 31, 32, 33,
34, 35, 36, 40, 41, 42, 43, 44, 45, 46, 50, 51, 52, 53, 54, 55, 56, 60, 61, 62, 63, 64, 65, 66
Natural numbers in base 7. [1,2; A7093]

M0512 1, 2, 3, 4, 5, 6, 10, 15, 45, 120
Maximal iterated binomial coefficients. Ref AMM 87 725 80. [1,2; A6543]

M0513 0, 1, 2, 3, 4, 5, 7, 8, 9, 10, 11, 12, 14, 15, 16, 18, 19, 21, 22, 23, 24, 25, 26, 29, 30,
31, 32, 33, 35, 37, 38, 40, 42, 43, 44, 45, 46, 47, 49, 51, 52, 53, 54, 56, 57, 58, 60, 63, 64
$\{m + n, m \in$ M1242 , $n \in$ M2614$\}$. Ref IAS 5 382 37. [1,3; A2855, N0183]

M0514 1, 2, 3, 4, 5, 7, 8, 9, 10, 11, 13, 14, 15, 16, 17, 19, 21, 22, 23, 25, 26, 27, 29, 31, 32,
33, 34, 35, 37, 38, 39, 41, 43, 44, 45, 46, 47, 49, 50, 51, 52, 53, 55, 57, 58, 59, 61, 62, 63
Deficient numbers: $\sigma(n) < 2n$. See Fig M0062. Ref UPNT B2. [1,2; A5100]

M0515 0, 2, 3, 4, 5, 7, 8, 9, 10, 12, 13, 14, 15, 16, 18, 19, 20, 21, 23, 24, 25, 26, 28, 29, 30,
31, 33, 34, 35, 36, 38, 39, 40, 41, 42, 44, 45, 46, 47, 49, 50, 51, 52, 54, 55, 56, 57, 59, 60
Wythoff game. Ref CMB 2 189 59. [0,2; A1967, N0184]

M0516 1, 2, 3, 4, 5, 7, 8, 9, 10, 12, 13, 14, 15, 17, 18, 19, 20, 22, 23, 24, 25, 26, 33, 34, 35,
36, 37, 39, 43, 44, 45, 46, 47, 49, 50, 51, 52, 59, 60, 62, 63, 64, 65, 66, 68, 69, 71, 73
No 6-term arithmetic progression. Ref MOC 33 1354 79. [0,2; A5838]

M0517 1, 2, 3, 4, 5, 7, 8, 9, 11, 13, 16, 17, 19, 23, 25, 27, 29, 31, 32, 37, 41, 43, 47, 49, 53,
59, 61, 64, 67, 71, 73, 79, 81, 83, 89, 97, 101, 103, 107, 109, 113, 121, 125, 127, 128, 131
Prime powers. Ref AS1 870. [1,2; A0961, N0185]

M0518 1, 1, 2, 3, 4, 5, 7, 8, 10, 12, 14, 16, 19, 21, 24, 27, 30, 33, 37, 40, 44, 48, 52, 56, 61,
65, 70, 75, 80, 85, 91, 96, 102, 108, 114, 120, 127, 133, 140, 147, 154, 161, 169, 176, 184
Partitions into at most 3 parts. See Fig M0663. Ref RS4 2. AMM 86 687 79. [0,3; A1399,
N0186]

M0519   1, 2, 3, 4, 5, 7, 9, 11, 13, 15, 18, 21, 24, 27, 30, 34, 38, 42, 46, 50, 55, 60, 65, 70,
75, 82, 89, 96, 103, 110, 119, 128, 137, 146, 155, 166, 177, 188, 199, 210, 223, 236, 249
Partitions of $5n$ into powers of 5. Ref rkg. [0,2; A5706]

M0520   2, 3, 4, 5, 7, 9, 11, 13, 16, 17, 19, 23, 24, 25, 29, 30, 31, 37, 40, 41, 42, 43, 47, 49,
53, 54, 56, 59, 60, 61, 66, 67, 70, 71, 72, 73, 78, 79, 81, 83, 84, 88, 89, 90, 96, 97, 101
A 2-way classification of integers. Cf. M4065. Ref CMB 2 89 59. Robe92 22. [1,1; A0028,
N0187]

M0521   1, 2, 3, 4, 5, 7, 9, 12, 15, 19, 24, 31, 40, 52, 67, 86, 110, 141, 181, 233, 300, 386,
496, 637, 818, 1051, 1351, 1737, 2233, 2870, 3688, 4739, 6090, 7827, 10060, 12930
From a nim-like game. Ref rkg. [0,2; A3413]

M0522   1, 2, 3, 4, 5, 7, 10, 11, 12, 14, 15, 18, 24, 25, 26, 28, 29, 31, 33, 35, 38, 39, 42, 43,
46, 49, 50, 53, 56, 59, 63, 64, 67, 68, 75, 81, 82, 87, 89, 91, 92, 94, 96, 106, 109, 120, 124
Values of x in M4363. Ref MES 41 144 12. [1,2; A2504, N0188]

M0523   1, 2, 3, 4, 5, 7, 10, 11, 17, 22, 23, 41, 47, 59, 89, 107, 167, 263, 347, 467, 683, 719,
1223, 1438, 1439, 2879, 3767, 4283, 6299, 10079, 11807, 15287, 21599, 33599, 45197
Smallest number of complexity n. Ref FQ 27 16 89. [1,2; A5520]

M0524   1, 1, 1, 2, 3, 4, 5, 7, 10, 13, 16, 21, 28, 35, 43, 55, 70, 86, 105, 130, 161, 196, 236,
287, 350, 420, 501, 602, 722, 858, 1016, 1206, 1431, 1687, 1981, 2331, 2741, 3206, 3740
Partitions with an even number of even parts. [0,4; A6950]

M0525   1, 2, 3, 4, 5, 7, 10, 13, 19, 28, 37, 55, 82, 109, 163, 244, 325, 487, 730, 973, 1459,
2188, 2917, 4375, 6562, 8749, 13123, 19684, 26245
Positions where M0456 increases. Ref HO85a 225. [1,2; A7601]

M0526   1, 1, 1, 1, 2, 3, 4, 5, 7, 10, 14, 19, 26, 36, 50, 69, 95, 131, 181, 250, 345, 476, 657,
907, 1252, 1728, 2385, 3292, 4544, 6272, 8657, 11949, 16493, 22765, 31422, 43371
$a(n) = a(n-1) + a(n-4)$. Ref BR72 120. [0,5; A3269]

M0527   1, 2, 3, 4, 5, 7, 11, 13, 21, 23, 41, 43, 71, 94, 139, 211, 215, 431, 863
Smallest number of complexity n. Ref BS71. [0,2; A3037]

M0528   1, 1, 1, 1, 2, 3, 4, 5, 7, 11, 16, 22, 30, 43, 62, 88, 124, 175, 249, 354, 502, 710,
1006, 1427, 2024, 2870, 4068, 5767, 8176, 11593, 16436, 23301, 33033, 46832, 66398
Compositions into squares. Ref BIT 19 301 79. [0,5; A6456]

M0529   1, 2, 3, 4, 5, 8, 9, 10, 12, 13, 16, 17, 18, 20, 25, 26, 27, 29, 32, 34, 36, 37, 40, 41,
45, 48, 49, 50, 52, 53, 58, 61, 64, 65, 68, 72, 73, 74, 75, 80, 81, 82, 85, 89, 90, 97, 98, 100
Sums of 2 squares or 3 times a square. Ref SW91. [1,2; A5792]

M0530   2, 3, 4, 5, 8, 9, 13, 16, 17, 24, 25, 35, 44, 63, 64, 91, 97, 128, 193, 221, 259, 324,
353, 391, 477, 702, 929, 1188, 1269, 1589, 1613, 2017, 2309, 2623, 3397, 4064, 4781
Related to iterates of bi-unitary totient function. Ref UM 10 349 76. [1,1; A5424]

M0531   2, 3, 4, 5, 9, 16, 17, 41, 83, 113, 137, 257, 773, 977, 1657, 2048, 2313, 4001, 5725,
   7129, 11117, 17279, 19897, 22409, 39283, 43657, 55457
Related to iterates of unitary totient function. Ref MOC 28 302 74. [1,1; A3271]

M0532   1, 2, 3, 4, 5, 10, 11, 12, 13, 14, 15, 20, 21, 22, 23, 24, 25, 30, 31, 32, 33, 34, 35, 40,
   41, 42, 43, 44, 45, 50, 51, 52, 53, 54, 55, 100, 101, 102, 103, 104, 105, 110, 111, 112, 113
Natural numbers in base 6. [1,2; A7092]

M0533   1, 2, 3, 4, 5, 11, 21, 36, 57, 127, 253, 463, 793, 1717, 3433, 6436, 11441, 24311,
   48621, 92379, 167961, 352717, 705433, 1352079, 2496145, 5200301, 10400601
Euler characteristics of polytopes. Ref JCT A17 346 74. [1,2; A6481]

M0534   1, 1, 2, 3, 4, 5, 12, 21, 32, 45, 120, 231, 384, 585, 1680, 3465, 6144, 9945, 30240,
   65835, 122880, 208845, 665280, 1514205, 2949120, 5221125, 17297280, 40883535
Quadruple factorial numbers $n!!!!$: $a(n)=na(n-4)$. Ref SpOl87 23. [0,3; A7662]

M0535   1, 1, 2, 3, 4, 6, 6, 9, 10, 12, 10, 22, 12, 18, 24, 27, 16, 38, 18, 44, 36, 30, 22, 78, 36,
   36, 50, 66, 28, 104, 30, 81, 60, 48, 72, 158, 36, 54, 72, 156, 40, 156, 42, 110, 152, 66
Mu-atoms of period n on continent of Mandelbrot set. Ref Man82 183. Pen91 138. rpm.
[1,3; A6874]

M0536   1, 1, 2, 3, 4, 6, 6, 11, 10, 18, 16, 20, 24, 26, 20, 45, 40, 38, 34, 62, 46, 54, 50, 84,
   50, 102, 78, 104, 98, 90, 70, 189, 82, 130, 84, 120, 112, 130, 120, 232, 152, 234, 132, 130
Shifts left under g.c.d.-convolution with itself. Ref BeSl94. [0,3; A7464]

M0537   1, 2, 3, 4, 6, 6, 12, 15, 20, 30, 30, 60, 60, 84, 105, 140, 210, 210, 420, 420, 420,
   420, 840, 840, 1260, 1260, 1540, 2310, 2520, 4620, 4620, 5460, 5460, 9240, 9240, 13860
Largest order of permutation of length n. Ref BSMF 97 187 69. [1,2; A0793, N0190]

M0538   2, 3, 4, 6, 6, 13, 10, 24, 22, 45, 30, 158, 74, 245, 368, 693, 522, 2637, 1610, 7341
Necklaces. Ref IJM 8 269 64. [1,1; A2729, N0191]

M0539   1, 2, 3, 4, 6, 7, 8, 9, 10, 11, 12, 13, 15, 16, 17, 18, 19, 21, 22, 23, 24, 25, 26, 27, 28,
   30, 31, 32, 33, 34, 36, 37, 38, 40, 41, 42, 43, 44, 46, 47, 48, 49, 50, 51, 52, 53, 55, 56, 57
Related to Fibonacci representations. Ref FQ 11 385 73. [1,2; A3247]

M0540   1, 2, 3, 4, 6, 7, 8, 9, 11, 12, 13, 14, 16, 17, 18, 19, 21, 22, 23, 24, 25, 27, 28, 29, 30,
   32, 33, 34, 35, 37, 38, 39, 40, 42, 43, 44, 45, 46, 48, 49, 50, 51, 53, 54, 55, 56, 58, 59, 60
A Beatty sequence: $[n(\sqrt{5} - 1]$. Cf. M3795. Ref CMB 2 189 59. [1,2; A1961, N0192]

M0541   0, 2, 3, 4, 6, 7, 8, 9, 11, 12, 13, 15, 16, 17, 19, 20, 21, 23, 24, 25, 26, 28, 29, 30, 32,
   33, 34, 36, 37, 38, 39, 41, 42, 43, 45, 46, 47, 49, 50, 51, 52, 54, 55, 56, 58, 59, 60, 62, 63
Wythoff game. Ref CMB 2 188 59. [0,2; A1959]

M0542   1, 2, 3, 4, 6, 7, 8, 9, 11, 12, 14, 16, 18, 19, 21, 22, 23, 24, 27, 28, 31, 32, 33, 36, 38,
   42, 43, 44, 46, 47, 48, 49, 54, 56, 57, 59, 62, 63, 64, 66, 67, 69, 71, 72, 76, 77, 79, 81, 83
Not the sum of two distinct squares. Ref FQ 13 319 75. [1,2; A4144]

M0543 0, 2, 3, 4, 6, 7, 9, 10, 12, 13, 14, 16, 17, 19, 20, 21, 23, 24, 26, 27, 28, 30, 31, 33, 34, 36, 37, 38, 40, 41, 43, 44, 45, 47, 48, 50, 51, 53, 54, 55, 57, 58, 60, 61, 62, 64, 65, 67
Wythoff game. Ref CMB 2 188 59. [0,2; A1953, N0193]

M0544 0, 1, 2, 3, 4, 6, 7, 9, 15, 22, 28, 30, 46, 60, 63, 127, 153, 172, 303, 471, 532, 865, 900, 1366, 2380, 3310, 4495, 6321, 7447, 10198, 11425, 21846, 24369, 27286, 28713
$x^n + x + 1$ is irreducible over $GF(2)$. Ref IFC 16 502 70. [1,3; A2475, N0194]

M0545 0, 1, 2, 3, 4, 6, 7, 11, 18, 34, 38, 43, 55, 64, 76, 94, 103, 143, 206, 216, 306, 324, 391, 458, 470, 827, 1274, 3276, 4204, 5134, 7559, 12676, 26459
$3.2^n - 1$ is prime. Ref MOC 23 874 69. Rie85 384. Cald94. [1,3; A2235, N0195]

M0546 2, 3, 4, 6, 8, 9, 10, 12, 16, 18, 20, 24, 30, 32, 36, 40, 48, 60, 64, 72, 80, 84, 90, 96, 100, 108, 120, 128, 144, 160, 168, 180, 192, 200, 216, 224, 240, 256, 288, 320, 336
Number of divisors of highly composite numbers. Ref RAM1 87. [1,1; A2183, N0196]

M0547 0, 1, 2, 3, 4, 6, 8, 9, 11, 12, 16, 17, 18, 19, 22, 24, 25, 27, 32, 33, 34, 36, 38, 41, 43, 44, 48, 49, 50, 51, 54, 57, 59, 64, 66, 67, 68, 72, 73, 75, 76, 81, 82, 83, 86, 88, 89, 96, 97
Of the form $x^2 + 2y^2$. Ref EUL (1) 1 421 11. L1 59. [1,3; A2479, N0197]

M0548 1, 2, 3, 4, 6, 8, 9, 12, 15, 16, 20, 24, 25, 30, 35, 36, 42, 48, 49, 56, 63, 64, 72, 80, 81, 90, 99, 100, 110, 120, 121, 132, 143, 144, 156, 168, 169, 182, 195, 196, 210, 224, 225
$[\sqrt{n}]$ divides n. Ref AMM 82 854 75. jos. [1,2; A6446]

M0549 1, 2, 3, 4, 6, 8, 9, 12, 15, 16, 21, 24, 24, 32, 36, 36, 45, 48, 48, 60, 66, 64, 75, 84, 81, 96, 105, 96, 120, 128, 120, 144, 144, 144, 171, 180, 168, 192, 210, 192, 231, 240, 216
Degree of rational porism of n-gon. Ref BCMS 39 103 47. [3,2; A2348, N0198]

M0550 0, 0, 0, 1, 2, 3, 4, 6, 8, 10, 12, 15, 16, 19, 22, 25, 27, 30, 32, 35, 37, 40, 42, 45, 48, 51, 54, 57, 60, 63, 66, 69, 72, 75, 78, 81, 84, 87, 90, 93, 96, 99, 102, 105, 108, 111, 114
Maximal splittance of graph with n nodes. Ref COMB 1 284 81. [0,5; A7183]

M0551 2, 3, 4, 6, 8, 10, 12, 15, 18, 21, 24, 28, 32, 36, 40, 45, 50
Restricted partitions. Ref CAY 2 277. [2,1; A1972, N0199]

M0552 1, 2, 3, 4, 6, 8, 10, 12, 15, 18, 21, 24, 28, 32, 36, 40, 46, 52, 58, 64, 72, 80, 88, 96, 106, 116, 126, 136, 148, 160, 172, 184, 199, 214, 229, 244, 262, 280, 298, 316, 337, 358
Partitions of $4n$ into powers of 4. Ref rkg. [0,2; A5705]

M0553 1, 2, 3, 4, 6, 8, 10, 12, 16, 18, 20, 24, 30, 36, 42, 48, 60, 72, 84, 90, 96, 108, 120, 144, 168, 180, 210, 216, 240, 288, 300, 336, 360, 420, 480, 504, 540, 600, 630, 660, 720
Highly abundant numbers: where $\sigma(n)$ increases. Ref TAMS 56 467 44. AS1 842. [1,2; A2093, N0200]

M0554 1, 2, 3, 4, 6, 8, 10, 13, 16, 20, 24, 28, 33, 38, 44, 50, 57, 64, 72, 80, 88, 97, 106, 116, 126, 137, 148, 160, 172, 185, 198, 212, 226, 241, 256, 272, 288, 304, 321, 338, 356
$a(n) = a(n-1) + [\sqrt{a(n-1)}]$. [0,2; A2984]

M0555   1, 2, 3, 4, 6, 8, 10, 13, 17, 21, 27, 30, 37, 47, 57, 62, 75, 87, 102, 116
Correlations of length n. Ref JCT A30 29 81. [1,2; A5434]

M0556   1, 1, 2, 3, 4, 6, 8, 10, 14, 17, 22, 27, 33, 41, 49, 59, 71, 83, 99, 115, 134, 157, 180,
    208, 239, 272, 312, 353, 400, 453, 509, 573, 642, 717, 803, 892, 993, 1102, 1219, 1350
Partitions of n into Fibonacci parts (with a single type of 1). Cf. M1045. [0,3; A3107]

M0557   1, 2, 3, 4, 6, 8, 11, 13, 16, 18, 26, 28, 36, 38, 47, 48, 53, 57, 62, 69, 72, 77, 82, 87,
    97, 99, 102, 106, 114, 126, 131, 138, 145, 148, 155, 175, 177, 180, 182, 189, 197, 206
Ulam numbers: next is uniquely the sum of 2 earlier terms. See Fig M0557. Ref SIAR 6
348 64. AB71 249. JCT A12 39 72. PC 2 13-7 74. UPNT C4. [1,2; A2858, N0201]

||||||||||||||||||||||||||||||||||||||||||||||||||||||||||||||||||||||||||||||||||||||||||||||||||||||||||||

Figure M0557.   ULAM SEQUENCES.

    M0557 shows the **Ulam numbers**. We start with $a_1 = 1$, $a_2 = 2$. Then having found
$a_1, \ldots, a_n$, we choose a_{n+1} to be the smallest number that can be written uniquely in the
form $a_i + a_j$ with $1 \le i < j \le n$. Many variations are possible. For example, starting with
$a_1 = 1$ and taking a_{n+1} to be the smallest number that is not the sum of **consecutive** terms of
$a_1, \ldots, a_n$ leads to M0972. The partial sums give M2633. See also M0634, M0689, M0794,
M1112, M2303, etc., and Fig. M0436.

||||||||||||||||||||||||||||||||||||||||||||||||||||||||||||||||||||||||||||||||||||||||||||||||||||||||||||

M0558   1, 2, 3, 4, 6, 8, 11, 14, 18, 23, 29, 36, 44, 54, 66, 79, 95, 113, 133, 157, 184, 216,
    250, 290, 335, 385, 442, 505, 576, 656, 743, 842, 951, 1070, 1204, 1351, 1514, 1691
Partitions of n into partition numbers. [1,2; A7279]

M0559   1, 2, 3, 4, 6, 8, 11, 14, 18, 24, 32, 43, 54, 68, 86, 110, 142, 185, 239, 307, 393, 503,
    645, 830, 1069, 1376, 1769, 2272, 2917, 3747, 4816, 6192, 7961, 10233, 13150, 16897
From a nim-like game. Ref rkg. [0,2; A3412]

M0560   1, 2, 3, 4, 6, 8, 11, 15, 21, 28, 39, 53, 99, 137, 186
Regions in certain maps. Ref HM85 311. [1,2; A6683]

M0561   1, 2, 3, 4, 6, 8, 11, 15, 21, 29, 40, 55, 76, 105, 145, 200, 276, 381, 526, 726, 1002,
    1383, 1909, 2635, 3637, 5020, 6929, 9564, 13201, 18221, 25150, 34714, 47915, 66136
Expansion of $(1 + x + x^2 + x^3 + x^4)/(1 - x - x^4)$. Ref rkg. [0,2; A3411]

M0562   1, 1, 1, 2, 3, 4, 6, 8, 12, 16, 22, 29, 41, 53, 71, 93, 125, 160, 211, 270, 354, 450,
    581, 735, 948, 1191, 1517, 1902, 2414, 3008, 3791, 4709, 5909, 7311, 9119, 11246
Symmetric plane partitions of n. Ref SAM 50 261 71. [0,4; A5987]

M0563   1, 2, 3, 4, 6, 8, 13, 18, 30, 46, 78, 126, 224, 380, 687, 1224, 2250, 4112, 7685,
    14310, 27012
Necklaces with n beads, allowing turning over. See Fig M3860. Ref IJM 5 662 61. [0,2;
A0029, N0202]

M0564 1, 2, 3, 4, 6, 8, 14, 20, 36, 60, 108, 188, 352, 632, 1182, 2192, 4116, 7712, 14602, 27596, 52488, 99880, 190746, 364724, 699252, 1342184, 2581428, 4971068, 9587580
n-bead necklaces with 2 colors; binary irreducible polynomials of degree n. See Fig M3860. Cf. M0116. Ref IJM 5 662 61. GO67 172. NAT 261 463 76. [0,2; A0031, N0203]

M0565 1, 2, 3, 4, 6, 9, 11, 15, 19, 25, 31, 41, 49, 61, 75, 91, 110, 134, 157, 189, 222, 264, 308, 363, 420, 489, 566, 654, 751, 866, 985, 1130, 1283, 1462, 1655, 1877, 2115, 2387
Oscillates under partition transform. Cf. M0630. Ref BeSl94. EIS § 2.7. [1,2; A7210]

M0566 1, 1, 2, 3, 4, 6, 9, 12, 16, 22, 29, 38, 50, 64, 82, 105, 132, 166, 208, 258, 320, 395, 484, 592, 722, 876, 1060, 1280, 1539, 1846, 2210, 2636, 3138, 3728, 4416, 5222, 6163
Expansion of Π $(1+x^{2k})/(1-x^{2k-1})$. Ref CAY 9 128. HO85a 241. [0,3; A1935, N0204]

M0567 1, 2, 3, 4, 6, 9, 12, 16, 22, 31, 40, 52, 68, 90, 121, 152, 192, 244, 312, 402, 523, 644, 796, 988, 1232, 1544, 1946, 2469, 2992, 3636, 4432, 5420, 6652, 8196, 10142
$a(n) = a(n-1) + a(n-1 -$ number of odd terms so far$)$. Ref rgw. [1,2; A7604]

M0568 1, 2, 3, 4, 6, 9, 12, 18, 27, 36, 54, 81, 108, 162, 243, 324, 486, 729, 972, 1458, 2187, 2916, 4374, 6561, 8748, 13122, 19683, 26244, 39366, 59049, 78732, 118098
$a(3n) = 3^n$, $a(3n+1) = 4.3^{n-1}$, $a(3n+2) = 2.3^n$. Ref CMB 8 627 65. JRM 4 168 71. FQ 27 16 89. [1,2; A0792, N0205]

M0569 1, 2, 3, 4, 6, 9, 13, 19, 27, 38, 54, 77, 109, 154, 218, 309, 437, 618, 874, 1236, 1748, 2472, 3496, 4944, 6992, 9888, 13984, 19777, 27969, 39554, 55938, 79108, 111876
$a(n+1) = [\sqrt{(2a(n)(a(n)+1))}]$. Ref MMAG 43 143 70; 64 168 91. AMM 95 705 88. [1,2; A1521, N0206]

M0570 1, 1, 2, 3, 4, 6, 9, 13, 19, 27, 38, 54, 77, 109, 155, 219, 310, 438, 621, 877, 1243, 1755, 2486, 3510, 4973, 7021, 9947, 14043, 19894, 28086, 39789, 56173, 79579, 112347
$a(2n) = [17.2^n/14]$, $a(2n+1) = [12.2^n/7]$. Ref KN1 3 207. [0,3; A3143]

M0571 1, 1, 1, 2, 3, 4, 6, 9, 13, 19, 28, 41, 60, 88, 129, 189, 277, 406, 595, 872, 1278, 1873, 2745, 4023, 5896, 8641, 12664, 18560, 27201, 39865, 58425, 85626, 125491
$a(n) = a(n-1) + a(n-3)$. Ref LA62 13. FQ 2 225 64. JA66 91. MMAG 41 15 68. [0,4; A0930, N0207]

M0572 2, 3, 4, 6, 9, 14, 21, 31, 47, 70, 105, 158, 237, 355, 533, 799, 1199, 1798, 2697, 4046, 6069, 9103, 13655, 20482, 30723, 46085, 69127, 103691, 155536, 233304, 349956
Josephus problem. Ref SC68 374. JNT 26 207 87. [1,1; A5428]

M0573 2, 3, 4, 6, 9, 14, 22, 35, 55, 89, 142, 230, 373, 609, 996, 1637, 2698, 4461, 7398, 12301, 20503, 34253, 57348
From sequence of numbers with abundancy n. Ref MMAG 59 87 86. [2,1; A5579]

M0574 2, 3, 4, 6, 9, 14, 22, 35, 56, 90, 145, 234, 378, 611, 988, 1598, 2585, 4182, 6766, 10947, 17712, 28658; 46369, 75026, 121394, 196419, 317812, 514230
Expansion of $(2-x-2x^2) / (1-x)(1-x+x^2)$. Ref CMB 4 32 61 (divided by 3). [3,1; A0381, N1692]

M0575   2, 3, 4, 6, 9, 14, 23, 38
Pairwise relatively prime polynomials of degree n. Ref IFC 13 615 68. [1,1; A1115, N0209]

M0576   1, 2, 3, 4, 6, 11, 22, 43, 79, 137, 231, 397, 728, 1444, 3018, 6386, 13278, 26725, 51852, 97243, 177671, 320286, 579371, 1071226, 2053626, 4098627, 8451288
$\Sigma C(n,k^2)$, $k = 0 \ldots n$. Ref hwg. [0,2; A3099]

M0577   1, 2, 3, 4, 6, 12, 15, 20, 30, 60, 84, 105, 140, 210, 420, 840, 1260, 1540, 2310, 2520, 4620, 5460, 9240, 13860, 16380, 27720, 30030, 32760, 60060, 120120, 180180
Largest order of permutation of length n. Ref BSMF 97 187 69. [1,2; A2809, N0210]

M0578   1, 1, 2, 3, 4, 6, 16, 16, 30
Point-symmetric tournaments with $2n+1$ nodes. Ref CMB 13 322 70. [1,3; A2087, N0211]

M0579   0, 1, 2, 3, 4, 7, 6, 12, 12, 23, 10, 51, 12, 75, 50, 144, 16, 324, 18, 561, 156, 1043, 22, 2340, 80, 4119, 540, 8307, 28, 17521, 30, 32928, 2096, 65567, 366, 135432, 36
Non-seed mu-atoms of period n in Mandelbrot set. Ref Man82 183. Pen91 138. rpm. [1,3; A6875]

M0580   1, 2, 3, 4, 7, 8, 11, 12, 18, 24, 30, 41, 42, 55, 72, 78, 97, 98, 108, 114, 139, 140, 155, 192, 198, 215, 264, 281, 282, 335, 408, 431, 432, 438, 517, 576, 582, 685, 828, 857
$a(n) = a(n-1) +$ sum of prime factors of $a(n-1)$. Ref MMAG 48 57 75. [1,2; A3508]

M0581   2, 3, 4, 7, 8, 15, 24, 60, 168, 480, 1512, 4800, 15748, 28672, 65528, 122880, 393192, 1098240, 4124736, 15605760, 50328576, 149873152, 371226240, 1710858240
$a(n) = \sigma(a(n-1))$. Ref rgw. [0,1; A7497]

M0582   1, 2, 3, 4, 7, 8, 16, 31, 127, 256, 8191, 65536, 131071, 524287
n and $n+1$ are prime powers. [1,2; A6549]

M0583   1, 1, 2, 3, 4, 7, 9, 13, 17, 25, 32, 43
Arrangements of pennies in rows. Ref PCPS 47 686 51. QJMO 23 153 72. rkg. [0,3; A5576]

M0584   1, 1, 1, 2, 3, 4, 7, 11, 18, 25, 32, 39, 71, 110, 181, 252, 323, 394, 465, 536, 1001, 1537, 2538, 3539, 4540, 5541, 6542, 7543, 8544, 9545, 18089, 27634, 45723, 63812
Denominators of approximations to e. Cf. M0686. Ref GKP 122. [1,4; A6259]

M0585   1, 2, 3, 4, 7, 12, 22, 30, 32, 61, 65, 115, 161, 189, 296, 470, 598, 841, 904, 1856, 2158, 2416, 1925, 3462, 2130, 3749, 6546, 11201, 2159, 2360, 5186, 6071, 8664, 14735
Worst case of Euclid's algorithm. Ref FQ 25 210 87. STNB 3 51 91. [1,2; A6537]

M0586   1, 2, 3, 4, 7, 13, 24, 44, 83, 157, 297, 567, 1085, 2086, 4019, 7766, 15039, 29181, 56717, 110408, 215225, 420076, 820836, 1605587, 3143562, 6160098, 12080946
Landau's approximation to population of $x^2 + y^2$. Ref MOC 18 79 64. [0,2; A0690, N0212]

M0587 1, 2, 3, 4, 8, 9, 11, 12, 13, 14, 16, 17, 18, 21, 23, 26, 29, 34, 36, 37, 38, 47, 48, 49, 51, 53, 54, 56, 62, 63, 66, 67, 68, 69, 73, 74, 77, 79, 82, 83, 91, 99, 101, 102, 103, 107
$4.n^2 + 25$ is prime. Ref KK71 1. [1,2; A2971]

M0588 1, 2, 3, 4, 8, 9, 16, 27, 32, 64, 81, 128, 243, 256, 512, 729, 1024, 2048, 2187, 4096, 6561, 8192, 16384, 19683, 32768, 59049, 65536, 131072, 177147, 262144, 524288
Powers of 2 or 3. Ref RAM2 78. [0,2; A6899]

M0589 1, 2, 3, 4, 8, 10, 14, 20, 22, 26, 30, 38, 39, 49, 54, 58, 70, 81, 84, 87, 102, 111, 140, 159, 207, 224, 328, 358, 360, 447, 484, 908, 1083, 1242, 1461, 1705
$10.3^n - 1$ is prime. Ref MOC 26 997 72. [1,2; A5542]

M0590 0, 2, 3, 4, 8, 14, 25, 47, 86, 164, 314, 603, 1159, 2271, 4456, 8748, 17182, 33761, 66919, 132679, 263087
A generalized Conway-Guy sequence. Ref MOC 50 312 88. [0,2; A6755]

M0591 1, 1, 1, 2, 3, 4, 8, 15, 16, 24
The coding-theoretic function $A(n,8,7)$. See Fig M0240. Ref PGIT 36 1336 90. [8,4; A5853]

M0592 1, 2, 3, 4, 9, 10, 12, 14, 19, 23, 24, 36, 38, 39, 48, 62, 93, 106, 120, 134, 150, 196, 294, 317, 586, 597
Unique period lengths of primes. Cf. M2890. Ref JRM 18 24 85. [1,2; A7498]

M0593 1, 2, 3, 4, 9, 27, 512, 134217728
An exponential function on partitions (next term is 2^{512}). Ref AMM 76 830 69. [1,2; A1144, N0214]

M0594 2, 3, 4, 9, 28, 225, 6076, 1361025, 8268226876, 11253255215681025, 9304446720552777233254 6876, 10470531358708673960627431922039587436 81025
$a(n+2) = (a(n) - 1)a(n+1) + 1$. Ref MFM 111 122 91. [1,1; A7704]

M0595 1, 2, 3, 4, 10, 11, 12, 13, 14, 20, 21, 22, 23, 24, 30, 31, 32, 33, 34, 40, 41, 42, 43, 44, 100, 101, 102, 103, 104, 110, 111, 112, 113, 114, 120, 121, 122, 123, 124, 130, 131
Natural numbers in base 5. [1,2; A7091]

M0596 1, 1, 2, 3, 4, 10, 18, 28, 80, 162, 280, 880, 1944, 3640, 12320, 29160, 58240, 209440, 524880, 1106560, 4188800, 11022480, 24344320, 96342400, 264539520
Triple factorial numbers $n!!!: a(n) = na(n-3)$. Ref SpO187 23. [0,3; A7661]

M0597 1, 0, 2, 3, 4, 11, 17, 29, 49, 85, 144
A partition function. Ref JNSM 9 103 69. [0,3; A2098, N0215]

M0598 2, 3, 4, 12, 20, 55, 127, 371, 1037, 3249
Related to hexaflexagrams. Ref JRM 8 186 76. [1,1; A7499]

M0599 1, 0, 1, 2, 3, 4, 15, 32, 89, 266, 797, 2496, 8012, 26028, 85888, 286608, 965216, 3278776, 11221548, 38665192, 134050521, 467382224, 1638080277, 5768886048
3-connected planar maps with n edges. Ref JCT B32 41 82. [6,4; A5645]

M0600 1, 2, 3, 4, 22, 30, 12, 128, 147, 132, 548, 516, 552
Expansion of a modular function. Ref PLMS 9 384 59. [−2,2; A6709]

M0601 1, 0, 2, 3, 4, 30, 66, 0, 496, 1512, 1800, 51480, 487752, 4633200, 50605296,
620703720, 8278947840, 118504008000, 1811156124096, 29452505385600
Expansion of $(1 − x)^x$. [0,3; A7114]

M0602 2, 3, 4, 40, 210, 1477, 11672, 104256, 1036050, 11338855, 135494844,
1755206648, 24498813794, 366526605705, 5851140525680, 99271367764480
From ménage polynomials. Ref R1 197. [2,1; A0033, N0216]

M0603 1, 1, 2, 3, 5, 1, 13, 7, 17, 11, 89, 1, 233, 29, 61, 47, 1597, 19, 37, 41, 421, 199,
28657, 23, 3001, 521, 53, 281, 514229, 31, 557, 2207, 19801, 3571, 141961, 107, 73, 113
Primitive prime factor of Fibonacci number $F(n)$. Ref FQ 1(3) 15 63. [1,3; A1578, N0217]

M0604 1, 2, 3, 5, 4, 7, 6, 9, 13, 8, 10, 19, 14, 12, 29, 16, 21, 22, 37, 18, 27, 20, 43, 33, 34,
28, 49, 24, 61, 32, 67, 30, 73, 45, 57, 44, 40, 36, 50, 42, 52, 101, 63, 85, 109, 91, 74, 54
Inverse of the sum-of-divisors function. Ref BA8 85. [1,2; A2192, N0218]

M0605 1, 1, 2, 3, 5, 4, 8, 8, 1, 0
Number of '$(n,2)$'-sequences of length $2n$. Ref SoGo94. [1,3; A7281]

M0606 1, 2, 3, 5, 5, 5, 7, 5, 6, 8, 8, 9, 11, 11, 12, 15, 14, 14, 15, 14, 14, 15, 15, 14, 16
Nontrivial disconnected complements of Steinhaus graphs on n nodes. Ref DM 37 167 81.
[7,2; A3660]

M0607 2, 3, 5, 5, 7, 7, 7, 11, 9, 9, 11, 13, 11, 11, 15, 13, 13, 13, 17, 15, 19, 15, 19, 17, 21,
17, 19, 17, 17, 19, 21, 25, 19, 19, 23, 25, 23, 21, 23, 21, 21, 29, 23, 25, 23, 27, 29, 23
x such that $p = x^2 − 2y^2$. Cf. M0139. Ref CU04 1. L1 55. [2,1; A2334, N0219]

M0608 1, 2, 3, 5, 5, 8, 13, 13, 13, 26, 13, 91, 13, 106
Periods of patterns of growth. Ref SU70. [1,2; A6447]

M0609 1, 2, 3, 5, 6, 5, 8, 9, 11, 10, 7, 15, 15, 14, 17, 24, 24, 21, 13, 19, 27, 25, 29, 26, 44,
44, 29, 46, 39, 46, 27, 42, 47, 47, 54, 35, 41, 60, 51, 37, 48, 45, 49, 50, 49, 53
Harmonic means of divisors of harmonic numbers. See Fig M4299. Cf. M4185. Ref AMM
61 95 54. [1,2; A1600, N0220]

M0610 2, 3, 5, 6, 6, 6, 7, 8, 10, 13, 13, 13, 14, 17, 17, 17, 18, 19, 20, 22, 23, 27, 29, 29, 29,
31, 32, 35, 36, 37, 40, 43, 46, 48, 50, 53, 55, 57, 60, 60, 61, 63, 66, 66, 68, 71, 74, 77
Related to lattice points in circles. Ref MOC 20 306 66. [1,1; A0036, N0221]

M0611 2, 3, 5, 6, 7, 2, 10, 11, 3, 13, 14, 15, 17, 2, 19, 5, 21, 22, 23, 6, 26, 3, 7, 29, 30, 31,
2, 33, 34, 35, 37, 38, 39, 10, 41, 42, 43, 11, 5, 46, 47, 3, 2, 51, 13, 53, 6, 55, 14, 57, 58
Remove squares! Ref NCM 4 168 1878. [1,1; A2734, N0222]

M0612 1, 2, 3, 5, 6, 7, 8, 9, 10, 12, 13, 14, 16, 17, 18, 19, 20, 21, 23, 24, 25, 27, 28, 30, 31, 32, 34, 35, 36, 37, 38, 39, 41, 42, 43, 45, 46, 47, 48, 49, 50, 52, 53, 54, 55, 56, 57, 59, 60
Related to Fibonacci representations. Ref FQ 11 385 73. [1,2; A3251]

M0613 2, 3, 5, 6, 7, 8, 10, 11, 12, 13, 14, 15, 17, 18, 19, 20, 21, 22, 23, 24, 26, 27, 28, 29, 30, 31, 32, 33, 34, 35, 37, 38, 39, 40, 41, 42, 43, 44, 45, 46, 47, 48, 50, 51, 52, 53, 54, 55
The non-squares: $a(n) = n + [\, \tfrac{1}{2} + \sqrt{n}\,]$. Ref MMAG 63 53 90. [1,1; A0037, N0223]

M0614 2, 3, 5, 6, 7, 8, 10, 11, 13, 14, 15, 17, 19, 21, 22, 23, 24, 26, 27, 29, 30, 31, 32, 33, 34, 35, 37, 38, 39, 40, 41, 42, 43, 46, 47, 51, 53, 54, 55, 56, 57, 58, 59, 61, 62, 65, 66, 67
Contain primes to odd powers only. Ref AMM 73 139 66. [1,1; A2035, N0224]

M0615 1, 2, 3, 5, 6, 7, 9, 10, 11, 13, 14, 15, 16, 18, 19, 20, 22, 23, 24, 26, 27, 28, 29, 31, 32, 33, 35, 36, 37, 39, 40, 41, 42, 44, 45, 46, 48, 49, 50, 52, 53, 54, 56, 57, 58, 59, 61, 62
A Beatty sequence. Cf. M3327. Ref CMB 2 188 59. [1,2; A1955, N0225]

M0616 1, 2, 3, 5, 6, 7, 9, 11, 12, 13
First row of 2-shuffle of spectral array $W(\sqrt{2})$. Ref FrKi94. [1,2; A7071]

M0617 1, 2, 3, 5, 6, 7, 10, 11, 13, 14, 15, 17, 19, 21, 22, 23, 26, 29, 30, 31, 33, 34, 35, 37, 38, 39, 41, 42, 43, 46, 47, 51, 53, 55, 57, 58, 59, 61, 62, 65, 66, 67, 69, 70, 71, 73, 74, 77
Square-free numbers. Ref NZ66 251. [1,2; A5117]

M0618 2, 3, 5, 6, 7, 11, 13, 14, 17, 19, 21, 22, 23, 29, 31, 33, 37, 38, 41, 43, 46, 47, 53, 57, 59, 61, 62, 67, 69, 71, 73, 77, 83, 86, 89, 93, 94, 97, 101, 103, 107, 109, 113, 118, 127
$Q(\sqrt{n})$ is unique factorization domain. Ref BA4 1. BS66 422. ST70 296. [1,1; A3172]

M0619 2, 3, 5, 6, 7, 11, 13, 17, 19, 21, 29, 33, 37, 41, 57, 73.
Real quadratic Euclidean fields (a finite sequence). Ref LE56 2 57. AMM 75 948 68. ST70 294. HW1 213. [1,1; A3174]

M0620 2, 3, 5, 6, 7, 19, 21, 23, 31, 37, 38, 44, 69, 73
Least positive primitive roots. Ref RS9 XLIV. [1,1; A2229, N0226]

M0621 1, 2, 3, 5, 6, 8, 9, 10, 15, 16, 17, 19, 26, 27, 29, 30, 31, 34, 37, 49, 50, 51, 53, 54, 56, 57, 58, 63, 65, 66, 67, 80, 87, 88, 89, 91, 94, 99
No 4-term arithmetic progression. Ref MOC 33 1354 79. [0,2; A5837]

M0622 1, 2, 3, 5, 6, 8, 10, 11, 13, 15
First column of inverse Stolarsky array. Ref PAMS 117 318 93. [1,2; A7067]

M0623 1, 2, 3, 5, 6, 8, 10, 11, 14, 16, 17, 18, 19, 21, 22, 24, 25, 29, 30, 32, 33, 34, 35, 37, 40, 41, 43, 45, 46, 47
A self-generating sequence. Ref UPNT E31. [1,2; A5243]

M0624 0, 1, 2, 3, 5, 6, 8, 10, 12, 15, 16, 18, 21, 23, 26, 28, 31, 34, 38, 41, 44, 47, 50, 54,
57, 61, 65, 68
Maximal edges in n-node graph of girth 5. Ref bdm. [1,3; A6856]

M0625 1, 1, 2, 3, 5, 6, 8, 10, 13, 15, 18, 21, 25, 28, 32, 36, 41, 45, 50
Restricted partitions. Ref CAY 2 277. PJM 86 1 60. [0,3; A1971, N0227]

M0626 1, 2, 3, 5, 6, 8, 12, 14, 15, 17, 20, 21, 24, 27, 33, 38, 41, 50, 54, 57, 59, 62, 66, 69,
71, 75, 77, 78, 80, 89, 90, 99, 101, 105, 110, 111, 117, 119, 131, 138, 141, 143, 147, 150
$n^2 + n + 1$ is prime. Ref CU23 1 245. LINM 3 209 29. L1 46. [1,2; A2384, N0228]

M0627 1, 2, 3, 5, 6, 9, 11, 15, 18, 23, 27, 34, 39, 47, 54, 64, 72, 84, 94, 108, 120, 136, 150,
169, 185, 206, 225, 249, 270, 297, 321, 351, 378, 411, 441, 478, 511, 551, 588, 632, 672
Partitions of n into at most 4 parts. Ref RS4 2. [1,2; A1400, N0229]

M0628 1, 1, 2, 3, 5, 6, 10, 11, 16, 19, 26, 27, 40, 41, 53, 61, 77, 78, 104, 105, 134, 147,
175, 176, 227, 233, 275, 294, 350, 351, 438, 439, 516, 545, 624, 640, 774, 775, 881, 924
Shifts one place under Moebius transform: $a(n + 1) = \Sigma\, a(k)$, $k \mid n$. Ref JRAM 278 334 75.
AcMaSc 2 109 82. [1,3; A3238]

M0629 2, 3, 5, 6, 10, 11, 17, 21, 27, 33, 46, 53, 68, 82, 104, 123, 154, 179, 221, 262, 314,
369, 446, 515, 614, 715, 845, 977, 1148, 1321, 1544, 1778, 2060, 2361, 2736, 3121
Mock theta numbers. Ref TAMS 72 495 52. [1,1; A0039, N0230]

M0630 1, 2, 3, 5, 6, 10, 12, 17, 22, 29, 36, 48, 58, 73, 91, 111, 134, 165, 197, 236, 283,
335, 395, 468, 547, 639, 747, 866, 1001, 1160, 1334, 1530, 1757, 2007, 2286, 2606, 2958
Oscillates under partition transform. Cf. M0565. Ref BeSl94. EIS § 2.7. [1,2; A7211]

M0631 1, 2, 3, 5, 6, 12, 14, 26, 37, 62, 90, 159, 234, 392, 618, 1013, 1598, 2630, 4182,
6830, 10962, 17802, 28658, 46548, 75031, 121628, 196455, 318206, 514230, 832722
Inverse Moebius transform of Fibonacci numbers. Ref EIS § 2.7. [1,2; A7435]

M0632 2, 3, 5, 7, 1, 3, 7, 9, 3, 9, 1, 7, 1, 3, 7, 3, 9, 1, 7, 1, 3, 9, 3, 9, 7, 1, 3, 7, 9, 3, 7, 1, 7,
9, 9, 1, 7, 3, 7, 3, 9, 1, 1, 3, 7, 9, 1, 3, 7, 9, 3, 9, 1, 1, 7, 3, 9, 1, 7, 1, 3, 3, 7, 1, 3, 7, 1, 7, 7
Final digits of primes. Ref AS1 870. [1,1; A7652]

M0633 2, 3, 5, 7, 2, 4, 8, 10, 5, 11, 4, 10, 5, 7, 11, 8, 14, 7, 13, 8, 10, 16, 11, 17, 16, 2, 4, 8,
10, 5, 10, 5, 11, 13, 14, 7, 13, 10, 14, 11, 17, 10, 11, 13, 17, 19, 4, 7, 11, 13, 8, 14, 7, 8, 14
Sum of digits of nth prime. Ref rgw. [1,1; A7605]

M0634 2, 3, 5, 7, 8, 9, 13, 14, 18, 19, 24, 25, 29, 30, 35, 36, 40, 41, 46, 51, 56, 63, 68, 72,
73, 78, 79, 83, 84, 89, 94, 115, 117, 126, 153, 160, 165, 169, 170, 175, 176, 181, 186, 191
$a(n)$ is smallest number which is uniquely $a(j) + a(k)$, $j < k$. See Fig M0557. Ref
Ulam60 IX. EXPM 1 57 92. [1,1; A1857, N0231]

M0635 2, 3, 5, 7, 8, 10, 12, 13, 15, 16, 18, 20, 21, 23, 24, 26, 28, 29, 31, 33, 34, 36, 37, 39,
41, 42, 44, 46, 47, 49, 50, 52, 54, 55, 57, 58, 60, 62, 63, 65, 67, 68, 70, 71, 73, 75, 76
Related to Fibonacci representations. Ref FQ 11 386 73. [1,1; A3258]

M0636 1, 2, 3, 5, 7, 8, 10, 12, 13, 18, 20, 27, 28, 33, 37, 42, 45, 47, 55, 58, 60, 62, 63, 65, 67, 73, 75, 78, 80, 85, 88, 90, 92, 102, 103, 105, 112, 115, 118, 120, 125, 128, 130, 132
$4 . n^2 + 1$ is prime. Ref Krai24 1 11. KK71 1. OG72 116. [1,2; A1912, N0232]

M0637 0, 0, 0, 1, 2, 3, 5, 7, 9, 11, 13, 15, 17, 19, 22, 25, 28, 31, 34, 37, 40, 43, 46, 49, 52, 55, 58, 61, 64, 67
Low discrepancy sequences in base 3. Ref JNT 30 68 88. [1,5; A5357]

$$\text{G.f.: } x^3 (1 + x^3 + x^{11}) / (1 - x)^2 .$$

M0638 1, 2, 3, 5, 7, 9, 12, 15, 18, 22, 26, 30, 35, 40, 45, 51, 57, 63, 70, 77, 84, 92, 100, 108, 117, 126, 135, 145, 155, 165, 176, 187, 198, 210, 222, 234, 247, 260, 273, 287, 301
Expansion of $1 / (1 - x)^2 (1 - x^3)$. Ref TI68 126 (divided by 2). [0,2; A1840, N0233]

M0639 1, 2, 3, 5, 7, 9, 12, 15, 18, 23, 28, 33, 40, 47, 54, 63, 72, 81, 93, 105, 117, 132, 147, 162, 180, 198, 216, 239, 262, 285, 313, 341, 369, 402, 435, 468, 508, 548, 588, 635, 682
Partitions of $3n$ into powers of 3. Ref rkg. [0,2; A5704]

M0640 1, 2, 3, 5, 7, 10, 11, 13, 14, 18, 21, 22, 31, 42, 67, 70, 71, 73, 251, 370, 375, 389, 407, 518, 818, 865, 1057, 1602, 2211, 3049
$39 . 2^n + 1$ is prime. Ref PAMS 9 674 58. Rie85 381. [1,2; A2269, N0234]

M0641 1, 2, 3, 5, 7, 10, 12, 17, 18, 23, 25, 30, 32, 33, 38, 40, 45, 47, 52, 58, 70, 72, 77, 87, 95, 100, 103, 107, 110, 135, 137, 138, 143, 147, 170, 172, 175, 177, 182, 192, 205, 213
$6n - 1$, $6n + 1$ are twin primes. Ref AMM 58 338 51. LE56 69. [1,2; A2822, N0235]

M0642 1, 2, 3, 5, 7, 10, 13, 18, 23, 30, 37, 47, 57, 70, 84, 101, 119, 141, 164, 192, 221, 255, 291, 333, 377, 427, 480, 540, 603, 674, 748, 831, 918, 1014, 1115, 1226, 1342, 1469
Partitions of n into at most 5 parts. Ref RS4 2. [1,2; A1401, N0237]

M0643 1, 2, 3, 5, 7, 10, 13, 18, 24, 35, 50, 75, 109, 161, 231, 336, 482, 703, 1020, 1498, 2188, 3214, 4694, 6877, 10039, 14699, 21487, 31489, 46097, 67582, 98977, 145071
Twopins positions. Ref GU81. [6,2; A5691]

M0644 1, 1, 1, 2, 3, 5, 7, 10, 14, 19, 26, 35, 47, 62, 82, 107, 139, 179, 230, 293
Stacks, or planar partitions of n. Ref PCPS 47 686 51. QJMO 23 153 72. [1,4; A1522, N0238]

M0645 1, 1, 2, 3, 5, 7, 10, 14, 20, 27, 37, 49, 66, 86, 113, 146, 190, 242, 310, 392, 497, 623, 782, 973, 1212, 1498, 1851, 2274, 2793, 3411, 4163, 5059, 6142, 7427, 8972, 10801
Representations of the symmetric group. Ref CJM 4 383 52. [2,3; A0701, N0239]

M0646 0, 2, 3, 5, 7, 10, 14, 20, 29, 43, 65, 100, 156, 246, 391, 625, 1003, 1614, 2602, 4200, 6785, 10967, 17733, 28680, 46392, 75050, 121419, 196445, 317839, 514258
nth Fibonacci number $+ n$. Ref HO70 96. [0,2; A2062, N0240]

M0647 1, 2, 3, 5, 7, 10, 14, 20, 30, 45, 69, 104, 157, 236, 356, 540, 821, 1252, 1908, 2909, 4434, 6762, 10319, 15755, 24066, 36766, 56176, 85837, 131172, 200471, 306410
Numbers of Twopins positions. Ref GU81. [5,2; A5688]

M0648 1, 2, 3, 5, 7, 10, 15, 22, 32, 47, 69, 101, 148, 217, 318, 466, 683, 1001, 1467, 2150, 3151, 4618, 6768, 9919, 14537, 21305, 31224, 45761, 67066, 98290, 144051, 211117
Expansion of $(1+x)(1+x^2)/(1-x-x^3)$. Ref rkg. [0,2; A3410]

M0649 1, 2, 3, 5, 7, 10, 16, 26, 36, 50, 71, 101, 161, 257, 417, 677, 937, 1297, 1801, 2501, 3551, 5042, 7172, 10202, 16262, 25922, 41378, 66050, 107170, 173890, 282310, 458330
$a(n) = 1 + a([n/2]) a(\lceil n/2 \rceil)$. Ref clm. [1,2; A5468]

M0650 1, 2, 3, 5, 7, 11, 13, 15, 17, 19, 23, 29, 31, 33, 35, 37, 41, 43, 47, 51, 53, 59, 61, 65, 67, 69, 71, 73, 77, 79, 83, 85, 87, 89, 91, 95, 97, 101, 103, 107, 109, 113, 115, 119, 123
n and $\phi(n)$ are relatively prime. Ref JIMS 12 75 48. AS1 840. [1,2; A3277]

M0651 2, 3, 5, 7, 11, 13, 17, 19, 23, 29, 31, 37, 41, 43, 47, 53, 59, 60, 61, 67, 71, 73, 79, 83, 89, 97, 101, 103, 107, 109, 113, 127, 131, 137, 139, 149, 151, 157, 163, 167, 168, 173
Orders of simple groups. Ref ATLAS. [1,1; A5180]

M0652 2, 3, 5, 7, 11, 13, 17, 19, 23, 29, 31, 37, 41, 43, 47, 53, 59, 61, 67, 71, 73, 79, 83, 89, 97, 101, 103, 107, 109, 113, 127, 131, 137, 139, 149, 151, 157, 163, 167, 173, 179
The prime numbers. Ref AS1 870. [1,1; A0040, N0241]

M0653 2, 3, 5, 7, 11, 13, 17, 19, 23, 31, 37, 41, 59, 61, 67, 83, 89, 97, 101, 103, 107, 109, 127, 131, 137, 139, 149, 167, 197, 199, 227, 241, 269, 271, 281, 293, 347, 373, 379, 421
$2^p - 1$ has at most 2 prime factors. Ref CUNN. [1,1; A6514]

M0654 2, 3, 5, 7, 11, 13, 17, 19, 31, 37, 41, 61, 73, 97, 101, 109, 151, 163, 181, 193, 241, 251, 257, 271, 401, 433, 487, 541, 577, 601, 641, 751, 769, 811, 1153, 1201, 1297
A restricted class of primes. Ref Krai24 1 53. [1,1; A2200, N0242]

M0655 2, 3, 5, 7, 11, 13, 17, 23, 25, 29, 37, 41, 43, 47, 53, 61, 67, 71, 77, 83, 89, 91, 97, 107, 115, 119, 121, 127, 131, 143, 149, 157, 161, 173, 175, 179, 181, 193, 209, 211, 221
Generated by a sieve. Ref PC 2 13-6 74. [1,1; A3309]

M0656 2, 3, 5, 7, 11, 13, 17, 23, 37, 47, 61, 73, 83, 101, 103, 107, 131, 137, 151, 173, 181, 233, 241, 257, 263, 271, 277, 283, 293, 311, 313, 331, 347, 367, 373, 397, 443, 461, 467
n and $6n+1$ are prime. Ref JNT 27 63 87. Robe92 83. [1,1; A7693]

M0657 2, 3, 5, 7, 11, 13, 17, 31, 37, 71, 73, 79, 97, 101, 107, 113, 131, 149, 151, 157, 167, 179, 181, 191, 199, 311, 313, 337, 347, 353, 359, 373, 383, 389, 701, 709, 727, 733, 739
Palindromic primes: reversal is prime. Ref Well86 134. [1,1; A7500]

M0658 2, 3, 5, 7, 11, 13, 17, 31, 37, 71, 73, 79, 97, 113, 131, 199, 311, 337, 373, 733, 919, 991, 11111111111111111111, 1111111111111111111111111
Every permutation of digits is a prime. Ref MMAG 47 233 74. rcs. [1,1; A3459]

M0659 2, 3, 5, 7, 11, 13, 17, 107, 197, 3293, 74057, 1124491, 1225063003, 48403915086083
$a(n) = \min (p \pm q > 1 : pq = \Pi a(k), k = 1..n-1)$. Ref jhc. [1,1; A3681]

M0660   2, 3, 5, 7, 11, 13, 19, 23, 29, 31, 37, 43, 47, 53, 59, 61, 67, 71, 79, 101
Higgs' primes: $a(n+1)$ = next prime such that $a(n+1)-1 \mid (a(1)...a(n))^2$. Ref AMM 100 233 93. [1,1; A7459]

M0661   1, 2, 3, 5, 7, 11, 13, 19, 23, 29, 33, 43, 47, 59, 65, 73, 81, 97, 103, 121, 129, 141, 151, 173, 181, 201, 213, 231, 243, 271, 279, 309, 325, 345, 361, 385, 397, 433, 451, 475
Fractions in Farey series of order n (1+M1008). Ref AMM 95 699 88. [1,2; A5728]

M0662   1, 1, 2, 3, 5, 7, 11, 14, 20, 26, 35, 44, 58, 71, 90, 110, 136, 163, 199, 235, 282, 331, 391, 454, 532, 612, 709, 811, 931, 1057, 1206, 1360, 1540, 1729, 1945, 2172, 2432, 2702
Partitions of n into at most 6 parts. Ref CAY 10 415. RS4 2. [0,3; A1402, N0243]

M0663   1, 1, 2, 3, 5, 7, 11, 15, 22, 30, 42, 56, 77, 101, 135, 176, 231, 297, 385, 490, 627, 792, 1002, 1255, 1575, 1958, 2436, 3010, 3718, 4565, 5604, 6842, 8349, 10143, 12310
Partitions of n. See Fig M0663. Ref RS4 90. R1 122. AS1 836. [0,3; A0041, N0244]

$$\text{G.f.:} \quad \prod_{n=1}^{\infty} (1 - x^n)^{-1}.$$

Figure M0663.   PARTITIONS.

The k-th entry in the n-th row of the following **partition triangle** (beginning the numbering at 1) gives $p(n, k)$, the number of ways of partitioning n into exactly k parts ([RS4], [C1 307]).

```
1
1  1
1  1  1
1  2  1  1
1  2  2  1  1
1  3  3  2  1  1
1  3  4  3  2  1  1
      ......
```

Many sequences arise from this table. For example the sum of the first three columns gives M0518. The row sums give the **partition numbers** $p(n)$, M0663, the total number of partitions of n into integer parts. The partitions of the first few integers are as follows:

$$p(1) \ = \ 1: \quad 1$$
$$p(2) \ = \ 2: \quad 2, 1^2$$
$$p(3) \ = \ 3: \quad 3, 21, 1^3$$
$$p(4) \ = \ 5: \quad 4, 31, 2^2, 21^2, 1^4$$
$$p(5) \ = \ 7: \quad 5, 41, 32, 31^2, 2^2 1, 21^3, 1^5$$
$$p(6) \ = \ 11: \quad 6, 51, 42, 41^2, 3^2, 321, 31^3, 2^3, 2^2 1^2, 21^4, 1^6$$

It is easy to write down generating functions for partition problems. For example the number of ways of partitioning n into part of sizes $a_1, a_2, a_3, ...$ is given by the coefficient of x^n in the expansion of

$$\frac{1}{(1 - x^{a_1})(1 - x^{a_2})(1 - x^{a_3}) \cdots}.$$

See [HW1 Chap. 19], [Wilf90].

M0664   1, 2, 3, 5, 7, 11, 16, 26, 40, 65, 101, 163, 257, 416, 663, 1073, 1719, 2781, 4472,
   7236, 11664, 18873, 30465, 49293, 79641, 128862, 208315, 337061, 545071
Twopins positions. Ref GU81. [4,2; A5685]

M0665   2, 3, 5, 7, 11, 17, 23, 31, 47, 53, 71, 107, 127, 191, 383, 431, 647, 863, 971, 1151,
   2591, 4373, 6143, 6911, 8191, 8747, 13121, 15551, 23327, 27647, 62207, 73727, 139967
Class 1+ primes. Ref UPNT A18. [1,1; A5105]

M0666   1, 1, 2, 3, 5, 7, 11, 17, 25, 38, 57, 86, 129, 194, 291, 437, 656, 985, 1477, 2216,
   3325, 4987, 7481, 11222, 16834, 25251, 37876, 56815, 85222, 127834, 191751, 287626
$[3^n/2^n]$. Ref JIMS 2 40 36. L1 82. MMAG 63 8 90. [0,3; A2379, N0245]

M0667   2, 3, 5, 7, 11, 19, 29, 47, 71, 127, 191, 379, 607, 1087, 1903, 3583, 6271, 11231
Smallest number with addition chain of length n. Ref KN1 2 458. [1,1; A3064]

M0668   2, 3, 5, 7, 11, 19, 43, 53, 79, 107, 149
Least positive prime primitive roots. Ref RS9 XLV. [1,1; A2231, N0246]

M0669   2, 3, 5, 7, 11, 31, 379, 1019, 1021, 2657, 3229, 4547, 4787, 11549, 13649, 18523,
   23801, 24029
1 + product of primes up to p is prime. Ref JRM 21 276 89. Cald94. [1,1; A5234]

M0670   2, 3, 5, 7, 11, 101, 131, 151, 181, 191, 313, 353, 373, 383, 727, 757, 787, 797, 919,
   929, 10301, 10501, 10601, 11311, 11411, 12421, 12721, 12821, 13331, 13831, 13931
Palindromic primes. Ref B1 228. [1,1; A2385, N0247]

M0671   1, 2, 3, 5, 7, 12, 17, 29, 41, 70, 99, 169, 239, 408, 577, 985, 1393, 2378, 3363,
   5741, 8119, 13860, 19601, 33461, 47321, 80782, 114243, 195025, 275807, 470832
$a(2n+1)=a(2n)+a(2n-1)$, $a(2n)=a(2n-1)+a(2n-2)-a(2n-5)$. Ref JALG 20
173 72. [0,2; A2965]

M0672   2, 3, 5, 7, 13, 17, 19, 31, 61, 89, 107, 127, 521, 607, 1279, 2203, 2281, 3217, 4253,
   4423, 9689, 9941, 11213, 19937, 21701, 23209, 44497, 86243, 110503, 132049, 216091
Mersenne primes (p such that 2^p-1 is prime). Ref CUNN. [1,1; A0043, N0248]

M0673   2, 3, 5, 7, 13, 17, 19, 37, 73, 97, 109, 163, 193, 257, 433, 487, 577, 769, 1153,
   1297, 1459, 2593, 2917, 3457, 3889, 10369, 12289, 17497, 18433, 39367, 52489, 65537
Class 1− primes. Ref UPNT A18. [1,1; A5109]

M0674   1, 1, 1, 2, 3, 5, 7, 13, 20, 35, 55, 96, 156, 267, 433, 747, 1239, 2089, 3498, 5912
Paraffins with n carbon atoms. Ref JACS 54 1544 32. [1,4; A0627, N0249]

M0675   2, 3, 5, 7, 13, 23, 43, 83, 163, 317, 631, 1259, 2503, 5003, 9973, 19937, 39869,
   79699, 159389, 318751, 637499, 1274989, 2549951, 5099893, 10199767, 20399531
Bertrand primes: $a(n)$ is largest prime $< 2a(n-1)$. Ref NZ66 189. [1,1; A6992]

M0676   1, 2, 3, 5, 7, 14, 11, 66, 127, 992, 5029, 30899, 193321, 1285300, 8942561,
   65113125, 494605857, 3911658640, 32145949441, 274036507173, 2419502677445
Reversion of g.f. for Bell numbers. Cf. M1484. [1,2; A7311]

M0677   1, 1, 1, 2, 3, 5, 7, 14, 21, 40, 61, 118, 186, 365
Achiral trees with n nodes. Ref TET 32 356 76. [1,4; A5629]

M0678   1, 2, 3, 5, 7, 17, 31, 89, 127, 521, 607, 1279, 2281, 3217, 4423, 9689, 19937,
   21701, 23209, 44497, 86243
$x^n + x^k + 1$ is irreducible (mod 2) for some k. Ref IFC 15 68 69. MOC 56 819 91. [1,2; A1153, N0250]

M0679   2, 3, 5, 7, 23, 67, 89, 4567, 78901, 678901, 23456789, 45678901, 9012345678901,
   789012345678901, 567890123456789012345678901123
Primes with consecutive digits. Ref JRM 10 33 77. rcs. [1,1; A6510]

M0680   1, 2, 3, 5, 7, 26, 27, 53, 147, 401
$2.10^n - 1$ is prime. Ref PLC 2 567 71. [1,2; A2957]

M0681   2, 3, 5, 7, 2411
Ramanujan number $\tau(p)$ is divisible by p. Cf. M5153. Ref Robe92 275. [1,1; A7659]

M0682   2, 3, 5, 8, 9, 10, 11, 12, 18, 19, 22, 26, 28, 30, 31, 33, 35, 36, 38, 39, 40, 41, 44, 46,
   47, 48, 50, 52, 54, 55, 56, 58, 61, 62, 66, 67, 68, 69, 71, 72, 74, 76, 77, 80, 82, 83, 91, 92
Elliptic curves. Ref JRAM 212 23 63. [1,1; A2153, N0251]

M0683   0, 2, 3, 5, 8, 11, 12, 14, 18, 20, 21, 27, 29, 30, 32, 35, 44, 45, 48, 50, 53, 56, 59, 62,
   66, 72, 75, 77, 80, 83, 84, 93, 98, 99, 101, 107, 108, 110, 116, 120, 125, 126, 128, 131
Of the form $2x^2 + 3y^2$. Ref EUL (1) 1 425 11. [1,2; A2480, N0252]

M0684 1, 2, 3, 5, 8, 11, 16, 21, 29, 40, 51, 67, 88, 109, 138, 167, 207, 258, 309, 376, 443, 531, 640, 749, 887, 1054, 1221, 1428, 1635, 1893, 2202, 2511, 2887, 3330, 3773, 4304
$a(n) = a(n-1) + a(n-1 -$ number of even terms so far). Ref drh. [1,2; A6336]

M0685 0, 1, 2, 3, 5, 8, 11, 16, 23, 31, 43, 58, 74, 95, 122, 151, 186, 229, 274, 329, 394, 460, 537, 626, 722, 832, 953, 1080, 1223, 1383, 1552, 1737, 1940, 2153, 2389, 2648
Taylor series from Ramanujan's Lost Notebook. Ref LNM 899 44 81. [0,3; A6304]

M0686 1, 2, 3, 5, 8, 11, 19, 30, 49, 68, 87, 106, 193, 299, 492, 685, 878, 1071, 1264, 1457, 2721, 4178, 6899, 9620, 12341, 15062, 17783, 20504, 23225, 25946, 49171, 75117
Numerators of approximations to e. Cf. M0584. Ref GKP 122. [1,2; A6258]

M0687 1, 1, 2, 3, 5, 8, 12, 18, 26, 38, 53, 75, 103, 142, 192, 260, 346, 461, 607, 797, 1038, 1348, 1738, 2234, 2856, 3638, 4614, 5832, 7342, 9214, 11525, 14369
Arrangements of n pennies in contiguous rows, each touching 2 in row below. Ref PCPS 47 686 51. QJMO 23 153 72. MMAG 63 6 90. [1,3; A1524, N0253]

M0688 1, 1, 1, 1, 2, 3, 5, 8, 12, 18
Dimension of primitive Vassiliev knot invariants of order n (next term at least 27). Ref BAMS 28 281 93. BarN94. [0,5; A7478]

M0689 1, 2, 3, 5, 8, 13, 17, 26, 34, 45, 54, 67, 81, 97, 115, 132, 153, 171, 198, 228, 256, 288, 323, 357, 400, 439, 488, 530, 581, 627, 681, 732, 790, 843, 908, 963, 1029, 1085
A self-generating sequence. See Fig M0557. Ref MMAG 63 6 90. [1,2; A1149, N0254]

M0690 1, 1, 2, 3, 5, 8, 13, 20, 34, 53, 88, 143, 236, 387, 641, 1061, 1763, 2937, 4903, 8202, 13750, 23095
From sequence of numbers with abundancy n. Ref MMAG 59 88 86; 63 5 90. [1,3; A5347]

M0691 1, 1, 2, 3, 5, 8, 13, 21, 34, 55, 89, 144, 232, 375, 606, 979, 1582, 2556, 4130, 6673, 10782, 17421, 28148, 45480, 73484, 118732, 191841, 309967, 500829, 809214, 1307487
Dying rabbits: $a(n+13) = a(n+12) + a(n+11) - a(n)$. Ref FQ 2 108 64. [0,3; A0044, N0255]

M0692 1, 1, 2, 3, 5, 8, 13, 21, 34, 55, 89, 144, 233, 377, 610, 987, 1597, 2584, 4181, 6765, 10946, 17711, 28657, 46368, 75025, 121393, 196418, 317811, 514229, 832040, 1346269
Fibonacci numbers: $F(n) = F(n-1) + F(n-2)$. See Fig M0692. Ref HW1 148. HO69. [0,3; A0045, N0256]

Figure M0692. FIBONACCI AND LUCAS NUMBERS.

The **Fibonacci numbers**, M0692, are defined by $F_0 = F_1 = 1, F_n = F_{n-1} + F_{n-2}$ $(n \geq 2)$. They are illustrated by the Fibonacci tree:

which grows according to the rules that every mature branch sprouts a new branch at the end of each year, and new branches take a year to reach maturity. At the end of the n-th year there are F_n branches. These numbers have generating function

$$F_0 + F_1 x + F_2 x^2 + \cdots = \frac{1}{1 - x - x^2} \, ,$$

and

$$F_n = \frac{1}{\sqrt{5}} \left\{ \left(\frac{1 + \sqrt{5}}{2} \right)^{n+1} - \left(\frac{1 - \sqrt{5}}{2} \right)^{n+1} \right\} .$$

The **Lucas numbers** L_n, M2341, are similarly defined by $L_0 = 1, L_1 = 3, L_n = L_{n-1} + L_{n-2}$ $(n \geq 2)$, with

$$L_0 + L_1 x + L_2 x^2 + \cdots = \frac{1 + 2x}{1 - x - x^2} \, ,$$

$$L_n = \left(\frac{1 + \sqrt{5}}{2} \right)^{n+1} + \left(\frac{1 - \sqrt{5}}{2} \right)^{n+1} .$$

They are illustrated by the following tree, which grows according to the same rules as the Fibonacci tree, except that in the first year two new branches are formed instead of one:

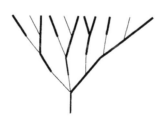

M0693 1, 1, 2, 3, 5, 8, 13, 21, 34, 55, 91, 149, 245, 404, 666, 1097, 1809, 2981, 4915, 8104, 13360, 22027, 36316, 59875, 98716, 162755, 268338, 442414, 729417, 1202605 $\lceil e^{(n-1)/2} \rceil$. Ref MMAG 63 5 90. [0,3; A5181]

M0694 1, 0, 1, 1, 2, 3, 5, 8, 13, 21, 35, 55, 93, 149, 248, 403, 670, 1082
Related to counting fountains of coins. Ref AMM 95 706 88. [1,5; A5170]

M0695 1, 2, 3, 5, 8, 13, 22, 37, 63, 108, 186, 322, 559, 973, 1697, 2964, 5183, 9071, 15886, 27835, 48790, 85545, 150021, 263136, 461596, 809812, 1420813, 2492945
Numbers of Twopins positions. Ref GU81. [3,2; A5683]

M0696 1, 1, 1, 2, 3, 5, 8, 14, 21, 39, 62, 112, 189, 352, 607, 1144, 2055, 3883, 7154, 13602
Necklaces with n beads. Ref IJM 5 663 61. HW84 88. [1,4; A0046, N0257]

M0697 1, 1, 1, 2, 3, 5, 8, 14, 23, 39, 65, 110, 184, 310, 520, 876, 1471, 2475, 4159, 6996, 11759
Paraffins with n carbon atoms. Ref JACS 54 1105 32. [1,4; A0621, N0258]

M0698 1, 1, 1, 2, 3, 5, 8, 14, 23, 41, 69, 122, 208, 370, 636
Achiral planted trees with n nodes. Ref TET 32 356 76. [0,4; A5627]

M0699 1, 1, 2, 3, 5, 8, 14, 24, 43, 77
Balanced ordered trees with n nodes. Ref RSA 5 115 94. [1,3; A7059]

M0700 1, 1, 1, 1, 2, 3, 5, 8, 15, 26, 47, 82
Distributive lattices on n nodes. Ref pdl. [0,5; A6982]

M0701 1, 2, 3, 5, 8, 15, 26, 48, 87, 161, 299, 563, 1066, 2030, 3885, 7464, 14384, 27779, 53782, 104359, 202838, 394860, 769777, 1502603, 2936519, 5744932
Integers $\leq 2^n$ of form $x^2 - 2y^2$. Ref MOC 20 560 66. [0,2; A0047, N0259]

M0702 1, 2, 3, 5, 8, 15, 27, 54, 110, 238, 526, 1211, 2839, 6825, 16655, 41315, 103663, 263086, 673604, 1739155, 4521632, 11831735, 31134338, 82352098, 218837877
Stable forests with n nodes. Ref LNM 403 84 74. [1,2; A6544]

M0703 1, 2, 3, 5, 8, 21, 29, 79, 661, 740, 19161, 19901, 118666, 138567, 3167140, 3305707, 29612796, 32918503, 62531299, 595700194, 658231493, 1253931687
Convergents to fifth root of 5. Ref AMP 46 116 1866. L1 67. hpr. [1,2; A2363, N0260]

M0704 2, 3, 5, 9, 14, 17, 26, 27, 33, 41, 44, 50, 51, 53, 69, 77, 80, 81, 84, 87, 98, 99, 101, 105, 122, 125, 129
A self-generating sequence. Ref UPNT E31. [1,1; A5244]

M0705 1, 2, 3, 5, 9, 15, 25, 43, 73, 123, 209, 355, 601, 1019, 1729, 2931, 4969, 8427, 14289, 24227, 41081, 69659, 118113, 200275, 339593, 575819, 976369, 1655555
$a(n) = a(n-1) + 2a(n-3)$. Ref DT76. [1,2; A3476]

M0706 1, 2, 3, 5, 9, 15, 25, 45, 75
Codes for rooted trees on n nodes. Ref JCT B29 142 80. [1,2; A5517]

M0707 1, 2, 3, 5, 9, 15, 26, 44, 78, 136, 246, 432, 772, 1382, 2481
Number of integers with addition chain of length n. Ref KN1 2 459. [1,2; A3065]

M0708 1, 1, 1, 2, 3, 5, 9, 15, 26, 45, 78, 135, 234, 406, 704, 1222, 2120, 3679, 6385,
11081, 19232, 33379, 57933, 100550, 174519, 302903, 525734, 912493, 1583775
Fountains of n coins. Ref AMM 95 705 88; 95 840 88. [0,4; A5169]

M0709 0, 1, 2, 3, 5, 9, 16, 28, 49, 86, 151, 265, 465, 816, 1432, 2513, 4410, 7739, 13581,
23833, 41824, 73396, 128801, 226030, 396655, 696081, 1221537, 2143648, 3761840
$a(n)=2a(n-1)-a(n-2)+a(n-3)$. Ref LAA 62 130 84. [0,3; A5314]

M0710 1, 1, 2, 3, 5, 9, 16, 28, 50, 89, 159, 285, 510, 914, 1639, 2938, 5269, 9451, 16952,
30410, 54555, 97871, 175586, 315016, 565168, 1013976, 1819198, 3263875, 5855833
Partitions into powers of ½, or binary rooted trees. Ref PEMS 11 224 59. prs. PGIT 17 309
71. IFC 21 482 72. DM 65 150 87. [1,3; A2572, N0261]

M0711 1, 1, 1, 2, 3, 5, 9, 16, 28, 51, 93, 170, 315, 585, 1091, 2048, 3855, 7280, 13797,
26214, 49929, 95325, 182361, 349520, 671088, 1290555, 2485504, 4793490, 9256395
Necklaces with n beads. Ref IJM 5 663 61. JCT A15 31 73. NAT 261 463 76. [1,4; A0048,
N0262]

M0712 1, 1, 1, 2, 3, 5, 9, 16, 28, 51, 93, 170, 315, 585, 1092, 2048, 3855, 7281, 13797,
26214, 49932, 95325, 182361, 349525, 671088, 1290555, 2485513, 4793490, 9256395
$[2^{n-1}/n]$. [1,4; A6788]

M0713 1, 2, 3, 5, 9, 16, 29, 52, 94, 175, 327, 616, 1169, 2231, 4273, 8215, 15832, 30628,
59345, 115208, 224040, 436343, 850981, 1661663, 3248231, 6356076, 12448925
Ramanujan's approximation to population of x^2+y^2. Ref MOC 18 79 64. [0,2; A0691,
N0263]

M0714 0, 0, 2, 3, 5, 9, 16, 29, 53, 98, 181, 341, 640, 1218, 2321, 4449, 8546, 16482,
31845, 61707, 119760, 232865, 453511, 884493, 1727125, 3376376, 6607207
Integers $\le 2^n$ of form $3x^2+4y^2$. Ref MOC 20 567 66. [0,3; A0049, N0264]

M0715 1, 2, 3, 5, 9, 16, 29, 54, 97, 180, 337, 633, 1197, 2280, 4357, 8363, 16096, 31064,
60108, 116555, 226419, 440616, 858696, 1675603, 3273643, 6402706, 12534812
Integers $\le 2^n$ of form x^2+y^2. Ref MOC 20 560 66. [0,2; A0050, N0265]

M0716 2, 3, 5, 9, 17, 33, 64, 126, 249, 495, 984, 1962, 3913, 7815
Weighted voting procedures. Ref LNM 686 70 78. NA79 100. MSH 84 48 83. [1,1;
A5257]

M0717 2, 3, 5, 9, 17, 33, 65, 129, 257, 513, 1025, 2049, 4097, 8193, 16385, 32769, 65537,
131073, 262145, 524289, 1048577, 2097153, 4194305, 8388609, 16777217, 33554433
2^n+1. Ref BA9. [0,1; A0051, N0266]

M0718   1, 1, 1, 1, 2, 3, 5, 9, 18, 35, 75, 159, 355, 802, 1858, 4347, 10359, 24894, 60523,
148284, 366319, 910726, 2278658, 5731580, 14490245, 36797588, 93839412
Quartic trees with n nodes. Ref JACS 55 680 33. BS65 201. TET 32 356 76. BA76 28.
LeMi91. [0,5; A0602, N0267]

M0719   1, 2, 3, 5, 9, 18, 42
Weights of threshold functions. Ref MU71 268. [2,2; A3218]

M0720   1, 1, 2, 3, 5, 9, 32, 56, 144, 320, 1458, 3645, 9477
Largest determinant of (0,1)-matrix of order n. Cf. M1291. Ref ZAMM 42 T21 62. MZT
83 127 64. AMM 79 626 72. MS78 54. [1,3; A3432]

M0721   1, 2, 3, 5, 10, 11, 26, 32, 39, 92, 116, 134, 170, 224, 277, 332, 370, 374, 640, 664,
820, 1657, 1952, 1969
$25.4^n + 1$ is prime. Ref PAMS 9 674 58. Rie85 381. [1,2; A2263, N0269]

M0722   1, 1, 2, 3, 5, 10, 13
Equidistant permutation arrays. Ref ANY 555 303 89. [1,3; A5677]

M0723   1, 2, 3, 5, 10, 18, 35, 63, 126, 231
From Radon's theorem. Ref MFM 73 12 69. [1,2; A2661, N0270]

M0724   1, 2, 3, 5, 10, 19, 42, 57, 135, 171, 341, 313, 728, 771, 1380, 1393, 2397, 1855,
3895, 3861, 6006, 5963, 8878, 7321, 12675, 12507, 17577, 17277, 23780, 16831, 31496
Nodes in regular n-gon with all diagonals drawn. Cf. M3833. Ref PoRu94. [1,2; A7569]

M0725   1, 1, 1, 2, 3, 5, 10, 21, 43, 97, 215, 503, 1187, 2876, 7033, 17510, 43961, 111664,
285809, 737632, 1915993, 5008652, 13163785, 34774873, 92282214, 245930746
n-node trees with a forbidden limb. Ref HA73 297. [1,4; A2991]

M0726   1, 1, 1, 2, 3, 5, 10, 24, 69, 384
Linear geometries on n points. See Fig M1197. Ref BSMB 19 424 67. JCT A49 28 88.
[0,4; A1200, N0271]

M0727   2, 3, 5, 10, 27, 119, 1113, 29375, 2730166
Threshold functions of n variables. Ref PGEC 19 821 70. MU71 38. [0,1; A0617, N0272]

M0728   1, 2, 3, 5, 10, 28, 154, 3520, 1551880, 267593772160, 7160642690122633501504,
46613457941460641338430989649193052264116096
$a(n+1) = (1 + a(0)^2 + \cdots + a(n)^2)/(n+1)$ (not always integral!). Ref AMM 95 704 88.
[0,2; A3504]

M0729   2, 3, 5, 10, 30, 210, 16353
Inequivalent monotone Boolean functions (or Dedekind numbers). Cf. M0817. Ref PAMS
21 677 69. Wels71 181. MU71 38. [0,1; A3182]

M0730 2, 3, 5, 11, 16, 38, 54, 130, 184, 444, 628, 1516, 2144, 5176, 7320, 17672, 24992,
60336, 85328, 206000, 291328, 703328, 994656, 2401312, 3395968, 8198592, 11594560
$a(2n)=a(2n-1)+2a(2n-2)$, $a(2n+1)=a(2n)+a(2n-1)$. Ref AMM 72 1024 65.
[1,1; A1882, N0273]

M0731 2, 3, 5, 11, 23, 29, 41, 53, 83, 89, 113, 131, 173, 179, 191, 233, 239, 251, 281, 293,
359, 419, 431, 443, 491, 509, 593, 641, 653, 659, 683, 719, 743, 761, 809, 911, 953, 1013
Germain primes: p and $2p+1$ are prime. Ref AS1 870. Robe92 83. [1,1; A5384]

M0732 1, 1, 1, 1, 2, 3, 5, 11, 24, 55, 136, 345, 900, 2412, 6563, 18127, 50699, 143255,
408429, 1173770, 3396844, 9892302, 28972080, 85289390, 252260276, 749329719
Steric planted trees with n nodes. Ref JACS 54 1544 32. TET 32 356 76. BA76 44.
LeMi91. [0,5; A0628, N0274]

M0733 1, 1, 1, 2, 3, 5, 11, 26, 81, 367, 2473, 32200, 939791, 80570391, 30341840591,
75749670168872, 2444729709746709953, 229838686181445202099305
$a(n) = a(n-1) + a(n-2) a(n-3)$. Ref rpm. [0,4; A6888]

M0734 1, 2, 3, 5, 11, 31, 127, 709, 5381, 52711, 648391, 9737333, 174440041,
3657500101, 88362852307, 2428096940717
R. G. Wilson's primeth recurrence: $a(n+1) = a(n)$-th prime. Ref rgw. [0,2; A7097]

M0735 1, 1, 1, 1, 1, 2, 3, 5, 11, 37, 83, 274, 1217, 6161, 22833, 165713, 1249441,
9434290, 68570323, 1013908933, 11548470571, 142844426789, 2279343327171
Somos-5 sequence. Ref MINT 13(1) 41 91. [0,6; A6721]

$$a(n) = (a(n-1)a(n-4)+a(n-2)a(n-3))/a(n-5).$$

M0736 2, 3, 5, 11, 47, 923, 409619, 83763206255, 3508125906290858798171,
6153473687096578758448522809275077520433167
Hamilton numbers. Ref PTRS 178 288 1887. LU91 496. [1,1; A0905, N0275]

M0737 2, 3, 5, 12, 14, 11, 13, 20, 72, 19, 42, 132, 84, 114, 29, 30, 110, 156, 37, 156, 420,
210, 156, 552, 462, 72, 53, 420, 342, 59
Shuffling $2n$ cards. Ref SIAR 3 296 61. [1,1; A2139, N0276]

M0738 0, 0, 0, 0, 0, 0, 2, 3, 5, 12, 22, 47, 94, 201, 417, 907, 1948, 4289, 9440, 21063,
47124, 106377, 240980, 549272, 1256609, 2888057, 6660347, 15416623, 35794121
Trees by stability index. Ref LNM 403 51 74. [1,7; A3428]

M0739 0, 1, 0, 0, 1, 2, 3, 5, 12, 36, 110, 326, 963, 2964, 9797, 34818, 130585, 506996,
2018454, 8238737, 34627390, 150485325, 677033911, 3147372610, 15066340824
From a differential equation. Ref AMM 67 766 60. [0,6; A0997, N0277]

M0740 1, 2, 3, 5, 13, 83, 2503, 976253, 31601312113, 2560404986164794683,
2025231131890379524787222304798003
From a continued fraction. Ref AMM 63 711 56. [0,2; A1685, N0278]

M0741   2, 3, 5, 13, 89, 233, 1597, 28657, 514229, 433494437, 2971215073,
99194853094755497, 1066340417491710595814752169
Prime Fibonacci numbers. Cf. M2309. Ref MOC 50 251 88. [1,1; A5478]

M0742   1, 2, 3, 5, 13, 610
Fibonacci tower: $a(n) = F(a(n-1)+1)$ (there is no room for next term). Ref SIAC 22
751 93. [0,2; A6985]

M0743   1, 2, 3, 5, 16, 231, 53105, 2820087664, 7952894429824835871,
63248529811938901240357985099443351745
$a(n) = a(n-1)^2 - a(n-2)^2$. Ref EUR 27 6 64. FQ 11 432 73. [0,2; A1042, N0279]

M0744   2, 3, 5, 19, 97, 109, 317, 353, 701, 9739
$(12^n - 1)/11$ is prime. Ref CUNN. MOC 61 928 93. [1,1; A4064]

M0745   1, 2, 3, 6, 2, 0, 1, 10, 0, 2, 10, 6, 7, 14, 0, 10, 12, 0, 6, 0, 9, 4, 10, 0, 18, 2, 0, 6, 14,
18, 11, 12, 0, 0, 22, 0, 20, 14, 6, 22, 0, 0, 23, 26, 0, 18, 4, 0, 14, 2, 0, 20, 0, 0, 0, 12, 3, 30
Glaisher's χ numbers. Ref QJMA 20 151 1884. [1,2; A2171, N0280]

M0746   0, 2, 3, 6, 5, 11, 14, 22, 30, 47, 66, 99, 143, 212, 308, 454, 663, 974, 1425, 2091,
3062, 4490, 6578, 9643, 14130, 20711, 30351, 44484, 65192, 95546, 140027, 205222
$a(n) = a(n-2)+a(n-3)+a(n-4)$. Ref IDM 8 64 01. FQ 6(3) 68 68. [0,2; A1634,
N0281]

M0747   2, 3, 6, 7, 8, 9, 12, 15, 18, 19, 20, 21, 22, 23, 24, 25, 26, 27, 30, 33, 36, 39, 42, 45,
48, 51, 54, 55, 56, 57, 58, 59, 60, 61, 62, 63, 64, 65, 66, 67, 68, 69, 70, 71, 72, 73, 74, 75
Unique monotone sequence with $a(a(n)) = 3n$. Ref jpropp. [1,1; A3605]

M0748   1, 2, 3, 6, 7, 9, 18, 25, 27, 54, 73, 97, 129, 171, 231, 313, 327, 649, 703, 871, 1161,
2223, 2463, 2919, 3711, 6171, 10971, 13255, 17647, 23529, 26623, 34239, 35655, 52527
'3x+1' record-setters (iterations). See Fig M2629. Ref GEB 400. ScAm 250(1) 12 84.
CMWA 24 96 92. [1,2; A6877]

M0749   1, 2, 3, 6, 7, 10, 14, 15, 21, 30, 35, 42, 70, 105, 210, 221, 230, 231, 238, 247, 253,
255, 266, 273, 285, 286, 299, 322, 323, 330, 345, 357, 374, 385, 390, 391, 399, 418, 429
A specially constructed sequence. Ref AMM 74 874 67. [0,2; A2038, N0282]

M0750   2, 3, 6, 7, 10, 19, 31, 34, 46, 79, 106, 151, 211, 214, 274, 331, 394, 631, 751, 919,
991, 1054, 1486, 1654
Extreme values of Dirichlet series. Ref PSPM 24 279 73. [1,1; A3421]

M0751   1, 2, 3, 6, 7, 11, 14, 17, 33, 42, 43, 63, 65, 67, 81, 134, 162, 206, 211, 366, 663,
782, 1305, 1411, 1494, 2297, 2826, 3230, 3354, 3417, 3690, 4842, 5802, 6937, 7967
$9.2^n + 1$ is prime. Ref PAMS 9 674 58. Rie85 381. Cald94. [1,2; A2256, N0283]

M0752   0, 2, 3, 6, 8, 10, 22, 35, 42, 43, 46, 56, 91, 102, 106, 142, 190, 208, 266, 330, 360,
382, 462, 503, 815
$33.2^n - 1$ is prime. Ref MOC 23 874 69. [1,2; A2240, N0284]

M0753   1, 2, 3, 6, 8, 13, 18, 29, 40, 58, 79, 115, 154, 213, 284, 391, 514, 690, 900, 1197,
1549, 2025, 2600, 3377, 4306, 5523, 7000, 8922, 11235, 14196, 17777, 22336, 27825
Generalized partition function. Ref KN75 293. [1,2; A4101]

M0754   1, 1, 2, 3, 6, 8, 13, 19, 30, 41, 59, 80, 113, 149, 202, 264, 350, 447, 578, 730, 928,
1155, 1444, 1777, 2193, 2667, 3249, 3915, 4721, 5635, 6728, 7967, 9432, 11083, 13016
Certain partially ordered sets of integers. Ref P4BC 123. [0,3; A3405]

M0755   1, 2, 3, 6, 8, 18, 23
Triangulations. Ref WB79 337. [0,2; A5508]

M0756   1, 2, 3, 6, 9, 10, 11, 12, 28, 29, 30, 53, 56, 57, 80, 82, 104, 105, 107, 129, 130, 132,
154, 155, 157, 179, 180, 182, 204, 205, 207, 229, 230, 232, 254, 255, 257, 279, 280, 282
Next term is uniquely the sum of 3 earlier terms. Ref AB71 249. [1,2; A7086]

M0757   0, 0, 1, 2, 3, 6, 9, 14, 20, 29, 42, 58, 79, 108, 145, 191, 252, 329, 427, 549, 704,
894, 1136, 1427, 1793, 2237, 2789, 3450, 4268, 5248, 6447, 7880, 9619, 11691, 14199
Mixed partitions of n. Ref JNSM 9 91 69. [1,4; A2096, N0286]

M0758   1, 0, 1, 1, 1, 2, 3, 6, 9, 15, 45, 59, 188, 399, 827, 2472, 5073, 14153, 35489, 85726
Minimal 3-polyhedra with n edges. Ref md. [6,6; A6868]

M0759   1, 1, 1, 2, 3, 6, 9, 16, 23, 35, 51, 72
Achiral trees. Ref JRAM 278 334 75. [1,4; A3244]

M0760   1, 1, 1, 2, 3, 6, 9, 19, 30, 61, 99, 208
Partially achiral trees. Ref JRAM 278 334 75. [1,4; A3243]

M0761   1, 2, 3, 6, 9, 26, 53, 146, 369, 1002
Necklaces. Ref IJM 2 302 58. [1,2; A2076, N0288]

M0762   0, 2, 3, 6, 10, 11, 21, 30, 48, 72, 110, 171, 260, 401, 613, 942, 1445, 2216, 3401,
5216, 8004, 12278, 18837, 28899, 44335, 68018, 104349, 160089, 245601, 376791
A Fielder sequence. Ref FQ 6(3) 68 68. [1,2; A1635, N0289]

M0763   0, 2, 3, 6, 10, 17, 21, 38, 57, 92, 143, 225, 351, 555, 868, 1366, 2142, 3365, 5282,
8296, 13023, 20451, 32108, 50417, 79160, 124295, 195159, 306431, 481139, 755462
A Fielder sequence. Ref FQ 6(3) 68 68. [1,2; A1636, N0290]

M0764   0, 2, 3, 6, 10, 17, 28, 46, 75, 122, 198, 321, 520, 842, 1363, 2206, 3570, 5777,
9348, 15126, 24475, 39602, 64078, 103681, 167760, 271442, 439203, 710646, 1149850
$a(n) = a(n-1) + a(n-2) + 1$. Ref JA66 96. MOC 15 397 61. [0,2; A1610, N0291]

M0765   1, 1, 2, 3, 6, 10, 19, 33, 60, 104
Dimension of space of weight systems of chord diagrams. Ref BarN94. [0,3; A7473]

M0766   0, 1, 1, 2, 3, 6, 10, 19, 33, 62, 110, 204
Partially achiral planted trees. Ref JRAM 278 334 75. [1,4; A3237]

M0767 1, 2, 3, 6, 10, 19, 35, 62, 118, 219, 414, 783, 1497, 2860, 5503, 10593, 20471,
39637, 76918, 149501, 291115, 567581, 1108022, 2165621, 4237085, 8297727
Related to population of numbers of form $x^2 + y^2$. Ref MOC 18 84 64. [1,2; A0693,
N0292]

M0768 0, 1, 0, 1, 1, 2, 3, 6, 10, 19, 35, 67, 127, 248, 482, 952, 1885, 3765, 7546, 15221,
30802, 62620, 127702, 261335, 536278, 1103600, 2276499, 4706985, 9752585
Series-reduced planted trees with n nodes. See Fig M0320. Ref AMA 101 150 59. dgc.
JAuMS A20 502 75. [1,6; A1678, N0293]

M0769 1, 2, 3, 6, 10, 20, 35, 70, 126, 252, 462, 924, 1716, 3432, 6435, 12870, 24310,
48620, 92378, 184756, 352716, 705432, 1352078, 2704156, 5200300, 10400600
Central binomial coefficients: $C(n, [n/2])$. Ref RS3. AS1 828. JCT 1 299 66. [1,2; A1405,
N0294]

M0770 1, 2, 3, 6, 10, 20, 36, 72, 135, 272, 528, 1048, 2080, 4160, 8242
Nonperiodic autocorrelation functions of length n. Ref LNM 686 332 78. [1,2; A6606]

M0771 1, 2, 3, 6, 10, 20, 36, 72, 136, 272, 528, 1056, 2080, 4160, 8256, 16512, 32896,
65792, 131328, 262656, 524800, 1049600, 2098176, 4196352, 8390656, 16781312
Binary grids: $2^{n-2} + 2^{[n/2]-1}$. Ref TYCM 9 267 78. [1,2; A5418]

M0772 1, 1, 2, 3, 6, 10, 20, 36, 72, 137, 274, 543
Restricted hexagonal polyominoes with n cells. Ref PEMS 17 11 70. [1,3; A2215, N0295]

M0773 1, 1, 1, 2, 3, 6, 10, 20, 36, 72, 137, 275, 541, 1098, 2208, 4521, 9240, 19084,
39451, 82113, 171240, 358794, 753460, 1587740, 3353192, 7100909, 15067924
Shifts left 2 places under Euler transform. Ref BeSl94. EIS § 2.7. [1,4; A7562]

M0774 1, 2, 3, 6, 10, 20, 37, 74, 143, 284
Self-complementary 2-colored necklaces with $2n$ beads. Ref PJM 110 210 84. [1,2; A7148]

M0775 1, 2, 3, 6, 10, 20, 37, 76, 152, 320, 672, 1454, 3154, 6959, 15439, 34608, 77988,
176985, 403510, 924683, 2127335, 4913452, 11385955, 26468231, 61700232
Binary forests with n nodes. Ref MW63. [1,2; A3214]

M0776 1, 2, 3, 6, 10, 20, 37, 76, 153, 329, 710, 1601, 3658, 8599, 20514, 49905, 122963,
307199, 775529, 1977878, 5086638, 13184156, 34402932, 90328674, 238474986
Forests with n nodes. Ref JCT B27 116 79. [1,2; A5195]

M0777 1, 1, 0, 1, 1, 2, 3, 6, 10, 21, 39, 82, 167, 360, 766, 1692, 3726, 8370, 18866, 43029,
98581, 227678, 528196, 1232541, 2888142, 6798293, 16061348, 38086682, 90607902
Trimmed trees with n nodes. Ref AMM 80 874 73. HA73 297. klm. [1,6; A2988]

M0778 1, 1, 1, 2, 3, 6, 10, 22, 45, 102, 226, 531, 1253, 3044, 7456, 18604, 46798, 119133,
305567, 790375, 2057523, 5390759, 14200122, 37598572, 100005401, 267131927
n-node trees with a forbidden limb. Ref HA73 297. [1,4; A2992]

M0779 1, 1, 2, 3, 6, 11, 12, 18, 28, 42, 48, 68
The coding-theoretic function $A(n,6,5)$. See Fig M0240. Ref PGIT 36 1335 90. [6,3; A4038]

M0780 1, 2, 3, 6, 11, 14, 29, 44, 64, 65, 74, 92, 106, 127
From a nonlinear recurrence. Ref PC 4 42-13 76. [1,2; A5211]

M0781 1, 2, 3, 6, 11, 18, 31, 54, 91, 154, 263, 446, 755, 1282, 2175, 3686, 6251, 10602, 17975, 30478, 51683, 87634, 148591, 251958, 427227, 724410, 1228327, 2082782
Expansion of $1/(1-x)(1-x-2x^3)$. Ref DT76. [0,2; A3479]

M0782 2, 3, 6, 11, 19, 28, 40, 56, 72
Additive bases. Ref SIAA 1 384 80. [2,1; A4135]

M0783 1, 1, 1, 2, 3, 6, 11, 20, 36, 64, 108, 179, 292, 464, 727, 1124, 1714, 2585, 3866, 5724, 8418, 12290, 17830, 25713, 36898, 52664, 74837, 105873, 149178, 209364
Trees in an n-node wheel. Ref HA72. [1,4; A2985]

M0784 1, 0, 1, 2, 3, 6, 11, 20, 37, 68, 125, 230, 423, 778, 1431, 2632, 4841, 8904, 16377, 30122, 55403, 101902, 187427, 344732, 634061, 1166220, 2145013, 3945294, 7256527
Tribonacci numbers: $a(n) = a(n-1) + a(n-2) + a(n-3)$. Ref FQ 5 211 67. [0,4; A1590, N0296]

M0785 1, 1, 2, 3, 6, 11, 20, 40, 77, 148, 285, 570, 1120, 2200, 4323, 8498, 16996, 33707, 66844, 132568, 262936, 521549, 1043098, 2077698, 4138400, 8243093, 16419342
Stern's sequence: $a(n+1)$ is sum of $1+[n/2]$ preceding terms. Ref JRAM 18 100 1838. MSH 84 48 83. [1,3; A5230]

M0786 1, 2, 3, 6, 11, 21, 40, 77, 149, 289, 563, 1099, 2152, 4222, 8299, 16339, 32217, 63612, 125753, 248870, 493015, 977576, 1940042, 3853117, 7658211, 15231219
Directed site animals on hexagonal lattice. Ref JPA 15 L282 82. [1,2; A6861]

M0787 1, 1, 2, 3, 6, 11, 22, 42, 84, 165, 330, 654, 1308, 2605, 5210, 10398, 20796, 41550, 83100, 166116, 332232, 664299, 1328598, 2656866, 5313732, 10626810, 21253620
Narayana-Zidek-Capell numbers: $a(2n) = 2a(2n-1)$, $a(2n+1) = 2a(2n) - a(n)$. Ref CRP 267 32 68. CMB 13 108 70. NA79 100. MSH 84 48 83. [2,3; A2083, N0297]

M0788 1, 1, 2, 3, 6, 11, 22, 43, 86, 171, 342, 683, 1366, 2731, 5462, 10923, 21846, 43691, 87382, 174763, 349526, 699051, 1398102, 2796203, 5592406, 11184811, 22369622
$a(2n) = 2a(2n-1)$, $a(2n+1) = 2a(2n) - 1$. Ref GTA91 603. [0,3; A5578]

M0789 1, 1, 1, 2, 3, 6, 11, 22, 44, 90, 187, 392, 832, 1778, 3831, 8304, 18104, 39666, 87296, 192896, 427778, 951808, 2124135, 4753476, 10664458, 23981698, 54045448
Shifts 2 places left when convolved with itself. Ref BeSl94. [0,4; A7477]

M0790 0, 1, 1, 1, 2, 3, 6, 11, 23, 46, 98, 207, 451, 983, 2179, 4850, 10905, 24631, 56011, 127912, 293547, 676157, 1563372, 3626149, 8436379, 19680277, 46026618, 107890609
Wedderburn-Etherington numbers: binary rooted trees with n endpoints inequivalent under reflections in vertical axes. Ref C1 55. [0,5; A1190, N0298]

$$\text{G.f.:}\quad A(x) = x + \tfrac{1}{2}A^2(x) + \tfrac{1}{2}A(x^2).$$

M0791 1, 1, 1, 1, 2, 3, 6, 11, 23, 47, 106, 235, 551, 1301, 3159, 7741, 19320, 48629, 123867, 317955, 823065, 2144505, 5623756, 14828074, 39299897, 104636890
Trees with n nodes. See Fig M0791. Ref R1 138. HA69 232. [0,5; A0055, N0299]

||||||||||||||||||||||||||||||||||||||||||||||||||||||||||||||||||||||||||||||||||||||||||||||||||||||||||||||||||||||||||||||

Figure M0791. TREES.

   A **tree** is a connected graph containing no closed paths. A **rooted tree** has a distinguished node called the root. A **planted** tree is a rooted tree in which the root has degree 1. M0791 and M1180 give the numbers t_n and r_n of trees and rooted trees with n nodes, respectively, as shown in the following diagrams:

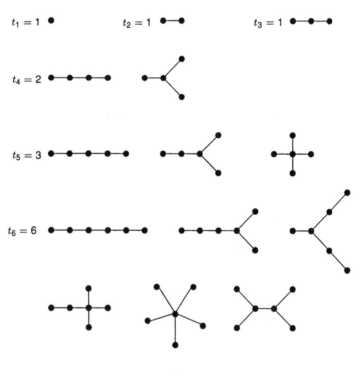

(a) Trees

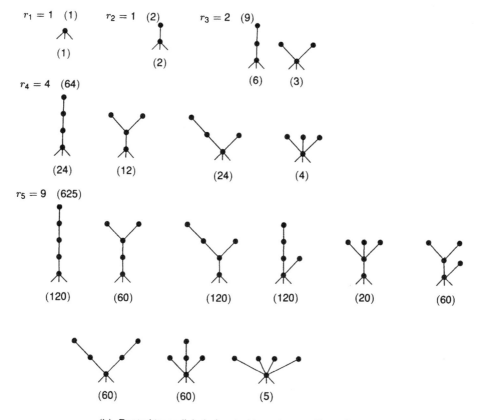

(b) Rooted trees (labeled rooted trees in parentheses)

The generating function for rooted trees,

$$r(x) = \sum_{n=1}^{\infty} r_n x^n = x + x^2 + 2x^3 + 4x^4 + 9x^5 + \cdots$$

satisfies

$$r(x) = x \exp[r(x) + \frac{1}{2}r(x^2) + \frac{1}{3}r(x^3) + \cdots].$$

The generating function for trees,

$$t(x) = x + x^2 + x^3 + 2x^4 + 3x^5 + 6x^6 + \cdots,$$

is then given by

$$t(x) = r(x) - \frac{1}{2}r^2(x) + \frac{1}{2}r(x^2).$$

See [HA69 187], [HP73 57]. The numbers of labeled trees with n nodes (n^{n-2}, M3027) and labeled rooted trees (n^{n-1}, M1946, shown in Fig. (b)) are much simpler.

M0792   1, 2, 3, 6, 11, 23, 48, 114, 293, 869, 2963, 12066, 58933, 347498, 2455693,
20592932, 202724920
n-node graphs of girth 5. Ref bdm. [1,2; A6787]

M0793   1, 1, 1, 2, 3, 6, 11, 24, 47, 103, 214, 481, 1030, 2337, 5131, 11813, 26329, 60958,
137821, 321690, 734428, 1721998, 3966556, 9352353, 21683445, 51296030, 119663812
$a(n) = \Sigma a(k) a(n-k)$, $k = 1 \ldots [n/2]$. Ref DUMJ 31 517 64. [1,4; A0992, N0300]

M0794   2, 3, 6, 12, 18, 24, 48, 54, 96, 162, 192, 216, 384, 486, 768, 864, 1458, 1536, 1944,
3072, 3456, 4374, 6144, 7776, 12288, 13122, 13824, 17496, 24576, 31104, 39366, 49152
MU-numbers: next term is uniquely the product of 2 earlier terms. See Fig M0557. Ref
Pick92 359. [1,1; A7335]

M0795   0, 0, 1, 2, 3, 6, 12, 23, 44, 85, 164, 316, 609, 1174, 2263, 4362, 8408, 16207,
31240, 60217, 116072, 223736, 431265, 831290, 1602363, 3088654, 5953572, 11475879
Tetranacci numbers: $a(n) = a(n-1) + a(n-2) + a(n-3) + a(n-4)$. Ref FQ 8 7 70. [0,4;
A1630, N0301]

M0796   1, 1, 1, 2, 3, 6, 12, 25, 52, 113, 247, 548, 1226, 2770, 6299, 14426, 33209, 76851,
178618, 416848, 976296, 2294224, 5407384, 12780394, 30283120, 71924647
Rooted identity trees with n nodes. Ref JAuMS A20 502 75. [1,4; A4111]

M0797   1, 2, 3, 6, 12, 26, 59, 146, 368
Series-parallel networks with n nodes. Ref ICM 1 646 50. [2,2; A1677, N0302]

M0798   1, 1, 1, 2, 3, 6, 12, 27, 65, 175, 490, 1473, 4588, 14782, 48678, 163414, 555885,
1913334, 6646728, 23278989, 82100014, 291361744, 1039758962, 3729276257
Projective plane trees with n nodes. Ref LNM 406 348 74. rwr. Fel92. [1,4; A6082]

M0799   1, 1, 1, 1, 2, 3, 6, 12, 28, 68
Minimal blocks with n nodes. Ref NBS B77 56 73. [1,5; A3317]

M0800   2, 3, 6, 13, 28, 62
Alkyls with n carbon atoms. Ref ZFK 93 437 36. [1,1; A0646, N0303]

M0801   1, 2, 3, 6, 13, 30, 72, 178, 450, 1158, 3023, 7986, 21309, 57346, 155469, 424206,
1164039, 3210246, 8893161, 24735666, 69051303, 193399578, 543310782, 1530523638
Sums of successive Motzkin numbers. Cf. M1184. Ref JCT B29 82 80. [1,2; A5554]

$$(n + 1)\, a(n) = 2n\, a(n - 1) + (3n - 9)\, a(n - 2).$$

M0802   1, 1, 2, 3, 6, 13, 35, 116
Connected weighted linear spaces of total weight n. Ref BSMB 22 234 70. [1,3; A2877,
N0304]

M0803   0, 1, 2, 3, 6, 14, 15, 39, 201, 249, 885, 1005, 1254, 1635
$4.3^n + 1$ is prime. Ref MOC 26 996 72. Cald94. [1,3; A5537]

M0804 0, 1, 2, 3, 6, 14, 30, 77, 196, 525, 1414, 3960, 11056, 31636, 90818, 264657, 774146, 2289787, 6798562, 20354005, 61164374, 184985060, 561433922, 1712696708
2-connected planar maps with n edges. Ref trsw. [1,3; A6444]

M0805 0, 1, 1, 1, 2, 3, 6, 14, 34, 95, 280, 854, 2694, 8714, 28640, 95640, 323396, 1105335, 3813798, 13269146, 46509358, 164107650, 582538732, 2079165208
Planar trees with n nodes. Ref JRAM 278 334 75. LA91. [0,5; A2995]

M0806 2, 3, 6, 14, 36, 94, 250, 675, 1838, 5053, 14016, 39169, 110194, 311751, 886160, 2529260, 7244862, 20818498, 59994514, 173338962, 501994070, 1456891547
Fixed triangular polyominoes with n cells. Ref RE72 97. dhr. [1,1; A1420, N0305]

M0807 2, 3, 6, 14, 40, 152, 784, 5168, 40576, 363392, 3629824, 39918848, 479005696, 6227028992, 87178307584, 1307674400768, 20922789953536, 355687428227072
$n! + 2^n$. Ref MMAG 64 141 91. [0,1; A7611]

M0808 0, 1, 2, 3, 6, 15, 36, 114, 396, 1565, 6756, 31563, 154370, 785113, 4099948, 21870704, 118624544, 652485364, 3631820462, 20426666644, 115949791342
2-connected planar maps with n edges. Ref SIAA 4 174 83. trsw. [1,3; A6403]

M0809 1, 2, 3, 6, 15, 63, 567, 14755, 1366318
Threshold functions of n variables. Ref PGEC 19 821 70. MU71 38. [0,2; A1529, N0306]

M0810 1, 2, 3, 6, 15, 85
Maximal tight voting schemes on n points. Ref Loeb94b. [0,2; A7364]

M0811 1, 1, 2, 3, 6, 16, 35, 90, 216, 768, 2310, 7700, 21450, 69498, 292864, 1153152, 4873050, 16336320, 64664600, 249420600, 1118939184, 5462865408, 28542158568
Largest irreducible character of symmetric group of degree n. Ref LI50 265. MOC 14 110 60. [1,3; A3040]

M0812 1, 2, 3, 6, 16, 42, 151, 596, 2605, 12098, 59166, 297684, 1538590, 8109078, 43476751, 236474942, 1302680941, 7256842362, 40832979283, 231838418310
Nonseparable planar maps with n edges. Ref SIAA 4 174 83. CJM 35 434 83. trsw. [2,2; A6402]

M0813 1, 2, 3, 6, 16, 122, 8003, 18476850, 190844194212235, 1923037112470381326001440866
Row sums of Fibonacci-Pascal triangle. Ref FQ 13 281 75. [0,2; A6449]

M0814 2, 3, 6, 17, 112, 8282
Unate Boolean functions of n variables. Ref MU71 38. [0,1; A3183]

M0815 2, 3, 6, 18, 206, 7888299
Boolean functions of n variables. Ref JSIAM 12 294 64. [1,1; A0614, N0307]

M0816 2, 3, 6, 20, 150, 3287, 244158, 66291591, 68863243522
Threshold functions of n variables. Ref PGEC 19 821 70. MU71 38. [0,1; A2078, N0308]

M0817 2, 3, 6, 20, 168, 7581, 7828354, 2414682040998, 56130437228687557907788
Monotone Boolean functions of n variables (or Dedekind numbers). See Fig M2051. Ref HA65 188. BI67 63. C1 273. Wels71 181. MU71 214. AN87 38. dhw. [0,1; A0372, N0309]

M0818 2, 3, 6, 21, 231, 26796, 359026206, 64449908476890321,
    20768953513397694604776113701866681
$a(n+1) = a(n) (a(n)+1) / 2$. Ref JRM 12 111 79. [0,1; A7501]

M0819 2, 3, 6, 22, 402, 1228158, 400507806843728
Irreducible Boolean functions of n variables. Ref JSIAM 11 827 63. MU71 38. HA65 149. [0,1; A0616, N0310]

M0820 2, 3, 6, 30, 75, 81, 115, 123, 249, 362, 384, 462, 512, 751, 822
$n.2^n - 1$ is prime. Ref MOC 23 875 69. [1,1; A2234, N0311]

M0821 2, 3, 7, 8, 9, 10, 16, 17, 18, 19, 20, 21, 29, 30, 31, 32, 33, 34, 35, 36, 46, 47, 48, 49,
    50, 51, 52, 53, 54, 55, 67, 68, 69, 70, 71, 72, 73, 74, 75, 76, 77, 78, 92, 93, 94, 95, 96, 97
Skip 1, take 2, skip 3, etc. Cf. M3241. Ref HO85a 177. [1,1; A7607]

M0822 2, 3, 7, 8, 10, 16, 18, 19, 40, 48, 55, 90, 96, 98, 190, 398, 456, 502, 719, 1312,
    1399, 1828
$57.2^n + 1$ is prime. Ref PAMS 9 675 58. Rie85 382. [1,1; A2274, N0313]

M0823 1, 2, 3, 7, 8, 12, 20, 23, 27, 35, 56, 62, 68, 131, 222
$2.3^n - 1$ is prime. Ref MOC 26 997 72. [1,2; A3307]

M0824 1, 1, 2, 3, 7, 9, 20, 26, 54, 74, 137, 184, 356, 473, 841, 1154, 2034, 2742, 4740,
    6405, 10874, 14794, 24515, 33246, 54955, 74380, 120501, 163828, 263144, 356621
Set-like molecular species of degree n. Ref AAMS 15(896) 94. [0,3; A7649]

M0825 1, 2, 3, 7, 10, 13, 18, 27, 37, 51, 74, 157, 271, 458, 530, 891
$21.2^n - 1$ is prime. Ref MOC 23 874 69. [1,2; A2238, N0314]

M0826 1, 2, 3, 7, 10, 13, 25, 26, 46, 60, 87, 90, 95, 145, 160, 195, 216, 308, 415, 902,
    1128, 3307, 6748, 7747, 8348, 11193, 27243
$7.4^n + 1$ is prime. Ref PAMS 9 674 58. Rie85 381. Cald94. [1,2; A2255, N0315]

M0827 1, 2, 3, 7, 11, 19, 43, 67, 163.
Imaginary quadratic fields with unique factorization (a finite sequence). Ref ST70 295. LNM 751 226 79. HW1 213. [1,2; A3173]

M0828 1, 2, 3, 7, 11, 25, 39, 89, 139, 317, 495, 1129, 1763, 4021, 6279, 14321, 22363,
51005, 79647, 181657, 283667, 646981, 1010295, 2304257, 3598219, 8206733
Subsequences of [1,...,n] in which each even number has an odd neighbor. Ref GuMo94.
[1,2; A7481]

$$a(n) = 3\,a(n-2) + 2\,a(n-4).$$

M0829 1, 1, 1, 2, 3, 7, 11, 26, 41, 97, 153, 362, 571, 1351, 2131, 5042, 7953, 18817,
29681, 70226, 110771, 262087, 413403, 978122, 1542841, 3650401, 5757961, 13623482
$a(n) = (1 + a(n-1)\,a(n-2))/a(n-3)$. Ref MAG 69 263 85. [0,4; A5246]

M0830 0, 1, 2, 3, 7, 12, 18, 32, 59, 81, 105, 132, 228, 265, 284, 304, 367, 389, 435, 483,
508, 697, 726, 944, 1011, 1045, 1080, 1115, 1187, 1454, 1494, 1617, 1659, 1788, 1921
$n!$ has a square number of digits. Ref GA78 55. rgw. [0,3; A6488]

M0831 1, 0, 1, 1, 2, 3, 7, 12, 27, 55, 127, 284, 682
Centered trees with n nodes. Ref CAY 9 438. [1,5; A0676, N0316]

M0832 2, 3, 7, 13, 17, 23, 37, 43, 47, 53, 67, 73, 83, 97, 103, 107, 113, 127, 137, 157, 163,
167, 173, 193, 197, 223, 227, 233, 257, 263, 277, 283, 293, 307, 313, 317, 337, 347, 353
Inert rational primes in $Q(\sqrt{5})$. Ref Hass80 498. [1,1; A3631]

M0833 2, 3, 7, 13, 21, 31
A problem in (0,1)-matrices. Ref AMM 81 1113 74. [2,1; A3509]

M0834 1, 2, 3, 7, 13, 28
Independence number of De Bruijn graph of order n. [1,2; A6946]

M0835 1, 1, 1, 2, 3, 7, 13, 31, 65, 154, 347, 824, 1905, 4512, 10546, 24935, 58476,
138002, 323894, 763172, 1790585, 4213061, 9878541
Filaments with n square cells. Ref PLC 1 337 70. [0,4; A2013, N0317]

M0836 1, 1, 2, 3, 7, 13, 31, 66, 159
Arborescences of type $(n, 1)$. Ref DM 5 197 73. [1,3; A3120]

M0837 1, 2, 3, 7, 13, 35, 85, 257, 765, 2518
Binary sequences of period $2n$ with n 1's per period. Ref JAuMS A33 14 82. CN 40 89 83.
[1,2; A6840]

M0838 2, 3, 7, 13, 97, 193, 18817, 37633, 708158977, 1416317953,
1002978273411373057, 2005956546822746113
$a(2n) = (a(2n-1)+1)\,a(2n-2)-1$, $a(2n+1) = 2a(2n)-1$. Ref ACA 55 311 90. FQ 31
37 93. [2,1; A6695]

M0839 1, 1, 2, 3, 7, 14, 32, 72, 171, 405, 989, 2426, 6045, 15167, 38422, 97925, 251275, 648061, 1679869, 4372872, 11428365, 29972078, 78859809, 208094977, 550603722
Alkyl derivatives of acetylene with n carbon atoms. Ref JACS 55 253 33. LNM 303 255 72. BA76 28. [1,3; A0642, N0318]

M0840 1, 1, 2, 3, 7, 14, 36, 81, 221, 538, 1530, 3926, 11510, 30694, 92114, 252939
Meanders in which first bridge is 3. See Fig M4587. Ref SFCA91 293. [3,3; A6660]

M0841 1, 2, 3, 7, 14, 38, 107, 410, 1897, 12172, 105071, 1262180, 20797002
Triangle-free graphs on n vertices. Ref bdm. [1,2; A6785]

M0842 1, 1, 2, 3, 7, 14, 54, 224, 2038, 32728
2-graphs with n nodes. Ref LNM 885 70 81. [1,3; A6627]

M0843 1, 2, 3, 7, 15, 27, 255, 447, 639, 703, 1819, 4255, 4591, 9663, 20895, 26623, 31911, 60975, 77671, 113383, 138367, 159487, 270271, 665215, 704511, 1042431
'3x+1' record-setters (values). See Fig M2629. Ref GEB 400. ScAm 250(1) 12 84. CMWA 24 94 92. [1,2; A6884]

M0844 2, 3, 7, 15, 34, 78, 182, 429, 1019, 2433, 5830, 14004, 33694, 81159, 195635, 471819, 1138286, 2746794, 6629290, 16001193, 38624911, 93240069, 225087338
Sum of Fibonacci and Pell numbers. [0,1; A1932, N0319]

M0845 1, 1, 2, 3, 7, 15, 35, 81, 195
n-level expressions. Ref AMM 80 876 73. [1,3; A3006]

M0846 1, 1, 2, 3, 7, 16, 54, 243, 2038, 33120, 1182004, 87723296, 12886193064, 3633057074584, 1944000150734320, 1967881448329407496
Euler graphs or 2-graphs with n nodes. Ref JSIAM 28 877 75. [1,3; A2854, N0321]

M0847 0, 0, 1, 1, 2, 3, 7, 18, 41, 123, 367, 1288, 4878
Alternating prime knots with n crossings. See Fig M0847. Ref TAIT 1 345. LE70 343. JK85 11. mt. [1,5; A2864, N0322]

M0848 2, 3, 7, 18, 47, 123, 322, 843, 2207, 5778, 15127, 39603, 103682, 271443, 710647, 1860498, 4870847, 12752043, 33385282, 87403803, 228826127, 599074578
$a(n) = 3a(n-1) - a(n-2)$. Ref FQ 9 284 71. MMAG 48 209 75. MAG 69 264 85. [1,1; A5248]

M0849 2, 3, 7, 19, 31, 37, 79, 97, 139, 157, 199, 211, 229, 271, 307, 331, 337, 367, 379, 439, 499, 547, 577, 601, 607, 619, 661, 691, 727, 811, 829, 877, 937, 967, 997, 1009
n and $2n-1$ are prime. Cf. M2492. Ref AS1 870. [1,1; A5382]

M0850 2, 3, 7, 19, 56, 174, 561, 1859, 6292, 21658, 75582, 266798, 950912, 3417340, 12369285, 45052515, 165002460, 607283490, 2244901890, 8331383610, 31030387440
Sums of adjacent Catalan numbers. Ref dek. [0,1; A5807]

|||||||||||||||||||||||||||||||||||||||||||||||||||||||||||||||||||||||||||||||||||||||||||||||||||||||||||||||

Figure M0847.   KNOTS.

The figure shows M0847: 0, 0, 1, 1, 2, 3, 7, …, the number of 'prime' knots with n crossings. (The product of two knots is formed by tying them on the same piece of string. A prime knot is one that is not a product.) Only thirteen terms of this sequence are known. M0851, which begins in the same way, gives the number of knots in which the over and under crossings alternate.

The classification of knots was begun in the nineteenth century by P. G. Tait [TA1 1 334] and C. N. Little. In 1967 J. H. Conway [LE70 329] introduced a new notation for attacking the problem, and remarked: "Little tells us that the enumeration of the knots in his 1900 paper took him six years — from 1893 to 1899 — the notation we shall describe made this just one afternoon's work!" The most recent results have been obtained by M. Thistlethwaite [JK85 11], [mt]. For further information see [Kauf87], [Atiy90], [Jone91].

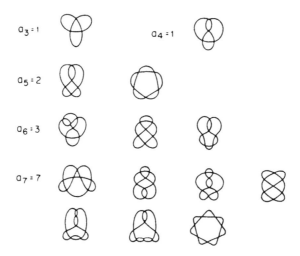

|||||||||||||||||||||||||||||||||||||||||||||||||||||||||||||||||||||||||||||||||||||||||||||||||||||||||||||||

M0851   0, 0, 1, 1, 2, 3, 7, 21, 49, 165, 552, 2176, 9988
Prime knots with n crossings. See Fig M0847. Ref TAIT 1 345. LE70 343. JK85 11. mt.
MMAG 61 7 88. [1,5; A2863, N0323]

M0852   1, 1, 2, 3, 7, 21, 135, 2470, 175428
Self-dual threshold functions of n variables. Ref PGEC 19 821 70. MU71 38. [1,3; A1532, N0324]

M0853   0, 0, 1, 1, 2, 3, 7, 22, 155, 3411, 528706, 1803416167, 953476947989903,
171951574286680922296 1802, 16395186225292360779521443 18816050685207
$a(n) = a(n-1).a(n-2)+1$. Ref rgw. [0,5; A7660]

M0854   1, 2, 3, 7, 23, 29, 157, 1307, 1669, 1879, 2089
Related to arithmetic progressions of primes. Ref UPNT A5. [0,2; A5115]

M0855   2, 3, 7, 23, 41, 71, 191, 409, 2161, 5881, 36721, 55441, 71761, 110881, 760321
Least positive primitive roots. Ref RS9 XLIV. [1,1; A2230, N0325]

M0856   2, 3, 7, 23, 43, 67, 83, 103, 127, 163, 167, 223, 227, 283, 367, 383, 443, 463, 467,
    487, 503, 523, 547, 587, 607, 643, 647, 683, 727, 787, 823, 827, 863, 883, 887, 907
Primes dividing all Fibonacci sequences. Ref FQ 2 38 64. [1,1; A0057, N0326]

M0857   1, 1, 1, 1, 2, 3, 7, 23, 59, 314, 1529, 8209, 83313, 620297, 7869898, 126742987,
    1687054711, 47301104551, 1123424582771, 32606721084786, 1662315215971057
Somos-4 sequence: $a(n) = (a(n-1)a(n-3) + a(n-2)^2)/a(n-4)$. Ref MINT 13(1) 41
91. [0,5; A6720]

M0858   2, 3, 7, 23, 89, 113, 523, 887, 1129, 1327, 9551, 15683, 19609, 31397, 155921,
    360653, 370261, 492113, 1349533, 1357201, 2010733, 4652353, 17051707, 20831323
Increasing gaps between primes (lower end). Cf. M2485. Ref Krai24 1 14. MOC 52 222
89. [1,1; A2386, N0327]

M0859   1, 1, 2, 3, 7, 23, 164, 3779, 619779, 2342145005, 1451612289057674,
    3399886472013047316638149, 4935316984175079105557291745555191750431
$a(n) = a(n-1)a(n-2) + a(n-3)$. Ref rkg. [0,3; A1064, N0328]

M0860   1, 2, 3, 7, 23, 167, 7223, 13053767, 42600227803223,
    45369485222168737744001767
Representation requires n squares with greedy algorithm. Ref Lem82. jos. [1,2; A6892]

M0861   2, 3, 7, 29, 71, 127, 271, 509, 1049, 6389, 6883, 10613
$(6^n - 1)/5$ is prime. Ref CUNN. MOC 61 928 93. [1,1; A4062]

M0862   1, 1, 1, 2, 3, 7, 33
n-dimensional perfect lattices. Ref PRS A418 56 88. [1,4; A4026]

M0863   2, 3, 7, 43, 13, 53, 5, 6221671, 38709183810571, 139, 2801, 11, 17, 5471,
    52662739, 23003, 30693651606209, 37, 1741, 1313797957, 887, 71, 7127, 109, 23, 97
Euclid-Mullin sequence: $a(n+1)$ is smallest prime in $\Pi a(i) + 1$. Ref BAMS 69 737 63.
GN75. BICA 8 26 93. [0,1; A0945, N0329]

M0864   2, 3, 7, 43, 139, 50207, 340999, 2365347734339, 4680225641471129,
    1368845206580129, 889340324577880670089824574922371
Euclid-Mullin sequence: $a(n+1)$ is largest prime in $\Pi a(i) + 1$. Ref GN75. PAMS 90 43
84. BICA 8 27 93. [0,1; A0946, N0330]

M0865   2, 3, 7, 43, 1807, 3263443, 10650056950807, 113423713055421844361000443,
    12864938683278671740537145998360961546653259485195807
Sylvester's sequence: $a(n+1) = a(n)^2 - a(n) + 1$. Ref CJM 15 475 63. AMM 70 403 63.
FQ 11 430 73. [0,1; A0058, N0331]

M0866 2, 3, 7, 127, 170141183460469231731687303715884105727
$a(n+1) = 2^{a(n)} - 1$. Ref SI64 91. [0,1; A7013]

M0867 1, 2, 3, 8, 10, 12, 14, 17, 23, 24, 27, 28, 37, 40, 41, 44, 45, 53, 59, 66, 70, 71, 77,
80, 82, 87, 90, 97, 99, 102, 105, 110, 114, 119, 121, 124, 127, 133, 136, 138, 139, 144
$(2n)^4 + 1$ is prime. Ref MOC 21 246 67. BIT 13 371 73. [1,2; A0059, N0332]

M0868 1, 2, 3, 8, 10, 54, 42, 944, 5112, 47160, 419760, 4297512, 47607144, 575023344,
7500202920, 105180931200, 1578296510400, 25238664189504
From nth derivative of x^x. Ref AMM 95 705 88. [1,2; A5727]

M0869 2, 3, 8, 11, 19, 87, 106, 193, 1264, 1457, 2721, 23225, 25946, 49171, 517656,
566827, 1084483, 13580623, 14665106, 28245729, 410105312, 438351041, 848456353
Numerators of convergents to e. Cf. M2343. Ref LE77 240. [1,1; A7676]

M0870 2, 3, 8, 12, 15, 27, 48, 89, 137
$2.25^n - 1$ is prime. Ref PLC 2 568 71. [1,1; A2958]

M0871 2, 3, 8, 13, 20, 31, 32, 53, 76, 79, 80, 117, 176, 181, 182, 193, 200, 283, 284, 285,
286, 293, 440, 443, 468, 661, 678, 683, 684, 1075, 1076, 1087, 1088, 1091, 1092
Related to Liouville's function. Ref IAS 12 408 40. [0,1; A2053, N0333]

M0872 1, 2, 3, 8, 13, 24, 37, 66, 107, 186, 303, 516, 849, 1436, 2377, 3998, 6639, 11134,
18531, 31024, 51701, 86464, 144205, 241018, 402163, 671906, 1121463, 1873244
A counter moving problem. Ref BA62 38. [1,2; A4138]

M0873 1, 1, 2, 3, 8, 14, 42, 79, 252, 494
P-graphs with $2n$ edges. Ref AEQ 31 54 86. [1,3; A7165]

M0874 1, 1, 2, 3, 8, 14, 42, 81, 262, 538, 1828, 3926, 13820, 30694, 110954, 252939,
933458, 2172830, 8152860, 19304190, 73424650, 176343390, 678390116, 1649008456
Meandric numbers: ways a river can cross a road n times. See Fig M4587. Ref PH88.
SFCA91 289. [1,3; A5316]

M0875 1, 2, 3, 8, 15, 24, 49, 128, 189, 480, 1023, 1536, 4095, 6272, 10125, 32768, 65025,
96768, 262143, 491520, 583443, 2095104, 4190209, 6291456, 15728625, 33546240
Generalized Euler Φ function. Ref MOC 28 1168 74. [1,2; A3473]

M0876 1, 1, 2, 3, 8, 15, 48, 105, 384, 945, 3840, 10395, 46080, 135135, 645120, 2027025,
10321920, 34459425, 185794560, 654729075, 3715891200, 13749310575, 81749606400
Double factorials $n!!$: $a(n)=n.a(n-2)$. Cf. M1878, M3002. Ref MOC 24 231 70. [0,3;
A6882]

M0877 1, 2, 3, 8, 15, 52, 126, 568, 1782, 10436, 42471, 323144, 1706562, 16866856,
115640460, 1484714416, 13216815036, 220426128584, 2548124192970
Alternating sign matrices. Ref LNM 1234 292 86. SFCA92 2 32. [1,2; A5162]

M0878 1, 2, 3, 8, 16, 50, 133, 440, 1387, 4752, 16159, 56822
Necklaces with n red and n blue beads. Ref MMAG 60 90 87. PLM 110 210 84. [1,2; A5648]

M0879 1, 2, 3, 8, 18, 44, 115, 294, 783
Folding a strip of n rectangular stamps. See Fig M4587. Ref SPH 7 203 37. [1,2; A2369, N0334]

M0880 1, 1, 2, 3, 8, 18, 47, 123, 338, 935, 2657, 7616, 22138, 64886, 191873, 571169,
1711189, 5153883, 15599094, 47415931, 144692886, 443091572, 1361233280
Alkyl derivatives of acetylene with n carbon atoms. Ref BA76 44. [1,3; A5957]

M0881 1, 2, 3, 8, 19, 27, 100, 227, 781, 1008, 3805, 4813, 148195, 153008, 760227,
913235, 2586697, 24193508, 147747745, 615184488, 762932233, 1378116721
Convergents to cube root of 4. Ref AMP 46 106 1866. L1 67. hpr. [1,2; A2356, N0335]

M0882 1, 2, 3, 8, 19, 64, 225, 928, 3441
Cyclic neofields of order n. Ref LNM 824 189 80. GTA85 100. [4,2; A6609]

M0883 1, 2, 3, 8, 19, 65, 84, 485, 1054, 24727, 50508, 125743, 176251, 301994,
16785921, 17087915, 85137581, 272500658, 357638239, 630138897, 9809721694
Convergents to $\log_2 3$. Ref rkg. [0,2; A5663]

M0884 1, 1, 2, 3, 8, 22, 3, 1
Solutions to knights on an $n \times n$ board. See Fig M3224. Cf. M3224. Ref GA78 194. [1,3; A6076]

M0885 1, 2, 3, 8, 22, 62
Stable unicyclic graphs with n nodes. Ref LNM 403 84 74. klm. [3,2; A6545]

M0886 2, 3, 8, 22, 77, 285, 1259, 5863, 29322, 151308
A subclass of $2n$-node trivalent planar graphs without triangles. Ref JCT B45 309 88. [6,1; A6796]

M0887 1, 1, 2, 3, 8, 24, 108, 640, 4492, 36336, 329900, 3326788, 36846288, 444790512,
5811886656, 81729688428, 1230752346368, 19760413251956, 336967037143596
2-colored patterns on an $n \times n$ board. Ref MES 37 61 07. clm. [1,3; A2619, N0336]

M0888 1, 1, 2, 3, 8, 27, 224, 6075, 1361024, 8268226875, 11253255215681024,
9304446720552777232332546875, 10470531358708673960627431922039587436 81024
$a(n) = (a(n-1)+1).a(n-2)$. Ref MFM 111 122 91. [0,3; A6277]

M0889 2, 3, 8, 29, 148, 1043, 11984, 229027, 6997682, 366204347, 30394774084,
4363985982959, 994090870519508, 393850452332173999, 249278602955869472540
Line-self-dual nets with n nodes. Ref CCC 2 32 77. rwr. JGT 1 295 77. [1,1; A4106]

M0890 2, 3, 8, 30, 144, 840, 5760, 45360, 403200, 3991680, 43545600, 518918400,
6706022400, 93405312000, 1394852659200, 22230464256000, 376610217984000
$n! + (n-1)!$. Ref CJM 22 26 70. [1,1; A1048, N0337]

M0891 1, 1, 2, 3, 8, 34, 377, 17711, 9227465, 225851433717, 2880067194370816120,
898923707008479989274290850145
$F(F(n))$, where F is a Fibonacci number. [0,3; A5370]

M0892 1, 0, 2, 3, 8, 40, 20, 651, 1160, 10872, 53102, 213235, 2765684, 4784988,
42740178, 433914495, 15184673712, 207885336400, 2790244125426, 48660847539651
Expansion of $(1-x)^{\sin x}$. [0,3; A7119]

M0893 1, 1, 2, 3, 8, 51, 1538, 599871, 19417825808, 1573273218577214751,
124442887685693556895657990772138
From a continued fraction. Ref AMM 63 711 56. [0,3; A1686, N0338]

M0894 2, 3, 8, 63, 3968, 15745023, 247905749270528,
61457260521381894004129398783
$a(n) = a(n-1)^2 - 1$. Ref FQ 11 430 73. [0,1; A3096]

M0895 1, 1, 1, 2, 3, 9, 9, 71, 96, 1325, 6843, 54922, 417975, 3586117, 32531983,
316599861, 3274076017, 35914014266, 416386323306, 5088908019824
Reversion of g.f. for Euler numbers. Cf. M1492. [1,4; A7316]

M0896 1, 1, 2, 3, 9, 10, 19, 20, 32, 84, 85, 161, 212, 214, 260, 521, 818, 820, 1189, 1406,
1415, 2005, 2375, 3351, 5698, 6122, 6141, 6600, 6623, 7270
Coprime chains with largest member p. Ref PAMS 16 809 65. [1,3; A3140]

M0897 1, 1, 2, 3, 9, 19, 71, 249, 1058, 4705, 22380
Irreducible polyhedral graphs with n nodes. Ref Dil92. [4,3; A6866]

M0898 1, 1, 2, 3, 9, 20, 75, 262, 1117, 4783, 21971, 102249, 489077, 2370142, 11654465,
57916324, 290693391, 1471341341, 7504177738, 38532692207, 199076194985
Symmetric dissections of a polygon. Ref BAMS 54 359 48. AEQ 18 388 78. [0,3; A1004,
N0339]

M0899 1, 1, 0, 1, 2, 3, 9, 28, 97, 378, 1601, 7116
A subclass of $2n$-node trivalent planar graphs without triangles. Ref JCT B45 309 88. [4,5;
A6797]

M0900 1, 2, 3, 10, 11, 12, 13, 20, 21, 22, 23, 30, 31, 32, 33, 100, 101, 102, 103, 110, 111,
112, 113, 120, 121, 122, 123, 130, 131, 132, 133, 200, 201, 202, 203, 210, 211, 212, 213
Natural numbers in base 4. [1,2; A7090]

M0901 2, 3, 10, 12, 13, 20, 21, 22, 23, 24, 25, 26, 27, 28, 29, 30, 31, 32, 33, 34, 35, 36, 37, 38, 39, 200, 201, 202, 203, 204, 205, 206, 207, 208, 209, 210, 211, 212, 213, 214, 215
Numbers beginning with t. Ref EUR 48 56 88. [1,1; A6092]

M0902 1, 2, 3, 10, 25, 140, 588, 5544, 39204, 622908, 7422987
Alternating sign matrices. Ref LNM 1234 292 86. [1,2; A5158]

M0903 1, 2, 3, 10, 25, 176, 721, 6406, 42561, 436402, 3628801, 48073796, 479001601, 7116730336, 88966701825, 1474541093026, 20922789888001
Permutations of length n with equal cycles. Ref JCT A35 201 83. [1,2; A5225]

M0904 1, 2, 3, 10, 27, 98, 350, 1402, 5743, 24742, 108968, 492638, 2266502, 10600510, 50235931
Signed trees with n nodes. Ref AMA 101 154 59. LeMi91. [1,2; A0060, N0340]

M0905 1, 0, 2, 3, 10, 27, 126, 593
Bipartite polyhedral graphs with n faces. Ref Dil92. [6,3; A7029]

M0906 1, 1, 1, 2, 3, 10, 1382, 420, 10851, 438670, 7333662, 51270780, 7090922730, 2155381956, 94997844116, 68926730208040
Numerators of Bernoulli numbers. Ref DA63 2 208. [0,4; A2443, N0341]

M0907 1, 2, 3, 11, 22, 26, 101, 111, 121, 202, 212, 264, 307, 836, 1001, 1111, 2002, 2285, 2636, 10001, 10101, 10201, 11011, 11111, 11211, 20002, 20102, 22865, 24846, 30693
Square is a palindrome. Ref JRM 20 69 88; 22 124 90. [1,2; A2778, N0342]

M0908 1, 2, 3, 11, 27, 37, 41, 73, 77, 116, 154, 320, 340, 399, 427, 872, 1477
$n! + 1$ is prime. Ref JRM 19 198 87. [1,2; A2981]

M0909 1, 2, 3, 11, 29, 122, 479, 2113, 9369, 43392, 203595, 975563, 4736005, 23296394, 115811855, 581324861, 2942579633, 15008044522, 77064865555, 398150807179
Dissections of a polygon. Ref AEQ 18 388 78. [3,2; A3455]

M0910 2, 3, 11, 36, 119, 393, 1298, 4287, 14159, 46764, 154451, 510117, 1684802, 5564523, 18378371, 60699636, 200477279, 662131473, 2186871698, 7222746567
$a(n) = 3a(n-1) + a(n-2)$. Ref FQ 15 292 77. [0,1; A6497]

M0911 1, 2, 3, 11, 69, 701, 10584, 222965, 6253604, 225352709, 10147125509, 558317255704, 36859086001973, 2875567025409598, 261713458398275391
$a(n) = n(n-1)a(n-1)/2 + a(n-2)$. [0,2; A1052, N0343]

M0912 1, 1, 2, 3, 12, 10, 60, 105, 280, 252, 2520, 2310, 27720, 25740, 24024, 45045, 720720, 680680, 12252240, 11639628, 11085360, 10581480, 232792560, 223092870
L.c.m. of $C(n,0)$, $C(n,1)$, ..., $C(n,n)$. [0,3; A2944, N0344]

M0913 1, 1, 2, 3, 12, 52, 456, 6873, 191532, 9733032, 903753248, 1541108311046
Nontransitive prime tournaments. Ref DUMJ 37 332 70. [1,3; A2638, N0345]

M0914 1, 1, 2, 3, 14, 129, 25298, 420984147, 269425140741515486,
4774958509020952887348253156297 7121
Convergents to Cahen's constant: $a(n+2) = a(n)^2 a(n+1) + a(n)$. Ref MFM 111 122
91. [0,3; A6279]

M0915 1, 1, 2, 3, 16, 10, 114, 462, 496, 2952, 16920, 31680, 130008, 1707576, 14259504,
138375720, 1652311680, 22238105280, 321916019904, 4959460972224
Expansion of $(1+x)^{1-x}$. [0,3; A7120]

M0916 1, 0, 2, 3, 16, 80, 440, 3171, 24680, 218952, 2170018, 23566675, 279907076,
3603250716, 49968204078, 742893013695, 11785962447792, 198748512229968
Expansion of $(1+x)^{\sin x}$. [0,3; A7118]

M0917 1, 1, 1, 2, 3, 16, 135, 3315, 158830
Pseudolines with a marked cell. Ref JCT A37 257 84. KN91. [1,4; A6247]

M0918 1, 2, 3, 16, 423, 5337932, 198815685282
(5,4)-graphs. Ref PE79. [3,2; A5273]

M0919 1, 0, 2, 3, 20, 90, 594, 4200, 34544, 316008, 3207240, 35699400, 432690312,
5672581200, 79991160144, 1207367605080, 19423062612480, 331770360922560
Expansion of $(1+x)^x$. [0,3; A7113]

M0920 1, 1, 1, 2, 3, 20, 242, 6405, 316835
Primitive sorting networks on n elements. Ref KN91. [1,4; A6246]

M0921 1, 1, 2, 3, 24, 5, 720, 105, 2240, 189, 3628800, 385, 479001600, 19305, 896896,
2027025, 20922789888000, 85085, 6402373705728000, 8729721, 47297536000
n-ϕ-torial: Πk, $(k,n) = 1$. Ref AMM 60 422 53. [1,3; A1783, N0346]

M0922 0, 0, 2, 3, 24, 130, 930, 7413, 66752, 667476, 7342290, 88107415, 1145396472,
16035550518, 240533257874, 3848532125865, 65425046139840, 1177650830516968
Even permutations of length n with no fixed points. Ref AMM 79 394 72. [1,3; A3221]

M0923 1, 2, 3, 26, 13, 1074, 1457, 61802, 7929, 4218722, 6385349, 934344762,
5065189307, 141111736466, 235257551943, 23219206152074, 97011062913167
Sums of logarithmic numbers. Ref TMS 31 78 63. jos. [0,2; A2748, N0347]

M0924 2, 3, 56, 43265728
Invertible Boolean functions of n variables. Ref JSIAM 12 297 64. [1,1; A0656, N0348]

M0925 1, 2, 3, 128, 150, 252, 332, 338, 374, 510, 600, 702, 758, 810, 878, 906, 908, 960,
978, 998, 1020, 1088, 1200, 1208, 1212, 1244, 1260, 1272, 1478, 1530, 1542, 1550, 1590
Not of the form $p + 2^x + 2^y$. Ref jos. [1,2; A6286]

M0926 1, 2, 3, 130, 131, 132, 133, 120, 121, 122, 123, 110, 111, 112, 113, 100, 101, 102,
103, 230, 231, 232, 233, 220, 221, 222, 223, 210, 211, 212, 213, 200, 201, 202, 203, 330
Integers written in base -4. Ref KN1 2 189. [1,2; A7608]

M0927 2, 3, 251, 9843019, 121174811
n consecutive primes in arithmetic progression. [2,1; A6560]

SEQUENCES BEGINNING . . ., 2, 4, . . .

M0928 2, 4, 0, 0, 8, 8, 0, 0, 10, 8, 0, 0, 8, 16, 0, 0, 16, 12, 0, 0, 16, 8, 0, 0, 10, 24, 0, 0, 24, 16, 0, 0, 16, 16, 0, 0, 8, 24, 0, 0, 32, 16, 0, 0, 24, 16, 0, 0, 18, 28, 0, 0, 24, 32, 0, 0, 16, 8, 0
Theta series of b.c.c. lattice w.r.t. long edge. Ref JCP 6532 85. [0,1; A4025]

M0929 2, 4, 0, 2, 10, 32, 38, 140, 496, 1186, 3178, 16792, 82038, 289566
Nonattacking superqueens. Ref GA91 240. [4,1; A7631]

M0930 0, 2, 4, 1, 3, 6, 5, 2, 8, 4, 10, 9, 1, 8, 5, 11, 12, 10, 2, 4, 9, 13, 6, 11, 8, 16, 5, 13, 17, 18, 15, 2, 4, 11, 6, 19, 17, 13, 16, 10, 1, 3, 20, 12, 22, 18, 17, 22, 23, 11, 2, 16, 19, 13, 8
x such that $p = x^2 + 7y^2$. Cf. M0197. Ref CU04 1. L1 55. [7,2; A2344, N0349]

M0931 2, 4, 2, 4, 4, 0, 6, 4, 0, 4, 4, 4, 2, 4, 0, 4, 8, 0, 4, 0, 2, 8, 4, 0, 4, 4, 0, 4, 4, 4, 2, 8, 0, 0, 4, 0, 8, 4, 4, 4, 0, 0, 6, 4, 0, 4, 8, 0, 4, 4, 0, 8, 0, 0, 0, 8, 6, 4, 4, 0, 4, 4, 0, 0, 4, 4, 8, 4
Theta series of square lattice w.r.t. edge. See Fig M3218. Ref SPLAG 106. [0,1; A4020]

M0932 1, 2, 4, 2, 12, 12, 12, 8, 8, 10, 60, 60, 84, 84, 84, 16, 18, 180, 20
Periods for game of Third One Lucky. Ref WW 487. [1,2; A6018]

M0933 0, 2, 4, 2, 24, 38, 44, 278, 336, 718, 3116, 2642, 10296, 33802, 16124, 136762, 354144, 24478, 1721764, 3565918, 1476984, 20783558, 34182196, 35553398
Imaginary part of $(1 + 2i)^n$. Cf. M2880. Ref FQ 15 235 77. [0,2; A6496]

M0934 2, 4, 3, 2, 3, 1, 2, 4, 3, 1, 2, 3, 2, 4, 3, 2, 3, 1, 2, 3, 2, 4, 3, 1, 2, 4, 3, 2, 3, 1, 2, 4, 3, 1, 2, 3, 2, 4, 3, 1, 2, 4, 3, 2, 3, 1, 2, 3, 2, 4, 3, 2, 3, 1, 2, 4, 3, 1, 2, 3, 2, 4, 3, 2, 3, 1, 2, 3, 2
A square-free quaternary sequence. Ref SA81 10. [1,1; A5681]

M0935 1, 2, 4, 3, 6, 7, 8, 16, 18, 25, 32, 11, 64, 31, 128, 10, 256
The game of Sym. Ref WW 441. [0,2; A6016]

M0936 1, 2, 4, 3, 6, 10, 12, 4, 8, 18, 6, 11, 20, 18, 28, 5, 10, 12, 36, 12, 20, 14, 12, 23, 21, 8, 52, 20, 18, 58, 60, 6, 12, 66, 22, 35, 9, 20, 30, 39, 54, 82, 8, 28, 11, 12, 10, 36, 48, 30
Multiplicative order of 2 mod $2n + 1$. Ref MAG 4 266 08. MOD 10 226 61. SIAR 3 296 61. [0,2; A2326, N0350]

M0937 1, 2, 4, 4, 6, 8, 8, 8, 13, 12, 12, 16, 14, 16, 24, 16, 18, 26, 20, 24, 32, 24, 24, 32, 31, 28, 40, 32, 30, 48, 32, 32, 48, 36, 48, 52, 38, 40, 56, 48, 42, 64, 44, 48, 78, 48, 48, 64, 57
Generalized divisor function. Ref PLMS 19 111 19. [0,2; A2131, N0351]

G.f. of Moebius transf.: $(1 + x + x^2) / (1 - x^2)^2$.

M0938 1, 1, 2, 4, 4, 6, 8, 8, 12, 14
Generalized tangent numbers. Ref MOC 21 690 67. [1,3; A0061, N0352]

M0939 1, 1, 2, 4, 4, 6, 16, 16, 30, 88
Related to the enumeration of symmetric tournaments. Ref CMB 13 322 70. [1,3; A2086, N0353]

M0940 2, 4, 4, 8, 6, 4, 12, 8, 8, 12, 8, 8, 14, 16, 4, 16, 16, 8, 20, 8, 8, 20, 20, 16, 18, 8, 12, 24, 16, 12, 20, 24, 8, 28, 16, 8, 32, 20, 16, 16, 18, 20, 24, 24, 16, 24, 24, 8, 40, 20, 12, 40
Theta series of f.c.c. lattice w.r.t. edge. Ref JCP 83 6526 85. [0,1; A5884]

M0941 1, 2, 4, 4, 9, 6, 12, 12, 17, 10, 28, 12, 25, 30, 32, 16, 45, 18, 52, 44, 41, 22, 76, 40, 49, 54, 76, 28, 105, 30, 80, 72, 65, 82, 132, 36, 73, 86, 140, 1, 4, 8, 20, 21, 56, 60, 96, 105
Σ gcd $\{k, n-k\}$, $k = 1 \ .. \ n-1$. Ref mlb. [2,2; A6579]

M0942 2, 4, 5, 6, 4, 3, 4, 4, 4, 3, 4, 5, 6, 8, 6, 5, 8, 8, 8, 5, 11, 10, 11, 12, 10, 9, 10, 10, 10, 6, 12, 11, 12, 13, 11, 10, 11, 11, 11, 8, 14, 13, 14, 15, 13, 12, 13, 13, 13, 9, 15, 14, 15, 16
Number of letters in n (in French). [1,1; A7005]

M0943 0, 1, 2, 4, 5, 6, 7, 8, 10, 11, 12, 13, 15, 16, 17, 18, 20, 21, 22, 23, 25, 26, 27, 28, 29, 31, 32, 33, 34, 36, 37, 38, 39, 41, 42, 43, 44, 46, 47, 48, 49, 50, 52, 53, 54, 55, 57, 58, 59
Wythoff game. Ref CMB 2 189 59. [0,3; A1963, N0354]

M0944 1, 2, 4, 5, 6, 7, 9, 10, 12, 13, 14, 15, 17, 18, 20, 22, 23, 25, 26, 27, 28, 30, 31, 33, 34, 35, 36, 38, 39, 40, 41, 43, 44, 46, 47, 48, 49, 51, 52, 54, 56, 57, 59, 60, 61, 62, 64, 65
Related to Fibonacci representations. Ref FQ 11 385 73. [1,2; A3233]

M0945 0, 1, 2, 4, 5, 6, 7, 10, 11, 12, 13, 15, 16, 17, 18, 22, 23, 24, 25, 27, 28, 29, 30, 33, 34, 35, 36, 38, 39, 40, 41, 46, 47, 48, 49, 51, 52, 53, 54, 57, 58, 59, 60, 62, 63, 64, 65, 69
Longest chain of subgroups in S_n. Ref CALG 14 1730 86. pjc. [1,3; A7238]

$\lceil 3n/2 \rceil - b(n) - 1$, where $b(n) = \#1$'s in binary expansion of n.

M0946 1, 2, 4, 5, 6, 8, 9, 10, 12, 13, 15, 16, 17, 19, 20, 21, 23, 24, 25, 27, 28, 30, 31, 32, 34, 35, 36, 38, 39, 40, 42, 43, 45, 46, 47, 49, 50, 51, 53, 54, 56, 57, 58, 60, 61, 62, 64, 65
A Beatty sequence: $[n(1+\sqrt{3})/2]$. See Fig M1332. Cf. M2622. Ref DM 2 338 72. [1,2; A3511]

M0947 1, 2, 4, 5, 6, 8, 9, 10, 12, 13, 15, 16, 17, 19, 20, 21, 23, 24, 25, 27, 28, 30, 31, 32, 34, 35, 36, 38, 39, 41, 42, 43, 45, 46, 47, 49, 50, 51, 53, 54, 56, 57, 58, 60, 61, 62, 64, 65
A Beatty sequence: $[n(1+1/e)]$. See Fig M1332. Cf. M2621. Ref CMB 3 21 60. [1,2; A6594]

M0948 1, 2, 4, 5, 6, 8, 9, 11, 12, 13, 15, 16, 18, 19, 20, 22, 23, 25, 26, 27, 29, 30, 32, 33, 34, 36, 37, 38, 40, 41, 43, 44, 45, 47, 48, 50, 51, 52, 54, 55, 57, 58, 59, 61, 62, 64, 65, 66
A Beatty sequence: $[n/(e-2)]$. Ref CMB 3 21 60. [1,2; A0062, N0355]

M0949 1, 2, 4, 5, 6, 8, 9, 11, 13, 14, 15, 17, 18, 20, 22, 23, 24, 26, 28, 29, 31, 32, 34, 35, 36
A density problem involving linear forms. Ref ZA77 108. [0,2; A4059]

M0950 1, 2, 4, 5, 6, 9, 10, 11, 12, 14, 15, 16, 19, 20, 22, 23, 24, 25, 26, 27, 28, 29, 33, 34, 35, 36, 37, 39, 40, 42
Størmer numbers. Ref AMM 56 518 49. TO51 2. CoGu95. [1,2; A5528]

M0951 1, 2, 4, 5, 6, 9, 16, 17, 30, 54, 57, 60, 65, 132, 180, 320, 696, 782, 822, 897, 1252, 1454
$2.3^n + 1$ is prime. Ref MOC 26 996 72. Cald94. [1,2; A3306]

M0952 1, 2, 4, 5, 6, 12, 18, 20, 30, 46, 60, 62, 72, 89, 105, 113, 117, 119, 120, 241, 483, 633, 647, 654, 1074, 1752, 1806, 3050, 3609, 3611, 3612, 5459, 5460, 7976, 7999, 8005
Numerators of worst case for Engel expansion. Ref STNB 3 52 91. [0,2; A6539]

M0953 1, 2, 4, 5, 7, 3, 0, 9, 3, 9, 6, 1, 5, 5, 1, 7, 3, 2, 5, 9, 6, 6, 6, 8, 0, 3, 3, 6, 6, 4, 0, 3, 0, 5, 0, 8, 0, 9, 3, 9, 3, 0, 9, 9, 9, 3, 0, 6, 8, 7, 7, 9, 8, 1, 1, 0, 4, 6, 1, 7, 3, 0, 1, 4, 3, 6, 0, 7, 4
Decimal expansion of fifth root of 3. [1,2; A5532]

M0954 1, 2, 4, 5, 7, 8, 9, 10, 11, 13, 14, 16, 17, 18, 19, 20, 22, 23, 25, 26, 28, 29, 31, 32, 34, 35, 36, 37, 38, 40, 41, 43, 44, 45, 46, 47, 49, 50, 52, 53, 55, 56, 58, 59, 61, 62, 63, 64
If n appears, $3n$ doesn't. [1,2; A7417]

M0955 1, 2, 4, 5, 7, 8, 9, 11, 12, 14, 15, 16, 18, 19, 21, 22, 24, 25, 26, 28, 29, 31, 32, 33, 35, 36, 38, 39, 41, 42, 43, 45, 46, 48, 49, 50, 52, 53, 55, 56, 57, 59, 60, 62, 63, 65, 66, 67
A Beatty sequence: $[n.\sqrt{2}]$. Cf. M2534. Ref CMB 2 188 59. FQ 10 487 72. GKP 77. [1,2; A1951, N0356]

M0956 0, 1, 2, 4, 5, 7, 8, 9, 14, 15, 16, 18, 25, 26, 28, 29, 30, 33, 36, 48, 49, 50, 52, 53, 55, 56, 57, 62
No 4-term arithmetic progression. Ref UPNT E10. [0,3; A5839]

M0957 1, 2, 4, 5, 7, 8, 10, 11, 13, 14, 16, 17, 19, 20, 22, 23, 25, 26, 28, 29, 31, 32, 34, 35, 37, 38, 40, 41, 43, 44, 46, 47, 49, 50, 52, 53, 55, 56, 58, 59, 61, 62, 64, 65, 67, 68, 70, 71
$a(n) = a(n-1) + a(n-2) - a(n-3)$. Ref FQ 6(3) 261 68. [0,2; A1651, N0357]

M0958 1, 2, 4, 5, 7, 8, 10, 11, 13, 14, 16, 17, 19, 20, 22, 24, 25, 27, 28, 30, 31, 33, 34, 36, 37, 39, 40, 42, 43, 45, 46, 48, 49, 51, 52, 54, 55, 57, 58, 60, 62, 63, 65, 66, 68, 69, 71, 72
$a(n) = (4-n).a(n-1) + 2.a(n-2) + (n-3).a(n-3)$. Ref FQ 11 386 73. [1,2; A3253]

M0959 1, 2, 4, 5, 7, 8, 10, 13, 15, 16, 20, 23, 25, 28, 31, 32, 37, 39, 40, 47, 52, 55, 58, 60, 63, 64, 71, 79, 80, 85, 87, 92, 95, 100, 103, 111, 112, 119, 124, 127, 128, 130, 135, 143
Not the sum of three nonzero squares. [1,2; A4214]

M0960 2, 4, 5, 7, 8, 10, 25, 53, 62, 134, 574, 2431, 13147, 27167, 229073, 315416, 435474, 771789, 1522716, 3853889, 7878986, 7922488, 8844776, 9182596, 9388467
Modified Engel expansion of 3/7. Ref FQ 31 37 93. [1,1; A6693]

M0961 2, 4, 5, 7, 8, 11, 13, 16, 17, 19, 31, 37, 41, 47, 53, 61, 71, 79, 113, 313, 353, 503, 613, 617, 863
Prime Lucas numbers. Ref MOC 50 251 88. [1,1; A1606, N0358]

M0962 1, 2, 4, 5, 7, 9, 10, 12, 14, 16, 17, 19, 21, 23, 25, 26, 28, 30, 32, 34, 36, 37, 39, 41, 43, 45, 47, 49, 50, 52, 54, 56, 58, 60, 62, 64, 65, 67, 69, 71, 73, 75, 77, 79, 81, 82, 84, 86
Connell sequence: 1 odd, 2 even, 3 odd, ... Ref AMM 67 380 60. Pick91 276. [1,2; A1614, N0359]

$$a(n) = 2n - [(1 + \sqrt{(8n-7)}) / 2].$$

M0963 0, 1, 2, 4, 5, 7, 9, 11, 12, 15, 16, 18, 20, 22, 24, 26
Rotation distance between trees. Ref CoGo87 135. JAMS 1 654 88. [0,3; A5152]

M0964 0, 1, 2, 4, 5, 7, 9, 12, 13, 15, 17, 20, 22, 25, 28, 32, 33, 35, 37, 40, 42, 45, 48, 52, 54, 57, 60, 64, 67, 71, 75, 80, 81, 83, 85, 88, 90, 93, 96, 100, 102, 105, 108, 112, 115, 119
1's in binary expansion of 0, ..., n. Ref SIAR 4 21 62. CMB 8 481 65. ANY 175 177 70. SIAC 18 1189 89. [0,3; A0788, N0360]

M0965 2, 4, 5, 7, 10, 12, 13, 15, 18, 20, 23, 25, 26, 28, 31, 33, 34, 36, 38, 39, 41, 44, 46, 47, 49, 52, 54, 57, 59, 60, 62, 65, 67, 68, 70, 72, 73, 75, 78, 80, 81, 83, 86, 88, 89, 91, 93
Sum of 2 terms is never a Fibonacci number. Complement of M2517. Ref DM 22 202 78. [1,1; A5653]

M0966 1, 2, 4, 5, 7, 12, 14, 15, 23
From reversals of n-tuples of natural numbers. Ref AMM 96 57 89. [1,2; A7062]

M0967 1, 2, 4, 5, 7, 17, 25, 40, 63, 99, 156, 249, 397
Spiral sieve using Fibonacci numbers. Ref FQ 12 395 74. [1,2; A5620]

M0968 0, 1, 2, 4, 5, 8, 9, 10, 13, 16, 17, 18, 20, 25, 26, 29, 32, 34, 36, 37, 40, 41, 45, 49, 50, 52, 53, 58, 61, 64, 65, 68, 72, 73, 74, 80, 81, 82, 85, 89, 90, 97, 98, 100, 101, 104, 106
Sums of 2 squares. See Fig M3218. Ref EUL (1) 1 417 11. KNAW 53 872 50. SPLAG 106. [0,3; A1481, N0361]

M0969 1, 2, 4, 5, 8, 9, 10, 14, 15, 16, 17, 18, 20, 26, 27, 28, 29, 30, 32, 33, 34, 36, 40, 44, 47, 50, 51, 52, 53, 54, 56, 57, 58, 60, 62, 63, 64, 66, 68, 72, 80, 83, 86, 87, 88, 89, 92, 93
If n appears so do $2n$, $3n+2$, $6n+3$. Ref AMM 90 40 83. [1,2; A5658]

M0970 1, 2, 4, 5, 8, 9, 12, 14, 17, 18, 23, 24, 27, 30, 34, 35, 40, 41, 46, 49, 52, 53, 60, 62, 65, 68, 73, 74, 81, 82, 87, 90, 93, 96, 104, 105, 108, 111, 118, 119, 126, 127, 132, 137
$\Sigma [(n-k)/k]$, $k = 1 . . 6n-1$. Ref DVSS 2 281 1884. [2,2; A2541, N0362]

M0971 1, 2, 4, 5, 8, 10, 12, 14, 15, 16, 19, 20, 21, 24, 25, 27, 28, 32, 33, 34, 37, 38, 40, 42, 43, 44, 46, 47, 48, 51, 53, 54, 56, 57, 58, 59, 61
A self-generating sequence. Ref UPNT E30. [1,2; A5242]

M0972   1, 2, 4, 5, 8, 10, 14, 15, 16, 21, 22, 25, 26, 28, 33, 34, 35, 36, 38, 40, 42, 46, 48, 49, 50, 53, 57, 60, 62, 64, 65, 70, 77, 80, 81, 83, 85, 86, 90, 91, 92, 100, 104, 107, 108, 116
Segmented numbers, or prime numbers of measurement. See Fig M0557. Cf. M2633. Ref AMM 75 80 68; 82 922 75. UPNT E30. [1,2; A2048, N0363]

M0973   2, 4, 5, 8, 12, 7, 24, 16, 10, 36, 13, 20, 44, 15, 56, 32, 18, 68, 21, 48, 76, 23, 88, 28, 26, 100, 29, 96, 108, 31, 120, 64, 34, 132, 37, 40, 140, 39, 152, 144, 42, 164, 45, 52, 172
Earliest sequence with $a(a(n)) = 4n$. Ref clm. [1,1; A7379]

M0974   2, 4, 5, 8, 12, 19, 30, 48, 77, 124, 200, 323, 522, 844, 1365, 2208, 3572, 5779, 9350, 15128, 24477, 39604, 64080, 103683, 167762, 271444, 439205, 710648, 1149852
$a(n) = a(n-1) + a(n-2) - 1$. Ref JA66 97. [0,1; A1612, N0364]

M0975   1, 2, 4, 5, 10, 11, 13, 14, 28, 29, 31, 32, 37, 38, 40, 41, 82, 83, 85, 86, 91, 92, 94, 95, 109, 110, 112, 113, 118, 119, 121, 122, 244, 245, 247, 248, 253, 254, 256, 257, 271
$a(n) - 1$ in ternary $= n - 1$ in binary. Ref JLMS 11 263 36. [1,2; A3278]

M0976   1, 2, 4, 5, 10, 14, 17, 31, 41, 73, 80, 82, 116, 125, 145, 157, 172, 202, 224, 266, 289, 293, 463, 1004
$15.2^n - 1$ is prime. Ref MOC 23 874 69. Rie85 384. [1,2; A2237, N0365]

M0977   1, 0, 1, 1, 1, 2, 4, 5, 10, 19, 36, 68, 138, 277, 581, 1218, 2591, 5545, 12026, 26226, 57719, 127685, 284109, 634919, 1425516, 3212890, 7269605, 16504439, 37592604
Centered boron trees with n nodes. Ref CAY 9 451. rcr. [1,6; A0675, N0366]

M0978   1, 1, 2, 4, 5, 14, 14, 39, 42, 132, 132, 424, 429, 1428, 1430, 4848, 4862, 16796, 16796, 58739, 58786, 208012, 208012, 742768, 742900, 2674426, 2674440, 9694416
Symmetrical dissections of an n-gon. Ref GU58. [5,3; A0063, N0367]

M0979   1, 2, 4, 5, 14, 24, 29, 36, 46, 80
$2.7^n - 1$ is prime. Ref PLC 2 569 71. [1,2; A2959]

M0980   0, 0, 1, 2, 4, 6, 3, 10, 25, 12, 42, 8, 40, 202, 21
Solutions of $x + y = z$ from $\{1, 2, ..., n\}$. Ref GU71. [1,4; A2849, N0368]

M0981   1, 2, 4, 6, 7, 10, 11, 12, 22, 23, 25, 26, 27, 30, 36, 38, 42, 43, 44, 45, 50, 52, 54, 58, 59, 70, 71, 72, 74, 75, 76, 78, 86, 87, 91, 102, 103, 106, 107, 108, 110, 116, 118, 119, 122
Elliptic curves. Ref JRAM 212 25 63. [1,2; A2158, N0369]

M0982   1, 1, 2, 4, 6, 7, 10, 12, 16, 19
Maximal number of 3-tree rows in n-tree orchard problem. See Fig M0982. Ref GR72 22. GMD 2 399 74. [3,3; A3035]

Figure M0982. ORCHARD PROBLEM.

> "Your aid I want, Nine trees to plant,
> In rows just half a score, And let there be,
> In each row, three — Solve this: I ask no more!"

J. Jackson, Rational Amusement for Winter Evenings,
Longman, London, 1821.

In other words, plant 9 trees so that there are 10 rows of 3. M0982 gives the maximal number of rows possible with n trees, for $n \leq 12$. The next term is conjectured to be 19 — see [GMD 2 397 74]. Here are the solutions for 7, 8, 9 and 11 trees. (Much less is known about the 4-trees-in-a-row problem, M0290.)

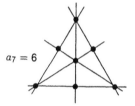

$a_7 = 6$

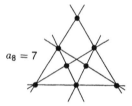

$a_8 = 7$

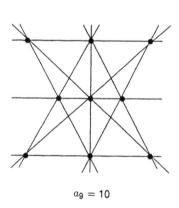

$a_9 = 10$

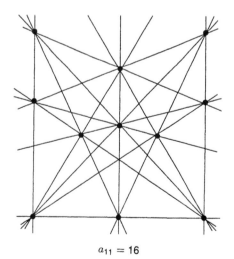

$a_{11} = 16$

M0983   0, 2, 4, 6, 8, 1, 3, 5, 7, 9, 11, 31, 33, 53, 55, 75, 77, 97, 99, 101, 301, 303, 503, 505,
705, 707, 907, 909, 119, 121, 321, 323, 523, 525, 725, 727, 927, 929, 139, 141, 341, 343
Add 2, then reverse digits! Ref Robe92 15. [0,2; A7396]

M0984   2, 4, 6, 8, 9, 10, 12, 14, 15, 17, 19, 21, 23, 24, 25, 27, 29, 31, 33, 34, 35, 37, 39, 40,
42
Related to Fibonacci representations. Ref FQ 11 385 73. [1,1; A3254]

M0985   2, 4, 6, 8, 10, 12, 14, 16, 18, 20, 22, 24, 26, 28, 30, 32, 34, 36, 38, 40, 42, 44, 46,
48, 50, 52, 54, 56, 58, 60, 62, 64, 66, 68, 70, 72, 74, 76, 78, 80, 82, 84, 86, 88, 90, 92, 94
The even numbers. [1,1; A5843]

M0986   1, 2, 4, 6, 8, 10, 12, 16, 18, 20, 22, 24, 28, 30, 32, 36, 40, 42, 44, 46, 48, 52, 54, 56,
58, 60, 64, 66, 70, 72, 78, 80, 82, 84, 88, 90, 92, 96, 100, 102, 104, 106, 108, 110, 112
Values taken by reduced totient function $\psi(n)$. Cf. M0298. Ref NADM 17 305 1898. L1 7.
[1,2; A2174, N0370]

M0987   1, 2, 4, 6, 8, 10, 12, 16, 18, 20, 22, 24, 28, 30, 32, 36, 40, 42, 44, 46, 48, 52, 54, 56,
58, 60, 64, 66, 70, 72, 78, 80, 82, 84, 88, 92, 96, 100, 102, 104, 106, 108, 110, 112, 116
Values of totient function. See Fig M0500. Cf. M0299. Ref BA8 64. [1,2; A2202, N0371]

M0988   1, 2, 4, 6, 8, 10, 14, 15, 18, 22, 24, 27, 31, 33, 37, 40, 44, 47, 51, 53, 57, 63, 65, 68,
73, 75, 81, 85, 87, 91, 97, 100, 104, 108, 112, 115, 121, 125, 129, 134, 136, 142, 146, 148
Sum of nearest integer to $(n-k)/k$, $k = 1 \,.\,.\, n-1$. Ref mlb. [2,2; A6586]

M0989   0, 1, 2, 4, 6, 8, 11, 13
Least number of edges in graph containing all trees on n nodes. Ref rlg. [1,3; A4401]

M0990   0, 1, 2, 4, 6, 8, 12, 14, 16, 24, 26, 28, 32, 40, 48, 52, 54, 56, 64, 72, 80, 96, 100,
104, 108, 110, 112, 128, 136, 144, 160, 176, 192, 200, 204, 208, 216, 218, 220, 224, 240
Partitioning integers to avoid arithmetic progressions of length 3. Ref PAMS 102 771 88.
[0,3; A6998]

$$a(n) = a([2n/3]) + a([(2n+1)/3]).$$

M0991   1, 2, 4, 6, 8, 12, 16, 18, 20, 24, 28, 30, 32, 36, 40, 42, 48, 54, 56, 60, 64, 66, 72, 78,
80, 84, 88, 90, 96, 100, 104, 108, 112, 120, 126, 128, 132, 140, 144, 150, 156, 160, 162
Practical numbers (first definition): all $k < n$ are sums of proper divisors of n. Ref HO73
113. [1,2; A5153]

M0992   1, 2, 4, 6, 8, 12, 16, 18, 24, 32, 36, 48, 54, 64, 72, 96, 108, 128, 144, 162, 192, 216,
256, 288, 324, 384, 432, 486, 512, 576, 648, 768, 864, 972, 1024, 1152, 1296, 1458, 1536
$\phi(n)$ divides n. Ref rgw. [1,2; A7694]

M0993   1, 2, 4, 6, 8, 12, 16, 24, 32, 36, 48, 64, 72, 96, 120, 128, 144, 192, 216, 240, 256,
288, 384, 432, 480, 512, 576, 720, 768, 864, 960, 1024, 1152, 1296, 1440, 1536, 1728
Jordan-Pólya numbers. Ref JCT 5 25 68. [1,2; A1013, N0372]

M0994 1, 2, 4, 6, 8, 14, 18, 20, 22, 34, 36, 44, 52, 72, 86, 96, 112, 114, 118, 132, 148, 154, 180, 210, 220, 222, 234, 248, 250, 282, 288, 292, 320, 336, 354, 382, 384, 394, 456, 464
Increasing gaps between primes. Cf. M0858. Ref MOC 52 222 89. UPNT A8. [1,2; A5250]

M0995 2, 4, 6, 8, 14, 26
Snake-in-the-box problem in n-dimensional cube. Ref AMM 77 63 70. [1,1; A0937, N0373]

M0996 1, 2, 4, 6, 9, 10, 13, 15, 19, 19, 21, 23, 27, 28, 31, 34, 39, 38, 39, 40, 43, 44, 47, 50
Bipartite Steinhaus graphs on n nodes. Ref DyWh94. [1,2; A3661]

M0997 2, 4, 6, 9, 12, 15, 19, 23, 27, 31, 35, 40, 45, 50, 55, 60, 65, 70, 75, 80, 86
Problimes (first definition). Ref AMM 80 677 73. [1,1; A3066]

M0998 0, 0, 1, 2, 4, 6, 9, 12, 16, 20, 25, 30, 36, 42, 49, 56, 64, 72, 81, 90, 100, 110, 121, 132, 144, 156, 169, 182, 196, 210, 225, 240, 256, 272, 289, 306, 324, 342, 361, 380, 400
Quarter-squares: $[n/2] . \lceil n/2 \rceil$. Ref AMS 26 304 55. GKP 99. [0,4; A2620, N0374]

M0999 1, 2, 4, 6, 9, 12, 18, 26, 41, 62, 96, 142, 212, 308, 454, 662, 979, 1438, 2128, 3126, 4606, 6748, 9910, 14510, 21298, 31212, 45820, 67176, 98571, 144476
Twopins positions. Ref GU81. [8,2; A5690]

M1000 1, 2, 4, 6, 9, 12, 20, 24, 30, 35, 44, 50
Consistent sets in tournaments. Ref BW78 195. [2,2; A5779]

M1001 1, 2, 4, 6, 9, 13, 17, 22, 28, 35, 43, 51, 60, 70, 81, 93, 106, 120, 135, 151, 167, 184, 202, 221, 241, 262, 284, 307, 331, 356, 382, 409, 437, 466, 496, 527, 559, 591, 624, 658
Subwords of length n in word generated by $a \rightarrow aab$, $b \rightarrow b$. Ref jos. [0,2; A6697]

M1002 1, 2, 4, 6, 9, 13, 18, 24, 31, 39, 50, 62, 77, 93, 112, 134, 159, 187, 218, 252, 292, 335, 384, 436, 494, 558, 628, 704, 786, 874, 972, 1076, 1190, 1310, 1440, 1580, 1730
Denumerants. Ref R1 152. [0,2; A0064, N0375]

$$\text{G.f.:} \quad 1 / (1 - x)^2 (1 - x^2) (1 - x^5) (1 - x^{10}).$$

M1003 1, 1, 2, 4, 6, 9, 14, 19, 27, 37, 49, 64, 84, 106, 134, 168, 207, 253, 309, 371, 445, 530, 626, 736, 863, 1003, 1163, 1343, 1543, 1766, 2017, 2291, 2597, 2935, 3305, 3712
Certain triangular arrays of integers. Ref P4BC 112. [0,3; A3402]

M1004 1, 2, 4, 6, 9, 14, 22, 36, 57, 90, 139, 214, 329, 506, 780, 1200, 1845, 2830, 4337, 6642, 10170, 15572, 23838, 36486, 55828, 85408, 130641, 199814, 305599
Twopins positions. Ref GU81. [7,2; A5687]

M1005 1, 2, 4, 6, 9, 15, 25, 40, 64, 104, 169, 273, 441, 714, 1156, 1870, 3025, 4895, 7921, 12816, 20736, 33552, 54289, 87841, 142129, 229970, 372100, 602070, 974169, 1576239
$a(n) = a(n-1)+a(n-3)+a(n-4)$. Ref FQ 16 113 78. [0,2; A6498]

M1006 1, 2, 4, 6, 10, 12, 16, 18, 22, 28, 30, 36, 40, 42, 46, 52, 58, 60, 66, 70, 72, 78, 82, 88, 96, 100, 102, 106, 108, 112, 126, 130, 136, 138, 148, 150, 156, 162, 166, 172, 178
Primes minus 1. Ref EUR 40 28 79. [1,2; A6093]

M1007 1, 2, 4, 6, 10, 12, 16, 20, 22, 24, 28, 32, 36, 40, 42, 44, 46, 48, 52, 56, 60, 64, 68, 72, 76, 80, 82, 84, 86, 88, 90, 92, 94, 96, 100, 104, 108, 112, 116, 120, 124, 128, 132, 136
$a(2n) = a(n) + a(n+1)$, $a(2n+1) = 2a(n+1)$. Ref DAM 24 93 89. [0,2; A5942]

M1008 0, 1, 2, 4, 6, 10, 12, 18, 22, 28, 32, 42, 46, 58, 64, 72, 80, 96, 102, 120, 128, 140, 150, 172, 180, 200, 212, 230, 242, 270, 278, 308, 324, 344, 360, 384, 396, 432, 450, 474
Sum of totient function. Cf. M0299. Ref SYL 4 103. L1 7. GKP 138. [0,3; A2088, N0376]

M1009 1, 2, 4, 6, 10, 12, 18, 22, 30, 34, 42, 48, 58, 60, 78, 82, 102, 108, 118, 132, 150, 154, 174, 192, 210, 214, 240, 258, 274, 282, 322, 330, 360, 372, 402, 418, 442, 454, 498
Generated by a sieve. Ref RLM 11 27 57. ADM 37 51 88. [1,2; A2491, N0377]

M1010 1, 2, 4, 6, 10, 14, 16, 20, 24, 26, 36, 40, 54, 56, 66, 74, 84, 90, 94, 110, 116, 120, 124, 126, 130, 134, 146, 150, 156, 160, 170, 176, 180, 184, 204, 206, 210, 224, 230, 236
$n^2 + 1$ is prime. [1,2; A5574]

M1011 1, 2, 4, 6, 10, 14, 20, 26, 36, 46, 60, 74, 94, 114, 140, 166, 202, 238, 284, 330, 390, 450, 524, 598, 692, 786, 900, 1014, 1154, 1294, 1460, 1626, 1828, 2030, 2268, 2506
Binary partitions (partitions of $2n$ into powers of 2): $a(n) = a(n-1) + a([n/2])$. Ref FQ 4 117 66. PCPS 66 376 69. AB71 400. BIT 17 387 77. [0,2; A0123, N0378]

$$\text{G.f.: } (1-x)^{-1} \prod_{n=0}^{\infty} (1 - x^{2\uparrow n})^{-1}.$$

M1012 1, 2, 4, 6, 10, 14, 21, 29, 41, 55, 76, 100, 134, 175, 230, 296, 384, 489, 626, 791, 1001, 1254, 1574, 1957, 2435, 3009, 3717, 4564, 5603, 6841, 8348, 10142, 12309, 14882
$-1 +$ number of partitions of n. Ref IBMJ 4 475 60. KU64. AS1 836. [2,2; A0065, N0379]

M1013 2, 4, 6, 10, 14, 24, 30, 58, 70
Minimal trivalent graph of girth n. Ref FI64 94. bdm. JCT B29 91 80. [2,1; A0066, N0380]

M1014 1, 2, 4, 6, 10, 16, 25, 38, 57, 80, 113, 156, 210, 278, 362, 462, 586, 732, 904, 1106, 1344, 1616, 1931, 2288, 2690, 3150, 3671, 4248, 4896, 5612, 6407, 7290, 8267, 9332
Taylor series from Ramanujan's Lost Notebook. Ref LNM 899 44 81. [0,2; A6305]

M1015 2, 4, 6, 10, 16, 26, 44, 76, 132, 234, 420, 761, 1391, 2561, 4745, 8841, 16551, 31114, 58708, 111136, 211000, 401650, 766372, 1465422, 2807599, 5388709, 10359735
$\Sigma[2^k / k]$, $k = 1 .. n$. [1,1; A0801, N0381]

M1016 1, 2, 4, 6, 10, 18, 33, 60, 111, 205, 385, 725, 1374, 2610, 4993, 9578, 18426, 35568, 68806, 133411, 259145, 504222, 982538, 1917190, 3745385, 7324822
Integers $\leq 2^n$ of form $x^2 + 2y^2$. Ref MOC 20 560 66. [0,2; A0067, N0382]

M1017   1, 2, 4, 6, 10, 22, 38, 102, 182
Balanced symmetric graphs. Ref DM 15 384 76. [1,2; A5194]

M1018   0, 1, 2, 4, 6, 11, 18, 31, 54, 97, 172, 309, 564, 1028, 1900, 3512, 6542, 12251, 23000, 43390, 82025, 155611, 295947, 564163, 1077871, 2063689, 3957809, 7603553
Number of primes $\leq 2^n$. Ref rgw. [0,3; A7053]

M1019   1, 2, 4, 6, 11, 18, 32, 52, 88, 142, 236, 382, 629, 1018, 1664, 2692, 4383, 7092, 11520, 18640, 30232, 48916, 79264, 128252, 207705, 336074, 544084
Twopins positions. Ref GU81. [6,2; A5684]

M1020   1, 1, 2, 4, 6, 11, 19, 33, 55, 95, 158, 267, 442, 731, 1193, 1947
Planar partitions of n. Ref MA15 2 332. [1,3; A0786, N0383]

M1021   1, 2, 4, 6, 11, 19, 34, 63, 117, 218, 411, 780, 1487, 2849, 5477, 10555, 20419, 39563, 76805, 149360, 290896, 567321, 1107775, 2165487, 4237384, 8299283
Related to population of numbers of form $x^2 + y^2$. Ref MOC 18 84 64. [1,2; A0694, N0384]

M1022   1, 2, 4, 6, 12, 16, 24, 36, 48, 60, 64, 120, 144, 180, 192, 240, 360, 576, 720, 840, 900, 960, 1024, 1260, 1296, 1680, 2520, 2880, 3072, 3600, 4096, 5040, 5184, 6300, 6480
Minimal numbers: n is smallest number with this number of divisors. Cf. M1026. Ref AMM 75 725 68. Robe92 86. [1,2; A7416]

M1023   1, 0, 1, 2, 4, 6, 12, 18, 32, 50, 88, 134, 232, 364, 604, 966, 1596, 2544, 4180, 6708, 10932, 17622, 28656, 46206, 75020, 121160, 196384, 317432, 514228, 831374, 1346268
Moebius transform of Fibonacci numbers. Ref EIS § 2.7. [1,4; A7436]

M1024   1, 0, 2, 4, 6, 12, 22, 36, 62, 104, 166, 268, 426, 660, 1022, 1564, 2358, 3540, 5266, 7756, 11362, 16524, 23854, 34252, 48890, 69368, 97942, 137588, 192314, 267628
Representation degeneracies for Raymond strings. Ref NUPH B274 544 86. [0,3; A5303]

$$\text{G.f.:}\quad \Pi\,(1 - x^k)^{-c(k)}, \quad c(k) = 0, 2, 4, 3, 4, 2, 4, 2, 4, 2, \dots.$$

M1025   2, 4, 6, 12, 24, 36, 48, 60, 120, 180, 240, 360, 720, 840, 1260, 1680, 2520, 5040, 7560, 10080, 15120, 20160, 25200, 27720, 45360, 50400, 55440, 83160, 110880, 166320
Highly composite (or super abundant) numbers: where $d(n)$ increases. Ref RAM1 87. TAMS 56 468 44. HO73 112. Well86 128. [1,1; A2182, N0385]

M1026   1, 2, 4, 6, 16, 12, 64, 24, 36, 48, 1024, 60, 4096, 192, 144, 120, 65536, 180, 262144, 240, 576, 3072, 4194304, 360, 1296, 12288, 900, 960, 268435456, 720
Smallest number with n divisors. Ref AS1 840. [1,2; A5179]

M1027   1, 2, 4, 6, 16, 20, 24, 28, 34, 46, 48, 54, 56, 74, 80, 82, 88, 90, 106, 118, 132, 140, 142, 154, 160, 164, 174, 180, 194, 198, 204, 210, 220, 228, 238, 242, 248, 254, 266, 272
$n^4 + 1$ is prime. Ref MOC 21 246 67. [1,2; A0068, N0386]

M1028 2, 4, 6, 16, 20, 36, 54, 60, 96, 124, 150, 252, 356, 460, 612, 654, 664, 698, 702, 972
$17.2^n - 1$ is prime. Ref MOC 22 421 68. Rie85 384. [1,1; A1774, N0387]

M1029 1, 1, 2, 4, 6, 19, 20, 107, 116, 567, 640
Atomic species of degree n. Ref JCT A50 279 89. [1,3; A5227]

M1030 2, 4, 6, 30, 32, 34, 36, 40, 42, 44, 46, 50, 52, 54, 56, 60, 62, 64, 66, 2000, 2002,
    2004, 2006, 2030, 2032, 2034, 2036, 2040, 2042, 2044, 2046, 2050, 2052, 2054, 2056
The 'eban' numbers (the letter 'e' is banned!). See Fig M2629. [1,1; A6933]

M1031 1, 2, 4, 7, 8, 11, 13, 14, 16, 19, 21, 22, 25, 26, 28, 31, 32, 35, 37, 38, 41, 42, 44, 47,
    49, 50, 52, 55, 56, 59, 61, 62, 64, 67, 69, 70, 73, 74, 76, 79, 81, 82, 84, 87, 88, 91, 93, 94
Odious numbers: odd number of 1's in binary expansion. Ref CMB 2 86 59. Robe92 22.
[0,2; A0069, N0388]

M1032 1, 2, 4, 7, 8, 12, 13, 17, 20, 26, 28, 35, 37, 44, 48, 57, 60, 70, 73, 83, 88, 100, 104,
    117, 121, 134, 140, 155, 160, 176, 181, 197, 204, 222, 228, 247, 253, 272, 280, 301, 308
The coding-theoretic function $A(n,4,3)$. See Fig M0240. Ref PGIT 36 1335 90. [4,2;
A1839, N0389]

M1033 2, 4, 7, 9, 12, 14, 16, 19, 21, 24, 26, 28, 31, 33, 36, 38, 41, 43, 45, 48
A Beatty sequence. Ref FQ 10 487 72. [1,1; A3151]

M1034 1, 2, 4, 7, 9, 17, 25, 46, 51, 83, 158, 233, 365
Spiral sieve using Fibonacci numbers. Ref FQ 12 395 74. [1,2; A5625]

M1035 1, 2, 4, 7, 10, 12, 16, 17, 32, 36, 42, 57, 72, 73, 98, 102, 104, 129, 159, 164, 174,
    189, 199, 221, 224, 255, 286, 287, 347, 372, 378, 403, 428, 443, 444, 469, 494, 529, 560
Next term is uniquely the sum of 3 earlier terms. Ref AB71 249. [1,2; A7087]

M1036 1, 2, 4, 7, 10, 12, 18, 40, 44, 45, 1850, 11604, 11616, 11617, 2132568, 17001726,
    17001743, 17001744, 6660587898, 64431061179, 64431061180, 64431061181
Earliest monotonic sequence fixed under reversion. Ref BeSl94. [1,2; A7303]

M1037 2, 4, 7, 10, 13, 17, 21, 25, 29, 34, 39, 44, 49, 54, 59, 64, 69, 74, 79, 84, 90
Problimes (second definition). Ref AMM 80 677 73. [1,1; A3067]

M1038 2, 4, 7, 11, 15, 19, 23, 28, 33, 38, 43, 48, 53, 58, 63, 68, 73, 79, 85, 91, 97
Problimes (third definition). Ref AMM 80 677 73. [1,1; A3068]

M1039 2, 4, 7, 11, 16, 21, 28, 35
Integral points in a quadrilateral. Ref JRAM 226 22 67. [1,1; A2789, N0390]

M1040 0, 1, 2, 4, 7, 11, 16, 22, 27, 34
Solution to Berlekamp's switching game on $n \times n$ board. See Fig M1040. Ref DM 74 265 89. [1,3; A5311]

||||||||||||||||||||||||||||||||||||||||||||||||||||||||||||||||||||||||||||||||||||||||||||||||||||||||||||

Figure M1040. LIGHT-BULB GAME.

In the first author's office at Murray Hill is a game built more than 25 years ago by Elwyn Berlekamp (a photograph can be seen in [DM 74 264 89]). It consists of a 10×10 array of light-bulbs, with an individual switch on the back for each bulb. On the front are 20 switches that complement each row and column. Let S be the set of light bulbs that are turned on at the start. One then attempts to minimize the number that are on by throwing row and column switches. Call this number $f(S)$. The problem, first solved in the above reference, is to determine maximal value of $f(S)$ over all choices of S. The answer is 34. Extremal configurations of light-bulbs for an $n \times n$ board for $3 \leq n \leq 10$ are shown below. This is M1040. No further terms are known.

||||||||||||||||||||||||||||||||||||||||||||||||||||||||||||||||||||||||||||||||||||||||||||||||||||||||||||

M1041 1, 2, 4, 7, 11, 16, 22, 29, 37, 46, 56, 67, 79, 92, 106, 121, 137, 154, 172, 191, 211, 232, 254, 277, 301, 326, 352, 379, 407, 436, 466, 497, 529, 562, 596, 631, 667, 704, 742
Central polygonal numbers (the Lazy Caterer's sequence): $\frac{1}{2} n(n-1) + 1$. See Fig M1041. Ref MAG 30 150 46. HO50 22. FQ 3 296 65. [1,2; A0124, N0391]

Figure M1041. SLICING A PANCAKE.

M1041 gives the maximal number of pieces that can be obtained by slicing a pancake with n cuts. The n-th term is $a_n = \tfrac{1}{2}n(n + 1) + 1$, the triangular numbers (M2535) plus 1.

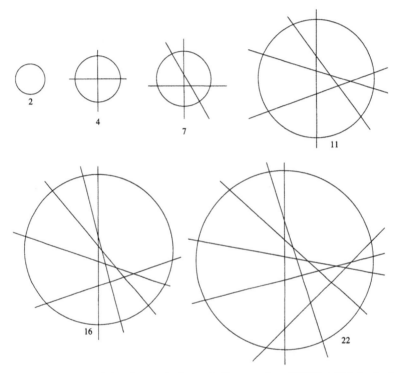

The number of n-sided polygons in the n-th diagram gives M2937, which is dually the number of n points in general position that can be chosen from among the intersection points of n lines in general position in the plane. Comtet [C1 274] calls such sets of points **clouds**. M1100 gives the maximal number of pieces that can be obtained with n slices of a cake, and M1594 is the corresponding sequence for a doughnut ([GA61] has a nice illustration of the third term in M1594).

M1042 1, 2, 4, 7, 11, 16, 22, 30, 42, 61, 91, 137, 205, 303, 443, 644, 936, 1365, 1999,
   2936, 4316, 6340, 9300, 13625, 19949, 29209, 42785, 62701, 91917, 134758, 197548
Twopins positions. Ref FQ 16 85 78. GU81. [6,2; A5689]

M1043 1, 2, 4, 7, 11, 16, 23, 31, 41, 53, 67, 83, 102, 123, 147, 174, 204, 237, 274, 314,
   358, 406, 458, 514, 575, 640, 710, 785, 865, 950, 1041, 1137, 1239, 1347, 1461, 1581
Expansion of $1 / (1-x)^2 (1-x^2)(1-x^3)$. Ref CAY 2 278. JACS 53 3084 31. AMS 26 304
55. [0,2; A0601, N0392]

M1044   1, 1, 1, 1, 2, 4, 7, 11, 16, 23, 34, 52, 81, 126, 194, 296, 450, 685, 1046, 1601, 2452,
  3753, 5739, 8771, 13404, 20489, 31327, 47904, 73252, 112004, 171245, 261813, 400285
Binary words not containing ..01110... Ref FQ 16 85 78. [0,5; A5253]

M1045   1, 2, 4, 7, 11, 17, 25, 35, 49, 66, 88, 115, 148, 189, 238, 297, 368, 451, 550, 665,
  799, 956, 1136, 1344, 1583, 1855, 2167, 2520, 2920, 3373, 3882, 4455, 5097, 5814, 6617
Partitions of n into Fibonacci parts (with 2 types of 1). Cf. M0556. [0,2; A7000]

M1046   1, 2, 4, 7, 11, 17, 25, 36, 50, 70, 94, 127, 168, 222, 288, 375, 480, 616, 781, 990,
  1243, 1562, 1945, 2422, 2996
Graphical partitions with n nodes. Ref CN 21 683 78. [3,2; A4250]

M1047   0, 1, 2, 4, 7, 11, 17, 26, 40, 61, 92, 139, 209, 314, 472, 709, 1064, 1597, 2396,
  3595, 5393, 8090, 12136, 18205, 27308, 40963, 61445, 92168, 138253, 207380, 311071
Partitioning integers to avoid arithmetic progressions of length 3. Ref PAMS 102 771 88.
[0,3; A6999]

$$a(n) = [(3a(n-1)+2)/2].$$

M1048   1, 1, 1, 1, 2, 4, 7, 11, 17, 27, 44, 72, 117, 189, 305, 493, 798, 1292, 2091, 3383,
  5473, 8855, 14328, 23184, 37513, 60697, 98209, 158905, 257114, 416020, 673135
$\Sigma C(n-2k,2k)$, $k = 0 .. n$. Ref FQ 7 341 69; 16 85 78. [0,5; A5252]

M1049   1, 1, 2, 4, 7, 11, 18, 27, 41, 60, 87, 122, 172, 235, 320, 430, 572, 751, 982, 1268,
  1629, 2074, 2625, 3297, 4123, 5118, 6324, 7771, 9506, 11567, 14023, 16917, 20335
Certain triangular arrays of integers. Ref P4BC 118. [0,3; A3403]

M1050   1, 2, 4, 7, 11, 19, 29, 46, 70, 106, 156, 232, 334, 482, 686, 971, 1357, 1894, 2612,
  3592, 4900, 6656, 8980, 12077, 16137, 21490, 28476, 37600, 49422, 64763, 84511
4-line partitions of n decreasing across rows. Ref MOC 26 1004 72. [1,2; A3292]

G.f.: $\Pi (1 - x^k)^{-c(k)}$, $c(k)=1,1,2,2,2,2,....$

M1051   1, 2, 4, 7, 12, 18, 27, 38, 53, 71, 94, 121, 155, 194, 241, 295, 359, 431, 515, 609,
  717, 837, 973, 1123, 1292, 1477, 1683, 1908, 2157, 2427, 2724, 3045, 3396, 3774, 4185
Expansion of $1 / (1-x)^2(1-x^2)(1-x^3)(1-x^4)$. Ref AMS 26 304 55. [0,2; A2621,
N0394]

M1052   1, 2, 4, 7, 12, 18, 28, 39, 55, 74, 100, 127, 167, 208, 261, 322, 399, 477, 581, 686,
  820, 967, 1142, 1318, 1545, 1778, 2053, 2347, 2697, 3048, 3486, 3925, 4441, 4986, 5610
$a(n+1)=1+a([n/1])+a([n/2])+ \cdots +a([n/n])$. Ref MAZ 4 173 68. rcr. [1,2; A3318]

M1053   1, 2, 4, 7, 12, 19, 29, 42, 60, 83, 113, 150, 197, 254, 324, 408, 509, 628, 769, 933,
  1125, 1346, 1601, 1892, 2225, 2602, 3029, 3509, 4049, 4652, 5326, 6074, 6905, 7823
A partition function. Ref AMS 26 304 55. [0,2; A2622, N0395]

M1054 1, 2, 4, 7, 12, 19, 30, 45, 67, 97, 139, 195, 272, 373, 508, 684, 915, 1212, 1597,
2087, 2714, 3506, 4508, 5763, 7338, 9296, 11732, 14742, 18460, 23025, 28629, 35471
Partitions of n into parts of 2 kinds. Ref RS4 90. RCI 199. FQ 9 332 71. HO85a 6. [0,2; A0070, N0396]

M1055 1, 2, 4, 7, 12, 20, 33, 53, 85, 133, 210, 322, 505, 759, 1192, 1748, 2782, 3931,
6476, 8579, 15216, 17847, 36761, 33612, 93961, 47282, 262987, 16105, 827382, 571524
Site percolation series for square lattice. Ref JPA 21 3821 88. [0,2; A6731]

M1056 0, 0, 1, 2, 4, 7, 12, 20, 33, 54, 88, 143, 232, 376, 609, 986, 1596, 2583, 4180, 6764,
10945, 17710, 28656, 46367, 75024, 121392, 196417, 317810, 514228, 832039, 1346268
Fibonacci numbers -1. Ref R1 155. AENS 79 203 62. FQ 3 295 65. [0,4; A0071, N0397]

M1057 0, 1, 1, 2, 4, 7, 12, 20, 33, 54, 90, 148, 244, 403, 665, 1096, 1808, 2980, 4914,
8103, 13359, 22026, 36315, 59874, 98715, 162754, 268337, 442413, 729416, 1202604
$[e^{(n-1)/2}]$. Ref rkg. [0,4; A5182]

M1058 1, 2, 4, 7, 12, 21, 34, 56, 90, 143, 223, 348, 532, 811, 1224, 1834, 2725, 4031,
5914, 8638, 12540, 18116, 26035, 37262, 53070, 75292, 106377, 149738, 209980
Planar partitions of n decreasing across rows. Ref MOC 26 1004 72. [1,2; A3293]

$$\text{G.f.: } \Pi \, (1 - x^k)^{-c(k)} , \quad c(k)=1,1,2,2,3,3,4,4,5,5,....$$

M1059 0, 1, 1, 1, 2, 4, 7, 12, 21, 37, 65, 114, 200, 351, 616, 1081, 1897, 3329, 5842,
10252, 17991, 31572, 55405, 97229, 170625, 299426, 525456, 922111, 1618192
$a(n)=a(n-1)+a(n-2)+a(n-4)$. Ref BR72 112. FQ 16 85 78. LAA 62 113 84. [0,5; A5251]

M1060 1, 2, 4, 7, 12, 21, 38, 68, 124, 229, 428, 806, 1530, 2919, 5591, 10750, 20717,
40077, 77653, 150752, 293161, 570963, 1113524, 2174315, 4250367, 8317036
Related to population of numbers of form x^2+y^2. Ref MOC 18 85 64. [1,2; A0709, N0398]

M1061 2, 4, 7, 12, 21, 38, 71, 136, 265, 522, 1035, 2060, 4109, 8206, 16399, 32784,
65553, 131090, 262163, 524308, 1048597, 2097174, 4194327, 8388632, 16777241
$2^n + n + 1$. Ref clm. [0,1; A5126]

M1062 1, 1, 2, 4, 7, 12, 22, 39, 70, 126, 225, 404, 725, 1299, 2331, 4182, 7501, 13458,
24145, 43316, 77715, 139430, 250152, 448808, 805222, 1444677, 2591958, 4650342
Restricted partitions. Ref PEMS 11 224 59. IFC 21 481 72. [2,3; A2573, N0399]

M1063 1, 1, 2, 4, 7, 12, 22, 41, 72, 137, 254, 476, 901, 1716, 3274, 6286, 12090, 23331,
45140, 87511, 169972, 330752, 644499, 1257523, 2456736, 4804666, 9405749
Integers $\leq 2^n$ of form x^2+4y^2. Ref MOC 20 560 66. [0,3; A0072, N0400]

M1064 1, 2, 4, 7, 12, 23, 39, 67, 118, 204, 343, 592, 1001, 1693, 2857, 4806
Number of bases for symmetric functions of n variables. Ref dz. [1,2; A7323]

M1065 2, 4, 7, 13, 15, 18, 19, 20, 21, 22, 23, 25, 28, 29, 30, 35, 38, 40, 43, 44, 45, 48, 49,
   50, 51, 54, 55, 56, 57, 58, 59, 60, 63, 65, 66, 71, 72, 74, 75, 79, 81, 84, 85, 87, 91, 93, 94
Elliptic curves. Ref JRAM 212 23 63. [1,1; A2152, N0401]

M1066 1, 1, 2, 4, 7, 13, 17, 30, 60, 107, 197, 257, 454, 908, 1619
A jumping problem. Ref DO64 259. [1,3; A2466, N0402]

M1067 1, 2, 4, 7, 13, 22, 40, 70, 126, 225, 411, 746, 1376, 2537, 4719, 8799, 16509,
   31041, 58635, 111012, 210870, 401427, 766149, 1465019, 2807195, 5387990, 10358998
Partial sums of M0116. Ref JSIAM 12 288 64. [1,2; A1036, N0403]

M1068 1, 1, 2, 4, 7, 13, 23, 41, 72, 127, 222, 388, 677, 1179, 2052, 3569, 6203, 10778,
   18722, 32513, 56455, 98017, 170161, 295389, 512755, 890043, 1544907, 2681554
Expansion of reciprocal of a determinant. Ref dhl. hpr. [0,3; A3116]

M1069 1, 2, 4, 7, 13, 23, 46, 88, 186, 395, 880, 1989, 4644, 10934, 26210, 63319, 154377,
   378443, 933022, 2308956, 5735371
States of a dynamic storage system. Ref CJN 25 391 82. [0,2; A5595]

M1070 1, 1, 2, 4, 7, 13, 24, 42, 76, 137, 245, 441
Restricted partitions. Ref PEMS 11 224 59. [3,3; A2574, N0404]

M1071 1, 2, 4, 7, 13, 24, 43, 77, 139, 249, 443, 786, 1400, 2486, 4395, 7758, 13732,
   24251, 42710, 75154, 132487, 233173, 409617, 719157, 1264303, 2219916, 3892603
Indefinitely growing n-step self-avoiding walks on Manhattan lattice. Ref JPA 22 3119 89.
[1,2; A6745]

M1072 1, 1, 2, 4, 7, 13, 24, 43, 78, 141, 253, 456
Partitions of n into parts 1/2, 3/4, 7/8, etc. Ref PEMS 11 224 59. [1,3; A2843, N0405]

M1073 1, 2, 4, 7, 13, 24, 44, 77, 139, 250, 450, 788, 1403, 2498, 4447, 7782, 13769,
   24363, 43106, 75396, 132865, 234171, 412731, 721433, 1267901, 2228666, 3917654
n-step self-avoiding walks on Manhattan lattice. Ref JPA 22 3117 89. [1,2; A6744]

M1074 0, 0, 1, 1, 2, 4, 7, 13, 24, 44, 81, 149, 274, 504, 927, 1705, 3136, 5768, 10609,
   19513, 35890, 66012, 121415, 223317, 410744, 755476, 1389537, 2555757, 4700770
Tribonacci numbers: $a(n) = a(n-1) + a(n-2) + a(n-3)$. Ref FQ 1(3) 71 63; 5 211 67.
[0,5; A0073, N0406]

M1075 1, 2, 4, 7, 13, 24, 44, 84, 161, 309, 594, 1164, 2284, 4484, 8807, 17305, 34301,
   68008, 134852, 267420, 530356, 1051905, 2095003, 4172701, 8311101, 16554194
Conway-Guy sequence: $a(n+1) = 2a(n) - a(n - [\frac{1}{2} + \sqrt{(2n)}])$. Ref ADM 12 143 82. JRM
15 149 83. CRP 296 345 83. MSH 84 59 83. [1,2; A5318]

M1076 1, 2, 4, 7, 13, 24, 46, 88, 172, 337, 667, 1321, 2629, 5234, 10444, 20842, 41638, 83188, 166288, 332404, 664636, 1328935, 2657533, 5314399, 10628131, 21254941
$a(n+1)=2a(n)-a(n-[\frac{1}{2}n+1])$. Ref LNM 686 70 78. NA79 100. MOC 50 298 88. DM 80 122 90. [1,2; A5255]

M1077 1, 1, 2, 4, 7, 13, 25, 43, 83, 157, 296, 564, 1083, 2077, 4006, 7733, 14968, 29044, 56447, 109864, 214197, 418080, 816907, 1598040, 3129063, 6132106
Odd integers $\leq 2^n$ of form x^2+y^2. Ref MOC 18 84 64. [1,3; A0074, N0407]

M1078 0, 1, 2, 4, 7, 14, 23, 42, 76, 139, 258, 482, 907, 1717, 3269, 6257, 12020, 23171, 44762, 86683, 168233, 327053, 636837, 1241723, 2424228, 4738426
Integers $\leq 2^n$ of form $2x^2+3y^2$. Ref MOC 20 563 66. [0,3; A0075, N0408]

M1079 0, 0, 1, 2, 4, 7, 14, 24, 43, 82, 149, 284, 534, 1015, 1937, 3713, 7136, 13759, 26597, 51537, 100045, 194586, 378987, 739161, 1443465, 2821923, 5522689
Integers $\leq 2^n$ of form $4x^2+4xy+5y^2$. Ref MOC 20 567 66. [0,4; A0076, N0409]

M1080 1, 2, 4, 7, 14, 26, 59, 122, 284, 647, 1528, 3602, 8679, 20882, 50824, 124055, 304574, 750122, 1855099, 4600202, 11442086
States of a dynamic storage system. Ref CJN 25 391 82. [0,2; A5594]

M1081 0, 0, 1, 0, 1, 2, 4, 7, 14, 27, 52, 100, 193, 372, 717, 1382, 2664, 5135, 9898, 19079, 36776, 70888, 136641, 263384, 507689, 978602, 1886316, 3635991, 7008598, 13509507
Tetranacci numbers: $a(n)=a(n-1)+a(n-2)+a(n-3)+a(n-4)$. Ref FQ 8 7 70. [0,6; A1631, N0410]

M1082 1, 1, 1, 1, 2, 4, 7, 14, 28, 61, 131, 297, 678, 1592, 3770, 9096, 22121, 54451, 135021, 337651, 849698, 2152048, 5479408, 14022947, 36048514, 93061268
n-node trees with a forbidden limb. Ref HA73 297. [1,5; A2989]

M1083 1, 1, 2, 4, 7, 14, 29, 60, 127, 275, 598, 1320, 2936, 6584, 14858, 33744, 76999, 176557, 406456, 939241, 2177573, 5064150, 11809632, 27610937, 64705623
Boron trees with n nodes. Ref CAY 9 450. rcr. [1,3; A0671, N0411]

M1084 1, 2, 4, 7, 15, 20, 48, 65, 119, 166
Number of basic invariants for cyclic group of order and degree n. Ref IOWA 55 290 48. [1,2; A2956]

M1085 1, 2, 4, 8, 7, 5, 10, 11, 13, 8, 7, 14, 19, 20, 22, 26, 25, 14, 19, 29, 31, 26, 25, 41, 37, 29, 40, 35, 43, 41, 37, 47, 58, 62, 61, 59, 64, 56, 67, 71, 61, 50, 46, 56, 58, 62, 70, 68, 73
Sum of digits of 2^n. Ref EUR 26 12 63. [0,2; A1370, N0414]

M1086 1, 2, 4, 8, 9, 10, 12, 16, 17, 18, 20, 24, 25, 26, 28, 32, 33, 34, 36, 40, 41, 42, 44, 48, 49, 50, 52, 56, 57, 58, 60, 64, 65, 66, 68, 72, 73, 74, 76, 80, 81, 82, 84, 88, 89, 90, 92, 96
Hurwitz-Radon function at powers of 2. Cf. M0161. Ref LA73a 131. [0,2; A3485]

$$G.f.: \quad (1+x+2x^2+4x^3) \,/\, (1-x)(1-x^4).$$

M1087 2, 4, 8, 10, 12, 14, 18, 32, 48, 54, 72, 148, 184, 248, 270, 274, 420
$5.2^n - 1$ is prime. Ref MOC 22 421 68. [1,1; A1770, N0415]

M1088 2, 4, 8, 12, 16, 20, 26, 32, 40, 44, 54, 64
Restricted postage stamp problem. Ref LNM 751 326 82. [1,1; A6638]

M1089 2, 4, 8, 12, 16, 20, 26, 32, 40, 46, 54, 64, 72
Postage stamp problem. Ref CJN 12 379 69. AMM 87 208 80. SIAA 1 383 80. [1,1; A1212, N0972]

M1090 0, 0, 2, 4, 8, 12, 18, 24, 32, 40, 50, 60, 72, 84, 98, 112, 128, 144, 162, 180, 200, 220, 242, 264, 288, 312, 338, 364, 392, 420, 450, 480, 512, 544, 578, 612, 648, 684, 722
$[n^2/2]$. [0,3; A7590]

M1091 1, 2, 4, 8, 12, 18, 27, 36, 48, 64, 80, 100, 125, 150, 180, 216, 252, 294, 343, 392, 448, 512, 576, 648, 729, 810, 900, 1000, 1100, 1210, 1331, 1452, 1584, 1728, 1872, 2028
Expansion of $(1+x^2) / (1-x)^2(1-x^3)^2$. Ref FQ 16 116 78. [0,2; A6501]

M1092 1, 2, 4, 8, 12, 18, 27, 45, 75, 125, 200, 320, 512, 832, 1352, 2197, 3549, 5733, 9261, 14994, 24276, 39304, 63580, 102850, 166375, 269225, 435655, 704969, 1140624
Restricted combinations. Ref FQ 16 116 78. [0,2; A6500]

M1093 1, 2, 4, 8, 12, 32, 36, 40, 24, 48, 160, 396, 2268, 704, 312, 72, 336, 216, 936, 144, 624, 1056, 1760, 360, 2560, 384, 288, 1320, 3696, 240, 768, 9000, 432, 7128, 4200, 480
Smallest k such that $\phi(x) = k$ has exactly n solutions. Ref AS1 840. [2,2; A7374]

M1094 1, 2, 4, 8, 13, 21, 31, 45, 66, 81, 97, 123, 148, 182, 204, 252, 290, 361, 401, 475, 565, 593, 662, 775, 822, 916, 970, 1016, 1159, 1312, 1395, 1523, 1572, 1821, 1896, 2029
A B_2 sequence. Ref UPNT E28. MOC 60 837 93. [1,2; A5282]

M1095 1, 1, 2, 4, 8, 13, 24, 42, 76, 140, 257, 483, 907, 1717, 3272, 6261, 12027, 23172, 44769, 86708, 168245, 327073, 636849, 1241720, 2424290, 4738450
Integers $\leq 2^n$ of form $x^2 + 6y^2$. Ref MOC 20 563 66. [0,3; A0077, N0417]

M1096 1, 2, 4, 8, 14, 18, 28, 40, 52, 70, 88, 104, 140
Generalized divisor function. Ref PLMS 19 111 19. [4,2; A2132, N0418]

M1097 2, 4, 8, 14, 24, 36, 54, 76, 104, 136
Straight binary strings of length n. Ref DO86. [1,1; A5598]

M1098 1, 2, 4, 8, 14, 26, 43, 74, 120, 197, 311, 495, 768, 1189, 1811, 2748, 4116, 6136, 9058, 13299, 19370, 28069, 40399, 57856, 82374, 116736, 164574, 231007, 322749
n-step spirals on hexagonal lattice. Ref JPA 20 492 87. [1,2; A6777]

M1099 2, 4, 8, 15, 12, 27, 24, 36, 90, 96, 245, 288, 368, 676, 1088, 2300, 1596, 1458, 3344, 3888, 5360, 8895, 11852, 25971, 23360, 38895, 35540, 35595, 36032, 53823
Smallest number such that nth iterate of Chowla function is 0. Ref MOC 25 924 71. [1,1; A2954]

M1100   1, 2, 4, 8, 15, 26, 42, 64, 93, 130, 176, 232, 299, 378, 470, 576, 697, 834, 988, 1160, 1351, 1562, 1794, 2048, 2325, 2626, 2952, 3304, 3683, 4090, 4526, 4992, 5489
Cake numbers: slicing a cake with n slices: $C(n+1,3)+n+1$. See Fig M1041. Ref MAG 30 150 46. FQ 3 296 65. [0,2; A0125, N0419]

M1101   1, 1, 2, 4, 8, 15, 26, 45, 71, 110, 168, 247
Achiral rooted trees. Ref JRAM 278 334 75. [1,3; A3241]

M1102   1, 2, 4, 8, 15, 27, 47, 79, 130, 209, 330, 512, 784, 1183, 1765, 2604, 3804, 5504, 7898, 11240
Stacks, or planar partitions of n. Ref PCPS 47 686 51. QJMO 23 153 72. [1,2; A1523, N0420]

M1103   1, 2, 4, 8, 15, 27, 47, 80, 134, 222, 365, 597, 973, 1582, 2568, 4164, 6747, 10927, 17691, 28636, 46346, 75002, 121369, 196393, 317785, 514202, 832012, 1346240
A nonlinear binomial sum. Ref FQ 3 295 65. [1,2; A0126, N0421]

$$\text{G.f.:}\quad (1 - x + x^3) / (x^2 + x - 1)(x - 1)^2.$$

M1104   2, 4, 8, 15, 28, 50, 90, 156
Percolation series for square lattice. Ref SSP 10 921 77. [1,1; A6808]

M1105   1, 2, 4, 8, 15, 28, 50, 90, 156, 274, 466, 804, 1348, 2300, 3804, 6450, 10547, 17784, 28826, 48464, 77689, 130868, 207308, 350014, 548271, 931584, 1433966
Bond percolation series for square lattice. Ref JPA 21 3820 88. [0,2; A6727]

M1106   1, 2, 4, 8, 15, 28, 51, 92, 165, 294, 522, 924, 1632, 2878, 5069, 8920, 15686, 27570, 48439, 85080, 149405, 262320, 460515, 808380, 1418916, 2490432
Twopins positions. Ref GU81. [5,2; A5682]

M1107   1, 1, 2, 4, 8, 15, 29, 53, 98, 177, 319, 565, 1001, 1749, 3047, 5264, 9054, 15467, 26320, 44532, 75054, 125904, 210413, 350215, 580901, 960035, 1581534, 2596913
n-node trees of height at most 3. Ref IBMJ 4 475 60. KU64. [1,3; A1383, N0422]

$$\text{G.f.:}\quad \Pi\,(1 - x^k)^{-p(k)}, \text{ where } p(k) = \text{number of partitions of } k.$$

M1108   0, 0, 0, 1, 1, 2, 4, 8, 15, 29, 56, 108, 208, 401, 773, 1490, 2872, 5536, 10671, 20569, 39648, 76424, 147312, 283953, 547337, 1055026, 2033628, 3919944, 7555935
Tetranacci numbers: $a(n) = a(n-1) + a(n-2) + a(n-3) + a(n-4)$. Ref AMM 33 232 26. FQ 1(3) 74 63. [0,6; A0078, N0423]

M1109   1, 2, 4, 8, 15, 38, 74
Coins needed for ApSimon's mints problem. Ref AMM 101 359 94. [1,2; A7673]

M1110 1, 2, 4, 8, 15, 240, 15120, 672, 8400, 100800, 69300, 4950, 17199000, 22422400, 33633600, 201801600, 467812800, 102918816000
Denominators of coefficients for numerical differentiation. Cf. M2651. Ref PHM 33 11 42. BAMS 48 922 42. [1,2; A2546, N0424]

M1111 1, 1, 1, 1, 2, 4, 8, 16, 20, 40
The coding-theoretic function $A(n,4)$. See Fig M0240. Ref PGIT 36 1338 90. [1,5; A5864]

M1112 1, 2, 4, 8, 16, 21, 42, 51, 102, 112, 224, 235, 470, 486, 972, 990, 1980, 2002, 4004, 4027, 8054, 8078, 16156, 16181, 32362, 32389, 64778, 64806, 129612, 129641, 259282
A self-generating sequence. See Fig M0557. Ref AMM 75 80 68. SI64a. MMAG 63 15 90. [1,2; A1856, N0425]

M1113 2, 4, 8, 16, 22, 24, 28, 36, 42, 44, 48, 56, 62, 64, 68, 76, 82, 84, 88, 96, 102, 104, 108, 116, 122, 124, 128, 136, 142, 144, 148, 156, 162, 164, 168, 176, 182, 184, 188, 196
First differences are periodic. Ref TCPS 2 219 1827. [0,1; A2081, N0426]

M1114 1, 2, 4, 8, 16, 23, 28, 29, 31, 35, 43, 50, 55, 56, 58, 62, 70, 77, 82, 83, 85, 89, 97, 104, 109, 110, 112, 116, 124, 131, 136, 137, 139, 143, 151, 158, 163, 164, 166, 170, 178
$a(n+1) = a(n) + $ digital root of $a(n)$. Ref Robe92 65. [1,2; A7612]

M1115 1, 2, 4, 8, 16, 23, 28, 38, 49, 62, 70, 77, 91, 101, 103, 107, 115, 122, 127, 137, 148, 161, 169, 185, 198, 216, 225, 234, 243, 252, 261, 270, 279, 297, 306, 315, 324, 333, 342
$a(n) = a(n-1) + $ sum of digits of $a(n-1)$. Ref Robe92 65. [0,2; A4207]

M1116 1, 2, 4, 8, 16, 24, 36, 46, 56, 64, 72, 80, 88, 96
Subwords of length n in Rudin-Shapiro word. Ref jos. [0,2; A5943]

M1117 2, 4, 8, 16, 30, 56, 100, 172, 290, 480, 780, 1248, 1970, 3068, 4724, 7200, 10862, 16240, 24080
Representation degeneracies for Raymond strings. Ref NUPH B274 548 86. [2,1; A5305]

M1118 1, 2, 4, 8, 16, 30, 57, 88, 163, 230, 386, 456, 794, 966, 1471, 1712, 2517, 2484, 4048, 4520, 6196, 6842, 9109, 9048, 12951, 14014, 17902, 19208, 24158, 21510, 31931
Join n equal points around circle in all ways, count regions. Cf. M3411. Ref WP 10 62 72. PoRu94. [1,2; A6533]

M1119 1, 2, 4, 8, 16, 31, 57, 99, 163, 256, 386, 562, 794, 1093, 1471, 1941, 2517, 3214, 4048, 5036, 6196, 7547, 9109, 10903, 12951, 15276, 17902, 20854, 24158, 27841, 31931
$C(n,4) + C(n,3) + \cdots + C(n,0)$. Ref MAG 30 150 46. FQ 3 296 65. [0,2; A0127, N0427]

M1120 1, 2, 4, 8, 16, 31, 58, 105, 185, 319, 541, 906, 1503, 2476, 4058, 6626, 10790, 17537, 28464, 46155, 74791, 121137, 196139, 317508, 513901, 831686, 1345888
A nonlinear binomial sum. Ref FQ 3 295 65. [1,2; A0128, N0428]

$$\text{G.f.: } (1 - 2x + x^2 + x^3) / (1 - x - x^2)(1 - x)^3.$$

M1121 1, 2, 4, 8, 16, 31, 61, 115, 213, 388, 691, 1218, 2110, 3617, 6113, 10238, 16945, 27802, 45180, 72838, 116479, 184936, 291556, 456694, 710907, 1100192, 1693123
n-step spirals on hexagonal lattice. Ref JPA 20 492 87. [1,2; A6775]

M1122 0, 0, 0, 0, 1, 1, 2, 4, 8, 16, 31, 61, 120, 236, 464, 912, 1793, 3525, 6930, 13624, 26784, 52656, 103519, 203513, 400096, 786568, 1546352, 3040048, 5976577, 11749641
Pentanacci numbers: $a(n+1)=a(n)+...+a(n-4)$. Ref FQ 5 260 67. [0,7; A1591, N0429]

M1123 1, 1, 2, 4, 8, 16, 31, 62, 120, 236, 454, 904
Partially achiral rooted trees. Ref JRAM 278 334 75. [1,3; A3240]

M1124 1, 2, 4, 8, 16, 32, 52, 100, 160, 260, 424
Words of length n in a certain language. Ref DM 40 231 82. [0,2; A7055]

M1125 1, 0, 2, 4, 8, 16, 32, 60, 114, 212
Fermionic string states. Ref CU86. [0,3; A5309]

M1126 1, 2, 4, 8, 16, 32, 63, 120, 219, 382, 638, 1024, 1586, 2380, 3473, 4944, 6885, 9402, 12616, 16664, 21700, 27896, 35443, 44552, 55455, 68406, 83682, 101584, 122438
$\Sigma C(n,k)$, $k = 0 \,.\,. \,5$. Ref MIS 4(3) 32 75. [0,2; A6261]

M1127 1, 2, 4, 8, 16, 32, 63, 124, 244, 480, 944, 1856, 3649, 7174, 14104, 27728, 54512, 107168, 210687, 414200, 814296, 1600864, 3147216, 6187264, 12163841
A probability difference equation. Ref AMM 32 369 25. [5,2; A1949, N0430]

M1128 0, 0, 0, 0, 0, 1, 1, 2, 4, 8, 16, 32, 63, 125, 248, 492, 976, 1936, 3840, 7617, 15109, 29970, 59448, 117920, 233904, 463968, 920319, 1825529, 3621088, 7182728, 14247536
Hexanacci numbers: $a(n+1)=a(n)+...+a(n-5)$. Ref FQ 5 260 67. [0,8; A1592, N0431]

M1129 1, 2, 4, 8, 16, 32, 64, 128, 256, 512, 1024, 2048, 4096, 8192, 16384, 32768, 65536, 131072, 262144, 524288, 1048576, 2097152, 4194304, 8388608, 16777216, 33554432
Powers of 2. Ref BA9. MOC 23 456 69. AS1 1016. [0,2; A0079, N0432]

M1130 0, 0, 1, 2, 4, 8, 16, 32, 64, 130, 264, 538, 1104, 2272, 4692, 9730, 20236, 42208, 88288, 185126, 389072, 819458, 1729296, 3655936, 7742124
Generalized Fibonacci numbers. Ref FQ 27 120 89. [1,4; A6211]

M1131 1, 1, 1, 1, 1, 2, 4, 8, 16, 33, 69, 146, 312, 673, 1463, 3202, 7050, 15605, 34705, 77511, 173779, 390966, 882376, 1997211, 4532593, 10311720, 23512376, 53724350
Generalized Catalan numbers: $a(n+2) = a(n+1)+a(n)+a(n-1) + \Sigma\ a(k)a(n-k)$, $k=0..n$. Ref DM 26 264 79. BeSl94. [2,6; A4149]

M1132   0, 1, 2, 4, 8, 16, 34, 72, 154, 336, 738, 1632, 3640, 8160, 18384, 41616, 94560, 215600, 493122, 1130976, 2600388, 5992560, 13838306, 32016576, 74203112
Generalized Fibonacci numbers. Ref FQ 27 120 89. [1,3; A6210]

M1133   1, 1, 1, 1, 2, 4, 8, 16, 34, 72, 157, 343
Modular lattices on n nodes. Ref pdl. [0,5; A6981]

M1134   0, 0, 0, 0, 1, 1, 1, 2, 4, 8, 16, 34, 72, 158, 348, 784, 1777, 4080, 9425, 21965, 51456, 121300, 287215, 683268, 1631532, 3910235, 9401000, 22670058, 54813780
Trees by stability index. Ref LNM 403 50 74. [1,8; A3427]

M1135   1, 2, 4, 8, 16, 36, 80
Orthogonal lattices in dimension n. Ref PCPS 76 23 74. SC80 65. [1,2; A7669]

M1136   1, 2, 4, 8, 16, 36, 85, 239
Weighted linear spaces of total weight n. Ref BSMB 22 234 70. [1,2; A2876, N0433]

M1137   1, 2, 4, 8, 16, 77, 145, 668, 1345, 6677, 13444, 55778, 133345, 666677, 1333444, 5567777, 12333445, 66666677, 133333444, 556667777, 1233334444, 5566667777
RATS: Reverse Add Then Sort! See Fig M2629. Ref AMM 96 425 89. [1,2; A4000]

M1138   1, 2, 4, 8, 17, 35, 71, 152, 314, 628, 1357, 2725, 5551, 12212, 24424, 48848, 108807, 218715, 438531, 878162, 1867334, 3845668, 7802447, 16705005, 34511011
Powers of 2 written in base 9. Ref EUR 14 13 51. [0,2; A1357, N0434]

M1139   1, 1, 1, 2, 4, 8, 17, 36, 78, 171, 379, 888, 1944
Distinct values taken by $2\uparrow 2\uparrow \cdots \uparrow 2$ (with n 2's). Ref AMM 80 875 73. jql. [1,4; A2845, N0435]

M1140   1, 1, 1, 2, 4, 8, 17, 36, 79, 175, 395, 899, 2074, 4818, 11291, 26626, 63184, 150691, 361141, 869057, 2099386, 5088769, 12373721, 30173307, 73771453
Rooted trimmed trees with n nodes. Ref AMM 80 874 73. HA73 297. klm. [1,4; A2955]

M1141   1, 1, 1, 2, 4, 8, 17, 37, 82, 185, 423, 978, 2283, 5373, 12735, 30372, 72832, 175502, 424748, 1032004, 2516347, 6155441, 15101701, 37150472, 91618049
Generalized Catalan numbers: $a(n+1)=a(n)+a(n-1)+\Sigma\, a(k)\, a(n-1-k)$, $k=0..n-1$.
See Fig M1141. Ref Wate78. DM 26 264 79. JCT B29 89 80. [0,4; A4148]

Figure M1141.   RNA MOLECULES.

M1141 arises in enumerating secondary structures of RNA molecules. The 17 structures with 6 nucleotides are as follows (after [Wate78]).

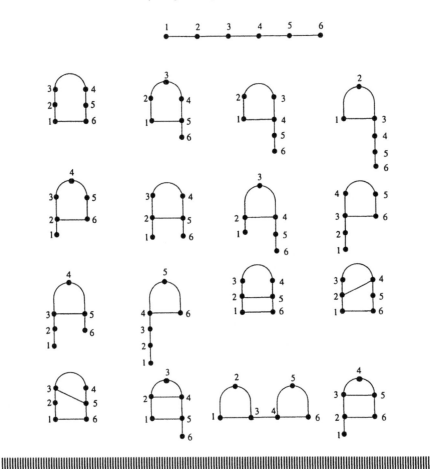

M1142   1, 1, 1, 1, 2, 4, 8, 17, 37, 85, 196, 469, 1134, 2799, 6975, 17628, 44903, 115497, 299089, 780036, 2045924, 5396078, 14299878, 38067356, 101748748, 272995157
Stable trees with *n* nodes. Ref LNM 403 50 74. [1,5; A3426]

M1143   1, 1, 1, 2, 4, 8, 17, 38, 87, 203, 482, 1160, 2822, 6929, 17149, 42736, 107144, 270060, 683940, 1739511
Leftist trees with *n* leaves. Ref IPL 25 228 87. [1,4; A6196]

M1144 1, 0, 1, 2, 4, 8, 17, 38, 88, 210, 511, 1264, 3165, 8006, 20426, 52472, 135682, 352562, 920924, 2414272, 6356565, 16782444, 44470757, 118090648, 314580062
Percolation series for directed square lattice. Ref SSP 10 921 77. JPA 21 3200 88. [0,4; A6461]

M1145 1, 1, 2, 4, 8, 17, 38, 89, 208
n-level expressions. Ref AMM 80 876 73. [1,3; A3007]

M1146 1, 1, 1, 2, 4, 8, 17, 39, 89, 211, 507, 1238, 3057, 7639, 19241, 48865, 124906, 321198, 830219, 2156010, 5622109, 14715813, 38649152, 101821927, 269010485
Quartic planted trees with n nodes. Ref JACS 53 3042 31. FI50 41.397. TET 32 356 76. [0,4; A0598, N0436]

M1147 1, 1, 2, 4, 8, 17, 39, 90, 213
n-level expressions. Ref AMM 80 876 73. [1,3; A3008]

M1148 1, 2, 4, 8, 18, 40, 91, 210, 492, 1165, 2786, 6710, 16267, 39650, 97108, 238824, 589521
Sums of Fermat coefficients. Ref MMAG 27 143 54. [1,2; A0967, N0437]

M1149 1, 2, 4, 8, 18, 44, 117, 351, 1230, 5069, 25181, 152045, 1116403, 9899865, 104980369
Square-free graphs on n vertices. Ref bdm. [1,2; A6786]

M1150 1, 2, 4, 8, 18, 44, 122, 362, 1162, 3914, 13648
Even sequences with period $2n$. Ref IJM 5 664 61. [0,2; A0117, N0438]

M1151 1, 1, 2, 4, 8, 19, 44, 112, 287, 763
n-node connected graphs with at most one cycle. Ref R1 150. rkg. [1,3; A5703]

M1152 1, 1, 1, 2, 4, 8, 19, 48, 126, 355, 1037, 3124, 9676, 30604, 98473, 321572
Triangular cacti. Ref HP73 73. LeMi91. [0,4; A3081]

M1153 2, 4, 8, 19, 53, 209
Hierarchical linear models on n factors allowing 2-way interactions; or graphs with $\leq n$ nodes. Cf. M1253. Ref clm. [1,1; A6897]

M1154 1, 2, 4, 8, 19, 67, 331, 2221, 19577
Codes for rooted trees on n nodes. Ref JCT B29 142 80. [1,2; A5518]

M1155 2, 4, 8, 20, 52, 152, 472, 1520, 5044, 17112, 59008, 206260, 729096, 2601640, 9358944, 33904324, 123580884, 452902072, 1667837680, 6168510256
Representations of 0 as $\Sigma \pm k$, $k = -n \ . \ . \ n$. Ref CMB 11 292 68. [0,1; A0980, N0439]

M1156 1, 2, 4, 8, 20, 56, 180, 596, 2068, 7316, 26272
Even sequences with period $2n$. Ref IJM 5 664 61. [0,2; A0116, N0440]

M1157   1, 1, 2, 4, 8, 20, 58, 177
2-connected planar maps. Ref SIAA 4 174 83. [3,3; A6407]

M1158   2, 4, 8, 20, 100, 2116, 1114244, 68723671300, 1180735735906024030724,
170141183460507917357914971986913657860
$\Sigma 2 \uparrow C(n,k)$, $k = 0 .. n$. Ref GO61. [0,1; A1315, N0441]

M1159   1, 2, 4, 8, 21, 52, 131, 316, 765, 1846, 4494
Related to partitions. Ref AMM 76 1036 69. [0,2; A2040, N0442]

M1160   1, 1, 2, 4, 8, 22, 52, 140, 366, 992
Necklaces. Ref IJM 2 302 58. [1,3; A2075, N0443]

M1161   2, 4, 8, 24, 84, 328, 1372, 6024, 27412, 128228, 613160, 2985116, 14751592,
73825416, 373488764, 1907334616, 9820757380, 50934592820, 265877371160
Energy function for square lattice. Ref PHA 28 925 62. DG74 386. [1,1; A2908, N0444]

M1162   1, 2, 4, 8, 27, 76, 295
Planar maps without loops or isthmuses. Ref SIAA 4 174 83. [2,2; A6399]

M1163   1, 2, 4, 8, 29, 92, 403
Planar maps without loops or isthmuses. Ref SIAA 4 174 83. [2,2; A6398]

M1164   1, 2, 4, 8, 121, 151, 212, 242, 484, 656, 757, 29092, 48884, 74647, 75457, 76267,
92929, 93739, 848848, 1521251, 2985892, 4022204, 4219124, 4251524, 4287824
Palindromic in bases 3 and 10. Ref JRM 18 169 85. [1,2; A7633]

M1165   1, 2, 4, 9, 10, 12, 27, 37, 38, 44, 48, 78, 112, 168, 229, 297, 339, 517, 522, 654,
900, 1518, 2808, 2875, 3128, 3888, 4410, 6804, 7050, 7392
$15.2^n + 1$ is prime. Ref PAMS 9 674 58. Rie85 381. [1,2; A2258, N0445]

M1166   2, 4, 9, 10, 22, 26, 40, 50, 54, 55, 78, 115, 123, 154, 155, 209, 288, 220, 221, 292,
301, 378, 494, 494, 551, 715, 670, 786, 805, 803, 1079, 966, 1190, 1222, 1274, 1274
Binomial coefficients with many divisors. Ref MSC 39 277 76. [1,1; A5733]

M1167   1, 2, 4, 9, 16, 20, 30, 42, 49, 64
Related to Ramsey numbers. Ref GTA91 547. [1,2; A6474]

M1168   2, 4, 9, 16, 29, 47, 77, 118, 181, 267, 392, 560, 797, 1111, 1541, 2106, 2863, 3846,
5142, 6808, 8973, 11733, 15275, 19753, 25443, 32582, 41569, 52770, 66757
Bipartite partitions. Ref PCPS 49 72 53. ChGu56 1. [0,1; A0291, N0447]

M1169   1, 0, 2, 4, 9, 16, 35, 63, 129, 234, 445, 798, 1458, 2568, 4561, 7924, 13770, 23584,
40301, 68097, 114646, 191336, 317893, 524396, 861054, 1405130, 2282651, 3688254
Related to solid partitions of n. Ref MOC 24 956 70. [0,3; A5980]

M1170 1, 2, 4, 9, 17, 33, 61, 112, 202, 361, 639, 1123, 1961, 3406, 5888, 10137, 17389, 29733, 50693, 86204, 146246, 247577, 418299, 705479, 1187857, 1997018, 3352636
Les Marvin sequence: $F(n)+(n-1)F(n-1)$. Ref JRM 10 230 77. [1,2; A7502]

M1171 1, 1, 2, 4, 9, 18, 42, 96, 229, 549, 1347, 3326, 8330, 21000, 53407, 136639, 351757, 909962, 2365146, 6172068, 16166991, 42488077, 112004630, 296080425
Carbon trees with n carbon atoms. Ref CAY 9 454. ZFK 93 437 36. BA76 28. [1,3; A0678, N0448]

M1172 1, 1, 2, 4, 9, 19, 42, 89, 191, 402, 847, 1763, 3667, 7564, 15564, 31851, 64987, 132031, 267471, 539949, 1087004, 2181796, 4367927, 8721533, 17372967, 34524291
n-node trees of height at most 4. Ref IBMJ 4 475 60. KU64. [1,3; A1384, N0449]

M1173 1, 2, 4, 9, 19, 48, 117, 307, 821, 2277
Minimal triangle graphs. Ref MOC 21 249 67. [4,2; A0080, N0450]

M1174 1, 2, 4, 9, 20, 45, 105, 249, 599, 1463, 3614, 9016, 22695, 57564, 146985, 377555, 974924, 2529308, 6589734, 17234114, 45228343, 119069228, 314368027, 832193902
Esters with n carbon atoms. Ref JACS 56 157 34. BA76 28. [2,2; A0632, N0451]

M1175 1, 2, 4, 9, 20, 46, 105, 242, 557, 1285, 2964, 6842, 15793, 36463, 84187, 194388, 448847, 1036426, 2393208, 5526198, 12760671, 29466050, 68041019, 157115917
Staircase polyominoes with n cells. Ref DM 8 31 74; 8 219 74. Fla91. [1,2; A6958]

M1176 1, 1, 2, 4, 9, 20, 46, 105, 246, 583, 1393, 3355, 8133, 19825, 48554, 119412, 294761
Sums of Fermat coefficients. Ref MMAG 27 143 54. [1,3; A0968, N0452]

M1177 1, 1, 2, 4, 9, 20, 47, 108, 252, 582, 1345, 3086, 7072, 16121, 36667, 83099, 187885, 423610, 953033, 2139158, 4792126, 10714105, 23911794, 53273599
n-node trees of height at most 5. Ref IBMJ 4 475 60. KU64. [1,3; A1385, N0453]

M1178 1, 1, 2, 4, 9, 20, 47, 111, 270, 664, 1659
Distinct values taken by $3 \uparrow 3 \uparrow ... \uparrow$ (n 3's). Ref AMM 80 874 73. [1,3; A3018]

M1179 1, 1, 2, 4, 9, 20, 48, 114, 282, 703, 1787
Distinct values taken by $4 \uparrow 4 \uparrow ... \uparrow 4$ (n 4's). Ref AMM 80 874 73. [1,3; A3019]

M1180 1, 1, 2, 4, 9, 20, 48, 115, 286, 719, 1842, 4766, 12486, 32973, 87811, 235381, 634847, 1721159, 4688676, 12826228, 35221832, 97055181, 268282855, 743724984
Rooted trees with n nodes. See Fig M0791. Ref R1 138. HA69 232. [1,3; A0081, N0454]

$$\text{G.f.: } \Pi\,(1-x^{n+1})^{-a(n)}.$$

M1181 1, 1, 2, 4, 9, 20, 50, 124, 332, 895
Graphs with no isolated vertices. Ref LNM 952 101 82. [2,3; A6648]

M1182 1, 2, 4, 9, 20, 51, 125, 329, 862, 2311, 6217, 16949, 46350, 127714, 353272, 981753, 2737539, 7659789, 21492286, 60466130
Rings and branches with n edges. Ref FI50 41.399. [1,2; A2861, N0455]

M1183 1, 2, 4, 9, 21, 51, 127, 322, 826, 2135, 5545, 14445, 37701, 98514, 257608, 673933, 1763581, 4615823, 12082291, 31628466, 82798926, 216761547, 567474769
$(F(2n)+F(n+1))/2$, where $F(n)$ is a Fibonacci number. Ref CJN 25 391 82. [0,2; A5207]

M1184 1, 1, 2, 4, 9, 21, 51, 127, 323, 835, 2188, 5798, 15511, 41835, 113634, 310572, 853467, 2356779, 6536382, 18199284, 50852019, 142547559, 400763223, 1129760415
Motzkin numbers. Ref BAMS 54 359 48. JSIAM 18 254 69. JCT A23 292 77. [0,3; A1006, N0456]

$$\text{G.f.:}\ \ (1 - x - (1-2x-3x^2)^{1/2}) \ / \ 2x^2.$$

M1185 1, 2, 4, 9, 21, 52, 129, 332, 859, 2261, 5983, 15976, 42836, 115469, 312246, 847241, 2304522, 6283327, 17164401, 46972357, 128741107, 353345434, 970999198
Paraffins with n carbon atoms. Ref JACS 56 157 34. BA76 28. [1,2; A0636, N0457]

M1186 1, 2, 4, 9, 21, 55, 151, 447, 1389, 4502, 15046, 51505, 179463, 634086, 2265014, 8163125, 29637903, 108282989, 397761507, 1468063369, 5441174511, 20242989728
Unit interval graphs. Ref TAMS 272 423 82. rwr. [1,2; A5217]

M1187 1, 1, 2, 4, 9, 21, 56, 148, 428, 1305, 4191, 14140, 50159, 185987, 720298, 2905512, 12180208
Graphs with n nodes and n edges. Ref R1 146. SS67. [2,3; A1430, N0458]

M1188 1, 1, 2, 4, 9, 21, 56, 155, 469, 1480
Projective plane trees with n nodes. Ref LNM 406 348 74. [1,3; A6080]

M1189 1, 1, 2, 4, 9, 22, 59, 167, 490, 1486, 4639, 14805, 48107, 158808, 531469, 1799659, 6157068, 21258104, 73996100, 259451116, 951695102, 3251073303
Scores in n-person round-robin tournament. Ref CMB 7 135 64. MO68 68. [1,3; A0571, N0459]

M1190 1, 2, 4, 9, 23, 63, 177, 514, 1527, 4625, 14230, 44357, 139779, 444558, 1425151, 4600339, 14939849, 48778197, 160019885, 527200711
Related to series-parallel networks. Ref SAM 21 92 42. [1,2; A1573, N0460]

M1191 1, 1, 2, 4, 9, 23, 63, 188
Mixed Husimi trees with n nodes. Ref PNAS 42 535 56. [1,3; A0083, N0461]

M1192 1, 1, 1, 2, 4, 9, 23, 65, 199, 654, 2296, 8569, 33825, 140581, 612933, 2795182, 13298464, 65852873, 338694479, 1805812309, 9963840219, 56807228074
Shifts 2 places left under binomial transform. Ref BeSl94. EIS § 2.7. [0,4; A7476]

M1193 1, 2, 4, 9, 24, 30, 99, 154, 189, 217, 1183, 1831, 2225, 3385, 14357, 30802, 31545, 40933, 103520, 104071, 149689, 325852, 1094421, 1319945, 2850174, 6957876
Index of primes where largest gap occurs. Cf. M0858. Ref MOC 52 222 89. Rei85 85. [0,2; A5669]

M1194 1, 2, 4, 9, 24, 76, 279, 1156, 5296
Rhyme schemes. Ref ANY 319 463 79. [1,2; A5001]

M1195 1, 1, 2, 4, 9, 24, 81, 274
2-connected planar maps. Ref SIAA 4 174 83. [3,3; A6406]

M1196 1, 2, 4, 9, 26, 95
Path sums of n-point graphs. Ref CN 21 259 78. [1,2; A4252]

M1197 1, 1, 1, 2, 4, 9, 26, 101, 950
Matroids (or geometries) with n points. See Fig M1197. Ref SAM 49 127 70. MOC 27 155 73. [0,4; A2773, N0462]

‖‖‖

Figure M1197. GEOMETRIES.

The numbers of topologies are shown in Fig. M2817. The following are some other geometrical sequences. A **linear space** is a system of (abstract) points and lines such that any two points lie on a unique line, and every line contains at least two points. A **geometry** or **matroid** is a system of points, lines, planes, ... with an analogous definition. The illustration shows M1197, the number of geometries with n points (for $n \geq 2$). The $*$ denotes 5 points in general position in 4-dimensional space. The numbers of geometries with 9 or more points are not known. The planar figures in the illustration form M0726. M0292 is a related sequence. References: [BSM 19 424 67], [JM2 49 127 70], [Wels76], [Aign77 133].

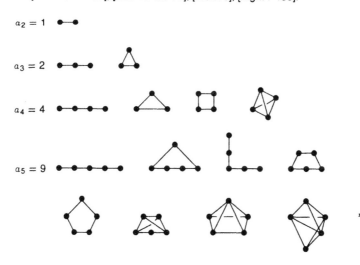

M1198   1, 1, 2, 4, 9, 27, 81, 256, 1024, 4096, 16384, 78125, 390625, 1953125, 10077696,
60466176, 362797056, 2176782336, 13841287201, 96889010407, 678223072849
Max of k^{n-k}, $k = 0 .. n$. Ref TO72 231. [1,3; A3320]

M1199   0, 0, 0, 1, 2, 4, 9, 38, 308, 4937, 158022
Coding a recurrence. Ref FQ 15 313 77. [0,5; A5204]

M1200   1, 2, 4, 10, 17, 50, 170, 184, 194, 209
$8.3^n - 1$ is prime. Ref MOC 26 997 72. [1,2; A5541]

M1201   1, 2, 4, 10, 20, 48, 104, 282, 496, 1066, 2460, 6128, 12840, 29380, 74904, 212728,
368016, 659296, 1371056, 2937136
Permutations with no 3-term arithmetic progression. Ref JLMS 11 263 36. AMM 82 76 75.
[1,2; A3407]

M1202   2, 4, 10, 22, 40, 76, 138, 238, 408, 682, 1112, 1792, 2844, 4444, 6872, 10510,
15896, 23834
Representation degeneracies for Raymond strings. Ref NUPH B274 548 86. [3,1; A5306]

M1203   0, 1, 2, 4, 10, 24, 55, 128, 300, 700, 1632, 3809, 8890, 20744, 48406
Permutations according to distance. Ref AENS 79 207 62. [0,3; A2525, N0463]

M1204   2, 4, 10, 24, 60, 156, 410, 1092
An expansion into products. Ref MOC 26 271 72. [1,1; A6575]

M1205   1, 1, 2, 4, 10, 24, 66, 174, 504, 1406, 4210, 12198, 37378, 111278, 346846,
1053874, 3328188, 10274466, 32786630, 102511418, 329903058, 1042277722
Folding a strip of n labeled stamps. See Fig M4587. Equals 2.M1420. Ref CJM 2 397 50.
JCT 5 151 68. MOC 22 198 68. [1,3; A0682, N0464]

M1206   1, 1, 2, 4, 10, 24, 66, 176, 493, 1361
Folding a piece of wire of length n. See Fig M4587. Ref AMM 44 51 37. [0,3; A1997,
N0465]

M1207   1, 2, 4, 10, 24, 66, 180, 522, 1532, 4624, 14136, 43930, 137908, 437502, 1399068,
4507352, 14611576, 47633486, 156047204, 513477502, 1696305720, 5623993944
Series-parallel networks. Ref SAM 21 87 42. R1 142. AAP 4 123 72. [1,2; A0084, N0466]

M1208   1, 1, 2, 4, 10, 24, 67
Hexagonal n-element polyominoes whose graph is a path. Ref LNM 303 216 72. [1,3;
A3104]

M1209   2, 4, 10, 25, 64, 166
Alkyls with n carbon atoms. Ref ZFK 93 437 36. [1,1; A0645, N0467]

M1210 1, 2, 4, 10, 25, 64, 172, 472, 1319, 3750, 10796, 31416, 92276, 273172, 814246, 2441688, 7360877, 22295746, 67819576, 207083944, 634512581, 1950301202
Esters with n carbon atoms. Ref BA76 44. [2,2; A5958]

M1211 1, 1, 2, 4, 10, 25, 70, 196, 574, 1681, 5002, 14884, 44530, 133225, 399310, 1196836, 3589414, 10764961, 32291602, 96864964, 290585050, 871725625
Folding a piece of wire of length n. See Fig M4587. Ref AMM 44 51 37. GMJ 15 146 74. [0,3; A1998, N0468]

M1212 1, 2, 4, 10, 25, 70, 196, 588, 1764, 5544, 17424, 56628, 184041, 613470, 2044900, 6952660, 23639044, 81662152, 282105616, 987369656, 3455793796, 12228193432
$C(n).C([(2n+1)]/2)$. Ref JCT A43 1 86. [0,2; A5817]

M1213 1, 2, 4, 10, 25, 76, 251, 968
Caskets of order n. Ref TCS 81 31 91. [1,2; A6901]

M1214 1, 1, 1, 1, 1, 2, 4, 10, 25, 87, 313, 1357, 6244, 30926, 158428
4-connected simplicial polyhedra with n nodes. Ref ADM 41 230 89. Dil92. [3,6; A7021]

M1215 1, 1, 2, 4, 10, 26, 75, 215
Generalized ballot numbers. Ref clm. [0,3; A6123]

M1216 1, 1, 2, 4, 10, 26, 75, 225, 711, 2311, 7725, 26313, 91141, 319749, 1134234, 4060128, 14648614, 53208998, 194423568, 714130372, 2635256408, 9764995800
n-element posets which are unions of 2 chains. Ref AMM 88 294 81. [0,3; A6251]

M1217 1, 2, 4, 10, 26, 76, 231, 756, 2556, 9096, 33231, 126060, 488488, 1948232, 7907185, 32831370, 138321690, 593610420, 2579109780, 11377862340, 50726936820
Young tableaux of height 6. Ref BFK94. [1,2; A7579]

M1218 1, 1, 2, 4, 10, 26, 76, 232, 750, 2494, 8524, 29624, 104468, 372308, 1338936, 4850640, 17685270, 64834550, 238843660, 883677784, 3282152588, 12233309868
Connected unit interval graphs with n nodes. Ref rwr. [1,3; A7123]

M1219 1, 2, 4, 10, 26, 76, 232, 763, 2611, 9415, 35135, 136335, 544623, 2242618, 9463508, 40917803, 180620411, 813405580, 3728248990, 17377551032, 82232982872
Young tableaux of height 7. Ref BFK94. [1,2; A7578]

M1220 1, 2, 4, 10, 26, 76, 232, 764, 2619, 9486, 35596, 139392, 562848, 2352064, 10092160, 44546320, 201158620, 930213752, 4387327088, 21115314916
Young tableaux of height 8. Ref BFK94. [1,2; A7580]

M1221 1, 1, 2, 4, 10, 26, 76, 232, 764, 2620, 9496, 35696, 140152, 568504, 2390480, 10349536, 46206736, 211799312, 997313824, 4809701440, 23758664096
Self-conjugate permutations on n letters: $a(n)=a(n-1)+(n-1)a(n-2)$. Ref LU91 1 221. R1 86. MU60 6. DUMJ 35 659 68. JCT A21 162 76. [0,3; A0085, N0469]

M1222 1, 1, 2, 4, 10, 26, 80, 246, 810, 2704, 9252, 32066, 56360
Rooted plane trees with n nodes. Ref JRAM 278 334 75. [1,3; A3239]

M1223 1, 2, 4, 10, 27, 74, 202, 548, 1490, 4052, 11013, 29937, 81377, 221207, 601302,
     1634509, 4443055, 12077476, 32829985, 89241150, 242582598, 659407867
Nearest integer to $\cosh(n)$. Ref AMP 3 33 1843. MNAS 14(5) 14 25. HA26. LF60 93. [0,2;
A2459, N0470]

M1224 1, 2, 4, 10, 27, 92, 369, 1807, 10344, 67659, 491347, 3894446, 33278992,
     304256984, 2960093835, 30523315419, 332524557107, 3816805831381
Interval graphs with n nodes. Ref TAMS 272 422 82. pjh. [1,2; A5975]

M1225 1, 1, 1, 2, 4, 10, 28, 127
Permutation arrays of period n. Ref LNM 952 404 82. [1,4; A6841]

M1226 1, 2, 4, 10, 28, 130
Superpositions of cycles. Ref AMA 131 143 73. [3,2; A3223]

M1227 1, 2, 4, 10, 29, 86, 266
Triangulations. Ref WB79 337. [0,2; A5505]

M1228 0, 1, 0, 1, 2, 4, 10, 29, 90, 295, 1030, 3838, 15168, 63117, 275252, 1254801,
     5968046, 29551768, 152005634, 810518729, 4472244574, 25497104007, 149993156234
From a differential equation. Ref AMM 67 766 60. [0,5; A0995, N0471]

M1229 1, 2, 4, 10, 30, 98, 328, 1140, 4040, 14542, 53060, 195624, 727790, 2728450,
     10296720, 39084190, 149115456
n-step walks on hexagonal lattice. Ref JPA 6 352 73. [1,2; A3289]

M1230 1, 1, 1, 2, 4, 10, 30, 100, 380, 1600, 7400, 37400, 204600, 1205600, 7612000,
     51260000, 366784000, 2778820000, 22222332000, 187067320000, 1653461480000
Shifts 2 places left when e.g.f. is squared. Ref BeSl94. [0,4; A7558]

M1231 1, 2, 4, 10, 30, 106, 426, 1930, 9690
Balanced labeled graphs. Ref DM 15 384 76. [1,2; A5193]

M1232 2, 4, 10, 31, 43, 121, 424, 853
Stopping times. Ref MOC 54 393 90. [1,1; A7177]

M1233 1, 1, 2, 4, 10, 31, 120, 578, 3422, 24504, 208744
Elementary sequences of length n. Ref DAM 44 261 93. [1,3; A5268]

M1234 1, 1, 2, 4, 10, 31, 127, 711, 5621, 64049, 1067599
Sub-Fibonacci sequences of length n. Ref DAM 44 261 93. [1,3; A5269]

M1235 2, 4, 10, 32, 122, 544, 2770, 15872, 101042, 707584, 5405530, 44736512,
     398721962, 3807514624, 38783024290, 419730685952, 4809759350882
Expansion of $2(1 + \sin x) / \cos x$. Cf. M1492. Ref AMM 65 534 58. DKB 262. C1 261.
[2,1; A1250, N0472]

M1236 1, 2, 4, 10, 34, 114, 475
Planar maps without faces or vertices of degree 1. Ref SIAA 4 174 83. [2,2; A6397]

M1237 1, 1, 2, 4, 10, 34, 154, 874, 5914, 46234, 409114, 4037914, 43954714, 522956314,
6749977114, 93928268314, 1401602636314, 22324392524314, 378011820620314
Left factorials: $!n = \Sigma k!$, $k = 0 \ldots n$. Ref BALK 1 147 71. MR 44 #3945. [0,3; A3422]

M1238 1, 2, 4, 10, 36, 132, 616
Planar maps without faces or vertices of degree 1. Ref SIAA 4 174 83. [2,2; A6396]

M1239 1, 2, 4, 10, 36, 202, 1828, 27338, 692004, 30251722, 2320518948, 316359580362,
77477180493604
Related to binary partition function. Ref RSE 65 190 59. NMT 10 65 62. PCPS 66 376 69.
AB71 400. BIT 17 388 77. [0,2; A2577, N0473]

M1240 1, 2, 4, 10, 37, 138
Rooted planar maps. Ref CJM 15 542 63. [2,2; A0087, N0474]

M1241 2, 4, 10, 46, 1372, 475499108
Boolean functions of n variables. Ref JSIAM 12 294 64. [1,1; A0613, N0475]

M1242 0, 1, 2, 4, 11, 15, 18, 23, 37, 44, 57, 78, 88, 95, 106, 134, 156, 205, 221, 232, 249,
310, 323, 414, 429, 452, 550, 576, 639, 667, 715, 785, 816, 837, 946, 1003, 1038, 1122
Of form $(p^2 - 49)/120$ where p is prime. Ref IAS 5 382 37. [1,3; A2382, N0476]

M1243 1, 1, 2, 4, 11, 16, 49, 72, 214, 319, 947, 1408, 4187, 6223, 18502, 27504, 81769,
121552, 361379, 537196
Spanning trees in third power of cycle. Ref FQ 23 258 85. [1,3; A5822]

M1244 1, 2, 4, 11, 19, 56, 96, 296, 554, 1593, 3093
Permutation groups of degree n. Ref JPC 33 1069 29. LE70 169. [1,2; A0638, N0477]

M1245 1, 2, 4, 11, 23, 64, 134, 373, 781, 2174, 4552, 12671, 26531, 73852, 154634,
430441, 901273, 2508794, 5253004, 14622323, 30616751, 85225144, 178447502
Solution to a diophantine equation. Ref TR July 1973 p. 74. jos. [0,2; A6452]

M1246 1, 2, 4, 11, 28, 77, 209, 573, 1576, 4340, 11964, 33004, 91080, 251407, 694065,
1916306, 5291223, 14610468, 40344380, 111406090, 307637516, 849517917
Irreducible positions of size n in Montreal solitaire. Ref JCT A60 55 92. [1,2; A7048]

M1247 1, 2, 4, 11, 28, 78, 213, 598, 1670, 4723
Fixed points in trees. Ref PCPS 85 410 79. [1,2; A5200]

M1248 1, 2, 4, 11, 28, 91, 311
Triangulations. Ref WB79 336. [0,2; A5503]

M1249 1, 2, 4, 11, 31, 83, 227, 616, 1674, 4550, 12367, 33617, 91380, 248397, 675214,
1835421, 4989191, 13562027, 36865412, 100210581, 272400600, 740461601
$a(n)$ terms of harmonic series exceed n. See Fig M4299. Ref AMM 78 870 71. SI74 181.
MMAG 65 308 92. [0,2; A2387, N1385]

M1250   1, 2, 4, 11, 31, 102, 342, 1213, 4361, 16016, 59348, 222117
Graphical partitions with n points. Ref CN 21 684 78. [1,2; A4251]

M1251   1, 1, 1, 2, 4, 11, 33, 116, 435, 1832, 8167, 39700, 201785, 1099449, 6237505,
   37406458
Sequences of refinements of partitions of n into 1^n. Ref JLMS 9 565 75. [1,4; A2846, N0478]

M1252   1, 2, 4, 11, 33, 142, 822, 6910
Planar graphs with n nodes. Ref WI72 162. ST90. [1,2; A5470]

M1253   1, 1, 2, 4, 11, 34, 156, 1044, 12346, 274668, 12005168, 1018997864,
   165091172592, 50502031367952, 29054155657235488, 31426485969804308768
Number of graphs with n nodes. See Fig M1253. Ref MIT 17 22 55. MAN 174 68 67.
HA69 214. [0,3; A0088, N0479]

||||||||||||||||||||||||||||||||||||||||||||||||||||||||||||||||||||||||||||||||||||||||||||||||||||||||||

Figure M1253.   GRAPHS.

   M1253 gives the number of graphs with n nodes, shown here. For generating functions
see [R1 145], [HP73 84], [C1 264]. See also Fig. M3032. The table contains a large number of
related sequences. For example M1657 counts connected graphs. Unless identified otherwise,
graphs are normally **unlabeled**. A graph with n nodes may be **labeled** by attaching the numbers from 1 to n to the nodes (see for example Fig. M1141). Labeled objects are much easier
to count than unlabeled ones.

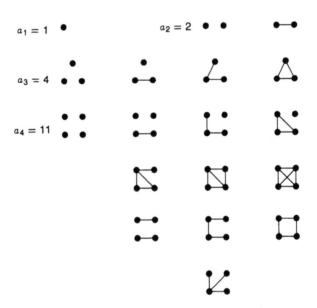

|||||||||||||||||||||||||||||||||||||||||||||||||||||||||||||||||||||||||||||||||||||||||||||||||||||||||||

M1254 1, 2, 4, 11, 67, 2279, 2598061, 3374961778892, 569518350449261 4029263279, 16217557574922386301420536972254869595782763547561
Representation requires n triangular numbers with greedy algorithm. Ref Lem00. jos. [1,2; A6894]

M1255 1, 2, 4, 12, 30, 88
Zero-entropy permutations of length n. Ref eco. [1,2; A6948]

M1256 0, 0, 2, 4, 12, 32, 108, 336, 1036, 3120, 9540, 29244
Permutations according to distance. Ref AENS 79 213 62. [0,3; A2528, N0480]

M1257 1, 2, 4, 12, 33, 102, 312, 1010
Symmetric trivalent maps with n nodes. Ref WB79 337. [3,2; A5028]

M1258 1, 2, 4, 12, 34, 111, 360, 1226
Rooted planar 2-trees with n nodes. Ref MAT 15 121 68. [1,2; A1895, N0481]

M1259 1, 1, 2, 4, 12, 36, 152, 624, 3472, 18256, 126752, 814144, 6781632, 51475776, 500231552, 4381112064, 48656756992, 482962852096, 6034272215552
Expansion of $e^x/\cos x$. [0,3; A3701]

M1260 1, 2, 4, 12, 39, 202, 1219, 9468, 83435, 836017, 9223092, 111255228, 1453132944, 20433309147, 307690667072, 4940118795869, 84241805734539
n-gons. Ref AMM 67 349 60. AMA 131 143 73. [3,2; A0940, N0482]

M1261 1, 2, 4, 12, 48, 200, 1040, 5600, 33600
Sorting numbers. Ref PSPM 19 173 71. [0,2; A2871, N0483]

M1262 1, 1, 2, 4, 12, 56, 456, 6880, 191536, 9733056, 903753248, 154108311168, 48542114686912, 28401423719122304, 31021002160355166848
Tournaments with n nodes. Ref MO68 87. HP73 245. [1,3; A0568, N0484]

M1263 1, 2, 4, 12, 60, 444, 4284, 50364, 695484, 11017404, 196811964, 3912703164, 85662309564, 2047652863164, 53059407256764, 1481388530277564
$1 + \Sigma 2^k k!$, $k = 1 .. n$. [-1,2; A4400]

M1264 1, 2, 4, 12, 60, 780, 47580, 37159980, 1768109008380, 65702897157329640780, 116169884340604934905464739377180
$a(n+1) = a(n)(a(n-1)+1)$. Ref FQ 25 208 87. [0,2; A5831]

M1265 2, 4, 12, 80, 3984, 37333248, 25626412338274304
Boolean functions of n variables. Ref HA65 147. MU71 38. [0,1; A3180]

M1266 1, 2, 4, 12, 81, 1684, 123565, 33207256, 34448225389
Self-dual functions of n variables. Ref PGEC 19 821 70. MU71 38. [1,2; A2080, N0485]

M1267 0, 1, 2, 4, 12, 81, 2646, 1422564
Self-dual monotone Boolean functions of n variables. Ref Wels71 181. Loeb94. Loeb94a.
[0,3; A1206, N0486]

M1268 1, 1, 2, 4, 12, 108, 10476, 108625644, 11798392680793836,
13920206856860156878594694949658348
A nonlinear recurrence. Ref FQ 11 436 73. [0,3; A1696, N0487]

M1269 1, 2, 4, 13, 41, 226, 1072, 9374, 60958, 723916, 5892536, 86402812, 837641884,
14512333928, 162925851376, 3252104882056, 41477207604872
Terms in a skew determinant. Ref RSE 21 354 1896. MU06 3 282. [1,2; A2771, N0488]

M1270 1, 1, 0, 1, 1, 2, 4, 13, 42, 308
Connected linear spaces with n points. Ref BSMB 19 228 67. [0,6; A1548, N0489]

M1271 1, 1, 1, 2, 4, 13, 46, 248, 1516, 13654, 142873, 2156888, 38456356, 974936056
Alternating sign matrices. Ref LNM 1234 292 86. SFCA92 2 32. [1,4; A5164]

M1272 0, 0, 0, 2, 4, 14, 34, 98, 270, 768, 2192, 6360, 18576, 54780, 162658, 486154,
1461174, 4413988, 13393816, 40807290
Paraffins with n carbon atoms. Ref JACS 54 1105 32. [1,4; A0622, N0490]

M1273 0, 0, 0, 0, 0, 2, 4, 14, 38, 89, 234, 579, 1466
n-node forests not determined by their spectra. Ref LNM 560 89 76. [1,6; A6611]

M1274 0, 1, 2, 4, 14, 38, 118, 338, 1006, 2990, 8974, 26862, 80510, 241390, 723934,
2171046, 6512910, 19536974, 58608782, 175821710, 527470318, 1582385678
Shifts left under XOR-convolution with itself. Ref BeSl94. [0,3; A7462]

M1275 1, 1, 1, 2, 4, 14, 38, 216, 600, 6240, 9552, 319296, 519312, 28108560, 176474352,
3998454144, 43985078784, 837126163584, 12437000028288, 237195036797184
Expansion of $1/(1 - \log(1+x))$. Ref PO54 9. [0,4; A6252]

M1276 1, 1, 1, 2, 4, 14, 46, 224
Column of Kempner tableau. There is a simple 2-D recurrence. Ref STNB 11 41 81. [1,4; A5437]

M1277 2, 4, 14, 47, 184, 761, 3314, 14997, 69886, 333884, 1626998, 8067786, 40580084,
206734083, 1064666724, 5536480877, 29036188788, 153450351924, 816503772830
Connected planar maps with n edges. Ref DM 36 205 81. [1,1; A6443]

M1278 2, 4, 14, 52, 194, 724, 2702, 10084, 37634, 140452, 524174, 1956244, 7300802,
27246964, 101687054, 379501252, 1416317954, 5285770564, 19726764302
$a(n) = 4a(n-1) - a(n-2)$. Ref FQ 11 29 73. MMAG 48 209 75. [0,1; A3500]

M1279 1, 2, 4, 14, 52, 248, 1416, 9172, 66366, 518868, 4301350, 37230364, 333058463,
3057319072, 28566583950, 273298352168, 2645186193457, 25931472185976
Connected planar maps with n edges. Ref DM 36 224 81. SIAA 4 169 83. trsw. [0,2;
A6385]

M1280 1, 2, 4, 14, 54, 332, 2246, 18264, 164950, 1664354, 18423144, 222406776, 2905943328, 40865005494, 615376173184, 9880209206458, 168483518571798
n-gons. Ref AMM 67 349 60. [3,2; A0939, N0491]

M1281 1, 2, 4, 14, 57, 312, 2071, 15030, 117735, 967850, 8268816, 72833730, 658049140, 6074058060, 57106433817, 545532037612, 5284835906037
Planar maps with n edges. Ref DM 36 224 81. SIAA 4 169 83. trsw. [0,2; A6384]

M1282 1, 1, 1, 2, 4, 14, 62
Necklace permutations. Ref AMM 81 340 74. [1,4; A3322]

M1283 1, 2, 4, 14, 72, 316, 1730, 9728
Isolated reformed permutations. Ref GN93. [2,2; A7712]

M1284 2, 4, 14, 96, 1146, 19996, 456774, 12851768, 429005426
Fanout-free functions of n variables. Ref CACM 23 705 76. PGEC 27 315 78. [0,1; A5737]

M1285 2, 4, 14, 104, 1882, 94572, 15028134, 8378070864, 17561539552946
Threshold functions of n variables. Ref PGEC 19 821 70. MU71 38. [0,1; A0609, N0492]

M1286 2, 4, 14, 128, 3882, 412736, 151223522, 189581406208, 820064805806914, 12419746847290729472, 6685900833067943215516802
A binomial coefficient sum. Ref PGEC 14 322 65. [0,1; A1527, N0493]

M1287 1, 2, 4, 14, 222, 616126, 200253952527184
Boolean functions of n variables. Ref HA65 153. MU71 38. [0,2; A0370, N0494]

M1288 0, 0, 0, 0, 0, 2, 4, 15, 36, 108, 276, 770, 2036, 5586, 15072, 41370, 113184, 312488, 863824, 2401344, 6692368, 18724990, 52531788, 147824963, 417006316
Identity connected unit interval graphs with n nodes. Ref rwr. [1,6; A7122]

M1289 0, 0, 0, 0, 2, 4, 15, 36, 108, 276, 771, 2044, 5622, 15204
Identity unit interval graphs. Ref TAMS 272 425 82. rwr. [1,5; A5219]

M1290 1, 1, 2, 4, 15, 102, 4166, 402631, 76374899, 27231987762, 18177070202320, 22801993267433275, 54212469444212172845, 246812697326518127351384
Series-reduced labeled graphs with n nodes. Ref JCT B19 282 75. [0,3; A3514]

M1291 1, 2, 4, 16, 48, 160, 576, 4096, 14336, 73728, 327680, 2985984, 14929920, 77635584
Largest determinant of $(+1, -1)$-matrix of order n. Cf. M0244. Ref ZAMM 42 T21 62. MZT 83 127 64. AMM 79 626 72. MS78 54. [1,2; A3433]

M1292 1, 2, 4, 16, 56, 256, 1072, 6224, 33616, 218656, 1326656, 9893632, 70186624, 574017536, 4454046976, 40073925376, 347165733632, 3370414011904
Degree n permutations of order dividing 4. Ref CJM 7 159 55. [1,2; A1472, N0495]

E.g.f.: $(1 + x + x^3)\exp(x(4 + 2x + x^3)/4)$.

M1293   1, 2, 4, 16, 56, 256, 1072, 11264, 78976, 672256, 4653056, 49810432, 433429504,
   4448608256, 39221579776, 1914926104576, 29475151020032, 501759779405824
Degree n permutations of order a power of 2. Ref CJM 7 159 55. [1,2; A5388]

M1294   1, 2, 4, 16, 63, 328, 1933, 12653
Trivalent maps. Ref WB79 337. [3,2; A5027]

M1295   1, 1, 2, 4, 16, 80, 520, 3640, 29120, 259840, 2598400, 28582400, 343235200,
   4462057600, 62468806400, 936987251200, 14991796019200, 254860532326400
Expansion of $\exp(-x^3/3) / (1-x)$. Ref R1 85. [0,3; A0090, N0496]

M1296   1, 2, 4, 16, 89, 579, 3989, 28630, 210847, 1584308
Q-graphs with $2n$ edges. Ref AEQ 31 63 86. [1,2; A7171]

M1297   2, 4, 16, 256, 65536, 4294967296, 18446744073709551616,
   340282366920938463463374607431768211456
$2\!\uparrow\!2^n$. Ref MOC 23 456 69. FQ 11 429 73. [0,1; A1146, N0497]

M1298   2, 4, 17, 19, 5777, 5779
A predictable Pierce expansion. Ref FQ 22 333 84. [0,1; A6276]

M1299   2, 4, 18, 648, 3140062, 503483766022188
Balanced colorings of n-cube. Ref JALC 1 266 92. [1,1; A6853]

M1300   1, 2, 4, 24, 128, 880, 7440
Sorting numbers. Ref PSPM 19 173 71. [0,2; A2875, N0498]

M1301   1, 2, 4, 24, 1104, 2435424, 11862575248704, 28144138306230580975686 1824,
   15841850420004711107538836924188411800321048574349 0304
A slowly converging series. Ref AMM 54 138 47. FQ 11 432 73. [0,2; A1510, N0499]

M1302   1, 2, 4, 44, 164, 616
N-free graphs. Ref QJMO 38 166 87. [0,2; A7596]

M1303   1, 2, 4, 54, 5337932, 3253511965960720
(6,5)-graphs. Ref PE79. [4,2; A5274]

M1304   2, 4, 60, 1276, 41888, 1916064, 116522048, 9069595840, 878460379392
Related to Latin rectangles. Ref BCMS 33 125 41. [2,1; A1625, N0500]

M1305   1, 0, 2, 4, 76, 109875
Self-dual Boolean functions of n variables. Ref PGEC 11 284 62. MU71 38. [1,3; A6688]

M1306   2, 4, 94, 96, 98, 400, 402, 404, 514, 516, 518, 784, 786, 788, 904, 906, 908, 1114,
   1116, 1118, 1144, 1146, 1148, 1264, 1266, 1268, 1354, 1356, 1358, 3244, 3246, 3248
Even numbers not the sum of a pair of twin primes. Ref Well86 132. [1,1; A7534]

M1307 1, 2, 4, 104, 272, 3104, 79808, 631936, 1708288, 7045156352, 1413417032704,
6587672324096, 37378439704576, 66465881481076736, 80812831866241024
Numerators of expansion of sinh x/sin x. Cf. M2294. Ref MMAG 31 189 58. [0,2; A0965, N0501]

M1308 1, 2, 4, 118, 132, 140, 152, 208, 240, 242, 288, 290, 306, 378, 392, 426, 434, 442,
508, 510, 540, 542, 562, 596, 610, 664, 680, 682, 732, 782, 800, 808, 866, 876, 884, 892
$n^8 + 1$ is prime. [1,2; A6314]

M1309 1, 2, 4, 243, 198815685282
(7,6)-graphs. Ref PE79. [5,2; A5275]

M1310 1, 2, 4, 65536
Ackermann function $A(n)$ (the next term is very large!). Ref SIAC 20 160 91. [0,2; A6263]

SEQUENCES BEGINNING . . ., 2, 5, . . .

M1311 2, 5, 0, 2, 9, 0, 7, 8, 7, 5, 0, 9, 5, 8, 9, 2, 8, 2, 2, 2, 8, 3, 9, 0, 2, 8, 7, 3, 2, 1, 8, 2, 1,
5, 7, 8, 6, 3, 8, 1, 2, 7, 1, 3, 7, 6, 7, 2, 7, 1, 4, 9, 9, 7, 7, 3, 3, 6, 1, 9, 2, 0, 5, 6
Decimal expansion of Feigenbaum reduction parameter. Ref JPA 12 275 79. MOC 57 438 91. [1,1; A6891]

M1312 0, 2, 5, 3, 3, 1, 3, 5, 3, 1, 5, 3, 1, 3, 3, 5, 3, 1, 5, 3, 3, 1, 3, 5, 1, 3, 5, 3, 1, 3, 3, 5, 3,
1, 5, 3, 3, 1, 3, 5, 3, 1, 5, 3, 1, 3, 3, 5, 1, 3, 5, 3, 3, 1, 3, 5, 1, 3, 5, 3, 1, 3, 3, 5, 3, 1, 5, 3, 3
A continued fraction. Ref JNT 11 213 79. [0,2; A4200]

M1313 0, 2, 5, 3, 15, 140, 5, 56
Queens problem. Ref SL26 49. [1,2; A2565, N0502]

M1314 1, 2, 5, 4, 12, 6, 9, 23, 11, 27, 34, 22, 10, 33, 15, 37, 44, 28, 80, 19, 81, 14, 107, 89,
64, 16, 82, 60, 53, 138, 25, 114, 148, 136, 42, 104, 115, 63, 20, 143, 29, 179, 67, 109
Related to Størmer numbers. Ref AMM 56 526 49. [2,2; A2314, N0503]

M1315 2, 5, 5, 3, 2, 3, 3, 4, 6, 3, 7, 10, 10, 8, 7, 8, 8, 9, 11, 3, 7, 10, 10, 8, 7, 8, 8, 9, 11, 7,
9, 12, 12, 10, 9, 10, 10, 11, 13, 6, 8, 11, 11, 9, 8, 9, 9, 10, 12, 5, 7, 10, 10, 8, 7, 8, 8, 9
Number of letters in n (in Hungarian). [1,1; A7292]

M1316 1, 1, 2, 5, 5, 16, 7, 50, 34, 45, 8, 301, 9, 53, 104
Transitive groups of degree n. Ref BAMS 2 143 1896. LE70 178. CALG 11 870 83. gb. jmckay. [1,3; A2106, N0504]

M1317 2, 5, 6, 7, 10, 12, 14, 15, 20, 21, 22, 23, 25, 26, 30, 31, 34, 36, 37, 38, 39, 41, 42,
45, 46, 47, 49, 50, 52, 53, 54, 55, 57, 58, 60, 62, 66, 69, 70, 71, 72, 73, 74, 76, 78, 79, 84
Elliptic curves. Ref JRAM 212 24 63. [1,1; A2157, N0505]

M1318 1, 2, 5, 6, 8, 12, 18, 30, 36, 41, 66, 189, 201, 209, 276, 353, 408, 438, 534, 2208, 2816, 3168, 3189, 3912, 20909, 34350, 42294, 42665, 44685, 48150, 55182
$3.2^n + 1$ is prime. Ref KN1 2 614. Rie85 381. Cald94. [1,2; A2253, N0506]

M1319 1, 2, 5, 6, 9, 10, 10, 15, 15, 16, 18, 24, 18, 26
Asymmetric families of palindromic squares. Ref JRM 22 130 90. [7,2; A7573]

M1320 2, 5, 6, 9, 10, 13, 17, 20, 21, 24, 28, 32, 35, 36, 39, 43, 47, 50, 51, 54, 58, 62, 65, 66, 69, 73, 77, 80, 81, 84, 88, 92, 95, 96, 99, 103, 107, 110, 111, 114, 118, 122, 125, 126
$a(n)$ is smallest number not $= a(j) + a(k)$, $j < k$. Ref GU94. [1,1; A3664]

M1321 2, 5, 6, 10, 8, 16, 10, 19, 16, 22, 14, 34, 16, 28, 28, 36, 20, 45, 22, 48, 36, 40, 26, 68, 34, 46, 44, 62, 32, 80, 34, 69, 52, 58, 52, 100, 40, 64, 60, 98, 44, 104, 46, 90, 84, 76
Subgroups of dihedral group: $\sigma(n) + d(n)$. Ref TYCM 23 150 92. [1,1; A7503]

M1322 2, 5, 6, 10, 13, 14, 21, 22, 29, 30, 33, 34, 37, 38, 41, 42, 46, 57, 58, 61, 65, 66, 69, 70, 73, 77, 78, 82, 85, 86, 93, 94, 101, 102, 105, 106, 109, 110, 113, 114, 118, 122, 129
n, $n + 1$ are square-free. Ref Halm91 28. [1,1; A7674]

M1323 1, 2, 5, 6, 10, 14, 21, 22, 27, 32, 42, 48, 59, 70
Related to recurrences over rings. Ref MSC 33 10 73. [1,2; A5984]

M1324 1, 2, 5, 6, 11, 13, 17, 22, 27, 29, 37, 44, 44, 55
Generalized divisor function. Ref PLMS 19 112 19. [3,2; A2133, N0507]

M1325 2, 5, 6, 14, 21, 26, 141, 278, 281, 306, 345
$(2^{2n+1} - 2^{n+1} + 1)/5$ is prime. Ref CUNN. [1,1; A6596]

M1326 1, 2, 5, 6, 14, 21, 29, 30, 54, 90, 134, 155, 174, 230, 234, 251, 270, 342, 374, 461, 494, 550, 666, 750, 810, 990, 1890, 2070, 2486, 2757, 2966, 3150, 3566, 3630, 4554
Related to lattice points in spheres. Ref MOC 20 306 66. [1,2; A0092, N0508]

M1327 1, 2, 5, 7, 8, 13, 21, 40, 75, 113, 146, 281, 425
Spiral sieve using Fibonacci numbers. Ref FQ 12 395 74. [1,2; A5624]

M1328 2, 5, 7, 9, 11, 12, 13, 15, 19, 23, 27, 29, 35, 37, 41, 43, 45, 49, 51, 55, 61, 67, 69, 71, 79, 83, 85, 87, 89, 95, 99, 107, 109, 119, 131, 133, 135, 137, 139, 141, 145, 149, 153
$a(n)$ is smallest number which is uniquely $a(j) + a(k)$, $j < k$ (periodic mod 126). Ref JCT A12 31 72. JCT A60 124 92. GU94. [1,1; A7300]

M1329 1, 2, 5, 7, 9, 11, 12, 15, 16, 19
First column of spectral array $W(\sqrt{2})$. Ref FrKi94. [1,2; A7069]

M1330 2, 5, 7, 9, 12, 14, 17, 19, 21, 24, 26, 28, 31, 33, 36, 38, 40, 43, 45, 47, 49, 51, 54, 56, 58, 61, 63, 66, 68, 70
Related to Fibonacci representations. Ref FQ 11 385 73. [1,1; A3256]

M1331   2, 5, 7, 10, 12, 14, 17, 19, 22, 24, 26, 29, 31, 34, 36, 39, 41, 43, 46, 48
Related to Pellian representations of numbers. Ref FQ 10 487 72. [1,1; A3153]

M1332   2, 5, 7, 10, 13, 15, 18, 20, 23, 26, 28, 31, 34, 36, 39, 41, 44, 47, 49, 52, 54, 57, 60,
   62, 65, 68, 70, 73, 75, 78, 81, 83, 86, 89, 91, 94, 96, 99, 102, 104, 107, 109, 112, 115, 117
A Beatty sequence: $[n\tau^2]$. See Fig M1332. Cf. M2322. Ref CMB 2 191 59. AMM 72 1144
65. FQ 11 385 73. [1,1; A1950, N0509]

||||||||||||||||||||||||||||||||||||||||||||||||||||||||||||||||||||||||||||||||||||||||||||||||||||||||||||||||||||||||||

Figure M1332.   BEATTY SEQUENCES.

If α and β are positive irrational numbers such that $\alpha^{-1} + \beta^{-1} = 1$, then it is a remarkable
fact that the **Beatty sequences**

$$[\alpha], [2\alpha], [3\alpha], \ldots \text{ and } [\beta], [2\beta], [3\beta], \ldots$$

are disjoint, and together contain all the positive integers! For example if $\alpha = (1 + \sqrt{5})/2$ we
obtain M2322 and M1332. The table contains a number of other Beatty sequences. The pair
M0946, M2622 resulting from $\alpha = (1 + \sqrt{3})/2$ is surprisingly similar to the pair M0947, M2621
obtained from $\alpha = 1 + e^{-1}$: the first 39 terms of M0946 coincide with those of M0947.

||||||||||||||||||||||||||||||||||||||||||||||||||||||||||||||||||||||||||||||||||||||||||||||||||||||||||||||||||||||||||

M1333   1, 2, 5, 7, 11, 14, 20, 24, 30, 35
Consistent arcs in a tournament (equals $C(n,2)$ − M2334). Ref CMB 12 263 69. MSH 37
23 72. MR 46 15(87) 73. [2,2; A1225, N0510]

M1334   0, 0, 1, 2, 5, 7, 11, 15, 21
Maximum triangles formed from n lines. Ref GA83 171. [1,4; A6066]

M1335   2, 5, 7, 12, 13, 23, 19, 31, 30, 45, 33, 67, 43, 65, 65, 84, 61, 107, 69, 123, 97, 115,
   85, 175, 110, 147, 133, 179, 111, 223, 129, 215, 175, 203, 179, 302, 159, 235, 215, 315
Inverse Moebius transform of primes. Ref EIS § 2.7. [1,1; A7445]

M1336   0, 1, 2, 5, 7, 12, 15, 22, 26, 35, 40, 51, 57, 70, 77, 92, 100, 117, 126, 145, 155, 176,
   187, 210, 222, 247, 260, 287, 301, 330, 345, 376, 392, 425, 442, 477, 495, 532, 551, 590
Generalized pentagonal numbers: $n(3n-1)/2$, $n = 0, \pm 1, \pm 2, \ldots$ Ref NZ66 231. AMM 76
884 69. HO70 119. [0,3; A1318, N0511]

M1337   1, 2, 5, 7, 12, 18, 26, 35, 50, 67, 88, 116, 149, 191, 245, 306, 381, 477, 585, 718
Weighted count of partitions with distinct parts. Ref ADV 61 160 86. [1,2; A5895]

M1338   2, 5, 7, 12, 19, 31, 50, 81, 131, 212, 343, 555, 898, 1453, 2351, 3804, 6155, 9959,
   16114, 26073, 42187, 68260, 110447, 178707, 289154, 467861, 757015, 1224876
$a(n) = a(n-1) + a(n-2)$. Ref FQ 3 129 65. BR72 52. [0,1; A1060, N0512]

M1339   1, 2, 5, 7, 14, 13, 27, 26, 39, 38, 65, 50, 90, 75, 100, 100, 152, 111, 189, 148, 198,
   185, 275, 196, 310, 258, 333, 294, 434, 292, 495, 392, 490, 440, 588, 438, 702, 549, 684
Moebius transform of triangular numbers. Ref EIS § 2.7. [1,2; A7438]

M1340 1, 1, 2, 5, 7, 19, 26, 71, 97, 265, 362, 989, 1351, 3691, 5042, 13775, 18817, 51409, 70226, 191861, 262087, 716035, 978122, 2672279, 3650401, 9973081, 13623482
$a(2n) = a(2n-1) + a(2n-2)$, $a(2n+1) = 2a(2n) + a(2n-1)$. Ref MQET 1 10 16. NZ66 181. [0,3; A2531, N0513]

M1341 2, 5, 7, 26, 265, 1351, 5042, 13775, 18817, 70226, 716035, 3650401
Related to Genocchi numbers. Ref ANN 36 645 35. [0,1; A2317, N0514]

M1342 2, 5, 7, 197, 199, 7761797, 7761799
A predictable Pierce expansion. Ref FQ 22 334 84. [0,1; A6275]

M1343 2, 5, 7, 257, 521, 97, 911, 673, 530713, 27961, 58367, 2227777, 79301, 176597, 142111, 67280421310721, 45957792327018709121, 33388093, 870542161121
Largest factor of $n^n + 1$. Ref rgw. [1,1; A7571]

M1344 1, 2, 5, 8, 11, 14, 18, 22, 27, 31, 36, 41, 46, 52, 58, 64, 70, 76, 82, 89, 96, 103, 110, 117, 125, 132, 140, 148, 156, 164, 172, 181, 189, 198, 207, 216, 225, 234, 243, 252, 262
$[n^{3/2}]$. Ref PCPS 47 214 51. [1,2; A0093, N0515]

M1345 1, 2, 5, 8, 12, 16, 20, 24, 29, 34, 39, 44, 49, 54, 59, 64, 70, 76, 82, 88, 94, 100, 106, 112, 118, 124, 130, 136, 142, 148, 154, 160, 167, 174, 181, 188, 195, 202, 209, 216, 223
Binary entropy: $a(n) = n + \min \{ a(k) + a(n-k) : 1 \le k \le n-1 \}$. Ref KN1 3 374. [1,2; A3314]

M1346 2, 5, 8, 12, 17, 22, 28, 34, 41, 48, 56, 65, 74, 84, 94, 105, 116, 128, 140, 153, 166, 180, 194, 209, 224, 240, 257, 274, 292, 310, 329, 348, 368, 388, 409, 430, 452, 474, 497
The square sieve. Ref JRM 4 288 71. [1,1; A2960]

M1347 2, 5, 8, 13, 16, 21, 26, 35
Ramsey numbers. Ref CMB 8 579 65. [2,1; A0789, N0516]

M1348 1, 2, 5, 8, 13, 18, 25, 32, 41, 50, 61, 72, 85, 98, 113, 128, 145, 162, 181, 200, 221, 242, 265, 288, 313, 338, 365, 392, 421, 450, 481, 512, 545, 578, 613, 648, 685, 722, 761
$\lceil n^2/2 \rceil$. [1,2; A0982, N0517]

M1349 0, 1, 2, 5, 8, 14, 20, 30, 40, 55, 70, 91, 112, 140, 168, 204, 240, 285, 330, 385, 440, 506, 572, 650, 728, 819, 910, 1015, 1120, 1240, 1360, 1496, 1632, 1785, 1938, 2109
$C(n+3,3)/4$, n odd; $n(n+2)(n+4)/24$, n even. Ref Lie92. [0,3; A6918]

M1350 1, 2, 5, 8, 14, 21, 32, 45, 65, 88, 121, 161, 215, 280, 367, 471, 607, 771, 980, 1232, 1551, 1933, 2410, 2983, 3690, 4536, 5574, 6811, 8317, 10110, 12276
Trees of diameter 4. Ref IBMJ 4 476 60. KU64. [5,2; A0094, N0518]

M1351 1, 2, 5, 8, 17, 24, 46, 64, 107, 147, 242, 302, 488, 629, 922, 1172, 1745, 2108, 3104, 3737
Partitions of $2n$ with all subsums different from n. Ref ADM 43 164 89. [1,2; A6827]

M1352 2, 5, 8, 18, 29, 57, 96, 183, 318, 603, 1080, 2047, 3762
Polytopes. Ref GR67 424. [3,1; A0943, N0519]

M1353 1, 0, 1, 1, 2, 5, 8, 21, 42, 96, 222, 495, 1177, 2717, 6435, 15288, 36374, 87516,
210494, 509694, 1237736, 3014882, 7370860, 18059899, 44379535, 109298070
Partitions of points on a circle. Ref BAMS 54 359 48. [0,5; A1005, N0520]

M1354 1, 2, 5, 9, 9, 2, 1, 0, 4, 9, 8, 9, 4, 8, 7, 3, 1, 6, 4, 7, 6, 7, 2, 1, 0, 6, 0, 7, 2, 7, 8, 2, 2,
8, 3, 5, 0, 5, 7, 0, 2, 5, 1, 4, 6, 4, 7, 0, 1, 5, 0, 7, 9, 8, 0, 0, 8, 1, 9, 7, 5, 1, 1, 2, 1, 5, 5, 2, 9
Decimal expansion of cube root of 2. Ref SMA 18 175 52. [1,2; A2580, N0521]

M1355 2, 5, 9, 10, 11, 16, 17, 19, 21, 22, 23, 25, 26, 27, 29, 33, 34, 35, 37, 41, 43, 45, 46,
47, 49, 50, 51, 52, 53, 55, 58, 59, 61, 64, 65, 66, 67, 69, 70, 71, 73, 75, 76, 77, 79, 81, 82
$\sigma(x) = n$ has no solution. Ref AS1 840. [1,1; A7369]

M1356 0, 2, 5, 9, 14, 20, 27, 35, 44, 54, 65, 77, 90, 104, 119, 135, 152, 170, 189, 209, 230,
252, 275, 299, 324, 350, 377, 405, 434, 464, 495, 527, 560, 594, 629, 665, 702, 740, 779
$n(n+3)/2$. Ref AS1 797. [0,2; A0096, N0522]

M1357 1, 1, 0, 2, 5, 9, 14, 20, 69, 125, 209, 329, 923, 1715, 3002, 5004, 12869, 24309,
43757, 75581, 184755, 352715, 646645, 1144065, 2704155, 5200299, 9657699
Euler characteristics of polytopes. Ref JCT A17 346 74. [1,4; A6482]

M1358 0, 1, 2, 5, 9, 14, 78, 81, 141, 189, 498
$2^{2n+1} + 2^{n+1} + 1$ is prime. Ref CUNN. [1,3; A6599]

M1359 2, 5, 9, 15, 25, 33, 393, 12231
Stopping times. Ref MOC 54 392 90. [1,1; A7176]

M1360 2, 5, 9, 17, 27, 40, 55, 73, 117, 143
Solutions to a linear inequality. Ref JRAM 227 47 67. [3,1; A2797, N0524]

M1361 1, 2, 5, 9, 17, 28, 47, 73, 114, 170, 253, 365, 525, 738, 1033, 1422, 1948, 2634,
3545, 4721, 6259, 8227, 10767, 13990, 18105, 23286, 29837, 38028, 48297, 61053
Partitions of n into parts of 2 kinds. Ref RS4 90. RCI 199. [0,2; A0097, N0525]

M1362 1, 2, 5, 9, 18, 31, 57, 92, 159
Allomorphic polyhedra with n nodes. Ref JRM 4 123 71. md. [4,2; A2883, N0526]

M1363 0, 1, 2, 5, 9, 21, 44, 103, 232, 571, 1368, 3441
Total diameter of unlabeled trees with n nodes. Ref IBMJ 4 476 60. [1,3; A1851, N0527]

M1364 1, 1, 2, 5, 9, 22, 62, 177, 560, 1939
Series-reduced star graphs with n edges. Ref JMP 7 1585 66. [3,3; A2935, N0528]

M1365 1, 1, 2, 5, 9, 24, 70, 222
2-connected maps without faces of degree 2. Ref SIAA 4 174 83. [3,3; A6405]

M1366   2, 5, 10, 13, 17, 26, 29, 37, 41, 53, 58, 61, 65, 73, 74, 82, 85, 89, 97, 101, 106, 109, 113, 122, 130, 137, 145, 149, 157, 170, 173, 181, 185, 193, 197, 202, 218, 226, 229, 233
Fundamental unit of $Q(\sqrt{n})$ has norm -1. Ref BU89 236. [1,1; A3654]

M1367   1, 2, 5, 10, 15, 25, 37, 52, 67, 97, 117
Generalized divisor function. Ref PLMS 19 112 19. [6,2; A2134, N0530]

M1368   1, 2, 5, 10, 16, 24, 33, 44, 56, 70, 85, 102, 120, 140, 161, 184, 208, 234, 261, 290, 320, 352, 385, 420, 456, 494, 533, 574, 616, 660, 705, 752, 800, 850, 901, 954, 1008
Series-reduced planted trees with n nodes, $n-3$ endpoints. Ref jr. [6,2; A1859, N0531]

$$\text{G.f.:} \quad (1 + x^2 + 2x^3 - x^4) / (1 - x)^2 (1 - x^2).$$

M1369   2, 5, 10, 17, 24, 35
Length of uncrossed knight's path on $n \times n$ board. See Fig M1369. Ref JRM 2 157 69. [3,1; A3192]

Figure M1369.   UNCROSSED KNIGHT"S TOURS.

M1369 gives the maximal length of an knight's tour on an $n \times n$ board that does not cross itself:

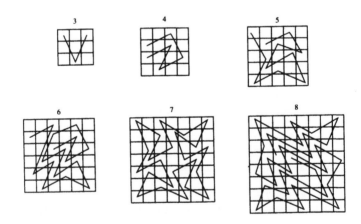

M1370   2, 5, 10, 17, 28, 41, 58, 77, 100, 129, 160, 197, 238, 281, 328, 381, 440, 501, 568, 639, 712, 791, 874, 963, 1060, 1161, 1264, 1371, 1480, 1593, 1720, 1851, 1988, 2127
Sum of first n primes. Ref JRM 14 205 81. [1,1; A7504]

M1371   2, 5, 10, 18, 31, 52, 86, 141, 230, 374, 607, 984, 1594, 2581, 4178, 6762, 10943, 17708, 28654, 46365, 75022, 121390, 196415, 317808, 514226, 832037, 1346266
Total preorders. Ref MSH 53 20 76. [3,1; A6327]

$$\text{G.f.:}\ \ (2 + x)\ /\ (1 - x)\ (1 - x - x^2).$$

M1372   1, 2, 5, 10, 18, 32, 55, 90, 144, 226, 346, 522, 777, 1138, 1648, 2362, 3348, 4704, 6554, 9056, 12425, 16932, 22922, 30848, 41282, 54946, 72768, 95914, 125842, 164402
Coefficients of an elliptic function. Ref CAY 9 128. MOC 29 852 75. [0,2; A1936, N0532]

$$\text{G.f.:}\ \ \Pi\,(1 - x^k)^{-c(k)},\ c(k) = 2,2,2,0,2,2,2,0,....$$

M1373   1, 2, 5, 10, 19, 33, 57, 92, 147, 227, 345, 512, 752, 1083, 1545, 2174, 3031, 4179, 5719, 7752, 10438, 13946, 18519, 24428, 32051, 41805, 54265, 70079, 90102, 115318
Partitions of n into parts of 2 kinds. Ref RS4 90. RCI 199. [0,2; A0098, N0533]

M1374   1, 2, 5, 10, 20, 24, 26, 41, 53, 130, 149, 205, 234, 287, 340, 410, 425, 480, 586, 840, 850, 986, 1680, 1843, 2260, 2591, 3023, 3024, 3400, 3959, 3960, 5182, 5183, 7920
Related to lattice points in circles. Ref MOC 20 306 66. [1,2; A0099, N0534]

M1375   1, 2, 5, 10, 20, 35, 62, 102, 167, 262, 407, 614, 919, 1345, 1952, 2788, 3950, 5524, 7671, 10540, 14388, 19470, 26190, 34968, 46439, 61275, 80455, 105047, 136541
Partitions of n into parts of 2 kinds. Ref RS4 90. RCI 199. [0,2; A0710, N0535]

M1376   1, 2, 5, 10, 20, 36, 65, 110, 185, 300, 481, 752, 1165, 1770, 2665, 3956, 5822, 8470, 12230, 17490, 24842, 35002, 49010, 68150, 94235, 129512, 177087, 240840
Partitions of n into parts of 2 kinds. Ref RS4 90. RCI 199. [0,2; A0712, N0536]

M1377   1, 2, 5, 10, 20, 38, 71, 130, 235, 420, 744, 1308, 2285, 3970, 6865, 11822, 20284, 34690, 59155, 100610, 170711, 289032, 488400, 823800, 1387225, 2332418, 3916061
Convolved Fibonacci numbers. Ref RCI 101. FQ 15 118 77. [0,2; A1629, N0537]

$$\text{G.f.:}\ \ (1 - x - x^2)^{-2}.$$

M1378   2, 5, 10, 20, 40, 86, 192, 440, 1038, 2492, 6071, 14960, 37198, 93193, 234956, 595561, 1516638, 3877904, 9950907, 25615653, 66127186, 171144671, 443966370
Nearest integer to exponential integral of n. Ref PTRS 160 384 1870. PHM 33 757 42. FMR 1 267. [1,1; A2460, N0538]

M1379   0, 1, 2, 5, 10, 20, 41, 86, 182, 393, 853
Percolation series for directed hexagonal lattice. Ref JPA 16 3146 83. [2,3; A6836]

M1380   1, 2, 5, 10, 22, 40, 75, 130, 230, 382, 636, 1022, 1645, 2570, 4002, 6110, 9297, 13910, 20715, 30462, 44597, 64584, 93085, 132990, 189164, 266990, 375186, 523784
Expansion of a modular function. Ref PLMS 9 386 59. [0,2; A2512, N0539]

$$\text{G.f.:}\ \ \Pi\,(1 - x^k)^{-c(k)},\ c(k) = 2,2,2,4,2,2,2,4,....$$

M1381   2, 5, 10, 23, 45, 94, 179, 358, 672, 1292, 2382, 4470, 8132, 14937, 26832, 48507
Dimension of nth compound of a certain space. Ref Chir93. [1,1; A7182]

M1382   1, 1, 2, 5, 10, 24, 63, 165, 467, 1405, 4435, 14775, 51814, 190443, 732472, 2939612
Graphs by nodes and edges. Ref R1 146. SS67. [0,3; A1431, N0540]

M1383   2, 5, 10, 25, 56, 139, 338, 852
Alcohols with n carbon atoms. Ref BER 8 1545 1875. [3,1; A2094, N0541]

M1384   1, 1, 2, 5, 10, 28, 86, 285
Connected planar graphs without vertices of degree 1. Ref SIAA 4 174 83. [3,3; A6401]

M1385   1, 1, 2, 5, 10, 29, 96, 339
2-connected maps without faces of degree 2. Ref SIAA 4 174 83. [3,3; A6404]

M1386   0, 1, 2, 5, 10, 40, 40, 106, 5627, 14501, 330861, 658110
From the powers that be. Ref AMM 83 805 76. [2,3; A4143]

M1387   2, 5, 11, 14, 26, 41, 89, 101, 194, 314, 341, 689, 1091, 1154, 1889, 2141, 3449, 3506, 5561, 6254, 8126, 8774, 10709, 13166, 15461, 24569
Extreme values of Dirichlet series. Ref PSPM 24 278 73. [1,1; A3420]

M1388   2, 5, 11, 17, 23, 29, 41, 47, 53, 59, 71, 83, 89, 101, 107, 113, 131, 137, 149, 167, 173, 179, 191, 197, 227, 233, 239, 251, 257, 263, 269, 281, 293, 311, 317, 347, 353, 359
Primes of form $3n - 1$. Ref AS1 870. [1,1; A3627]

M1389   2, 5, 11, 17, 29, 37, 53, 67, 83, 101, 127, 149, 173, 197, 227, 257, 293, 331, 367, 401, 443, 487, 541, 577, 631, 677, 733, 787, 853, 907, 967, 1031, 1091, 1163, 1229, 1297
First prime between n^2 and $(n+1)^2$. Ref HW1 19. [1,1; A7491]

M1390   0, 0, 1, 2, 5, 11, 20, 37, 63, 110, 174, 283, 435, 671, 1001, 2160, 3127, 4442, 6269, 8739, 12109, 16597, 22618, 30576
Arrangements of pennies in rows. Ref PCPS 47 686 51. QJMO 23 153 72. rkg. [1,4; A5575]

M1391   1, 1, 1, 1, 2, 5, 11, 21, 37, 64, 113, 205, 377, 693, 1266, 2301, 4175, 7581, 13785, 25088, 45665, 83097, 151169, 274969, 500162, 909845, 1655187, 3011157, 5477917
$\Sigma C(n-k,3k)$, $k = 0 .. n$. Ref BR72 113. [0,5; A3522]

M1392   0, 0, 0, 1, 2, 5, 11, 21, 39, 73, 129, 226, 388, 659, 1100, 1821
Asymmetrical planar partitions of n. Ref MA15 2 332. [1,5; A0785, N0542]

M1393   1, 2, 5, 11, 23, 45, 87, 160, 290, 512, 889, 1514, 2547, 4218, 6909, 11184, 17926, 28449, 44772, 69862, 108205, 166371, 254107, 385617, 581729, 872535, 1301722
Column-strict plane partitions of n. Ref SAM 50 260 71. [0,2; A5986]

M1394 1, 2, 5, 11, 23, 47, 94, 185, 360, 694, 1328, 2526, 4781, 9012, 16929, 31709, 59247, 110469, 205606, 382087, 709108, 1314512, 2434364, 4504352, 8328253
Compositions. Ref R1 155. ARS 31 28 91. [3,2; A0100, N0543]

$$\text{G.f.: } (1 - x - x^2)^{-1} (1 - x - x^2 - x^3)^{-1}.$$

M1395 2, 5, 11, 23, 47, 191, 383, 6143, 786431, 51539607551, 824633720831, 26388279066623, 108086391056891903, 55340232221128654847
Primes of form $3.2^n - 1$. Ref Rie85 384. [0,1; A7505]

M1396 0, 0, 1, 2, 5, 11, 25, 56, 126, 283, 636, 1429, 3211, 7215, 16212, 36428, 81853, 183922, 413269, 928607, 2086561, 4688460, 10534874, 23671647, 53189708
$a(n) = 2a(n-1) + a(n-2) - a(n-3)$. [0,4; A6054]

M1397 1, 1, 2, 5, 11, 25, 66, 172, 485, 1446, 4541, 15036, 52496, 192218, 737248
Graphs by nodes and edges. Ref R1 146. SS67. [0,3; A1432, N0544]

M1398 2, 5, 11, 26, 56, 122, 287, 677, 1457, 3137, 6833, 14885, 35015, 82370, 194300, 458330, 986390, 2122850, 4570610, 9840770, 21435122, 46689890, 101709206
$a(n) = 1 + a([n/2]) a(\lceil n/2 \rceil)$. Ref clm. [1,1; A5469]

M1399 1, 2, 5, 11, 26, 59, 137, 314, 725, 1667, 3842, 8843, 20369, 46898, 108005, 248699, 572714, 1318811, 3036953, 6993386, 16104245, 37084403, 85397138
$a(n) = a(n-1) + 3a(n-2)$. Ref FQ 11 52 73. [0,2; A6138]

M1400 1, 2, 5, 11, 26, 68, 177, 497, 1476, 4613, 15216, 52944, 193367, 740226
Number of graphs with n edges. Ref R1 146. SS67. MAN 174 68 67. [1,2; A0664, N0545]

M1401 1, 2, 5, 11, 27, 62, 152, 373
From the graph reconstruction problem. Ref LNM 952 101 82. [4,2; A6652]

M1402 1, 1, 1, 2, 5, 11, 28, 74, 199, 551, 1553, 4436, 12832, 37496, 110500, 328092, 980491, 2946889, 8901891, 27011286, 82300275, 251670563, 772160922, 2376294040
Steric planted trees with n nodes. Ref JACS 54 1105 32. TET 32 356 76. BA76 44. [0,4; A0625, N0546]

M1403 0, 1, 2, 5, 11, 31, 77, 214, 576, 1592, 4375, 12183, 33864, 94741, 265461, 746372
Proper rings with n edges. Ref FI50 41.399. [1,3; A2862, N0547]

M1404 1, 1, 2, 5, 11, 33, 117, 431
Connected planar graphs without vertices of degree 1. Ref SIAA 4 174 83. [3,3; A6400]

M1405 2, 5, 11, 38, 174, 984, 6600, 51120, 448560, 4394880, 47537280, 562464000, 7224940800, 100111334400, 1488257971200, 23625316915200, 398840682240000
$\Sigma (n+k)! C(2,k)$, $k = 0 .. 2$. Ref CJM 22 26 70. [-1,1; A1344, N0548]

M1406 2, 5, 11, 41, 89, 179, 359, 509, 719, 1019, 1031, 1229, 1409, 1451, 1481, 1511,
  1811, 1889, 1901, 1931, 2459, 2699, 2819, 3449, 3491, 3539, 3821, 3911, 5081, 5399
n, $2n+1$, $4n+3$ all prime. Ref SIAC 15 378 86. tm. [1,1; A7700]

M1407 1, 1, 2, 5, 12, 17, 63, 143, 492, 635, 2397, 3032, 93357, 96389, 478913, 575302,
  1629517, 15240955, 93075247, 387541943, 480617190, 868159133, 2216935456
Convergents to cube root of 4. Ref AMP 46 106 1866. L1 67. hpr. [1,3; A2355, N0549]

M1408 1, 0, 2, 5, 12, 24, 56, 113, 248, 503, 1043, 2080, 4169, 8145, 15897, 30545, 58402,
  110461, 207802, 387561, 718875, 1324038, 2425473, 4416193, 7999516, 14411507
Related to solid partitions of n. Ref MOC 24 956 70. [0,3; A2836, N0550]

M1409 1, 2, 5, 12, 27, 59, 127, 269, 563, 1167, 2400, 4903, 9960, 20135, 40534, 81300,
  162538, 324020, 644282, 1278152, 2530407, 5000178, 9863763, 19427976, 38211861
Compositions. Ref R1 155. ARS 31 28 91. [4,2; A0102, N0551]

$$\text{G.f.:}\quad (1 - x - x^2 - x^3)^{-1} (1 - x - x^2 - x^3 - x^4)^{-1}.$$

M1410 1, 2, 5, 12, 28, 63, 139, 303, 653, 1394, 2953, 6215, 13008, 27095, 56201, 116143,
  239231, 491326, 1006420, 2056633, 4193706, 8534653, 17337764, 35162804, 71205504
Compositions. Ref ARS 31 28 91. [5,2; A6979]

M1411 1, 2, 5, 12, 28, 64, 143, 315, 687, 1485, 3186, 6792, 14401, 30391, 63872, 133751,
  279177, 581040, 1206151, 2497895, 5161982, 10646564, 21919161, 45052841
Compositions. Ref ARS 31 28 91. [6,2; A6980]

M1412 2, 5, 12, 28, 64, 144, 320, 704, 1536, 3328, 7168, 15360, 32768, 69632, 147456,
  311296, 655360, 1376256, 2883584, 6029312, 12582912, 26214400, 54525952
$(n+5)2^n$. [-1,1; A3416]

M1413 0, 1, 2, 5, 12, 29, 70, 169, 408, 985, 2378, 5741, 13860, 33461, 80782, 195025,
  470832, 1136689, 2744210, 6625109, 15994428, 38613965, 93222358, 225058681
Pell numbers: $a(n)=2a(n-1)+a(n-2)$. Ref FQ 4 373 66. BPNR 43. Robe92 224. [0,3;
A0129, N0552]

M1414 2, 5, 12, 29, 71, 177, 448, 1147, 2960, 7679, 19989, 52145, 136214, 356121,
  931540, 2437513, 6379403, 16698113, 43710756, 114427391, 299560472, 784236315
$(F(2n+1)+F(2n-1)+F(n+3)-2)/2$. Ref CJN 25 391 82. [1,1; A5593]

M1415 1, 2, 5, 12, 30, 74, 188, 478, 1235, 3214, 8450, 22370, 59676, 160140, 432237,
  1172436, 3194870, 8741442, 24007045, 66154654, 182864692, 506909562, 1408854940
Powers of rooted tree enumerator. Ref R1 150. [1,2; A0106, N0553]

M1416 1, 2, 5, 12, 30, 76, 196, 512, 1353, 3610, 9713, 26324, 7199, 196938, 542895,
  1503312, 4179603, 11662902, 32652735, 9165540, 258215664, 728997192
Generalized ballot numbers. Ref JSIAM 18 254 69. JCT A23 293 77. [1,2; A2026, N0554]

M1417   1, 1, 2, 5, 12, 30, 79, 227, 710, 2322, 8071, 29503, 112822, 450141
Connected line graphs with n nodes. Ref HP73 221. [1,3; A3089]

M1418   1, 2, 5, 12, 31, 80, 210, 555, 1479, 3959, 10652, 28760, 77910, 211624, 576221,
1572210, 4297733, 11767328, 32266801, 88594626, 243544919, 670228623
Paraffins with n carbon atoms. Ref JACS 56 157 34. BA76 28. [1,2; A0635, N0555]

M1419   1, 2, 5, 12, 32, 94, 289, 910, 2934, 9686, 32540, 110780
Balancing weights. Ref JCT 7 132 69. [1,2; A2838, N0556]

M1420   1, 2, 5, 12, 33, 87, 252, 703, 2105, 6099, 18689, 55639, 173423, 526937, 1664094,
5137233, 16393315, 51255708, 164951529, 521138861, 1688959630, 5382512216
Folding a strip of n labeled stamps. See Fig M4587. Equals ½ M1205. Ref CJM 2 397 50.
JCT 5 151 68. MOC 22 198 68. [3,2; A0560, N0557]

M1421   1, 1, 2, 5, 12, 33, 90, 261, 766, 2312, 7068, 21965, 68954, 218751, 699534,
2253676, 7305788, 23816743, 78023602, 256738751, 848152864, 2811996972
Series-reduced planted trees with n nodes. Equals ½ M1207. Ref CAY 3 246. jr. MW63.
[1,3; A0669, N0558]

M1422   1, 1, 2, 5, 12, 33, 98, 305, 1002, 3424, 12016, 43230, 158516, 590621, 2230450,
8521967, 32889238, 128064009
Two-colored trees with n nodes. Ref JAuMS A20 503 75. [1,3; A4114]

M1423   0, 0, 0, 1, 1, 2, 5, 12, 34, 130, 525, 2472, 12400, 65619, 357504
n-node triangulations of sphere, with no node of degree 3. Ref MOC 21 252 67. JCT B45
309 88. [1,6; A0103, N0559]

M1424   1, 1, 2, 5, 12, 35, 107, 363, 1248, 4460, 16094, 58937, 217117, 805475, 3001211
n-celled polyominoes without holes. Ref PA67. JRM 2 182 69. [1,3; A0104, N0560]

M1425   1, 1, 2, 5, 12, 35, 108, 369, 1285, 4655, 17073, 63600, 238591, 901971, 3426576,
13079255, 50107909, 192622052, 742624232, 2870671950, 11123060678, 43191857688
Polyominoes with n cells. See Fig M1845. Ref AB71 363. RE72 97. DM 36 202 81. [1,3;
A0105, N0561]

M1426   1, 1, 2, 5, 12, 37, 123, 446, 1689, 6693, 27034, 111630, 467262, 1981353,
8487400, 36695369, 159918120, 701957539, 3101072051, 13779935438, 61557789660
Restricted hexagonal polyominoes with n cells. Ref GMJ 15 146 74. [1,3; A2216, N0562]

M1427   1, 1, 1, 2, 5, 12, 37, 128, 457, 1872, 8169, 37600, 188685, 990784, 5497741,
32333824, 197920145, 1272660224, 8541537105, 59527313920, 432381471509
Expansion of exp(sinh x). [0,4; A3724]

M1428   1, 1, 2, 5, 12, 41, 53, 306, 665, 15601, 31867, 79335, 111202, 190537, 10590737,
10781274, 53715833, 171928773, 225644606, 397573379, 6189245291, 6586818670
Convergents to $\log_2 3$. Ref rkg. [0,3; A5664]

M1429 1, 2, 5, 12, 53, 171, 566, 737, 4251, 4988, 9239, 41944, 428679, 7329487, 7758166, 115943811, 123701977, 239645788, 731522646953, 731762292741
Convergents to cube root of 5. Ref AMP 46 107 1866. L1 67. hpr. [1,2; A2358, N0563]

M1430 2, 5, 13, 17, 29, 37, 41, 53, 61, 73, 89, 97, 101, 109, 113, 137, 149, 157, 173, 181, 193, 197, 229, 233, 241, 257, 269, 277, 281, 293, 313, 317, 337, 349, 353, 373, 389, 397
Primes congruent to 1 or 2 modulo 4. Ref AS1 872. [1,1; A2313, N0564]

M1431 1, 2, 5, 13, 19, 32, 53, 89, 139, 199, 293, 887, 1129, 1331, 5591, 8467, 9551, 15683, 19609, 31397, 155921, 360653, 370261, 492113, 1349533, 1357201, 2010733
Increasing gaps between prime-powers. Ref DVSS 2 255 1884. vm. [1,2; A2540, N0565]

M1432 1, 2, 5, 13, 29, 34, 89, 169, 194, 233, 433, 610, 985, 1325, 1597, 2897, 4181, 5741, 6466, 7561, 9077, 10946, 14701, 28657, 33461, 37666, 43261, 51641, 62210, 75025
Markoff numbers. Ref EM 6 19 60. AMM 90 39 83. UPNT D12. jhc. [1,2; A2559, N0566]

M1433 1, 1, 1, 2, 5, 13, 33, 80, 184, 402, 840
Expansion of bracket function. Ref FQ 2 254 64. [0,4; A1659, N0567]

M1434 1, 2, 5, 13, 33, 81, 193, 449, 1025, 2305, 5121, 11265, 24577, 53249, 114689, 245761, 524289, 1114113, 2359297, 4980737, 10485761, 22020097, 46137345
$n2^{n-1} + 1$. Ref MMAG 63 15 90. [0,2; A5183]

M1435 1, 2, 5, 13, 33, 81, 193, 449, 1089, 2673, 6561, 15633, 37249, 88209, 216513, 531441, 1266273, 3017169, 7189057, 17537553, 43046721, 102568113, 244390689
Ways to add n ordinals. Ref MMAG 63 15 90. [1,2; A5348]

M1436 2, 5, 13, 33, 83, 205, 495, 1169, 2707, 6169, 13889, 30993, 68701, 151469, 332349, 725837, 1577751, 3413221, 7349029, 15751187, 33616925, 71475193
Binomial transform of primes. Ref EIS § 2.7. [1,1; A7443]

M1437 1, 1, 1, 2, 5, 13, 33, 85, 199, 445, 947, 1909, 3713, 7006
Maximal planar degree sequences with n nodes. Ref Dil92. [3,4; A7020]

M1438 0, 0, 1, 2, 5, 13, 33, 89, 240, 657, 1806, 5026, 13999, 39260, 110381, 311465, 880840, 2497405
n-node connected graphs with one cycle. Ref R1 150. SS67. [1,4; A1429, N0568]

M1439 1, 2, 5, 13, 34, 89, 233, 610, 1597, 4181, 10946, 28657, 75025, 196418, 514229, 1346269, 3524578, 9227465, 24157817, 63245986, 165580141, 433494437, 1134903170
Bisection of Fibonacci sequence: $a(n) = 3a(n-1) - a(n-2)$. Cf. M0692. Ref R1 39. FQ 9 283 71. [0,2; A1519, N0569]

M1440 1, 2, 5, 13, 34, 90, 239, 635, 1689, 4494, 11960, 31832, 84727, 225524, 600302, 1597904, 4253371, 11321838, 30137079, 80220557
2-dimensional directed animals of size n. Ref JPA 19 3265 86. [0,2; A6801]

M1441 1, 2, 5, 13, 35, 95, 260, 714, 1965, 5415, 14934, 41206, 113730, 313958, 866801, 2393315, 6608473, 18248017, 50389350, 139144906, 384237186
Irreducible positions of size n in Montreal solitaire. Ref JCT A60 56 92. [1,2; A7075]

M1442 1, 2, 5, 13, 35, 95, 262, 727, 2033, 5714, 16136, 45733, 130046, 370803, 1059838, 3035591, 8710736, 25036934, 72069134, 207727501, 599461094, 1731818878
Partially labeled rooted trees with n nodes (invert M1180). Ref R1 134. [1,2; A0107, N0570]

M1443 1, 2, 5, 13, 35, 96, 267, 750, 2123, 6046, 17303, 49721, 143365, 414584, 1201917, 3492117, 10165779, 29643870, 86574831, 253188111, 741365049, 2173243128
Directed animals of size n. Inverse of M1184. Ref AAM 9 340 88. [1,2; A5773]

$$\text{G.f.:} \quad \tfrac{1}{2}\left((1+x)/(1-3x)\right)^{1/2} - \tfrac{1}{2}.$$

M1444 2, 5, 13, 35, 97, 275, 793, 2315, 6817, 20195, 60073, 179195, 535537, 1602515, 4799353, 14381675, 43112257, 129271235, 387682633, 1162785755, 3487832977
$2^n + 3^n$. Ref KN1 1 92. [0,1; A7689]

M1445 1, 1, 2, 5, 13, 36, 102, 296, 871, 2599
Nonisentropic binary rooted trees with n nodes. Ref rkg. [1,3; A2844, N0571]

M1446 1, 0, 1, 1, 2, 5, 13, 36, 109, 359, 1266, 4731, 18657, 77464, 337681, 1540381, 7330418, 36301105, 186688845, 995293580, 5491595645, 31310124067, 184199228226
From a differential equation. Ref AMM 67 766 60. [0,5; A0994, N0572]

M1447 1, 2, 5, 13, 37, 108, 325, 993, 3070, 9564, 29979, 94392, 298311, 945592, 3005021, 9570559, 30539044, 97611676, 312462096, 1001554565, 3214232129
Paraffins with n carbon atoms. Ref BA76 44. [1,2; A5961]

M1448 1, 1, 2, 5, 13, 37, 111, 345, 1105, 3624, 12099, 41000, 140647, 487440, 1704115, 6002600
Rooted triangular cacti. Ref HP73 73. LeMi91. [0,3; A3080]

M1449 1, 2, 5, 13, 38, 116, 382, 1310, 4748, 17848, 70076, 284252, 1195240, 5174768, 23103368, 105899656, 498656912, 2404850720, 11879332048, 59976346448
$a(n)=a(n-1)+n\,a(n-2)$. Ref R1 86 (divided by 2). [1,2; A1475, N0573]

M1450 1, 2, 5, 13, 38, 149, 703, 4132
Connected trivalent bipartite graphs with $2n$ nodes. Ref OR76 135. [4,2; A6823]

M1451 1, 1, 2, 5, 13, 41, 145, 604, 2938, 16947
General partition graphs on n vertices. Ref DM 113 258 93. [1,3; A7269]

M1452 0, 1, 2, 5, 13, 44, 191, 1229, 13588, 288597
Disconnected graphs with n nodes. Ref TAMS 78 459 55. SS67. [1,3; A0719, N0574]

M1453 1, 2, 5, 14, 36, 98, 273, 768, 2197, 6360, 18584, 54780, 162672, 486154, 1461197, 4413988, 13393855, 40807290, 124783669, 382842018, 1178140280, 3635626680
Secondary alcohols with n carbon atoms. Ref BA76 44. [3,2; A5955]

M1454 2, 5, 14, 38, 107
Domino n-tuples. Ref JRM 7 324 74. [1,1; A6574]

M1455 1, 1, 2, 5, 14, 38, 120, 353, 1148, 3527, 11622, 36627, 121622, 389560, 1301140, 4215748, 13976335, 46235800, 155741571, 512559185, 1732007938, 5732533570
Folding a strip of n blank stamps. See Fig M4587. Ref ScAm 209(3) 262 63. JCT 5 151 68. CBUL (2) 3 36 75. [1,3; A1011, N0576]

M1456 1, 2, 5, 14, 39, 109
Paraffins with n carbon atoms. Ref ZFK 93 437 36. [1,2; A0641, N0575]

M1457 1, 1, 2, 5, 14, 40, 128, 369, 1214, 3516, 12776, 40534, 137404, 463232, 1602348, 5216253, 17753898, 58597316, 212150928, 710453534, 2366853608, 8584498376
Shifts left under l.c.m.-convolution with itself. Ref BeSl94. [0,3; A7463]

M1458 1, 2, 5, 14, 41, 122, 365, 1094, 3281, 9842, 29525, 88574, 265721, 797162, 2391485, 7174454, 21523361, 64570082, 193710245, 581130734, 1743392201
$(3^n + 1)/2$. Ref BPNR 60. Ribe91 53. HM94. [0,2; A7051]

M1459 1, 1, 2, 5, 14, 42, 132, 429, 1430, 4862, 16796, 58786, 208012, 742900, 2674440, 9694845, 35357670, 129644790, 477638700, 1767263190, 6564120420, 24466267020
Catalan numbers: $C(n) = C(2n,n)/(n+1)$. See Fig M1459. Ref AMM 72 973 65. RCI 101. C1 53. PLC 2 109 71. MAG 61 211 88. [0,3; A0108, N0577]

$$\text{G.f.:} \quad \frac{1 - (1 - 4x)^{1/2}}{2x}.$$

Figure M1459. CATALAN NUMBERS.

The **Catalan numbers**, defined by

$$C_n = \frac{1}{n+1}\binom{2n}{n} = \frac{1}{n+1}C(2n,n)$$

(M1459), are probably the most frequently occurring combinatorial numbers after the binomial coefficients. [GO4] lists over 240 references. See also [MAG 45 199 61], [AMM 72 973 65], [PLC 2 109 71], [C1], [Stan86] and [GKP]. Some of the dozens of interpretations of C_n are:

(i) The number of ways of dissecting a convex polygon of $n + 2$ sides into n triangles by drawing $n - 1$ nonintersecting diagonals (Fig. (a)).

(ii) The number of ways of completely parenthesizing a product of $n + 1$ letters (so that there are two factors inside each set of parentheses):

$n = 1$ (ab); $n = 2$ $a(bc)$, $(ab)c$;
$n = 3$ $(ab)(cd)$, $a((bc)d)$, $((ab)c)d$, $a(b(cd))$, $(a(bc))d$.

(iii) The number of planar binary trees with n nodes (Fig. (b)).

(iv) The number of planar rooted trees with n nodes (Fig. (c)).

(v) In an election with two candidates A and B, each receiving n votes, C_n is the number of ways the votes can come in so that A is never behind B [Fell60 1 71], [C1 1 94].

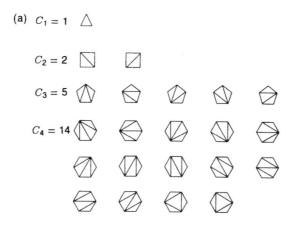

(a) $C_1 = 1$

$C_2 = 2$

$C_3 = 5$

$C_4 = 14$

(b) $C_1 = 1$

$C_2 = 2$

$C_3 = 5$

(c) $C_1 = 1$

$C_2 = 2$

$C_3 = 5$

M1460 1, 2, 5, 14, 43, 142, 494, 1780, 6563, 24566, 92890, 353740, 1354126, 5204396,
20066492, 77575144, 300572963, 1166868646, 4537698722, 17672894044
$(2^n + C(2n,n))/2$. Ref pcf. [0,2; A5317]

M1461 1, 2, 5, 14, 43, 142, 499, 1850, 7193, 29186, 123109, 538078, 2430355, 11317646,
54229907, 266906858, 1347262321, 6965034370, 36833528197, 199037675054
Switchboard problem with n subscribers: $a(n) = 2.a(n-1) + (n-2).a(n-2)$. Ref JCT
A21 162 1976. [0,2; A5425]

$$\text{G.f.: } \exp(2x + \tfrac{1}{2}x^2).$$

M1462 1, 1, 2, 5, 14, 43, 143, 509, 1922, 7651, 31965, 139685, 636712, 3020203,
14878176, 75982829, 401654560, 2194564531, 12377765239, 71980880885
Bessel numbers (Taylor expansion of n stages of c.f. in reference). Ref EJC 11 422 90.
[0,3; A6789]

M1463 1, 0, 1, 2, 5, 14, 44, 152
Partition function for square lattice. Ref AIP 9 279 60. [0,4; A2890, N0578]

M1464 1, 2, 5, 14, 45, 191, 871
Planar maps without loops. Ref SIAA 4 174 83. [1,2; A6391]

M1465 1, 1, 2, 5, 14, 46, 166, 652, 2780, 12644, 61136, 312676, 1680592, 9467680,
55704104, 341185496, 2170853456, 14314313872, 97620050080, 687418278544
The partition function $G(n,3)$. Ref CMB 1 87 58. [0,3; A1680, N0579]

$$\text{E.g.f.: } \exp (x + x^2 / 2 + x^3 / 6).$$

M1466 2, 5, 14, 46, 178, 800, 4094, 23536, 150178, 1053440, 8057774, 66750976,
595380178, 5688903680, 57975175454, 627692271616, 7195247514178
Entringer numbers. Ref NAW 14 241 66. DM 38 268 82. [0,1; A6216]

M1467 1, 0, 1, 2, 5, 14, 47, 186, 894, 5249
Partition graphs on n vertices. Ref DM 113 258 93. [1,4; A7268]

M1468 1, 2, 5, 14, 49, 240, 1259
Planar maps without loops. Ref SIAA 4 174 83. [1,2; A6390]

M1469 1, 1, 1, 2, 5, 14, 50, 233, 1249, 7595, 49566, 339722, 2406841, 17490241
Simplicial polyhedra with n nodes. Ref MOC 21 252 67. GR67 424. JCT 7 157 69. Dil92.
[3,4; A0109, N0580]

M1470 1, 1, 2, 5, 14, 51, 267, 2328, 56092
Number of groups of order 2^n. Ref HS64. JALG 129 136 90. JALG 143 219 91. [0,3;
A0679, N0581]

M1471 1, 1, 2, 5, 14, 58, 238, 1516, 9020, 79892, 635984, 7127764, 70757968, 949723600, 11260506056, 175400319992, 2416123951952, 42776273847184
Extreme points of set of $n \times n$ symmetric doubly-stochastic matrices. Ref JCT 8 422 70. EJC 1 180 80. [0,3; A6847]

M1472 0, 2, 5, 15, 32, 90, 189, 527, 1104, 3074, 6437, 17919, 37520, 104442, 218685, 608735, 1274592, 3547970, 7428869, 20679087, 43298624, 120526554, 252362877
Solution to a diophantine equation. Ref TR July 1973 p. 74. jos. [0,2; A6451]

M1473 1, 0, 0, 1, 2, 5, 15, 32, 99, 210, 650, 1379, 4268, 9055, 28025, 59458, 184021, 390420, 1208340, 2563621, 7934342, 16833545, 52099395, 110534372, 342101079
A ternary continued fraction. Ref TOH 37 441 33. [0,5; A0962, N0582]

M1474 1, 2, 5, 15, 41, 124, 369, 1132, 3491
Graphs with no isolated vertices. Ref LNM 952 101 82. [3,2; A6649]

M1475 1, 2, 5, 15, 48, 166, 596, 2221, 8472
From trees with valency ≤3. Ref QJMO 38 182 87. [1,2; A6570]

M1476 1, 2, 5, 15, 48, 167, 602, 2256, 8660, 33958, 135292, 546422, 2231462, 9199869, 38237213, 160047496, 674034147, 2854137769, 12144094756, 51895919734
N-free posets (generated by unions and sums) with n nodes. Ref PAMS 45 298 74. DM 75 97 89. [1,2; A3430]

M1477 1, 2, 5, 15, 49, 169, 602, 2191, 8095, 30239, 113906, 431886, 1646177, 6301715, 24210652, 93299841, 360490592, 1396030396, 5417028610
Permutations by inversions. Ref NET 96. DKB 241. MMAG 61 28 88. rkg. [2,2; A1892, N0583]

M1478 1, 1, 2, 5, 15, 49, 180, 701, 2891, 12371, 54564, 246319
Matched trees with n nodes. Ref DM 88 97 91. [1,3; A5751]

M1479 1, 2, 5, 15, 51, 187, 715, 2795, 11051, 43947, 175275, 700075, 2798251, 11188907, 44747435, 178973355, 715860651, 2863377067, 11453377195, 45813246635
$(3.2^{n-1} + 2^{2n-1} + 1)/3$. Ref JGT 17 625 93. [0,2; A7581]

M1480 1, 2, 5, 15, 51, 188, 731, 2950, 12235, 51822, 223191, 974427, 4302645, 19181100, 86211885, 390248055, 1777495635, 8140539950, 37463689775
Binomial transform of Catalan numbers. Cf. M1459. Ref EIS § 2.7. [1,2; A7317]

M1481 1, 1, 2, 5, 15, 51, 196, 827, 3795, 18755, 99146, 556711, 3305017, 20655285, 135399720, 927973061, 6631556521, 49294051497, 380306658250, 3039453750685
The partition function $G(n,4)$. Ref CMB 1 87 58. [0,3; A1681, N0584]

M1482 1, 2, 5, 15, 52, 200, 825, 3565, 15900, 72532, 336539, 1582593, 7524705, 36111810, 174695712, 851020367, 4171156249, 20555470155, 101787990805
Reversion of g.f. for partition numbers. Cf. M0663. [1,2; A7312]

M1483 1, 2, 5, 15, 52, 200, 827, 3596, 16191, 74702, 350794, 1669439, 8029728, 38963552, 190499461, 937550897, 4641253152, 23096403422, 115475977145
Reversion of g.f. for primes. [1,2; A7296]

M1484 1, 1, 2, 5, 15, 52, 203, 877, 4140, 21147, 115975, 678570, 4213597, 27644437, 190899322, 1382958545, 10480142147, 82864869804, 682076806159, 5832742205057
Bell or exponential numbers: $a(n+1) = \Sigma a(k)C(n,k)$. See Fig M4981. Ref MOC 16 418 62. AMM 71 498 64. PSPM 19 172 71. GO71. [0,3; A0110, N0585]

$$\text{E.g.f.: } \exp(e^x - 1).$$

M1485 1, 1, 1, 1, 2, 5, 15, 53, 213, 961, 4808, 26405, 157965, 1022573, 7122441, 53118601, 422362118, 3566967917, 31887812715, 300848966213, 2987359924149
Shifts 3 places left under exponentiation. Ref BeSl94. [1,5; A7548]

M1486 1, 1, 1, 2, 5, 15, 53, 222, 1078, 5994, 37622
Lattices on n nodes. Ref jrs. pdl. [0,4; A6966]

M1487 1, 1, 2, 5, 15, 56, 250, 1328, 8069, 54962, 410330, 3317302, 28774874, 266242936, 2616100423, 27205605275, 298569256590, 3449309394415
Connected interval graphs with n nodes. Ref TAMS 272 422 82. pjh. [1,3; A5976]

M1488 1, 2, 5, 16, 48, 164, 599, 1952
Triangulations of the disk. Ref PLMS 14 759 64. [0,2; A5497]

M1489 1, 2, 5, 16, 52, 208, 911
Inverse semigroups of order n. Ref PL65. MAL 2 2 67. SGF 14 71 77. [1,2; A1428, N0586]

M1490 1, 2, 5, 16, 54, 180, 595, 1964
Paths on square lattice. Ref ARS 6 168 78. [3,2; A6191]

M1491 1, 1, 2, 5, 16, 60, 261, 1243, 6257, 32721, 175760, 963900, 5374400, 30385256, 173837631, 1004867079, 5861610475, 34469014515, 204161960310, 1217145238485
Dissecting a polygon into n quadrilaterals. Ref DM 11 387 75. [1,3; A5036]

M1492 1, 1, 1, 2, 5, 16, 61, 272, 1385, 7936, 50521, 353792, 2702765, 22368256, 199360981, 1903757312, 19391512145, 209865342976, 2404879675441
Euler numbers: expansion of sec x + tan x. See Fig M4019. Ref JDM 7 171 1881. JO61 238. NET 110. DKB 262. C1 259. [0,4; A0111, N0587]

M1493 1, 2, 5, 16, 62, 276, 1377, 7596, 45789, 298626, 2090910, 15621640, 123897413, 1038535174, 9165475893, 84886111212, 822648571314, 8321077557124
Partitional matroids on n elements. Ref SMH 9 249 74. [0,2; A5387]

$$\text{E.g.f.: } \exp((x - 1)e^x + 2x + 1).$$

M1494 0, 1, 2, 5, 16, 62, 344, 2888, 42160, 1130244, 57349936, 5479605296, 979383162528, 326346868386848, 202753892253143616, 23531407879734441560
Strength 1 Eulerian graphs with n nodes, 2 of odd degree. Ref rwr. [1,3; A7124]

M1495 1, 1, 2, 5, 16, 63, 318, 2045, 16999, 183231, 2567284, 46749427, 1104891746, 33823827452
Partially ordered sets with n elements. See Fig M1495. Ref CN 8 180 73. C1 60. CRP 314 691 92. ORD 9 203 92. [0,3; A0112, N0588]

|||||||||||||||||||||||||||||||||||||||||||||||||||||||||||||||||||||||||||||||||||||||||||||||||||||||||||||||||||

Figure M1495. PARTIALLY ORDERED SETS.

   M1495 gives the number of **partially ordered sets** (or **posets**) on n points. Only 14 terms are known. See also Fig. M3032.

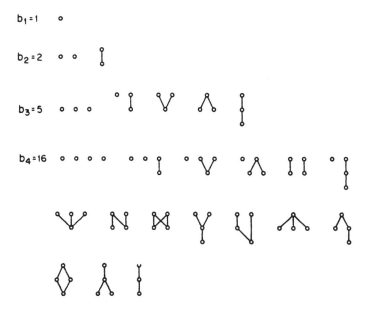

|||||||||||||||||||||||||||||||||||||||||||||||||||||||||||||||||||||||||||||||||||||||||||||||||||||||||||||||||||

M1496 1, 2, 5, 16, 64, 312, 1812, 12288, 95616, 840960, 8254080, 89441280, 1060369920, 13649610240, 189550368000, 2824077312000, 44927447040000
$\Sigma k!(n-k)!$, $k = 0 \ldots n$. Ref EJC 14 352 93. [0,2; A3149]

M1497 1, 2, 5, 16, 65, 326, 1957, 13700, 109601, 986410, 9864101, 108505112, 1302061345, 16926797486, 236975164805, 3554627472076, 56874039553217
$a(n) = na(n-1) + 1$. Ref R1 16. TMS 31 79 63. [0,2; A0522, N0589]

M1498 1, 1, 1, 2, 5, 16, 66, 343, 2167, 16193, 140919, 1414947, 16258868, 211935996,
  3105828560, 50748310068, 918138961643, 18287966027343, 399145502051200
Shifts left 2 places under Stirling-2 transform. Ref BeSl94. EIS § 2.7. [0,4; A7469]

M1499 1, 2, 5, 16, 66, 352, 2431, 21760, 252586
Totally symmetric plane partitions. Ref LNM 1234 292 86. [0,2; A5157]

M1500 1, 2, 5, 16, 67, 368, 2630, 24376, 293770, 4610624, 94080653, 2492747656,
  85827875506, 3842929319936, 223624506056156, 16901839470598576
Alternating sign matrices. Ref LNM 1234 292 86. BoHa92. [1,2; A5163]

M1501 1, 2, 5, 16, 67, 374, 2825, 29212, 417199, 8283458, 229755605, 8933488744,
  488176700923, 37558989808526, 4073773336877345, 623476476706836148
Sum of Gaussian binomial coefficients $[n,k]$ for $q = 2$. Ref TU69 76. GJ83 99. ARS A17
328 84. [0,2; A6116]

M1502 1, 2, 5, 16, 67, 435
Circuits of nullity n. Ref AIEE 51 311 32. [1,2; A2631, N0590]

M1503 1, 2, 5, 16, 73, 538
Circuits of rank n. Ref AIEE 51 313 32. [1,2; A2632, N0591]

M1504 1, 1, 2, 5, 16, 78, 588, 8047, 205914, 10014882, 912908876, 154636289460,
  48597794716736, 28412296651708628, 31024938435794151088
n-node graphs without endpoints. Ref rwr. [1,3; A4110]

M1505 2, 5, 17, 13, 37, 41, 101, 61, 29, 197, 113, 257, 181, 401, 97, 53, 577, 313, 677, 73,
  157, 421, 109, 89, 613, 1297, 137, 761, 1601, 353
Primes associated with Størmer numbers. Ref TO51 vi. CoGu95. [1,1; A5529]

M1506 2, 5, 17, 37, 101, 197, 257, 401, 577, 677, 1297, 1601, 2917, 3137, 4357, 5477,
  7057, 8101, 8837, 12101, 13457, 14401, 15377, 15877, 16901, 17957, 21317, 22501
Primes of form $n^2 + 1$. Ref EUL (1) 3 22 17. OG72 116. [1,1; A2496, N0592]

M1507 2, 5, 17, 41, 461, 26833, 26849, 26863, 26881, 26893, 26921, 616769, 616793,
  616829, 616843, 616871, 617027, 617257, 617363, 617387, 617411, 617447, 617467
Where prime race $4n - 1$ vs. $4n + 1$ is tied. Ref rgw. [1,1; A7351]

M1508 1, 2, 5, 17, 55, 186, 635, 2199, 7691, 27101, 96061
Spheroidal harmonics. Ref MES 52 75 24. [0,2; A2692, N0593]

M1509 1, 2, 5, 17, 62, 275, 1272, 6225, 31075, 158376, 816229, 4251412, 23319056,
  117998524, 627573216, 3355499036, 18025442261, 97239773408, 526560862829
Dissections of a polygon. Ref AEQ 18 388 78. [3,2; A3456]

M1510 1, 1, 2, 5, 17, 67, 352, 1969, 13295, 97619, 848354, 7647499, 82862683,
  897904165, 11226063188, 146116260203, 2089038231953, 30230018309161
An equivalence relation on permutations. Ref AMM 82 87 75. [0,3; A3510]

M1511 1, 2, 5, 17, 71, 357
Trivalent planar multigraphs with $2n$ nodes. Ref BA76 92. [1,2; A5966]

M1512 2, 5, 17, 71, 388
Trivalent graphs with $2n$ nodes. Ref BA76 92. [1,1; A5967]

M1513 1, 1, 2, 5, 17, 73, 388, 2461, 18155, 152531, 1436714, 14986879, 171453343,
2134070335, 28708008128, 415017867707, 6416208498137, 105630583492969
$a(n) = n.a(n-1) - (n-1)(n-2)a(n-3)/2$. Ref CAY 9 190. PLMS 17 29 17. EDMN 34
1 44. AMM 79 519 72. [0,3; A2135, N0594]

M1514 1, 2, 5, 18, 66, 266, 1111, 4792, 21124, 94888, 432415, 1994828
Rooted identity matched trees with n nodes. Ref DM 88 97 91. [1,2; A5753]

M1515 1, 2, 5, 18, 75, 414, 2643, 20550, 180057, 1803330, 19925541, 242749602,
3218286195, 46082917278, 710817377715, 11689297807734, 205359276208113
Extreme points of set of $n \times n$ symmetric doubly-substochastic matrices. Ref EJC 1 180
80. [0,2; A6848]

M1516 1, 1, 2, 5, 18, 88, 489, 3071
Triangulations. Ref WB79 336. [0,3; A5500]

M1517 1, 2, 5, 18, 100, 1242, 43425, 4925635, 1678993887, 1613875721946,
4293014800909806, 315749445343642595 07, 6444833270876996597 71857
Strength 2 Eulerian graphs with n nodes. Ref rwr. [1,2; A7127]

M1518 1, 2, 5, 18, 102, 848, 12452, 265759, 10454008, 598047612, 63620448978,
9974635937844, 2905660724913768, 1268590412128132389
Self-converse oriented graphs with n nodes. Ref LNM 686 264 78. [1,2; A5639]

M1519 1, 2, 5, 18, 107, 1008, 13113, 214238, 4182487, 94747196
Phylogenetic trees with n labels. Ref ARS A17 179 84. [1,2; A5805]

M1520 2, 5, 18, 113, 1450, 40069, 2350602, 286192513, 71213783666, 35883905263781,
36419649682706466, 74221659280476136241, 30319350595387164556 2970
Hierarchical linear models on n factors allowing 2-way interactions; or labeled graphs on
$\leq n$ nodes. Ref clm. [1,1; A6896]

$$1 + C(n,1) + C(n,2)^2 + C(n,3)\ 2^3 + C(n,4)\ 2^6 + \cdots + C(n,n)\ 2^{n(n-1)/2}.$$

M1521 1, 2, 5, 19, 85, 509, 4060, 41301, 510489, 7319447, 117940535, 2094480864,
40497138011, 845480228069, 18941522184590, 453090162062723
Trivalent connected graphs with $2n$ nodes. Ref HA69 195. BW78 429. GTA91 1020. JGT
7 465 83. [1,2; A2851, N0595]

M1522 1, 2, 5, 19, 132, 3107, 623615
Semigroups of order n with 1 idempotent. Ref MAL 2 2 67. SGF 14 71 77. [1,2; A2786,
N0596]

M1523 1, 1, 1, 2, 5, 20, 85, 520, 3145, 26000, 204425, 2132000, 20646925, 260104000, 2993804125, 44217680000, 589779412625, 9993195680000, 151573309044625
Expansion of exp(arcsin x). Ref AMM 28 114 21. JO61 150. jos. [0,4; A6228]

M1524 1, 2, 5, 20, 87, 616, 4843, 44128
Ménage permutations. Ref SMA 22 233 56. R1 195. BE71 162. [3,2; A2484, N0597]

M1525 1, 1, 2, 5, 20, 88, 632, 8816
n-node digraphs with same converse as complement. Ref HA73 200. [1,3; A3069]

M1526 0, 1, 2, 5, 20, 101, 743, 7350, 91763
Trivalent graphs of girth exactly 4 and $2n$ nodes. Ref gr. [2,3; A6924]

M1527 1, 2, 5, 20, 107, 826, 7703, 81231, 914973
Irreducible polyhedral graphs with n faces. Ref md. [4,2; A6867]

M1528 1, 2, 5, 20, 115, 790, 6217, 55160, 545135, 5938490, 70686805, 912660508, 12702694075, 189579135710, 3019908731105
Permutations of length n by rises. Ref DKB 263. [2,2; A0130, N0598]

M1529 1, 2, 5, 20, 132, 1452, 26741
Plane partitions associated with Weak Macdonald Conjecture. Ref INV 53 222 79. [1,2; A6366]

M1530 1, 2, 5, 20, 132, 1452, 26741, 826540
Alternating sign matrices. Ref LNM 1234 292 86. [0,2; A5159]

M1531 1, 2, 5, 20, 179, 4082, 218225, 25316720, 6135834479, 3047003143022, 3067545380897645, 6223557209578656620, 2536038487880235826879
Certain subgraphs of a directed graph (binomial transform of M1986). Ref DM 14 119 76. [1,2; A5331]

M1532 1, 2, 5, 20, 180
Hierarchical models with linear terms forced. Ref BF75 34. clm. aam. [1,2; A6602]

M1533 1, 2, 5, 20, 230, 26795, 359026205, 64449908476890320, 2076895351339769460477611370186680
Representation requires n triangular numbers with greedy algorithm. Ref Lem00. jos. [1,2; A6893]

M1534 1, 2, 5, 20, 350, 140, 1050, 300, 57750, 38500, 250250, 45500, 2388750, 367500, 318750, 42500, 1088106250, 128012500, 960093750, 101062500, 105761906250
Denominators of Van der Pol numbers. Cf. M5262. Ref JRAM 260 35 73. [0,2; A3163]

M1535 2, 5, 21, 61, 214, 669, 2240, 7330, 24695, 83257, 284928, 981079, 3410990, 11937328, 42075242, 149171958, 531866972, 1905842605, 6861162880
Asymmetrical dissections of n-gon. Ref GU58. [7,1; A0131, N0599]

M1536   1, 2, 5, 21, 106, 643, 4547, 36696, 332769, 3349507
Permutations of length n without 3-sequences. Ref BAMS 51 748 45. ARS 1 305 76. [1,2; A2628, N0600]

M1537   0, 1, 1, 1, 2, 5, 21, 233, 10946, 5702887, 139583862445, 1779979416004714189, 5555654042242926944040015791808
$F(F(n))$, where F is a Fibonacci number. Ref FQ 15 122 77. [0,5; A7570]

M1538   1, 1, 2, 5, 22, 138, 1579, 33366, 1348674, 105925685, 15968704512, 4520384306832, 2402814904220039, 2425664021535713098
Connected graphs with n nodes, $n(n-1)/4$ edges. Ref SS67. [2,3; A1437, N0601]

M1539   0, 0, 1, 2, 5, 22, 181, 5814, 1488565, 12194330294
Coding Fibonacci numbers. Ref FQ 15 312 77. PAMS 63 30 77. [1,4; A5203]

M1540   2, 5, 23, 110, 527, 2525, 12098, 57965, 277727, 1330670, 6375623, 30547445, 146361602, 701260565, 3359941223, 16098445550, 77132286527, 369562987085
$a(n) = 5a(n-1) - a(n-2)$. Ref MMAG 48 209 75. [0,1; A3501]

M1541   2, 5, 23, 113, 719, 5039, 40289, 362867, 3628789, 39916787, 479001599, 6227020777, 87178291199, 1307674367953, 20922789887947, 355687428095941
Largest prime $\leq n!$. Ref rgw. [2,1; A6990]

M1542   1, 2, 5, 24, 23, 76, 249, 168, 599, 1670, 1026, 3272, 8529, 5232
Expansion of a modular function. Ref PLMS 9 384 59. [-3,2; A2507, N0602]

M1543   2, 5, 24, 1430
Switching networks. Ref JFI 276 326 63. [1,1; A0895]

M1544   0, 1, 2, 5, 26, 677, 458330, 210066388901, 44127887745906175987802, 1947270476915296449559703445493848930452791205
$a(n) = a(n-1)^2 + 1$. Ref FQ 11 429 73. [0,3; A3095]

M1545   1, 2, 5, 27, 923, 909182, 1046593950039, 1168971346319460027570137, 1730152138254248421873938035330598736473956767 1241
Denominators of convergents to Lehmer's constant. Cf. M3034. Ref DUMJ 4 334 38. jww. [0,2; A2795, N0603]

M1546   2, 5, 29, 199, 2309, 30029, 510481, 9699667, 223092870, 6469693189, 200560490057, 7420738134751, 304250263527209, 13082761331669941
Largest prime $\leq \Pi p(k)$. Ref rgw. [1,1; A7014]

M1547   2, 5, 30, 2288, 67172352, 144115192303714304
Boolean functions of n variables. Ref HA65 153. [1,1; A0133, N0604]

M1548   1, 1, 0, 2, 5, 32, 234, 3638, 106147, 6039504, 633754161, 120131932774, 41036773627286, 25445404202438466, 28918219616495404542
n-node connected graphs without points of degree 2. Ref rwr. [1,4; A5636]

M1549 1, 2, 5, 34, 985, 1151138, 1116929202845, 1480063770341062927127746,
1846425204836010506550936273411258268076151412465
Continued fraction for Lehmer's constant. Ref DUMJ 4 334 38. jos. [0,2; A2665, N0605]

M1550 1, 1, 2, 5, 34, 2136, 7013488, 1788782616656, 53304527811667897504
Relations with 3 arguments on *n* nodes. Ref MAN 174 69 67. HP73 231. DM 6 384 73.
[1,3; A0665, N0606]

M1551 1, 2, 5, 45, 22815, 2375152056927, 2233176271342403475345148513527359103
$a(n+1)=(1+a(0)^3+\cdots+a(n)^3)/(n+1)$ (not always integral!). Ref AMM 95 704 88.
[0,2; A5166]

M1552 2, 5, 52, 88, 96, 120, 124, 146, 162, 188, 206, 210, 216, 238, 246, 248, 262, 268,
276, 288, 290, 292, 304, 306, 322, 324, 326, 336, 342, 372, 406, 408, 426, 430, 448, 472
Untouchable numbers: impossible values for sum of aliquot parts of *n*. Ref AS1 840.
UPNT B10. [1,1; A5114]

M1553 2, 5, 55, 9999, 3620211523, 25838201785967533906,
3408847366605453091140558218322023440765
Denominator of Egyptian fraction for $e-2$. Ref hpr. [0,1; A6525]

M1554 2, 5, 71, 369119, 415074643
Primes *p* that divide sum of all primes $\leq p$. Ref JRM 14 205 82. [1,1; A7506]

M1555 2, 5, 197, 776, 1797, 467613464999866416197,
102249460387306384473056172738577521087843948916391508591105797
Numerators of a continued fraction. Ref NBS B80 288 76. [0,1; A6271]

SEQUENCES BEGINNING . . ., 2, 6, . . .

M1556 2, 6, 2, 10, 2, 10, 14, 10, 6, 10, 18, 2, 6, 14, 22, 14, 22, 26, 18, 14, 2, 30, 26, 30, 2,
26, 18, 10, 34, 26, 22, 18, 10, 34, 14, 34, 38, 2, 6, 30, 34, 14, 42, 38, 10, 22, 42, 38, 26
Glaisher's χ numbers. Ref QJMA 20 152 1884. [1,1; A2172, N0607]

M1557 1, 1, 2, 6, 2, 60, 2, 42, 6, 30, 1, 660, 3, 364, 30, 42, 2, 1020, 1, 570, 42, 22, 1,
106260, 10, 390, 6, 1092, 1, 1740, 10, 1302, 66, 34, 70, 11220, 1, 1406, 78, 3990, 1
From polynomial identities. Ref BAMS 81 108 75. ACA 29 246 76. [1,3; A5729]

M1558 1, 2, 6, 4, 30, 4, 84, 24, 90, 20, 132, 8, 5460, 840, 360, 48, 1530, 4, 1596, 168,
1980, 1320, 8280, 80, 81900, 6552, 1512, 112, 3480, 80, 114576, 7392, 117810, 7140
Denominators of Cauchy numbers. Ref C1 294. [0,2; A6233]

M1559 1, 2, 6, 4, 30, 12, 84, 24, 90, 20, 132, 24, 5460, 840, 360, 16, 1530, 180, 7980, 840,
13860, 440, 1656, 720, 81900, 6552, 216, 112, 3480, 240, 114576, 7392, 117810, 2380
Denominators of Cauchy numbers. Cf. M3790. Ref MT33 136. C1 294. [0,2; A2790,
N0608]

M1560 2, 6, 6, 5, 1, 4, 4, 1, 4, 2, 6, 9, 0, 2, 2, 5, 1, 8, 8, 6, 5, 0, 2, 9, 7, 2, 4, 9, 8, 7, 3, 1, 3, 9, 8, 4, 8, 2, 7, 4, 2, 1, 1, 3, 1, 3, 7, 1, 4, 6, 5, 9, 4, 9, 2, 8, 3, 5, 9, 7, 9, 5, 9, 3, 3, 6, 4, 9, 2
Decimal expansion of $2 \uparrow \sqrt{2}$. Ref PSPM 28 16 76. [0,1; A7507]

M1561 2, 6, 6, 8, 12, 6, 12, 18, 6, 14, 18, 12, 18, 18, 12, 12, 30, 18, 14, 24, 6, 30, 30, 12, 24, 24, 18, 24, 30, 12, 26, 42, 24, 12, 30, 18, 24, 48, 18, 36, 24, 18, 36, 30, 24, 26, 48, 18
Theta series of b.c.c. lattice w.r.t. short edge. Ref JCP 83 6526 85. [0,1; A5869]

M1562 1, 1, 2, 6, 6, 10, 6, 210, 2, 30, 42, 110, 6, 546, 2, 30, 462, 170, 6, 51870, 2, 330, 42, 46, 6, 6630, 22, 30, 798, 290, 6, 930930, 2, 102, 966, 10, 66, 1919190, 2, 30, 42, 76670, 6
Euler-Maclaurin expansion of polygamma function. Ref AS1 260. [2,3; A6955]

M1563 0, 1, 1, 2, 6, 6, 27, 20, 130, 124, 598, 640
Atomic species of degree n. Ref JCT A50 279 89. [0,4; A5226]

M1564 2, 6, 8, 5, 4, 5, 2, 0, 0, 1, 0, 6, 5, 3, 0, 6, 4, 4, 5, 3, 0, 9, 7, 1, 4, 8, 3, 5, 4, 8, 1, 7, 9, 5, 6, 9, 3, 8, 2, 0, 3, 8, 2, 2, 9, 3, 9, 9, 4, 4, 6, 2, 9, 5, 3, 0, 5, 1, 1, 5, 2, 3, 4, 5, 5, 5, 7, 2, 1
Decimal expansion of Khintchine's constant. Ref MOC 14 371 60. VA91 164. [1,1; A2210, N0609]

M1565 2, 6, 8, 8, 56, 24, 168, 240, 608, 920, 5680, 6104, 18416, 43008, 148152, 325608, 980840, 2421096, 7336488, 19769312, 58192608, 164776248, 502085760, 1427051544
Symmetries in unrooted (1,3) trees on $2n$ vertices. Ref GTA91 849. [1,1; A3610]

M1566 1, 1, 2, 6, 8, 13, 29, 44, 66, 122, 184, 269, 448, 668, 972, 1505, 2205
Expansion of a modular function. Ref PLMS 9 386 59. [-6,3; A2511, N0610]

M1567 0, 2, 6, 8, 18, 20, 24, 26, 54, 56, 60, 62, 72, 74, 78, 80, 162, 164, 168, 170, 180, 182, 186, 188, 216, 218, 222, 224, 234, 236, 240, 242, 486, 488, 492, 494, 504, 506, 510
$a(2n) = 3a(n)$, $a(2n+1) = 3a(n) + 2$. Ref TCS 65 213 89. TCS 98 186 92. [0,2; A5823]

M1568 1, 2, 6, 8, 20, 12, 42, 32, 54, 40, 110, 48
Related to patterns on an $n \times n$ board. Ref MES 37 61 07. [1,2; A2618, N0611]

M1569 2, 6, 8, 90, 288, 840, 17280, 28350, 89600, 598752, 87091200, 63063000, 301771008000, 5003856000, 6199345152, 976924698750, 3766102179840000
Cotesian numbers. Ref QJMA 46 63 14. [1,1; A2176, N0612]

M1570 2, 6, 9, 12, 15, 18, 21, 24, 27, 31, 34, 37, 40, 43, 46, 49, 53, 56, 59, 62, 65, 68, 71, 75, 78, 81, 84, 87, 90, 93, 97, 100, 103, 106, 109, 112, 115, 119, 122, 125, 128, 131
Zeros of Bessel function of order 0. Ref BA6 171. AS1 409. [1,1; A0134, N0613]

M1571 2, 6, 9, 13, 15, 19, 22, 26, 30, 33, 37, 39, 43, 46, 50, 53, 57, 59, 63, 66, 70, 74, 77, 81, 83, 87, 90, 94, 96, 100, 103, 107, 111, 114, 118, 120, 124, 127, 131, 134, 138
A self-generating sequence. Ref FQ 10 49 72. [1,1; A3145]

M1572   2, 6, 9, 18, 22, 32, 46, 58, 77, 97, 114, 135, 160, 186, 218
Van der Waerden numbers $W(2;2,n-1)$. Ref GU70 31. DM 28 135 79. [0,1; A2886, N0614]

M1573   1, 2, 6, 9, 23, 38, 90, 147, 341, 564, 1294, 2148, 4896, 8195, 18612, 31349, 70983,
  120357, 271921, 463712, 1045559, 1792582
Axially symmetric polyominoes with n cells. Ref DM 36 203 81. [4,2; A6746]

M1574   2, 6, 10, 14, 18, 26, 30, 38
Conference matrices of these orders exist. Ref ASB 82 15 68. MS78 56. [1,1; A0952, N0615]

M1575   1, 2, 6, 10, 18, 40, 46, 86, 118, 170
$n.3^n - 1$ is prime. [1,2; A6553]

M1576   1, 2, 6, 10, 19, 28, 44, 60, 85, 110
Paraffins. Ref BER 30 1919 1897. [1,2; A5993]

M1577   1, 2, 6, 11, 23, 38, 64, 95, 141, 194
Paraffins. Ref BER 30 1921 1897. [1,2; A5999]

M1578   0, 0, 0, 0, 1, 2, 6, 11, 24, 42, 81, 138, 250, 419, 732, 1214, 2073, 3414, 5742, 9411,
  15664, 25586, 42273, 68882, 113202, 184131, 301428, 489654, 799273, 1297118
$F(n+1) - 2^{[(n+1)/2]} - 2^{[n/2]} + 1$. Ref rkg. [0,6; A5673]

M1579   2, 6, 11, 26, 51, 106, 111, 113, 261
Stopping times. Ref MOC 54 393 90. [1,1; A7186]

M1580   2, 6, 12, 18, 30, 22, 42, 60, 54, 66, 46, 90, 58, 62, 120, 126, 150, 98, 138, 94, 210,
  106, 162, 174, 118, 198, 240, 134, 142, 270, 158, 330, 166, 294, 276, 282, 420, 250, 206
Largest inverse of totient function. Cf. M0987. Ref BA8 64. [1,1; A6511]

M1581   0, 2, 6, 12, 20, 30, 42, 56, 72, 90, 110, 132, 156, 182, 210, 240, 272, 306, 342, 380,
  420, 462, 506, 552, 600, 650, 702, 756, 812, 870, 930, 992, 1056, 1122, 1190, 1260, 1332
The pronic numbers: $n(n+1)$. Ref D1 2 232. [0,2; A2378, N0616]

M1582   1, 2, 6, 12, 20, 30, 43
A problem in (0,1) matrices. Ref AMM 81 1113 74. [1,2; A5991]

M1583   1, 2, 6, 12, 20, 34, 56, 88, 136, 208, 314, 470, 700, 1038, 1534, 2262, 3330, 4896,
  7192, 10558, 15492, 22724, 33324, 48860, 71630, 105002, 153912, 225594, 330650
Key permutations of length n. Ref CJN 14 152 71. [1,2; A3274]

M1584   1, 2, 6, 12, 22, 34, 52, 74, 102, 134, 176, 222, 280, 344, 416, 496, 592, 694, 814,
  942, 1082, 1232, 1404, 1584, 1784, 1996, 2226, 2468, 2738, 3016
Words of length n in a certain language. Ref TCS 71 399 90. [1,2; A5819]

M1585 1, 2, 6, 12, 24, 40, 72, 126, 240, 272
Maximal kissing number of n-dimensional lattice. See Fig M2209. Ref SPLAG 15. [0,2; A1116, N0617]

M1586 2, 6, 12, 24, 48, 60, 192, 120, 180, 240, 3072, 360, 12288, 960, 720, 840, 196608, 1260, 786432, 1680, 2880, 15360, 12582912, 2520, 6480, 61440, 6300, 6720, 805306368
Smallest number with $2n$ divisors. Ref AS1 840. B1 23. [1,1; A3680]

M1587 2, 6, 12, 30, 60, 120, 210, 420, 840, 1260, 2520, 2520, 5040, 9240, 13860, 27720, 32760, 55440, 65520, 120120, 180180, 360360, 360360, 720720, 720720, 942480
Maximal periods of shift registers. Ref LU79 134. [0,1; A5417]

M1588 2, 6, 12, 31, 72, 178
Alkyls with n carbon atoms. Ref ZFK 93 437 36. [1,1; A0650, N0618]

M1589 1, 2, 6, 12, 60, 20, 140, 280, 2520, 2520, 27720, 27720, 360360, 360360, 360360, 720720, 12252240, 4084080, 77597520, 15519504, 5173168, 5173168, 118982864
Denominators of harmonic numbers. See Fig M4299. Cf. M2885. Ref KN1 1 615. [1,2; A2805, N0619]

M1590 1, 2, 6, 12, 60, 60, 420, 840, 2520, 2520, 27720, 27720, 360360, 360360, 360360, 720720, 12252240, 12252240, 232792560, 232792560, 232792560, 232792560
Least common multiple of 1,2, ..., n. Ref MSC 39 275 76. [1,2; A3418]

M1591 2, 6, 12, 60, 120, 360, 2520, 5040, 55440, 720720, 1441440, 4324320, 21621600, 367567200, 6983776800, 13967553600, 321253732800, 2248776129600
Superior highly composite numbers. Ref RAM1 87. [1,1; A2201, N0620]

M1592 1, 2, 6, 12, 60, 168, 360, 720, 2520, 20160, 29120, 443520
Order of largest group with n conjugacy classes. Ref CJM 20 457 68. AN71. ISJM 51 305 85; 56 188 86. [1,2; A2319, N0621]

M1593 1, 2, 6, 13, 21, 24, 225, 615, 17450, 23228, 57774, 221361, 522377, 793040, 1706305, 8664354, 19037086, 51965160, 56870701, 124645388
Pierce expansion for Euler's constant. Ref FQ 22 332 84. jos. [0,2; A6284]

M1594 1, 2, 6, 13, 24, 40, 62, 91, 128, 174, 230, 297, 376, 468, 574, 695, 832, 986, 1158, 1349, 1560, 1792, 2046, 2323, 2624, 2950, 3302, 3681, 4088, 4524, 4990, 5487, 6016
Slicing a torus with n cuts: $(n^3 + 3n^2 + 8n)/6$. See Fig M1041. Ref GA61. Pick91 373. [0,2; A3600]

M1595 1, 2, 6, 13, 24, 42, 73, 125, 204, 324
Partitions into non-integral powers. Ref PCPS 47 215 51. [1,2; A0135, N0622]

M1596 1, 2, 6, 13, 32, 92
Maximum terms in disjunctive normal form with n variables. Ref DIA 18 3 71. CRV 13 415 72. [1,2; A3039]

M1597 1, 0, 1, 2, 6, 13, 40, 100, 291, 797, 2273, 6389
Functional digraphs with n nodes. Ref MAN 143 110 61. [0,4; A1373, N0623]

M1598 0, 0, 0, 2, 6, 14, 24, 46, 88, 162, 300, 562, 1056
Sets with a congruence property. Ref MFC 15 58 65. [3,4; A2703, N0624]

M1599 0, 2, 6, 14, 30, 62, 126, 254, 510, 1022, 2046, 4094, 8190, 16382, 32766, 65534,
131070, 262142, 524286, 1048574, 2097150, 4194302, 8388606, 16777214, 33554430
$2^n - 2$. Ref VO11 31. DA63 2 212. R1 33. [1,2; A0918, N0625]

M1600 1, 1, 2, 6, 14, 31, 73, 172, 400, 932, 2177, 5081, 11854, 27662, 64554
Permutations of length n within distance 2. Ref AENS 79 207 62. [0,3; A2524, N0626]

M1601 1, 2, 6, 14, 33, 70, 149, 298, 591, 1132, 2139, 3948, 7199, 12894, 22836, 39894,
68982, 117948, 199852, 335426, 558429, 922112, 1511610, 2460208, 3977963, 6390942
Expansion of $\Pi(1-x^k)^{-k-1}$. Ref SAM 273 71. DM 75 94 89. [0,2; A5380]

M1602 1, 2, 6, 14, 36, 89, 229, 599, 1609
From the graph reconstruction problem. Ref LNM 952 101 82. [3,2; A6653]

M1603 1, 2, 6, 14, 37, 92, 236, 596, 1517, 3846, 9770, 24794, 62953, 159800, 405688,
1029864, 2614457, 6637066, 16849006, 42773094, 108584525, 275654292, 699780452
Hamiltonian cycles on $P_4 \times P_n$. Ref ARS 33 87 92. [2,2; A6864]

$$a(n) = 2a(n-1) + 2a(n-2) - 2a(n-3) + a(n-4).$$

M1604 1, 2, 6, 14, 38, 97, 260, 688, 1856, 4994, 13550, 36791, 100302, 273824, 749316,
2053247, 5635266, 15484532, 42599485, 117312742, 323373356, 892139389
Glycols with n carbon atoms. Ref JACS 56 157 34. BA76 28. [2,2; A0634, N0627]

M1605 2, 6, 14, 38, 97, 264, 728, 2084, 6100
From the graph reconstruction problem. Ref LNM 952 101 82. [3,1; A6654]

M1606 1, 2, 6, 15, 40, 104, 273, 714, 1870, 4895, 12816, 33552, 87841, 229970, 602070,
1576239, 4126648, 10803704, 28284465, 74049690, 193864606, 507544127
$F(n).F(n+1)$. Ref FQ 6 82 68. BR72 17. [0,2; A1654, N0628]

M1607 1, 2, 6, 15, 60, 260, 1820, 12376, 136136, 1514513, 27261234, 488605194,
14169550626, 411591708660, 19344810307020, 908637119420910
Central Fibonomial coefficients. Ref FQ 6 82 68. BR72 74. [0,2; A3268]

M1608 1, 1, 2, 6, 15, 84, 330, 1812, 9978
From the game of Mousetrap. Ref GN93. [1,3; A7709]

M1609 2, 6, 16, 30, 54, 84, 128, 180, 250, 330
Paraffins. Ref BER 30 1920 1897. [1,1; A5996]

M1610 1, 1, 1, 1, 1, 2, 6, 16, 36, 71, 128, 220, 376, 661, 1211, 2290, 4382, 8347, 15706,
29191, 53824, 99009, 182497, 337745, 627401, 1167937, 2174834, 4046070, 7517368
$\Sigma C(n-k,4k)$, $k = 0 \ldots n$. [0,6; A5676]

M1611 2, 6, 16, 43
Largest subset of $3 \times 3 \times \cdots$ cube with no 3 points collinear. Ref DM 4 326 73. [1,1;
A3142]

M1612 1, 2, 6, 16, 45, 126, 357, 1016, 2907, 8350, 24068, 69576, 201643, 585690,
1704510, 4969152, 14508939, 42422022, 124191258, 363985680, 1067892399
From expansion of $(1+x+x^2)^n$. Ref C1 78. [1,2; A5717]

M1613 2, 6, 16, 46, 140, 464, 1580, 5538, 19804, 71884, 264204, 980778, 3671652,
13843808
n-step walks on hexagonal lattice. Ref JPA 6 352 73. [2,1; A3291]

M1614 1, 2, 6, 16, 50, 144, 462, 1392, 4536, 14060, 46310, 146376, 485914, 1557892,
5202690, 16861984, 56579196, 184940388, 622945970, 2050228360, 6927964218
Folding a strip of n labeled stamps. See Fig M4587. Equals $2n$.M1420. Ref MOC 22 198
68. JCT 5 151 68. [1,2; A0136, N0630]

M1615 1, 1, 2, 6, 16, 50, 165, 554, 1908, 6667, 23556
Self-dual planar graphs with $2n$ edges. Ref JCT 7 157 69. Dil92. [3,3; A2841, N0631]

M1616 1, 1, 2, 6, 16, 52, 170, 579, 1996, 7021, 24892, 89214, 321994, 1170282, 4277352,
15715249, 57999700, 214939846, 799478680, 2983699498, 11169391168, 41929537871
Triangulations of an n-gon. Ref LNM 406 345 74. DM 11 387 75. AEQ 18 387 78. [1,3;
A3446]

M1617 1, 1, 2, 6, 16, 59, 265, 1544, 10778, 88168
Connected regular graphs of degree 4 with n nodes. Ref OR76 135. [5,3; A6820]

M1618 1, 0, 0, 1, 1, 1, 2, 6, 17, 44, 112, 304, 918, 3040, 10623, 38161, 140074, 528594,
2068751, 8436893, 35813251, 157448068, 713084042, 3315414747, 15805117878
From a differential equation. Ref AMM 67 766 60. [0,7; A0996, N0632]

M1619 2, 6, 17, 46, 122, 321, 842, 2206, 5777, 15126, 39602, 103681, 271442, 710646,
1860497, 4870846, 12752042, 33385281, 87403802, 228826126, 599074577
$F(2n+1)+F(2n-1)-1$. Equals M3867 + 1. Ref CJN 25 391 82. [1,1; A5592]

M1620 0, 0, 2, 6, 18, 46, 146, 460, 1436, 4352, 13252, 40532
Permutations according to distance. Ref AENS 79 213 62. [0,3; A2529, N0633]

M1621   1, 2, 6, 18, 50, 142, 390, 1086, 2958, 8134, 22050, 60146, 162466, 440750,
1187222, 3208298, 8622666, 23233338, 62329366, 167558310, 448848582
n-step walks on square lattice. Equals ½ M3448. Ref AIP 9 354 60. [0,2; A2900, N0634]

M1622   2, 6, 18, 52, 166
Triangulations. Ref WB79 337. [0,1; A5507]

M1623   1, 2, 6, 18, 55, 174, 566, 1868, 6237, 21050, 71666, 245696, 847317, 2937116,
10226574, 35746292, 125380257, 441125966, 1556301578, 18155586993
$2n + 2$-celled polygons with perimeter n on square lattice. Ref JSP 58 480 90. [1,2; A6725]

M1624   1, 2, 6, 18, 57, 186, 622, 2120, 7338, 25724, 91144, 325878, 1174281, 4260282,
15548694, 57048048, 210295326, 778483932, 2892818244, 10786724388
Fine's sequence: relations of valence ≥ 1 on an n-set (inversion of Catalan numbers). Ref
IFC 16 352 70. JCT A23 90 77. DM 19 101 77; 75 97 89. [2,2; A0957, N0635]

M1625   1, 2, 6, 18, 58, 186, 614, 2034, 6818, 22970
Series-parallel numbers. Ref R1 142. [1,2; A0137, N0636]

M1626   0, 1, 2, 6, 18, 60, 184, 560, 1695, 5200, 15956, 48916
Permutations according to distance. Ref AENS 79 213 62. [0,3; A2527, N0637]

M1627   1, 2, 6, 18, 60, 200, 700, 2450, 8820, 31752, 116424, 426888, 1585584, 5889312,
22084920, 82818450, 312869700, 1181952200, 4491418360, 17067389768
Walks on square lattice. Ref GU90. [0,2; A5566]

M1628   1, 2, 6, 18, 60, 200, 760
Bishops on an $n \times n$ board. Ref LNM 560 212 76. [0,2; A5631]

M1629   1, 2, 6, 18, 60, 204, 734, 2694, 10162, 38982, 151920, 599244, 2389028, 9608668,
38945230, 158904230, 652178206, 2690598570
Two-colored trees with n nodes. Ref JAuMS A20 503 75. [1,2; A4113]

M1630   1, 1, 2, 6, 18, 64, 227, 856, 3280, 12885, 51342, 207544, 847886, 3497384,
14541132, 60884173, 256480895, 1086310549, 4623128656, 19759964149
Disconnected N-free posets with n nodes. Ref DM 75 97 89. [1,3; A7454]

M1631   1, 1, 2, 6, 18, 68, 282, 1309
Bordered triangulations of sphere with n nodes. Ref ADM 41 231 89. [5,3; A6674]

M1632   1, 2, 6, 18, 68, 313, 1592
Planar maps without faces of degree 1. Ref SIAA 4 174 83. [1,2; A6389]

M1633   1, 2, 6, 18, 74, 393, 2282
Planar maps without faces of degree 1. Ref SIAA 4 174 83. [1,2; A6388]

M1634 1, 1, 2, 6, 18, 75, 295, 1575, 7196, 48993, 230413, 2164767, 8055938, 139431149,
70125991, 14201296057, 77573062280, 2389977322593, 28817693086263
Expansion of $(1 + x)^{\exp x}$. [0,3; A7116]

M1635 1, 1, 2, 6, 18, 90, 540, 3780, 31500, 283500, 2835000, 31185000, 372972600,
4848643800, 67881013200, 1018215198000, 16294848570000, 277012425690000
Expansion of $\exp(-x^4/4) / (1-x)$. Ref R1 85. [0,3; A0138, N0638]

M1636 1, 2, 6, 19, 61, 196, 629, 2017, 6466, 20727, 66441, 212980, 682721, 2188509,
7015418, 22488411, 72088165, 231083620, 740754589, 2374540265, 7611753682
Board-pile polyominoes with n cells. Ref JCT 6 103 69. AB71 363. JSP 58 477 90. [1,2;
A1169, N0639]

$$\text{G.f.:}\quad x(1-x)^3 / (1-5x+7x^2-4x^3).$$

M1637 1, 2, 6, 19, 63, 216, 756, 2684, 9638, 34930, 127560, 468837, 1732702, 6434322,
23993874, 89805691, 337237337, 1270123530, 4796310672, 18155586993
$2n$-celled polygons on square lattice. Ref JSP 58 480 90. [2,2; A6724]

M1638 1, 2, 6, 19, 63, 216, 760, 2723, 9880, 36168, 133237, 492993, 1829670, 6804267,
25336611, 94416842, 351989967, 1312471879, 4894023222, 18248301701
Board-pair-pile polyominoes with n cells. Ref AB71 363. [1,2; A1170, N0640]

M1639 1, 2, 6, 19, 63, 216, 760, 2725, 9910, 36446, 135268, 505861, 1903890, 7204874,
27394666, 104592937, 400795844, 1540820542, 5940738676, 22964779660
Fixed polyominoes with n cells. Ref AB71 363. RE72 97. DM 36 202 81. [1,2; A1168,
N0641]

M1640 1, 2, 6, 19, 66, 236, 868, 3235, 12190, 46252, 176484, 676270, 2600612,
10030008, 38781096, 150273315, 583407990, 2268795980, 8836340260, 34461678394
Necklaces with n red, 1 pink and $n - 1$ blue beads. Ref MMAG 60 90 87. [1,2; A5654]

M1641 0, 0, 0, 0, 2, 6, 20, 60, 176, 510, 1484, 4314, 12624, 37126, 109864
Chiral planted trees with n nodes. Ref TET 32 356 76. [0,5; A5628]

M1642 0, 0, 0, 0, 2, 6, 20, 60, 176, 512, 1488, 4326, 12648, 37186, 109980, 327216,
979020, 2944414, 8897732, 27004290, 82287516
Paraffins with n carbon atoms. Ref JACS 54 1105 32. [1,5; A0620, N0642]

M1643 1, 2, 6, 20, 65, 216, 728, 2472, 8451, 29050, 100298, 347568, 1208220, 4211312,
14712960, 51507280, 180642391, 634551606, 2232223626, 7862669700
Quadrinomial coefficients. Ref C1 78. [1,2; A5726]

M1644 1, 2, 6, 20, 68, 232, 792, 2704, 9232, 31520, 107616, 367424, 1254464, 4283008,
14623104, 49926400, 170459392, 581984768, 1987020288, 6784111616, 23162405888
$a(n) = 4a(n-1) - 2a(n-2)$. Ref GK90 86. [0,2; A6012]

M1645 1, 2, 6, 20, 70, 252, 924, 3432, 12870, 48620, 184756, 705432, 2704156,
   10400600, 40116600, 155117520, 601080390, 2333606220, 9075135300, 35345263800
Central binomial coefficients: $C(2n,n)$. See Fig M1645. Ref RS3. AS1 828. [0,2; A0984, N0643]

IIIIIIIIIIIIIIIIIIIIIIIIIIIIIIIIIIIIIIIIIIIIIIIIIIIIIIIIIIIIIIIIIIIIIIIIIIIIIIIIIIIIIIIIIIIIIIIII

Figure M1645. PASCAL'S TRIANGLE.

```
                         1
                      1     1
                   1     2     1
                1     3     3     1
             1     4     6     4     1
          1     5    10    10     5     1
       1     6    15    20    15     6     1
    1     7    21    35    35    21     7     1
 1     8    28    56    70    56    28     8     1
```

The k-th entry in the n-th row (if we begin the numbering at 0) is the **binomial coefficient**

$$C(n,k) = \frac{n(n-1)(n-2)\cdots(n-k+1)}{1.2.3\cdots k} = \frac{n!}{k!(n-k)!},$$

which is the number of ways of choosing k things out of n. These numbers occur in the **binomal theorem**:

$$(x+y)^n = \sum_{k=0}^{n} C(n,k)x^{n-k}y^k.$$

Many sequences arise from this triangle. The diagonals give M0472, M2535, M3382, M3853, M4142, etc., and reading down the columns we see M1645, M2848, M3500, M2225, etc. Sequences M0082, M4084, etc. can also be seen in the triangle. For a bigger version of the triangle see [RS3], [AS1 828], and for the extension of the triangle to negative values of n see [RCI 2], [GKP 164].

IIIIIIIIIIIIIIIIIIIIIIIIIIIIIIIIIIIIIIIIIIIIIIIIIIIIIIIIIIIIIIIIIIIIIIIIIIIIIIIIIIIIIIIIIIIIIIIII

M1646 1, 1, 2, 6, 20, 71, 259, 961, 3606, 13640, 51909, 198497, 762007, 2934764,
   11333950, 43874857, 170193528, 661386105, 2574320659, 10034398370, 39163212165
Permutations by inversions. Ref NET 96. DKB 241. KN1 3 15. MMAG 61 28 88. [1,3; A0707, N0644]

M1647 1, 2, 6, 20, 71, 263, 1005
Column-convex polyominoes with perimeter $2n+2$. Ref DE87. JCT A48 12 88. [1,2; A6027]

M1648 1, 2, 6, 20, 76, 312, 1384, 6512, 32400, 168992, 921184, 5222208, 30710464,
   186753920, 1171979904, 7573069568, 50305536256, 342949298688, 2396286830080
$a(n)=2(a(n-1)+(n-1)a(n-2))$. Ref JCT A21 162 76. LU91 1 221. [0,2; A0898, N0645]

M1649 1, 1, 2, 6, 20, 86, 662, 8120, 171526, 5909259, 348089533, 33883250874,
5476590066777, 1490141905609371, 666003784522738152, 509204473666338077658
Self-dual signed graphs with *n* nodes. Ref CCC 2 31 77. rwr. JGT 1 295 77. [1,3; A4104]

M1650 1, 2, 6, 20, 90, 544, 5096, 79264, 2208612, 113743760, 10926227136,
1956363435360, 652335084592096, 405402273420996800, 470568642161119963904
Symmetric reflexive relations on *n* nodes. See Fig M3032. Ref MIT 17 21 55. MAN 174 70
67. JGT 1 295 77. [0,2; A0666, N0646]

M1651 1, 2, 6, 20, 91, 506
Trivalent multigraphs with $2n$ nodes. Ref BA76 92. [1,2; A5965]

M1652 1, 1, 2, 6, 20, 99, 646, 5918
Connected planar graphs with *n* nodes. Ref WI72 162. ST90. [1,3; A3094]

M1653 1, 2, 6, 21, 65, 221, 771, 2769, 10250, 39243, 154658, 628635, 2632420,
11353457, 50411413, 230341716
Graphs by nodes and edges. Ref R1 146. SS67. [3,2; A1434, N0647]

M1654 0, 0, 0, 0, 2, 6, 21, 75, 411
n-node bipartite graphs not determined by their spectra. Ref LNM 560 89 76. [1,5; A6612]

M1655 1, 1, 2, 6, 21, 94, 512, 3485
Connected topologies with *n* nodes. Ref jaw. CN 8 180 73. [0,3; A1928, N0648]

M1656 1, 2, 6, 21, 94, 540, 4207, 42110, 516344, 7373924, 118573592, 2103205738,
40634185402, 847871397424, 18987149095005, 454032821688754
Trivalent graphs with $2n$ nodes. Ref JGT 7 464 83. [2,2; A5638]

M1657 1, 1, 1, 2, 6, 21, 112, 853, 11117, 261080, 11716571, 1006700565, 164059830476,
50335907869219, 29003487462848061, 3139738114276 1241960
Connected graphs with *n* nodes. See Fig M1253. Ref TAMS 78 459 55. SS67. CCC 2 199
77. [0,4; A1349, N0649]

M1658 0, 0, 0, 0, 0, 0, 0, 0, 2, 6, 22, 67, 213, 744, 2609, 9016, 31426, 110381
Perfect squared rectangles of order *n*. Ref GA61 207. cjb. [1,9; A2839, N0650]

M1659 1, 2, 6, 22, 90, 394, 1806, 8558, 41586, 206098, 1037718, 5293446, 27297738,
142078746, 745387038, 3937603038, 20927156706, 111818026018, 600318853926
Royal paths in a lattice: twice M2898. Ref CRO 20 12 73. DM 9 341 74; 26 271 79. SIAD
5 499 92. [1,2; A6318]

M1660 1, 2, 6, 22, 91, 408, 1938, 9614, 49335, 260130, 1402440, 7702632, 42975796,
243035536, 1390594458, 8038677054, 46892282815, 275750636070, 1633292229030
$2 \cdot C(3n, 2n+1)/n(n+1)$. Ref CJM 15 257 63. AB71 363. [1,2; A0139, N0651]

M1661 1, 2, 6, 22, 92, 422, 2074, 10754, 58202, 326240, 1882960, 11140560, 67329992, 414499438, 2593341586, 16458756586, 105791986682, 687782586844, 4517543071924
Baxter permutations of length $2n - 1$. Ref MAL 2 25 67. JCT A24 393 78. FQ 27 166 89. [1,2; A1181, N0652]

$$\sum_{k=1}^{n} C(n+1,k-1)\,C(n+1,k)\,C(n+1,k+1) \;/\; C(n+1,1)\,C(n+1,2).$$

M1662 1, 2, 6, 22, 94, 454, 2430, 14214, 89918, 610182, 4412798, 33827974, 273646526, 2326980998, 20732504062, 192982729350, 1871953992254, 18880288847750
Values of Bell polynomials. Ref PSPM 19 173 71. [0,2; A1861, N0653]

E.g.f.: $\exp 2\,(e^x - 1)$.

M1663 2, 6, 22, 98, 522, 3262, 23486, 191802, 1753618, 17755382, 197282022, 2387112466, 31249472282, 440096734638, 6635304614542, 106638824162282
$a(n) = (n+1).a(n-1) + (2-n).a(n-2)$. Ref DM 55 272 85. [2,1; A6183]

M1664 2, 6, 22, 101, 546, 3502, 25586, 214062, 1987516, 20599076, 232482372, 2876191276, 38228128472, 549706132536, 8408517839416, 137788390312712
Terms in a bordered skew determinant. Ref RSE 21 354 1896. MU06 4 278. [2,1; A2772, N0654]

M1665 1, 2, 6, 22, 101, 573, 3836, 29228, 250749, 2409581, 25598186, 296643390, 3727542188, 50626553988, 738680521142
Kendall-Mann numbers: maximal inversions in permutation of n letters. Ref DKB 241. PGEC 19 1226 70. Aign77 28. [1,2; A0140, N0655]

M1666 1, 1, 2, 6, 23, 103, 513, 2761, 15767, 94359, 586590, 3763290, 24792705, 167078577, 1148208090, 8026793118, 56963722223, 409687815151, 2981863943718
Permutations with subsequences of length ≤3. Ref JCT A53 281 90. [0,3; A5802]

M1667 1, 1, 2, 6, 23, 107, 586, 3690, 26245, 207997, 1817090, 17345358, 179595995, 2004596903, 23992185226, 306497734962, 4162467826729, 59882101858777
Length of standard paths in composition poset. Ref BeBoDu 93. [0,3; A7555]

M1668 1, 2, 6, 23, 109, 618, 4096, 31133, 267219, 2557502
Matrices with 2 rows. Ref PLMS 17 29 17. [3,2; A2136, N0656]

M1669 1, 1, 2, 6, 24, 1, 720, 3, 80, 12, 3628800, 2, 479001600, 360, 8, 45, 20922789888000, 40, 6402373705728000, 6, 240, 1814400, 1124000727777607680000
Smarandache quotients: $a(n)!/n$, n in M0453. Ref AMM 25 210 18. [1,3; A7672]

M1670 1, 1, 2, 6, 24, 44, 80, 144, 264, 484, 888, 1632
Rook polynomials. Ref JAuMS A28 375 79. [0,3; A4306]

M1671 1, 1, 2, 6, 24, 78, 230, 675, 2069, 6404, 19708, 60216, 183988
Permutations of length n within distance 3. Ref AENS 79 213 62. [0,3; A2526, N0657]

M1684   1, 1, 1, 1, 2, 6, 27, 177, 1680, ...

M1672   2, 6, 24, 80, 450, 2142, 17696, 112464, 1232370, 9761510, 132951192,
1258797696, 20476388114, 225380451870, 4261074439680, 53438049741152
Logarithmic numbers. Ref TMS 31 77 63. jos. [1,1; A2742, N0658]

M1673   1, 2, 6, 24, 89, 371, 1478, 6044, 24302, 98000, 392528, 1570490
Dissections of a polygon. Ref AEQ 18 388 78. [5,2; A3450]

M1674   1, 2, 6, 24, 116, 648, 4088, 28640, 219920, 1832224, 16430176, 157554048,
1606879040, 17350255744, 197553645440, 2363935624704, 29638547505408
Binomial transform of M2900. Ref EIS § 2.7. [0,2; A7405]

M1675   1, 1, 2, 6, 24, 120, 720, 5040, 40320, 362880, 3628800, 39916800, 479001600,
6227020800, 87178291200, 1307674368000, 20922789888000, 355687428096000
Factorial numbers $n!$. See Fig M4730. Ref AS1 833. MOC 24 231 70. [0,3; A0142,
N0659]

M1676   1, 2, 6, 25, 107, 509, 2468, 12258, 61797, 315830, 1630770, 8498303, 44629855,
235974495, 1255105304, 6710883952, 36050676617, 194478962422, 1053120661726
Dissections of a polygon. Ref AEQ 18 388 78. [3,2; A3454]

M1677   1, 2, 6, 25, 114, 591, 3298, 19532, 120687, 771373, 5061741, 33943662,
231751331, 1606587482, 11283944502
Directed trees with n nodes. Ref LeMi91. [1,2; A6965]

M1678   1, 1, 2, 6, 25, 135, 892, 6937, 61886
Labeled n-node trees with unlabeled end-points. Ref JCT 6 63 69. [2,3; A1258, N0660]

M1679   1, 2, 6, 26, 135, 875
Semigroups by number of idempotents. Ref MAL 2 2 67. [1,2; A2788, N0661]

M1680   1, 2, 6, 26, 147, 892, 5876, 40490
Triangulations of the disk. Ref PLMS 14 759 64. [0,2; A2710, N0662]

M1681   1, 1, 2, 6, 26, 152, 1144, 10742, 122772, 1673856, 26780972, 496090330,
10519217930, 252851833482, 6832018188414, 205985750827854, 6885220780488694
Shifts left under Stirling-2 transform. Ref BeS194. EIS § 2.7. [0,3; A3659]

M1682   1, 2, 6, 26, 164, 1529, 21439
Van Lier sequences of length n. Ref DAM 27 218 90. [1,2; A5272]

M1683   1, 1, 2, 6, 26, 166, 1626, 25510, 664666, 29559718, 2290267226, 314039061414,
77160820913242
From the binary partition function. Ref RSE 65 190 59. PCPS 66 376 69. AB71 400. [0,3;
A2449, N0664]

M1684   1, 1, 1, 1, 2, 6, 27, 177, 1680, 23009, 455368
Sub-Fibonacci sequences of length n. Ref DAM 44 261 93. [1,5; A5270]

M1685 1, 2, 6, 27, 192, 2280, 47097
Regular sequences of length n. Ref DAM 27 218 90. [1,2; A3513]

M1686 0, 2, 6, 28, 180, 662, 7266, 24568
Second order Euler numbers. Ref JRAM 79 69 1875. FMR 1 75. [1,2; A2435, N0665]

M1687 1, 2, 6, 28, 212, 2664, 56632, 2052656, 127902864, 13721229088,
2544826627424, 815300788443072, 452436459318538048, 434188323928823259776
Sum of Gaussian binomial coefficients $[n,k]$ for $q = 3$. Ref TU69 76. GJ83 99. ARS A17
328 84. [0,2; A6117]

M1688 1, 2, 6, 28, 244, 2544, 35600, 659632
$3 \times (2n + 1)$ zero-sum arrays. Ref JAuMS 7 25 67. AMM 85 365 78. [0,2; A2047, N0666]

M1689 1, 2, 6, 30, 42, 30, 66, 2730, 6, 510, 798, 330, 138, 2730, 6, 870, 14322, 510, 6,
1919190, 6, 13530, 1806, 690, 282, 46410, 66, 1590, 798, 870, 354, 56786730, 6, 510
Denominators of Bernoulli numbers. Ref AS1 260, 810. [0,2; A6954]

M1690 1, 2, 6, 30, 156, 1455, 11300
Permutation groups of degree n. Ref JCT A50 279 89. [1,2; A5432]

M1691 1, 2, 6, 30, 210, 2310, 30030, 510510, 9699690, 223092870, 6469693230,
200560490130, 7420738134810, 304250263527210, 13082761331670030
Primorial numbers: product of first n primes. Ref FMR 1 50. BPNR 4. [0,2; A2110,
N0668]

M1692 1, 1, 2, 6, 30, 240, 3120, 65520, 2227680, 122522400, 10904493600,
1570247078400, 365867569267200, 137932073613734400, 84138564904377984000
Product of Fibonacci numbers. Ref BR72 69. [1,3; A3266]

M1693 1, 1, 1, 1, 2, 6, 30, 390, 32370, 81022110, 79098077953830,
2499603048957386233742790, 6399996109983215106481566902449146981585570
From a continued fraction. Ref AMM 63 711 56. [0,5; A1684, N0669]

M1694 2, 6, 30, 630, 34650, 47297250, 309560501250, 618190773193743750,
665374338505449540154687 50, 6557284423409512500603061274936953125 0
Multiplicative encoding of partition triangle. See Fig M1722. Ref C1 307. Sloa94. [1,1;
A7280]

M1695 1, 0, 2, 6, 31, 180, 1255, 9949, 89162
From the game of Mousetrap. Ref GN93. [1,3; A7710]

M1696 1, 1, 2, 6, 31, 302, 5984, 243668, 20286025, 3424938010, 1165948612902,
797561675349580, 1094026876269892596, 300584736573545 6265830
Acyclic digraphs with n nodes. Ref LNM 622 36 77. [0,3; A3087]

M1697 2, 6, 32, 314, 4892, 104518, 2814520, 91069042
Fanout-free functions of n variables. Ref CACM 23 705 76. PGEC 27 315 78. [1,1; A5736]

M1698 2, 6, 32, 346, 6572, 176678, 5511738
Fanout-free functions of n variables. Ref PGEC 27 315 78. [1,1; A5742]

M1699 1, 1, 2, 6, 32, 353, 8390, 436399, 50468754
Linear spaces with n points. Ref BSMB 19 421 67. [1,3; A1199, N0670]

M1700 1, 1, 1, 2, 6, 33, 286, 4420, 109820
Alternating sign matrices. Ref LNM 1234 292 86. [1,4; A5161]

M1701 2, 6, 34, 198, 1154, 6726, 39202, 228486, 1331714, 7761798, 45239074,
263672646, 1536796802, 8957108166, 52205852194, 304278004998, 1773462177794
$a(n) = 6a(n-1) - a(n-2)$. Ref B1 198. MMAG 48 209 75. [0,1; A3499]

M1702 2, 6, 34, 250, 972, 15498, 766808, 5961306, 54891535, 2488870076
Coefficients for numerical integration. Ref MOC 6 217 52. [1,1; A2685, N0671]

M1703 1, 1, 2, 6, 36, 240, 1800, 15120, 141120, 1693440
Sets of lists. Ref PSPM 19 172 71. [0,3; A2868, N0673]

M1704 1, 1, 2, 6, 36, 240, 1800, 16800, 191520, 2328480
Lists of sets. Ref PSPM 19 172 71. [0,3; A2869, N0674]

M1705 1, 2, 6, 36, 270, 2520, 28560, 361200, 5481000, 88565400, 1654052400,
32885455680, 721400359680, 17024709461760
Maximal coefficient in $(x + x^2 + x^4 + x^8 + ...)^n$. Ref Shal94. [1,2; A7657]

M1706 1, 1, 2, 6, 36, 876, 408696, 83762796636, 3508125906207095591916,
6153473687096578758444501468336878666616349966
Hypothenusal numbers. Ref PTRS 178 288 1887. LU91 1 496. [0,3; A1660, N0675]

M1707 2, 6, 38, 390, 6062, 134526
Colored graphs. Ref CJM 22 596 70. [2,1; A2031, N0676]

M1708 2, 6, 38, 526, 12022, 376430, 14821942
Fanout-free functions of n variables. Ref PGEC 27 315 78. [1,1; A5738]

M1709 2, 6, 38, 558, 14102, 493230, 21734582
Fanout-free functions of n variables. Ref PGEC 27 315 78. [1,1; A5740]

M1710 2, 6, 38, 684, 50224, 13946352, 14061131152, 50947324188128
Switching classes of digraphs. Ref rcr. [1,1; A6536]

M1711   2, 6, 38, 942, 325262
Boolean functions. Ref TO72 129. [1,1; A5530]

M1712   2, 6, 40, 1992, 18666624, 12813206169137152
Boolean functions of n variables. Ref HA65 153. [1,1; A0612, N0677]

M1713   1, 2, 6, 42, 1806, 3263442, 10650056950806, 113423713055421844361000442,
   12864938683278671740537145998360961546653259485195806
$a(n+1) = a(n)^2 + a(n)$. Ref HO85a 94. [0,2; A7018]

M1714   1, 2, 6, 42, 4094, 98210640
Self-complementary Boolean functions of n variables. Ref PGEC 12 561 63. PJM 110 220
84. [1,2; A0610, N0678]

M1715   2, 6, 46, 522, 7970, 152166, 3487246, 93241002, 2849229890, 97949265606,
   3741386059246, 157201459863882, 7205584123783010, 357802951084619046
Generalized Euler numbers. Ref MOC 21 693 67. [0,1; A1587, N0679]

M1716   1, 2, 6, 48, 720, 23040, 1451520, 185794560, 47377612800, 24257337753600,
   24815256521932800, 50821645356918374400, 208114637736580743168000
Order of orthogonal group $O(n, GF(2))$. Ref AMM 76 158 69. [1,2; A3053]

M1717   1, 1, 1, 1, 1, 1, 1, 2, 6, 56, 528, 6193, 86579, 1425518, 27298230, 601580875,
   15116315766, 429614643062, 13711655205087, 488332318973594
Von Staudt-Clausen representation of Bernoulli numbers. Ref MOC 21 678 67. [1,7;
A0146, N0680]

M1718   1, 1, 2, 6, 60, 2880, 2246400, 135862272000, 10376834265907200000,
   775401153743482383237120000000000
An operational recurrence. Ref FQ 1(1) 31 63. [1,3; A1577, N0681]

M1719   1, 2, 6, 66, 946, 8646
n divides $2^n + 2$. Ref HO73 142. [1,2; A6517]

M1720   2, 6, 70, 700229
Switching networks. Ref JFI 276 326 63. [1,1; A0896, N0682]

M1721   1, 2, 6, 74, 169112, 39785643746726
Balanced Boolean functions of n variables. Ref PGEC 9 265 60. PJM 110 220 84. [1,2;
A0721, N0683]

M1722   2, 6, 90, 47250, 66852843750, 2806877704512541816406250,
   121693589658270389851935478170253711859753338623046 8750
Multiplicative encoding of Pascal triangle: $\Pi \, p(i+1) \uparrow C(n,i)$. See Fig M1722. Ref AS1
828. Sloa94. [0,1; A7188]

|||||||||||||||||||||||||||||||||||||||||||||||||||||||||||||||||||||||||||||||||||||||||||||||||||||||||||||||||||||||||||||||||||||||||||||||||||||

Figure M1722.   MULTIPLICATIVE ENCODING OF ARRAYS.

A triangular array

$$
\begin{array}{ccccccc}
& & & t_{00} & & & \\
& & t_{01} & & t_{11} & & \\
& t_{02} & & t_{12} & & t_{22} & \\
t_{03} & & t_{13} & & t_{23} & & t_{33} \\
\cdots & & \cdots & & & & \cdots
\end{array}
$$

of nonnegative integers may be **multiplicatively encoded** by the sequence whose n-th term is

$$\prod_{k=0}^{n} p_k^{t_{n,k}},$$

where $p_1 = 2$, $p_2 = 3$, $p_3 = 5$, ... are the prime numbers. For example Pascal's triangle (see Fig. M1645) becomes M1722: $2^1 = 2$, $2^1 3^1 = 6$, $2^1 3^2 5^1 = 90$, $2^1 3^3 5^3 7^1 = 90$, $2^1 3^4 5^6 5^7 4 11^1 = 47250$, ... M1694, M1724, M1725, M1726 are also of this type.

|||||||||||||||||||||||||||||||||||||||||||||||||||||||||||||||||||||||||||||||||||||||||||||||||||||||||||||||||||||||||||||||||||||||||||||||||||||

M1723 1, 1, 1, 2, 6, 156, 7013488, 29288387523484992,
234431745534048922731115019069056
3-plexes. Ref DM 6 384 73. [1,4; A3189]

M1724 2, 6, 270, 478406250, 265749227134771782159805297851562 50
Multiplicative encoding of Stirling numbers of 2nd kind. See Fig M1722. Ref R1 48. AS1 835. Sloa94. [1,1; A7190]

M1725 2, 6, 540, 1240029000000,
10891955767228539785770886673520561523437500000000000000000000000000
Multiplicative encoding of Stirling numbers of 1st kind. See Fig M1722. Ref R1 48. AS1 833. Sloa94. [1,1; A7189]

M1726 2, 6, 810, 121096582031250, 1490116119384765625
Multiplicative encoding of Eulerian number triangle. Ref R1 215. GKP 254. [1,1; A7338]

SEQUENCES BEGINNING . . ., 2, 7, . . .

M1727 2, 7, 1, 8, 2, 8, 1, 8, 2, 8, 4, 5, 9, 0, 4, 5, 2, 3, 5, 3, 6, 0, 2, 8, 7, 4, 7, 1, 3, 5, 2, 6, 6, 2, 4, 9, 7, 7, 5, 7, 2, 4, 7, 0, 9, 3, 6, 9, 9, 9, 5, 9, 5, 7, 4, 9, 6, 6, 9, 6, 7, 6, 2, 7, 7, 2, 4, 0
Decimal expansion of e. Ref MOC 4 14 50; 23 679 69. [1,1; A1113, N0684]

M1728 0, 0, 1, 2, 7, 7, 11, 15
Nonabelian groups with n conjugacy classes. Ref CJM 20 457 68. AN71. [1,4; A3061]

M1729 2, 7, 8, 10, 22, 52, 58, 76, 130, 143, 331, 332, 980, 1282, 1655
$8.3^n + 1$ is prime. Ref MOC 26 996 72. [1,1; A5538]

M1730 1, 1, 2, 7, 8, 37, 40, 200, 258, 1039, 1500
Permutation groups of degree n. Ref JPC 33 1069 29. LE70 169. [1,3; A0637, N0685]

M1731 2, 7, 9, 11, 13, 15, 16, 17, 19, 21, 25, 29, 33, 37, 39, 45, 47, 53, 61, 69, 71, 73, 75, 85, 89, 101, 103, 117, 133, 135, 137, 139, 141, 143, 145, 147, 151, 155, 159, 163, 165
$a(n)$ is smallest number which is uniquely $a(j) + a(k)$, $j < k$. Ref GU94. [1,1; A3668]

M1732 2, 7, 9, 17, 19, 20, 26, 28, 37, 43, 63, 65, 91, 124, 126, 182, 215, 217, 254, 342, 344, 422, 511, 513, 614, 635, 651, 728, 730, 813, 999, 1001, 1330, 1332, 1521, 1588
$x^3 + ny^3 = 1$ is solvable. Ref MA71 674. [1,1; A5988]

M1733 1, 2, 7, 9, 43, 52, 303, 355, 658, 4303, 9264, 50623, 414248, 1293367, 4294349, 18470763, 41235875, 265886013, 1104779927, 4685005721, 5789785648
Convergents to cube root of 3. Ref AMP 46 105 1866. L1 67. [1,2; A2353, N0686]

M1734 2, 7, 10, 13, 18, 23, 28, 31, 34, 39, 42, 45, 50, 53, 56, 61, 66, 71, 74, 77, 82, 87, 92, 95, 98, 103, 108, 113, 116, 119, 124, 127, 130, 135, 138, 141, 146, 151, 156, 159
Not representable by M2482. Ref FQ 10 499 72. BR72 67. [1,1; A3158]

M1735 2, 7, 11, 15, 20, 24, 28, 32, 37, 41, 45, 50, 54, 58, 63, 67, 71, 76, 80, 84, 88, 93, 97, 101, 106, 110, 114, 119, 123, 127, 131, 136, 140, 144, 149, 153, 157, 162, 166, 170, 174
Wythoff game. Ref CMB 2 188 59. [0,1; A1960, N0687]

M1736 1, 2, 7, 11, 101, 111, 1001, 2201, 10001, 10101, 11011, 100001, 101101, 110011, 1000001, 1001001, 1100011, 10000001, 10011001, 10100101, 11000011
Cube is a palindrome. Cf. M4583. Ref JRM 3 97 70. [1,2; A2780, N0688]

M1737 0, 0, 0, 1, 0, 0, 2, 7, 12, 14, 36, 95, 226, 501, 1056, 2377, 5448
Self-avoiding walks on square lattice. Ref JCT A13 181 72. [4,7; A6143]

M1738 0, 1, 2, 7, 12, 30, 54, 127, 226, 508, 930, 2046, 3780, 8182, 15288, 32767, 61680, 131042, 248346, 524284
Free subsets of multiplicative group of $GF(2^n)$. Ref SFCA92 2 15. [1,3; A7230]

M1739 2, 7, 13, 18, 23, 28, 34, 39, 44, 49, 54, 60, 65, 70, 75, 81, 86, 91, 96, 102, 107, 112, 117, 123, 128, 133, 138, 143, 149, 154, 159, 164, 170, 175, 180, 185, 191, 196, 201, 206
Wythoff game. Ref CMB 2 189 59. [0,1; A1966, N0689]

M1740 1, 2, 7, 14, 29, 48, 79, 116, 169, 230
Paraffins. Ref BER 30 1921 1897. [1,2; A5998]

M1741 1, 2, 7, 14, 32, 58, 110, 187, 322, 519, 839, 1302, 2015, 3032, 4542, 6668, 9738, 14006, 20036, 28324, 39830, 55473, 76875, 105692, 144629, 196585, 266038, 357952
Trees of diameter 5. Ref IBMJ 4 476 60. KU64. [6,2; A0147, N0690]

M1742 2, 7, 15, 28, 45, 69, 98, 136, 180, 235, 297, 372
Dissections of a polygon. Ref AEQ 18 387 78. [5,1; A3452]

M1743   1, 2, 7, 15, 28, 45, 70, 100, 138
Partitions into non-integral powers. Ref PCPS 47 214 51. [2,2; A0148, N0691]

M1744   2, 7, 16, 30, 50, 77, 112, 156, 210, 275, 352, 442, 546, 665, 800, 952, 1122, 1311,
   1520, 1750, 2002, 2277, 2576, 2900, 3250, 3627, 4032, 4466, 4930, 5425, 5952, 6512
$n(n+1)(n+5)/6$. Ref AS1 797. [1,1; A5581]

M1745   2, 7, 17, 19, 23, 29, 47, 59, 61, 97, 109, 113, 131, 149, 167, 179, 181, 193, 223,
   229, 233, 257, 263, 269, 313, 337, 367, 379, 383, 389, 419, 433, 461, 487, 491, 499, 503
$1/n$ has period $n-1$. Ref Krai24 1 61. HW1 115. [1,1; A6883, N1823]

M1746   1, 2, 7, 17, 49, 134, 368, 1017, 2806, 7743, 21374, 59015, 162942, 449916,
   1242352, 3430578, 9473170, 26159353, 72237232, 199478805, 550850090
Irreducible positions of size n in Montreal solitaire. Ref JCT A60 55 92. [3,2; A7049]

M1747   2, 7, 18, 28, 182, 845, 904, 5235, 36028, 74713, 526624, 977572, 4709369,
   9959574, 96696762, 7724076630, 35354759457, 138217852516, 642742746639
Increasing blocks of digits of e. Ref MOC 4 14 50; 23 679 69. [1,1; A1114, N0692]

M1748   1, 2, 7, 18, 52, 133, 330, 762, 1681
Vertex-degree sequences of n-faced polyhedral graphs. Ref md. [4,2; A6869]

M1749   1, 1, 2, 7, 18, 60, 196, 704, 2500, 9189, 33896, 126759, 476270, 1802312,
   6849777, 26152418, 100203194, 385221143
One-sided polyominoes with n cells. Ref GO65 105. wfl. [1,3; A0988, N0693]

M1750   1, 1, 2, 7, 18, 64, 226, 856, 3306, 13248, 53794, 222717
Restricted hexagonal polyominoes with n cells. Ref PEMS 17 11 70. [1,3; A2214, N0694]

M1751   1, 2, 7, 20, 54, 148, 403, 1096, 2980, 8103, 22026, 59874, 162754, 442413,
   1202604, 3269017, 8886110, 24154952, 65659969, 178482300, 485165195, 1318815734
$[e^n]$. Ref MNAS 14(5) 14 25. FW39. FMR 1 230. [0,2; A0149, N0695]

M1752   1, 1, 2, 7, 20, 58, 174, 519, 1550, 4634, 13884, 41616, 124824, 374390, 1123288,
   3369297, 10107324, 30320434, 90961626, 272878138, 818632094, 2455888346
Shifts left under OR-convolution with itself. Ref BeSl94. [0,3; A7460]

M1753   1, 2, 7, 20, 66, 212, 715, 2424, 8398, 29372, 104006, 371384, 1337220, 4847208,
   17678835, 64821680, 238819350, 883629164, 3282060210, 12233125112
Symmetrical dissections of a $2n$-gon. Ref GU58. MAT 15 121 68. [3,2; A0150, N0696]

M1754   1, 2, 7, 21, 57, 162, 452, 1255, 3474, 9621, 26604, 73531, 203166, 561242,
   1550216, 4281502, 11824338, 32654467, 90177615, 249028277, 687692923
Irreducible positions of size n in Montreal solitaire. Ref JCT A60 55 92. [6,2; A7050]

M1755   0, 2, 7, 21, 59, 163, 447, 1223, 3343, 9135, 24959, 68191, 186303, 508991,
   1390591, 3799167, 10379519, 28357375, 77473791, 211662335, 578272255
Tower of Hanoi with cyclic moves only. Ref IPL 13 118 81. GKP 18. [0,2; A5666]

M1756 2, 7, 22, 54, 118, 230, 418, 710, 1150
Homogeneous primitive partition identities of degree 6 with largest part n. Ref DGS94.
[4,1; A7344]

M1757 1, 2, 7, 22, 75, 250, 886, 3150
E-trees with at most 2 colors. Ref AcMaSc 2 109 82. [1,2; A7141]

M1758 1, 1, 2, 7, 22, 96, 380, 1853, 8510, 44940, 229836, 1296410, 7211116, 43096912,
    256874200, 1617413773, 10226972110, 67542201972, 449809389740, 3104409032126
Symmetric irreducible diagrams with $2n$ nodes. Ref JCT A24 361 78. [1,3; A4300]

M1759 1, 0, 2, 7, 23, 88, 414, 2371, 16071, 125672, 1112082, 10976183, 119481295,
    1421542640, 18348340126, 255323504931, 3809950977007, 60683990530224
$-1 + \Sigma (k-1)!C(n,k)$, $k = 1 .. n$. Ref CJM 22 26 70. [0,3; A1338, N0697]

M1760 1, 1, 2, 7, 23, 114, 625, 3974
n-celled solid polyominoes without holes. Ref GO65. [1,3; A6986]

M1761 1, 2, 7, 23, 115, 694, 5282, 46066, 456454, 4999004, 59916028, 778525516,
    10897964660, 163461964024, 2615361578344, 44460982752488, 800296985768776
Ways of placing n nonattacking rooks on $n \times n$ board. See Fig M0180. Ref LU91 1 222.
LNM 560 201 76. rcr. [2,2; A0903, N0698]

M1762 0, 1, 2, 7, 23, 122, 888, 11302, 262322, 11730500, 1006992696, 164072174728,
    50336940195360, 29003653625867536
n-node graphs without isolated nodes. Equals first differences of M1253. Ref AJM 49 453
27. HA69 214. JGT 16 133 92. [0,3; A2494, N0699]

M1763 1, 2, 7, 24, 82, 280, 956, 3264, 11144, 38048, 129904, 443520, 1514272, 5170048,
    17651648, 60266496, 205762688, 702517760, 2398545664, 8189147136, 27959497216
$a(n)=4a(n-1)-2a(n-2)$. Ref MOC 29 220 75. DM 75 95 89. [0,2; A3480]

M1764 2, 7, 24, 92, 388
Vacuously transitive relations on n nodes. Ref DM 4 194 73. [2,1; A3041]

M1765 1, 1, 2, 7, 24, 95, 388, 1650, 7183, 31965, 144502, 662241
Planted identity matched trees with n nodes. Ref DM 88 97 91. [1,3; A5754]

M1766 1, 1, 2, 7, 24, 96, 388, 1667, 7278, 32726
Skeins with vertical symmetry. Ref AEQ 31 54 86. [0,3; A7162]

M1767 2, 7, 25, 89, 317, 1129, 4021, 14321, 51005, 181657, 646981, 2304257, 8206733,
    29228713, 104099605, 370756241, 1320467933, 4702916281, 16749684709
Subsequences of $[1,...,2n+1]$ in which each even number has an odd neighbor. Ref
GuMo94. [0,1; A7484]

$$a(n) = 3\,a(n-1) + 2\,a(n-2).$$

M1768 1, 1, 1, 2, 7, 25, 108, 492, 2431, 12371, 65169, 350792, 1926372, 10744924, 60762760, 347653944, 2009690895, 11723100775, 68937782355, 408323229930
Dissections of a polygon. Ref DM 11 387 75. LeMi91. [0,4; A5034]

M1769 1, 2, 7, 26, 97, 362, 1351, 5042, 18817, 70226, 262087, 978122, 3650401, 13623482, 50843527, 189750626, 708158977, 2642885282, 9863382151, 36810643322
$a(n) = 4a(n-1) - a(n-2)$. Ref NCM 4 167 1878. MMAG 40 78 67. FQ 7 239 69. [0,2; A1075, N0700]

M1770 1, 2, 7, 26, 107, 458, 2058, 9498, 44947, 216598, 1059952, 5251806, 26297238, 132856766, 676398395, 3466799104, 17873808798, 92630098886, 482292684506
Oriented rooted trees with n nodes. Ref R1 138. [1,2; A0151, N0701]

$$\text{G.f.:} \quad \prod_{n=1}^{\infty} (1 - x^{2\,a(n)})^{-1}.$$

M1771 1, 2, 7, 26, 107, 468, 2141, 10124, 49101, 242934, 1221427, 6222838, 32056215, 166690696, 873798681, 4612654808, 24499322137, 130830894666, 702037771647
Generalized Fibonacci numbers. Ref LNM 622 186 77. [0,2; A6603]

M1772 1, 2, 7, 26, 108, 434, 1765, 7086, 28384, 113092, 449582, 1783092, 7062611
Dissections of a polygon. Ref AEQ 18 387 78. [4,2; A3447]

M1773 1, 1, 2, 7, 26, 111, 562, 3151, 19252, 128449, 925226
Forests of least height with n nodes. Ref JCT 5 97 68. jr. [0,3; A1862, N0702]

M1774 1, 2, 7, 26, 114, 512, 2427, 11794, 58787, 298188
P-graphs with $2n$ edges. Ref AEQ 31 56 86. [1,2; A7168]

M1775 1, 1, 2, 7, 26, 153, 1134, 11050
High-dimensional polyominoes with n cells. Ref wfl. [1,3; A5519]

M1776 1, 2, 7, 27, 110, 460, 1948, 8296, 35400, 151056, 643892, 2740216, 11639416, 49340080, 208727176, 881212272, 3713043152, 15615663008, 65555425780
Convex polygons of length $2n$ on square lattice. Ref TCS 34 179 84. [2,2; A5768]

M1777 1, 1, 2, 7, 27, 118, 537, 2570, 12587, 63173
Blobs with $2n + 1$ edges. Ref AEQ 31 56 86. [0,3; A7166]

M1778 1, 2, 7, 28, 120, 528, 2344, 10416, 46160, 203680, 894312, 3907056, 16986352, 73512288, 316786960, 1359763168, 5815457184, 24788842304, 105340982248
Convex polygons of length $2n$ on square lattice. Ref TCS 34 179 84. JPA 21 L472 88. [2,2; A5436]

$$a(n) = (2n+11)4^n - 4(2n+1)C(2n,n).$$

M1779   1, 2, 7, 28, 122, 558, 2641, 12822, 63501, 319554, 1629321, 8399092, 43701735, 229211236, 1210561517, 6432491192, 34364148528, 184463064936, 994430028087
Column-convex polyominoes with perimeter $2n$. Ref JCT A48 29 88. JPA 23 2323 90. [2,2; A5435]

M1780   1, 2, 7, 28, 124, 588, 2938, 15268, 81826, 449572, 2521270, 14385376, 83290424, 488384528, 2895432660, 17332874364, 104653427012, 636737003384, 3900770002646
Polygons of length $2n$ on square lattice. Ref JCP 31 1333 59. JPA 21 L167 88. [2,2; A2931, N0703]

M1781   1, 2, 7, 29, 196, 1788, 21994
Hamiltonian circuits on n-octahedron. Ref JCT B19 2 75. [2,2; A3437]

M1782   1, 2, 7, 30, 143, 728, 3876, 21318, 120175, 690690, 4032015, 23841480, 142498692, 859515920, 5225264024, 31983672534, 196947587823, 1219199353190
Convolution of M2926 with itself: $2C(3n+2,n)/(3n+2)$. Ref dek. [0,2; A6013]

M1783   0, 1, 2, 7, 30, 157, 972, 6961, 56660, 516901, 5225670, 57999271, 701216922, 9173819257, 129134686520, 1946194117057, 31268240559432, 533506283627401
$a(n)=n.a(n-1)+a(n-2)$. Ref EUR 20 15 57. [0,3; A1053, N0704]

M1784   1, 2, 7, 31, 147, 999, 6495, 44619, 327482, 2417860
P-graphs with $2n$ edges. Ref AEQ 31 54 86. [1,2; A7164]

M1785   1, 2, 7, 31, 162, 973, 6539, 48410, 390097, 3389877, 31534538, 312151125, 3271508959, 36149187780, 419604275375, 5100408982825, 64743452239424
Exponentiation of e.g.f. for primes. [0,2; A7446]

M1786   1, 2, 7, 31, 164, 999, 6841, 51790, 428131, 3827967, 36738144, 376118747, 4086419601, 46910207114, 566845074703, 7186474088735, 95318816501420
Sorting numbers. Ref PSPM 19 173 71. [0,2; A2872, N0705]

$$\text{E.g.f.: } \exp((e^{2x}-3)/2+e^x).$$

M1787   0, 0, 0, 0, 1, 2, 7, 31, 168, 1025, 7013, 52495, 425213, 3696032, 34291937, 338161526, 3532185118, 38963334652, 452704892533, 5526901638291
Reduced interval graphs with n nodes. Ref TAMS 272 422 82. pjh. [1,6; A5977]

M1788   1, 2, 7, 32, 168, 970, 5984, 38786, 261160, 1812630, 12895360, 93638634
Polygons of length $4n$ on Manhattan lattice. Ref JPA 18 1013 85. [1,2; A6781]

M1789   1, 2, 7, 32, 177, 1122, 7898, 60398, 494078, 4274228, 38763298, 366039104, 3579512809, 36091415154, 373853631974, 3966563630394, 42997859838010
Hoggatt sequence. Ref FQ 27 167 89. FA90. [0,2; A5362]

M1790   1, 2, 7, 32, 178, 1160, 8653, 72704, 679798, 7005632, 78939430, 965988224, 12762344596, 181108102016, 2748049240573, 44405958742016, 761423731533286
$a(n+1)=(n+1)a(n)+\Sigma a(k)a(n-k)$. Ref dek. [1,2; A6014]

M1791 0, 1, 2, 7, 32, 181, 1214, 9403, 82508, 808393, 8743994, 103459471, 1328953592, 18414450877, 273749755382, 4345634192131, 73362643649444, 1312349454922513
$a(n) = n.a(n-1) + (n-2)a(n-2)$. Ref R1 188. [0,3; A0153, N0706]

$$\text{E.g.f.: } (1-x)^{-3} e^{-x}.$$

M1792 1, 1, 2, 7, 32, 181, 1232, 9787, 88832, 907081, 10291712, 128445967, 1748805632, 25794366781, 409725396992, 6973071372547, 126585529106432
Expansion of $1/(1 - \sinh x)$. Ref ARS 10 138 80. [0,3; A6154]

M1793 0, 1, 1, 2, 7, 32, 184, 1268, 10186, 93356, 960646, 10959452, 137221954, 1870087808, 27548231008, 436081302248, 7380628161076, 132975267434552
Stochastic matrices of integers. Ref DUMJ 35 659 68. [0,4; A0987, N0707]

M1794 1, 2, 7, 33, 192
Permutations of length n with n in second orbit. Ref C1 258. [2,2; A6595]

M1795 1, 2, 7, 34, 209, 1546, 13327, 130922, 1441729, 17572114, 234662231, 3405357682, 53334454417, 896324308634, 16083557845279, 306827170866106
$a(n) = 2n.a(n-1) - (n-1)^2 a(n-2)$. Ref SE33 78. [0,2; A2720, N0708]

M1796 1, 2, 7, 34, 257, 2606, 32300, 440564, 6384634
Polyhedra with n nodes. Ref GR67 424. UPG B15. Dil92. [4,2; A0944, N0709]

M1797 2, 7, 35, 219, 1594, 12935, 113945, 1070324, 10586856, 109259633, 1168384157, 12877168147, 145656436074, 1685157199175, 19886174611045
Two-rowed truncated monotone triangles. Ref JCT A42 277 86. Zei93. [1,1; A6947]

M1798 1, 1, 2, 7, 35, 228, 1834, 17382, 195866, 2487832, 35499576, 562356672, 9794156448, 186025364016, 3826961710272, 84775065603888, 2011929826983504
Coefficients of iterated exponentials. Ref SMA 11 353 45. [0,3; A0154, N0710]

M1799 1, 2, 7, 35, 228, 1834, 17582, 195866, 2487832, 35499576, 562356672, 9794156448, 186025364016, 3826961710272, 84775065603888, 2011929826983504
Expansion of $\ln(1 + \ln(1 + x))$. [0,2; A3713]

M1800 1, 0, 1, 2, 7, 36, 300, 3218, 42335, 644808
Circular diagrams with n chords. Ref BarN94. [0,4; A7474]

M1801 1, 2, 7, 36, 317, 5624, 251610, 33642660, 14685630688
$n \times n$ binary matrices. Ref CPM 89 217 64. SLC 19 79 88. [0,2; A2724, N0711]

M1802 2, 7, 37, 216, 1780, 32652
Semigroups of order n with 2 idempotents. Ref MAL 2 2 67. SGF 14 71 77. [2,1; A2787, N0712]

M1803 1, 2, 7, 37, 266, 2431, 27007, 353522, 5329837, 90960751, 1733584106, 36496226977, 841146804577, 21065166341402, 569600638022431
$a(n) = (2n-1)a(n-1) + a(n-2)$. Ref RCI 77. [0,2; A1515, N0713]

M1804 1, 1, 2, 7, 38, 291, 2932, 36961, 561948, 10026505, 205608536, 4767440679,
123373203208, 3525630110107, 110284283006640, 3748357699560961
Forests of labeled trees with n nodes. Ref JCT 5 96 68. SIAD 3 574 90. [0,3; A1858,
N0714]

M1805 1, 1, 2, 7, 40, 357, 4824, 96428, 2800472, 116473461
n-element partial orders contained in linear order. Ref nbh. [0,3; A6455]

M1806 1, 2, 7, 41, 346, 3797, 51157, 816356, 15050581, 314726117, 7359554632,
190283748371, 5389914888541, 165983936096162, 5521346346543307
Planted binary phylogenetic trees with n labels. Ref LNM 884 196 81. [1,2; A6677]

M1807 1, 1, 2, 7, 41, 376, 5033, 92821, 2257166, 69981919, 2694447797, 126128146156,
7054258103921, 464584757637001, 35586641825705882, 3136942184333040727
Hammersley's polynomial $p_n(1)$. Ref MASC 14 4 89. [0,3; A6846]

M1808 1, 2, 7, 42, 429, 7436, 218348, 10850216, 911835460, 129534272700,
31095744852375, 12611311859677500, 8639383518297652500
Robbins numbers: $\Pi(3k+1)!/(n+k)!$, $k = 0 .. n-1$. Ref MINT 13(2) 13 91. JCT A66
17 94. [1,2; A5130]

M1809 1, 2, 7, 42, 582, 21480, 2142288, 575016219, 415939243032, 816007449011040,
4374406209970747314, 6453983693872074973 9356
Antisymmetric relations on n nodes. Ref PAMS 4 494 53. MIT 17 23 55. [1,2; A1174,
N0715]

M1810 0, 1, 2, 7, 44, 361, 3654, 44207, 622552, 10005041, 180713290, 3624270839,
79914671748, 1921576392793, 50040900884366, 1403066801155039
Modified Bessel function $K_n(1)$. Ref AS1 429. [0,3; A0155, N0716]

M1811 0, 1, 2, 7, 44, 447, 6749, 142176, 3987677, 143698548, 6470422337,
356016927083, 23503587609815, 1833635850492653, 166884365982441238
$a(n) = n(n-1)a(n-1)/2 + a(n-2)$. [0,3; A1046, N0717]

M1812 1, 2, 7, 44, 529, 12278, 565723, 51409856, 9371059621, 3387887032202,
2463333456292207, 3557380311703796564, 10339081666350180289849
Sum of Gaussian binomial coefficients $[n,k]$ for $q = 4$. Ref TU69 76. GJ83 99. ARS A17
328 84. [0,2; A6118]

M1813 2, 7, 52, 2133, 2590407, 3374951541062, 56951835044791 16640376509,
16217557574922386301420514191523784895639577710480
Free binary trees of height n. Ref JCIS 17 180 92. [1,1; A5588]

M1814 1, 1, 2, 7, 56, 2212, 2595782, 3374959180831, 56951835044892390 67484387,
16217557574922386301420531277071365103168734284282
Planted 3-trees of height n. Ref RSE 59(2) 159 39. CMB 11 87 68. JCIS 17 180 92. [0,3;
A2658, N0718]

M1815   2, 7, 60, 13733
Switching networks. Ref JFI 276 326 63. [1,1; A0892, N0719]

M1816   0, 2, 7, 63, 1234, 55447, 5598861
$n \times n$ binary matrices with no 2 adjacent 1's. Ref rhh. [0,2; A6506]

M1817   2, 7, 97, 18817, 708158977, 1002978273411373057,
   2011930833870518011412817828051050497
$a(n) = 2a(n-1)^2 - 1$. Ref D1 1 399. TCS 65 219 89. [0,1; A2812, N0720]

M1818   1, 2, 7, 111, 308063, 100126976263592
Boolean functions of n variables. Ref HA65 153 (divided by 2). [1,2; A0157, N0721]

M1819   2, 7, 124, 494298
Switching networks. Ref JFI 276 326 63. [1,1; A0889, N0722]

M1820   1, 2, 7, 160, 332381, 2751884514766, 2726229327962814088790659 87,
   36418399108354015676266835934360038942509313109 90279692
Free idempotent monoid on n letters. Ref Loth83 32. [0,2; A5345]

M1821   2, 7, 1172, 36325278240, 1827297478706355168798634830 6336
Invertible Boolean functions of n variables. Ref PGEC 13 530 64. [1,1; A0653, N0723]

SEQUENCES BEGINNING . . ., 2, 8, . . .

M1822   0, 1, 2, 8, 3, 1, 1, 1, 1, 7, 1, 1, 2, 1, 1, 1, 2, 7, 1, 2, 2, 1, 1, 1, 3, 7, 1, 3, 2, 1, 1, 1, 4,
   7, 1, 4, 2, 1, 1, 1, 5, 7, 1, 5, 2, 1, 1, 1, 6, 7, 1, 6, 2, 1, 1, 1, 7, 7, 1, 7, 2, 1, 1, 1, 8, 7, 1, 8, 2
Continued fraction for $e/4$. Ref KN1 2 601. [1,3; A6085]

M1823   2, 8, 9, 10, 11, 15, 19, 21, 22, 25, 26, 27, 28, 30, 31, 34, 40, 42, 45, 46, 47, 50, 55,
   57, 58, 59, 62, 64, 65, 66, 70, 74, 75, 78, 79, 80, 84, 86, 94, 96, 97, 98, 100, 101, 103, 106
Numbers with an even number of partitions. Ref JLMS 1 226 26. MOC 21 470 67. ASI
836. [1,1; A1560, N0724]

M1824   2, 8, 10, 8, 16, 16, 10, 24, 16, 8, 32, 24, 18, 24, 16, 24, 32, 32, 16, 32, 34, 16, 48,
   16, 16, 56, 32, 24, 32, 40, 26, 48, 48, 16, 32, 32, 32, 56, 48, 24, 64, 32, 26, 56, 16, 40, 64
Theta series of cubic lattice w.r.t. edge. Ref SPLAG 107. [0,1; A5876]

M1825   1, 1, 2, 8, 10, 24, 53, 74, 153, 280, 436, 793, 1322, 2085, 3510, 5648, 8796
Expansion of a modular function. Ref PLMS 9 385 59. [-6,3; A2510, N0725]

M1826   2, 8, 12, 8, 16, 24, 20, 32, 18, 24, 40, 48, 28, 48, 60, 32, 32, 56
Frequency of nth largest distance in $N \times N$ grid, $N > n$. Ref ReSk94. [1,1; A7543]

M1827   2, 8, 12, 16, 26, 24, 28, 48, 36, 40, 64, 48, 62, 80, 60, 64, 96, 96, 76, 112, 84, 88,
   156, 96, 114, 144, 108, 144, 160, 120, 124, 208, 168, 136, 192, 144, 148, 248, 192, 160
Theta series of D_4 lattice w.r.t. edge. [0,1; A5880]

M1828 1, 1, 1, 2, 8, 16, 384, 768, 3072, 6144, 61440, 983040
Denominators of an asymptotic expansion. Cf. M3817. Ref SIAD 3 575 90. [0,4; A6573]

M1829 1, 2, 8, 18, 55, 138, 470, 1164, 4055, 10140, 35609, 89782, 316513, 803040
Deficit in peeling rinds. Ref GTA85 727. [1,2; A5675]

M1830 1, 2, 8, 19, 41, 78, 134, 218
Partitions into non-integral powers. Ref PCPS 47 214 51. [3,2; A0158, N0726]

M1831 0, 2, 8, 20, 40, 70, 112, 168, 240, 330, 440, 572, 728, 910, 1120, 1360, 1632, 1938,
  2280, 2660, 3080, 3542, 4048, 4600, 5200, 5850, 6552, 7308, 8120, 8990, 9920, 10912
Series expansion for rectilinear polymers on square lattice. Ref JPA 12 2137 79. [0,2;
A7290]

$$\text{G.f.:}\quad 2\,x\,(1-x)^{-4}.$$

M1832 1, 2, 8, 20, 75, 210, 784, 2352, 8820, 27720, 104544, 339768, 1288287, 4294290,
  16359200, 55621280, 212751396, 734959368, 2821056160, 9873696560, 38013731756
Walks on square lattice. Ref GU90. [1,2; A5559]

M1833 0, 0, 2, 8, 20, 80, 350, 1232, 5768, 31040, 142010, 776600, 4874012, 27027728,
  168369110, 1191911840, 7678566800, 53474964992, 418199988338
Degree n permutations of order exactly 3. Ref CJM 7 159 55. [1,3; A1471, N0727]

M1834 2, 8, 20, 152, 994, 7888, 70152, 695760, 7603266, 90758872, 1174753372,
  16386899368, 245046377410, 3910358788256, 66323124297872, 1191406991067168
From ménage polynomials. Ref R1 197. [3,1; A0159, N0728]

M1835 1, 2, 8, 21, 48, 99, 186
Partitions into non-integral powers. Ref PCPS 47 214 51. [4,2; A0160, N0729]

M1836 2, 8, 22, 50, 110, 226, 464, 938, 1888, 3794, 7598, 15208, 30438, 60890, 121792,
  243606, 487238, 974488, 1948998, 3898034, 7796078, 15592168, 31184358, 62368754
$a(n) = \min\{\, a(k-1) + 2^k(n + a(n-k))\,\}$, $k = 1 .. n$. Ref CK90. [1,1; A6696]

M1837 0, 2, 8, 22, 52, 112, 228, 442, 832, 1516, 2720, 4754, 8264, 14000, 23824, 39318,
  66052, 106282, 177884, 277936, 469384, 703924, 1225052, 1718226, 3203156, 3974696
First moment of site percolation series for square lattice. Ref JPA 21 3821 88. [0,2; A6732]

M1838 0, 2, 8, 22, 52, 114, 240, 494, 1004, 2026, 4072, 8166, 16356, 32738, 65504,
  131038, 262108, 524250, 1048536, 2097110, 4194260, 8388562, 16777168, 33554382
Second-order Eulerian numbers: $2^{n+1} - 2n - 2$. Ref JCT A24 28 78. GKP 256. [1,2;
A5803]

M1839 0, 2, 8, 24, 60, 136, 288, 582, 1132, 2138, 3940, 7114, 12632, 22080, 38160,
  65056, 110172, 184032, 306968, 503650, 831408, 1340338, 2201840, 3479116, 5733312
Second perpendicular moment of site percolation series for square lattice. Ref JPA 21 3821
88. [0,2; A6734]

M1840 0, 2, 8, 24, 62, 148, 330, 710, 1464, 2962, 5814, 11288, 21406, 40364, 74570, 137602, 249088, 451868, 804766, 1440580, 2529686, 4482584, 7775166, 13664146
First moment of bond percolation series for square lattice. Ref JPA 21 3820 88. [0,2; A6728]

M1841 0, 2, 8, 24, 64, 156, 358, 786, 1664, 3434, 6902, 13656, 26464, 50772, 95754, 179442, 331294, 609496, 1106106, 2004852, 3586874, 6423028, 11351274, 20126538
Second perpendicular moment of bond percolation series for square lattice. Ref JPA 21 3820 88. [0,2; A6730]

M1842 1, 2, 8, 24, 78, 232, 720, 2152, 6528, 19578, 58944, 176808, 531128, 1593288, 4781952, 14345792, 43043622, 129130584, 387411144, 1162232520, 3486755688
Conjugacy classes in $GL(n,3)$. Ref wds. [0,2; A6952]

M1843 2, 8, 24, 85, 286, 1008, 3536, 12618, 45220, 163504
Perforation patterns for punctured convolutional codes (2,1). Ref SFCA92 1 9. [3,1; A7223]

M1844 1, 0, 2, 8, 26, 80, 268, 944, 3474, 13072, 49672, 191272, 744500, 2924680, 11596284, 46364456
Magnetization for diamond lattice. Ref JMP 6 297 65. JPA 6 1511 73. DG74 420. [0,3; A2930, N0730]

M1845 1, 1, 2, 8, 29, 166, 1023, 6922, 48311, 346543, 2522572, 18598427
3-dimensional polyominoes with n cells. See Fig M1845. Ref FQ 3 19 65. cjb. RE72 108. wfl. [1,3; A0162, N0731]

||||||||||||||||||||||||||||||||||||||||||||||||||||||||||||||||||||||||||||||||||||||||||||||||||||||||||||||||||||||

Figure M1845. POLYOMINOES. A **polyomino** with n cells is a connected set of n squares cut from a square grid [GO65]. The basic sequence is M1425, the number of **free** polyominoes with n cells, i.e. allowing pieces to be rotated and turned over. (The sequences refer to free polyominoes unless stated otherwise.) Only the first 23 terms are known. Polyominoes can also be formed from triangles (M2374), hexagons (M2682), cubes (M1845): The table contains a large number of other sequences arising in the enumeration of polyominoes of various types.

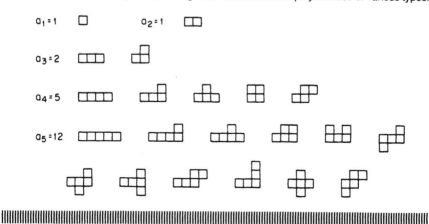

||||||||||||||||||||||||||||||||||||||||||||||||||||||||||||||||||||||||||||||||||||||||||||||||||||||||||||||||||||||

M1846 2, 8, 31, 88, 199, 384, 659, 1056, 1601, 2310, 3185, 4364, 5693, 7360, 9287, 11494, 14189, 17258, 20517, 24526, 28967, 33736, 38917, 45230, 51797, 59180, 66831
Sum of next n primes. [0,1; A7468]

M1847 1, 2, 8, 31, 139, 724
Trivalent graphs with $2n$ nodes. Ref pam. [0,2; A3175]

M1848 2, 8, 32, 54, 114, 414, 1400, 1850, 2848, 4874
$n.3^n + 1$ is prime. Ref JRM 21 191 89. [1,1; A6552]

M1849 1, 2, 8, 32, 136, 592, 2624, 11776, 53344, 243392, 1116928, 5149696, 23835904, 110690816, 515483648, 2406449152, 11258054144, 52767312896, 247736643584
$na(n) = 2(2n-1)a(n-1) + 4(n-1)a(n-2)$. Ref FQ 27 434 89. [0,2; A6139]

M1850 2, 8, 32, 144, 708, 3696, 20296
Figure 8's with $2n$ edges on the square lattice. Ref JPA 3 23 70. [2,1; A3304]

M1851 1, 1, 2, 8, 33, 194, 1196, 8196, 58140, 427975, 3223610, 24780752, 193610550, 1534060440, 12302123640, 99699690472, 815521503060, 6725991120004
Dissections of a polygon. Ref DM 11 388 75. [1,3; A5040]

M1852 2, 8, 34, 136, 538, 2080, 7970, 30224, 113874
Series-parallel numbers. Ref R1 142. [2,1; A0163, N0732]

M1853 1, 2, 8, 34, 152, 714, 3472, 17318, 88048
Magnetization for square lattice. Ref PHA 22 934 56. [0,2; A2928, N0733]

M1854 0, 0, 0, 0, 0, 0, 1, 2, 8, 35, 168, 999, 6340, 43133, 305271, 2231377
Noncircumscribable simplicial polyhedra with n nodes. Ref Dil92. [1,8; A7034]

M1855 2, 8, 35, 205, 1224, 8169, 58980, 440312, 3424506, 27412679, 224376048
Number of twin primes $< 10^n$. Ref BPNR 202. [1,1; A7508]

M1856 0, 2, 8, 36, 184, 1110, 7776, 62216, 559952, 5599530, 61594840, 739138092, 9608795208, 134523132926, 2017846993904, 32285551902480
Transpositions needed to generate permutations of length n. Ref CJN 13 155 70. [1,2; A1540, N0734]

M1857 2, 8, 38, 192, 1002, 5336, 28814, 157184, 864146, 4780008, 26572086, 148321344, 830764794, 4666890936, 26283115038, 148348809216, 838944980514
$2\Sigma C(n-1,k)C(n+k,k)$, $k = 0 .. n-1$. Ref AMM 43 29 36. [1,1; A2003, N0735]

M1858 2, 8, 38, 212, 1370, 10112, 84158, 780908, 8000882, 89763320, 1094915222, 14431179908, 204423631178, 3097603939952, 50001759773870, 856665220770332
$\Sigma(n+2)!C(n,k)$, $k = 0 .. n$. Ref CJM 22 26 70. [0,1; A1340, N0736]

M1859 1, 2, 8, 40, 165, 712, 2912, 11976, 48450, 195580, 784504, 3139396
Dissections of a polygon. Ref AEQ 18 387 78. [5,2; A3445]

M1860   2, 8, 40, 208, 1120, 6200, 35236
Figure 8's with $2n$ edges on the square lattice. Ref JPA 3 24 70. [4,1; A3305]

M1861   2, 8, 40, 240, 1680, 13440, 120960, 1209600, 13305600, 159667200, 2075673600,
   29059430400, 435891456000, 6974263296000, 118562476032000, 2134124568576000
$n!/3$. Ref TOH 42 152 36. [3,1; A2301, N0737]

M1862   1, 2, 8, 42, 262, 1828, 13820, 110954, 933458, 8152860, 73424650, 678390116,
   6405031050, 61606881612, 602188541928, 5969806669034, 59923200729046
Closed meandric numbers: ways a loop can cross a road $2n$ times. See Fig M4587. Ref
PH88. SFCA91 291. jar. vrp. [1,2; A5315]

M1863   1, 2, 8, 42, 268, 1994, 16852, 158778, 1644732, 18532810, 225256740,
   2933174842, 40687193548, 598352302474, 9290859275060, 151779798262202
Sorting numbers. Ref PSPM 19 173 71. [0,2; A2874, N0738]

$$\text{E.g.f.:} \quad \exp((e^{3x} - 4) / 3 + e^x).$$

M1864   1, 1, 2, 8, 42, 296, 2635
Polyhedra with n nodes and n faces. Ref JCT 7 157 69. [4,3; A2856, N0739]

M1865   0, 1, 2, 8, 43, 283, 1946, 14010, 103274, 776784
Q-graphs rooted at a polygon. Ref AEQ 31 63 86. [1,3; A7169]

M1866   1, 2, 8, 44, 308, 2612, 25988, 296564, 3816548, 54667412, 862440068,
   14857100084, 277474957988, 5584100659412, 120462266974148, 2772968936479604
Expansion of $(2 - e^x)^{-2}$. Ref C1 294. [1,2; A5649]

M1867   1, 2, 8, 44, 310, 2606, 25202, 272582, 3233738, 41454272, 567709144,
   8230728508, 125413517530, 1996446632130, 33039704641922, 566087847780250
Hoggatt sequence. Ref FQ 27 167 89. FA90. [0,2; A5363]

M1868   2, 8, 44, 436, 7176, 484256
Self-converse relations on n points. Ref MAT 13 157 66. [1,1; A2500, N0740]

M1869   0, 1, 2, 8, 44, 490, 14074, 1349228
Threshold functions of n variables. Ref PGEC 19 823 70. [0,3; A2833, N0741]

M1870   1, 1, 1, 2, 8, 45, 416, 6657, 189372, 9695869, 902597327, 154043277297,
   48535481831642, 28400190511772276, 31020581422991798557
$[2^{n(n-1)/2}/n!]$. Ref HP73 246. [1,4; A3091]

M1871   1, 1, 2, 8, 46, 322, 2546, 21870, 199494
Irreducible meanders. See Fig M4587. Ref SFCA91 299. [0,3; A6664]

M1872   1, 1, 2, 8, 46, 332, 2874, 29024, 334982, 4349492, 62749906, 995818760,
   17239953438, 323335939292, 6530652186218, 141326092842416, 3262247252671414
Expansion of $(1-x)e^x/(2-e^x)$. Ref Stan89. [0,3; A5840]

M1873 1, 2, 8, 46, 352, 3362, 38528, 515086, 7869952, 135274562, 2583554048,
54276473326, 1243925143552, 30884386347362, 825787662368768
Alternating 3-signed permutations. Ref EhRe94. [0,2; A7289]

$$\text{G.f.: } (\sin 2x + \cos x) \, / \cos 3x \,.$$

M1874 1, 2, 8, 48, 80, 96, 128, 240, 288, 480, 1008, 1200, 1296, 1440, 1728, 2592, 2592,
4800, 5600, 6480, 8640, 11040, 12480, 14976, 19008, 19200, 22464, 24320, 24576
Values of $\phi(n) = \phi(n+1)$. Cf. M2999. Ref AMM 56 22 49. MI72. [1,2; A3275]

M1875 2, 8, 48, 98, 350, 440, 2430, 2430, 13310, 13454, 17575, 212380
Every sequence of 3 numbers $> a(n)$ contains a prime $> p(n)$. Ref AMM 79 1087 72. [2,1; A3032]

M1876 1, 2, 8, 48, 256, 1280, 6912, 39424, 212992, 1118208
Norm of a matrix. Ref AMM 86 843 79. [1,2; A4141]

M1877 0, 1, 2, 8, 48, 328, 2335, 17133, 128244, 975547
Q-graphs with $2n$ edges. Ref AEQ 31 63 86. [1,3; A7170]

M1878 1, 2, 8, 48, 384, 3840, 46080, 645120, 10321920, 185794560, 3715891200,
81749606400, 1961990553600, 51011754393600, 1428329123020800
Double factorials: $(2n)!! = 2^n n!$. Ref AMM 55 425 48. MOC 24 231 70. [0,2; A0165, N0742]

M1879 1, 2, 8, 48, 450, 5751, 90553
Trivalent graphs of girth exactly 5 and $2n$ nodes. Ref gr. [5,2; A6925]

M1880 1, 1, 2, 8, 48, 480, 5760, 92160, 1658880, 36495360, 1021870080, 30656102400,
1103619686400, 44144787456000, 1854081073152000, 85287729364992000
$a(n) = (p(n)-1) \, a(n-1)$, where $p(n)$ is the nth prime. [1,3; A5867]

M1881 1, 2, 8, 50, 416, 4322, 53888, 783890, 13031936, 243733442, 5064892768
From Fibonacci sums. Ref FQ 5 48 67. [0,2; A0557, N0743]

M1882 1, 1, 2, 8, 50, 418, 4348, 54016, 779804, 12824540, 236648024, 4841363104,
108748223128, 2660609220952, 70422722065040, 2005010410792832
$a(n) = (2n-1) a(n-1) - (n-1) a(n-2)$. Ref AJM 2 94 1879. LU91 1 223. [0,3; A2801, N0744]

M1883 2, 8, 50, 442
Essential complementary partitions of n-set. Ref PGCT 18 562 71. [2,1; A7334]

M1884 1, 2, 8, 50, 872, 55056, 14330784, 14168055824, 51063045165248,
667216816957658368, 31770676467810344050944
Strength 3 Eulerian graphs with n nodes. Ref rwr. [1,2; A7128]

M1885 1, 2, 8, 52, 472, 5504, 78416, 1320064, 25637824, 564275648, 13879795712, 377332365568, 11234698041088, 363581406419456, 12707452084972544
Related to series-parallel networks. Ref AAP 4 127 72. [1,2; A6351]

M1886 1, 2, 8, 54, 533, 6944, 111850, 2135740, 47003045, 1168832808
Degree sequences of n-node graphs. Ref Stan89. [1,2; A5155]

M1887 2, 8, 54, 556, 8146
Coefficients of Gandhi polynomials. Ref DUMJ 41 311 74. [2,1; A5440]

M1888 1, 1, 2, 8, 56, 608, 9440, 198272
Genocchi medians. Ref STNB 11 85 41. [1,3; A5439]

M1889 1, 2, 8, 58, 444, 4400, 58140, 785304, 12440064, 238904904, 4642163952, 101180433024, 2549865473424, 64728375139872, 1797171220690560
Maximal Eulerian numbers of second kind. Ref GKP 256. [1,2; A7347]

M1890 1, 2, 8, 58, 612, 8374, 140408, 2785906, 63830764, 1658336270
Phylogenetic trees with n labels. Ref ARS A17 179 84. [1,2; A5804]

M1891 1, 2, 8, 60, 320, 1980, 10512, 60788, 320896, 1787904, 9381840, 51081844
Folding a $2 \times n$ strip of stamps. See Fig M4587. Ref CJN 14 77 71. [0,2; A1415, N0745]

M1892 1, 2, 8, 60, 672, 9953, 184557, 4142631, 109813842, 3373122370, 118280690398, 4678086540493, 206625802351035, 10107719377251109, 543762148079927802
Even graphs with n edges. Ref CJM 8 410 56. dgc. [1,2; A1188, N0746]

M1893 1, 2, 8, 61, 822, 17914, 571475, 24566756, 1346167320, 90729050427, 7341861588316, 700870085606926, 77858914606919461, 9954018225212149326
Labeled interval graphs with n nodes. Ref TAMS 272 422 82. pjh. [1,2; A5215]

M1894 1, 1, 2, 8, 62, 908, 24698, 1232944, 112018190
Primitive sorting networks on n elements. Ref KN91. [1,3; A6245]

M1895 2, 8, 64, 736, 10624, 183936, 3715072
Fanout-free functions of n variables. Ref PGEC 27 1183 78. [1,1; A5612]

M1896 1, 2, 8, 64, 832, 15104, 352256, 10037248, 337936384, 13126565888
Phylogenetic trees with n labels. Ref JACM 23 705 76. PGEC 27 315 78. LNM 829 122 80. [1,2; A5640]

M1897 1, 1, 2, 8, 64, 1024, 32768, 2097152, 268435456, 68719476736, 35184372088832, 36028797018963968, 73786976294838206464, 302231454903657293676544
$2^{n(n-1)/2}$. Ref HA69 178. [0,3; A6125]

M1898 1, 2, 8, 64, 1120, 42176, 3583232, 666124288, 281268665344, 260766671206400,
549874114073747456, 2547649010961476288512, 268544167244050088788829568
Sum of Gaussian binomial coefficients [n,k] for $q = 5$. Ref TU69 76. GJ83 99. ARS A17
329 84. [0,2; A6119]

M1899 2, 8, 72, 1152, 26304, 773376, 27792384
Fanout-free functions of n variables. Ref PGEC 27 315 78. [1,1; A5615]

M1900 2, 8, 75, 8949, 119646723, 15849841722437093,
708657580163382065836292133774995
Continued cotangent for e. Ref DUMJ 4 339 38. jos. [1,1; A2668, N0748]

M1901 1, 2, 8, 96, 4608, 798720, 361267200
Folding a $2 \times 2 \times \cdots \times 2$ n-dimensional map. See Fig M4587. Ref CJN 14 77 71. [0,2;
A1417, N0750]

M1902 1, 1, 2, 8, 96, 10368, 108615168, 11798392572168192,
139202068568601556987554268864512
$a(n+1) = a(n)(a(0) + \cdots + a(n))$. [0,3; A1697, N0751]

M1903 2, 8, 96, 43008, 187499658240
Hamiltonian cycles on n-cube. Ref GA86 24. [1,1; A6069]

M1904 2, 8, 112, 5856
Reachable configurations on n circles. Ref CACM 31 1231 88. [1,1; A5787]

M1905 1, 2, 8, 152, 5024, 247616, 16845056, 1510219136, 172781715968,
24607783918592, 4275324219846656, 8908279475571834880
Reversion of o.g.f. for tangent numbers. Cf. M2096. [1,2; A7314]

M1906 2, 8, 214, 10740500
Switching networks. Ref JFI 276 326 63. [1,1; A0893, N0752]

M1907 1, 2, 8, 496, 9088, 12032, 12004352, 4139008, 51347456
Coefficients of Green function for cubic lattice. Ref PTRS 273 593 73. [0,2; A3301]

M1908 2, 8, 502, 547849868
Switching networks. Ref JFI 276 326 63. [1,1; A0890, N0753]

M1909 2, 8, 23040, 24, 1857945600, 326998425600, 714164561510400, 64,
8391719263571804160000, 25510826561258284646400000
Group generated by perfect shuffles of $2n$ cards. Ref AAM 4 177 83. [1,1; A7346]

SEQUENCES BEGINNING . . ., 2, 9, . . .

M1910 1, 2, 9, 4, 28, 18, 118, 80, 504, 466, 1631, 2160, 5466, 7498
Expansion of a modular function. Ref PLMS 9 384 59. [−3,2; A2508, N0754]

M1911 1, 1, 1, 2, 9, 4, 95, 414, 49, 10088, 55521, 13870, 2024759, 15787188, 28612415, 616876274, 7476967905, 32522642896, 209513308607, 4924388011050
Expansion of $\exp(x\,e^{-x})$. [0,4; A3725]

M1912 2, 9, 9, 7, 9, 2, 4, 5, 8
Decimal expansion of speed of light (m/sec). Ref FiFi87. Lang91. [9,1; A3678]

M1913 1, 0, 1, 1, 2, 9, 9, 50, 267, 413, 2180, 17731, 50533, 110176, 1966797, 9938669, 8638718, 278475061, 2540956509, 9816860358, 27172288399, 725503033401
Expansion of $\exp(1-e^x)$. Ref JIA 76 153 50. FQ 7 448 69. JM1 96 45 83. [0,5; A0587, N0755]

M1914 2, 9, 10, 19, 37, 39, 75, 76, 77, 149, 151, 152, 155, 299, 303, 309, 597, 605, 607, 619, 1195, 1211, 1213, 1214, 1237, 2389, 2421, 2427, 2475, 4779, 4843, 4853, 4854
Positions of remoteness 3 in Beans-Don't-Talk. Ref MMAG 59 267 86. [1,1; A5695]

M1915 2, 9, 10, 42, 79, 252, 582, 1645, 4106, 11070, 28459, 75348, 195898
From sum of $1/F(n)$. Ref FQ 16 169 78. [1,1; A6172]

M1916 1, 2, 9, 18, 118
Abstract n-dimensional crystallographic point groups. Ref PCPS 47 650 51. ACA 22 605 67. JA73 73. Enge93 1020. [0,2; A6226]

M1917 1, 0, 2, 9, 20, 30, 66, 0, 7216, 155736, 2447640, 40095000, 696155448, 13193809200, 269899395024, 5951688692040, 140573490904320, 3543930826470720
Expansion of $(1-x-x^2)^x$. [0,3; A7115]

M1918 1, 2, 9, 20, 149, 467, 237385, 237852, 1426645, 7371077, 8797722, 16168799, 24966521, 66101841, 91068362, 157170203, 3863153234, 4020323437
Convergents to cube root of 6. Ref AMP 46 107 1866. L1 67. hpr. [1,2; A2360, N0756]

M1919 1, 2, 9, 20, 670
Cutting numbers of graphs. Ref GU70 149. [1,2; A2888, N0757]

M1920 2, 9, 24, 50, 90, 147, 224, 324
Paraffins. Ref BER 30 1922 1897. [1,1; A6002]

M1921 0, 1, 2, 9, 24, 130, 720, 8505, 35840, 412776, 3628800, 42030450, 479001600, 7019298000, 82614884352, 1886805545625, 20922789888000, 374426276224000
Dimensions of representations by Witt vectors. Ref CRP 312 488 91. [1,3; A6973]

M1922 2, 9, 25, 55, 105, 182, 294, 450, 660, 935, 1287, 1729, 2275, 2940, 3740, 4692, 5814, 7125, 8645, 10395, 12397, 14674, 17250, 20150, 23400, 27027, 31059, 35525
$n(n+1)(n+2)(n+7)/24$. Ref AS1 797. [1,1; A5582]

M1923 0, 1, 2, 9, 28, 101, 342, 1189, 4088, 14121, 48682, 167969, 579348, 1998541, 6893822, 23780349, 82029808, 282961361, 976071762, 3366950329, 11614259468
$a(n)=2a(n-1)+5a(n-2)$. Ref MQET 1 11 16. [0,3; A2532, N0758]

M1924 1, 2, 9, 28, 185, 846, 7777, 47384, 559953, 4264570, 61594841, 562923252,
9608795209, 102452031878, 2017846993905, 24588487650736, 548854382342177
Logarithmic numbers. Ref TMS 31 78 63. jos. [1,2; A2747, N0759]

M1925 1, 2, 9, 31, 109, 339, 1043, 2998, 8406, 22652, 59521, 151958, 379693, 927622,
2224235, 5236586, 12130780, 27669593, 62229990, 138095696, 302673029
Bipartite partitions. Ref PCPS 49 72 53. ChGu56 1. [0,2; A2774, N0760]

M1926 1, 2, 9, 32, 121, 450, 1681, 6272, 23409, 87362, 326041, 1216800, 4541161,
16947842, 63250209, 236052992, 880961761, 3287794050, 12270214441, 45793063712
Stacking bricks. Ref GKP 360. [0,2; A6253]

M1927 2, 9, 34, 119, 401, 1316, 4247, 13532, 42712
Partially labeled rooted trees with n nodes. Ref R1 134. [2,1; A0524, N0761]

M1928 0, 2, 9, 35, 132, 494, 1845, 6887, 25704, 95930, 358017, 1336139, 4986540,
18610022, 69453549, 259204175, 967363152, 3610248434, 13473630585, 50284273907
From the solution to a Pellian. Ref AMM 56 175 49. [0,2; A1571, N0762]

M1929 2, 9, 36, 142, 558, 2189, 8594, 33796
Value of an urn. Ref DM 5 307 73. [1,1; A3125]

M1930 1, 0, 1, 2, 9, 36, 154, 684, 3128, 14666, 70258, 342766
Polygons of length $4n$ on L-lattice. Ref JPA 18 1013 85. [1,4; A6782]

M1931 1, 2, 9, 36, 190, 980, 5705, 33040, 204876, 1268568, 8209278, 53105976,
354331692, 2364239592, 16140234825, 110206067400
A binomial coefficient summation. Ref AMM 81 170 74. [1,2; A3161]

M1932 1, 2, 9, 37, 183, 933, 5314
Relations on an infinite set. Ref MAN 174 67 67. [0,2; A0663, N0763]

M1933 2, 9, 38, 143, 546, 2066, 7752, 29070, 108968
Perforation patterns for punctured convolutional codes (2,1). Ref SFCA92 1 9. [4,1;
A7224]

M1934 1, 2, 9, 38, 161, 682, 2889, 12238, 51841, 219602, 930249, 3940598, 16692641,
70711162, 299537289, 1268860318, 5374978561, 22768774562, 96450076809
$a(n) = 4a(n-1) + a(n-2)$. Ref TH52 282. [0,2; A1077, N0764]

M1935 1, 2, 9, 40, 355, 11490, 7758205, 549758283980
Precomplete Post functions. Ref SMD 10 619 69. JCT A14 6 73. [1,2; A2825, N0765]

M1936 1, 2, 9, 43, 212, 1115, 6156, 34693, 199076
A subclass of $2n$-node trivalent planar graphs without triangles. Ref JCT B45 309 88. [7,2;
A6795]

M1937 1, 0, 1, 2, 9, 44, 265, 1854, 14833, 133496, 1334961, 14684570, 176214841,
2290792932, 32071101049, 481066515734, 7697064251745, 130850092279664
Subfactorial or rencontres numbers (permutations of n elements with no fixed points):
expansion of $1/(1-x)e^x$. See Fig M1937. Ref R1 65. DB1 168. RY63 23. MOC 21 502
67. C1 182. [0,4; A0166, N0766]

||||||||||||||||||||||||||||||||||||||||||||||||||||||||||||||||||||||||||||||||||||||||||||||||||||||||||||||||||||||||||

Figure M1937. DERANGEMENTS.

M1937 gives the number of **derangements** of n objects, i.e. those permutations in which
every object is moved from its original position. These are also called **subfactorial** or **rencontres** numbers. Here are the first few such permutations:

$$a_2 = 1 \quad \begin{matrix} 1\,2 \\ 2\,1 \end{matrix}$$

$$a_3 = 2 \quad \begin{matrix} 1\,2\,3 & 1\,2\,3 \\ 2\,3\,1 & 3\,1\,2 \end{matrix}$$

$$a_4 = 9 \quad \begin{matrix} 1\,2\,3\,4 & 1\,2\,3\,4 & 1\,2\,3\,4 \\ 2\,1\,4\,3 & 2\,3\,4\,1 & 2\,4\,1\,3 \\ 1\,2\,3\,4 & 1\,2\,3\,4 & 1\,2\,3\,4 \\ 3\,1\,4\,2 & 3\,4\,1\,2 & 3\,4\,2\,1 \\ 1\,2\,3\,4 & 1\,2\,3\,4 & 1\,2\,3\,4 \\ 4\,1\,2\,3 & 4\,3\,1\,2 & 4\,3\,2\,1 \end{matrix}$$

Also

$$a_n = na_{n-1} + (-1)^n, \quad \sum_{n=0}^{\infty} a_n \frac{x^n}{n!} = \frac{1}{(1-x)e^x}.$$

References: [R1 57], [C1 180], [GKP 194], [Wilf90 42].

||||||||||||||||||||||||||||||||||||||||||||||||||||||||||||||||||||||||||||||||||||||||||||||||||||||||||||||||||||||||||

M1938 0, 0, 0, 1, 2, 9, 49, 306, 2188, 17810, 162482, 1642635, 18231462, 220420179,
2883693795, 40592133316, 611765693528, 9828843229764, 167702100599524
Modified Bessel function $K_n(2)$. Ref AS1 429. hpr. [0,5; A0167, N0767]

M1939 1, 2, 9, 52, 365, 3006, 28357, 301064, 3549177, 45965530, 648352001,
9888877692, 162112109029, 2841669616982, 53025262866045, 1049180850990736
Expansion of $x \exp (x/(1-x))$. Ref ARS 10 142 80. [0,2; A6152]

M1940 1, 2, 9, 54, 378, 2916, 24057, 208494, 1876446, 17399772, 165297834,
1602117468, 15792300756, 157923007560, 1598970451545, 16365932856990
$2.3^n(2n)!/n!(n+2)!$. Ref CJM 15 254 63; 33 1039 81. JCT 3 121 67. [0,2; A0168,
N0768]

M1941 1, 2, 9, 54, 450, 4500, 55125, 771750, 12502350
Expansion of an integral. Ref C1 167. [1,2; A1757, N0769]

M1942   2, 9, 56, 705, 19548, 1419237, 278474976, 148192635483, 213558945249402, 836556995284293897, 896297565838112393\ 7708, 26440451619023468566266\ 6051
Nets with n nodes. Ref CCC 2 32 77. rwr. JGT 1 295 77. [1,1; A4103]

M1943   1, 2, 9, 58, 506, 5462, 70226, 1038578
Hoggatt sequence. Ref FA90. [0,2; A5364]

M1944   1, 2, 9, 60, 525, 5670, 72765, 1081080, 18243225
Expansion of an integral. Ref C1 167. [1,2; A1193, N0770]

$$\text{E.g.f.:} \quad (1-x)\,(1-2\,x)^{-3/2}.$$

M1945   1, 2, 9, 61, 551, 6221, 84285, 1332255, 24066691, 489100297, 11044268633, 274327080611, 7433424980943, 218208342366093, 6898241919264181
Expansion of $1/(2-x-e^x)$. Ref ARS 10 138 80. [0,2; A6155]

M1946   1, 2, 9, 64, 625, 7776, 117649, 2097152, 43046721, 1000000000, 25937424601, 743008370688, 23298085122481, 793714773254144, 29192926025390625
n^{n-1}. See Fig M0791. Ref BA9. R1 128. [1,2; A0169, N0771]

M1947   2, 9, 69, 567, 5112
Strongly self-dual planar maps with $2n$ edges. Ref SMS 4 321 85. [1,1; A6849]

M1948   1, 1, 2, 9, 76, 1095, 25386, 910161, 49038872, 3885510411
Connected labeled topologies with n points. Ref MSM 11 243 74. [0,3; A6059]

M1949   0, 2, 9, 76, 1145, 27486, 962017, 46176824, 2909139921, 232731193690, 23040388175321, 2764846581038532, 395373061088510089, 66422674262869694966
$a(n+1)=(n^2-1)\,a(n)+n+1$. Ref rkg. [1,2; A6041]

M1950   1, 2, 9, 82, 1313, 32826, 1181737, 57905114, 3705927297, 300180111058, 30018011105801, 3632179343801922, 523033825507476769, 88392716510763573962
$a(n+1)=n^2 a(n)+1$. Ref rkg. [1,2; A6040]

M1951   1, 1, 2, 9, 88, 1802, 75598, 6421599, 1097780312, 376516036188, 258683018091900, 355735062429124915, 978786413996934006272
Number of full sets of size n. Ref PAMS 13 828 62. C1 123. [1,3; A1192, N0772]

M1952   1, 2, 9, 88, 2111, 118182, 16649389, 5547079988, 4671840869691, 9326302435784002, 471000399781522102\ 49, 5640200352649980315528\ 48
Sum of Gaussian binomial coefficients $[n,k]$ for $q=6$. Ref TU69 76. GJ83 99. ARS A17 329 84. [0,2; A6120]

M1953   1, 1, 2, 9, 96, 2500, 162000, 26471025, 11014635520, 11759522374656, 32406091200000000, 2316276860430802500\ 00, 431150066170386038784\ 0000
$C(n,0).C(n,1).\cdots.C(n,n)$. Ref AS1 828. [0,3; A1142, N0773]

M1954 1, 2, 9, 114, 6894
Hierarchical models with linear terms forced. Ref BF75 34. clm. aam. [1,2; A6126]

M1955 1, 2, 9, 272, 589185
Perfect matchings in n-cube. Ref AML 1 46 88. [1,2; A5271]

M1956 2, 9, 443, 11211435
Switching networks. Ref JFI 276 324 63. [1,1; A0883, N0774]

M1957 1, 2, 9, 2193, 5782218987645,
22356722575362383325389316291986782893945666485024 1
$a(n+1) = (1 + a(0)^4 + \cdots + a(n)^4)/(n+1)$ (not always integral!). Ref AMM 95 704 88.
[0,2; A5167]

SEQUENCES BEGINNING ..., 2, 10, ..., ..., 2, 11, ...

M1958 1, 0, 0, 2, 10, 4, 40, 92, 352, 724, 2680, 14200, 73712, 365596, 2279184,
14772512, 95815104, 666090624, 4968057848, 39029188884
Ways of placing n nonattacking queens on $n \times n$ board. See Fig M0180. Ref PSAM 10 93
60. Well71 238. CACM 18 653 75. AMM 101 637 94. [1,4; A0170, N0775]

M1959 0, 0, 0, 0, 0, 0, 0, 2, 10, 8, 60, 119, 415, 826, 2470, 5246, 14944, 32347
n-node trees not determined by their spectra. Ref LNM 560 91 76. [1,8; A6610]

M1960 1, 2, 10, 11, 12, 20, 21, 22, 100, 101, 102, 110, 111, 112, 120, 121, 122, 200, 201,
202, 210, 211, 212, 220, 221, 222, 1000, 1001, 1002, 1010, 1011, 1012, 1020, 1021, 1022
Natural numbers in base 3. [1,2; A7089]

M1961 2, 10, 12, 21, 102, 111, 122, 201, 212, 1002, 1011, 1101, 1112, 1121, 1202, 1222,
2012, 2021, 2111, 2122, 2201, 2221, 10002, 10022, 10121, 10202, 10211, 10222
Primes in ternary. Ref EUR 23 23 60. [1,1; A1363, N0776]

M1962 1, 1, 2, 10, 22, 60, 158, 439, 1229, 3525, 10178, 29802, 87862, 261204, 781198,
2350249, 7105081, 21577415, 65787902, 201313311, 618040002, 1903102730
Carbon trees with n carbon atoms. Ref BA76 44. [1,3; A5962]

M1963 2, 10, 28, 60, 110, 182, 280, 408, 570, 770, 1012, 1300, 1638, 2030, 2480, 2992,
3570, 4218, 4940, 5740, 6622, 7590, 8648, 9800, 11050, 12402, 13860, 15428, 17110
From the enumeration of corners. Ref CRO 6 82 65. [0,1; A6331]

$$\text{G.f.:}\quad (2 + 2x) / (1 - x)^4.$$

M1964 0, 1, 2, 10, 28, 106, 344, 1272, 4592, 17692, 69384, 283560, 1191984, 5171512,
23087168, 105883456, 498572416, 2404766224, 11878871456, 59975885856
Symmetric permutations. Ref LU91 1 222. JRM 7 181 74. LNM 560 201 76. [2,3; A0900,
N0777]

M1965 1, 2, 10, 32, 227
Geometric n-dimensional crystal classes. Ref JA73 73. BB78 52. Enge93 1020. [0,2; A4028]

M1966 1, 2, 10, 36, 145, 560, 2197, 8568, 33490, 130790, 510949, 1995840, 7796413, 30454814, 118965250, 464711184, 1815292333, 7091038640, 27699580729
Product of Fibonacci and Pell numbers. Ref FQ 3 213 65. [0,2; A1582, N0779]

M1967 1, 2, 10, 37, 162, 674, 2871, 12132, 51436, 217811, 922780, 3908764, 16558101, 70140734, 297121734, 1258626537, 5331629710, 22585142414, 95672204155
Sum of cubes of Fibonacci numbers. Ref BR72 18. [1,2; A5968]

M1968 1, 1, 2, 10, 43, 346
Sub-Hamiltonian graphs with n nodes. Ref ST90. [2,3; A5144]

M1969 0, 0, 1, 2, 10, 45, 210, 1002, 4883, 23797, 116518, 571471
Restricted hexagonal polyominoes with n cells. Equals M2682 − M1426. Ref BA76 75. [1,4; A5963]

M1970 0, 2, 10, 46, 224, 1202, 7120, 46366, 329984, 2551202, 21306880, 191252686, 1836652544, 18793429202, 204154071040, 2346705139006, 28459289083904
Entringer numbers. Ref NAW 14 241 66. DM 38 268 82. [0,2; A6213]

M1971 1, 2, 10, 56, 346, 2252, 15184, 104960, 739162, 5280932, 38165260, 278415920, 2046924400, 15148345760, 112738423360, 843126957056, 6332299624282
Franel numbers: $\Sigma C(n,k)^3$, $k = 0..n$. Ref R1 193. JCT A52 77 89. [0,2; A0172, N0781]

M1972 1, 2, 10, 70, 588, 5544, 56628, 613470, 6952660, 81662152, 987369656, 12228193432, 154532114800, 1986841476000, 25928281261800, 342787130211150
Product of successive Catalan numbers. Ref JCT A43 1 86. [0,2; A5568]

$$(n + 1) \, n \, a(n) \;\; = \;\; 4 \, (2n - 1) \, (2n - 3) \, a(n - 1).$$

M1973 2, 10, 74, 518, 3934, 29914
$(2n + 1)$-step walks on diamond lattice. Ref PCPS 58 100 62. [0,1; A1395, N0782]

M1974 1, 1, 2, 10, 74, 706, 8162, 110410, 1708394, 29752066, 576037442, 12277827850, 285764591114, 7213364729026, 196316804255522, 5731249477826890
A problem of configurations. Ref CJM 4 25 52. PRV D18 1949 78. [0,3; A0698, N0783]

M1975 2, 10, 74, 730, 9002, 133210, 2299754, 45375130
Generalized weak orders on n points. Ref ARC 39 147 82. [1,1; A4123]

M1976 1, 2, 10, 74, 782, 10562, 175826, 3457742
Hoggatt sequence. Ref FA90. [0,2; A5365]

M1977 1, 2, 10, 83, 690, 6412, 61842, 457025
Planar 2-trees with n nodes. Ref JLMS 6 592 73. [3,2; A3093]

M1978 0, 1, 2, 10, 83, 946, 13772, 244315, 5113208, 123342166, 3369568817,
  102831001120, 3467225430308, 128006254663561, 5135734326127862
Planted binary phylogenetic trees with n labels. Ref LNM 884 196 81. [0,3; A6679]

M1979 1, 1, 2, 10, 104, 1816, 47312, 1714000, 82285184
Generalized Euler numbers of type 2^n. Ref JCT A53 266 90. [0,3; A5799]

M1980 1, 2, 10, 104, 3044, 291968, 96928992, 112282908928, 458297100061728,
  6666621572153927936, 349390545493499839161856
Relations on n nodes. See Fig M3032. Ref PAMS 4 494 53. MIT 17 19 55. MAN 174 66
67. JGT 1 295 77. [0,2; A0595, N0784]

M1981 0, 0, 0, 0, 2, 10, 110, 1722, 51039
n-node graphs not determined by their spectra. Ref LNM 560 85 76. [1,5; A6608]

M1982 2, 10, 114, 1842, 37226, 902570, 25530658, 825345250, 30016622298,
  1212957186330, 53916514446482, 2614488320210258, 137345270749953610
Cascade-realizable functions of n variables. Ref PGEC 24 683 75. [1,1; A5613]

M1983 2, 10, 114, 2154, 56946, 1935210, 80371122, 3944568042, 223374129138,
  14335569726570, 1028242536825906, 81514988432370666, 70775780569723777714
Disjunctively-realizable functions of n variables. Ref PGEC 24 687 75. [1,1; A5616]

M1984 1, 2, 10, 116, 3652, 285704, 61946920, 33736398032, 51083363186704,
  194585754101247008, 2061787082699360148640, 54969782721182164414355264
Sum of Gaussian binomial coefficients $[n,k]$ for $q = 7$. Ref TU69 76. GJ83 99. ARS A17
329 84. [0,2; A6121]

M1985 2, 10, 122, 2554, 75386, 2865370, 133191386
Fanout-free functions of n variables. Ref PGEC 27 315 78. [1,1; A5617]

M1986 1, 1, 2, 10, 122, 3346, 196082, 23869210, 5939193962, 2992674197026,
  3037348468846562, 6189980791404487210, 25285903982959247885402
Upper triangular (0,1)-matrices. Ref DM 14 119 76. [0,3; A5321]

M1987 1, 1, 1, 2, 10, 140, 5880, 776160, 332972640, 476150875200, 2315045555222400,
  38883505145515430400, 228580573348427009149440
Products of Catalan numbers. Ref UM 45 81 94. [0,4; A3046]

M1988 1, 1, 2, 10, 148, 7384
Intertwining numbers. Ref clm. [1,3; A4065]

M1989   1, 2, 10, 152, 7736, 1375952, 877901648, 2046320373120, 17658221702361472,
569773219836965265152, 69280070663388783890248448
Labeled Eulerian digraphs with n nodes. Ref CN 40 215 83. [1,2; A7080]

M1990   0, 2, 10, 162, 6218, 739198, 292320730, 393805101318, 1834614855993394,
30008091277676005830, 1747116355298560745383906
Connected strength 3 Eulerian graphs with n nodes. Ref rwr. [1,2; A7131]

M1991   1, 1, 2, 10, 208, 615904, 200253951911058
Nondegenerate Boolean functions of n variables. Ref PGEC 14 323 65. MU71 38. [0,3;
A1528, N0785]

M1992   2, 10, 268, 195472, 104310534400, 29722161121961969778688,
2413441860555924454205324333893477339897004032
Stable matchings. Ref GI89 25. [1,1; A5154]

M1993   1, 1, 2, 10, 280, 235200, 173859840000, 98238542885683200000000,
3216937102767405756074510254080000000000000000000
$a(n) = $ Catalan number $\times \, \Pi a(k)$, $k = 0 \ldots n - 1$. Ref JCT B23 188 77. [1,3; A3047]

M1994   2, 10, 2104, 13098898366
Switching networks. Ref JFI 276 324 63. [1,1; A0884, N0786]

M1995   2, 10, 3866, 297538923922, 675089708540070294583609203589639922
Essentially n-ary operations in a certain 3-element algebra. Ref Berm83. [0,1; A7158]

M1996   2, 11, 23, 24, 26, 33, 47, 49, 50, 59, 73, 74, 88, 96, 97, 107, 121, 122, 146, 169,
177, 184, 191, 193, 194, 218, 239, 241, 242, 249, 289, 297, 299, 311, 312, 313, 337, 338
Sum of squares of n consecutive integers is a square. Ref MMAG 37 218 64. AMM 101
439 94. [1,1; A1032, N0787]

M1997   1, 2, 11, 32, 50, 132, 380, 368, 1135
No-3-in-line problem on $n \times n$ grid. Ref GK68. Wels71 124. LNM 403 7 74. [2,2; A0755,
N0788]

M1998   2, 11, 35, 85, 175, 322, 546, 870, 1320, 1925, 2717, 3731, 5005, 6580, 8500,
10812, 13566, 16815, 20615, 25025, 30107, 35926, 42550, 50050, 58500, 67977, 78561
Stirling numbers of first kind. See Fig M4730. Ref AS1 833. DKB 226. [1,1; A0914,
N0789]

M1999   2, 11, 36, 91, 196, 378, 672, 1122, 1782, 2717, 4004, 5733, 8008, 10948, 14688,
19380, 25194, 32319, 40964, 51359, 63756, 78430, 95680, 115830, 139230, 166257
Coefficients of Chebyshev polynomials. Ref AS1 797. [1,1; A5583]

$$\text{G.f.: } (2 - x) \, / \, (1 - x)^6.$$

M2000 1, 1, 2, 11, 38, 946, 4580, 202738, 3786092, 261868876, 1992367192, 2381255244240
Related to zeros of Bessel function. Ref MOC 1 406 45. [1,3; A0175, N0790]

M2001 2, 11, 46, 128, 272, 522, 904, 1408, 2160, 3154
Generalized tangent numbers. Ref MOC 21 690 67. [1,1; A0176, N0791]

M2002 1, 2, 11, 48, 208, 858, 3507, 14144, 56698, 226100, 898942, 3565920, 14124496
Dissections of a polygon. Ref AEQ 18 386 78. [4,2; A3442]

M2003 1, 2, 11, 62, 406, 3046, 25737, 242094
Permutations of length n with one 3-sequence. Ref BAMS 51 748 45. ARS 1 305 76. [3,2; A2629, N0792]

M2004 2, 11, 64, 426, 3216, 27240, 256320, 2656080, 30078720, 369774720, 4906137600, 69894316800, 1064341555200, 17255074636800, 296754903244800
3rd differences of factorial numbers. Ref JRAM 198 61 57. [0,1; A1565, N0793]

M2005 0, 0, 0, 2, 11, 77, 499, 3442, 24128, 173428, 1262464, 9307494
3-dimensional polyominoes with n cells. Ref CJN 18 367 75. [1,4; A6766]

M2006 1, 2, 11, 92, 1157, 19142, 403691, 10312304
Hoggatt sequence. Ref FA90. [0,2; A5366]

M2007 2, 11, 101, 13, 137, 9091, 9901, 909091, 5882353, 52579, 27961, 8779, 99990001, 1058313049, 121499449, 9091, 69857, 21993833369, 999999000001
Largest factor of $10^n + 1$. Ref CUNN. [0,1; A3021]

M2008 2, 11, 101, 1009, 10007, 100003, 1000003, 10000019, 100000007, 1000000007, 10000000019, 100000000003, 1000000000039, 10000000000037, 100000000000031
Smallest n-digit prime. Ref JRM 22 278 90. [1,1; A3617]

M2009 2, 11, 123, 1364, 15127, 167761, 1860498, 20633239, 228826127, 2537720636, 28143753123, 312119004989, 3461452808002, 38388099893011, 425730551631123
$a(n) = 11a(n-1) + a(n-2)$. Ref RCI 139. [0,1; A1946, N0794]

M2010 1, 2, 11, 148, 5917, 617894, 195118127, 162366823096, 409516908802369, 2724882133766162378, 54969878431787791720019, 292592984952707262305175132
Sum of Gaussian binomial coefficients $[n,k]$ for $q = 8$. Ref TU69 76. GJ83 99. ARS A17 329 84. [0,2; A6122]

M2011 1, 2, 11, 172, 8603
Unilateral digraphs with n nodes. Ref HP73 218. [1,2; A3088]

M2012 2, 11, 590, 7644658
Switching networks. Ref JFI 276 324 63. [1,1; A0886, N0795]

M2013 1, 2, 12, 5, 8, 1, 7, 11, 10, 12
Coefficients of a modular function. Ref GMJ 8 29 67. [-2,2; A5760]

M2014 1, 2, 12, 8, 240, 96, 4032, 1152, 34560, 7680, 101376, 18432, 50319360
Denominators of Bernoulli polynomials. Ref NO24 459. [0,2; A1898, N0749]

M2015 2, 12, 8, 720, 288, 60480, 17280, 3628800, 89600, 95800320, 17418240,
   2615348736000, 402361344000, 4483454976000, 98402304, 32011868528640000
Numerators of coefficients for numerical integration. Cf. M3737. Ref SAM 22 49 43. [1,1;
A2209, N0796]

M2016 2, 12, 12, 1120, 3360, 6720, 6720, 172972800
State assignments for n-state machine. Ref Woo68 263. [2,1; A7041]

M2017 2, 12, 24, 720, 160, 60480, 24192, 3628800, 1036800, 479001600, 788480,
   2615348736000, 475517952000, 31384184832000, 689762304000
Denominators of logarithmic numbers. Cf. M5066. Ref SAM 22 49 43. PHM 38 336 47.
MOC 20 465 66. [1,1; A2207, N0797]

M2018 2, 12, 32, 110, 310, 920
Alkyls with n carbon atoms. Ref ZFK 93 437 36. [1,1; A0647, N0798]

M2019 2, 12, 40, 101, 216, 413, 728, 1206, 1902, 2882, 4224, 6019, 8372, 11403, 15248,
   20060, 26010, 33288, 42104, 52689, 65296, 80201, 97704, 118130, 141830, 169182
Quadrinomial coefficients. Ref C1 78. [2,1; A5719]

M2020 0, 2, 12, 46, 144, 402, 1040, 2548, 5992, 13632, 30220, 65486, 139404, 291770,
   602908, 1229242, 2482792, 4959014, 9836840, 19323246, 37773464, 73182570
Second perpendicular moment of site percolation series for hexagonal lattice. Ref JPA 21
3822 88. [0,2; A6742]

M2021 0, 2, 12, 48, 160, 480, 1344, 3584, 9216, 23040, 56320, 135168, 319488, 745472,
   1720320, 3932160, 8912896, 20054016, 44826624, 99614720, 220200960, 484442112
$C(n,2).2^{n-1}$. Ref AS1 801. [1,2; A1815, N0799]

M2022 2, 12, 50, 180, 606, 1924, 5910, 17564, 51186, 146180
Susceptibility for square lattice. Ref DG72 136. [1,1; A3493]

M2023 2, 12, 52, 232, 952, 3888, 15504, 61333
Perforation patterns for punctured convolutional codes (2,1). Ref SFCA92 1 9. [5,1;
A7225]

M2024 0, 2, 12, 54, 206, 712, 2294, 7024, 20656, 58842, 163250, 443062, 1180156,
   3092964, 7993116, 20401250, 51502616, 128748512, 319010540, 784179992
Second perpendicular moment of bond percolation series for hexagonal lattice. Ref JPA 21
3822 88. [0,2; A6738]

M2025   2, 12, 56, 240, 990, 4004
Closed meanders. See Fig M4587. Ref SFCA91 292. [1,1; A6659]

M2026   1, 1, 2, 12, 57, 366, 2340, 16252, 115940, 854981, 6444826, 49554420,
   387203390, 3068067060, 24604111560, 199398960212, 1631041938108
Dissecting a polygon into n hexagons. Ref DM 11 388 75. [1,3; A5038]

M2027   1, 2, 12, 58, 300, 1682, 10332, 69298, 505500, 3990362, 33925452, 309248938
Quasi-alternating permutations of length n. Equals 2.M4188. Ref NET 113. C1 261. [2,2;
A1758, N0800]

M2028   0, 2, 12, 60, 280, 1260, 5544, 24024, 102960, 437580, 1847560, 7759752,
   32449872, 135207800, 561632400, 2326762800, 9617286240, 39671305740
Apéry numbers: $n.C(2n,n)$. Ref MINT 1 195 78. JNT 20 92 85. [0,2; A5430]

M2029   0, 2, 12, 60, 292, 1438, 7180, 36566
Colored series-parallel networks. Ref R1 159. [1,2; A1574, N0801]

M2030   0, 2, 12, 70, 408, 2378, 13860, 80782, 470832, 2744210, 15994428, 93222358,
   543339720, 3166815962, 18457556052, 107578520350, 627013566048, 3654502875938
$a(n)=6a(n-1)-a(n-2)$. Bisection of M1413. Ref NCM 4 166 1878. ANN 30 72 28.
AMM 75 683 68. [0,2; A1542, N0802]

M2031   2, 12, 70, 442, 3108, 24216, 208586, 1972904, 20373338, 228346522,
   2763259364, 35927135944
Permutations of length n by length of runs. Ref DKB 262. [3,1; A1251, N0803]

M2032   1, 2, 12, 71, 481, 3708, 32028
Permutations of length n with two 3-sequences. Ref BAMS 51 748 45. ARS 1 305 76. [4,2;
A2630, N0804]

M2033   1, 2, 12, 72, 240, 2400, 907200, 4233600, 25401600, 1371686400
Related to numerical integration formulas. Ref MOC 11 198 57. [1,2; A2670, N0805]

M2034   0, 1, 2, 12, 72, 600, 5760, 65520, 846720, 12337920
From a Fibonacci-like differential equation. Ref FQ 27 306 89. [0,3; A5443]

M2035   1, 2, 12, 72, 720, 7200, 100800, 1411200, 25401600, 457228800, 10059033600,
   221298739200, 5753767219200, 149597947699200, 4487938430976000
$C(n,[n/2]).(n+1)!$. Ref PSPM 19 172 71. [0,2; A2867, N0806]

M2036   1, 2, 12, 72, 1440, 7200, 302400, 4233600, 101606400, 914457600, 100590336000
Coefficients for step-by-step integration. Ref JACM 11 231 64. [0,2; A2397, N0807]

M2037   2, 12, 84, 640, 5236, 45164
Closed meanders with 2 components. See Fig M4587. Ref SFCA91 292. [2,1; A6657]

M2038 2, 12, 92, 800, 7554, 75664, 792448, 8595120, 95895816, 1095130728, 12753454896
Hamiltonian rooted triangulations with n internal nodes. Ref DM 6 167 73. [0,1; A3123]

M2039 2, 12, 120, 252, 240, 132, 32760, 12, 8160, 14364, 6600, 276, 65520, 12, 3480, 85932, 16320, 12, 69090840, 12, 541200, 75852, 2760, 564, 2227680, 132, 6360, 43092
From asymptotic expansion of harmonic numbers. Ref AS1 259. [1,1; A6953]

M2040 1, 2, 12, 120, 1680, 30240, 665280, 17297280, 518918400, 17643225600, 670442572800, 28158588057600, 1295295050649600, 64764752532480000
$(2n)!/n!$. Ref MOC 3 168 48. [0,2; A1813, N0808]

M2041 1, 2, 12, 120, 3400, 306016, 98563520, 112894101120, 459097587148864, 6670310734264082432, 349450667631321436169216
Connected strength 3 Eulerian graphs with n nodes, 2 of odd degree. Ref rwr. [1,2; A7132]

M2042 1, 2, 12, 128, 1872, 37600, 990784, 32333824, 1272660224, 59527313920, 3252626013184, 204354574172160, 14594815769038848, 1174376539738169344
Expansion of sin(sin x). [0,2; A3712]

M2043 1, 1, 2, 12, 146, 3060, 101642, 5106612, 377403266, 40299722580
Connected labeled partially ordered sets with n points. Ref jaw. CN 8 180 73. MSM 11 243 74. [0,3; A1927, N0809]

M2044 1, 2, 12, 152, 3472, 126752, 6781632, 500231552, 48656756992, 6034272215552, 929327412759552, 174008703107274752, 38928735228629389312
Expansion of cosh x / cos x. Ref MMAG 34 37 60. [0,2; A0795, N0810]

M2045 0, 2, 12, 176, 6416, 745920, 293075904, 394099077120, 1835009281314048, 30009926711011488256, 174714636716450426961817 6
Strength 3 Eulerian graphs with n nodes, 2 of odd degree. Ref rwr. [1,2; A7129]

M2046 1, 2, 12, 183, 8884, 1495984, 872987584, 1787227218134
Labeled mating digraphs with n nodes. Ref RE89. [1,2; A6023]

M2047 1, 2, 12, 240, 16800, 4233600, 3911846400, 13425456844800, 172785629592576000, 8400837310791045120000, 15521050981925103321907 20000
Π $C(2k,k)$, $k = 1..n$. Ref UM 45 81 94. [0,2; A7685]

M2048 1, 2, 12, 286, 33592, 23178480, 108995910720
Strict sense ballot numbers: n candidates, k-th candidate gets k votes. Ref MMJ 1 81 52. AMS 36 241 65. PLIS 26 87 81. clm. [1,2; A3121]

M2049 1, 1, 2, 12, 288, 34560, 24883200, 125411328000, 5056584744960000, 1834933472251084800000, 6658606584104736522240000000
Super factorials: product of first n factorials. Ref FMR 1 50. RY63 53. GKP 231. [0,3; A0178, N0811]

M2050   2, 12, 360, 75600, 174636000, 5244319080000, 2677277333530800000,
54079050769003260000000, 579344523873625579898552724000000
Chernoff sequence: $\Pi\, p(k)^{n-k+1}$. Ref Pick92 353. [1,1; A6939]

M2051   1, 2, 12, 576, 161280, 812851200, 61479419904000, 108776032459082956800,
5524751496156892842531225600, 9982437658213039871725064756920320000
Latin squares of order n. See Fig M2051. Equals $n!.(n-1)!$.M3690. Ref R1 210. RY63
53. JCT 3 98 67. DM 11 94 75. bdm. [1,2; A2860, N0812]

||||||||||||||||||||||||||||||||||||||||||||||||||||||||||||||||||||||||||||||||||||||||||||||||||||||||||||||||||||||||||||

Figure M2051.   HARD SEQUENCES.

The following sequences are known to be hard to extend, or are connected with hard problems.

(1) Sequences M2051 and M3690, the number of **Latin squares** of order n, are known only for $n \le 10$.

(2) Sequences M0729 and M0817, the numbers of **monotone Boolean functions** of n variables, or **Dedekind's problem**.

(3) It would be very nice if someone would show that the laminated lattice Λ_9 is the densest lattice packing in 9 dimensions, and hence that the next terms in M3201 and M2209 are respectively 512 and 1 [SPLAG], [CoSl94]. What about the analogous Hermite constant when nonlattice packings are allowed? Only the first two terms are known: 1, 4/3 [Hale94].

(4) Similar questions can be asked about the kissing number problem [SPLAG]. Is the next term in M1585 equal to 336? The maximal kissing number in n dimensions when nonlattice arrangements are permitted is known to be 2, 6, 12 in dimensions 1, 2, 3; and 240, 196560 in dimensions 8, 24; no other values are known [SPLAG 23]. It is known that this sequence is strictly different from M1585 by the time we reach 9 dimensions.

(5) In spite of the expenditure of years of computer time, we still don't know enough terms of

$$1, 1, 1, 1, 0, 1, 1, 4, 0, ?$$

the number of **projective planes** of order n, for $n \ge 2$, to place it in the table [HuPi73], [CJM 41 1117 89], [DM 92 187 91].

(6) There are many other sequences (not as difficult as the preceding) where it would be nice to know more terms, for example M1495 (see Fig. M1495), M2817, M1197.

(7) M3736 gives the number of inequivalent **Hadamard matrices** of order $4n$ (see Fig. M3736). It is believed, but not proved, that this sequence is never zero, i.e. that there always exists a Hadamard matrix of order $4n$ [SeYa92].

(8) The $n = 3$ term of M5197 is equal to 1 only if the **Poincaré conjecture** is true. This states that every simply connected compact 3-manifold without boundary is homeomorphic to the 3-sphere [Hirs76], [ANN 74 391 61], [JDG 17 357 82].

(9) Finally, consider the sequence of the zeros of the Riemann zeta function $\zeta(s) = \sum_{n=1}^{\infty} n^{-s}$, arranged in order of magnitude. The first 1.5×10^9 terms of this sequence have the form $s = \frac{1}{2} + i\alpha$, α real. (M4924 gives the nearest integer to α for the first 40 zeros.) It is a famous hard problem, the 'Riemann hypothesis', to show that **all** zeros are of this form [Edwa74], [Odly95].

||||||||||||||||||||||||||||||||||||||||||||||||||||||||||||||||||||||||||||||||||||||||||||||||||||||||||||||||||||||||||||

M2052 2, 12, 1112, 3112, 132112, 1113122112, 311311222112, 13211321322112,
   11131221131211113222112, 31131122211311123113322112
Describe the previous term! Ref CoGo87 176. [1,1; A6751]

M2053 1, 2, 12, 2688, 1813091520
Hamiltonian cycles on n-cube. M1903 divided by 2^n. Ref GA86 24. [1,2; A3042]

M2054 2, 12, 2828, 8747130342
Switching networks. Ref JFI 276 324 63. [1,1; A0887, N0813]

M2055 2, 12, 3888, 297538935552, 67508970854007029458361039374535884
$(2 \uparrow (2 \uparrow n))(3 \uparrow (3 \uparrow n - 2 \uparrow n))$. Ref Berm83. [0,1; A7155]

M2056 2, 13, 19, 6173, 6299, 6353, 6389, 16057, 16369, 16427, 16883, 17167, 17203,
   17257, 18169, 18517, 18899, 20353, 20369, 20593, 20639, 20693, 20809, 22037, 22109
Where prime race among $10n + 1, ... , 10n + 9$ changes leader. Ref rgw. [1,1; A7355]

M2057 2, 13, 37, 73, 1021
Smallest prime of class $n +$. Ref UPNT A18. [1,1; A5113]

M2058 0, 1, 2, 13, 44, 205, 806, 3457, 14168, 59449, 246410, 1027861, 4273412,
   17797573, 74055854, 308289865, 1283082416, 5340773617, 22229288978
$a(n) = 2a(n-1) + 9a(n-2)$. Ref MQET 1 11 16. [0,3; A2534, N0814]

M2059 2, 13, 49, 140, 336, 714, 1386, 2508, 4290, 7007, 11011, 16744, 24752, 35700,
   50388, 69768, 94962, 127281, 168245, 219604, 283360, 361790, 457470, 573300
Coefficients of Chebyshev polynomials. Ref AS1 797. [1,1; A5584]

$$\text{G.f.: } (2 - x) / (1 - x)^7.$$

M2060 1, 2, 13, 73, 710
Arithmetic n-dimensional crystal classes. Ref SC80 34. BB78 52. Enge93 1021. [0,2;
A4027]

M2061 1, 2, 13, 76, 263, 2578, 36979, 33976, 622637, 11064338, 11757173, 255865444,
   1346255081, 3852854518, 116752370597, 3473755390832, 3610501179557
Denominators of convergents to $4/\pi$. Ref Beck71 131. [1,2; A7509]

M2062 1, 1, 0, 1, 2, 13, 80, 579, 4738, 43387, 439792, 4890741, 59216642, 775596313,
   10927434464, 164806435783, 2649391469058, 45226435601207, 817056406224416
Ménage numbers. Ref CJM 10 478 58. R1 197. [0,5; A0179, N0815]

M2063 1, 2, 13, 116, 1393, 20894, 376093, 7897952, 189550849, 5117872922,
   153536187661, 5066694192812, 182400990941233, 7113638646708086
Expansion of $e^{-x} / (1 - 3x)$. Ref R1 83. [0,2; A0180, N0816]

M2064 2, 13, 123, 1546, 24283, 457699, 10064848, 252945467
Generalized weak orders on n points. Ref ARC 39 147 82. [1,1; A4122]

M2065 1, 2, 13, 171, 3994, 154303, 9415189, 878222530
Transitive relations on n nodes. Ref FoMK91. [0,2; A6905]

M2066 1, 1, 0, 0, 2, 13, 199, 3773
Rigid tournaments with n nodes. Ref DM 11 65 75. [1,5; A3507]

M2067 1, 2, 13, 199, 9364, 1530843, 880471142, 1792473955306, 13026161682466252,
341247400399400765678, 32522568098548115377595264
Connected digraphs with n nodes. Ref HP73 124. [1,2; A3085]

M2068 2, 14, 21, 26, 33, 34, 38, 44, 57, 75, 85, 86, 93, 94, 98, 104, 116, 118, 122, 133,
135, 141, 142, 145, 147, 158, 171, 177, 189, 201, 202, 205, 213, 214, 217, 218, 230, 231
n and $n + 1$ have same number of divisors. Ref AS1 840. UPNT B18. [1,1; A5237]

M2069 2, 14, 72, 330, 1430, 6006, 24052, 100776, 396800, 1634380, 6547520
Partitions of a polygon by number of parts. Ref CAY 13 95. [5,1; A2058, N0817]

M2070 1, 0, 0, 2, 14, 90, 646, 5242, 47622, 479306, 5296790, 63779034, 831283558,
11661506218, 175203184374, 2806878055610, 47767457130566, 860568917787402
Hertzsprung's problem: kings on an $n \times n$ board. Ref IDM 26 121 19. AH21 1 271. AMS
38 1253 67. SIAD 4 279 91. [1,4; A2464, N0818]

M2071 1, 2, 14, 182, 3614, 99302, 3554894, 159175382
Quadratic invariants. Ref CJM 8 310 56. [0,2; A0807, N0819]

M2072 1, 1, 2, 14, 546, 16944
Comparative probability orderings on n elements. Ref ANP 4 670 76. [1,3; A5806]

M2073 2, 14, 2786, 21624372014, 101118475259126798441921131854786
A continued cotangent. Ref NBS B80 288 76. [1,1; A6266]

M2074 1, 0, 0, 2, 15, 36, 104, 312, 1050, 3312, 10734, 34518, 113210, 370236, 1220922,
4028696, 13364424, 44409312
Low temperature antiferromagnetic susceptibility for cubic lattice. Ref DG74 422. [0,4;
A7217]

M2075 2, 15, 60, 175, 420, 882, 1680, 2970, 4950, 7865, 12012, 17745, 25480, 35700,
48960, 65892, 87210, 113715, 146300, 185955, 233772, 290950, 358800, 438750
Rooted planar maps. Ref JCT B18 257 75. [1,1; A6470]

M2076 2, 15, 60, 469, 3660, 32958, 328920, 3614490, 43341822, 563144725,
7880897892, 118177520295, 1890389939000, 32130521850972, 578260307815920
From ménage polynomials. Ref R1 197. [4,1; A0181, N0820]

M2077 2, 15, 74, 409, 1951, 9765, 48827, 256347, 1220699, 6103515, 30517572,
160216158, 762939452, 3814697265, 19073486293, 101327896117, 476837158134
Free subsets of multiplicative group of $GF(5^n)$. Ref SFCA92 2 15. [1,1; A7232]

M2078 0, 2, 15, 84, 420, 1980, 9009, 40040, 175032, 755820, 3233230, 13728792, 57946200
Tree-rooted planar maps. Ref SE33 97. JCT B18 257 75. [1,2; A2740, N0821]

M2079 2, 15, 104, 770, 6264, 56196
Paths through an array. Ref EJC 5 52 84. [2,1; A6675]

M2080 1, 0, 1, 2, 15, 140, 1915
Tensors with n external gluons. Ref PRV D14 1549 76. [0,4; A5415]

M2081 0, 2, 15, 148, 1785, 26106, 449701, 8927192, 200847681
Total height of labeled trees with n nodes. Ref IBMJ 4 478 60. [1,2; A1854, N0822]

M2082 1, 2, 15, 150, 1707, 20910, 268616, 3567400, 48555069, 673458874, 9481557398, 135119529972, 1944997539623, 28235172753886, 412850231439153
Coefficients of Jacobi nome. Ref MOC 29 853 75. [0,2; A2103, N0823]

M2083 1, 2, 15, 316, 16885, 2174586, 654313415, 450179768312, 696979588034313, 2398044825254021110
n-node acyclic digraphs with 1 out-point. Ref HA73 254. [1,2; A3025]

M2084 2, 15, 825, 725, 1925, 2275, 425, 390, 330, 290, 770, 910, 170, 156, 132, 116, 308, 364, 68, 4, 30, 225, 12375, 10875, 28875, 25375, 67375, 79625, 14875, 13650, 2550
Successive integers produced by Conway's PRIMEGAME. Ref MMAG 56 28 83. CoGo87 4. Oliv93 21. [1,1; A7542]

M2085 2, 15, 1001, 215441, 95041567, 66238993967, 63009974049301, 877967704491685553, 173955570033393401009, 42138536059332405470076
Product of next n primes. [1,1; A7467]

M2086 1, 2, 16, 52, 160, 9232, 13120, 39364, 41524, 250504, 1276936, 6810136, 8153620, 27114424, 50143264, 106358020, 121012864, 593279152, 1570824736
'3x+1' records (values). See Fig M2629. Ref GEB 400. ScAm 250(1) 12 84. CMWA 24 94 92. [1,2; A6885]

M2087 0, 2, 16, 68, 220, 608, 1520, 3526, 7756, 16302, 33172, 65378, 126224, 237600, 441776, 802820, 1451932, 2563356, 4544304, 7818078, 13684784, 22938278, 39986208
Second moment of site percolation series for square lattice. Ref JPA 21 3821 88. [0,2; A6733]

M2088 0, 2, 16, 72, 252, 764, 2094, 5362, 12968, 30138, 67446, 147048, 311940, 649860, 1325234, 2668130, 5278066, 10346200, 19977010, 38329556, 72546986, 136785444
Second moment of bond percolation series for square lattice. Ref JPA 21 3820 88. [0,2; A6729]

M2089 2, 16, 88, 416, 1824, 7680, 31616, 128512, 518656, 2084864, 8361984, 33497088, 134094848, 536608768, 2146926592, 8588754944, 34357248000, 137433710592
Expansion of $2/(1-2x)^2(1-4x)$. Ref DKB 261. [1,1; A0431, N0824]

M2090 0, 2, 16, 96, 512, 2560, 12288, 57344, 262144, 1179648, 5242880, 23068672, 100663296, 436207616, 1879048192, 8053063680, 34359738368, 146028888064
$n.2^{2n-1}$. Ref LA56 518. [0,2; A2699, N0825]

M2091 2, 16, 130, 1424, 23682
Coefficients of Bell's formula. Ref NMT 10 65 62. [3,1; A2576, N0826]

M2092 2, 16, 134, 1164, 10982, 112354, 1245676, 14909340, 191916532, 2646066034, 38932027996
Permutations of length n by length of runs. Ref DKB 262. [4,1; A1252, N0827]

M2093 2, 16, 136, 1232, 12096, 129024, 1491840, 18627840
Generalized tangent numbers. Ref TOH 42 152 36. [3,1; A2302, N0828]

M2094 1, 2, 16, 192, 2816, 46592, 835584, 15876096, 315031552, 6466437120, 136383037440, 2941129850880, 64614360416256, 1442028424527872
$4^n.(3n)!/(n+1)!(2n+1)!$. Ref CRO 6 99 65. [1,2; A6335]

M2095 2, 16, 208, 3968, 109568, 4793344
Generalized weak orders on n points. Ref ARC 39 147 82. [1,1; A4121]

M2096 1, 2, 16, 272, 7936, 353792, 22368256, 1903757312, 209865342976, 29088885112832, 4951498053124096, 1015423886506852352
Tangent numbers. See Fig M4019. Ref MOC 21 672 67. [1,2; A0182, N0829]

M2097 1, 1, 2, 16, 768, 292864, 1100742656, 48608795688960, 29258366996258488320, 27303528066353552248799 2320, 44261486084874072183645699204710400
$C(n,2)!/(1^{n-1}.3^{n-2}\cdots(2n-3)^1)$. Ref EJC 5 359 84. PLIS 26 87 81. [1,3; A5118]

M2098 0, 2, 16, 980, 9332768
Complete Post functions of n variables. Ref ZML 7 198 61. PLMS 16 191 66. [1,2; A2543, N0830]

M2099 2, 17, 40, 5126, 211888, 134691268742, 28539643139633848, 244353369161294832262756 3638932102
A simple recurrence. Ref MMAG 37 167 64. [1,1; A0956, N0831]

M2100 1, 1, 2, 17, 62, 1382, 21844, 929569, 6404582, 443861162, 18888466084, 113927491862, 58870668456604, 8374643517010684, 689005380505609448
Numerators in expansion of $\tan(x)$. Ref RO00 329. FMR 1 74. [1,3; A2430, N0832]

M2101 2, 17, 131, 227, 733, 829, 929, 997, 1097, 1123, 1237, 1277, 1447, 1487, 1531, 1627, 1811, 1907, 1993, 2141, 2203, 2267, 2441, 2677, 2707, 3209, 3299, 3433, 3547
Where prime race among $7n+1, \dots, 7n+6$ changes leader. Ref rgw. [1,1; A7354]

M2102 2, 17, 167, 227, 362, 398
Extreme values of Dirichlet series. Ref PSPM 24 277 73. [1,1; A3419]

M2103 1, 2, 17, 219, 4783
n-dimensional space groups. Ref JA73 119. SC80 34. BB78 52. Enge93 1025. [0,2; A4029]

M2104 1, 2, 17, 230, 4895
n-dimensional space groups (including enantiomorphs). Ref JA73 119. BB78 52. Enge93 1025. [0,2; A6227]

M2105 1, 2, 17, 5777, 192900153617, 717790523757994658974359292468417
$a(n) = a(n-1)^3 + 3a(n-1)^2 - 3$. Ref D1 1 397. NBS B80 290 76. TCS 65 219 89. [0,2; A2814, N0833]

M2106 1, 2, 18, 99, 724, 4820, 33381, 227862, 1564198, 10714823, 73457064,
503438760, 3450734281, 23651386922, 162109796922, 1111115037483
Sum of fourth powers of Fibonacci numbers. Ref BR72 19. [1,2; A5969]

M2107 2, 18, 108, 540, 2430, 10206, 40824, 157464, 590490, 2165130, 7794468,
27634932, 96722262, 334807830, 1147912560, 3902902704, 13172296626
A traffic light problem: expansion of $2/(1-3x)^3$. Ref BIO 46 422 59. [0,1; A6043]

M2108 2, 18, 136, 1030, 7992, 63796, 522474, 4369840, 37179840, 320861342
n-step walks on f.c.c. lattice. Ref JPA 6 351 73. [2,1; A5544]

M2109 2, 18, 144, 1200, 10800, 105840, 1128960, 13063680, 163296000, 2195424000,
31614105600, 485707622400, 7933224499200, 137305808640000, 2510734786560000
$n!.C(n,2)$. Ref AS1 799. [2,1; A1804, N0834]

M2110 1, 2, 18, 164, 1810, 21252, 263844, 3395016, 44916498, 607041380, 8345319268,
116335834056, 1640651321764, 23365271704712, 335556407724360
$\Sigma C(n,k)^4$, $k = 0 \ .. \ n$. Ref JNT 25 201 87. [0,2; A5260]

M2111 1, 2, 18, 4608, 1800000000, 50779978334208000000000,
5608281336980602120684110607059901336780800000000
$\Pi \ k \uparrow (2 \uparrow (k-2))$, $k = 2 \ .. \ n$. Ref JCMCC 1 147 87. [1,2; A6262]

M2112 1, 2, 18, 5712, 5859364320
Hamiltonian paths on n-cube. Ref ScAm 228(4) 111 73. [1,2; A3043]

M2113 1, 2, 18, 39366, 23841243993846402,
8170984911139653687798259509222498825805303332039774701014
$\Pi (2 \uparrow (2 \uparrow k - 1) + 1) \uparrow C(n,k)$. Ref Berm83. [0,2; A7184]

M2114 2, 19, 23, 317, 1031
$(10^n - 1)/9$ is prime. Ref CUNN. [1,1; A4023]

M2115 2, 19, 29, 199, 569, 809, 1289, 1439, 2539, 3319, 3559, 3919, 5519, 9419, 9539,
9929, 11279, 11549, 13229, 14489, 17239, 18149, 18959, 19319, 22279, 24359, 27529
Supersingular primes of the elliptic curve $X_0(11)$. Ref LNM 504 267. [1,1; A6962]

M2116 1, 2, 20, 70, 112, 352, 1232, 22880, 183040
Denominators of coefficients of Green's function for cubic lattice. Cf. M4360. Ref PTRS 273 590 73. [0,2; A3283]

M2117 1, 0, 1, 2, 20, 104, 775, 6140, 55427
Hit polynomials. Ref RI63. [1,4; A1884, N0835]

M2118 2, 20, 110, 2600, 16150, 208012, 1376550, 74437200, 511755750, 7134913500, 50315410002, 1433226830360
Coefficients of Legendre polynomials. Ref MOC 3 17 48. [1,1; A1797, N0836]

M2119 2, 20, 142, 880, 5106, 28252, 152142, 799736, 4141426, 21133476, 106827054
Susceptibility for cubic lattice. Ref DG72 136. [1,1; A3490]

M2120 2, 20, 143, 986, 6764, 46367, 317810, 2178308, 14930351, 102334154, 701408732, 4807526975, 32951280098, 225851433716, 1548008755919
$a(n) = 7a(n-1) - a(n-2) + 5$. Ref DM 9 89 74. [0,1; A3481]

M2121 0, 1, 2, 20, 144, 1265, 12072, 126565, 1445100, 17875140, 238282730, 3407118041, 52034548064, 845569542593, 14570246018686, 265397214435860
Discordant permutations of length n. Ref SMA 20 23 54. KYU 10 13 56. [3,3; A0183, N0838]

M2122 0, 2, 20, 198, 1960, 19402, 192060, 1901198, 18819920, 186298002, 1844160100, 18255302998, 180708869880, 1788833395802, 17707625088140, 175287417485598
$a(n) = 10a(n-1) - a(n-2)$. Ref TH52 281. [0,2; A1078, N0839]

M2123 2, 20, 198, 2048, 22468, 264538, 3340962, 45173518, 652197968, 10024549190
Permutations of length n by length of runs. Ref DKB 262. [5,1; A1253, N0840]

M2124 2, 20, 210, 2520, 34650, 540540, 9459450, 183783600, 3928374450, 91662070500, 2319050383650, 63246828645000, 1849969737866250
Expansion of $2(1 + 3x)/(1 - 2x)^{7/2}$. Equals 2.M4736. Ref TOH 37 259 33. JO39 152. DB1 296. C1 256. [0,1; A0906, N0841]

M2125 2, 20, 402, 14440, 825502, 69055260, 7960285802, 1209873973712
Some special numbers. Ref FMR 1 77. [0,1; A2116, N0842]

M2126 1, 1, 2, 21, 32, 331, 433, 4351, 53621, 647221, 7673221, 8883233, 891132333, 8101532334101, 101118423351110001, 141220533351220001001
At each step, record how many 1s, 2s etc. have been seen. See Fig M2629. Ref jpropp. [1,3; A6920]

M2127 1, 2, 21, 44, 725, 1494, 2219, 10370, 22959, 33329, 722868, 756197, 2991459, 15713492, 18704951, 53123394, 71828345, 124951739, 321731823, 3664001792
Convergents to cube root of 7. Ref AMP 46 106 1866. L1 67. hpr. [1,2; A5484]

M2128 2, 22, 164, 1030, 5868, 31388, 160648, 795846, 3845020, 18211380, 84876152,
390331292, 1775032504
Rooted planar maps with n edges. Ref BAMS 74 74 68. WA71. JCT A13 215 72. [2,1;
A0184, N0843]

M2129 1, 2, 22, 170, 1366, 10922, 87382, 699050, 5592406, 44739242, 357913942,
2863311530, 22906492246, 183251937962, 1466015503702, 11728124029610
$(8^n + 2(-1)^n)/3$. Ref MMAG 63 29 90. [0,2; A7613]

M2130 2, 23, 37, 47, 53, 67, 79, 83, 89, 97, 113, 127, 131, 157, 163, 167, 173, 211, 223,
233, 251, 257, 263, 277, 293, 307, 317, 331, 337, 353, 359, 367, 373, 379, 383, 389, 397
Single, isolated or non-twin primes. Ref JRM 11 17 78. rgw. [1,1; A7510]

M2131 1, 2, 23, 44, 563, 3254, 88069, 11384, 1593269, 15518938, 31730711, 186088972,
3788707301, 5776016314, 340028535787, 667903294192, 10823198495797
Numerator of $\Sigma 1/(n(2k-1))$, $k = 1 \ .. \ n$. Ref RO00 313. FMR 1 89. [1,2; A2428,
N0844]

M2132 1, 1, 2, 24, 11, 1085, 2542, 64344, 56415, 4275137, 10660486, 945005248,
6010194555, 147121931021, 88135620922, 23131070531152, 120142133444319
Sums of logarithmic numbers. Ref TMS 31 77 63. jos. [1,3; A2743, N0845]

M2133 1, 2, 24, 48, 5760, 11520, 35840, 215040, 51609600, 103219200, 13624934400
Denominators of coefficients for numerical differentiation. Cf. M3151. Ref PHM 33 13 42.
[1,2; A2552, N0846]

M2134 2, 24, 72, 144, 1584, 32544, 30528, 188928, 4030848, 12029184, 66104064,
524719872, 2364433920, 28794737664, 194617138176, 962354727936, 6901447938048
Symmetries in unrooted (1,4) trees on $3n - 1$ vertices. Ref GTA91 849. [1,1; A3614]

M2135 2, 24, 140, 1232, 11268, 115056, 1284360, 15596208, 204710454, 2888897032,
43625578836, 702025263328, 11993721979336, 216822550325472, 4135337882588880
From ménage polynomials. Ref R1 197. [5,1; A0185, N0847]

M2136 0, 2, 24, 180, 1120, 6300, 33264, 168168, 823680, 3938220, 18475600, 85357272,
389398464, 1757701400, 7862853600, 34901442000, 153876579840, 674412197580
Apéry numbers: $n^2 C(2n,n)$. Ref SE33 93. MINT 1 195 78. [0,2; A2736, N0848]

M2137 2, 24, 272, 3424, 46720, 676608, 10251520, 160900608
Almost trivalent maps. Ref PLC 1 292 70. [0,1; A2006, N0849]

M2138 0, 2, 24, 312, 4720, 82800, 1662024, 37665152, 952401888, 26602156800,
813815035000, 27069937855488, 972940216546896, 37581134047987712
Total height of rooted trees with n nodes. Ref JAuMS 10 281 69. [1,2; A1864, N0850]

M2139 2, 24, 420, 27720, 720720, 36756720, 5354228880, 481880599200, 72201776446800, 10685862914126400
Largest number divisible by all numbers < its nth root. Ref CHIBA 3 429 62. MR 30 213(1085) 65. PME 4 124 65. [2,1; A3102]

M2140 1, 0, 0, 2, 24, 552, 21280, 1073760, 70299264, 5792853248, 587159944704, 71822743499520, 10435273503677440, 1776780700509416448
$3 \times n$ Latin rectangles. Ref JMSJ 1(4) 241 50. R1 210. C1 183. [0,4; A0186, N0851]

M2141 1, 2, 24, 912, 87360, 19226880, 9405930240
Colored graphs. Ref CJM 22 596 70. [1,2; A2032, N0852]

M2142 1, 2, 24, 2640, 3230080, 48251508480, 9307700611292160, 240619834982494428379648, 85584720554148149511797587 9680
Labeled regular tournaments with $2n + 1$ nodes. Ref CN 40 215 83. [0,2; A7079]

M2143 1, 2, 24, 3852, 18534400
Permanent of projective plane of order n. Ref RY63 124. [1,2; A0794, N2248]

M2144 2, 24, 40320, 20922789888000, 2631308369336935301672180121600 00000
Invertible Boolean functions of n variables. Ref PGEC 13 530 64. [1,1; A0722, N0853]

M2145 2, 26, 50, 54, 126, 134, 246, 354, 362, 950
$11.2^n - 1$ is prime. Ref MOC 22 421 68. Rie85 384. [1,1; A1772, N0854]

M2146 2, 26, 938, 42800, 2130458
Sets with a congruence property. Ref MFC 15 316 65. [0,1; A2704, N0855]

M2147 2, 28, 27, 52, 136, 108, 162, 620, 486, 760, 1970, 1404, 1940, 6048, 4293, 6100, 15796, 10692, 14264, 40232, 27108, 36496, 93285, 61020, 79054, 211624, 137781
McKay-Thompson series of class 6D for Monster. Ref CALG 18 257 90. FMN94. [1,1; A7257]

M2148 2, 28, 168, 660, 2002, 5096, 11424, 23256, 43890, 77924, 131560, 212940, 332514, 503440, 742016, 1068144, 1505826, 2083692, 2835560, 3801028, 5026098
From the enumeration of corners. Ref CRO 6 82 65. [1,1; A6332]

M2149 2, 28, 182, 4760, 31654, 428260, 2941470, 163761840, 1152562950, 16381761396, 117402623338
Coefficients of Legendre polynomials. Ref MOC 3 17 48. [1,1; A1798, N0856]

M2150 2, 28, 236, 1852, 14622, 119964, 1034992
Permutations by number of sequences. Ref C1 261. [1,1; A1759, N0857]

M2151   2, 29, 541, 7919, 104729, 1299709, 15485863, 179424673, 2038074743,
22801763489
10^n-th prime. Ref GKP 111. rgw. [0,1; A6988]

M2152   0, 0, 0, 0, 0, 0, 0, 0, 0, 0, 1, 2, 30, 239, 2369, 22039, 205663
Non-Hamiltonian simplicial polyhedra with n nodes. Ref Dil92. [1,12; A7030]

M2153   2, 30, 3522, 1066590, 604935042, 551609685150, 737740947722562,
1360427147514751710, 3308161927353377294082
Generalized Euler numbers. Ref MOC 21 689 67. [0,1; A0187, N0858]

M2154   0, 1, 2, 31, 264, 2783, 30818, 369321
Hit polynomials. Ref JAuMS A28 375 79. [4,3; A4307]

M2155   2, 33, 242, 11605, 28374, 171893
$a(n),...,a(n)+n$ have same number of divisors. Ref Well86 176. rgw. [1,1; A6558]

M2156   1, 2, 34, 488, 9826, 206252, 4734304, 113245568, 2816649826, 72001228052,
1883210876284, 50168588906768, 1357245464138656, 37198352117916992
$\Sigma C(n,k)^5$, $k = 0 .. n$. [0,2; A5261]

M2157   1, 2, 34, 5678, 910111213141516171819202122232425262728293031323334 3536
Each term divides the next. Ref JRM 3 40 70. [1,2; A2782, N0859]

M2158   0, 0, 2, 36, 840, 29680, 1429920, 90318144, 7237943552, 717442928640
3-line Latin rectangles. Ref PLMS 31 336 28. BCMS 33 125 41. [1,3; A1626, N0860]

M2159   0, 1, 2, 36, 1200, 57000, 3477600, 257826240, 22438563840, 2238543216000,
251584613280000, 31431367287936000, 4319334744012288000
$n(n-1)^2(5n-10)!/(4n-6)!$. Ref JLMS 6 590 73. [1,3; A3092]

M2160   1, 2, 36, 6728, 12988816, 258584046700, 53060478020000000,
112202210500000000000000, 2444888766000000000000000000000000
Domino tilings of $2n \times 2n$ square. Ref PRV 124 1664 61. [0,2; A4003]

M2161   1, 1, 2, 37, 329, 1501, 31354, 1451967, 39284461, 737652869
Related to ménage numbers. Ref BCMS 39 83 47. [1,3; A1569, N0861]

M2162   2, 41, 130, 269, 458, 697, 986, 1325, 1714, 2153, 2642, 3181, 3770, 4409, 5098,
5837, 6626, 7465, 8354, 9293, 10282, 11321, 12410, 13549, 14738, 15977, 17266, 18605
$(5n+1)^2+4n+1$. Ref SI64a 323. [0,1; A7533]

M2163   2, 44, 56, 92, 104, 116, 140, 164, 204, 212, 260, 296, 332, 344, 356, 380, 392, 444,
452, 476, 524, 536, 564, 584, 588, 620, 632, 684, 692, 716, 744, 764, 776, 836, 860, 884
$\phi(x) = n$ has exactly 3 solutions. Ref AS1 840. [1,1; A7367]

M2164 1, 2, 44, 74, 76, 94, 156, 158, 176, 188, 198, 248, 288, 306, 318, 330, 348, 370,
382, 396, 452, 456, 470, 474, 476, 478, 560, 568, 598, 642, 686, 688, 690, 736, 774, 776
$n^{16} + 1$ is prime. [1,2; A6313]

M2165 1, 0, 2, 44, 1008, 34432, 1629280, 101401344, 8030787968, 788377273856
Related to Latin rectangles. Ref BCMS 33 125 41. [1,3; A1627, N0862]

M2166 2, 46, 3362, 515086, 135274562, 54276473326, 30884386347362,
23657073914466766, 23471059057478981762, 292793578518565951135406
Generalized tangent numbers. Ref MOC 21 690 67. [1,1; A0191, N0864]

M2167 2, 46, 7970, 3487246, 2849229890, 3741386059246, 7205584123783010,
19133892392367261646, 67000387673723462963330
Generalized Euler numbers. Ref MOC 21 689 67. [0,1; A0192, N0865]

M2168 1, 2, 47, 4720, 1256395, 699971370, 706862729265, 1173744972139740,
2987338986043236825, 1105245737952209385450, 570351058228012953756857 5
Trivalent labeled graphs with $2n$ nodes. Ref SIAA 4 192 83. [0,2; A5814]

M2169 0, 2, 48, 540, 4480, 31500, 199584, 1177176, 6589440, 35443980, 184756000,
938929992, 4672781568, 22850118200, 110079950400, 523521630000, 2462025277440
Apéry numbers: $n^3 . C(2n,n)$. Ref MINT 1 195 78. JNT 20 92 85. [0,2; A5429]

M2170 2, 48, 5824, 2887680, 5821595648
$n \times n$ invertible binary matrices A such that $A + I$ is invertible. Ref JSIAM 20 377 71. [2,1;
A2820, N0866]

M2171 1, 1, 2, 49, 629, 6961, 38366, 1899687, 133065253, 6482111309
Related to 3-line Latin rectangles. Ref BCMS 39 72 47. [1,3; A1568, N0867]

M2172 2, 50, 325, 1105, 5525, 27625, 71825, 138125, 160225, 801125, 2082925,
4005625, 5928325, 29641625, 77068225, 148208125, 243061325, 1215306625
Two square partitions. Ref JRM 11 1328 78; 18 70 85. [1,1; A7511]

M2173 2, 52, 142090700, 17844701940501123640681816160
Invertible Boolean functions of n variables. Ref PGEC 13 530 64. [1,1; A0654, N0868]

M2174 0, 2, 56, 16256, 1073709056, 4611686016279904256,
850705917302346158566202798210872 77056
Complete Post functions of n variables. Ref PLMS 16 191 66. [1,2; A2542, N0869]

M2175 2, 60, 660, 4290, 20020, 74256, 232560, 639540, 1586310, 3617900, 7696260,
15438150, 29451240, 53796160, 94607040, 160908264, 265670730, 427156860
From the enumeration of corners. Ref CRO 6 82 65. [1,1; A6333]

M2176 2, 60, 836, 9576, 103326, 1106820
Permutations by number of sequences. Ref C1 261. [1,1; A1760, N0870]

M2177 2, 83, 137, 293, 337, 443, 487, 523, 557, 743, 797, 1213, 1277, 1523, 1657, 1733,
1867, 1973, 2027, 2063, 2797, 2833, 2887, 4733, 5227, 5323, 5437, 5503, 5527, 5623
Where prime race among $5n + 1$, ... , $5n + 4$ changes leader. Ref rgw. [1,1; A7353]

M2178 1, 2, 88, 3056, 319616, 18940160, 283936226304
From higher order Bernoulli numbers. Ref NO24 462. [0,2; A1904, N0871]

M2179 2, 110, 2002, 20020, 136136, 705432, 2984520, 10786908, 34370050, 98768670,
260390130, 638110200, 1468635168, 3200871520, 6650874912, 13248113736
From the enumeration of corners. Ref CRO 6 82 65. [1,1; A6334]

M2180 2, 136, 22377984, 7686143354122719232,
35446079887598386374927067091514163232
Relations with 3 arguments on n nodes. Ref MAN 174 66 67. [1,1; A0662, N0872]

M2181 1, 2, 154, 2270394624
Invertible Boolean functions of n variables. Ref JACM 10 27 63. [1,2; A0725, N0873]

M2182 1, 2, 168, 32738, 20825760, 47942081642
(0,1)-matrices by 1-width. Ref DM 20 110 77. [1,2; A5020]

M2183 2, 264, 1015440, 90449251200, 169107043478365440,
626741682116507920359936, 4435711276305905572695127676467200
Eulerian circuits on $2n + 1$ nodes. Ref CN 40 221 83. [1,1; A7082]

M2184 2, 271, 2718281
Primes found in decimal expansion of e (next term has 85 digits). [1,1; A7512]

M2185 2, 324, 48869983488
2-tournaments on n nodes. Ref GTA91 1085. [4,1; A6475]

M2186 2, 523, 109, 79, 2, 13, 5, 127, 47, 17, 5, 127, 53, 17, 7, 67, 31, 37, 47, 37, 83, 11,
43, 19, 157, 2, 37, 5, 47, 5, 19, 67, 7, 29, 19, 53, 31, 73, 53, 29, 139, 13, 67, 83, 7, 47, 29
n consecutive primes with 1st and last having same digit sum. Ref JRM 7 293 74. rgw.
[1,1; A7513]

M2187 1, 1, 2, 720, 620448401733239439360000
$(n!)!$. Ref MOC 24 231 70. EUR 37 11 74. [0,3; A0197, N0874]

M2188 1, 2, 2248, 54103952, 9573516562048
Generalized Euler numbers of type 3^{2n}. Ref JCT A53 266 90. [0,2; A5800]

M2189 2, 32896, 40297527320597594793574
Relations with 4 arguments on n nodes. Ref OB66. [1,1; A1377, N0875]

M2190   2, 161038, 215326, 2568226, 3020626, 7866046, 9115426, 49699666, 143742226
Even pseudoprimes to base 2: $n \mid 2^n - 2$, n even. Ref rcs. [1,1; A6935]

M2191   2, 608981813029, 608981813357, 608981813707, 608981813717, 608981819119,
608981819273, 608981819437, 608981820869, 608981836423, 608981836481
Where prime race $3n - 1$ vs. $3n + 1$ changes leader. Ref rgw. [1,1; A7352]

SEQUENCES BEGINNING . . ., 3, 0, . . . TO . . ., 3, 3, . . .

M2192   1, 0, 0, 0, 0, 0, 0, 0, 3, 0, 0, 0, 0, 0, 24, 24, 54, 0, 0, 0, 252, 504, 900, 1152, 1452,
3312, 7344, 11484, 35856, 30132, 50184, 264, 113160, 175464, 712176, 319098
Zero-field low-temperature series for 3-state Potts model. Ref JPA 12 1608 79. [0,9; A7271]

M2193   3, 0, 0, 0, 3, 1, 5, 6, 5, 0, 1, 4, 7, 8, 0, 6, 7, 10, 7, 10, 4, 10, 6, 16, 1, 11, 20, 3, 18,
12, 9, 13, 18, 21, 14, 34, 27, 11, 27, 33, 36, 18, 5, 18, 5, 23, 39, 1, 10, 42, 28, 17, 20, 51, 8
$\pi = \Sigma a(n)/n!$. Ref rgw. [0,1; A7514]

M2194   1, 0, 0, 0, 0, 0, 3, 0, 0, 0, 18, 18, 42, 0, 135, 270, 477, 648, 1980, 2988, 4140,
14052, 21690, 52920, 55020, 201852, 162774, 914538, 555750, 3229524, 1188399
Zero-field low-temperature series for 3-state Potts model. Ref JPA 12 1608 79. [0,7; A7270]

M2195   0, 0, 0, 3, 0, 0, 1, 0, 0, 3, 0, 1, 2, 0, 0, 4, 0, 2, 2, 2, 2, 2, 1, 2, 1, 1, 0, 4, 0, 0, 0, 2, 1,
6, 2, 4, 1, 2, 1, 2, 0, 5, 2, 3, 1, 6, 0, 4, 0, 4, 2, 2, 2, 4, 0, 2, 0, 5, 2, 2, 2, 4, 0, 2, 1, 4, 3, 5, 2
Theta series of h.c.p. w.r.t. triangle between layers. Ref JCP 83 6530 85. [0,4; A5890]

M2196   3, 0, 1, 0, 2, 9, 9, 9, 5, 6, 6, 3, 9, 8, 1, 1, 9, 5, 2, 1, 3, 7, 3, 8, 8, 9, 4, 7, 2, 4, 4, 9, 3,
0, 2, 6, 7, 6, 8, 1, 8, 9, 8, 8, 1, 4, 6, 2, 1, 0, 8, 5, 4, 1, 3, 1, 0, 4, 2, 7, 4, 6, 1, 1, 2, 7, 1, 0, 8
$\text{Log}_2 10$. [0,1; A7524]

M2197   1, 0, 1, 1, 3, 0, 5, 2, 4, 0, 9, 1, 11, 0, 3, 4, 15, 0, 17, 3, 5, 0, 21, 2, 16, 0, 12, 5, 27, 0,
29, 8, 9, 0, 15, 4, 35, 0, 11, 6, 39, 0, 41, 9, 12, 0, 45, 4, 36, 0, 15, 11, 51, 0, 27, 10, 17, 0
Moebius transform applied twice to natural numbers. Ref EIS § 2.7. [1,5; A7431]

M2198   0, 0, 3, 0, 5, 3, 7, 8, 3, 15, 22, 15, 39, 35, 38, 72, 85, 111, 152, 175, 241, 308, 414,
551, 655, 897, 1164, 1463, 2001, 2538, 3286, 4296, 5503, 7259, 9357, 12147, 15910
From symmetric functions. Ref PLMS 23 310 23. [1,3; A2123, N0876]

M2199   0, 0, 0, 0, 0, 0, 0, 0, 1, 0, 3, 0, 6, 0, 10, 0, 15, 1, 21, 4, 28, 10, 36, 20, 45, 35, 56, 56,
71, 84, 93, 120, 126, 165, 175, 221, 246, 292, 346, 385, 483, 511, 666, 686, 906, 932
Strict 7th-order maximal independent sets in path graph. Ref YaBa94. [1,11; A7386]

M2200   0, 0, 0, 0, 0, 0, 1, 0, 3, 0, 6, 0, 10, 1, 15, 4, 21, 10, 28, 20, 37, 35, 50, 56, 70, 84,
101, 121, 148, 171, 217, 241, 315, 342, 451, 490, 638, 707, 896, 1022, 1256, 1473, 1765
Strict 5th-order maximal independent sets in path graph. Ref YaBa94. [1,9; A7385]

M2201 0, 0, 0, 0, 1, 0, 3, 0, 6, 1, 10, 4, 15, 10, 22, 20, 33, 35, 51, 57, 80, 90, 125, 141, 193, 221, 295, 346, 449, 539, 684, 834, 1045, 1283, 1600, 1967, 2451, 3012, 3752, 4612, 5738
Strict 3rd-order maximal independent sets in path graph. Ref YaBa94. [1,7; A7384]

M2202 1, 3, 0, 6, 3, 0, 0, 6, 0, 6, 0, 0, 6, 6, 0, 0, 3, 0, 0, 6, 0, 12, 0, 0, 0, 3, 0, 6, 6, 0, 0, 6, 0, 0, 0, 0, 6, 6, 0, 12, 0, 0, 0, 6, 0, 0, 0, 0, 6, 9, 0, 0, 6, 0, 0, 0, 0, 12, 0, 0, 0, 6, 0, 12, 3, 0, 0, 6
Theta series of hexagonal net w.r.t. node. Ref JMP 28 1654 87. [0,2; A5928]

M2203 3, 0, 6, 3, 0, 6, 6, 0, 18, 0, 0, 6, 6, 0, 24, 3, 0, 6, 12, 0, 24, 6, 0, 6, 15, 0, 18, 6, 0, 12, 18, 0, 24, 6, 0, 18, 18, 0, 36, 0, 0, 0, 12, 0, 36, 12, 0, 6, 21, 0, 48, 6, 0, 18, 12, 0, 36, 0, 0
Theta series of h.c.p. w.r.t. triangle between octahedra. Ref JCP 83 6529 85. [1,1; A5889]

M2204 3, 0, 6, 18, 40, 81, 201, 414, 916, 1899, 3973, 8059, 16402, 32561, 64520, 125986, 244448, 469195, 895077, 1692143, 3179406, 5929721, 10993373, 20250589, 37096872
Solid partitions. Ref PNISI 26 135 60. [2,1; A2043, N1710]

M2205 0, 0, 0, 0, 0, 3, 0, 8, 3, 15, 11, 27, 26, 49, 53, 88, 102, 156, 190, 275, 346, 484, 621, 851, 1105, 1495, 1956, 2625, 3451, 4608, 6076, 8088, 10684, 14195, 18772, 24912
Strict 1st-order maximal independent sets in cycle graph. Ref YaBa94. [1,6; A7391]

M2206 1, 3, 0, 9, 5, 7, 12, 6, 15, 13, 3, 9, 17, 4, 21, 3, 23, 16, 21, 25, 15, 20, 1, 5, 27, 18, 30, 12, 19, 27, 35, 9, 37, 25, 39, 15, 2, 30, 24, 10, 29, 21, 39, 31, 3, 43, 40, 45, 15, 47, 48
x such that $p = (x^2 + 11y^2)/4$. Cf. M0151. Ref CU04 1. L1 55. [3,2; A2346, N0877]

M2207 0, 1, 0, 3, 1, 0, 21, 34, 101, 249, 921, 2524, 5613, 8914, 6206
Percolation series for directed cubic lattice. Ref JPA 16 3146 83. [2,4; A6837]

M2208 3, 1, 1, 0, 3, 7, 5, 5, 2, 4, 2, 1, 0, 2, 6, 4, 3, 0, 2, 1, 5, 1, 4, 2, 3, 0, 6, 3, 0, 5, 0, 5, 6, 0, 0, 6, 7, 0, 1, 6, 3, 2, 1, 1, 2, 2, 0, 1, 1, 1, 6, 0, 2, 1, 0, 5, 1, 4, 7, 6, 3, 0, 7, 2, 0, 0, 2, 0, 2
Digits of pi in base eight. [1,1; A6941]

M2209 1, 3, 1, 1, 1, 3, 1, 1
Denominator of nth power of Hermite constant for dimension n. See Fig M2209. Cf. M3201. Ref Cass71 332. GrLe87 410. SPLAG 20. [1,2; A7362]

M2210 1, 1, 3, 1, 1, 1, 7, 7, 1, 1, 11, 1, 1, 1, 15, 1, 16, 1, 19, 1, 1, 23, 22, 1, 25, 27, 1, 1, 1, 31, 1, 1, 1, 35, 1, 1, 1, 39, 1, 1, 1, 43
Size of Doehlert-Klee design with n blocks. Ref DM 2 322 72. UM 12 263 77. [2,3; A5765]

M2211 1, 1, 1, 1, 3, 1, 1, 15, 1, 5, 21, 5, 1, 21, 1, 1, 231, 5, 1, 1365, 1, 55, 21, 1, 1, 663, 11, 5, 57, 5, 1, 15015, 1, 17, 483, 1, 11, 25935, 1, 5, 21, 935, 1, 7917, 1, 23, 19437, 5, 1, 3315
Euler-Maclaurin expansion of polygamma function. Ref AS1 260. [3,5; A6956]

M2212 1, 3, 1, 2, 2, 4, 2, 6, 1, 8, 2, 10, 2, 5, 4, 14, 3, 16, 2, 7, 4, 20, 4, 10, 5, 18, 4, 26, 2, 28, 8, 16, 7, 8, 6, 34, 8, 20, 4, 38, 3, 40, 8, 12, 10, 44, 8, 28, 5, 30, 10, 50, 9, 16, 8, 33, 13
Number of first n tetrahedral numbers prime to n. Ref AMM 41 585 34. [1,2; A2016, N0878]

II

Figure M2209.   SPHERE-PACKING PROBLEM.

What is the densest way to pack unit spheres in n-dimensional space? The answer is known only for $n = 1$ and 2, but for lattice packings it is known for $n \leq 8$ [SPLAG], [CoSl94], [Hale94]. The **Hermite constant** γ_n gives the minimal nonzero squared length in the densest lattice when the determinant is 1. The known values of γ_n^n are

$$1, \frac{4}{3}, 2, 4, 8, \frac{64}{3}, 64, 256,$$

whose numerators and denominators give M3201, M2209 [Cass71 332], [GrLe87 410], [SPLAG 20]. The densest lattices in 2 and 3 dimensions are respectively the hexagonal lattice shown in Fig. M2336 and the face-centered cubic (or f.c.c.) lattice (or fruit-stand packing) shown here. M1585 gives the maximal number of unit spheres that can touch another unit sphere in an n-dimensional lattice packing. This is known for $n \leq 9$. (See also Fig. M2051.)

IIIIIIIIIIIIIIIIIIIIIIIIIIIIIIIIIIIIIIIIIIIIIIIIIIIIIIIIIIIIIIIIIIIIIIIIIIIIIIIIIIIIIIIIIIIIIIIIIIIIIIIIIIIIIIIIIIIIIIIIIIIIIIIIIIII

M2213   1, 1, 1, 1, 1, 1, 1, 1, 3, 1, 3, 1, 2, 1, 1, 5, 1, 1, 2, 2, 3, 1, 2, 7, 5, 3, 1, 3, 1, 1, 4, 5, 3, 3, 4, 1, 3, 5, 1, 1, 5, 5, 1, 2, 3, 5, 1, 7, 3, 2, 5, 1, 4, 11, 1, 5, 4, 2, 1, 13, 1, 9, 2, 3, 3, 7, 2, 7
Classes per genus in quadratic field with discriminant $-n$. Ref BU89 224. [3,9; A3636]

M2214   3, 1, 3, 3, 6, 3, 6, 1, 9, 0, 12, 3, 6, 6, 12, 3, 6, 3, 12, 6, 12, 6, 12, 3, 15, 0, 9, 6, 18, 6, 18, 1, 12, 6, 12, 9, 18, 6, 18, 0, 18, 0, 12, 12, 18, 12, 12, 3, 21, 7, 24, 6, 18, 9, 12, 6, 18, 0
Theta series of f.c.c. lattice w.r.t. triangle. Ref JCP 83 6526 85. [0,1; A5885]

M2215   1, 1, 1, 1, 1, 3, 1, 3, 5, 3, 3, 7, 3, 5, 7, 3, 3, 5, 9, 7, 3, 5, 5, 15, 9, 19, 5, 13, 9, 9, 5, 19, 9, 5, 7, 15, 13, 9, 9, 15, 25, 13, 9, 27, 19, 15, 21, 7, 13, 11, 23, 9, 13, 13, 11, 33, 15, 25
Class numbers of quadratic fields. Ref MOC 28 1143 74. [1,6; A5474]

M2216 1, 1, 3, 1, 3, 5, 7, 1, 3, 5, 7, 9, 11, 13, 15, 1, 3, 5, 7, 9, 11, 13, 15, 17, 19, 21, 23, 25,
27, 29, 31, 1, 3, 5, 7, 9, 11, 13, 15, 17, 19, 21, 23, 25, 27, 29, 31, 33, 35, 37, 39, 41, 43, 45
Josephus problem. Ref GKP 10. [1,3; A6257]

M2217 0, 1, 3, 1, 3, 11, 9, 8, 27, 37, 33, 67, 117, 131, 192, 341, 459, 613, 999, 1483, 2013,
3032, 4623, 6533, 9477, 14311, 20829, 30007, 44544, 65657, 95139, 139625, 206091
Associated Mersenne numbers. Ref EUR 11 22 49. [0,3; A1351, N0879]

M2218 3, 1, 4, 1, 5, 9, 2, 6, 5, 3, 5, 8, 9, 7, 9, 3, 2, 3, 8, 4, 6, 2, 6, 4, 3, 3, 8, 3, 2, 7, 9, 5, 0,
2, 8, 8, 4, 1, 9, 7, 1, 6, 9, 3, 9, 9, 3, 7, 5, 1, 0, 5, 8, 2, 0, 9, 7, 4, 9, 4, 4, 5, 9, 2, 3, 0, 7, 8, 1
Decimal expansion of π. See Fig M2218. Ref MOC 16 80 62. [1,1; A0796, N0880]

||||||||||||||||||||||||||||||||||||||||||||||||||||||||||||||||||||||||||||||||||||||||||||||||||||||||||||||||||||||||||||

Figure M2218. DECIMAL EXPANSIONS.

The decimal expansions of some important real numbers ($\pi, e, \sqrt{2}$, the golden ratio $\tau = (1 + \sqrt{5})/2$, etc.) have been included. For example, M2218: 3, 1, 4, 1, 5, 9, ... gives the decimal expansion of π, the ratio of the circumference of a circle to its diameter. The offset (in this case 1) gives the number of digits before the decimal point.

||||||||||||||||||||||||||||||||||||||||||||||||||||||||||||||||||||||||||||||||||||||||||||||||||||||||||||||||||||||||||||

M2219 0, 1, 1, 3, 1, 4, 1, 7, 4, 6, 1, 10, 1, 8, 6, 15, 1, 13
Maundy cake values. Ref WW 28. [1,4; A6022]

M2220 1, 3, 1, 5, 1, 1, 4, 1, 1, 8, 1, 14, 1, 10, 2, 1, 4, 12, 2, 3, 2, 1, 3, 4, 1, 1, 2, 14, 3, 12, 1,
15, 3, 1, 4, 534, 1, 1, 5, 1, 1, 121, 1, 2, 2, 4, 10, 3, 2, 2, 41, 1, 1, 1, 3, 7, 2, 2, 9, 4, 1, 3, 7, 6
Continued fraction for cube root of 2. Ref JRAM 255 118 72. [1,2; A2945]

M2221 1, 3, 1, 5, 1, 5, 7, 5, 3, 5, 9, 1, 3, 7, 11, 7, 11, 13, 9, 7, 1, 15, 13, 15, 1, 13, 9, 5, 17,
13, 11, 9, 5, 17, 7, 17, 19, 1, 3, 15, 17, 7, 21, 19, 5, 11, 21, 19, 13, 1, 23, 5, 17, 19, 25, 13
From quadratic partitions of primes. Ref KK71 243. [5,2; A2972]

M2222 1, 1, 3, 1, 5, 3, 7, 1, 9, 5, 11, 3, 13, 7, 15, 1, 17, 9, 19, 5, 21, 11, 23, 3, 25, 13, 27, 7,
29, 15, 31, 1, 33, 17, 35, 9, 37, 19, 39, 5, 41, 21, 43, 11, 45, 23, 47, 3, 49, 25, 51, 13, 53
Remove 2s from n. Ref FQ 6 52 68. [1,3; A0265, N0881]

M2223 1, 3, 1, 5, 3, 15, 3, 20, 1, 1, 1, 32, 37, 22, 36, 8, 36, 10, 1, 7, 49, 48, 23, 77, 92, 81,
13, 95, 49, 1, 17, 95, 30, 96, 66, 132, 67, 107, 3, 50, 148, 25, 52, 175, 167, 109, 143, 201
Fermat remainders. Ref KAB 35 666 13. L1 10. [3,2; A2323, N0882]

M2224 0, 0, 0, 0, 0, 0, 0, 1, 0, 0, 1, 3, 1, 5, 4, 11, 20, 46
Trivalent 3-connected bipartite planar graphs with $4n$ nodes. Ref JCT B38 295 85. [2,12; A7085]

M2225 1, 1, 3, 1, 5, 7, 3, 17, 11, 23, 45, 1, 91, 89, 93, 271, 85, 457, 627, 287, 1541, 967,
2115, 4049, 181, 8279, 7917, 8641, 24475, 7193, 41757, 56143, 27371, 139657, 84915
$a(n) = -a(n-1)-2a(n-2)$. Ref JA66 82. AMM 79 772 72. [0,3; A1607, N0883]

M2226 0, 1, 1, 3, 1, 6, 1, 7, 4, 8, 1, 16, 1, 10, 9, 15, 1, 21, 1, 22, 11, 14, 1, 36, 6, 16, 13, 28,
1, 42, 1, 31, 15, 20, 13, 55, 1, 22, 17, 50, 1, 54, 1, 40, 33, 26, 1, 76, 8, 43, 21, 46, 1, 66, 17
Sum of aliquot parts of n. Ref AS1 840. [1,4; A1065, N0884]

M2227 0, 1, 1, 1, 3, 1, 6, 1, 10, 4, 12, 1, 33, 1, 29, 13, 64, 1, 100, 1, 156, 30, 187, 1, 443,
10, 476, 78, 877, 1, 1326, 1, 2098, 188, 2745, 36, 5203, 1, 6408, 477, 11084, 1, 15687, 1
Set-like atomic species of degree n. Ref AAMS 15(896) 94. [0,5; A7650]

M2228 0, 0, 1, 0, 3, 1, 6, 4, 11, 10, 20, 21, 36, 41, 64, 77, 113, 141, 199, 254, 350, 453,
615, 803, 1080, 1418, 1896, 2498, 3328, 4394, 5841, 7722, 10251, 13563, 17990, 23814
Strict 1st-order maximal independent sets in path graph. Ref YaBa94. [1,5; A7383]

M2229 1, 0, 1, 0, 3, 1, 8, 6, 19, 21, 48, 57, 117, 150, 268, 366, 609, 840, 1338, 1866, 2856,
4004, 5961, 8332, 12163, 16938, 24278, 33666, 47577, 65571, 91584, 125469, 173394
Representation degeneracies for Neveu-Schwarz strings. Ref NUPH B274 544 86. [0,5;
A5295]

M2230 1, 0, 1, 0, 3, 1, 8, 7, 37, 55, 220, 499, 1862, 5174, 18258, 57107, 198474
4-connected polyhedra with n nodes. Ref Dil92. [6,5; A7023]

M2231 1, 3, 1, 9, 5, 0, 7, 9, 1, 0, 7, 7, 2, 8, 9, 4, 2, 5, 9, 3, 7, 4, 0, 0, 1, 9, 7, 1, 2, 2, 9, 6, 4,
0, 1, 3, 3, 0, 3, 3, 4, 6, 9, 0, 1, 3, 1, 9, 3, 4, 1, 8, 6, 8, 1, 5, 0, 5, 8, 0, 7, 7, 9, 5, 9, 8, 0, 5, 3
Decimal expansion of fifth root of 4. [1,2; A5533]

M2232 3, 1, 9, 6, 29, 27, 99, 108
Percolation series for hexagonal lattice. Ref SSP 10 921 77. [0,1; A6803]

M2233 3, 1, 11, 43, 19, 683, 2731, 331, 43691, 174763, 5419, 2796203, 251, 87211, 59,
715827883, 67, 281, 1777, 22366891, 83, 2932031007403, 18837001, 283
Smallest primitive factor of $2^{2n+1} + 1$. Ref Krai24 2 85. CUNN. [0,1; A2185, N0885]

M2234 3, 1, 11, 43, 19, 683, 2731, 331, 43691, 174763, 5419, 2796203, 4051, 87211,
3033169, 715827883, 20857, 86171, 25781083, 22366891, 8831418697, 2932031007403
Largest primitive factor of $2^{2n+1} + 1$. Ref Krai24 2 85. CUNN. [0,1; A2589, N0886]

M2235 1, 1, 3, 1, 19, 25, 11, 161, 227, 681, 1019, 3057, 5075, 15225, 29291, 55105,
34243, 233801, 439259, 269201, 1856179, 3471385, 6219851, 1882337, 5647011
3^n reduced modulo 2^n. Ref JIMS 2 40 36. L1 82. [1,3; A2380, N0887]

M2236 0, 3, 2, 0, 3, 12, 0, 6, 0, 6, 0, 12, 6, 6, 12, 12, 3, 0, 2, 6, 0, 24, 0, 24, 6, 3, 0, 24, 6,
12, 12, 6, 0, 12, 0, 0, 18, 6, 12, 48, 0, 24, 0, 6, 0, 36, 0, 0, 6, 9, 14, 24, 6, 12, 12, 0, 0, 48, 0
Theta series of h.c.p. w.r.t. triangle between tetrahedra. Ref JCP 83 6529 85. [0,2; A5874]

M2237 3, 2, 1, 4, 3, 1, 3, 3, 2, 5, 4, 5, 2, 5, 1, 3, 3, 2, 5, 4, 3, 4, 5, 2, 5, 4, 3, 5, 4, 3, 2, 5, 3,
5, 5, 3, 4, 5, 2, 5, 4, 2, 5, 4, 7, 2, 5, 2, 5, 4, 2, 4, 5, 3, 5, 4, 5, 2, 5, 4, 2, 4, 4, 7, 2, 1, 4, 3, 3
Suspense numbers for Tribulations. Ref WW 502. [1,1; A6020]

M2238 1, 1, 1, 3, 2, 1, 5, 23, 25, 27, 49, 74, 62, 85
Generalized divisor function. Ref PLMS 19 111 19. [3,4; A2130, N0888]

M2239 3, 2, 1, 7, 4, 1, 1, 8, 5, 2, 9, 8, 2, 1, 6, 8, 5, 2, 3, 8, 5, 4, 8, 5, 9, 9, 7, 0, 9, 4, 3, 5, 2,
     2, 3, 3, 8, 5, 4, 3, 6, 6, 2, 0, 6, 2, 4, 8, 3, 7, 3, 4, 8, 7, 3, 1, 2, 3, 7, 5, 9, 2, 5, 6, 0, 6, 2, 2, 8
Mix digits of π and e. See Fig M5405. Ref EUR 13 11 50. [1,1; A1355, N0889]

M2240 1, 1, 3, 2, 1, 9, 5, 8, 3, 1, 19, 10, 7, 649, 15, 4, 1, 33, 17, 170, 9, 55, 197, 24, 5, 1,
     51, 26, 127, 9801, 11, 1520, 17, 23, 35, 6, 1, 73, 37, 25, 19, 2049, 13, 3482, 199, 161
Solution to Pellian: x such that $x^2 - ny^2 = 1$. Cf. M0046. Ref DE17. CAY 13 434. L1 55.
[1,3; A2350, N0890]

M2241 0, 0, 0, 0, 1, 1, 1, 1, 3, 2, 2, 3, 0, 4, 3, 1, 2, 2, 0, 1, 2
Non-cyclic simple groups with n conjugacy classes. Ref LA73. [1,9; A6379]

M2242 3, 2, 2, 5, 11, 59, 659, 38939, 25661459, 999231590939, 25641740502411581459,
     25622037156669717708454796390939
Knopfmacher expansion of ½: $a(n+1) = a(n-1)(a(n)+1) - 1$. Ref Knop. [0,1; A7567]

M2243 3, 2, 3, 6, 14, 36, 99, 286, 858, 2652, 8398, 27132, 89148, 297160, 1002915,
     3421710, 11785890, 40940460, 143291610, 504932340, 1790214660, 6382504440
Super ballot numbers: $6(2n)! / n!(n+2)!$. Ref JSC 14 181 92. [0,1; A7054]

M2244 1, 1, 3, 2, 5, 5, 4, 2, 9, 5, 8, 5, 13, 12, 8, 5, 17, 8, 6, 11, 14, 11, 23, 7, 23, 26, 11, 16,
     14, 15, 31, 10, 28, 16, 24, 15, 37, 9, 39, 16, 20, 27, 20, 31, 14, 43, 47, 23, 32, 20, 51, 17
Related to perfect powers. Ref FQ 8 268 70. [1,3; A1598, N0891]

M2245 1, 3, 2, 5, 7, 4, 9, 11, 6, 13, 15, 8, 17, 19, 10, 21, 23, 12, 25, 27, 14, 29, 31, 16, 33,
     35, 18, 37, 39, 20, 41, 43, 22, 45, 47, 24, 49, 51, 26, 53, 55, 28, 57, 59, 30, 61, 63, 32, 65
Expansion of $(1 + 3x + 2x^2 + 3x^3 + x^4)/(1 - x^3)^2$. Ref UPNT E17. jhc. [0,2; A6369]

M2246 3, 2, 5, 29, 11, 7, 13, 37, 32222189, 131, 136013303998782209, 31, 197, 19, 157,
     17, 8609, 1831129, 35977, 508326079288931, 487, 10253, 1390043
From Euclid's proof. Ref GN75. BICA 8 27 93. [0,1; A5265]

M2247 3, 2, 5, 29, 79, 68729, 3739, 6221191, 157170297801581,
     70724343608203457341903, 4631629768201473138715887759877
From Euclid's proof. Ref GN75. BICA 8 29 93. [0,1; A5266]

M2248 3, 2, 5, 29, 869, 756029, 571580604869, 326704387862983487112029,
     1067357570489267520408564952748713861262836088869
$a(n+1) = a(n)^2 + a(n) + 1$. Ref GN75. [0,1; A5267]

M2249 1, 3, 2, 6, 4, 9, 5, 12, 7, 15, 8, 18, 10, 21, 11, 24, 13, 27, 14, 30, 16, 33, 17, 36, 19,
     39, 20, 42, 22, 45, 23, 48, 25, 51, 26, 54, 28, 57, 29, 60, 31, 63, 32, 66, 34, 69, 35, 72, 37
Expansion of $(1 + 3x + x^2 + 3x^3 + x^4)/(1 - x^2)(1 - x^4)$. Ref UPNT E17. jhc. [0,2; A6368]

M2250 0, 1, 3, 2, 6, 7, 5, 4, 12, 13, 15, 14, 10, 11, 9, 8, 24, 25, 27, 26, 30, 31, 29, 28, 20, 21, 23, 22, 18, 19, 17, 16, 48, 49, 51, 50, 54, 55, 53, 52, 60, 61, 63, 62, 58, 59, 57, 56, 40
Decimal equivalent of Gray code for n. Ref ScAm 227(2) 107 72. GA86 15. [1,3; A3188]

M2251 1, 3, 2, 6, 7, 5, 4, 13, 12, 14, 15, 11, 10, 8, 9, 24, 25, 27, 26, 30, 31, 29, 28, 21, 20, 22, 23, 19, 18, 16, 17, 52, 53, 55, 54, 50, 51, 49, 48, 57, 56, 58, 59, 63, 62, 60, 61, 44, 45
Nim-squares. Ref ONAG 52. [1,2; A6042]

M2252 1, 1, 3, 2, 7, 5, 13, 8, 29, 21, 55, 34, 115, 81, 209, 128, 465, 337, 883
From Pascal's triangle mod 2. Ref FQ 30 35 92. [1,3; A6921]

M2253 0, 1, 3, 2, 7, 6, 4, 5, 15, 14, 12, 13, 8, 9, 11, 10, 31, 30, 28, 29, 24, 25, 27, 26, 16, 17, 19, 18, 23, 22, 20, 21, 63, 62, 60, 61, 56, 57, 59, 58, 48, 49, 51, 50, 55, 54, 52, 53, 32
$a(n)$ is Gray-coded into n. Ref ScAm 227(2) 107 72. GA86 15. [1,3; A6068]

M2254 0, 0, 1, 1, 3, 2, 7, 6, 19, 16, 51, 45, 141, 126
Interval schemes. Ref TAMS 272 409 82. [1,5; A5213]

M2255 1, 0, 3, 2, 10, 14, 40, 74, 176, 358, 798, 1670, 3626, 7638, 16366, 34462, 73230, 153830, 324896, 680514, 1430336, 2987310, 6253712, 13025954, 27176052, 56465878
Convex polygons of length $2n$ on honeycomb. Ref JPA 21 L472 88. [3,3; A6743]

M2256 1, 0, 1, 0, 3, 2, 12, 14, 54, 86, 274, 528, 1515, 3266, 8854, 20422, 53786, 129368, 336103, 830148, 2145020
Dyck paths of knight moves. Ref DAM 24 218 89. [0,5; A5220]

M2257 1, 0, 3, 2, 12, 18, 65, 138, 432, 1074, 3231, 8718, 25999, 73650, 220215, 643546, 1937877, 5783700, 17564727, 53222094, 163009086, 499634508, 1542392088
$2n$-step polygons on honeycomb. Ref JPA 22 1379 89. [3,3; A6774]

M2258 1, 0, 3, 2, 12, 24, 80, 222, 687, 2096, 6585, 20892, 67216, 218412
Low temperature antiferromagnetic susceptibility for honeycomb lattice. Ref DG74 422. [0,3; A7214]

M2259 3, 2, 13, 12, 15, 14, 9, 8, 11, 10, 53, 52, 55, 54, 49, 48, 51, 50, 61, 60, 63, 62, 57, 56, 59, 58, 37, 36, 39, 38, 33, 32, 35, 34, 45, 44, 47, 46, 41, 40, 43, 42, 213, 212, 215, 214
Base -2 representation for $-n$ read as binary number. Ref GA86 101. [1,1; A5352]

M2260 0, 3, 2, 18, 98, 33282, 319994402, 354455304050635218, 362949532317927139026406479889080098
Partial quotients in c.f. expansion of $2C - 1$, where C is Cahen's constant. Ref MFM 111 123 91. [0,2; A6281]

M2261 1, 3, 2, 45, 72, 105, 6480, 42525, 22400, 56133, 32659200, 7882875
Coefficients of Chebyshev polynomials. Ref SAM 26 192 47. [1,2; A2680, N0892]

M2262 1, 1, 3, 2, 115, 11, 5887, 151, 259723, 15619, 381773117, 655177, 20646903199, 27085381, 467168310097, 2330931341
Values of an integral. Ref PHM 36 295 45. MOC 19 114 65. [1,3; A2297, N0893]

M2263 1, 1, 1, 1, 1, 3, 3, 1, 2, 1, 5, 1, 2, 3, 1, 7, 5, 3, 3, 1, 4, 5, 3, 4, 1, 5, 1, 5, 5, 3, 5, 7, 2, 5, 1, 11, 5, 4, 1, 13, 1, 9, 2, 3, 7, 2, 7, 5, 3, 1, 15, 3, 7, 3, 2, 13, 1, 11, 3, 7, 2, 4, 4, 5, 3, 19
Classes per genus in quadratic field with discriminant $-4n+1$. Cf. M0191. Ref BU89 224. [1,6; A3637]

M2264 0, 1, 3, 3, 1, 3, 5, 3, 7, 1, 9, 9, 5, 3, 9, 9, 3, 11, 1, 9, 11, 7, 15, 15, 13, 3, 15, 9, 11, 17, 5, 13, 7, 3, 15, 19, 3, 11, 9, 19, 21, 21, 13, 15, 21, 7, 3, 19, 23, 15, 21, 11, 17, 3, 9, 23
x such that $p = x^2 + 2y^2$. Cf. M0444. Ref CU04 1. L1 55. MOC 23 459 69. [2,3; A2332, N0894]

M2265 1, 3, 3, 1, 3, 6, 3, 0, 3, 6, 6, 3, 1, 6, 6, 0, 3, 9, 6, 3, 6, 6, 3, 0, 3, 9, 12, 4, 0, 12, 6, 0, 3, 6, 9, 6, 6, 6, 9, 0, 6, 15, 6, 3, 3, 12, 6, 0, 1, 9, 15, 6, 6, 12, 12, 0, 6, 6, 6, 9, 0, 12, 12, 0, 3
Nonnegative solutions of $x^2 + y^2 + z^2 = n$. Ref PNISI 13 39 47. [0,2; A2102, N0895]

M2266 1, 1, 1, 1, 3, 3, 1, 5, 3, 1, 7, 5, 3, 5, 3, 5, 5, 3, 7, 1, 11, 5, 13, 9, 3, 7, 5, 15, 7, 13, 11, 3, 3, 19, 3, 5, 19, 9, 3, 17, 9, 21, 15, 5, 7, 7, 25, 7, 9, 3, 21, 5, 3, 9, 5, 7, 25, 13, 5, 13, 3, 23
Class numbers $h(-p)$, $p = 4n - 1$. Ref MOC 23 458 69. [3,5; A2143, N0896]

M2267 3, 3, 2, 2, 3, 13, 1, 174, 1, 1, 1, 2, 2, 2, 1, 1, 1, 2, 2, 1
Continued fraction for Wirsing's constant. Ref KN1 2 350. [0,1; A7515]

M2268 1, 3, 3, 2, 48, 362, 49711, 13952
Numerators of coefficients in an asymptotic expansion. Cf. M3976. Ref JACM 3 14 56. [0,2; A2073, N0897]

M2269 3, 3, 3, 3, 4, 4, 5, 5, 5, 5, 5, 5, 5, 6, 6, 6, 7, 7, 7
Chromatic number of cycle with n nodes. Ref ADM 41 21 89. [3,1; A6671]

M2270 1, 1, 3, 3, 3, 3, 5, 7, 9, 9, 9, 9, 9, 9, 9, 9, 9, 9, 11, 13, 15, 17, 19, 21, 23, 25, 27, 27, 27, 27, 27, 27, 27, 27, 27, 27, 27, 27, 27, 27, 27, 27, 27, 27, 27, 27, 27, 27, 27, 27, 27
$a(3n+2) = 2a(n+1) + a(n)$, $a(3n+1) = a(n+1) + 2a(n)$, $a(3n) = 3a(n)$. Ref JRM 22 91 90. [1,3; A6166]

M2271 1, 3, 3, 3, 3, 9, 3, 1, 3, 9, 3, 9, 3, 9, 9, 0, 3, 9, 3, 9, 9, 9, 3, 3, 3, 9, 1, 9, 3, 27, 3, 0, 9, 9, 9, 9, 3, 9, 9, 9, 3, 0, 3, 9, 9, 9, 3, 3, 9, 3, 9, 9, 3, 27, 3, 9, 9, 0, 9, 27, 3, 9
Moebius transform applied thrice to sequence 1,0,0,0,.... Ref EIS § 2.7. [1,2; A7428]

M2272 3, 3, 3, 4, 5, 5, 6, 6, 7, 9, 9, 10, 11, 11, 12, 13, 14, 15, 16, 17, 17, 18, 19, 20, 21, 22, 22, 23, 23, 24, 26, 27, 28, 28, 30, 30, 31, 32, 33, 34, 35, 35, 36, 37, 37, 38, 39, 41, 42
Compressed primes. Ref AMM 74 43 67. [1,1; A2036, N0898]

M2273 3, 3, 3, 5, 3, 3, 5, 3, 3, 5, 3, 5, 7, 3, 3, 5, 7, 3, 5, 3, 3, 5, 3, 5, 7, 3, 5, 7, 3, 3, 5, 7, 3,
5, 3, 3, 5, 7, 3, 5, 3, 5, 7, 3, 5, 7, 19, 3, 5, 3, 3, 5, 3, 3, 5, 3, 5, 7, 13, 11, 13, 19, 3, 5, 3, 5
Smallest prime in decomposition of $2n$ into sum of two odd primes. Ref FVS 4(4) 7 27. L1
80. [3,1; A2373, N0899]

M2274 1, 0, 0, 1, 3, 3, 4, 3, 3, 1, 0, 0, 1, 0, 0, 1, 3, 3, 4, 6, 9, 10, 12, 12, 13, 12, 12, 13, 15,
15, 16, 15, 13, 12, 12, 13, 12, 12, 10, 9, 6, 4, 3, 3, 1, 0, 0
Related to binary expansion of numbers. Ref P5BC 573. [1,5; A5536]

M2275 3, 3, 4, 3, 3, 4, 6, 5, 6, 6, 6, 7
Minimal number of ordinary lines through n points. Ref MMAG 41 34 68. GR72 18. UPG
F12. [3,1; A3034]

M2276 1, 3, 3, 4, 7, 7, 7, 9, 9, 10, 13, 13, 13, 15, 15, 19, 19, 19, 19, 21, 21, 22, 27, 27, 27,
27, 27, 28, 31, 31, 31, 39, 39, 39, 39, 39, 39, 39, 39, 40, 43, 43, 43, 45, 45, 46, 55, 55, 55
Knuth's sequence: $a(n+1) = 1 + \min(2a[n/2], 3a[n/3])$. Cf. M2335. Ref GKP 78.
[0,2; A7448]

M2277 3, 3, 5, 4, 4, 3, 5, 5, 4, 3, 6, 6, 8, 8, 7, 7, 9, 8, 8, 6, 9, 9, 11, 10, 10, 9, 11, 11, 10, 6,
9, 9, 11, 10, 10, 9, 11, 11, 10, 6, 9, 9, 11, 10, 10, 9, 11, 11, 10, 5, 8, 8, 10, 9, 9, 8, 10, 10, 9
Number of letters in n. Ref EUR 37 11 74. [1,1; A5589]

M2278 3, 3, 5, 5, 7, 5, 7, 7, 11, 11, 13, 11, 13, 13, 17, 17, 19, 17, 19, 13, 23, 19, 19, 23, 23,
19, 29, 29, 31, 23, 29, 31, 29, 31, 37, 29, 37, 37, 41, 41, 43, 41, 43, 31, 47, 43, 37, 47
Largest prime in decomposition of $2n$ into sum of two odd primes. Ref FVS 4(4) 7 27. L1
80. [3,1; A2374, N0900]

M2279 1, 3, 3, 5, 7, 11, 17, 27, 43, 69, 111, 179, 289, 467, 755, 1221, 1975, 3195, 5169,
8363, 13531, 21893, 35423, 57315, 92737, 150051, 242787, 392837, 635623, 1028459
$a(n) = a(n-1) + a(n-2) - 1$. Ref FQ 5 288 67. [0,2; A1588, N0901]

M2280 3, 3, 5, 9, 21, 21, 81, 81, 81, 243, 243, 441, 1215, 1701, 1701, 6561, 6561, 6561,
45927, 45927, 45927, 137781, 137781, 229635, 1594323
Largest group of a tournament with n nodes. Ref MO68 81. [3,1; A0198, N0902]

M2281 3, 3, 6, 0, 6, 3, 6, 0, 3, 6, 6, 0, 6, 0, 6, 0, 9, 6, 0, 0, 6, 3, 6, 0, 6, 6, 6, 0, 0, 12, 0, 6,
3, 6, 0, 6, 6, 0, 0, 3, 6, 6, 0, 12, 0, 6, 0, 0, 6, 6, 0, 6, 0, 6, 0, 9, 6, 6, 0, 6, 0, 0, 0, 6, 9, 6, 0, 0
Theta series of planar hexagonal lattice w.r.t. deep hole. See Fig M2336. Ref JCP 83 6524
85. [0,1; A5882]

M2282 1, 3, 3, 6, 3, 9, 3, 10, 6, 9, 3, 18, 3, 9, 9, 15, 3, 18, 3, 18, 9, 9, 3, 30, 6, 9, 10, 18, 3,
27, 3, 21, 9, 9, 9, 36, 3, 9, 9, 30, 3, 27, 3, 18, 18, 9, 3, 45, 6, 18, 9, 18, 3, 30, 9, 30, 9, 9, 3
Inverse Moebius transform applied twice to all 1's sequence. Ref EIS § 2.7. [1,2; A7425]

M2283 3, 3, 6, 15, 26, 39, 192, 45
Percolation series for directed hexagonal lattice. Ref SSP 10 921 77. [0,1; A6807]

M2284 3, 3, 7, 4, 2, 30, 1, 8, 3, 1, 1, 1, 9, 2, 2, 1, 3, 22986, 2, 1, 32, 8, 2, 1, 8, 55, 1, 5, 2, 28, 1, 5, 1, 1501790, 1, 2, 1, 7, 6, 1, 1, 5, 2, 1, 6, 2, 2, 1, 2, 1, 1, 3, 1, 3, 1, 2, 4, 3, 1, 35657
An exotic continued fraction (for real root of $x^3 - 8x - 10$). Ref AB71 21. Robe92 227. [0,1; A2937, N0903]

M2285 1, 3, 3, 7, 6, 12, 13, 20, 21, 34, 36, 51, 58, 78, 89, 118, 131, 171, 197, 245, 279, 349, 398, 486, 557, 671, 767, 920, 1046, 1244, 1421, 1667, 1898, 2225, 2525, 2937, 3333
Mock theta numbers. Ref TAMS 72 495 52. [1,2; A0199, N0904]

M2286 3, 3, 7, 11, 28, 57, 155, 353, 1003, 2458, 7214, 18575, 55880, 149183
Meanders in which first bridge is 5. See Fig M4587. Ref SFCA91 293. [3,1; A6661]

M2287 1, 0, 3, 3, 9, 11, 26, 36, 71, 102, 183, 268, 450, 661, 1059, 1554
Representation degeneracies for Neveu-Schwarz strings. Ref NUPH B274 547 86. [0,3; A5296]

M2288 0, 1, 0, 1, 1, 3, 3, 9, 15, 38, 73
Bicentered hydrocarbons with n atoms. Ref BS65 201. [1,6; A0200, N0905]

M2289 1, 3, 3, 9, 21, 33, 1173, 13515, 113739, 532209, 6284379, 264830061, 5897799141, 104393462439, 1459983940203, 10308316834293, 308010522508395
$a(n) = -\Sigma \ (n+k)! a(k)/(2k)!$, $k = 1..n-1$. Ref UM 45 82 94. [1,2; A7683]

M2290 1, 0, 1, 1, 3, 3, 11, 18, 58, 139, 451, 1326, 4461, 14554, 49957, 171159, 598102
4-regular polyhedra with n nodes. Ref Dil92. [6,5; A7022]

M2291 3, 3, 12, 21, 55, 114, 273, 611, 1437, 3300, 7714, 17913
Percolation series for cubic lattice. Ref SSP 10 921 77. [0,1; A6804]

M2292 3, 3, 13, 19, 55, 61, 139, 139, 181, 181, 391, 439, 559, 619, 619, 829, 859, 1069
Smallest odd number expressible as $p + 2m^2$ in at least n ways, where p is 1 or prime. Ref MMAG 66 47 93. [1,1; A7697]

M2293 1, 0, 0, 1, 0, 3, 3, 15, 30, 101, 261, 807, 2308, 7065, 21171
Partition function for cubic lattice. Ref PCPS 47 425 51. [0,6; A2891, N0906]

M2294 1, 3, 3, 21, 9, 11, 21, 9, 1, 133, 693, 69, 7, 189, 3, 7161, 231, 7, 399, 63, 77, 3311, 4347, 987, 49, 33, 33, 627, 57, 59, 7161, 2079, 11, 10787, 207, 2343, 1463, 4389, 231
Denominators of expansion of $\sinh x / \sin x$. Cf. M1307. Ref MMAG 31 189 58. [0,2; A6656]

M2295 0, 1, 3, 3, 140, 420, 840, 840, 1081800, 75675600, 454053600, 2270268000, 9081072000, 27243216000, 54486432000, 54486432000, 524011612740029568000
State assignments for n-state machine. Ref HP81 308. [1,3; A6845]

M2296 1, 1, 3, 3, 217, 2951, 5973, 1237173, 52635599, 1126610929, 20058390573, 3920482183827, 256734635981833, 8529964147714967, 383670903748980603
Expansion of $e^{\sin x}$. [0,3; A7301]

SEQUENCES BEGINNING . . ., 3, 4, . . . TO . . ., 3, 7, . . .

M2297 3, 4, 3, 3, 5, 5, 5, 4, 3, 3, 5, 3, 3, 7, 7, 4, 7, 8, 9, 8, 7, 7, 7, 7, 5, 5, 3, 3, 9, 3, 3, 4, 8, 8, 5, 3, 3, 9, 3, 3, 13, 13, 13, 11, 11, 11, 8, 7, 5, 5, 5, 13, 9, 5, 5, 5, 7, 7, 5, 5, 5, 7, 4, 7
Least number not dividing $C(2n,n)$. Ref MOC 29 91 75. [1,1; A6197]

M2298 1, 1, 3, 4, 3, 4, 7, 7, 9, 7, 7, 12, 13, 12, 13, 16, 13, 13, 19, 16, 21, 19, 19, 21, 25, 21, 27, 28, 21, 27, 31, 28, 27, 28, 31, 36, 37, 31, 39, 37, 37, 36, 43, 39, 39, 39, 39, 48, 49, 43
Greatest minimal norm of sublattice of index n in hexagonal lattice. Ref BSW94. [1,3; A6984]

M2299 3, 4, 4, 4, 3, 3, 5, 4, 5, 4, 3, 6, 7, 8, 7, 7, 9, 8, 9, 7, 12, 13, 13, 13, 12, 12, 14, 13, 14, 6, 11, 12, 12, 12, 11, 11, 13, 12, 13, 7, 12, 13, 13, 13, 12, 12, 14, 13, 14, 6, 11, 12, 12, 12
Number of letters in n (in Dutch). [1,1; A7485]

M2300 1, 3, 4, 5, 6, 8, 9, 10, 11, 12, 13, 14, 15, 16, 17, 18, 19, 21, 22, 23, 24, 25, 26, 27, 28, 29, 30, 31, 32, 33, 34, 35, 36, 37, 38, 39, 40, 41, 42, 43, 44, 45, 46, 47, 48, 49, 50, 51
Not of form $[e^m]$, $m \geq 1$. Ref AMM 61 454 54. Robe92 11. [1,2; A3619]

$$a(n) = n + [\log(n + 1 + [\log(n + 1)])].$$

M2301 0, 1, 3, 4, 5, 6, 8, 9, 10, 11, 12, 14, 15, 16, 17, 19, 20, 21, 22, 24, 25, 26, 27, 29, 30, 31, 32, 33, 35, 36, 37, 38, 40, 41, 42, 43, 45, 46, 47, 48, 50, 51, 52, 53, 55, 56, 57, 58, 59
Wythoff game. Ref CMB 2 189 59. [0,3; A1965, N0907]

M2302 0, 1, 3, 4, 5, 6, 8, 9, 10, 12, 13, 14, 16, 17, 18, 19, 21, 22, 23, 25, 26, 27, 29, 30, 31, 33, 34, 35, 36, 38, 39, 40, 42, 43, 44, 46, 47, 48, 49, 51, 52, 53, 55, 56, 57, 59, 60, 61, 62
Wythoff game. Ref CMB 2 188 59. [0,3; A1957, N0908]

M2303 1, 3, 4, 5, 6, 8, 10, 12, 17, 21, 23, 28, 32, 34, 39, 43, 48, 52, 54, 59, 63, 68, 72, 74, 79, 83, 98, 99, 101, 110, 114, 121, 125, 132, 136, 139, 143, 145, 152, 161, 165, 172, 176
$a(n)$ is smallest number which is uniquely $a(j)+a(k)$, $j<k$. See Fig M0557. Ref Ulam60 IX. JCT A12 39 72. Pick92 358. GU94. [1,2; A2859, N0909]

M2304 3, 4, 5, 7, 8, 9, 11, 12, 13, 15, 16, 17, 19, 20, 21, 24, 25, 27, 28, 32, 33, 35, 36, 40, 44, 45, 48, 60, 84.
Cyclotomic fields with class number 1 (a finite sequence). Ref LNM 751 234 79. BPNR 259. [1,1; A5848]

M2305 3, 4, 5, 7, 8, 11, 13, 17, 19, 20, 26, 29, 32, 37, 38, 43, 49, 50, 56, 62, 67, 68, 71, 73, 86, 89, 91, 98, 103, 113, 116, 121, 127, 131, 133, 137, 140, 151, 158, 161, 169, 173
Generated by a sieve. Ref PC 2 13-6 74. [1,1; A3310]

M2306 1, 3, 4, 5, 7, 9, 11, 12, 13, 15, 16, 17, 19, 20, 21, 23, 25, 27, 28, 29, 31, 33, 35, 36, 37, 39, 41, 43, 44, 45, 47, 48, 49, 51, 52, 53, 55, 57, 59, 60, 61, 63, 64, 65, 67, 68, 69, 71
If n appears, $2n$ doesn't (the parity of number of 1s in binary expansion alternates). See Fig M0436. Ref FQ 10 501 72. AMM 87 671 80. [1,2; A3159]

M2307 1, 3, 4, 5, 7, 9, 14, 18, 24, 31, 43, 55, 72, 94, 123, 156, 200, 254, 324, 408, 513,
641, 804, 997, 1236, 1526, 1883, 2308, 2829, 3451, 4209, 5109, 6194, 7485, 9038, 10871
Conjugacy classes in alternating group A_n. Ref CJM 4 383 52. [2,2; A0702, N0910]

M2308 1, 3, 4, 5, 7, 10, 14, 20, 29, 43, 64, 95, 142, 212, 317, 475, 712, 1067, 1600, 2399,
3598, 5396, 8093, 12139, 18208, 27311, 40966, 61448, 92171, 138256, 207383, 311074
$a(n+1) = a(n) + [(a(n) - 1)/2]$. Ref PC 5 51-17 77. [0,2; A3312]

M2309 3, 4, 5, 7, 11, 13, 17, 23, 29, 43, 47, 83, 131, 137, 359, 431, 433, 449, 509, 569,
571, 2971, 4723, 5387
Prime Fibonacci numbers. Ref MOC 50 251 88. [1,1; A1605, N0911]

M2310 3, 4, 5, 8, 10, 7, 9, 18, 24, 14, 30, 19, 20, 44, 16, 27, 58, 15, 68, 70, 37, 78, 84, 11,
49, 50, 104, 36, 27, 19, 128, 130, 69, 46, 37, 50, 79, 164, 168, 87, 178, 90, 190, 97, 99
First occurrence of p in Fibonacci sequence. Ref JA66 7. MOC 20 618 66. BR72 25. [2,1;
A1602, N0912]

M2311 1, 3, 4, 5, 9, 15, 27, 50, 92, 171, 322, 610, 1161, 2220, 4260, 8201, 15828, 30622,
59362, 115287, 224260, 436871, 852161, 1664196, 3253531, 6366973, 12471056
An approximation to population of $x^2 + y^2$. Ref MOC 18 79 64. [0,2; A0692, N0913]

M2312 0, 1, 3, 4, 5, 11, 12, 13, 15, 16, 17, 19, 20, 21, 43, 44
Loxton-van der Poorten sequence: base 4 representation contains only -1, 0, $+1$. Ref
JRAM 392 57 88. TCS 98 188 92. [1,3; A6288]

M2313 1, 1, 3, 4, 6, 2, 4, 3, 6, 16, 14, 33, 31, 37, 51, 56, 54, 55, 53, 45, 55, 25, 23, 17, 8,
72, 79, 135, 137, 235, 237, 343, 369, 479, 463, 622, 624, 732, 792, 898, 900, 1056, 1058
Shifts left when Moebius transformation applied twice. Ref BeSl94. EIS § 2.7. [1,3;
A7551]

M2314 3, 4, 6, 5, 12, 8, 6, 12, 15, 10, 12, 7, 24, 20, 12, 9, 12, 18, 30, 8, 30, 24, 12, 25, 21,
36, 24, 14, 60, 30, 24, 20, 9, 40, 12, 19, 18, 28, 30, 20, 24, 44, 30, 60, 24, 16, 12, 56
First occurrence of n in Fibonacci sequence. Ref HM68. MOC 23 459 69. ACA 16 109 69.
BR72 25. [2,1; A1177, N0914]

M2315 1, 3, 4, 6, 6, 12, 8, 12, 12, 18, 12, 24, 14, 24, 24, 24, 18, 36, 20, 36, 32, 36, 24, 48,
30, 42, 36, 48, 30, 72, 32, 48, 48, 54, 48, 72, 38, 60, 56, 72, 42, 96, 44, 72, 72, 72, 48, 96
$n \Pi(1 + p^{-1})$, $p \mid n$. Ref NBS B67 62 63. [1,2; A1615, N0915]

M2316 1, 3, 4, 6, 7, 8, 10, 11, 12, 13, 15, 16, 17, 18, 19, 21, 22, 23, 24, 25, 26, 28, 29, 30,
31, 32, 33, 34, 36, 37, 38, 39, 40, 41, 42, 43, 45, 46, 47, 48, 49, 50, 51, 52, 53, 55, 56, 57
Add $n - 1$ to nth term of 'n appears n times' sequence. Cf. M0250. [1,2; A7401]

M2317 3, 4, 6, 7, 8, 10, 12, 13, 14, 15, 16, 18, 20, 22, 24, 25, 26, 27, 28, 29, 30, 31, 32, 34,
36, 38, 40, 42, 44, 46, 48, 49, 50, 51, 52, 53, 54, 55, 56, 57, 58, 59, 60, 61, 62, 63, 64, 66
Unique monotone sequence with $a(a(n)) = 2n$, $n \geq 2$. Ref clm. [2,1; A7378]

$$a(2^i + j) = 3.2^{(i-1)} + j, \ 0 \leq j < 2^{(i-1)} \ ; \ a(3.2^{(i-1)} + j) = 2^{(i+1)} + 2j, \ 0 \leq j < 2^{(i-1)}.$$

M2318 1, 3, 4, 6, 7, 8, 12, 13, 14, 15, 18, 20, 24, 28, 30, 31, 32, 36, 38, 39, 40, 42, 44, 48,
54, 56, 57, 60, 62, 63, 68, 72, 74, 78, 80, 84, 90, 91, 93, 96, 98, 102, 104, 108, 110, 112
Possible numbers of divisors of n. Ref BA8 85. AS1 840. [1,2; A2191, N0916]

M2319 1, 3, 4, 6, 7, 8, 13, 14, 15, 20, 28, 30, 36, 38, 39, 40, 44, 57, 62, 63, 68, 74, 78, 91,
93, 102, 110, 112, 121, 127, 133, 138, 150, 158, 160, 162, 164, 171, 174, 176, 183, 194
$\sigma(x) = n$ has unique solution. Ref AS1 840. [1,2; A7370]

M2320 0, 1, 3, 4, 6, 7, 9, 11, 13, 16, 18, 21, 24, 27, 30, 33, 36, 39, 42, 46, 50, 52
Maximal edges in n-node square-free graph. Ref bdm. [1,3; A6855]

M2321 3, 4, 6, 7, 12, 14, 30, 32, 33, 38, 94, 166, 324, 379, 469, 546, 974, 1963, 3507, 3610
$n! - 1$ is prime. Ref MOC 26 568 72; 38 640 82. Cald94. [1,1; A2982]

M2322 1, 3, 4, 6, 8, 9, 11, 12, 14, 16, 17, 19, 21, 22, 24, 25, 27, 29, 30, 32, 33, 35, 37, 38,
40, 42, 43, 45, 46, 48, 50, 51, 53, 55, 56, 58, 59, 61, 63, 64, 66, 67, 69, 71, 72, 74, 76, 77
A Beatty sequence: $[n.\tau]$. See Fig M1332. Cf. M1332. Ref CMB 2 191 59. AMM 72 1144
65. FQ 11 385 73. [1,2; A0201, N0917]

M2323 1, 3, 4, 6, 8, 9, 11, 12, 14, 16, 17, 19, 21, 22, 24, 25, 27, 29, 30, 32, 34, 35, 37, 38,
40, 42, 43, 45, 47, 48, 50, 51, 53, 55, 56, 58, 60, 61, 63, 64, 66, 68, 69, 71, 73, 74, 76, 77
$a(8i+j) = 13i + a(j)$. Ref FQ 1(4) 50 63. [1,2; A0202, N0918]

M2324 1, 3, 4, 6, 8, 9, 11, 13, 15, 17, 19, 20, 22, 26, 28, 30, 31, 33, 35, 37, 39, 41, 43, 45,
48, 50, 52, 54, 56, 58, 62, 64, 65, 67, 69, 71, 73, 75, 79, 81, 83, 85, 86, 90, 92, 94, 96, 98
Nonnegative solutions of $x^2 + y^2 = z$ in first n shells. Ref PURB 20 14 52. [0,2; A0592,
N0919]

M2325 1, 3, 4, 6, 8, 10, 11, 13, 15, 16, 18, 20, 22, 23, 25, 27, 29, 30, 32, 34, 35, 37, 39, 41,
42, 44, 46, 48, 50, 52, 53, 55, 57, 59, 60, 62, 64, 65, 67, 69
Related to Fibonacci representations. Ref FQ 11 385 73. [1,2; A3257]

M2326 1, 3, 4, 6, 9, 13, 21, 28, 59, 93, 122, 249, 385
Spiral sieve using Fibonacci numbers. Ref FQ 12 395 74. [1,2; A5626]

M2327 3, 4, 6, 11, 45, 906, 409182, 83762797735
Related to Hamilton numbers. Ref SYL 4 551. [1,1; A2090, N0920]

M2328 3, 4, 6, 12, 48, 924, 409620, 83763206256
Related to Hamilton numbers. Ref SYL 4 551. [1,1; A6719]

M2329 1, 3, 4, 7, 6, 12, 8, 15, 13, 18, 12, 28, 14, 24, 24, 31, 18, 39, 20, 42, 32, 36, 24, 60,
31, 42, 40, 56, 30, 72, 32, 63, 48, 54, 48, 91, 38, 60, 56, 90, 42, 96, 44, 84, 78, 72, 48, 124
$\sigma(n) =$ sum of the divisors of n. Ref AS1 840. [1,2; A0203, N0921]

M2330 0, 1, 3, 4, 7, 8, 10, 11, 15, 16, 18, 19, 22, 23, 25, 26, 31, 32, 34, 35, 38, 39, 41, 42,
46, 47, 49, 50, 53, 54, 56, 57, 63, 64, 66, 67, 70, 71, 73, 74, 78, 79, 81, 82, 85, 86, 88, 89
$a(n) = a([n/2]) + n$. Ref arw. [0,3; A5187]

M2331 3, 4, 7, 8, 11, 12, 15, 16, 19, 20, 24, 27, 28, 32, 35, 36, 40, 43, 48, 51, 52, 60, 64,
67, 72, 75, 84, 88, 91, 96, 99, 100, 112, 115, 120, 123, 132, 147, 148, 160, 163, 168, 180
Discriminants of orders of imaginary quadratic fields with 1 class per genus (a finite
sequence). Ref DI57 85. BS66 426. [1,1; A3171, N0922]

M2332 3, 4, 7, 8, 11, 15, 19, 20, 23, 24, 31, 35, 39, 40, 43, 47, 51, 52, 55, 56, 59, 67, 68,
71, 79, 83, 84, 87, 88, 91, 95, 103, 104, 107, 111, 115, 116, 119, 120, 123, 127, 131, 132
Discriminants of imaginary quadratic fields, negated. Ref Ribe72 97. [1,1; A3657]

M2333 3, 4, 7, 8, 11, 15, 19, 20, 24, 35, 40, 43, 51, 52, 67, 84, 88, 91, 115, 120, 123, 132,
148, 163, 168, 187, 195, 228, 232, 235, 267, 280, 312, 340, 372, 403, 408, 420, 427, 435
Discriminants of imaginary quadratic fields with 1 class per genus (a finite sequence). Ref
DI57 85. BS66 426. [1,1; A3644]

M2334 0, 1, 1, 3, 4, 7, 8, 12, 15, 20
Minimum arcs whose reversal yields transitive tournament. Equals $C(n,2)$ − M1333. Ref
CMB 12 263 69. MSH 37 23 72. MR 46 15(87) 73. [2,4; A3141]

M2335 1, 3, 4, 7, 9, 10, 13, 15, 19, 21, 22, 27, 28, 31, 39, 40, 43, 45, 46, 55, 57, 58, 63, 64,
67, 79, 81, 82, 85, 87, 91, 93, 94, 111, 115, 117, 118, 121, 127, 129, 130, 135, 136, 139
If n is present so are $2n+1$ and $3n+1$. Cf. M2276. Ref NAMS 18 960 71. GKP 78. [1,2;
A2977]

M2336 1, 3, 4, 7, 9, 12, 13, 16, 19, 21, 25, 27, 28, 31, 36, 37, 39, 43, 48, 49, 52, 57, 61, 63,
64, 67, 73, 75, 76, 79, 81, 84, 91, 93, 97, 100, 103, 108, 109, 111, 112, 117, 121, 124, 127
Numbers of form x^2+xy+y^2. See Fig M2336. Ref SPLAG 111. [1,2; A3136]

‖‖

Figure M2336. THETA SERIES OF HEXAGONAL LATTICE.

M4042 and M2336 are the coefficients and exponents in the theta series

$$1 + 6q + 6q^3 + 6q^4 + 12q^7 + 6q^9 + \cdots .$$

of the planar hexagonal lattice [SPLAG 111]. The exponents are also the numbers that can be
written in the form $a^2 + ab + b^2$. M0187, M2281 give the theta function with respect to other
points. (See also Fig. M2347.)

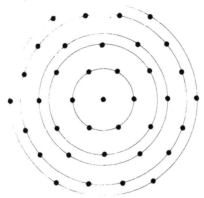

M2337   1, 3, 4, 7, 9, 12, 18, 22, 102, 112, 157, 162, 289
$10.3^n + 1$ is prime. Ref MOC 26 996 72. [1,2; A5539]

M2338   1, 1, 3, 4, 7, 9, 14, 19, 26, 34, 45, 59, 76, 96, 121, 153, 189, 234
Weighted count of partitions with odd parts. Ref ADV 61 160 86. [3,3; A5896]

M2339   3, 4, 7, 10, 11, 13, 15, 16, 21, 22, 27, 30, 35, 36, 41, 44, 50, 53, 55, 61, 69, 70, 75,
   78, 84, 87, 92, 93, 98, 101, 107, 112, 118, 121, 132, 135, 138, 141, 149, 150, 164, 166
$a(n)$ is smallest number which is uniquely $a(j) + a(k)$, $j < k$. Ref GU94. [1,1; A3669]

M2340   3, 4, 7, 10, 50
Graphs by cutting center. Ref GU70 149. [1,1; A2887, N0923]

M2341   1, 3, 4, 7, 11, 18, 29, 47, 76, 123, 199, 322, 521, 843, 1364, 2207, 3571, 5778,
   9349, 15127, 24476, 39603, 64079, 103682, 167761, 271443, 439204, 710647, 1149851
Lucas numbers (beginning at 1): $L(n) = L(n-1) + L(n-2)$. See Fig M0692. M0155.
Ref HW1 148. HO69. C1 46. [1,2; A0204, N0924]

M2342   1, 3, 4, 7, 11, 29, 40, 109, 912, 1021, 26437, 27458, 163727, 191185, 4369797,
   4560982, 40857653, 45418635, 86276288, 821905227, 908181515, 1730086742
Convergents to fifth root of 5. Ref AMP 46 116 1866. L1 67. hpr. [1,2; A2364, N0925]

M2343   1, 1, 3, 4, 7, 32, 39, 71, 465, 536, 1001, 8544, 9545, 18089, 190435, 208524,
   398959, 4996032, 5394991, 10391023, 150869313, 161260336, 312129649, 5155334720
Denominators of convergents to e. Cf. M0869. Ref LE77 240. [0,3; A7677]

M2344   1, 3, 4, 8, 9, 11, 12, 20, 21, 23, 24, 28, 29, 31, 32, 48, 49, 51, 52, 56, 57, 59, 60, 68,
   69, 71, 72, 76, 77, 79, 80, 112, 113, 115, 116, 120, 121, 123, 124, 132, 133, 135, 136, 140
Partial sums of M0162. [0,2; A6520]

M2345   1, 3, 4, 8, 9, 11, 13, 18, 19, 24, 27, 28, 29, 33, 35, 40, 43, 44, 51, 59, 61, 63, 67, 68,
   75, 83, 88, 91, 92, 93, 98, 100, 104, 107, 108, 109, 115, 120, 121, 123, 125, 126, 129, 131
Elliptic curves. Ref JRAM 212 24 63. [1,2; A2156, N0926]

M2346   1, 3, 4, 8, 10, 16, 20, 29, 35, 47, 56, 72, 84, 104, 120, 145, 165, 195, 220, 256, 286,
   328, 364, 413, 455, 511, 560, 624, 680, 752, 816, 897, 969, 1059, 1140, 1240, 1330, 1440
n-bead necklaces with 4 red beads. Ref JAuMS 33 12 82. AJG 22 266 85. [4,2; A5232]

M2347   0, 3, 4, 8, 11, 12, 16, 19, 20, 24, 27, 32, 35, 36, 40, 43, 44, 48, 51, 52, 56, 59, 64,
   67, 68, 72, 75, 76, 80, 83, 84, 88, 91, 96, 99, 100, 104, 107, 108, 115, 116, 120, 123, 128
Numbers represented by f.c.c. lattice. Ref SPLAG 116. [0,2; A4014]

M2348   1, 1, 3, 4, 8, 11, 18, 24, 36, 47, 66, 84, 113, 141, 183, 225, 284, 344, 425, 508, 617,
   729, 872, 1020, 1205, 1397, 1632, 1877, 2172, 2480, 2846, 3228, 3677
Expansion of a generating function. Ref CAY 10 415. [0,3; A1994, N0927]

M2349   1, 3, 4, 8, 11, 20, 27, 45, 61, 95, 128, 193, 257, 374, 497, 703, 927, 1287, 1683,
   2297
Factorization patterns of n. Ref ARS 25 77 88. [1,2; A6167]

M2350 1, 1, 3, 4, 8, 14, 25, 45, 82, 151, 282, 531, 1003, 1907, 3645, 6993, 13456, 25978, 50248, 97446, 189291, 368338, 717804, 1400699, 2736534, 5352182, 10478044
Integers $\le 2^n$ of form $x^2 + 3y^2$. Ref MOC 20 560 66. [0,3; A0205, N0928]

M2351 1, 3, 4, 8, 15, 27, 50, 92, 169, 311, 572, 1052, 1935, 3559, 6546, 12040, 22145, 40731, 74916, 137792, 253439, 466147, 857378, 1576964, 2900489, 5334831, 9812284
$a(n) = a(n-1) + a(n-2) + a(n-3)$. [1,2; A7486]

M2352 1, 3, 4, 8, 65536
Ackermann function (the next term cannot even be described here). Ref AMM 70 133 63. [0,2; A1695, N0929]

M2353 0, 1, 3, 4, 9, 10, 12, 13, 27, 28, 30, 31, 36, 37, 39, 40, 81, 82, 84, 85, 90, 91, 93, 94, 108, 109, 111, 112, 117, 118, 120, 121, 243, 244, 246, 247, 252, 253, 255, 256, 270, 271
Base 3 representation contains no 2. Ref UPNT E10. TCS 98 186 92. [0,3; A5836]

M2354 1, 1, 3, 4, 9, 12, 23, 31, 54, 73, 118, 159, 246, 329, 489, 651, 940, 1242, 1751, 2298, 3177, 4142, 5630, 7293, 9776, 12584, 16659, 21320, 27922, 35532, 46092, 58342
Expansion of $\Pi (1 - x^{2k+1})^{-1}(1 - x^{2k})^{-2}$. Ref PLMS 9 387 59. AMM 76 194 69. [0,3; A2513, N0931]

M2355 0, 0, 0, 1, 1, 3, 4, 9, 13, 26, 40, 74, 118, 210, 342, 595, 981, 1684, 2798, 4763
Paraffins with n carbon atoms. Ref JACS 54 1105 32. [1,6; A0624, N0932]

M2356 0, 1, 1, 3, 4, 9, 14, 27, 44
Dimension of space of Vassiliev knot invariants of order n. Ref BAMS 28 281 93. BarN94. [0,4; A7293]

M2357 1, 1, 3, 4, 9, 14, 27, 48, 93, 163, 315, 576, 1085
Sequences related to transformations on the unit interval. Ref JCT A15 39 73. [3,3; A2823, N0933]

M2358 0, 0, 1, 1, 3, 4, 9, 14, 28, 47, 89, 155, 286, 507, 924, 1652, 2993, 5373, 9707, 17460, 31501, 56714, 102256, 184183, 331981, 598091, 1077870, 1942071, 3499720
$a(n) = a(n-1) + 2a(n-2) - a(n-3)$. [0,5; A6053]

M2359 1, 1, 3, 4, 9, 16, 41, 78, 179, 382, 889, 1992, 4648, 10749, 25462, 59891, 142793, 340761, 819533, 1975109, 4784055, 11617982, 28316757, 69185852, 169516558
Symmetries in planted 3-trees on $n+1$ vertices. Ref GTA91 849. [1,3; A3611]

M2360 1, 3, 4, 10, 14, 34, 48, 116, 164, 396
First row of spectral array $W(\sqrt{2})$. Ref FrKi94. [1,2; A7068]

M2361 0, 1, 3, 4, 10, 15, 35
Valence of graph of maximal intersecting families of sets. Ref Loeb94a. Meye94. [1,3; A7007]

M2362   1, 0, 3, 4, 10, 18, 35, 64, 117, 210, 374, 660, 1157, 2016, 3495, 6032, 10370, 17766, 30343, 51680, 87801, 148830, 251758, 425064, 716425, 1205568, 2025675
Generalized Lucas numbers. Ref FQ 15 252 77. [1,3; A6490]

M2363   1, 1, 3, 4, 11, 15, 41, 56, 153, 209, 571, 780, 2131, 2911, 7953, 10864, 29681, 40545, 110771, 151316, 413403, 564719, 1542841, 2107560, 5757961, 7865521
$a(2n) = a(2n-1) + a(2n-2)$, $a(2n+1) = 2a(2n) + a(2n-1)$. Ref MQET 1 10 16. NZ66 181. [1,3; A2530, N0934]

M2364   1, 3, 4, 11, 16, 30, 50, 91, 157, 278, 485, 854, 1496, 2628, 4609, 8091, 14196, 24915, 43720, 76726, 134642, 236283, 414645, 727654, 1276941, 2240878, 3932464
A Fielder sequence. Ref FQ 6(3) 69 68. [1,2; A1641, N0935]

M2365   0, 0, 3, 4, 11, 16, 32, 49, 87, 137, 231, 369, 608, 978, 1595, 2574, 4179, 6754, 10944, 17699, 28655, 46355, 75023, 121379, 196416, 317796, 514227, 832024, 1346267
Strict (-1)st-order maximal independent sets in path graph. Ref YaBa94. [1,3; A7382]

M2366   0, 1, 0, 1, 1, 3, 4, 11, 20, 51, 108, 267, 619
Bicentered trees with n nodes. Ref CAY 9 438. [1,6; A0677, N0936]

M2367   1, 3, 4, 11, 21, 36, 64, 115, 211, 383, 694, 1256, 2276, 4126, 7479, 13555, 24566, 44523, 80694, 146251, 265066, 480406, 870689, 1578040, 2860046, 5183558, 9394699
A Fielder sequence. Ref FQ 6(3) 69 68. [1,2; A1642, N0937]

M2368   1, 3, 4, 11, 21, 42, 71, 131, 238, 443, 815, 1502, 2757, 5071, 9324, 17155, 31553, 58038, 106743, 196331, 361106, 664183, 1221623, 2246918, 4132721, 7601259
A Fielder sequence. Ref FQ 6(3) 69 68. [1,2; A1643, N0938]

M2369   0, 0, 1, 1, 3, 4, 11, 21, 55, 124, 327, 815, 2177, 5712, 15465, 41727, 114291, 313504, 866963, 2404251, 6701321, 18733340, 52557441, 147849031, 417080105
Reduced unit interval graphs. Ref TAMS 272 424 82. rwr. [1,5; A5218]

M2370   0, 1, 1, 3, 4, 11, 136, 283, 419, 1121, 1540, 38081, 39621, 117323, 156944, 431211, 5331476, 11094163, 16425639, 43945441, 60371080, 1492851361, 1553222441
A continued fraction. Ref IFC 13 623 68. [0,4; A1112, N0939]

M2371   0, 0, 1, 1, 3, 4, 12, 22, 61, 128, 335, 756, 1936, 4580, 11652, 28402, 72209, 179460, 457274, 1151725, 2945129
Dyck paths of knight moves. Ref DAM 24 218 89. [0,5; A5221]

M2372   1, 1, 3, 4, 12, 22, 71, 181, 618, 1957, 6966
Even sequences with period $2n$. Ref IJM 5 664 61. [0,3; A0206, N0940]

M2373   1, 0, 1, 1, 3, 4, 12, 23, 71, 187, 627, 1970, 6833, 23384, 82625, 292164
Cyclically-5-connected planar trivalent graphs with $2n$ nodes. Ref JCT B45 309 88. bdm. [10,5; A6791]

M2374 1, 1, 1, 3, 4, 12, 24, 66, 160, 448, 1186, 3334, 9235, 26166, 73983, 211297
Triangular polyominoes with n cells. See Fig M1845. Ref HA67 37. JRM 2 216 69. RE72
97. [1,4; A0577, N0941]

M2375 1, 1, 1, 3, 4, 12, 27, 82, 228, 733, 2282, 7528, 24834, 83898, 285357, 983244,
    3412420, 11944614, 42080170, 149197152, 531883768, 1905930975, 6861221666
Dissecting a polygon into n triangles. See Fig M2375. Ref BAMS 54 355 48. CMB 6 175
63. GU58. MAT 15 121 68. DM 11 387 75. [1,4; A0207, N0942]

||||||||||||||||||||||||||||||||||||||||||||||||||||||||||||||||||||||||||||||||||||||||||||||||||||||||

Figure M2375. DISSECTIONS.

    One interpretation of the Catalan numbers (Fig. M1459) is that they give the number of ways
of dissecting a polygon with $n + 2$ sides into n triangles. If two such dissections are regarded
as equivalent when one can be obtained from the other by a rotation or reflection, the number
of inequivalent dissections is given by M2375:

$a_3 = 1$

$a_4 = 1$

$a_5 = 1$

$a_6 = 3$

$a_7 = 4$

||||||||||||||||||||||||||||||||||||||||||||||||||||||||||||||||||||||||||||||||||||||||||||||||||||||||

M2376 1, 3, 4, 12, 27, 84, 247, 826, 2777, 9868
Hydrocarbons with n carbon atoms. Ref LNM 303 257 72. [1,2; A2986]

M2377 1, 1, 3, 4, 12, 28, 94, 298, 1044, 3658, 13164
Even sequences with period $2n$. Ref IJM 5 664 61. [0,3; A0208, N0943]

M2378 1, 3, 4, 13, 53, 690, 36571, 25233991, 922832284862, 23286741570717144243,
    214897569306958209736833319349467
$a(n) = a(n-1) a(n-2) + 1$. Ref EUR 19 13 57. FQ 11 436 73. [0,2; A1056, N0944]

M2379 1, 3, 4, 14, 30, 107, 318, 1106, 3671
Polyaboloes with n triangles. Ref GA78 151. [1,2; A6074]

M2380 1, 3, 4, 23, 27, 50, 227, 277, 504, 4309, 4813, 71691, 76504, 836731, 1749966,
2586697, 12096754, 147747745, 307592244, 1070524477, 2448641198, 3519165675
Convergents to cube root of 2. Ref AMP 46 105 1866. L1 67. hpr. [1,2; A2351, N0945]

M2381 3, 4, 25, 108, 735, 5248, 40824, 362000
Permutations of length n with spread 2. Ref JAuMS A21 489 76. [3,1; A4206]

M2382 1, 1, 3, 4, 28, 16, 256, 324, 3600, 3600, 129774, 63504, 3521232, 3459600,
60891840, 32626944, 8048712960, 3554067456, 425476094976, 320265446400
Permanent of 'coprime?' matrix. Ref JCT A23 253 77. [1,3; A5326]

M2383 1, 1, 3, 4, 125, 3, 16807, 256, 19683, 125, 2357947691, 144, 1792160394037,
16807, 1265625, 16777216, 2862423051509815793, 19683, 5480386857784802185939
Discriminant of nth cyclotomic polynomial. Ref BE68 91. MA77 27. [1,3; A4124]

M2384 1, 3, 5, 3, 9, 3, 51, 675, 5871
From discordant permutations. Ref KYU 10 11 56. [0,2; A2633, N0946]

M2385 3, 5, 3, 17, 3, 5, 3, 257, 3, 5, 3, 17, 3, 5, 3, 65537, 3, 5, 3, 17, 3, 5, 3, 97, 3, 5, 3, 17,
3, 5, 3, 641, 3, 5, 3, 17, 3, 5, 3, 257, 3, 5, 3, 17, 3, 5, 3, 193, 3, 5, 3, 17, 3, 5, 3, 257, 3, 5, 3
Smallest factor of $2^n + 1$. Ref AJM 1 239 1878. Krai24 2 85. CUNN. [1,1; A2586, N0947]

M2386 3, 5, 3, 17, 11, 13, 43, 257, 19, 41, 683, 241, 2731, 113, 331, 65537, 43691, 109,
174763, 61681, 5419, 2113, 2796203, 673, 4051, 1613, 87211, 15790321, 3033169, 1321
Largest factor of $2^n + 1$. Ref AJM 1 239 1878. Krai24 2 85. CUNN. [1,1; A2587, N0948]

M2387 1, 3, 5, 4, 10, 7, 15, 8, 20, 9, 18, 24, 31, 14, 28, 22, 42, 35, 33, 46, 53, 6, 36, 23, 2
Diagonal of an expulsion array. Ref CRUX 17(2) 44 91. [1,2; A7063]

M2388 3, 5, 6, 7, 8, 9, 10, 10, 11, 12
Rational points on curves of genus n over $GF(2)$. Ref CRP 296 401 83. HW84 51. [2,1;
A5527]

M2389 1, 3, 5, 6, 7, 11, 13, 14, 15, 17, 19, 20, 21, 22, 23, 27, 29, 31, 33, 35, 37, 38, 39, 41,
43, 44, 45, 46, 47, 51, 53, 54, 55, 56, 57, 59, 60, 61, 62, 65, 66, 67, 68, 69, 70, 71, 73, 77
Average of divisors of n is an integer. Ref AMM 55 616 48. [1,2; A3601]

M2390 0, 3, 5, 6, 7, 14, 25, 45, 84, 162, 310, 595, 1165, 2285, 4486, 8810, 17310, 34310,
68025, 134885, 267485
A generalized Conway-Guy sequence. Ref MOC 50 312 88. [0,2; A6754]

M2391 1, 3, 5, 6, 8, 9, 10, 12, 14, 16, 17, 24, 27, 31, 32, 33, 34, 36, 37, 41, 42, 46, 52, 62,
68, 69, 70, 73, 77, 78, 80, 82, 86, 88, 90, 92, 96, 97, 98, 99, 103, 108, 111, 114, 117, 119
Elliptic curves. Ref JRAM 212 23 63. [1,2; A2150, N0949]

M2392 1, 3, 5, 6, 8, 10, 11, 13, 15, 17, 18, 20, 22, 23, 25, 27, 29, 30, 32, 34
A Beatty sequence. Ref FQ 10 487 72. [1,2; A3152]

M2393 1, 3, 5, 6, 8, 10, 12, 13, 15, 17, 18, 20, 22, 24, 25, 27, 29, 30, 32, 34, 36, 37, 39, 41,
42, 44, 46, 48, 49, 51, 53, 54, 56, 58, 60, 61, 63, 65, 67, 68, 70, 72, 73, 75, 77, 79, 80
A Beatty sequence: $[n(e-1)]$. Ref CMB 3 21 60. [1,2; A0210, N0950]

M2394 1, 3, 5, 6, 8, 17, 25, 34, 67, 103, 134, 265, 405
Spiral sieve using Fibonacci numbers. Ref FQ 12 395 74. [1,2; A5623]

M2395 0, 3, 5, 6, 9, 10, 12, 15, 17, 18, 20, 23, 24, 27, 29, 30, 33, 34, 36, 39, 40, 43, 45, 46,
48, 51, 53, 54, 57, 58, 60, 63, 65, 66, 68, 71, 72, 75, 77, 78, 80, 83, 85, 86, 89, 90, 92
Evil numbers: even number of 1's in binary expansion. Ref CMB 2 86 59. TCS 98 188 92.
[0,2; A1969, N0952]

M2396 3, 5, 6, 9, 13, 20, 31, 49, 78, 125, 201, 324, 523, 845, 1366, 2209, 3573, 5780,
9351, 15129, 24478, 39605, 64081, 103684, 167763, 271445, 439206, 710649, 1149853
$a(n)=a(n-1)+a(n-2)-2$. Ref SMA 20 23 54. R1 233. JCT 7 292 69. [0,1; A0211,
N0953]

M2397 3, 5, 6, 11, 12, 14, 17, 18, 20, 29, 41, 44, 59, 62, 71, 92, 101, 107, 116, 137, 149,
164, 179, 191, 197, 212, 227, 239, 254, 269, 281, 311, 332, 347, 356, 419, 431, 452, 461
$\phi(n+2)=\phi(n)+2$. Ref AMM 56 22 49. AS1 840. [1,1; A1838]

M2398 1, 3, 5, 7, 7, 13, 19, 23, 31, 47, 61, 71, 73, 121, 121, 121, 121, 121, 121, 242, 484,
661, 661, 661, 1093, 1753, 1807, 3053, 3613, 3613, 3613, 5461, 5461, 8011, 8011, 8011
Denominators of worst case for Engel expansion. Ref STNB 3 52 91. [0,2; A6540]

M2399 1, 3, 5, 7, 8, 10, 12, 14, 16, 18, 20, 21, 23, 25, 27, 29, 31, 32, 34, 36, 38, 40, 42, 44,
45, 47, 49, 51, 52, 54, 56, 58, 60, 62, 64, 65, 67, 69, 71, 73, 75, 76, 78, 80, 82, 84, 86, 88
A self-generating sequence. Ref FQ 10 49 72. [1,2; A3144]

M2400 1, 3, 5, 7, 9, 11, 13, 15, 17, 19, 21, 23, 25, 27, 29, 31, 33, 35, 37, 39, 41, 43, 45, 47,
49, 51, 53, 55, 57, 59, 61, 63, 65, 67, 69, 71, 73, 75, 77, 79, 81, 83, 85, 87, 89, 91, 93, 95
The odd numbers. [1,2; A5408]

M2401 1, 3, 5, 7, 9, 11, 13, 15, 18, 21, 24, 27, 30, 33, 36, 39, 43, 47, 51, 54, 58, 63, 67, 71
Smallest square that contains all squares of sides 1, ..., n. Ref GA77 147. UPG D5. rkg.
[1,2; A5842]

M2402 1, 3, 5, 7, 9, 12, 15, 19, 23, 26, 29, 32, 35, 38, 41, 45, 49, 53, 57, 62, 67, 72, 77, 83,
89, 93, 97, 101, 105, 109, 113, 117, 121, 125, 129, 133, 137, 141, 145, 150, 155, 160, 165
Optimal cost of search tree. Ref SIAC 17 1213 88. [1,2; A7078]

M2403 0, 1, 3, 5, 7, 9, 15, 17, 21, 27, 31, 33, 45, 51, 63, 65, 73, 85, 93, 99, 107, 119, 127,
129, 153, 165, 189, 195, 219, 231, 255, 257, 273, 297, 313, 325, 341, 365, 381, 387, 403
Binary expansion is palindromic. [0,3; A6995]

M2404 1, 3, 5, 7, 9, 20, 31, 42, 53, 64, 75, 86, 97, 108, 110, 121, 132, 143, 154, 165, 176,
187, 198, 209, 211, 222, 233, 244, 255, 266, 277, 288, 299, 310, 312, 323, 334, 345, 356
Self or Colombian numbers. Ref KA59. AMM 81 407 74. GA88 116. [1,2; A3052]

M2405 1, 3, 5, 7, 9, 20, 42, 108, 110, 132, 198, 209, 222, 266, 288, 312, 378, 400, 468,
512, 558, 648, 738, 782, 804, 828, 918, 1032, 1098, 1122, 1188, 1212, 1278, 1300, 1368
Self numbers divisible by the sum of their digits. Ref KA67. [1,2; A3219]

M2406 1, 3, 5, 7, 9, 33, 99, 313, 585, 717, 7447, 9009, 15351, 32223, 39993, 53235,
53835, 73737, 585585, 1758571, 1934391, 1979791, 3129213, 5071705, 5259525
Palindromic in bases 2 and 10. Ref JRM 18 47 85. [1,2; A7632]

M2407 0, 1, 3, 5, 7, 10, 13, 16, 19, 22, 26, 29, 33, 37, 41, 45, 49, 53, 57, 62, 66, 70, 75, 80,
84, 89, 94, 98, 103, 108, 113, 118, 123, 128, 133, 139, 144, 149, 154, 160, 165, 170, 176
Smallest integer $\geq \log_2 n!$. Ref KN1 3 187. [1,3; A3070]

M2408 0, 1, 3, 5, 7, 10, 13, 16, 19, 22, 26, 30, 34, 38, 42, 46, 50, 54, 58, 62, 66, 71, 76, 81,
86, 91, 96, 101, 106, 111, 116, 121, 126
Number of comparisons for merge sort of n elements. Ref AMM 66 389 59. Well71 207.
KN1 3 187. TCS 98 193 92. [1,3; A1768, N0954]

M2409 3, 5, 7, 11, 13, 15, 17, 23, 25, 29, 31, 41, 47, 51, 53, 55, 59, 61, 83, 85, 89, 97, 101,
103, 107, 113, 115, 119, 121, 122, 123, 125
Related to iterates of $\phi(n)$. Ref UPNT B41. [1,1; A5239]

M2410 1, 3, 5, 7, 11, 13, 16, 18, 20, 22, 26, 28, 30, 32, 36, 38, 41
Related to Fibonacci representations. Ref FQ 11 385 73. [1,2; A3255]

M2411 3, 5, 7, 11, 13, 17, 19, 23, 29, 31, 41, 43, 47, 53, 61, 71, 73, 79, 83, 89, 97, 107,
109, 113, 127, 137, 139, 151, 163, 167, 173, 179, 181, 191, 193, 197, 199, 211, 223, 227
Regular primes. Ref BS66 430. [1,1; A7703]

M2412 3, 5, 7, 11, 13, 17, 19, 23, 29, 31, 105, 165, 195, 231, 255, 273, 285, 345, 357, 385,
399, 429, 435, 455, 465, 483, 561, 595, 609, 627, 651, 663, 665, 715, 741, 759, 805
Liouville function $\lambda(n)$ is negative. Ref JIMS 7 71 43. [1,1; A2556, N0955]

M2413 3, 5, 7, 11, 13, 17, 19, 23, 31, 43, 61, 79, 101, 127, 167, 191, 199, 313, 347, 701,
1709, 2617
$(2^n + 1)/3$ is prime. Ref MMAG 27 157 54. CUNN. rgw. [1,1; A0978, N0956]

M2414 1, 3, 5, 7, 11, 15, 19, 23, 27, 31, 39, 47, 55, 63, 71, 79, 87, 95, 103, 111, 127, 143,
159, 175, 191, 207, 223, 239, 255, 271, 287, 303, 319, 335, 351, 383, 415, 447, 479, 511
Tower of Hanoi with 5 pegs. Ref JRM 8 175 76. [1,2; A7665]

M2415 3, 5, 7, 11, 47, 71, 419, 4799
$(17^n - 1)/16$ is prime. Ref MOC 61 928 93. [1,1; A6034]

M2416 1, 3, 5, 7, 13, 15, 17, 27, 33, 35, 37, 45, 47, 57, 65, 67, 73, 85, 87, 95, 97, 103, 115, 117, 125, 135, 137, 147, 155, 163, 167, 177, 183, 193, 203, 207, 215, 217, 233, 235, 243
$n^2 + 4$ is prime. [0,2; A7591]

M2417 3, 5, 7, 13, 17, 31, 73, 127, 257, 307, 757, 1093, 1723, 2801, 3541, 5113, 8011, 8191, 10303, 17293, 19531, 28057, 30103, 30941, 65537, 86143, 88741, 131071, 147073
Primes of form $(p^x - 1)/(p^y - 1)$, p prime. Ref IJM 6 154 62. [1,1; A3424]

M2418 3, 5, 7, 13, 23, 17, 19, 23, 37, 61, 67, 61, 71, 47, 107, 59, 61, 109, 89, 103, 79, 151, 197, 101, 103, 233, 223, 127, 223, 191, 163, 229, 643, 239, 157, 167, 439, 239, 199, 191
Fortunate primes. Ref UPNT A2. [1,1; A5235]

M2419 1, 1, 3, 5, 7, 13, 23, 37, 63, 109, 183, 309, 527, 893, 1511, 2565, 4351, 7373, 12503, 21205, 35951, 60957, 103367, 175269, 297183, 503917, 854455, 1448821
$a(n) = a(n-1) + 2a(n-3)$. Ref DT76. [0,3; A3229]

M2420 3, 5, 7, 13, 23, 43, 281, 359, 487, 577
$(3^n + 1)/4$ is prime. Ref CUNN. [1,1; A7658]

M2421 1, 3, 5, 7, 15, 11, 13, 17, 19, 25, 23, 35, 29, 31, 51, 37, 41, 43, 69, 47, 65, 53, 81, 87, 59, 61, 85, 67, 71, 73, 79, 123, 83, 129, 89, 141, 97, 101, 103, 159, 107, 109, 121, 113
Inverse of Euler totient function. Ref BA8 64. [1,2; A2181, N0957]

M2422 1, 3, 5, 7, 15, 45, 95, 235
$4.3^n - 1$ is prime. Ref MOC 26 997 72. [1,2; A5540]

M2423 3, 5, 7, 17, 19, 37, 97, 113, 257, 401, 487, 631, 971, 1297, 1801, 19457, 22051, 28817, 65537, 157303, 160001
A special sequence of primes. Ref ACA 5 425 59. [1,1; A1259, N0958]

M2424 1, 3, 5, 7, 17, 29, 47, 61, 73, 83, 277, 317, 349, 419, 503, 601, 709, 829
From a Goldbach conjecture. Ref BIT 6 49 66. [1,2; A2092, N0959]

M2425 1, 3, 5, 7, 19, 21, 43, 81, 125, 127, 209, 211, 3225, 4543, 10179
$11.2^n + 1$ is prime. Ref PAMS 9 674 58. Rie85 381. [1,2; A2261, N0960]

M2426 3, 5, 7, 19, 29, 47, 59, 163, 257, 421, 937, 947, 1493, 1901
Prime NSW numbers. Cf. M4423. Ref BPNR 290. [1,1; A5850]

M2427 3, 5, 7, 31, 53, 97, 211, 233, 277, 367, 389, 457, 479, 547, 569, 613, 659, 727, 839, 883, 929, 1021, 1087, 1109, 1223, 1289, 1447, 1559, 1627, 1693, 1783, 1873, 2099, 2213
Prime self-numbers. Ref KA59. AMM 81 407 74. GA88 116. jos. [1,1; A6378]

M2428 1, 3, 5, 7, 32, 11, 13, 17, 19, 25, 23, 224, 29, 31, 128, 37, 41, 43, 115, 47, 119, 53, 81, 928, 59, 61, 256, 67, 71, 73, 79, 187, 83, 203, 89, 209, 235, 97, 101
Inverse of reduced totient function. Ref NADM 17 305 1898. L1 7. [1,2; A2396, N0961]

M2429 3, 5, 8, 9, 13, 15, 18, 19, 20, 21, 24, 28, 29, 31, 35, 37, 40, 47, 49, 51, 53, 56, 60, 61, 67, 69, 77, 79, 83, 84, 85, 88, 90, 92, 93, 95, 98, 100, 101, 104, 109, 111, 115, 120
Elliptic curves. Ref JRAM 212 25 63. [1,1; A2159, N0962]

M2430 1, 3, 5, 8, 10, 12, 14, 16, 18, 21, 23, 25, 27, 29, 32, 34, 36, 38, 40, 42, 45, 47, 49, 52, 54, 56, 58, 60, 62, 65, 67, 69, 71, 73, 76, 78, 80, 82, 84, 86, 89, 91, 93, 95, 97, 99
Not representable by truncated tribonacci sequence. Ref BR72 65. [1,2; A3265]

M2431 1, 3, 5, 8, 10, 14, 16, 20, 22, 26
Davenport-Schinzel numbers. Ref ARS 1 47 76. UPNT E20. [1,2; A5004]

M2432 1, 3, 5, 8, 10, 14, 16, 20, 23, 27, 29, 35, 37, 41, 45, 50, 52, 58, 60, 66, 70, 74, 76, 84, 87, 91, 95, 101, 103, 111, 113, 119, 123, 127, 131, 140, 142, 146, 150, 158, 160, 168
$\Sigma[n/k]$, $k = 1 . . n$. Ref MMAG 62 191 89. [1,2; A6218]

M2433 0, 1, 3, 5, 8, 11, 14, 17, 21, 25, 29, 33, 37, 41, 45, 49, 54, 59, 64, 69, 74, 79, 84, 89, 94, 99, 104, 109, 114, 119, 124, 129, 135, 141, 147, 153, 159, 165, 171, 177, 183, 189
$\Sigma\lceil\log_2 k\rceil$, $k = 1 . . n$. Ref AFI 32 519 68. KN1 3 184. TCS 98 192 92. [1,3; A1855, N0963]

M2434 1, 3, 5, 8, 11, 14, 17, 22, 24, 28, 33, 36, 40, 45, 48, 53, 57, 62, 66, 71, 74, 79, 86, 89, 93, 99, 102, 109, 114, 117, 122, 129, 133, 138, 143, 148, 152, 159, 164, 169, 175, 178
Sum of nearest integer to n/k, $k = 1 . . n$. Ref mlb. [2,2; A6591]

M2435 0, 0, 1, 3, 5, 8, 11, 14, 18, 22, 26, 30, 34, 38, 43, 48, 53, 58, 63, 68, 73, 78, 83, 89, 95, 101, 107, 113, 119, 125
Low discrepancy sequences in base 2. Ref JNT 30 68 88. [1,4; A5356]

M2436 3, 5, 8, 11, 15, 18, 23, 27, 32, 38, 42, 47, 53, 57, 63, 71, 75, 78, 90, 93, 98, 105, 113, 117, 123, 132, 137, 140, 147, 161, 165, 168, 176, 183, 188, 197, 206, 212, 215, 227
Generated by a sieve. Ref PC 2 13-6 74. [1,1; A3311]

M2437 1, 3, 5, 8, 11, 15, 19, 23, 27, 32, 36, 42, 47, 52, 58, 64, 70, 76, 83, 89, 96, 103, 110, 118, 125, 133, 140, 148, 156, 164, 173, 181, 190, 198, 207, 216, 225, 234, 244, 253, 263
Nearest integer to $n^{3/2}$. Ref BO47 46. LF60 17. AB71 177. [1,2; A2821, N0964]

M2438 1, 3, 5, 8, 11, 15, 19, 23, 28, 33, 38, 44, 50, 56, 62, 69, 76, 83, 90, 98, 106, 114, 122, 131, 140, 149, 158, 167, 177, 187, 197, 207, 217, 228, 239, 250, 261, 272, 284, 296
$a(n)$ is the last occurrence of n in M0257. See Fig M0436. Ref AMM 74 740 67. [1,2; A1463, N0965]

M2439 0, 1, 3, 5, 8, 12, 16, 21, 27, 33, 40, 48, 56, 65, 75, 85, 96, 108, 120, 133, 147, 161, 176, 192, 208, 225, 243, 261, 280, 300, 320, 341, 363, 385, 408, 432, 456, 481, 507, 533
$[n^2/3]$. [1,3; A0212, N0966]

M2440 3, 5, 8, 12, 17, 23, 30, 37, 45, 54
Integral points in a quadrilateral. Ref CRP 265 161 67. [1,1; A2579, N0967]

M2441 1, 1, 3, 5, 8, 12, 18, 24, 33, 43, 55, 69, 86, 104, 126, 150, 177, 207, 241, 277, 318, 362, 410, 462, 519, 579, 645, 715, 790, 870, 956, 1046, 1143, 1245, 1353, 1467, 1588
Expansion of $(1+x^3)/(1-x)(1-x^2)^2(1-x^3)$. Ref CAY 2 278. [0,3; A1973, N0969]

M2442 3, 5, 8, 20, 12, 9, 28, 11, 48, 39, 65, 20, 60, 15, 88, 51, 85, 52, 19, 95, 28, 60, 105, 120, 32, 69, 115, 160, 68, 25, 75, 175, 180, 225, 252, 189, 228, 40, 120, 29, 145, 280
x such that $p^2 = x^2 + y^2$, $x \le y$. Cf. M3430. Ref CU27 77. L1 60. [5,1; A2366, N0970]

M2443 0, 1, 3, 5, 9, 11, 14, 17, 25, 27, 30, 33, 38, 41, 45, 49, 65
Number of comparisons for sorting n elements by list merging. Ref KN1 3 184. TCS 98 193 92. [1,3; A3071]

M2444 3, 5, 9, 11, 15, 19, 25, 29, 35, 39, 45, 49, 51, 59, 61, 65, 69, 71, 79, 85, 95, 101, 121, 131, 139, 141, 145, 159, 165, 169, 171, 175, 181, 195, 199, 201, 205, 209, 219, 221
$(n^2 + 1)/2$ is prime. Ref EUL (1) 3 24 17. [1,1; A2731, N0971]

M2445 1, 3, 5, 9, 11, 15, 19, 27, 29, 33, 37, 45, 49, 57, 65, 81, 83, 87, 91, 99, 103, 111, 119, 135, 139, 147, 155, 171, 179, 195, 211, 243, 245, 249, 253, 261, 265, 273, 281, 297
Odd entries in first n rows of Pascal's triangle. Ref PAMS 62 19 77. SIAJ 32 717 77. JPA 21 1927 88. CG 13 59 89. [0,2; A6046]

M2446 0, 1, 3, 5, 9, 12, 16, 19, 25
Minimum comparisons in n-element sorting network. Last term unproved. Ref KN1 3 227. [1,3; A3075]

M2447 0, 1, 3, 5, 9, 12, 16, 19, 26, 31, 37, 41, 48, 53, 59, 63, 74, 82, 91, 97, 107, 114, 122, 127, 138, 146, 155, 161, 171, 178, 186, 191, 207, 219, 232, 241, 255, 265, 276, 283, 298
Comparisons in Batcher's parallel sort. Ref KN1 3 227. TCS 98 193 92. [1,3; A6282]

M2448 0, 1, 3, 5, 9, 12, 18, 21, 29, 34, 44, 48, 60, 67, 81, 85, 101, 110, 128, 134, 154, 165, 187, 192, 216, 229, 255, 263, 291, 306, 336, 341, 373, 390, 424, 434, 470, 489, 527, 534
Minimal multiplication-cost addition chains for n. Ref DM 23 115 78. [1,3; A5766]

$$a(2n) = a(n) + n^2 , \ a(2n+1) = a(n) + n(n+1).$$

M2449 1, 3, 5, 9, 13, 17, 25, 33, 41, 49, 65, 81, 97, 113, 129, 161, 193, 225, 257, 289, 321, 385, 449, 513, 577, 641, 705, 769, 897, 1025, 1153, 1281, 1409, 1537, 1665, 1793, 2049
Tower of Hanoi with 4 pegs. Ref JRM 8 172 76. [1,2; A7664]

M2450 3, 5, 9, 13, 19, 21, 30, 35
Modular postage stamp problem. Ref SIAA 1 384 80. [2,1; A4132]

M2451 0, 1, 3, 5, 9, 13, 20, 28, 40, 54, 75, 99, 133, 174, 229, 295, 383, 488, 625, 790, 1000, 1253, 1573, 1956, 2434, 3008, 3716, 4563, 5602, 6840, 8347
From a partition triangle. Ref AMM 100 288 93. [1,3; A7042]

M2452 1, 1, 3, 5, 9, 13, 22, 30, 45, 61, 85, 111
Expansion of a generating function. Ref CAY 10 414. [0,3; A1993, N0973]

M2453 1, 1, 3, 5, 9, 15, 25, 41, 67, 109, 177, 287, 465, 753, 1219, 1973, 3193, 5167, 8361,
13529, 21891, 35421, 57313, 92735, 150049, 242785, 392835, 635621, 1028457
$a(n) = a(n-1) + a(n-2) + 1$. Ref FQ 8 267 70. [0,3; A1595, N0974]

M2454 1, 1, 1, 3, 5, 9, 17, 31, 57, 105, 193, 355, 653, 1201, 2209, 4063, 7473, 13745,
25281, 46499, 85525, 157305, 289329, 532159, 978793, 1800281, 3311233, 6090307
Tribonacci numbers: $a(n) = a(n-1) + a(n-2) + a(n-3)$. Ref FQ 1(3) 72 63; 2 260 64.
[0,4; A0213, N0975]

M2455 3, 5, 9, 17, 35, 79, 209
Weights of threshold functions. Ref MU71 268. [2,1; A3217]

M2456 1, 1, 1, 1, 1, 1, 1, 3, 5, 9, 17, 41, 137, 769, 1925, 7203, 34081, 227321, 1737001,
14736001, 63232441, 702617001, 8873580481, 122337693603, 1705473647525
Somos-7 sequence. Ref MINT 13(1) 41 91. [0,8; A6723]

$$a(n) = (a(n-1)a(n-6) + a(n-2)a(n-5) + a(n-3)a(n-4)) / a(n-7).$$

M2457 1, 1, 1, 1, 1, 1, 3, 5, 9, 23, 75, 421, 1103, 5047, 41783, 281527, 2534423,
14161887, 232663909, 3988834875, 45788778247, 805144998681, 14980361322965
Somos-6 sequence. Ref MINT 13(1) 40 91. [0,7; A6722]

$$a(n) = (a(n-1)a(n-5) + a(n-2)a(n-4) + a(n-3)^2)/a(n-6).$$

M2458 1, 1, 3, 5, 10, 12, 24, 26, 43, 52, 78, 80, 133, 135, 189, 219, 295, 297, 428, 430,
584, 642, 804, 806, 1100, 1123, 1395, 1494, 1856, 1858, 2428, 2430, 2977, 3143, 3739
Shifts left when inverse Moebius transform applied twice. Ref BeSl94. EIS § 2.7. [1,3;
A7557]

M2459 0, 0, 0, 0, 1, 3, 5, 10, 13, 26, 25, 50, 49, 73, 81, 133, 109, 196, 169, 241, 241, 375,
289, 476, 421, 568, 529, 806, 577, 1001, 833, 1081, 1009, 1393, 1081, 1768, 1441, 1849
Genus of modular group Γ_n. Ref GU62 15. [2,6; A1767, N0976]

M2460 3, 5, 10, 14, 21, 26, 36, 43, 55, 64, 78, 88, 105, 117, 136, 150, 171, 186, 210, 227,
253, 272, 300, 320, 351, 373, 406, 430, 465, 490, 528, 555, 595, 624, 666, 696, 741
Related to Zarankiewicz's problem. Ref TI68 126. [3,1; A1841, N0977]

M2461 1, 3, 5, 10, 15, 29, 42, 72, 107, 170, 246, 383, 542, 810, 1145, 1662, 2311, 3305,
4537, 6363
Factorization patterns of n. Ref ARS 25 77 88. [1,2; A6168]

M2462 1, 3, 5, 10, 16, 29, 45, 75, 115, 181, 271, 413, 605, 895, 1291, 1866, 2648, 3760, 5260, 7352, 10160, 14008, 19140, 26085, 35277, 47575, 63753, 85175, 113175, 149938
2-line partitions of n. Ref DUMJ 31 272 64. [1,2; A0990, N0978]

$$\text{G.f.:} \quad \Pi \, (1 - x^n)^{-2} \, / \, (1 - x).$$

M2463 1, 3, 5, 10, 17, 31, 53, 92, 156, 265
Protruded partitions of n. Ref FQ 13 230 75. [1,2; A5403]

M2464 1, 1, 3, 5, 10, 19, 39
Connected planar graphs with n edges. Ref GA69 80. [1,3; A3055]

M2465 1, 3, 5, 10, 25, 64, 160, 390, 940, 2270, 5515, 13440, 32735, 79610, 193480, 470306, 1143585, 2781070, 6762990, 16445100, 39987325, 97232450, 236432060
Related to partitions (g.f. is inverse to M2329). Ref AMM 76 1034 69. [0,2; A2039, N0979]

M2466 3, 5, 10, 26, 96, 553, 5461, 100709, 3718354, 289725509, 49513793526, 19089032278261, 16951604697397302, 35231087224279091310
Cardinalities of Sperner families on 1,...,n. Ref DM 3 123 73. dek. KN1 4 Section 7.3. [1,1; A7695]

M2467 3, 5, 10, 27, 119
Positive threshold functions of n variables. Ref MU71 214. [1,1; A3187]

M2468 3, 5, 10, 30, 198
Positive pseudo-threshold functions. Ref MU71 214. [1,1; A3186]

M2469 1, 3, 5, 10, 30, 210
Antichains of subsets of an n-set. Ref MU71 214. AN87 38. clm. aam. [0,2; A6826]

M2470 3, 5, 10, 32, 382, 15768919
Boolean functions of n variables. Ref JACM 13 154 66. [1,1; A0214, N0980]

M2471 1, 3, 5, 11, 11, 19, 35, 47, 53, 95, 103, 179, 251, 299, 503, 743, 1019, 1319, 1439, 2939, 3359, 3959, 5387, 5387, 5879, 5879, 17747, 17747, 23399, 23399, 23399, 23399
Worst case of Euclid's algorithm. Ref FQ 25 210 87. STNB 3 51 91. [1,2; A6538]

M2472 3, 5, 11, 13, 19, 29, 37, 43, 53, 59, 61, 67, 83, 101, 107, 109, 131, 139, 149, 157, 163, 173, 179, 181, 197, 211, 227, 229, 251, 269, 277, 283, 293, 307, 317, 331, 347, 349
Primes $\equiv \pm 3$ (mod 8). Ref AS1 870. [1,1; A3629]

M2473 3, 5, 11, 13, 19, 29, 37, 53, 59, 61, 67, 83, 101, 107, 131, 139, 149, 163, 173, 179, 181, 197, 211, 227, 269, 293, 317, 347, 349, 373, 379, 389, 419, 421, 443, 461, 467, 491
Primes with 2 as primitive root. Ref Krai24 1 56. AS1 864. [1,1; A1122, N0981]

M2474 3, 5, 11, 13, 41, 89, 317, 337, 991, 1873, 2053, 2377, 4093, 4297, 4583, 6569, 13033, 15877
-1 + product of primes up to p is prime. Ref JRM 19 199 87. Cald94. [1,1; A6794]

M2475 1, 3, 5, 11, 16, 32, 47, 84, 124, 206, 299, 481, 687, 1058, 1506, 2255, 3163, 4638, 6444, 9258
Factorization patterns of n. Ref ARS 25 77 88. [1,2; A6169]

M2476 3, 5, 11, 17, 29, 41, 59, 71, 101, 107, 137, 149, 179, 191, 197, 227, 239, 269, 281, 311, 347, 419, 431, 461, 521, 569, 599, 617, 641, 659, 809, 821, 827, 857, 881, 1019
Lesser of twin primes. Cf. M3763. Ref AS1 870. [1,1; A1359, N0982]

M2477 3, 5, 11, 17, 31, 41, 59, 67, 83, 109, 127, 157, 179, 191, 211, 241, 277, 283, 331, 353, 367, 401, 431, 461, 509, 547, 563, 587, 599, 617, 709, 739, 773, 797, 859, 877, 919
Primes with prime subscripts. Ref JACM 22 380 75. [1,1; A6450]

M2478 1, 3, 5, 11, 17, 33, 50, 89, 135, 223, 332, 531, 776, 1194, 1730, 2591, 3700, 5429, 7660, 11035
Factorization patterns of n. Ref ARS 25 77 88. [1,2; A6170]

M2479 1, 3, 5, 11, 17, 34, 52, 94, 145, 244, 370, 603, 899, 1410, 2087, 3186, 4650, 6959, 10040, 14750, 21077, 30479, 43120, 61574, 86308, 121785, 169336, 236475, 326201
Factorization patterns of n. Ref ARS 25 77 88. [1,2; A6171]

M2480 1, 1, 3, 5, 11, 17, 39, 61, 139, 217, 495, 773, 1763, 2753, 6279, 9805, 22363, 34921, 79647, 124373, 283667, 442961, 1010295, 1577629, 3598219, 5618809
Subsequences of $[1,...,n]$ in which each odd number has an even neighbor. Ref GuMo94. [0,3; A7455]
$$a(n) = 3\,a(n-2) + 2\,a(n-4).$$

M2481 1, 3, 5, 11, 19, 29, 157, 163, 283, 379, 997
$2^n + 2^{(n+1)/2} + 1$ is prime. Ref CUNN xlviii. [1,2; A7671]

M2482 0, 1, 1, 3, 5, 11, 21, 43, 85, 171, 341, 683, 1365, 2731, 5461, 10923, 21845, 43691, 87381, 174763, 349525, 699051, 1398101, 2796203, 5592405, 11184811, 22369621
$a(n) = a(n-1) + 2a(n-2)$. Ref FQ 10 499 72; 26 306 88. JCT A26 149 79. [0,4; A1045, N0983]

M2483 1, 1, 1, 3, 5, 11, 22, 26, 42, 70, 112
The coding-theoretic function $A(n,6,6)$. See Fig M0240. Ref PGIT 36 1335 90. [6,4; A4039]

M2484 0, 1, 1, 3, 5, 11, 22, 47, 93, 193, 386, 793, 1586, 3238, 6476, 13167, 26333, 53381, 106762, 215955
Chvatal conjecture for radius of graph of maximal intersecting sets. Ref Loeb94a. Meye94. [1,4; A7008]

M2485 3, 5, 11, 29, 97, 127, 541, 907, 1151, 1361, 9587, 15727, 19661, 31469, 156007,
360749, 370373, 492227, 1349651, 1357333, 2010881, 4652507, 17051887, 20831533
Increasing gaps between primes (upper end). Cf. M0858. Ref MOC 18 649 64. [1,1;
A0101, N0984]

M2486 1, 1, 3, 5, 12, 30, 79, 227, 710, 2322, 8071, 29503, 112822, 450141
Connected graphs with n edges. Ref PRV 164 801 67. SS67. [1,3; A2905, N0985]

M2487 3, 5, 13, 17, 19, 31, 41, 47, 59, 61, 73, 83, 89, 97, 101, 103, 131, 139, 157, 167,
173, 181, 199, 223, 227, 229, 241, 251, 257, 269, 271, 283, 293, 307, 311, 313, 349, 353
Inert rational primes in $Q(\sqrt{-7})$. Ref Hass80 498. [1,1; A3625]

M2488 3, 5, 13, 17, 241, 257, 65281, 65537, 4294901761, 4294967297,
18446744069414584321, 18446744073709551617
An infinite coprime sequence. Ref MAG 48 420 64. jos. [0,1; A2716, N0986]

M2489 0, 1, 1, 3, 5, 13, 23, 59, 105, 269, 479, 1227, 2185, 5597, 9967, 25531, 45465,
116461, 207391, 531243, 946025, 2423293, 4315343, 11053979, 19684665, 50423309
$a(n) = 5a(n-2) - 2a(n-4)$. Ref JSC 10 599 90. [0,4; A5824]

M2490 1, 1, 3, 5, 13, 27, 66, 153, 377, 914, 2281, 5690, 14397, 36564, 93650, 240916,
623338, 1619346, 4224993, 11062046, 29062341, 76581151, 202365823, 536113477
Ethylene derivatives with n carbon atoms. Ref JACS 55 685 33; 56 157 34. LNM 303 255
72. BA76 28. [2,3; A0631, N0987]

M2491 0, 1, 1, 3, 5, 13, 27, 68, 160, 404, 1010, 2604, 6726, 17661, 46628, 124287,
333162, 898921, 2437254, 6640537, 18166568, 49890419, 137478389, 380031868
Forests of planted trees. Ref JCT B27 118 79. [1,4; A5198]

M2492 3, 5, 13, 37, 61, 73, 157, 193, 277, 313, 397, 421, 457, 541, 613, 661, 673, 733,
757, 877, 997, 1093, 1153, 1201, 1213, 1237, 1321, 1381, 1453, 1621, 1657, 1753, 1873
n and $(n+1)/2$ are prime. Cf. M0849. Ref AS1 870. [1,1; A5383]

M2493 0, 0, 0, 1, 1, 3, 5, 14, 27, 65, 142, 338, 773, 1832, 4296, 10231, 24296, 58128,
139132, 334350, 804441, 1940239, 4685806, 11335797, 27455949, 66585170
Symmetry sites in all planted 3-trees with n nodes. Ref DAM 5 157 83. CN 41 149 84. rwr.
[1,6; A7136]

M2494 1, 1, 3, 5, 14, 42, 150, 624
Connected planar graphs with n nodes. Ref SIAA 4 174 83. [1,3; A6395]

M2495 1, 3, 5, 15, 17, 51, 85, 255, 257, 771, 1285, 3855, 4369, 13107, 21845, 65535,
65537, 196611, 327685, 983055, 1114129, 3342387, 5570645, 16711935, 16843009
Pascal's triangle mod 2 converted to decimal. Ref GO61. FQ 15 183 77. MMAG 63 3 90.
[0,2; A1317, N0988]

M2496 1, 0, 3, 5, 15, 19, 58
Simple imperfect squared squares of order n. See Fig M4482. Ref cjb. [13,3; A2962]

M2497 1, 3, 5, 15, 23, 59, 93, 239, 375, 955, 1501, 3823, 6007, 15291, 24029, 61167,
96119, 244667, 384477, 978671, 1537911, 3914683, 6151645, 15658735, 24606583
Cellular automaton with $000,001,010,011,...,111 \rightarrow 0,1,1,0,0,1,1,1$. See Fig M2497.
Ref mlb. [1,2; A6977]

||||||||||||||||||||||||||||||||||||||||||||||||||||||||||||||||||||||||||||||||||||||||||||||||||||||||||||||||||||

Figure M2497. CELLULAR AUTOMATA.

M2497 is generated by a 1-dimensional cellular automaton [mlb]. Start with a single 1 in
the middle of an infinite string of 0's, and apply, from left to right, the rules

000	$\rightarrow$	0	100 $\rightarrow$	0
001	$\rightarrow$	1	101 $\rightarrow$	1
010	$\rightarrow$	1	110 $\rightarrow$	1
011	$\rightarrow$	0	111 $\rightarrow$	1 .

Similarly for M2642.

||||||||||||||||||||||||||||||||||||||||||||||||||||||||||||||||||||||||||||||||||||||||||||||||||||||||||||||||||||

M2498 1, 1, 3, 5, 15, 52, 213, 1002
Connected planar graphs with n nodes. Ref SIAA 4 174 83. [1,3; A6394]

M2499 3, 5, 16, 12, 15, 125, 24, 40, 75, 48, 80, 72, 84, 60, 32768, 192, 144, 524288, 384,
640, 9375, 168, 120, 300, 1536, 520, 576, 3072, 975, 2147483648, 336, 240, 1171875
Least side of n Pythagorean triples. Ref B1 114. [1,1; A6593]

M2500 1, 1, 1, 3, 5, 17, 41, 127, 365, 1119, 3413, 10685, 33561, 106827, 342129,
1104347, 3584649, 11701369, 38374065, 126395259
Related to series-parallel networks. Ref SAM 21 87 42. [1,4; A1572, N0989]

M2501 1, 1, 3, 5, 17, 44
Connected bipartite graphs with n nodes. Ref ST90. [2,3; A5142]

M2502 1, 3, 5, 17, 49, 161, 513, 1665, 5377, 17409, 56321, 182273, 589825, 1908737,
6176769, 19988481, 64684033, 209321985, 677380097, 2192048129, 7093616641
$F(n).2^n + 1$. Ref dsk. [0,2; A6483]

M2503 3, 5, 17, 257, 65537, 4294967297, 18446744073709551617,
340282366920938463463374607431768211457
Fermat numbers: $2\uparrow2^n + 1$. Ref HW1 14. [0,1; A0215, N0990]

M2504 1, 3, 5, 17, 65537
$(2\uparrow2\uparrow\cdots\uparrow2)$ $(n\times)$ $+ 1$. Ref BPNR 73. [0,2; A7516]

M2505 0, 1, 3, 5, 20, 168, 11748
Maximal self-dual antichains on n points. Ref Loeb94b. [1,3; A7363]

M2506 1, 3, 5, 21, 41, 49, 89, 133, 141, 165, 189, 293, 305, 395, 651, 665, 771, 801, 923, 953
$19.2^n - 1$ is prime. Ref MOC 22 421 68. Rie85 384. [1,2; A1775, N0991]

M2507 1, 3, 5, 24, 13, 22, 13, 5, 51
Coefficients of a modular function. Ref GMJ 8 29 67. [−3,2; A5761]

M2508 1, 1, 3, 5, 35, 63, 231, 429, 6435, 12155, 46189, 88179, 676039, 1300075, 5014575, 9694845, 300540195, 583401555, 2268783825, 4418157975, 34461632205
Numerators in expansion of $(1-x)^{-1/2}$. Ref PHM 33 13 42. MOC 3 17 48. [1,3; A1790, N0992]

M2509 1, 3, 5, 105, 81, 6765, 175747, 30375, 25219857, 142901109, 4548104883, 31152650265, 5198937484375
Permanent of Schur's matrix of order $2n+1$. Ref JNT 5 48 73. JAuMS A21 496 76. VA91 121. [0,2; A3112]

M2510 1, 1, 1, 3, 5, 691, 35, 3617, 43867, 1222277, 854513, 1181820455, 76977927, 23749461029, 8615841276005, 84802531453387, 90219075042845
Numerator of $2(2n+1)B_{2n}$. Ref EUL (1) 15 93 27. FMR 1 73. [1,4; A2427, N0993]

M2511 1, 3, 6, 2, 7, 13, 20, 12, 21, 11, 22, 10, 23, 9, 24, 8, 25, 43, 62, 42, 63, 41, 18, 42, 17, 43, 16, 44, 15, 45, 14, 46, 79, 113, 78, 114, 77, 39, 78, 38, 79, 37, 80, 36, 81, 35, 82
$a(n) = a(n-1) - n$ if positive & new, else $a(n) = a(n-1) + n$. Ref br. [1,2; A5132]

M2512 0, 3, 6, 4, 4, 2, 4, 6, 4, 2, 6, 4, 2, 4, 4, 6, 4, 2, 6, 4, 4, 2, 4, 6, 2, 4, 6, 4, 2, 4, 4, 6, 4, 2, 6, 4, 4, 2, 4, 6, 4, 2, 6, 4, 2, 4, 4, 6, 2, 4, 6, 4, 4, 2, 4, 6, 2, 4, 6, 4, 2, 4, 4, 6, 4, 2, 6, 4, 4
A continued fraction. Ref JNT 11 214 79. [0,2; A6464]

M2513 0, 3, 6, 5, 12, 2, 10, 9, 24, 11, 4, 14, 20, 15, 18, 17, 48, 26, 22, 21, 8, 23, 28, 38, 40, 27, 30, 29, 36, 50, 34, 33, 96, 35, 52, 62, 44, 39, 42, 41, 16, 74, 46, 45, 56, 47, 76, 86, 51
Earliest sequence with $a(a(a(n))) = 2n$. [0,2; A7479]

M2514 1, 3, 6, 6, 10, 16, 28, 28, 28, 28, 28, 28, 28, 28, 36, 40, 48, 48
Equiangular lines in n dimensions. Ref JALG 24 496 73. [1,2; A2853, N0994]

M2515 3, 6, 7, 7, 23, 62, 143, 44, 159, 46, 47, 174, 2239, 239, 719, 241, 5849, 2098, 2099, 43196, 14871, 19574, 35423, 193049, 2105, 36287, 1119, 284, 240479, 58782, 341087
All prime factors of $C(a(n),n)$ exceed n. Ref MOC 28 647 74; 59 256 92. [1,1; A3458]

M2516 0, 3, 6, 8, 9, 11, 12, 15, 17, 18, 19, 20, 21, 22, 25, 27, 29, 30, 31, 32, 33, 34, 35, 36, 37, 39, 41, 43, 44, 45, 46, 47, 48, 49, 51, 53, 54, 55, 56, 57, 59, 61, 63, 65, 66, 67, 68, 69
Numbers represented by hexagonal close-packing. Ref SPLAG 114. [0,2; A5870]

M2517 1, 3, 6, 8, 9, 11, 14, 16, 17, 19, 21, 22, 24, 27, 29, 30, 32, 35, 37, 40, 42, 43, 45, 48, 50, 51, 53, 55, 56, 58, 61, 63, 64, 66, 69, 71, 74, 76, 77, 79, 82, 84, 85, 87, 90, 92, 95, 97
Sum of 2 terms is never a Fibonacci number. Complement of M0965. Ref DM 22 202 78. [1,2; A5652]

M2518 1, 3, 6, 8, 9, 17, 25, 28, 79, 119, 132, 281, 437
Spiral sieve using Fibonacci numbers. Ref FQ 12 395 74. [1,2; A5622]

M2519 1, 3, 6, 8, 12, 18, 21, 27, 36, 38, 42, 48, 52, 60, 72, 78, 90, 108, 111, 117, 126, 132, 144, 162, 171, 189, 216, 218, 222, 228, 232, 240, 252, 258, 270, 288, 292, 300, 312, 320
Entries in first n rows of Pascal's triangle not divisible by 3. Ref JPA 21 1927 88. CG 13 59 89. TCS 98 188 92. [0,2; A6048]

M2520 1, 3, 6, 9, 9, 0, 27, 81, 162, 243, 243
Expansion of bracket function. Ref FQ 2 254 64. [3,2; A0748, N0995]

M2521 3, 6, 9, 12, 15, 18, 21, 23, 26, 29, 32, 35, 38, 41, 44, 47, 50, 53, 56, 59, 61, 64, 67, 70, 73, 76, 79, 81, 83, 86, 89, 92, 95, 98, 101, 104, 107, 110, 113, 116, 119, 121, 124
Related to Fibonacci representations. Ref FQ 11 386 73. [1,1; A3252]

M2522 1, 3, 6, 9, 13, 16, 21, 24, 29, 33, 38, 41, 48, 51, 56, 61, 67, 70, 77, 80, 87, 92, 97, 100, 109, 113, 118, 123, 130, 133, 142, 145, 152, 157, 162, 167, 177, 180, 185, 190, 199
$\Sigma \lceil n/k \rceil$, $k = 1..n$. Ref mlb. [2,2; A6590]

M2523 1, 3, 6, 9, 13, 17, 22, 27, 32, 37, 43, 49, 56, 63, 70, 77, 85, 93, 102, 111, 120, 129, 139, 149, 159, 169, 179, 189, 200, 211, 223, 235, 247, 259, 271, 283, 296, 309, 322, 335
$n + \Sigma \pi(k)$, $k = 1..n$. Ref IDM 7 136 1900. [1,2; A2815, N0996]

M2524 1, 3, 6, 9, 13, 17, 22, 27, 33, 39, 46, 53, 61, 69, 78, 87, 97, 107, 118, 129, 141, 153, 166, 179, 193, 207, 222, 237, 253, 269, 286, 303, 321, 339, 358, 377, 397, 417, 438, 459
$[(n^2 + 6n - 3)/4]$. Ref AMM 87 206 80. [1,2; A4116]

M2525 1, 3, 6, 9, 13, 17, 22, 27, 33, 40, 47, 56, 65
Postage stamp problem. Ref SIAA 1 383 80. [2,2; A4129]

M2526 3, 6, 9, 13, 17, 23, 29, 36, 43, 50, 59, 60, 79, 90, 101, 112, 123, 138
Minimal nodes in graceful graph with n edges. See Fig M2540. Ref AB71 306. WI78 29. [3,1; A4137]

M2527 1, 3, 6, 9, 13, 17, 24, 30, 36
Modular postage stamp problem. Ref SIAA 1 384 80. [2,2; A4131]

M2528 3, 6, 9, 13, 18, 24, 29, 37, 45, 51, 61, 70, 79, 93, 101, 113, 127
Maximal edges in $\hat{b}$-graceful graph with n nodes. Ref AB71 306. WI78 30. [3,1; A5488]

M2529 0, 0, 1, 3, 6, 9, 13, 18, 24, 31
Integral points in a quadrilateral. Ref CRP 265 161 67. [1,4; A2578, N0997]

M2530 3, 6, 9, 14, 18, 23, 28, 36
Ramsey numbers $R(3,n)$. Ref RY63 42. C1 288. bdm. [2,1; A0791, N0998]

M2531 1, 3, 6, 9, 15, 18, 27, 30, 45, 42, 66
Compositions into 3 relatively prime parts. Ref FQ 2 250 64. [3,2; A0741, N0999]

M2532 1, 3, 6, 9, 15, 20, 26, 34, 41
Leech's tree-labeling problem for n nodes. See Fig M2540. Ref AMM 100 946 93. [2,2; A7187]

M2533 1, 3, 6, 9, 15, 25, 34, 51, 73, 97, 132, 178, 226, 294, 376, 466, 582, 722, 872, 1062, 1282, 1522, 1812, 2147, 2507, 2937, 3422, 3947, 4557, 5243, 5978, 6825, 7763, 8771
A generalized partition function. Ref PNISI 17 237 51. [1,2; A2597, N1000]

M2534 3, 6, 10, 13, 17, 20, 23, 27, 30, 34, 37, 40, 44, 47, 51, 54, 58, 61, 64, 68, 71, 75, 78, 81, 85, 88, 92, 95, 99, 102, 105, 109, 112, 116, 119, 122, 126, 129, 133, 136, 139, 143
A Beatty sequence: $[n(2+\sqrt{2}]$. Cf. M0955. Ref CMB 2 188 59. FQ 10 487 72. GKP 77. [1,1; A1952, N1001]

M2535 1, 3, 6, 10, 15, 21, 28, 36, 45, 55, 66, 78, 91, 105, 120, 136, 153, 171, 190, 210, 231, 253, 276, 300, 325, 351, 378, 406, 435, 465, 496, 528, 561, 595, 630, 666, 703, 741
Triangular numbers $n(n+1)/2$. See Fig M2535. Ref D1 2 1. RS3. B1 189. AS1 828. [1,2; A0217, N1002]

$$\text{G.f.: } (1 - x)^{-3}.$$

Figure M2535. POLYGONAL NUMBERS.

The **polygonal** numbers have the form $P(r, s) = \frac{1}{2}r(rs - s + 2)$ [B1 189]. The figures show M2535: the **triangular** numbers $P(r, 1) = \frac{1}{2}r(r + 1)$; M3356: the **square** numbers $P(r, 2) = r^2$; and M3818: the **pentagonal** numbers $P(r, 3) = \frac{1}{2}r(3r - 1)$. Many similar sequences are in the table, including **hexagonal** (M4108), **heptagonal** (M4358), **octagonal** (M4493), etc., and **hex** (M4362), **star** (M4893) and **star-hex** (M5265) numbers.

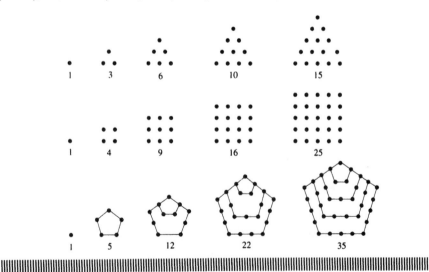

M2536   1, 3, 6, 10, 17, 25, 37, 51, 70, 92, 121, 153, 194, 240, 296, 358, 433, 515, 612, 718, 841, 975, 1129, 1295, 1484, 1688, 1917, 2163, 2438, 2732, 3058, 3406, 3789, 4197, 4644
3×3 matrices with row and column sums n. Ref MO78. NAMS 26 A-27 (763-05-13) 79. [2,2; A5045]

M2537   3, 6, 10, 21, 46, 108, 263, 658, 1674, 4305, 11146, 28980
From sum of $1/F(n)$. Ref FQ 15 46 77. [1,1; A5522]

M2538   0, 0, 1, 3, 6, 10, 30, 126, 448, 1296, 4140, 17380, 76296, 296088, 1126216, 4940040, 23904000, 110455936, 489602448, 2313783216, 11960299360, 61878663840
Degree n odd permutations of order 2. Ref CJM 7 167 55. [0,4; A1465, N1003]

M2539   1, 3, 6, 11, 4, 15, 2, 19, 38, 61, 32, 63, 26, 67, 24, 71, 18, 77, 16, 83, 12, 85, 164, 81, 170, 73, 174, 277, 384, 275, 162, 35, 166, 29, 168, 317, 468, 311, 148, 315, 142, 321
Cald's sequence: $a(n+1)=a(n)-p(n)$ if new and >0, else $a(n)+p(n)$ if new, otherwise 0, where p are primes. Ref JRM 7 318 74. PC 4 41-16 76. [1,2; A6509]

M2540   1, 3, 6, 11, 17, 25, 34, 44, 55, 72, 85, 106, 127, 151
Shortest Golomb ruler with n marks. See Fig M2540. Ref RE72 34. ScAm 253(6) 21 85; 254(3) 21 86. ARS 21 8 86. [2,2; A3022]

|||||||||||||||||||||||||||||||||||||||||||||||||||||||||||||||||||||||||||||||||||||||||||||||||||||||||||||

Figure M2540.   GOLOMB RULERS.

The problem is to label the nodes of a complete graph on n nodes with numbers taken from 0 to k so that the induced edge labels are all distinct (an edge with endpoints labeled i and j gets the label $|i - j|$), and k is minimized. The successive values for k for $n = 2, \ldots$ give M2540. For the connection with rulers see [RE72 34], [ScAm 253(6) 21 85]. Sequences M2526 and M2532 are closely related. Here are the optimal labelings for $n = 2, 3, 4, 5$, for which the corresponding values of k are $1, 3, 6, 11$.

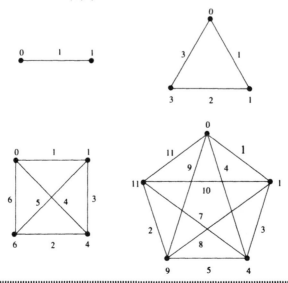

|||||||||||||||||||||||||||||||||||||||||||||||||||||||||||||||||||||||||||||||||||||||||||||||||||||||||||||

M2541 1, 3, 6, 11, 17, 26, 35, 45, 58, 73, 90, 106, 123, 146, 168, 193, 216, 243, 271, 302, 335, 365, 402, 437, 473, 516, 557, 600, 642, 687, 736, 782, 835, 886, 941, 999, 1050
Nonnegative solutions to $x^2 + y^2 \leq n$. Ref PNISI 13 37 47. [0,2; A0603, N1004]

M2542 1, 3, 6, 11, 17, 26, 36, 50, 65, 85, 106, 133, 161, 196, 232, 276, 321, 375, 430, 495, 561, 638, 716, 806, 897, 1001, 1106, 1225, 1345, 1480, 1616, 1768, 1921, 2091, 2262
Dissections of a polygon. Ref AEQ 18 388 78. [5,2; A3453]

$$\text{G.f.: } (1 + x - x^2) \,/\, (1 - x)^4 (1 + x)^2.$$

M2543 1, 1, 3, 6, 11, 18, 32, 48, 75, 111, 160
Multigraphs with 4 nodes. Ref HP73 88. [0,3; A3082]

M2544 1, 3, 6, 11, 19, 31, 43, 63, 80
Additive bases. Ref SIAA 1 384 80. [2,2; A4133]

M2545 1, 3, 6, 11, 19, 32, 48, 71, 101, 141, 188, 249, 322, 414, 518, 645, 791, 966
Restricted partitions. Ref CAY 2 278. [0,2; A1976, N1006]

M2546 0, 1, 3, 6, 11, 19, 32, 53, 87, 142, 231, 375, 608, 985, 1595, 2582, 4179, 6763, 10944, 17709, 28655, 46366, 75023, 121391, 196416, 317809, 514227, 832038, 1346267
$a(n) = a(n-1) + a(n-2) + 2$. Ref R1 233. LNM 748 151 79. [0,3; A1911, N1007]

M2547 1, 3, 6, 11, 20, 37, 70, 135, 264, 521, 1034, 2059, 4108, 8205, 16398, 32783, 65552, 131089, 262162, 524307, 1048596, 2097173, 4194326, 8388631, 16777240
$2^n + n$. [0,2; A6127]

M2548 1, 3, 6, 11, 24, 51, 130, 315, 834, 2195, 5934, 16107, 44368, 122643, 341802, 956635, 2690844, 7596483, 21524542, 61171659, 174342216, 498112275, 1426419858
n-bead necklaces with 3 colors. See Fig M3860. Ref R1 162. IJM 5 658 61. [0,2; A1867, N1008]

M2549 1, 3, 6, 11, 24, 69, 227, 753, 2451, 8004, 27138, 97806, 375313, 1511868, 6292884, 26826701, 116994453, 523646202, 2414394601, 11487130362, 56341183365
From a differential equation. Ref AMM 67 766 60. [0,2; A0998, N1009]

M2550 1, 3, 6, 12, 18, 30, 42, 60, 78, 108, 144, 204, 264, 342, 456, 618, 798, 1044, 1392, 1830, 2388, 3180, 4146, 5418, 7032, 9198, 11892, 15486
Ternary square-free words of length n. Ref QJMO 34 145 83. TCS 23 69 83. [0,2; A6156]

M2551 1, 1, 3, 6, 12, 20, 32, 49, 73, 102, 141, 190, 252, 325, 414, 521, 649, 795, 967
Restricted partitions. Ref CAY 2 278. [0,3; A1975, N1010]

M2552 1, 3, 6, 12, 20, 35, 54, 86, 128, 192, 275, 399, 556, 780, 1068, 1463, 1965, 2644, 3498, 4630, 6052, 7899, 10206, 13174, 16851
Expansion of $\Sigma \, n \, \Pi \, x/(1-x^k)$, $k = 1 \ldots n$. Ref clm. [1,2; A6128]

M2553 1, 1, 3, 6, 12, 21, 38, 63, 106
Corners. Ref DM 27 282 79. [0,3; A6330]

M2554 1, 3, 6, 12, 21, 40, 67, 117, 193, 319, 510, 818, 1274, 1983, 3032, 4610, 6915,
   10324, 15235, 22371, 32554, 47119, 67689, 96763, 137404, 194211, 272939, 381872
3-line partitions of n. Ref DUMJ 31 272 64. [1,2; A0991, N1011]

$$\text{G.f.:} \quad \prod_{k=1}^{\infty} (1 - x^k)^{-3} (1 - x)^2 (1 - x^2).$$

M2555 1, 3, 6, 12, 22, 42, 75, 135, 238, 416
Protruded partitions of n. Ref FQ 13 230 75. [1,2; A5404]

M2556 1, 3, 6, 12, 23, 45, 87, 171, 336, 666, 1320, 2628, 5233, 10443
Weighted voting procedures. Ref LNM 686 70 78. NA79 100. MSH 84 48 83. [1,2; A5256]

M2557 1, 3, 6, 12, 24, 33, 60, 99, 156, 276, 438, 597
Cluster series for honeycomb. Ref PRV 133 A315 64. DG72 225. [0,2; A3204]

M2558 1, 3, 6, 12, 24, 48, 90, 168, 318, 600, 1098, 2004, 3696, 6792, 12270, 22140,
   40224, 72888, 130650, 234012, 421176, 756624, 1348998, 2403840, 4299018
Susceptibility for honeycomb. Ref PHA 28 931 62. JPA 5 635 72. DG74 380. [0,2; A2910, N1012]

M2559 1, 3, 6, 12, 24, 48, 90, 174, 336, 648, 1218, 2328, 4416, 8388, 15780, 29892,
   56268, 106200, 199350, 375504, 704304, 1323996, 2479692, 4654464, 8710212
n-step walks on honeycomb. Ref JMP 2 61 61. JPA 5 659 72. [0,2; A1668, N1013]

M2560 1, 3, 6, 12, 24, 48, 96, 186, 360, 696, 1344, 2562, 4872, 9288, 17664, 33384,
   63120, 119280, 225072, 423630, 797400, 1499256, 2817216, 5286480, 9918768
Trails of length n on honeycomb. Ref JPA 18 576 85. [0,2; A6851]

M2561 3, 6, 12, 24, 48, 96, 192, 384, 768, 1536, 3072, 6144, 12288, 24576, 49152, 98304,
   196608, 393216, 786432, 1572864, 3145728, 6291456, 12582912, 25165824, 50331648
3.2^n. [0,1; A7283]

M2562 3, 6, 12, 24, 54, 138, 378, 1080, 3186, 9642, 29784, 93552, 297966, 960294,
   3126408, 10268688, 33989388, 113277582, 379833906, 1280618784
Energy function for hexagonal lattice. Ref DG74 386. [1,1; A7239]

M2563 1, 3, 6, 13, 23, 45, 78, 141, 239, 409
4-line partitions of n. Ref MES 52 115 24. DUMJ 31 272 64. [1,2; A2799, N1014]

M2564 1, 3, 6, 13, 24, 47, 83, 152, 263, 457, 768, 1292, 2118, 3462, 5564, 8888, 14016,
   21973, 34081, 52552, 80331, 122078, 184161, 276303, 411870, 610818, 900721
5-line partitions of n. Ref MES 52 115 24. DUMJ 31 272 64. [1,2; A1452, N1015]

M2565   1, 3, 6, 13, 24, 47, 86, 159, 285, 509
Protruded partitions of n. Ref FQ 13 230 75. [1,2; A5405]

M2566   1, 3, 6, 13, 24, 48, 86, 160, 282, 500, 859, 1479, 2485, 4167, 6879, 11297, 18334,
 29601, 47330, 75278, 118794, 186475, 290783, 451194, 696033, 1068745, 1632658
Planar partitions of n. See Fig M2566. Ref MA15 2 332. PCPS 63 1099 67. Andr76 241.
[1,2; A0219, N1016]

$$\text{G.f.:} \quad \prod_{k=1}^{\infty} (1 - x^k)^{-k}.$$

||||||||||||||||||||||||||||||||||||||||||||||||||||||||||||||||||||||||||||||||||||||||||||||||||||||||||||||||||||

Figure M2566.   PLANAR PARTITIONS.

M2566 gives the number of **planar partitions** of n:

$a_1 = 1$	1							
$a_2 = 3$	2	11	1					
			1					
$a_3 = 6$	3	21	2	111	11	1		
			1		1	1		
						1		
$a_4 = 13$	4	31	3	22	2	211	21	2
			1		2		1	1
								1
	1111		111	11	11	1		
			1	11	1	1		
					1	1		
						1		

||||||||||||||||||||||||||||||||||||||||||||||||||||||||||||||||||||||||||||||||||||||||||||||||||||||||||||||||||||

M2567   1, 3, 6, 13, 24, 49, 93, 190, 381, 803, 1703, 3755, 8401, 19338, 45275, 108229,
 262604, 647083, 1613941, 4072198, 10374138, 26663390, 69056163, 180098668
Random forests. Ref JCT B27 116 79. [1,2; A5196]

M2568   1, 3, 6, 13, 24, 52, 103, 222, 384, 832, 1648
Nim product $2^n.2^n$. Ref WW 444. [0,2; A6017]

M2569   1, 3, 6, 13, 25, 49, 91, 170, 309, 558
Protruded partitions of n. Ref FQ 13 230 75. [1,2; A5406]

M2570   1, 3, 6, 13, 25, 50, 93, 175, 320, 582
Protruded partitions of n. Ref FQ 13 230 75. [1,2; A5407]

M2571   1, 3, 6, 13, 25, 50, 94, 178, 328, 601
Protruded partitions of n. Ref FQ 13 230 75. [1,2; A5116]

M2572 1, 1, 3, 6, 13, 28, 60, 129, 277, 595, 1278, 2745, 5896, 12664, 27201, 58425,
   125491, 269542, 578949, 1243524, 2670964, 5736961, 12322413, 26467299, 56849086
Bisection of M0571. Ref EUL (1) 1 322 11. [0,3; A2478, N1017]

M2573 1, 3, 6, 13, 29, 70, 175, 449, 1164, 3035, 7931, 20748, 54301, 142143, 372114,
   974185, 2550425, 6677074, 17480779, 45765245, 119814936, 313679543, 821223671
Triangular anti-Hadamard matrices of order n. Ref LAA 62 117 84. [1,2; A5313]

M2574 3, 6, 14, 25, 53, 89, 167, 278, 480, 760
Restricted partitions. Ref JCT 9 373 70. [2,1; A2219, N1018]

M2575 1, 1, 3, 6, 14, 25, 56, 97, 198, 354, 672, 1170, 2207, 3762, 6786, 11675, 20524,
   34636, 60258, 100580, 171894, 285820, 480497, 791316, 1321346, 2156830, 3557353
$1 / \Pi (1 - kx^k)$. Ref gla. [0,3; A6906]

M2576 1, 3, 6, 14, 27, 58, 111, 223, 424, 817, 1527, 2870, 5279, 9710, 17622, 31877,
   57100, 101887, 180406, 318106, 557453, 972796, 1688797, 2920123, 5026410, 8619551
Functional determinants: Euler transform applied twice to all 1's sequence. Ref CAY 2
219. DM 75 93 89. EIS § 2.7. [1,2; A1970, N1019]

M2577 1, 1, 3, 6, 14, 27, 60, 117, 246, 490, 1002, 1998, 4053, 8088, 16284, 32559, 65330,
   130626, 261726, 523374, 1047690, 2095314, 4192479, 8384808, 16773552
Conjugacy classes in $GL(n,2)$. Ref wds. [0,3; A6951]

M2578 1, 3, 6, 14, 31, 70, 157, 353, 793, 1782, 4004
Distributive lattices. Ref MSH 53 19 76. MSG 121 121 76. [0,2; A6356]

M2579 1, 3, 6, 14, 33, 71, 150, 318, 665, 1375, 2830, 5798, 11825, 24039, 48742, 98606,
   199113, 401455, 808382, 1626038, 3267809, 6562295, 13169814, 26416318, 52962681
Expansion of $1/(1-2x)(1+x^2)(1-x-2x^3)$. Ref DT76. [0,2; A3477]

M2580 3, 6, 14, 36, 98, 276, 794, 2316, 6818, 20196, 60074, 179196, 535538, 1602516,
   4799354, 14381676, 43112258, 129271236, 387682634, 1162785756, 3487832978
$1^n + 2^n + 3^n$. Ref AS1 813. [0,1; A1550, N1020]

M2581 3, 6, 15, 24, 33, 48, 63, 90
Restricted postage stamp problem. Ref LNM 751 326 82. [1,1; A6639]

M2582 1, 3, 6, 15, 27, 63, 120, 252, 495, 1023, 2010, 4095, 8127, 16365, 32640, 65535,
   130788, 262143, 523770, 1048509, 2096127, 4194303, 8386440, 16777200, 33550335
$(\Sigma a(d), d \mid n) = 2^{n-1}$. Ref FQ 2 251 64. [1,2; A0740, N1021]

M2583 1, 0, 0, 0, 0, 0, 1, 1, 3, 6, 15, 29, 67, 139, 310, 667, 1480, 3244, 7241, 16104,
   36192, 81435, 184452, 418870, 955860, 2187664, 5025990, 11580130, 26765230
Asymmetric trees with n nodes. Ref JAuMS A20 502 75. HA69 232. ajs. [1,9; A0220,
N1022]

M2584 1, 1, 3, 6, 15, 31, 75, 164, 388, 887, 2092, 4884, 11599, 27443, 65509, 156427, 375263
Mappings from n points to themselves with in-degree ≤ 2. Ref SIAA 3 367 92. [0,3; A6961]

M2585 1, 1, 3, 6, 15, 33, 82, 194, 482, 1188, 2988, 7528, 19181, 49060, 126369, 326863, 849650, 2216862, 5806256, 15256265, 40210657, 106273050, 281593237, 747890675
Secondary alcohols with n carbon atoms. Ref JACS 53 3042 31; 54 2919 32. BA76 28. [3,3; A0599, N1023]

M2586 1, 1, 3, 6, 15, 33, 83, 202
Graphs with no isolated vertices. Ref LNM 952 101 82. [4,3; A6647]

M2587 1, 0, 1, 1, 3, 6, 15, 36, 91, 232, 603, 1585, 4213, 11298, 30537, 83097, 227475, 625992, 1730787, 4805595, 13393689, 37458330, 105089229, 295673994, 834086421
$(n+1)a(n)=(n-1)(2a(n-1)+3a(n-2))$. Ref JCT A23 293 77. JCP 67 5027 77. TAMS 272 406 82. JALG 93 189 85. [0,5; A5043]

M2588 1, 3, 6, 15, 41, 115, 345, 1103, 3664, 12763, 46415, 175652, 691001, 2821116, 11932174, 52211412
Graphs by nodes and edges. Ref R1 146. SS67. [2,2; A1433, N1024]

M2589 1, 0, 1, 3, 6, 15, 42
Occurrences of principal character. Ref SIAA 4 541 83. [0,4; A5368]

M2590 3, 6, 15, 46, 148, 522, 1869, 6910, 25767, 97256, 369127, 1409362
Positions in Mu Torere. Ref MMAG 60 90 87. [1,1; A5655]

M2591 1, 1, 1, 3, 6, 16, 43, 120, 339, 985, 2892, 8606, 25850, 78347, 239161, 734922, 2271085, 7054235, 22010418, 68958139, 216842102, 684164551, 2165240365
Shifts left when weigh-transform applied twice. Ref BeSl94. EIS § 2.7. [1,4; A7561]

M2592 1, 1, 3, 6, 16, 46, 126, 448, 1366, 5354, 18971
Sum of degrees of irreducible representations of A_n. Ref ATLAS. [1,3; A7002]

M2593 1, 1, 3, 6, 17, 44, 133, 404, 1319
Projective plane trees with n nodes. Ref LNM 406 348 74. [2,3; A6081]

M2594 3, 6, 17, 66, 327
Ramsey numbers. Ref BF72 175. hwg. [1,1; A3323]

M2595 1, 1, 3, 6, 18, 48, 156, 492, 1740, 6168, 23568, 91416, 374232, 1562640, 6801888, 30241488, 139071696, 653176992, 3156467520, 15566830368, 78696180768
$a(n)=a(n-1)+n.a(n-2)$. [0,3; A0932, N1025]

M2596 1, 1, 3, 6, 19, 47, 140, 374, 1082, 2998, 8574, 24130, 68876, 195587, 559076, 1596651, 4575978, 13122219, 37711998, 108488765, 312577827, 901531937
Unlabeled bisectable trees with $2n+1$ nodes. Ref COMB 4 177 84. [0,3; A7098]

M2597 1, 1, 3, 6, 19, 49, 163, 472, 1626, 5034, 17769, 57474, 206487, 688881, 2508195, 8563020
A binomial coefficient summation. Ref AMM 81 170 74. [1,3; A3162]

M2598 1, 1, 3, 6, 20, 50, 175, 490, 1764, 5292, 19404, 60984, 226512, 736164, 2760615, 9202050, 34763300, 118195220, 449141836, 1551580888, 5924217936, 20734762776
Walks on square lattice. Ref GU90. [0,3; A5558]

M2599 1, 1, 1, 3, 6, 24, 148, 1646, 34040, 1358852, 106321628, 16006173014, 4525920859198, 2404130854745735, 2426376196165902704
Graphs by nodes and edges. Ref R1 146. SS67. [1,4; A0717, N1027]

M2600 1, 1, 1, 3, 6, 26, 122
Classifications of n things. Ref CSB 4 2 79. [1,4; A5646]

M2601 0, 3, 6, 30, 360, 504, 44016, 204048, 8261760, 128422272, 1816480512, 76562054400, 124207469568
A partition function. Ref PRV 135 M4378 64. [1,2; A2164, N1028]

M2602 1, 3, 6, 38, 213, 1479, 11692, 104364, 1036809, 11344859, 135548466, 1755739218, 24504637741, 366596136399, 5852040379224, 99283915922264
From ménage numbers. Ref R1 198. [2,2; A0222, N1029]

M2603 1, 3, 6, 42, 618, 15990, 668526, 43558242, 4373213298, 677307561630, 162826875512646
Colored graphs. Ref CJM 22 596 70. rcr. [1,2; A2028, N1030]

M2604 3, 6, 44, 180, 1407, 10384, 92896
Hit polynomials. Ref RI63. [3,1; A1886, N1031]

M2605 1, 3, 6, 55, 66, 171, 595, 666, 3003, 5995, 8778, 15051, 66066, 617716, 828828, 1269621, 1680861, 3544453, 5073705, 5676765, 6295926, 35133153, 61477416
Palindromic triangular numbers. Ref JRM 6 146 73. [1,2; A3098]

M2606 1, 3, 7, 0, 3, 5, 9, 8, 9, 5
Decimal expansion of reciprocal of fine-structure constant. See Fig M2218. Ref RMP 59 1139 87. Lang91. [3,2; A5600]

M2607 1, 3, 7, 1, 2, 2, 1, 2, 4, 56, 1, 14, 2, 1, 1, 3, 5, 6, 2, 1, 1, 2, 1, 1, 8, 1, 2, 2, 1, 5, 1, 4, 1, 1, 3, 3, 1, 1, 3, 7, 4, 1, 10, 1, 2, 1, 8, 2, 4, 1, 1, 9, 2, 2, 2, 1, 2, 1, 1, 1, 92, 1, 26, 4, 31, 1
Continued fraction for fifth root of 4. [1,2; A3118]

M2608 0, 3, 7, 3, 9, 5, 5, 8, 1, 3, 6, 1, 9, 2, 0, 2, 2, 8, 8, 0, 5, 4, 7, 2, 8, 0, 5, 4, 3, 4, 6, 4, 1, 6, 4, 1, 5, 1, 1, 1, 6, 2, 9, 2, 4, 9
Decimal expansion of Artin's constant. Ref MOC 15 397 71. [1,2; A5596]

M2609 3, 7, 5, 31, 7, 127, 17, 73, 31, 89, 13, 8191, 127, 151, 257, 131071, 73, 524287, 41, 337, 683, 178481, 241, 1801, 8191, 262657, 127, 2089, 331, 2147483647, 65537, 599479
Largest factor of $2^n - 1$. Ref CUNN. [2,1; A5420]

M2610   1, 3, 7, 5, 93, 637, 1425, 22341
Related to Weber functions. Ref KNAW 66 751 63. [1,2; A1663, N1032]

M2611   1, 0, 0, 1, 3, 7, 8, 4, 0, 4
Decimal expansion of neutron-to-proton mass ratio. See Fig M2218. Ref RMP 59 1142 87.
Lang91. [1,5; A6834]

M2612   0, 3, 7, 8, 10, 14, 19, 20, 21, 23, 24, 27, 29, 31, 36, 37, 40, 45, 51, 52, 53, 54, 56,
57, 58, 61, 62, 64, 66, 67, 71, 73, 74, 76, 78, 81, 84, 86, 92, 93, 94, 97, 98, 102, 104, 107
Location of 0's when natural numbers are listed in binary. [0,2; A3607]

M2613   1, 3, 7, 8, 13, 17, 18, 21, 30, 31, 32, 38, 41, 43, 46, 47, 50, 55, 57, 68, 70, 72, 73,
75, 76, 83, 91, 93, 98, 99, 100, 105, 111, 112, 117, 119, 122, 123, 128, 129, 132, 133, 142
Reducible numbers. Ref AMM 56 525 49. TO51 94. [1,2; A2312, N1033]

M2614   0, 1, 3, 7, 8, 14, 29, 31, 42, 52, 66, 85, 99, 143, 161, 185, 190, 267, 273, 304, 330,
371, 437, 476, 484, 525, 603, 612, 658, 806, 913, 1015, 1074, 1197, 1261, 1340, 1394
Of form $(p^2 - 1)/120$ where p is prime. Ref IAS 5 382 37. [0,3; A2381, N1034]

M2615   1, 3, 7, 9, 7, 2, 9, 6, 6, 1, 4, 6, 1, 2, 1, 4, 8, 3, 2, 3, 9, 0, 0, 6, 3, 4, 6, 4, 2, 1, 6, 0, 1,
7, 6, 9, 2, 8, 5, 5, 6, 4, 9, 8, 7, 7, 7, 9, 7, 7, 6, 0, 6, 1, 2, 1, 7, 7, 2, 7, 3, 7, 6, 7, 4, 7, 9, 1, 5, 0
Decimal expansion of fifth root of 5. [1,2; A5534]

M2616   1, 3, 7, 9, 13, 15, 21, 25, 31, 33, 37, 43, 49, 51, 63, 67, 69, 73, 75, 79, 87, 93, 99,
105, 111, 115, 127, 129, 133, 135, 141, 151, 159, 163, 169, 171, 189, 193, 195, 201, 205
Lucky numbers. Ref MMAG 29 119 55. OG72 99. PC 2 13-7 74. UPNT C3. Well86 114.
[1,2; A0959, N1035]

M2617   3, 7, 9, 63, 63, 168, 322, 322, 1518, 1518, 1680, 10878, 17575, 17575, 17575,
17575, 17575, 17575, 70224, 70224, 97524, 97524, 97524, 97524, 224846, 224846
Every sequence of 4 numbers $> a(n)$ contains number with prime factor $> p(n)$. Ref AMM
79 1087 72. [3,1; A3033]

M2618   3, 7, 10, 14, 18, 21, 25, 28, 32, 36, 39, 43, 47, 50, 54, 57, 61, 65, 68, 72, 75, 79, 83,
86, 90, 94, 97, 101, 104, 108, 112, 115, 119, 123, 126, 130, 133, 137, 141, 144, 148
Related to a Beatty sequence. Ref FQ 11 385 73. [1,1; A3231]

M2619   3, 7, 10, 19, 32, 34, 37, 51, 81, 119, 122, 134, 157, 160, 161, 174, 221, 252, 254,
294, 305, 309, 364, 371, 405, 580, 682, 734, 756, 763, 776, 959, 1028, 1105, 1120, 1170
Related to lattice points in spheres. Ref MOC 20 306 66. [1,1; A0223, N1036]

M2620   3, 7, 11, 13, 47, 127, 149, 181, 619, 929, 3407, 10949
$(5^n - 1)/4$ is prime. Ref CUNN. MOC 61 928 93. [1,1; A4061]

M2621 3, 7, 11, 14, 18, 22, 26, 29, 33, 37, 40, 44, 48, 52, 55, 59, 63, 66, 70, 74, 78, 81, 85,
89, 92, 96, 100, 104, 107, 111, 115, 118, 122, 126, 130, 133, 137, 141, 145, 148, 152, 156
A Beatty sequence: $[n(e+1)]$. See Fig M1332. Cf. M0947. Ref CMB 3 21 60. [1,1;
A0572, N1037]

M2622 3, 7, 11, 14, 18, 22, 26, 29, 33, 37, 41, 44, 48, 52, 55, 59, 63, 67, 70, 74, 78, 82, 85,
89, 93, 97, 100, 104, 108, 111, 115, 119, 123, 126, 130, 134, 138, 141, 145, 149, 153, 156
A Beatty sequence: $[n(\sqrt{3}+2)]$. See Fig M1332. Cf. M0946. Ref DM 2 338 72. [1,1;
A3512]

M2623 0, 0, 3, 7, 11, 16, 22, 27, 33, 40, 46, 53, 60, 67, 74, 81, 89, 96, 104, 112, 120, 128,
136, 144, 153, 161, 169, 178, 187, 195, 204, 213, 222, 231, 240, 249, 258, 267, 276, 286
Nearest integer to $2n.\ln n$. Ref NBS B66 229 62. [0,3; A1618, N1038]

M2624 3, 7, 11, 19, 23, 31, 43, 47, 59, 67, 71, 79, 83, 103, 107, 127, 131, 139, 151, 163,
167, 179, 191, 199, 211, 223, 227, 239, 251, 263, 271, 283, 307, 311, 331, 347, 359, 367
Primes of form $4n+3$. Ref AS1 870. [1,1; A2145, N1039]

M2625 1, 3, 7, 11, 21, 39, 71, 131, 241, 443, 815, 1499, 2757, 5071, 9327, 17155, 31553,
58035, 106743, 196331, 361109, 664183, 1221623, 2246915, 4132721, 7601259
A Fielder sequence. Ref FQ 6(3) 69 68. [1,2; A1644, N1040]

M2626 1, 3, 7, 11, 26, 45, 85, 163, 304, 578, 1090, 2057, 3888, 7339, 13862, 26179,
49437, 93366, 176321, 332986, 628852, 1187596, 2242800, 4235569, 7998951
A Fielder sequence. Ref FQ 6(3) 69 68. [1,2; A1645, N1041]

M2627 3, 7, 11, 29, 47, 199, 521, 2207, 3571, 9349, 3010349, 54018521, 370248451,
6643838879, 119218851371, 5600748293801, 688846502588399, 32361122672259149
Prime Lucas numbers. Ref MOC 50 251 88. [1,1; A5479]

M2628 3, 7, 11, 47, 322, 9349, 1860498, 10749957122, 12360848947227307,
8212348881505319130910 3132, 6273762154066095121476727991896972575 45380
$L(L(n))$, where L is a Lucas number. [0,1; A5372]

M2629 1, 3, 7, 12, 18, 26, 35, 45, 56, 69, 83, 98, 114, 131, 150, 170, 191, 213, 236, 260,
285, 312, 340, 369, 399, 430, 462, 495, 529, 565, 602, 640, 679, 719, 760, 802, 845, 889
Sequence and first differences include all numbers. See Fig M2629. Ref GEB 73. [1,2;
A5228]

||||||||||||||||||||||||||||||||||||||||||||||||||||||||||||||||||||||||||||||||||||||||||||||||||||||||||||||||||||||||||||||||||||||||||||||||||||||

Figure M2629.   OUR FAVORITE SEQUENCES.

These may or may not make you popular at parties, but we like them a lot. Several of the sequences in Figs. M0436, M0557 could also have been included here.

(1) M2629: $1, 3, 7, 12, 18, \ldots$: every positive integer is either in the sequence itself or in the sequence of differences [GEB 73].

(2) The RATS sequence, M1137: 1, 2, 4, 8, 16, 77, 145, 668, ...: produced by the instructions "Reverse, Add, Then Sort". For example, after 668 we get

$$\begin{array}{r} 668 \\ \underline{866} \\ 1534 \end{array}$$

so the next term is 1345. Zeros are suppressed. J. H. Conway conjectures that no matter what number you start with, eventually the sequence either cycles or joins the ever-increasing sequence $\ldots, 1\,2\,3^n 4^4, 5^2\,6^n 7^4, 1\,2\,3^{n+1} 4^4, \ldots$ [AMM 96 425 89].

This is somewhat similar to the widely-studied '$3x + 1$' or Collatz sequence, where $a_{n+1} = a_n/2$ if a_n is even, or $a_{n+1} = 3a_n + 1$ if a_n is odd [AMM 92 3 85], [UPNT E16]. It is conjectured that every number eventually reaches the cycle 4, 2, 1, 4, 2, 1, ... M4323 gives the number of steps for n to reach 1. Sequences M0019, M0189, M0304, M0305, M0748, M0843, M2086, M3198, M3733, M4335 are related to this problem.

(3) Aronson's sequence, quoted in [HO85 44], M3406: 1, 4, 11, 16, 24, ..., whose definition is: "t is the first, fourth, eleventh, ... letter of this sentence"!

(4) M4780: 1, 11, 21, 1211, 111221, ..., in which the next term is obtained by describing the previous term (one 1, two 1's, one 2 two 1's, etc.). J. H. Conway's astonishing analysis of the asymptotic behavior of this sequence is well worth reading [CoGo87 176]. M4778, M4779, M2126 have similar descriptions.

(5) Everyone knows about the even numbers, M0985. Less well-known are the **eban** numbers, M1030: 2, 4, 6, 30, 32, 34, ... The reader unable to guess the rule can look it up in the table.

(6) M5100: the number of possible chess games after n moves, computed specially for this book by Ken Thompson. Finite, but we like it anyway!

||||||||||||||||||||||||||||||||||||||||||||||||||||||||||||||||||||||||||||||||||||||||||||||||||||||||||||||||||||||||||||||||||||||||||||||||||||||

M2630 1, 3, 7, 12, 18, 26, 35, 45, 57, 70, 84, 100, 117, 135, 155, 176, 198, 222, 247, 273, 301, 330, 360, 392, 425, 459, 495, 532, 570, 610, 651, 693, 737, 782, 828, 876, 925, 975
Fermat coefficients. Ref MMAG 27 141 54. [3,2; A0969, N1042]

M2631 1, 3, 7, 12, 19, 27, 37, 46
Queens of 3 colors on an $n \times n$ board. Ref MINT 12 66 90. [1,2; A6317]

M2632 3, 7, 12, 19, 30, 43, 49, 53, 70, 89, 112, 141, 172, 209, 250, 293, 301
Related to a highly composite sequence. Ref BSMF 97 152 69. [1,1; A2498, N1043]

M2633 1, 3, 7, 12, 20, 30, 44, 59, 75, 96, 118, 143, 169, 197, 230, 264, 299, 335, 373, 413, 455, 501, 549, 598, 648, 701, 758, 818, 880, 944, 1009, 1079, 1156, 1236, 1317, 1400
Prime numbers of measurement. See Fig M0557. Cf. M0972. Ref AMM 75 80 68; 82 922 75. UPNT E30. [1,2; A2049, N1044]

M2634 1, 3, 7, 13, 15, 21, 43, 63, 99, 109, 159, 211, 309, 343, 415, 469, 781, 871, 939, 1551, 3115, 3349, 5589, 5815, 5893, 7939, 8007, 11547, 12495, 35647
$9.2^n - 1$ is prime. Ref MOC 23 874 69. Rie85 384. Cald94. [1,2; A2236, N1045]

M2635 1, 3, 7, 13, 15, 25, 39, 55, 75, 85, 127, 1947, 3313, 4687, 5947, 13165, 23473, 26607
$5.2^n + 1$ is prime. Ref PAMS 9 674 58. Rie85 381. Cald94. [1,2; A2254, N1046]

M2636 1, 3, 7, 13, 19, 27, 39, 49, 63, 79, 91, 109, 133, 147, 181, 207, 223, 253, 289, 307, 349, 387, 399, 459, 481, 529, 567, 613, 649, 709, 763, 807, 843, 927, 949, 1009, 1093
Flavius' sieve. Ref MMAG 29 117 55. Bru65. [1,2; A0960, N1048]

M2637 3, 7, 13, 19, 31, 37, 43, 61, 67, 73, 79, 97, 103, 109, 127, 139, 151, 157, 163, 181, 193, 199, 211, 223, 229, 241, 271, 277, 283, 307, 313, 331, 337, 349, 367, 373, 379, 397
Primes of form $x^2 + xy + y^2$. [1,1; A7645]

M2638 1, 1, 3, 7, 13, 21, 31, 43, 57, 73, 91, 111, 133, 157, 183, 211, 241, 273, 307, 343, 381, 421, 463, 507, 553, 601, 651, 703, 757, 813, 871, 931, 993, 1057, 1123, 1191, 1261
Central polygonal numbers: $n^2 - n + 1$. Ref HO50 22. HO70 87. [0,3; A2061, N1049]

M2639 3, 7, 13, 21, 31, 48, 57, 73, 91
Additive bases. Ref SIAA 1 384 80. [2,1; A4136]

M2640 1, 3, 7, 13, 22, 34, 50, 70, 95, 125, 161, 203, 252, 308, 372, 444, 525, 615, 715, 825, 946, 1078, 1222, 1378, 1547, 1729, 1925, 2135, 2360, 2600, 2856, 3128, 3417, 3723
Expansion of $1 / (1-x)^3(1-x^2)$. Ref AMS 26 308 55. PGEC 22 1050 73. [0,2; A2623, N1050]

M2641 3, 7, 13, 31, 43, 73, 157, 211, 241, 307, 421, 463, 601, 757, 1123, 1483, 1723, 2551, 2971, 3307, 3541, 3907, 4423, 4831, 5113, 5701, 6007, 6163, 6481, 8011, 8191
Primes of form $n^2 + n + 1$. Ref LINM 3 209 29. L1 46. [1,1; A2383, N1051]

M2642 1, 3, 7, 13, 31, 49, 115, 215, 509, 775, 1805, 3359, 7985, 12659, 29655, 54909, 130759, 197581, 460383, 855793, 2038675, 3227319, 7562237, 14149127, 33304077
Cellular automaton with $000,001,010,011,...,111 \rightarrow 0,1,1,1,0,1,1,0$. See Fig M2497. Ref mlb. [1,2; A6978]

M2643 3, 7, 13, 71, 103, 541, 1019, 1367, 1627, 4177, 9011, 9551
$(3^n - 1)/2$ is prime. Ref CUNN. MOC 61 928 93. [1,1; A4060]

M2644 1, 1, 3, 7, 14, 18, 30, 35, 51, 65, 91, 105, 140
The coding-theoretic function $A(n,4,4)$. See Fig M0240. Ref TI68 126. PGIT 36 1335 90.
[4,3; A1843, N1052]

M2645 1, 3, 7, 14, 26, 46, 79, 133, 221, 364, 596, 972, 1581, 2567, 4163, 6746, 10926,
  17690, 28635, 46345, 75001, 121368, 196392, 317784, 514201, 832011, 1346239
From rook polynomials. Ref SMA 20 18 54. [0,2; A1924, N1053]

$$\text{G.f.:}\quad 1\ /\ (1 - x - x^2)\ (1 - x)^2.$$

M2646 3, 7, 15, 1, 292, 1, 1, 1, 2, 1, 3, 1, 14, 2, 1, 1, 2, 2, 2, 2, 1, 84, 2, 1, 1, 15, 3, 13, 1, 4,
  2, 6, 6, 99, 1, 2, 2, 6, 3, 5, 1, 1, 6, 8, 1, 7, 1, 2, 3, 7, 1, 2, 1, 1, 12, 1, 1, 1, 3, 1, 1, 8, 1, 1, 2
Continued fraction for π. See Fig M3097. Ref LE59. MFM 67 312 63. MOC 25 403 71.
[1,1; A1203, N1054]

M2647 3, 7, 15, 24, 36, 52, 70, 93, 121, 154
Postage stamp problem. Ref CJN 12 379 69. AMM 87 208 80. [1,1; A1213, N1340]

M2648 1, 3, 7, 15, 26, 51, 99, 191, 367, 708, 1365, 2631, 5071, 9775, 18842, 36319,
  70007, 134943, 260111, 501380, 966441, 1862875, 3590807, 6921503, 13341626
A Fielder sequence. Ref FQ 6(3) 70 68. [1,2; A1648, N1055]

M2649 1, 3, 7, 15, 26, 57, 106, 207, 403, 788, 1530, 2985, 5812, 11322, 22052, 42959,
  83675, 162993, 317491, 618440, 1204651, 2346534, 4570791, 8903409, 17342876
A Fielder sequence. Ref FQ 6(3) 70 68. [1,2; A1649, N1056]

M2650 3, 7, 15, 27, 41, 62, 85, 115, 150, 186, 229, 274, 323, 380, 443, 509, 577, 653, 733,
  818, 912, 1010, 1114, 1222, 1331, 1448, 1572, 1704, 1845, 1994, 2138, 2289, 2445
A number-theoretic function. Ref ACA 6 372 61. [2,1; A1276, N1057]

M2651 1, 3, 7, 15, 29, 469, 29531, 1303, 16103, 190553, 128977, 9061, 30946717,
  39646461, 58433327, 344499373, 784809203, 169704792667
Numerators of coefficients for numerical differentiation. Cf. M1110. Ref PHM 33 11 42.
BAMS 48 922 42. [3,2; A2545, N1058]

M2652 1, 3, 7, 15, 31, 59, 110, 198, 347, 592, 997, 1641, 2666, 4266, 6741, 10525, 16268,
  24882, 37717, 56683, 84504, 125031, 183716, 268125, 388873, 560647, 803723
n-step spirals on hexagonal lattice. Ref JPA 20 492 87. [1,2; A6778]

M2653 1, 3, 7, 15, 31, 60, 113, 207, 373, 663, 1167, 2038, 3537, 6107, 10499, 17983,
  30703, 52272, 88769, 150407, 254321, 429223, 723167, 1216490, 2043361, 3427635
Patterns in a dual ring. Ref MMAG 66 170 93. [1,2; A7574]

M2654 1, 3, 7, 15, 31, 62, 122, 235, 448, 842, 1572, 2904, 5341, 9743, 17718, 32009,
  57701, 103445, 185165, 329904, 587136, 1040674, 1843300, 3253020, 5738329
Site percolation series for hexagonal lattice. Ref JPA 21 3822 88. [0,2; A6739]

M2655   1, 3, 7, 15, 31, 63, 127, 255, 511, 1023, 2047, 4095, 8191, 16383, 32767, 65535,
131071, 262143, 524287, 1048575, 2097151, 4194303, 8388607, 16777215, 33554431
$2^n - 1$. See Fig M4981. Ref BA9. [1,2; A0225, N1059]

M2656   1, 1, 1, 3, 7, 15, 35, 87, 217, 547, 1417, 3735
Maximally stable towers of 2 x 2 LEGO blocks. Ref JRM 12 27 79. [1,4; A7576]

M2657   3, 7, 16, 31, 57, 97, 162, 257, 401, 608, 907, 1325, 1914, 2719, 3824, 5313, 7316,
9973, 13495, 18105, 24132, 31938, 42021, 54948, 71484, 92492, 119120, 152686
Bipartite partitions. Ref PCPS 49 72 53. ChGu56 1. [0,1; A0412, N1060]

M2658   1, 3, 7, 16, 33, 71, 141, 284, 552, 1067, 2020, 3803, 7043, 12957, 23566, 42536,
76068, 135093, 238001, 416591
Solid partitions of n, distinct along rows. Ref AB71 404. [1,2; A2936, N1061]

M2659   1, 3, 7, 16, 46, 138
Symmetric anti-Hadamard matrices of order n. Ref LAA 62 117 84. [1,2; A5312]

M2660   0, 1, 0, 3, 7, 16, 49, 104, 322, 683, 2114, 4485, 13881, 29450, 91147, 193378,
598500, 1269781, 3929940, 8337783, 25805227, 54748516, 169445269, 359496044
A ternary continued fraction. Ref TOH 37 441 33. [0,4; A0963, N1062]

M2661   1, 3, 7, 17, 31, 42, 54, 122, 143, 167, 211, 258, 414, 469, 525, 582, 640, 699, 759,
820, 882, 945, 1009, 1075, 1458, 1539, 1621
From a partition of the integers. Ref LNM 751 275 79. [1,2; A6628]

M2662   1, 3, 7, 17, 39, 85, 183, 389, 815, 1693, 3495, 7173, 14655, 29837, 60567, 122645,
247855, 500061, 1007495, 2027493, 4076191, 8188333, 16437623, 32978613, 66132495
Expansion of $1/(1-2x)(1-x-2x^3)$. Ref DT76. [0,2; A3478]

M2663   1, 3, 7, 17, 39, 96, 232, 583, 1474, 3797, 9864, 25947, 68738, 183612, 493471,
1334143, 3624800, 9893860, 27113492, 74577187, 205806860, 569678759, 1581243203
Random rooted forests. Ref JCT B27 117 79. [1,2; A5197]

M2664   1, 1, 3, 7, 17, 40, 102, 249, 631, 1594, 4074, 10443, 26981, 69923, 182158,
476141, 1249237, 3287448, 8677074, 22962118, 60915508, 161962845, 431536102
Tertiary alcohols with n carbon atoms. Ref JACS 53 3042 31; 54 2919 32. [4,3; A0600,
N1063]

M2665   1, 1, 3, 7, 17, 41, 99, 239, 577, 1393, 3363, 8119, 19601, 47321, 114243, 275807,
665857, 1607521, 3880899, 9369319, 22619537, 54608393, 131836323, 318281039
$a(n) = 2a(n-1) + a(n-2)$. Ref MQET 1 9 16. AMM 56 445 49. Robe92 224. [0,3;
A1333, N1064]

M2666   1, 1, 3, 7, 17, 42, 104, 259, 648, 1627, 4098, 10350, 26202, 66471, 168939,
430071, 1096451, 2799072, 7154189, 18305485, 46885179, 120195301, 308393558
Binary vectors with restricted repetitions. Ref PO74. [0,3; A3440]

M2667   1, 1, 3, 7, 18, 42, 109
Ammonium compounds with n carbon atoms. Ref JACS 56 157 34. [4,3; A0633, N1065]

M2668   1, 1, 3, 7, 18, 44, 117, 299
Connected graphs with one cycle. Ref R1 150. [3,3; A0226, N1066]

M2669   3, 7, 19, 25, 51, 109, 153, 213, 289, 1121, 1121, 1121, 3997, 7457, 12017, 12719,
   20299, 24503, 24503, 25817, 25817, 128755, 128755, 219207, 456929, 456929, 761619
Class numbers of quadratic fields. Ref MOC 24 437 70. [3,1; A1985, N1068]

M2670   3, 7, 19, 31, 41, 2687
$(14^n - 1)/13$ is prime. Ref MOC 61 928 93. [1,1; A6032]

M2671   1, 3, 7, 19, 47, 130, 343, 951, 2615, 7318, 20491, 57903, 163898, 466199,
   1328993, 3799624, 10884049, 31241170, 89814958, 258604642
Mappings from n points to themselves. See Fig M2671. Ref FI50 41.401. MAN 143 110
61. prs. JCT 12 18 72. [1,2; A1372, N1069]

‖‖

Figure M2671.   MAPPINGS ON AN n-SET.

Also called **functional digraphs**.

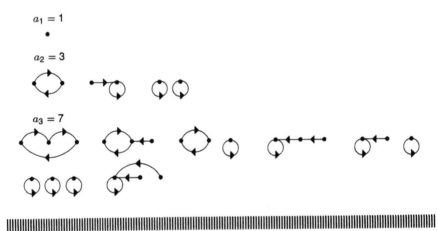

$a_1 = 1$

$a_2 = 3$

$a_3 = 7$

‖‖

M2672   1, 3, 7, 19, 49, 127, 321, 813, 2041, 5117, 12763, 31791, 78917, 195677, 484019
Expansion of layer susceptibility series for square lattice. Ref JPA 12 2451 79. [0,2;
A7288]

M2673   1, 1, 3, 7, 19, 51, 141, 393, 1107, 3139, 8953, 25653, 73789, 212941, 616227,
   1787607, 5196627, 15134931, 44152809, 128996853, 377379369, 1105350729
Expansion of $(1+x+x^2)^n$. Ref EUL (1) 15 59 27. FQ 7 341 69. Henr74 1 42. [0,3; A2426,
N1070]

$$\text{G.f.:}\quad 1 \,/\, (1 + x)^{1/2}(1 - 3x)^{1/2}.$$

M2674 3, 7, 19, 53, 147, 401, 1123, 3137, 8793, 24599, 69287, 194967, 550361, 1552645, 4393021, 12425121, 35213027, 99771855, 283162701
Expansion of critical exponent for walks on tetrahedral lattice. Ref JPA 14 443 81. [1,1; A7180]

M2675 1, 3, 7, 19, 53, 149, 419, 1191, 3403, 9755, 28077, 81097
Stable towers of 2 x 2 LEGO blocks. Ref JRM 12 27 79. [1,2; A7575]

M2676 1, 3, 7, 19, 57, 176
Triangulations. Ref WB79 337. [0,2; A5506]

M2677 1, 3, 7, 20, 52, 157
Critical connected topologies with n points. Ref JCT B15 193 73. [2,2; A3097]

M2678 1, 3, 7, 20, 55, 148, 403, 1097, 2981, 8103, 22026, 59874, 162755, 442413, 1202604, 3269017, 8886111, 24154953, 65659969, 178482301, 485165195, 1318815734
Nearest integer to e^n. Ref MNAS 14(5) 14 25. FW39. FMR 1 230. [0,2; A0227, N1071]

M2679 1, 1, 1, 3, 7, 20, 131, 815, 5142, 36800, 272093, 2077909, 16176607, 127997683, 1025727646, 8310377815, 68217725764, 560527576100, 4556993996246
Simplicial 4-clusters with n cells. Ref DM 40 216 82. [1,4; A7174]

M2680 0, 1, 1, 1, 3, 7, 21, 61, 187, 577, 1825, 5831, 18883, 61699, 203429, 675545, 2258291, 7592249, 25656477, 87096661, 296891287, 1015797379, 3487272317
Asymmetric permutation rooted trees with n nodes. Ref JSC 14 237 92. [0,5; A5355]

M2681 1, 1, 3, 7, 22, 66, 217, 715, 2438, 8398, 29414, 104006, 371516, 1337220, 4847637, 17678835, 64823110, 238819350, 883634026, 3282060210, 12233141908
$(C_n + C_{(n-1)/2})/2$. Ref QJMO 38 163 87. [1,3; A7595]

M2682 1, 1, 3, 7, 22, 82, 333, 1448, 6572, 30490, 143552, 683101
Hexagonal polyominoes with n cells. See Fig M1845. Ref CJM 19 857 67. RE72 97. [1,3; A0228, N1072]

M2683 3, 7, 23, 47, 1103, 2207, 2435423, 4870847, 11862575248703, 23725150497407, 281441383062305809756861823, 562882766124611619513723647
An infinite coprime sequence. Ref MAG 48 418 64. jos. [0,1; A2715, N1073]

M2684 3, 7, 23, 71, 311, 479, 1559, 5711, 10559, 18191, 31391, 422231, 701399, 366791, 3818929, 9257329
p is least nonresidue for $a(p)$. Ref PCPS 61 672 65. MNR 29 114 65. [2,1; A0229, N1074]

M2685 3, 7, 23, 89, 139, 199, 113, 1831, 523, 887, 1129, 1669, 2477, 2971, 4297, 5591, 1327, 9551, 30593, 19333, 16141, 15683, 81463, 28229, 31907, 19609, 35617, 82073
Lower prime of gap of $2n$ between primes. Cf. M3812. Ref MOC 52 222 89. [1,1; A0230]

M2686 3, 7, 23, 287, 291, 795
$13.2^n - 1$ is prime. Ref MOC 22 421 68. Rie85 384. [1,1; A1773, N1076]

M2687 1, 3, 7, 24, 74, 259, 891, 3176, 11326, 40942, 148646, 543515
Dissections of a polygon. Ref AEQ 18 388 78. [4,2; A3449]

M2688 1, 1, 1, 3, 7, 24, 93, 434, 2110, 10957, 58713, 321576, 1792133, 10131027,
57949430, 334970205, 1953890318, 11489753730, 68054102361, 405715557048
Simplicial 3-clusters with n cells. Ref DM 40 216 82. [1,4; A7172]

M2689 1, 3, 7, 24, 117, 663, 4824, 40367, 381554, 4001849, 46043780, 576018785,
7783281188, 112953364381, 1752128923245, 28930230194371, 506596534953769
2-diregular connected digraphs with n nodes. Ref JGT 11 477 87. [2,2; A5642]

M2690 1, 1, 3, 7, 25, 90, 350, 1701, 7770, 42525, 246730, 1379400, 9321312, 63436373,
420693273, 3281882604, 25708104786, 197462483400, 1709751003480
Largest Stirling numbers of second kind. Ref AS1 835. PSPM 19 172 71. [1,3; A2870,
N1077]

M2691 1, 3, 7, 27, 106, 681, 5972, 88963, 2349727, 117165818, 11073706216,
1968717966417, 654366802299848, 406048824479878828, 470960717141418629512
$\Sigma\ a(n)\ x^n\ /\ n\ =\ \log\ (1\ +\ \Sigma\ g(n)\ x^n)$, where $g(n)$ is # graphs on n nodes (M1253). Ref
HP73 91. [1,2; A3083]

M2692 1, 1, 3, 7, 31, 100, 331, 431, 2486, 2917, 5403, 24529, 250693, 4286310, 4537003,
67804352, 72341355, 140145707, 427797039119, 427937184826, 855734223945
Convergents to cube root of 5. Ref AMP 46 107 1866. L1 67. hpr. [1,3; A2357, N1078]

M2693 3, 7, 31, 127, 89, 8191, 131071, 524287, 178481, 2089, 2147483647, 616318177,
164511353, 2099863, 13264529, 20394401, 3203431780337, 2305843009213693951
Largest factor of Mersenne numbers. Ref CUNN. [1,1; A3260]

M2694 3, 7, 31, 127, 2047, 8191, 131071, 524287, 8388607, 536870911, 2147483647,
137438953471, 2199023255551, 8796093022207, 140737488355327
Mersenne numbers: $2^p - 1$, where p is prime. Ref HW1 16. [1,1; A1348, N1079]

M2695 3, 7, 31, 127, 2047, 8191, 131071, 524287, 8388607, 2147483647, 137438953471,
2199023255551, 576460752303423487, 2305843009213693951
Mersenne numbers with at most 2 prime factors. Ref CUNN. [1,1; A6515]

M2696 3, 7, 31, 127, 8191, 131071, 524287, 2147483647, 2305843009213693951,
618970019642690137449562111, 162259276829213363391578010288127
Mersenne primes (of form $2^p - 1$). Ref CUNN. [1,1; A0668, N1080]

M2697 3, 7, 31, 211, 2311, 509, 277, 27953, 703763, 34231, 200560490131, 676421,
11072701, 78339888213593, 13808181181, 18564761860301, 19026377261
Largest factor of 2.3.5.7... + 1. Ref SMA 14 26 48. Krai52 2. MOC 26 568 72. MMAG 48
93 75. [1,1; A2585, N1081]

M2698   3, 7, 31, 211, 2311, 30031, 510511, 9699691, 223092871, 6469693231,
200560490131, 7420738134811, 304250263527211, 13082761331670031
Euclid numbers: product of consecutive primes plus 1. Ref WAG91 35. [1,1; A6862]

M2699   1, 3, 7, 33, 67, 223, 663, 912, 1383, 3777
$n.4^n + 1$ is prime. Ref JRM 21 191 89. [1,2; A7646]

M2700   1, 1, 3, 7, 35, 155, 1395, 11811, 200787, 3309747, 109221651, 3548836819,
230674393235, 14877590196755, 1919209135381395, 246614610741341843
Gaussian binomial coefficient $[n, n/2]$ for $q = 2$. Ref TU69 76. GJ83 99. ARS A17 328 84.
[0,3; A6099]

M2701   1, 1, 3, 7, 41, 299, 6128
$n \times n$ binary matrices. Ref PGEC 22 1050 73. [0,3; A6383]

M2702   3, 7, 46, 4336, 134281216, 288230380379570176
Boolean functions of n variables. Ref JSIAM 11 827 63. HA65 143. [1,1; A0231, N1083]

M2703   3, 7, 47, 73, 79, 113, 151, 167, 239, 241, 353, 367, 457, 1367, 3041
$2^n - 2^{(n+1)/2} + 1$ is prime. Cf. M3098. Ref CUNN xlviii. [1,1; A7670]

M2704   1, 1, 3, 7, 47, 207, 2249, 14501, 216273, 1830449, 34662523, 362983263,
8330310559, 103938238111, 2801976629841, 40574514114061, 1256354802202337
Magic squares of order n. Ref C1 125. [0,3; A5650]

M2705   3, 7, 47, 2207, 4870847, 23725150497407, 562882766124611619513723647,
316837008400094222150776738483768236006420971486980607
$a(n) = a(n-1)^2 - 2$. Ref D1 1 397. HW1 223. FQ 11 432 73. TCS 65 219 89. [0,1; A1566,
N1084]

M2706   1, 1, 3, 7, 83, 109958
Self-dual Boolean functions of n variables. Ref PGEC 11 284 62. MU71 38. PJM 110 220
84. [1,3; A1531, N1085]

M2707   3, 7, 127, 170141183460469231731687303715884105727
$a(n+1) = 2^{a(n)} - 1$. Ref BPNR 81. [0,1; A5844]

M2708   3, 7, 137, 283, 883, 991, 1021, 1193, 3671
$(13^n - 1)/12$ is prime. Ref MOC 61 928 93. [1,1; A6031]

SEQUENCES BEGINNING . . ., 3, 8, . . . TO . . ., 3, 12, . . .

M2709   1, 1, 1, 0, 3, 8, 3, 56, 217, 64, 2951, 12672, 5973, 309376, 1237173, 2917888,
52635599, 163782656, 1126610929, 12716052480, 20058390573, 495644917760
Expansion of $e^{\sin x}$. Ref AMM 41 418 34. [0,5; A2017, N1086]

M2710 3, 8, 6, 20, 24, 16, 12, 24, 60, 10, 24, 28, 48, 40, 24, 36, 24, 18, 60, 16, 30, 48, 24,
   100, 84, 72, 48, 14, 120, 30, 48, 40, 36, 80, 24, 76, 18, 56, 60, 40, 48, 88, 30, 120, 48, 32
Pisano periods: period of Fibonacci numbers mod *n*. Ref HM68. MOC 23 459 69. ACA 16
109 69. Robe92 162. [2,1; A1175, N1087]

M2711 1, 1, 1, 1, 1, 1, 1, 3, 8, 9, 37, 121, 211, 695, 4889, 41241, 76301, 853513, 3882809,
   11957417, 100146415, 838216959, 13379363737, 411322824001, 3547404378125
First factor of prime cyclotomic fields. Ref MOC 24 217 70. [3,8; A0927, N1088]

M2712 3, 8, 10, 14, 15, 21, 24, 28, 35, 36, 45, 48, 52, 55, 63, 66, 78, 80, 91, 99, 105, 120,
   133, 136, 143, 153, 168, 171, 190, 195, 210, 224, 231, 248, 253, 255, 276, 288, 300, 323
Dimensions of simple Lie algebras. Ref JA62 146. BAMS 78 637 72. [1,1; A3038]

M2713 3, 8, 11, 14, 19, 24, 29, 32, 35, 40, 43, 46, 51, 54, 57, 62, 67, 72, 75, 78, 83, 88, 93,
   96, 99, 104, 109, 114, 117, 120, 125, 128, 131, 136, 139, 142, 147, 152, 157, 160
A self-generating sequence. Ref FQ 10 500 72. [1,1; A3157]

M2714 3, 8, 11, 16, 19, 21, 24, 29, 32, 37, 42, 45, 50, 53, 55, 58, 63, 66, 71, 74, 76, 79, 84,
   87, 92, 97, 100, 105, 108, 110, 113, 118, 121, 126, 129, 131, 134, 139, 142, 144, 147
Related to Fibonacci representations. Ref FQ 11 385 73. [1,1; A3234]

M2715 3, 8, 11, 16, 21, 24, 29, 32, 37, 42, 45, 50, 55, 58, 63, 66, 71, 76, 79, 84, 87, 92, 97,
   100, 105, 110, 113, 118, 121, 126, 131, 134, 139, 144, 147, 152, 155, 160, 165, 168, 173
From a 3-way splitting of positive integers: $[[n\tau^2]\tau]$. Cf. M3278. Ref Robe92 10. [1,1;
A3623]

M2716 3, 8, 12, 18, 26, 27, 38, 39, 54, 56, 57, 78, 80, 81, 84, 110, 114, 116, 117, 120, 158,
   162, 164, 165, 170, 171, 174, 222, 230, 234, 236, 237, 242, 243, 246, 255, 318, 326, 330
If *n* appears so do $2n+2$ and $3n+3$. [1,1; A5660]

M2717 1, 3, 8, 12, 24, 24, 48, 48, 72, 72, 120, 96, 168, 144, 192, 192, 288, 216, 360, 288,
   384, 360, 528, 384, 600, 504, 648, 576, 840, 576, 960, 768, 960, 864, 1152, 864, 1368
Moebius transform of squares. Ref EIS § 2.7. [1,2; A7434]

M2718 3, 8, 14, 14, 25, 24, 23, 22, 25, 59, 98, 97, 98, 97, 174, 176, 176, 176, 176, 291,
   290, 289, 740, 874, 873, 872, 873, 872, 871, 870, 869, 868, 867, 866, 2180, 2179, 2178
Related to gaps between primes. Ref MOC 13 122 59. SI64 35. [1,1; A0232, N1089]

M2719 3, 8, 14, 32, 62, 87, 169, 132, 367, 389, 510, 394, 512, 512
Binomial coefficients with many divisors. Ref MSC 39 275 76. [2,1; A5735]

M2720 3, 8, 15, 24, 35, 48, 63, 80, 99, 120, 143, 168, 195, 224, 255, 288, 323, 360, 399,
   440, 483, 528, 575, 624, 675, 728, 783, 840, 899, 960, 1023, 1088
Walks on square lattice. Ref GU90. [0,1; A5563]

$$\text{G.f.:} \quad (3-x)\,/\,(1-x)^3.$$

M2721   3, 8, 15, 26, 35, 52, 69, 89, 112, 146, 172, 212, 259, 302, 354, 418, 476, 548, 633,
714, 805, 902, 1012, 1127, 1254, 1382, 1524, 1678, 1841, 2010, 2188, 2382, 2584
Postage stamp problem. Ref CJN 12 379 69. AMM 87 208 80. [1,1; A1208, N1351]

M2722   1, 3, 8, 16, 30, 46, 64, 96, 126, 158
Generalized class numbers. Ref MOC 21 689 67. [1,2; A0233, N1090]

M2723   1, 3, 8, 16, 30, 50, 80, 120, 175, 245, 336, 448, 588, 756, 960, 1200, 1485, 1815,
2200, 2640, 3146, 3718, 4368, 5096, 5915, 6825, 7840, 8960, 10200, 11560, 13056
Expansion of $(1-x)^{-3}(1-x^2)^{-2}$. Ref AMS 26 308 55. [0,2; A2624, N1091]

M2724   1, 3, 8, 16, 32, 48, 64, 64
Minimal determinant of n-dimensional norm 3 lattice. Ref SPLAG 180. [0,2; A5103]

M2725   1, 3, 8, 16, 32, 55, 94, 147, 227, 332, 480, 668, 920, 1232, 1635
Restricted partitions. Ref CAY 2 279. [0,2; A1978, N1092]

M2726   1, 3, 8, 17, 33, 58, 97, 153, 233, 342, 489, 681, 930, 1245, 1641, 2130, 2730, 3456,
4330, 5370, 6602, 8048, 9738, 11698, 13963, 16563, 19538, 22923, 26763, 31098, 35979
Expansion of $1/(1-x)^3(1-x^2)^2(1-x^3)$. Ref AMS 26 308 55. [0,2; A2625, N1093]

M2727   1, 3, 8, 17, 34, 61, 105, 170, 267, 403, 594, 851, 1197, 1648, 2235, 2981, 3927,
5104, 6565, 8351, 10529, 13152, 16303, 20049, 24492, 29715, 35841, 42972, 51255
Expansion of $1/(1-x)^3(1-x^2)^2(1-x^3)(1-x^4)$. Ref AMS 26 308 55. [0,2; A2626,
N1094]

M2728   1, 3, 8, 18, 30, 43, 67, 90, 122, 161, 202, 260, 305, 388, 416, 450, 555, 624, 730,
750, 983, 1059, 1159, 1330, 1528, 1645, 1774, 1921, 2140, 2289, 2580, 2632, 2881, 3158
Magic integers. Ref ACC A34 634 78. [1,2; A4210]

M2729   1, 1, 3, 8, 18, 36, 66
The coding-theoretic function $A(n,4,5)$. See Fig M0240. Ref PGIT 36 1335 90. [5,3;
A4035]

M2730   1, 3, 8, 18, 37, 72, 136, 251, 445, 770
Partitions into non-integral powers. Ref PCPS 47 215 51. [1,2; A0234, N1095]

M2731   1, 3, 8, 18, 38, 74, 139, 249, 434, 734, 1215, 1967, 3132, 4902, 7567, 11523,
17345, 25815, 38045, 55535, 80377, 115379, 164389, 232539, 326774, 456286, 633373
Partitions of n into parts of 3 kinds. Ref RS4 122. [0,2; A0713, N1096]

M2732   1, 3, 8, 18, 38, 76, 147, 277, 509, 924, 1648, 2912, 5088, 8823, 15170, 25935,
44042, 74427, 125112, 209411, 348960, 579326, 958077, 1579098, 2593903, 4247768
n-node trees of height 3. Ref IBMJ 4 475 60. KU64. [4,2; A0235, N1097]

M2733 1, 3, 8, 18, 38, 76, 147, 277, 512, 932, 1676, 2984, 5269, 9239, 16104, 27926, 48210, 82900, 142055, 242665, 413376, 702408, 1190808, 2014608, 3401833, 5734251
$a(n) = a(n-1) + a(n-2) + F(n) - 1$. Ref BIT 13 93 73. [3,2; A6478]

M2734 0, 0, 0, 0, 0, 1, 3, 8, 19, 40
Unexplained difference between two partition g.f.s. Ref PCPS 63 1100 67. [1,7; A7326]

M2735 1, 3, 8, 19, 41, 81, 153
$4 \times n$ binary matrices. Ref PGEC 22 1050 73. [0,2; A6380]

M2736 1, 3, 8, 19, 42, 88, 176, 339, 633, 1150, 2040, 3544, 6042, 10128, 16720, 27219, 43746, 69483, 109160, 169758, 261504, 399272, 604560, 908248, 1354427, 2005710
Coefficients of elliptic function $\pi / 2K$. Ref QJMA 21 66 1885. [1,2; A2318, N1098]

M2737 3, 8, 20, 44, 80, 343, 399
Two consecutive residues. Ref MOC 24 738 70. [2,1; A0236, N1099]

M2738 1, 3, 8, 20, 47, 106, 230, 479, 973, 1924, 3712, 7021, 13034, 23780, 42732, 75703, 132360, 228664, 390611, 660296, 1105321, 1833358, 3014694, 4917036, 7958127
n-step spirals on hexagonal lattice. Ref JPA 20 492 87. [1,2; A6776]

M2739 1, 3, 8, 20, 48, 112, 256, 576, 1280, 2816, 6144, 13312, 28672, 61440, 131072, 278528, 589824, 1245184, 2621440, 5505024, 11534336, 24117248, 50331648
$(n+2).2^{n-1}$. Ref RSE 62 190 46. AS1 795. [0,2; A1792, N1100]

M2740 3, 8, 21, 54, 141, 372, 995, 2697, 7397, 20502, 57347
From sequence of numbers with abundancy n. Ref MMAG 59 87 86. [2,1; A5580]

M2741 1, 3, 8, 21, 55, 144, 377, 987, 2584, 6765, 17711, 46368, 121393, 317811, 832040, 2178309, 5702887, 14930352, 39088169, 102334155, 267914296, 701408733
Bisection of Fibonacci sequence: $a(n) = 3a(n-1) - a(n-2)$. Cf. M0692. Ref IDM 22 23 15. PLMS 21 729 70. FQ 9 283 71. [0,2; A1906, N1101]

M2742 1, 3, 8, 21, 56, 154, 434, 1252, 3675, 10954, 33044, 100676, 309569, 957424, 2987846
Percolation series for directed square lattice. Ref JPA 16 3146 83; 25 6609 92. [2,2; A6835]

M2743 0, 1, 3, 8, 22, 58, 158, 425, 1161, 3175, 8751, 24192, 67239
Total height of trees with n nodes. Ref IBMJ 4 475 60. [1,3; A1853, N1102]

M2744 1, 1, 3, 8, 22, 58, 160, 434, 1204, 3341, 9363, 26308, 74376, 210823, 599832, 1710803, 4891876, 14015505, 40231632, 115669419, 333052242, 960219974
Endpoints in planted trees with n nodes. Ref DM 12 364 75. [1,3; A3227]

M2745 1, 3, 8, 22, 65, 209, 732, 2780, 11377, 49863, 232768, 1151914, 6018785, 33087205, 190780212, 1150653920, 7241710929, 47454745803, 323154696184
$\Sigma(n-k+1)^k$, $k = 1 . . 6n$. Ref hwg. [1,2; A3101]

M2746 1, 1, 1, 3, 8, 23, 68, 215, 680, 2226, 7327
Triangulations of the disk. Ref PLMS 14 765 64. [0,4; A2712, N1103]

M2747 1, 3, 8, 23, 69, 208, 636, 1963, 6099, 19059, 59836, 188576, 596252, 1890548,
6008908, 19139155, 61074583, 195217253, 624913284, 2003090071, 6428430129
Paraffins with *n* carbon atoms. Ref BA76 44. [1,2; A5960]

M2748 1, 3, 8, 24, 75, 243, 808, 2742, 9458, 33062, 116868, 417022, 1500159, 5434563,
19808976, 72596742, 267343374, 988779258, 3671302176, 13679542632
A simple recurrence. Ref IFC 16 351 70. [1,2; A0958, N1104]

M2749 1, 3, 8, 24, 89, 415, 2372, 16072, 125673, 1112083, 10976184, 119481296,
1421542641, 18348340127, 255323504932, 3809950977008, 60683990530225
Logarithmic numbers. Ref TMS 31 78 63. CACM 13 726 70. [1,2; A2104, N1105]

$$\text{E.g.f.:} \quad -e^x \ln(1-x).$$

M2750 1, 1, 3, 8, 25, 72, 245, 772, 2692, 8925, 32065, 109890, 400023, 1402723,
5165327, 18484746, 68635477, 248339122, 930138521, 3406231198
Witt vector *2!/2!. Ref SLC 16 107 88. [1,3; A6177]

M2751 1, 1, 3, 8, 25, 77, 258, 871, 3049, 10834, 39207, 143609, 532193, 1990163,
7503471, 28486071, 108809503, 417862340, 1612440612, 6248778642, 24309992576
Shifts left when Euler transform applied twice. Ref BeSl94. EIS § 2.7. [1,3; A7563]

M2752 3, 8, 25, 89, 357, 1602, 7959, 43127
From descending subsequences of permutations. Ref JCT A53 99 90. [3,1; A6219]

M2753 1, 3, 8, 25, 108, 735, 4608, 40824, 362000
Permutations of length *n* with spread 1. Ref JAuMS A21 489 76. [2,2; A4205]

M2754 1, 1, 3, 8, 26, 84, 297, 1066
Mixed Husimi trees with *n* labeled nodes. Ref PNAS 42 535 56. [1,3; A0237, N1107]

M2755 1, 1, 3, 8, 26, 94, 435, 2564, 19983, 205729
Quasi-orders with *n* elements. Ref ErSt89. [0,3; A6870]

M2756 1, 1, 3, 8, 27, 91, 350, 1376, 5743, 24635, 108968, 492180, 2266502, 10598452,
50235931, 240872654, 1166732814, 5682001435, 48068787314, 139354922608
Oriented trees with *n* nodes. Ref R1 138. DM 88 97 91. [1,3; A0238, N1108]

M2757 1, 3, 8, 27, 131, 711, 5055, 41607, 389759, 4065605, 46612528, 581713045,
7846380548, 113718755478, 1762208816647, 29073392136390, 508777045979418
2-diregular digraphs with *n* nodes. Ref JGT 11 477 87. [2,2; A5641]

M2758 1, 3, 8, 28, 143, 933, 7150, 62310, 607445, 6545935, 77232740, 989893248,
13692587323, 203271723033, 3223180454138
Permutations of length *n* by rises. Ref DKB 264. [2,2; A0239, N1109]

M2759 1, 1, 3, 8, 31, 147, 853, 5824, 45741, 405845, 4012711, 43733976, 520795003, 6726601063, 93651619881, 1398047697152, 22275111534553, 377278848390249
$a(n) = n.a(n-1) - a(n-2) + 1 + (-1)^n$. [0,3; A3470]

M2760 1, 3, 8, 33, 164, 985, 6894, 55153, 496376, 4963761, 54601370, 655216441, 8517813732, 119249392249, 1788740883734, 28619854139745, 486537520375664
$a(n) = n.a(n-1) + (-1)^n$: nearest integer to $n!(1 + 1/e)$. [1,2; A1120, N1110]

M2761 1, 1, 3, 8, 36, 110, 666, 3250, 23436, 125198, 1037520, 7241272, 66360960, 503851928, 5080370400
Bishops on an $n \times n$ board. Ref LNM 560 212 76. [2,3; A5635]

M2762 1, 1, 1, 3, 8, 40, 211, 1406, 9754, 71591, 537699, 4131943, 32271490, 255690412, 2050376883, 16616721067, 135920429975, 1120999363012, 9313779465810
Simplicial 4-clusters with n cells. Ref DM 40 216 82. [1,4; A7175]

M2763 1, 0, 3, 8, 45, 264, 1855, 14832, 133497, 1334960, 14684571, 176214840, 2290792933, 32071101048, 481066515735, 7697064251744, 130850092279665
Expansion of $e^{-x}(1+x^3) / (1-x)(1-x^2)$. Ref R1 65. [1,3; A0240, N1111]

M2764 1, 0, 1, 3, 8, 48, 383, 6020
Hamiltonian graphs with n nodes. Ref CN 8 266 73. [1,4; A3216]

M2765 3, 8, 49, 3963
Switching networks. Ref JFI 276 324 63. [1,1; A0862, N1112]

M2766 1, 1, 3, 8, 50, 214, 2086, 11976, 162816, 1143576
From a Fibonacci-like differential equation. Ref FQ 27 309 89. [0,3; A5444]

M2767 3, 8, 178, 129054, 430903911398
Essentially n-ary operations in a certain 3-element algebra. Ref Berm83. [0,1; A7159]

M2768 1, 3, 9, 12, 16, 28, 49, 77, 121, 198, 324, 522, 841, 1363, 2209, 3572, 5776, 9348, 15129, 24477, 39601, 64078, 103684, 167762, 271441, 439203, 710649, 1149852
Restricted circular combinations. Ref FQ 16 115 78. [0,2; A6499]

M2769 3, 9, 14, 19, 24, 30, 35, 40, 45, 51, 56, 61, 66, 71, 77, 82, 87, 92, 98, 103, 108, 113, 119, 124, 129, 134, 140, 145, 150, 155, 161, 166, 171, 176, 181, 187, 192, 197, 202, 208
Wythoff game. Ref CMB 2 189 59. [0,1; A1968, N1113]

M2770 1, 3, 9, 15, 30, 45, 67, 99, 135, 175, 231, 306, 354, 465
Generalized divisor function. Ref PLMS 19 111 19. [3,2; A2127, N1114]

M2771 1, 3, 9, 17, 31, 53, 85, 133, 197, 293, 417, 593, 849, 1193, 1661, 2291, 3139, 4299
Number of elements in $Z[\sqrt{-2}]$ whose 'smallest algorithm' is $\leq n$. Ref JALG 19 290 71. hwl. [0,2; A6459]

M2772   0, 0, 0, 0, 1, 3, 9, 18, 36, 60, 100, 150, 225, 315, 441, 588
Crossing number of complete graph with n nodes. Dubious for $n \geq 11$. Ref GU60. AMM 80 53 73. [1,6; A0241, N1115]

M2773   3, 9, 19, 21, 55, 115, 193, 323, 611, 1081, 1571, 10771, 13067, 16321, 44881, 57887, 93167, 189947
From a Goldbach conjecture. Ref BIT 6 49 66. [1,1; A2091, N1116]

M2774   1, 3, 9, 19, 38, 66, 110, 170, 255, 365
Paraffins. Ref BER 30 1919 1897. [1,2; A5994]

M2775   1, 3, 9, 21, 9, 297, 2421, 12933, 52407, 145293, 35091, 2954097, 25228971, 142080669, 602217261, 1724917221, 283305033, 38852066421, 337425235479
$(n+1)^2 a(n+1) = (9n^2 + 9n + 3) a(n) - 27 n^2 a(n-1)$. [0,2; A6077]

M2776   0, 3, 9, 21, 39, 66, 102, 150, 210, 285, 375, 483, 609, 756, 924, 1116, 1332, 1575, 1845, 2145, 2475, 2838, 3234, 3666, 4134, 4641, 5187, 5775, 6405, 7080, 7800, 8568
$[n(n+2)(2n-1)/8]$. Ref JRM 7 151 75. [0,2; A7518]

M2777   1, 3, 9, 21, 47, 95, 186, 344, 620, 1078, 1835, 3045, 4967, 7947, 12534, 19470, 29879, 45285, 67924, 100820, 148301, 216199, 312690, 448738, 639464, 905024
Partitions of n into parts of 3 kinds. Ref RS4 122. [0,2; A0714, N1117]

M2778   3, 9, 21, 48, 105, 219, 459, 936
Percolation series for directed hexagonal lattice. Ref SSP 10 921 77. [1,1; A6813]

M2779   1, 3, 9, 21, 51, 117, 271, 607, 1363, 3013, 6643, 14491, 31495, 67965, 146115
Board of directors problem (identical to following sequence). Ref JRM 9 240 77. [1,2; A7517]

M2780   1, 3, 9, 21, 51, 117, 271, 607, 1363, 3013, 6643, 14491, 31495, 67965, 146115
Weighted voting procedures. Ref LNM 686 70 78. NA79 100. MSH 84 48 83. [1,2; A5254]

M2781   1, 3, 9, 21, 57, 123, 279, 549, 1209, 2127, 4689
Words of length n in a certain language. Ref DM 40 231 82. [0,2; A7056]

M2782   1, 1, 3, 9, 21, 81, 351, 1233, 5769, 31041, 142011, 776601, 4874013, 27027729, 168369111, 1191911841, 7678566801, 53474964993, 418199988339, 3044269834281
Degree n permutations of order dividing 3. Ref CJM 7 159 55. [1,3; A1470, N1118]

$$a(n) = a(n-1) + (n^2 - 3n + 2) a(n-3).$$

M2783   1, 3, 9, 21, 363, 2161, 4839
$n.10^n + 1$ is prime. Ref JRM 21 191 89. [1,2; A7647]

M2784   1, 3, 9, 22, 42, 84, 140, 231, 351, 551, 783
Generalized divisor function. Ref PLMS 19 111 19. [6,2; A2128, N1119]

M2785 1, 3, 9, 22, 48, 99, 194, 363, 657, 1155, 1977, 3312, 5443, 8787, 13968, 21894, 33873, 51795, 78345, 117312, 174033, 255945, 373353, 540486, 776848, 1109040
Coefficients of an elliptic function. Ref CAY 9 128. [0,2; A1937, N1120]

$$\text{G.f.:} \quad \Pi \, (1 - x^k)^{-c(k)} \, , \; c(k) = 3, 3, 3, 0, 3, 3, 3, 0, \dots$$

M2786 1, 3, 9, 22, 50, 104, 208, 394, 724, 1286, 2229, 3769, 6253, 10176, 16303, 25723, 40055, 61588, 93647, 140875, 209889, 309846, 453565, 658627, 949310, 1358589
Partitions of n into parts of 3 kinds. Ref RS4 122. [0,2; A0715, N1121]

M2787 1, 3, 9, 22, 51, 107, 217, 416, 775, 1393, 2446, 4185, 7028, 11569, 18749, 29908, 47083, 73157, 112396, 170783, 256972, 383003, 565961, 829410, 1206282, 1741592
Partitions of n into parts of 3 kinds. Ref RS4 122. [0,2; A0711, N1122]

M2788 1, 3, 9, 22, 51, 108, 221, 429, 810, 1479, 2640, 4599, 7868, 13209, 21843, 35581, 57222, 90882, 142769, 221910, 341649, 521196, 788460, 1183221, 1762462, 2606604
Partitions of n into parts of 3 kinds. Ref RS4 122. [0,2; A0716, N1123]

M2789 1, 3, 9, 22, 51, 111, 233, 474, 942, 1836, 3522, 6666, 12473, 23109, 42447, 77378, 140109, 252177, 451441, 804228, 1426380, 2519640, 4434420, 7777860, 13599505
Convolved Fibonacci numbers. Ref RCI 101. FQ 15 118 77. [0,2; A1628, N1124]

$$\text{G.f.:} \quad (1 - x - x^2)^{-3}.$$

M2790 3, 9, 23, 51, 103, 196, 348
3-covers of an n-set. Ref DM 81 151 90. [1,1; A5783]

M2791 1, 3, 9, 24, 61, 145, 333, 732, 1565, 3247, 6583, 13047, 25379, 48477, 91159, 168883, 308736, 557335, 994638, 1755909, 3068960, 5313318, 9118049
Terms in an n-th derivative. Ref CRP 278 250 74. C1 175. [1,2; A3262]

M2792 1, 3, 9, 25, 57, 145, 337, 793, 1921, 3849, 8835, 18889, 41473, 92305, 203211, 432699, 944313, 2027529, 4077769, 8745153, 18133305, 37898113, 80713737
Multilevel sieve: at k-th step, accept k numbers, reject k, accept k, ... Ref PC 4 43-15 76. [1,2; A5209]

M2793 1, 3, 9, 25, 59, 131, 277, 573, 1167, 2359, 4745, 9521, 19075, 38187, 76413, 152869, 305783, 611615, 1223281, 2446617, 4893291, 9786643, 19573349, 39146765
$[(7 . 2^{n+1} - 6n - 10)/3]$. Ref CRUX 13 331 87. [0,2; A5262]

M2794 1, 3, 9, 25, 60, 126, 238, 414, 675, 1045, 1551, 2223, 3094, 4200, 5580, 7276, 9333, 11799, 14725, 18165, 22176, 26818, 32154, 38250, 45175, 53001, 61803, 71659
$n(n+1)(n^2 - 3n + 6)/8$. Ref dsk. [1,2; A4255]

M2795 1, 3, 9, 25, 65, 161, 385, 897, 2049, 4609, 10241, 22529, 49153, 106497, 229377, 491521, 1048577, 2228225, 4718593, 9961473, 20971521, 44040193, 92274689
Cullen numbers: $n . 2^n + 1$. Ref SI64a 346. UPNT B20. [1,2; A2064, N1125]

M2796 3, 9, 25, 66, 168, 417, 1014, 2427
Percolation series for hexagonal lattice. Ref SSP 10 921 77. [1,1; A6809]

M2797 1, 3, 9, 25, 66, 168, 417, 1014, 2427, 5737, 13412, 31088, 71506, 163378, 371272,
839248, 1889019, 4235082, 9459687, 21067566, 46769977, 103574916, 228808544
Bond percolation series for hexagonal lattice. Ref JPA 21 3822 88. [0,2; A6735]

M2798 1, 3, 9, 25, 69, 186, 503, 1353, 3651, 9865, 26748, 72729, 198447, 543159,
1491402, 4107152, 11342826, 31408719, 87189987, 242603970, 676524372
Powers of rooted tree enumerator. Ref R1 150. [1,2; A0242, N1126]

M2799 1, 3, 9, 25, 69, 189, 518, 1422, 3915, 10813, 29964, 83304, 232323, 649845,
1822824, 5126520, 14453451, 40843521, 115668105, 328233969, 933206967
Column of Motzkin triangle. Ref JCT A23 293 77. [2,2; A5322]

M2800 1, 3, 9, 25, 70, 194, 537, 1485, 4104, 11338, 31318, 86498, 238885, 659713,
1821843, 5031071, 13893316, 38366206, 105947374, 292570493, 807923428
Irreducible positions of size n in Montreal solitaire. Ref JCT A60 56 92. [3,2; A7046]

M2801 0, 1, 3, 9, 25, 75, 231, 763, 2619, 9495, 35695, 140151, 568503, 2390479,
10349535, 46206735, 211799311, 997313823, 4809701439, 23758664095
Degree n permutations of order exactly 2. Equals M1221 − 1. Ref CJM 7 159 55. [1,3;
A1189, N1127]

M2802 1, 1, 3, 9, 25, 133, 631, 3857, 29505
Starters in cyclic group of order $2n + 1$. Ref DM 79 276 89. [1,3; A6204]

M2803 1, 3, 9, 26, 75, 214, 612, 1747, 4995
Partially labeled trees with n nodes. Ref R1 138. [2,2; A0243, N1128]

M2804 1, 3, 9, 26, 75, 216, 623, 1800, 5211, 15115, 43923
Directed animals of size n. Ref AAM 9 340 88. [2,2; A5774]

M2805 3, 9, 27, 78, 225, 633, 1785, 4944
Percolation series for cubic lattice. Ref SSP 10 921 77. [1,1; A6810]

M2806 1, 3, 9, 27, 81, 171, 243, 513, 729, 1539, 2187, 3249, 4617, 6561, 9747
n divides $2^n + 1$. Ref HO73 142. [1,2; A6521]

M2807 1, 3, 9, 27, 81, 243, 729, 2187, 6561, 19683, 59049, 177147, 531441, 1594323,
4782969, 14348907, 43046721, 129140163, 387420489, 1162261467, 3486784401
Powers of 3. Ref BA9. [0,2; A0244, N1129]

M2808 0, 1, 0, 0, 0, 1, 3, 9, 28, 85, 262, 827, 2651, 8626, 28507, 95393, 322938, 1104525,
3812367, 13266366, 46504495, 164098390, 582521687, 2079133141, 7457788295
Asymmetric planar trees with n nodes. Ref JSC 14 236 92. JSC 14 236 92. [0,7; A5354]

M2809 1, 3, 9, 28, 90, 297, 1001, 3432, 11934, 41990, 149226, 534888, 1931540,
7020405, 25662825, 94287120, 347993910, 1289624490, 4796857230, 17902146600
$3C(2n, n-1)/(n+2)$. Ref QAM 14 407 56. MOC 29 216 75. FQ 14 397 76. [1,2; A0245,
N1130]

M2810 1, 3, 9, 29, 98, 343, 1230, 4489, 16599, 61997, 233389, 884170, 3366951,
12876702, 49424984, 190297064, 734644291, 2842707951
Permutations by inversions. Ref NET 96. DKB 241. MMAG 61 28 88. rkg. [3,2; A1893,
N1132]

M2811 1, 3, 9, 29, 99, 351, 1275, 4707, 17577, 66197, 250953, 956385, 3660541,
14061141, 54177741, 209295261, 810375651, 3143981871, 12219117171, 47564380971
$\Sigma\, C(2k,k)$, $k = 0 .. n$. Ref FQ 15 204 77. [0,2; A6134]

M2812 1, 1, 1, 3, 9, 29, 105, 431, 1969, 9785, 52145, 296155, 1787385, 11428949,
77124569, 546987143, 4062341601, 31502219889, 254500383457, 2137863653811
Shifts 2 places left when binomial transform applied twice. Ref BeSl94. EIS § 2.7. [0,4;
A7472]

M2813 1, 1, 3, 9, 30, 103, 375, 1400, 5380, 21073, 83950, 338878, 1383576, 5702485,
23696081, 99163323, 417553252, 1767827220, 7520966100, 32135955585
Connected N-free posets with n nodes. Ref DM 75 97 89. [1,3; A7453]

M2814 3, 9, 30, 105, 378, 1386, 5148, 19305, 72930, 277134, 1058148, 4056234,
15600900, 60174900, 232676280, 901620585, 3500409330, 13612702950, 53017895700
$3C(2n-1, n)$. Ref DM 9 355 74. [1,1; A3409]

M2815 1, 3, 9, 30, 128, 675, 4231, 30969, 258689, 2428956, 25306287, 289620751,
3610490805
Number of primes $\leq n!$. Ref rwg. [2,2; A3604]

M2816 0, 1, 1, 3, 9, 32
Trivalent planar graphs with $2n$ nodes. Ref BA76 92. [1,4; A5964]

M2817 1, 1, 3, 9, 33, 139, 718, 4535
Topologies or unlabeled transitive digraphs with n nodes. See Fig M2817. Ref jaw. CN 8
180 73. [0,3; A1930, N1133]

||||||||||||||||||||||||||||||||||||||||||||||||||||||||||||||||||||||||||||||||||||||||||||||||||||||||||||||||||||||

Figure M2817. TOPOLOGIES. (See Fig. M3032.) Only 8 terms of this sequence are known.

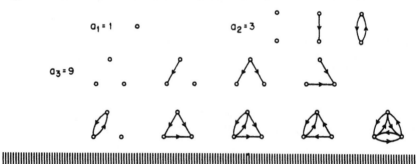

||||||||||||||||||||||||||||||||||||||||||||||||||||||||||||||||||||||||||||||||||||||||||||||||||||||||||||||||||||||

M2818   1, 3, 9, 33, 153, 873, 5913, 46233, 409113, 4037913, 43954713, 522956313, 6749977113, 93928268313, 1401602636313, 22324392524313, 378011820620313
Sum of $n!$, $n \geq 1$. [1,2; A7489]

M2819   1, 3, 9, 35, 178
Van der Waerden numbers. Ref Loth83 49. [1,2; A5346]

M2820   1, 3, 9, 35, 201, 1827
Coefficients of Bell's formula. Ref NMT 10 65 62. [2,2; A2575, N1134]

M2821   1, 3, 9, 37, 153, 951, 5473, 42729, 353937, 3455083, 30071001, 426685293, 4707929449, 59350096287, 882391484913, 15177204356401, 205119866263713
Sums of logarithmic numbers. Ref TMS 31 79 63. jos. [0,2; A2751, N1135]

M2822   1, 1, 1, 3, 9, 37, 177, 959, 6097, 41641, 325249, 2693691, 24807321, 241586893, 2558036145, 28607094455, 342232522657, 4315903789009, 57569080467073
Expansion of $e^{\tan x}$. Ref JO61 150. [0,4; A6229]

M2823   1, 3, 9, 42, 206, 1352, 10168
Regular semigroups of order n. Ref PL65. MAL 2 2 67. SGF 14 71 77. [1,2; A1427, N1136]

M2824   0, 1, 1, 3, 9, 45, 225, 1575, 11025, 99225, 893025, 9823275, 108056025, 1404728325, 18261468225, 273922023375, 4108830350625, 69850115960625
Expansion of $1 / (1-x)(1-x^2)^{1/2}$. Ref R1 87. [1,4; A0246, N1137]

M2825   1, 1, 1, 3, 9, 48, 504, 14188, 1351563
Threshold functions of n variables. Ref PGEC 19 821 70. MU71 38. [0,4; A1530, N1138]

M2826   3, 9, 54, 450, 4725, 59535, 873180, 14594580
Expansion of an integral. Ref C1 167. [2,1; A1194, N1139]

M2827   1, 3, 9, 89, 1705, 67774
Superpositions of cycles. Ref AMA 131 143 73. [3,2; A3225]

M2828   1, 3, 9, 93, 315, 3855, 13797, 182361, 9256395, 34636833, 1857283155, 26817356775, 102280151421, 1497207322929, 84973577874915, 4885260612740877
Fermat quotients: $(2^{p-1} - 1)/p$. Ref Well86 70. [0,2; A7663]

M2829   3, 10, 4, 5, 10, 2, 5, 3, 2, 3, 6, 6, 6, 3, 5, 6, 10, 5, 5, 10, 6, 6, 6, 2, 5, 8, 2, 6, 8, 4, 6, 6, 4, 5, 10, 2, 4, 7, 11, 5, 7, 9, 10, 7, 1, 6, 7, 11, 7, 10, 0, 6, 8, 9, 6, 4, 11, 7, 13, 2, 6, 4, 4
Iterations until $3n$ reaches 153 under x goes to sum of cubes of digits map. Ref Robe92 13. [1,1; A3620]

M2830   1, 3, 10, 13, 62, 75, 437, 512, 949, 6206, 13361, 73011, 597449, 1865358, 6193523, 26639450, 59472423, 383473988, 1593368375, 6756947488, 8350315863
Convergents to cube root of 3. Ref AMP 46 105 1866. L1 67. hpr. [1,2; A2354, N1140]

M2831   1, 1, 3, 10, 17, 38, 106, 253, 716, 1903, 5053, 13786, 39293, 107641, 302807,
860099, 2450684, 7038472, 20316895, 58849665, 171217429, 499926666, 1464276207
Symmetries in planted 4-trees on $n + 1$ vertices. Ref GTA91 849. [1,3; A3615]

M2832   1, 3, 10, 20, 39, 63, 100, 144, 205, 275
Paraffins. Ref BER 30 1920 1897. [1,2; A5997]

M2833   3, 10, 21, 44, 83
4-colorings of cyclic group of order n. Ref MMAG 63 212 90. [1,1; A7687]

M2834   3, 10, 21, 55, 78, 136, 171
Coefficients of period polynomials. Ref LNM 899 292 81. [3,1; A6308]

M2835   3, 10, 22, 40, 65, 98, 140, 192, 255, 330, 418, 520, 637, 770, 920, 1088, 1275,
1482, 1710, 1960, 2233, 2530, 2852, 3200, 3575, 3978, 4410, 4872, 5365, 5890, 6448
Coefficient of x^3 in $(1 - x - x^2)^{-n}$. Ref FQ 14 43 76. [1,1; A6503]

M2836   3, 10, 25, 56, 119, 246, 501, 1012, 2035, 4082, 8177, 16368, 32751, 65518,
131053, 262124, 524267, 1048554, 2097129, 4194280, 8388583, 16777190, 33554405
Expansion of $(3 - 2x) / (1 - 2x)(1 - x)^2$. Ref R1 76. DB1 296. C1 222. [0,1; A0247,
N1141]

M2837   0, 0, 0, 0, 1, 3, 10, 25, 63, 144, 327, 711, 1534, 3237, 6787, 14056, 28971, 59283,
120894, 245457, 497167, 1004256, 2025199, 4077007, 8198334, 16467597, 33052491
$2^{n-1} + 2^{[n/2]} + 2^{[(n-1)]/2} - F(n+2)$. Ref rkg. [0,6; A5674]

M2838   1, 1, 3, 10, 27, 79, 234, 686, 2036, 6080, 18224, 54920, 166245, 505201, 1541014,
4716540, 14480699, 44586619, 137648341, 425992838, 1321362034, 4107332002
Tertiary alcohols with n carbon atoms. Ref BA76 44. [4,3; A5956]

M2839   1, 3, 10, 30, 75, 161, 308, 540, 885, 1375, 2046, 2938, 4095, 5565, 7400, 9656,
12393, 15675, 19570, 24150, 29491, 35673, 42780, 50900, 60125, 70551, 82278, 95410
$n(n+1)(n^2 - 3n + 5)/6$. Ref dsk. [1,2; A6484]

M2840   1, 1, 3, 10, 30, 99, 335, 1144, 3978, 14000, 49742, 178296, 643856, 2340135,
8554275, 31429068, 115997970, 429874830, 1598952498, 5967382200, 22338765540
Dissections of a polygon. Ref DM 11 387 75. AEQ 18 386 78. [1,3; A3441]

M2841   1, 3, 10, 31, 97, 306, 961, 3020, 9489, 29809, 93648, 294204, 924269, 2903677,
9122171, 28658146, 90032221, 282844564, 888582403, 2791563950, 8769956796
Nearest integer to π^n. Ref PE57 1(Appendix) 1. FMR 1 122. [0,2; A2160, N1142]

M2842   3, 10, 31, 101, 311, 962, 3132, 10202, 31412, 96722, 299183, 925445, 3012985,
9809425, 31952665, 104080805, 320465225, 986713745, 3038231465, 9355145285
$a(n) = 1 + a([n/2]) \, a(\lceil n/2 \rceil)$. Ref clm. [1,1; A5510]

M2843 1, 1, 3, 10, 31, 101, 336, 1128, 3823, 13051, 44803, 154518, 534964, 1858156, 6472168, 22597760, 79067375, 277164295, 973184313, 3422117190, 12049586631
Quadrinomial coefficients. Ref C1 78. [0,3; A5725]

M2844 0, 1, 3, 10, 33, 109, 360, 1189, 3927, 12970, 42837, 141481, 467280, 1543321, 5097243, 16835050, 55602393, 183642229, 606529080, 2003229469, 6616217487
$a(n) = 3a(n-1) + a(n-2)$. Ref FQ 15 292 77. ARS 6 168 78. [0,3; A6190]

M2845 1, 3, 10, 33, 111, 379, 1312, 4596, 16266, 58082, 209010, 757259, 2760123, 10114131, 37239072, 137698584, 511140558, 1904038986, 7115422212, 26668376994
A simple recurrence. Ref IFC 16 351 70. [0,2; A1558, N1143]

M2846 1, 1, 3, 10, 33, 147
One-sided hexagonal polyominoes with n cells. Ref jm. [1,3; A6535]

M2847 1, 3, 10, 34, 116, 396, 1352, 4616, 15760, 53808, 183712, 627232, 2141504, 7311552, 24963200, 85229696, 290992384, 993510144, 3392055808, 11581202944
Order-consecutive partitions. Ref HM94. [0,2; A7052]

$$\text{G.f.:} \quad (1 - x) / (1 - 4x + 2x^2).$$

M2848 1, 3, 10, 35, 126, 462, 1716, 6435, 24310, 92378, 352716, 1352078, 5200300, 20058300, 77558760, 300540195, 1166803110, 4537567650, 17672631900
$C(2n+1, n+1)$. Ref RS3. [0,2; A1700, N1144]

M2849 1, 3, 10, 36, 136, 528, 2080, 8256, 32896, 131328, 524800, 2098176, 8390656, 33558528, 134225920, 536887296, 2147516416, 8590000128, 34359869440
$2^{n-1}(1 + 2^n)$. Ref JGT 17 625 93. [0,2; A7582]

M2850 1, 3, 10, 36, 137, 543, 2219, 9285, 39587, 171369, 751236, 3328218, 14878455, 67030785, 304036170, 1387247580, 6363044315, 29323149825, 135700543190
Restricted hexagonal polyominoes with n cells: reversion of M2741. Ref PEMS 17 11 70. rcr. [1,2; A2212, N1145]

M2851 1, 3, 10, 37, 151, 674, 3263, 17007, 94828, 562595, 3535027, 23430840, 163254885, 1192059223, 9097183602, 72384727657, 599211936355, 5150665398898
Expansion of $e \uparrow (e \uparrow x + 2x - 1)$. Ref JCT A24 316 78. SIAD 5 498 92. [0,2; A5493]

M2852 1, 1, 3, 10, 38, 154, 654, 2871, 12925, 59345, 276835, 1308320, 6250832, 30142360, 146510216, 717061938, 3530808798, 17478955570, 86941210950
Dissections of a polygon. Ref EDMN 32 6 40. BAMS 54 359 48. [0,3; A1002, N1146]

$$\text{Reversion of } x(1 - x - x^2).$$

M2853 1, 1, 3, 10, 38, 156, 692, 3256, 16200, 84496, 460592, 2611104, 15355232, 93376960, 585989952, 3786534784, 25152768128, 171474649344, 1198143415040
Symmetric permutations. Ref LU91 1 222. LNM 560 201 76. [0,3; A0902, N1147]

$$a(n) = 2.a(n-1) + (2n-4).a(n-2).$$

M2854 1, 3, 10, 39, 158, 674, 2944, 13191, 92154, 523706
P-graphs with vertical symmetry. Ref AEQ 31 54 86. [1,2; A7163]

M2855 1, 1, 3, 10, 39, 160, 702, 3177, 14830, 70678, 342860, 1686486, 8393681,
42187148, 213828802, 1091711076, 5609447942, 28982558389, 150496428594
Planted matched trees with n nodes (inverse convolution of M1770). Ref DM 88 97 91.
[1,3; A5750]

M2856 1, 0, 1, 3, 10, 40, 190, 1050, 6620, 46800, 365300, 3103100, 28269800,
271627200, 2691559000, 26495469000, 238131478000, 1394099824000
Expansion of $\cos(\ln(1+x))$. [0,4; A3703]

M2857 1, 1, 3, 10, 41, 196, 1057, 6322, 41393, 293608, 2237921, 18210094, 157329097,
1436630092, 13810863809, 139305550066, 1469959371233, 16184586405328
Forests with n nodes and height at most 1. Ref JCT 3 134 67; 5 102 68. C1 91. [0,3;
A0248, N1148]

$$\text{E.g.f.:}\quad \exp x\, e^x.$$

M2858 0, 1, 3, 10, 41, 206, 1237, 8660, 69281, 623530, 6235301, 68588312, 823059745,
10699776686, 149796873605, 2246953104076, 35951249665217, 611171244308690
$a(n) = n.a(n-1)+1$. Ref TMS 20 70 52. [0,3; A2627, N1149]

M2859 0, 0, 0, 0, 0, 0, 0, 0, 1, 3, 10, 42, 193, 966, 5215, 30170, 186234, 1222065,
8496274, 62395234, 482700052, 3923995651, 33444263516, 298233514595
Modified Bessel function $K_n(5)$. Ref AS1 429. [0,10; A0249, N1150]

M2860 1, 3, 10, 42, 204, 1127, 6924, 46704, 342167, 2700295, 22799218, 204799885,
1947993126, 19540680497, 206001380039, 2275381566909, 26261810071925
Exponentiation of Fibonacci numbers. Ref BeS194. [1,2; A7552]

M2861 1, 3, 10, 42, 216, 1320, 9360, 75600, 685440, 6894720, 76204800, 918086400,
11975040000, 168129561600, 2528170444800, 40537905408000, 690452066304000
$(2n+1).n!$. Ref UM 45 82 94. [0,2; A7680]

M2862 1, 1, 3, 10, 43, 223, 1364, 9643, 77545, 699954
Permutations with strong fixed points. Ref AMM 100 800 93. kw. [0,3; A6932]

M2863 1, 1, 3, 10, 43, 225, 1393, 9976, 81201, 740785, 7489051, 83120346, 1004933203,
13147251985, 185066460993, 2789144166880, 44811373131073, 764582487395121
$a(n+1)=n.a(n)+a(n-1)$. Ref EUR 22 15 59. [1,3; A1040, N1151]

M2864 1, 1, 1, 3, 10, 44, 238, 1650, 14512, 163341, 2360719, 43944974
Connected partially ordered sets with n elements. Ref NAMS 17 646 70. jaw. CN 8 180 73.
gm. [0,4; A0608, N1152]

M2865 0, 1, 1, 3, 10, 44, 274, 2518, 39159, 1087472, 56214536, 5422178367, 973901229150, 325367339922914, 202427527012666564, 235111320292288931449
Connected strength-1 Eulerian graphs with n nodes, 2 of odd degree. Ref rwr. [1,4; A7125]

M2866 1, 1, 3, 10, 45, 251, 1638, 12300, 104877, 1000135
From descending subsequences of permutations. Ref JCT A53 99 90. [1,3; A6220]

M2867 1, 1, 3, 10, 45, 256, 1743, 13840, 125625, 1282816, 14554683, 181649920, 2473184805, 36478744576, 579439207623, 9861412096000, 179018972217585
Expansion of $\ln(1+\sinh x)$. [0,3; A3704]

M2868 1, 3, 10, 45, 272, 2548, 39632, 1104306, 56871880, 5463113568, 978181717680, 326167542296048, 202701136710498400, 235284321080559981952
Symmetric reflexive relations on n nodes: ½ M1650. See Fig M3032. Ref MIT 17 21 55. MAN 174 70 67. JGT 1 295 77. [1,2; A0250, N1153]

M2869 1, 1, 3, 10, 45, 274
Sub-Eulerian graphs with n nodes. Ref ST90. [2,3; A5143]

M2870 1, 1, 3, 10, 47, 246, 1602, 11481, 95503, 871030, 8879558, 98329551, 1191578522, 15543026747, 218668538441, 3285749117475, 52700813279423
Sums of multinomial coefficients. Ref C1 126. [0,3; A5651]

$$\text{G.f.: } 1 \,/\, \Pi \,(1-x^k/k!).$$

M2871 1, 3, 10, 48, 312, 2520, 24480, 277200, 3588480, 52254720
From solution to a difference equation. Ref FQ 25 363 87. [0,2; A5921]

M2872 1, 1, 3, 10, 53, 265, 1700
Sorting numbers. Ref PSPM 19 173 71. [0,3; A2873, N1154]

M2873 0, 1, 1, 3, 10, 56, 468, 7123, 194066, 9743542, 900969091, 153620333545, 48432939150704, 28361824488394169, 30995890806033380784
Nonseparable graphs with n nodes. Ref JCT 9 352 70. CCC 2 199 77. JCT B57 294 93. [1,4; A2218, N1155]

M2874 1, 3, 10, 66, 792, 25506, 2302938, 591901884, 420784762014, 819833163057369, 4382639993148435207, 64588133532185722290294, 263857237581576280415666529
Signed graphs with n nodes. Ref CCC 2 31 77. rwr. JGT 1 295 77. [1,2; A4102]

M2875 1, 3, 10, 70, 708, 15224, 544152, 39576432, 5074417616, 1296033011648, 604178966756320, 556052774253161600, 95489532201976258566, 4
Self-converse digraphs with n nodes. Ref MAT 13 157 66. rwr. [1,2; A2499, N1156]

M2876 3, 10, 84, 10989, 363883, 82620, 137550709
Coefficients of period polynomials. Ref LNM 899 292 81. [3,1; A6311]

M2877 0, 0, 1, 3, 10, 93, 2521, 612696, 4019900977
Coding Fibonacci numbers. Ref FQ 15 315 77. [1,4; A5205]

M2878 0, 1, 0, 3, 10, 355, 6986, 297619, 15077658, 1120452771, 111765799882,
   15350524923547, 2875055248515242, 738416821509929731, 260316039943139322858
Labeled nonseparable bipartite graphs. Ref CJM 31 65 79. NR82. [1,4; A4100]

M2879 3, 10, 1297, 2186871697, 104585123175352403839929505297
Numerators of a continued fraction. Ref NBS B80 288 76. [0,1; A6273]

M2880 1, 1, 3, 11, 7, 41, 117, 29, 527, 1199, 237, 6469, 11753, 8839, 76443, 108691,
   164833, 873121, 922077, 2521451, 9653287, 6699319, 34867797, 103232189, 32125393
Real part of $(1+2i)^n$. Cf. M0933. Ref FQ 15 235 77. [0,3; A6495]

M2881 3, 11, 13, 31, 37, 41, 43, 53, 67, 71, 73, 79, 83, 89, 101, 103, 107, 127, 137, 139,
   151, 157, 163, 173, 191, 197, 199, 211, 227, 239, 241, 251, 271, 277, 281, 283, 293, 307
Short period primes. [1,1; A6559]

M2882 3, 11, 19, 43, 59, 67, 83, 107, 131, 139, 163, 179, 211, 227, 251, 283, 307, 331,
   347, 379, 419, 443, 467, 491, 499, 523, 547, 563, 571, 587, 619, 643, 659, 683, 691, 739
Primes $\equiv$ 3 (mod 8). Ref AS1 870. [1,1; A7520]

M2883 1, 0, 1, 3, 11, 20, 57, 108, 240, 472, 1013, 1959, 4083, 8052, 16315, 32496, 65519,
   130464, 262125, 523209, 1048353, 2095084, 4194281, 8384100, 16777120
Mu-molecules in Mandelbrot set whose seeds have period n. Ref Man82 183. Pen91 138.
rpm. [1,4; A6876]

M2884 3, 11, 23, 83, 131, 179, 191, 239, 251, 359, 419, 431, 443, 491, 659, 683, 719, 743,
   911, 1019, 1031, 1103, 1223, 1439, 1451, 1499, 1511, 1559, 1583, 1811, 1931, 2003
$p \equiv 3$ (mod 4) with $2p+1$ prime. Ref BAR 564 1894. D1 1 27. [1,1; A2515, N2039]

M2885 1, 3, 11, 25, 137, 49, 363, 761, 7129, 7381, 83711, 86021, 1145993, 1171733,
   1195757, 2436559, 42142223, 14274301, 275295799, 55835135, 18858053, 19093197
Numerators of harmonic numbers. See Fig M4299. Cf. M1589. Ref KN1 1 615. [1,2;
A1008, N1157]

M2886 1, 3, 11, 27, 101, 41, 7, 239, 73, 81, 451, 21649, 707, 53, 2629, 31, 17, 2071723,
   19, 111111111111111111111, 3541, 43, 23, 1111111111111111111111111, 511, 21401, 583
Smallest number with reciprocal of period n. Ref PC 1 4-13 73. CUNN xxxiv. [0,2; A3060]

M2887 1, 3, 11, 29, 74, 167, 367, 755, 1515, 2931, 5551, 10263, 18677, 33409, 59024,
   102984, 177915, 304458, 516939, 871180, 1458882, 2428548, 4021670, 6627515
Trees of diameter 6. Ref IBMJ 4 476 60. KU64. [7,2; A0251, N1158]

M2888 3, 11, 37, 101, 41, 7, 239, 73, 333667, 9091, 21649, 9901, 53, 909091, 31, 17,
2071723, 19, 111111111111111111, 3541, 43, 23, 11111111111111111111111111
Smallest primitive factor of $10^n - 1$. Ref CUNN. [1,1; A7138]

M2889 3, 11, 37, 101, 271, 37, 4649, 137, 333667, 9091, 513239, 9901, 265371653,
909091, 2906161, 5882353, 5363222357, 333667, 1111111111111111111, 27961
Largest factor of $10^n - 1$. Ref CUNN. [1,1; A5422]

M2890 3, 11, 37, 101, 333667, 9091, 9901, 909091, 1111111111111111111,
11111111111111111111111111, 99990001, 999999000001, 909090909090909091
Primes with unique period length. Ref JRM 18 24 85. [1,1; A7615]

M2891 1, 3, 11, 38, 126, 415, 1369, 4521
Paths on square lattice. Ref ARS 6 168 78. [3,2; A6189]

M2892 1, 3, 11, 39, 131, 423, 1331, 4119, 12611, 38343, 116051, 350199, 1054691,
3172263, 9533171, 28632279, 85962371, 258018183, 774316691, 2323474359
$2(3^n - 2^n) + 1$. Ref IJ1 11 162 69. [0,2; A2783, N1159]

M2893 1, 3, 11, 39, 139, 495, 1763, 6279, 22363, 79647, 283667, 1010295, 3598219,
12815247, 45642179, 162557031, 578955451, 2061980415, 7343852147, 26155517271
Subsequences of $[1,...,2n]$ in which each odd number has an even neighbor. Ref GuMo94.
[0,2; A7482]

$$a(n) = 3\,a(n-1) + 2\,a(n-2).$$

M2894 1, 1, 3, 11, 41, 153, 571, 2131, 7953, 29681, 110771, 413403, 1542841, 5757961,
21489003, 80198051, 299303201, 1117014753, 4168755811, 15558008491
$a(n) = 4a(n-1) - a(n-2)$. Ref EUL (1) 1 375 11. MMAG 40 78 67. [0,3; A1835, N1160]

M2895 1, 3, 11, 43, 171, 683, 2731, 10923, 43691, 174763, 699051, 2796203, 11184811,
44739243, 178956971, 715827883, 2863311531, 11453246123, 45812984491
$(2^{2n+1} + 1)/3$. Ref JGT 17 625 93. [0,2; A7583]

M2896 3, 11, 43, 683, 2731, 43691, 174763, 2796203, 715827883, 2932031007403,
768614336404564651, 201487636602438195784363
Primes of form $(2^p + 1)/3$. Ref MMAG 27 157 54. [1,1; A0979, N1161]

M2897 1, 3, 11, 44, 186, 814, 3652, 16689, 77359, 362671, 1716033, 8182213, 39267086,
189492795, 918837374, 4474080844, 21866153748, 107217298977, 527266673134
Fixed hexagonal polyominoes with n cells. Ref RE72 97. dhr. [1,2; A1207, N1162]

M2898 1, 1, 3, 11, 45, 197, 903, 4279, 20793, 103049, 518859, 2646723, 13648869,
71039373, 372693519, 1968801519, 10463578353, 55909013009, 300159426963
Schroeder's second problem: $(n+1)a(n+1) = 3(2n-1)a(n) - (n-2)a(n-1)$. Ref
EDMN 32 6 40. BAMS 54 359 48. RCI 168. C1 57. VA91 198. [1,3; A1003, N1163]

M2899 1, 1, 3, 11, 46, 207, 979, 4797, 24138, 123998, 647615, 3428493, 18356714,
99229015, 540807165, 2968468275, 16395456762, 91053897066, 508151297602
Modes of connections of $2n$ points. Ref LNM 686 326 78. [0,3; A6605]

M2900 1, 1, 3, 11, 49, 257, 1539, 10299, 75905, 609441, 5284451, 49134923, 487026929,
5120905441, 56878092067, 664920021819, 8155340557697, 104652541401025
Coincides with its 2nd order binomial transform. Ref DM 21 320 78. EIS § 2.7. [0,3;
A4211]

$$\text{Lgd.e.g.f.: } e^{2x}.$$

M2901 1, 3, 11, 49, 261, 1631, 11743, 95901, 876809, 8877691, 98641011, 1193556233,
15624736141, 220048367319, 3317652307271, 53319412081141, 909984632851473
$\Sigma(n+1)!C(n,k)$, $k = 0 \ .. \ n$. Ref CJM 22 26 70. Adam74 70. [0,2; A1339, N1164]

M2902 1, 3, 11, 50, 274, 1764, 13068, 109584, 1026576, 10628640, 120543840,
1486442880, 19802759040, 283465647360, 4339163001600, 70734282393600
Stirling numbers of first kind: $a(n+1)=(n+1)\ a(n)+n!$. See Fig M4730. Ref AS1 833.
DKB 226. [1,2; A0254, N1165]

M2903 1, 3, 11, 51, 299, 2163, 18731, 189171, 2183339, 28349043, 408990251
Chains in power set of n-set. Ref MMAG 64 29 91. [0,2; A7047]

M2904 3, 11, 53, 295, 1867
Triangulations. Ref WB79 336. [0,1; A5502]

M2905 1, 1, 3, 11, 53, 309, 2119, 16687, 148329, 1468457, 16019531, 190899411,
2467007773, 34361893981, 513137616783, 8178130767479, 138547156531409
$a(n)=n.a(n-1)+(n-1)a(n-2)$. Ref R1 188. DKB 263. MAG 52 381 68. FQ 18 228
80. [0,3; A0255, N1166]

$$\text{E.g.f.: } e^{-x}(1-x)^{-2}.$$

M2906 1, 1, 3, 11, 55, 330, 2345
Dimensions of subspaces of Jordan algebras. Ref LE70 309. [1,3; A1776, N1167]

M2907 1, 1, 3, 11, 56, 348, 2578, 22054, 213798, 2313638, 27627434, 360646314,
5107177312, 77954299144, 1275489929604, 22265845018412, 412989204564572
Stochastic matrices of integers. Ref DUMJ 35 659 68. [0,3; A0985, N1168]

M2908 1, 1, 3, 11, 57, 361, 2763, 24611, 250737, 2873041, 36581523, 512343611,
7828053417, 129570724921, 2309644635483, 44110959165011, 898621108880097
Generalized Euler numbers. Ref MOC 21 693 67. [0,3; A1586, N1169]

M2909 1, 0, 1, 3, 11, 60, 502, 7403, 197442, 9804368, 902818087, 153721215608,
48443044675155, 28363687700395422, 30996524108446916915
Bridgeless graphs with n nodes. Ref JCT B33 303 82. [1,4; A7146]

M2910 1, 0, 1, 3, 11, 61, 507, 7442, 197772, 9808209, 902884343, 153723152913,
   48443147912137, 28363697921914475, 30996525982586676021
n-node connected graphs without endpoints. Ref rwr. [1,4; A4108]

M2911 3, 11, 101, 131, 181, 313, 383, 10301, 11311, 13331, 13831, 18181, 30103, 30803,
   31013, 38083, 38183, 1003001, 1008001, 1180811, 1183811, 1300031, 1303031
Palindromic reflectable primes. Ref JRM 15 252 83. [1,1; A7616]

M2912 3, 11, 171, 43691, 2863311531, 12297829382473034411,
   2268549112806256423089164049545512140971
$(2 \uparrow (2^n + 1) + 1)/3$. Ref dsk. [1,1; A6485]

M2913 1, 3, 11, 173, 2757, 176275, 11278843, 2887207533, 739113849605,
   756849694787987, 775013348349049083, 3174453917988010255981
$a(n+2) = (4^{n+1} - 5) a(n) - 4a(n-2)$. Ref dhl. hpr. [1,2; A3115]

M2914 3, 11, 197, 129615, 430904428717
Spectrum of a certain 3-element algebra. Ref Berm83. [0,1; A7156]

M2915 3, 12, 15, 36, 138, 276, 4326, 21204, 65274, 126204, 204246, 1267356, 10235538,
   54791316, 212311746, 678889380, 4946455134, 20113372464
Specific heat for crystobalite lattice. Ref CJP 48 310 70. [0,1; A5392]

M2916 1, 3, 12, 28, 66, 126, 236, 396, 651, 1001
Paraffins. Ref BER 30 1919 1897. [1,2; A5995]

M2917 3, 12, 29, 57, 99, 157, 234, 333, 456, 606, 786, 998, 1245
Series-reduced planted trees with n nodes, $n-4$ endpoints. Ref jr. [9,1; A1860, N1171]

M2918 3, 12, 31, 65, 120, 203, 322, 486, 705, 990, 1353, 1807, 2366, 3045, 3860, 4828,
   5967, 7296, 8835, 10605, 12628, 14927, 17526, 20450, 23725, 27378, 31437, 35931
Quadrinomial coefficients. Ref C1 78. [2,1; A5718]

M2919 0, 3, 12, 45, 168, 627, 2340, 8733, 32592, 121635, 453948, 1694157, 6322680,
   23596563, 88063572, 328657725, 1226567328, 4577611587, 17083879020
$a(n) = 4a(n-1) - a(n-2)$. [0,2; A5320]

M2920 0, 0, 1, 3, 12, 45, 170, 651, 2520, 97502, 37854, 147070
Necklaces with n red, 1 pink and $n-3$ blue beads. Ref MMAG 60 90 87. [1,4; A5656]

M2921 1, 3, 12, 50, 27, 1323, 928, 1080, 48525, 3237113, 7587864, 23361540993,
   770720657, 698808195, 179731134720, 542023437008852, 3212744374395
Cotesian numbers. Ref QJMA 46 63 14. [2,2; A2179, N1172]

M2922 1, 3, 12, 52, 238, 1125, 5438, 26715, 132871, 667312, 3377906, 17210522,
   88169685, 453810095, 2345209383, 12162367228, 63270384303
n-node animals on f.c.c. lattice. Ref DU92 42. [1,2; A7198]

M2923 1, 1, 0, 1, 3, 12, 52, 241, 1173, 5929, 30880, 164796, 897380, 4970296, 27930828, 158935761, 914325657, 5310702819, 31110146416, 183634501753, 1091371140915
Simple triangulations of plane with n nodes. Ref CJM 15 268 63. [3,5; A0256, N1173]

M2924 1, 3, 12, 54, 260, 1310, 6821, 36413
Column-convex polyominoes with perimeter n. Ref DE87. JCT A48 12 88. [1,2; A6026]

M2925 1, 3, 12, 55, 273, 1425, 7695, 42576, 239925, 1371555, 7931817, 46310127, 272559558, 1615163592, 9627985773, 57688721354, 347228163630
n-node animals on f.c.c. lattice. Ref DU92 42. [1,2; A7199]

M2926 1, 1, 3, 12, 55, 273, 1428, 7752, 43263, 246675, 1430715, 8414640, 50067108, 300830572, 1822766520, 11124755664, 68328754959, 422030545335, 2619631042665
$C(3n,n)/(3n+1)$. See Fig M1645. Ref CMA 2 25 70. MAN 191 98 71. FQ 11 125 73. DM 9 355 74. [0,3; A1764, N1174]

M2927 1, 1, 3, 12, 56, 288, 1584, 9152, 54912, 339456, 2149888, 13891584, 91287552, 608583680, 4107939840, 28030648320, 193100021760
Rooted bicubic maps: $a(n)=(8n-4)a(n-1)/(n+2)$. Ref CJM 15 269 63. [0,3; A0257, N1175]

M2928 1, 3, 12, 56, 321, 2175, 17008, 150504, 1485465, 16170035, 192384876, 2483177808, 34554278857, 515620794591, 8212685046336
Permutations of length n by rises. Ref DKB 264. [2,2; A1277, N1176]

M2929 1, 3, 12, 58, 325, 2143, 17291
Commutative semigroups of order n. Ref PL65. MAL 2 2 67. SGF 14 71 77. [1,2; A1426, N1177]

M2930 1, 3, 12, 58, 335, 2261, 17465, 152020, 1473057, 15730705, 183571817, 2324298010, 31737207026, 464904410985, 7272666016725, 121007866402968
Sum of lengths of longest increasing subsequences of all permutations of n things. Ref MOC 22 390 68. [1,2; A3316]

M2931 1, 1, 3, 12, 60, 270, 1890, 14280, 128520, 1096200, 12058200, 139043520, 1807565760, 22642139520, 339632092800, 5237183952000, 89032127184000
Square permutations of n things. Ref JCT A17 156 74. [0,3; A3483]

M2932 1, 1, 3, 12, 60, 358, 2471, 19302, 167894, 1606137, 16733779, 188378402, 2276423485, 29367807524, 402577243425, 5840190914957, 89345001017415
Coefficients of iterated exponentials. Ref SMA 11 353 45. PRV A32 2342 85. [0,3; A0258, N1178]

M2933 1, 3, 12, 60, 360, 2520, 20160, 181440, 1814400, 19958400, 239500800, 3113510400, 43589145600, 653837184000, 10461394944000, 177843714048000
$n!/2$. Ref PEF 77 26 62. [2,2; A1710, N1179]

M2934   3, 12, 60, 420, 4620, 60060, 180180, 360360, 6126120, 116396280, 2677114440, 77636318760, 2406725881560, 89048857617720, 3651003162326520
A highly composite sequence. Ref BSMF 97 152 69. [1,1; A2497, N1180]

M2935   3, 12, 65, 480, 4851, 67256, 1281258, 33576120
Motifs in triangular window of side n. Ref grauzy. [1,1; A7017]

M2936   1, 1, 3, 12, 68, 483, 3946, 34485, 315810, 2984570, 28907970, 285601251
Hexagon trees. Ref GMJ 15 146 74. [1,3; A4127]

M2937   1, 0, 0, 1, 3, 12, 70, 465, 3507, 30016, 286884, 3026655, 34944085, 438263364, 5933502822, 86248951243, 1339751921865, 22148051088480, 388246725873208
Clouds with n points. See Fig M1041. Ref AMM 59 296 52. C1 276. [1,5; A1205, N1181]

$$a(n+1) = n\,a(n) + \tfrac{1}{2}\,n\,(n-1)\,a(n-2).$$

SEQUENCES BEGINNING . . ., 3, 13, . . ., . . ., 3, 14, . . .

M2938   3, 13, 17, 71, 43, 4733, 241, 757, 9091, 1806113, 20593, 1803647, 8108731, 39225301, 6700417, 2699538733, 465841, 10991220309223964380221, 222361
Largest factor of $n^n - 1$. Ref dsk. rgw. [2,1; A6486]

M2939   1, 3, 13, 27, 52791, 482427, 124996631
Numerators of an asymptotic expansion of an integral. Cf. 2305. Ref MOC 19 114 65. [0,2; A2304, N1182]

M2940   3, 13, 31, 43, 67, 71, 83, 89, 107, 151, 157, 163, 191, 197, 199, 227, 283, 293, 307, 311, 347, 359, 373, 401, 409, 431, 439, 443, 467, 479, 523, 557, 563, 569, 587, 599
Cyclic numbers: 10 is a quadratic residue modulo p and class of mantissa is 2. Ref Krai24 1 61. [1,1; A1914, N1183]

M2941   1, 3, 13, 57, 259, 1177, 5367, 24473, 111631, 509193
Worst case of a Jacobi symbol algorithm. Ref JSC 10 605 90. [0,2; A5827]

M2942   1, 3, 13, 63, 321, 1683, 8989, 48639, 265729, 1462563, 8097453, 45046719, 251595969, 1409933619, 7923848253, 44642381823, 252055236609, 1425834724419
$\Sigma C(n,k)\,C(n+k,k)$, $k = 0 \dots n$. Ref SIAR 12 277 70. [0,2; A1850, N1184]

M2943   1, 3, 13, 63, 326, 1761
Rooted planar maps. Ref CJM 15 542 63. [1,2; A0259, N1185]

M2944   1, 1, 3, 13, 63, 399, 3268, 33496, 412943
Trivalent graphs of girth exactly 3 and $2n$ nodes. Ref gr. [2,3; A6923]

M2945   0, 0, 1, 3, 13, 65, 397, 2819, 22831, 207605, 2094121, 23205383, 280224451, 3662810249, 51523391965, 776082247979, 12463259986087, 212573743211549
The game of Mousetrap with n cards. Ref QJMA 15 241 1878. GN93. [1,4; A2468, N1186]

M2946 1, 1, 3, 13, 68, 399, 2530, 16965, 118668, 857956, 6369883, 48336171,
373537388, 2931682810, 23317105140, 187606350645, 1524813969276
$2(4n+1)!/(n+1)!(3n+2)!$. Ref CJM 14 32 62. [0,3; A0260, N1187]

M2947 1, 1, 1, 3, 13, 70, 462, 3592, 32056, 322626, 3611890, 44491654, 597714474,
8693651092, 136059119332, 2279212812480, 40681707637888, 770631412413148
Stochastic matrices of integers. Ref DUMJ 35 659 68. [0,4; A1495, N1188]

M2948 1, 1, 3, 13, 71, 461, 3447, 29093, 273343, 2829325, 31998903, 392743957,
5201061455, 73943424413, 1123596277863, 18176728317413, 311951144828863
$a(n)=n!-\Sigma\ k!a(n-k)$. Ref CRP 275 569 72. C1 295. [1,3; A3319]

$$\text{G.f.:}\quad 1\ /\ \Sigma\ (k!\ x^k).$$

M2949 0, 1, 3, 13, 71, 465, 3539, 30637, 296967, 3184129, 37401155, 477471021,
6581134823, 97388068753, 1539794649171, 25902759280525, 461904032857319
$a(n)=n.a(n-1)+(n-3)a(n-2)$. Ref R1 188. [1,3; A0261, N1189]

$$\text{E.g.f.:}\quad e^{-x}\ (1-x)^{-4}.$$

M2950 1, 1, 3, 13, 73, 501, 4051, 37633, 394353, 4596553, 58941091, 824073141,
12470162233, 202976401213, 3535017524403, 65573803186921, 1290434218669921
Sets of lists: $a(n)=(2n-1)a(n-1)-(n-1)(n-2)a(n-2)$. Ref RCI 194. PSPM 19 172
71. TMJ 2 72 92. [0,3; A0262, N1190]

$$\text{E.g.f.:}\quad e^{x/(1-x)}.$$

M2951 1, 1, 3, 13, 75, 525, 4347, 41245, 441675, 5259885, 68958747, 986533053,
15292855019, 255321427725, 4567457001915
Ways to write 1 as ordered sum of n powers of ½, allowing repeats. Ref dek. Shal94. [1,3;
A7178]

M2952 1, 1, 3, 13, 75, 541, 4683, 47293, 545835, 7087261, 102247563, 1622632573,
28091567595, 526858348381, 10641342970443, 230283190977853, 5315654681981355
Preferential arrangements of n things: expansion of $1/(2-e^x)$. Ref CAY 4 113. PLMS 22
341 1891. AMM 69 7 62. PSPM 19 172 71. DM 48 102 84. [0,3; A0670, N1191]

M2953 1, 3, 13, 81, 673, 6993, 87193, 1268361, 21086113, 394368993, 8195230473
From solution to a difference equation. Ref FQ 25 83 87. [0,2; A5923]

M2954 1, 3, 13, 81, 721, 9153, 165313, 4244481, 154732801, 8005686273,
587435092993, 61116916981761, 9011561121239041, 1882834327457349633
Colored labeled n-node graphs with 2 interchangeable colors. Ref JCT 6 17 69. CJM 31 65
79. NR82. [1,2; A0684, N1192]

M2955 0, 3, 13, 83, 592, 4821, 43979, 444613, 4934720, 59661255, 780531033, 10987095719, 165586966816, 2660378564777, 45392022568023, 819716784789193
Ménage numbers. Ref LU91 1 495. [4,2; A0904, N1193]

$$a(n) = (1 + n) a(n - 1) + (2 + n) a(n - 2) + a(n - 3).$$

M2956 1, 1, 3, 13, 87, 841, 11643
Graded partially ordered sets with n elements. Ref JCT 6 17 69. [0,3; A1831, N1194]

M2957 3, 13, 87, 1053, 28576, 2141733, 508147108
Incidence matrices. Ref CPM 89 217 64. SLC 19 79 88. [1,1; A2725, N1195]

M2958 3, 13, 95, 1337, 38619
Hierarchical quadratic models on n factors (differences of M1520). Ref clm. [1,1; A6898]

M2959 1, 3, 13, 111, 1381, 25623, 678133, 26269735, 1447451707, 114973020921, 13034306495563
Tiered orders on n nodes. Ref DM 53 148 85. [1,2; A6860]

M2960 3, 13, 146, 40422
Switching networks. Ref JFI 276 324 63. [1,1; A1150, N1196]

M2961 1, 3, 13, 183, 33673, 1133904603, 1285739649838492213, 1653126447166808570252515315100129583
$a(n + 1) = a(n)^2 + a(n) + 1$. Ref DUMJ 4 325 38. FQ 11 436 73. [0,2; A2065, N1197]

M2962 1, 3, 13, 253, 218201, 61323543802, 5704059172637470075854, 1780598168152033955529170567877224451335939040
Egyptian fraction for square root of 2. [0,2; A6487]

M2963 3, 13, 308, 1476218
Switching networks. Ref JFI 276 324 63. [1,1; A0859, N1198]

M2964 3, 13, 781, 137257, 28531167061, 25239592216021, 51702516367896047761, 10991220309223964384021, 94911218181126872883431967753
$(p^p - 1)/(p - 1)$ where p is prime. Ref MOC 16 421 62. PSPM 19 174 71. [2,1; A1039, N1199]

M2965 3, 13, 1113, 3113, 132113, 1113122113, 311311222113, 13211321322113, 1113122113121113222113, 31131122211311123113322113
Describe the previous term! Ref CoGo87 176. VA91 4. [1,1; A6715]

M2966 3, 13, 51413, 951413, 2951413, 53562951413, 979853562951413
Primes in decimal expansion of π written backwards. Ref GA89a 84. [1,1; A7523]

M2967 3, 14, 39, 91, 173, 307, 502, 779
Partitions into non-integral powers. Ref PCPS 47 215 51. [3,1; A0263, N1200]

M2968 3, 14, 40, 90, 175, 308, 504, 780, 1155, 1650, 2288, 3094, 4095, 5320, 6800, 8568, 10659, 13110, 15960, 19250, 23023, 27324, 32200, 37700, 43875, 50778, 58464, 66990
$n(n+1)(n+2)(n+5)/12$. Ref LNM 1234 118 86. [0,1; A5701]

M2969 1, 3, 14, 42, 128, 334, 850, 2010, 4625, 10201, 21990, 46108, 94912, 191562, 380933, 746338, 1444676, 2763931, 5235309, 9822686, 18275648, 33734658, 61826344
Trees of diameter 7. Ref IBMJ 4 476 60. KU64. [8,2; A0550, N1201]

M2970 1, 3, 14, 60, 279, 1251
Related to Fibonacci numbers. Ref FQ 16 217 78. [0,2; A6502]

M2971 1, 3, 14, 70, 370, 2028, 11452, 66172, 389416, 2326202, 14070268, 86010680, 530576780, 3298906810, 20653559846, 130099026600, 823979294284, 5244162058026
Sum of spans of n-step polygons on square lattice. Ref JPA 21 L167 88. [1,2; A6772]

M2972 1, 3, 14, 78, 504, 3720, 30960, 287280, 2943360, 33022080, 402796800, 5308934400, 75203251200, 1139544806400, 18394619443200, 315149522688000
2nd differences of factorial numbers. Ref JRAM 198 61 57. [0,2; A1564, N1202]

M2973 1, 3, 14, 79, 494, 3294, 22952, 165127, 1217270, 9146746, 69799476, 539464358, 4214095612, 33218794236, 263908187100, 2110912146295, 16985386737830
2-line arrays. Ref FQ 11 124 73; 14 232 76. AEQ 31 52 86. [1,2; A3169]

M2974 1, 1, 3, 14, 80, 518, 3647, 27274, 213480, 1731652
Hamiltonian rooted maps with $2n$ nodes. Ref CJM 14 417 62. [1,3; A0264, N1203]

M2975 1, 1, 3, 14, 84, 594, 4719, 40898, 379236, 3711916, 37975756, 403127256
Dyck paths. Ref LNM 1234 118 86. [0,3; A5700]

$$\text{G.f.:} \quad {}_3F_2([1,1/2,3/2]\,;\,[3,4]\,;\,16x).$$

M2976 1, 1, 3, 14, 85, 626, 5387, 52882, 582149, 7094234, 94730611, 1374650042, 21529197077, 361809517954, 6492232196699, 123852300381986
n-term 2-sided generalized Fibonacci sequences. Ref FQ 27 355 89. SIAD 1 342 88. [1,3; A5189]

M2977 1, 1, 3, 14, 89, 716, 6967, 79524, 1041541, 15393100, 253377811, 4596600004, 91112351537, 1959073928124, 45414287553455, 1129046241331316
Shifts left when exponentiated twice. Ref BeSl94. [1,3; A7549]

M2978 3, 14, 95, 424, 3269, 21202, 178443, 1622798
Permutations of length n with 2 cycle lengths. Ref JCT A35 201 1983. [3,1; A5772]

M2979 1, 1, 3, 14, 97, 934, 11814, 188650, 3698399, 87133235, 2424143590, 78483913829, 2920947798710, 123676552368689, 5904927996501989
Shifts left when Stirling-2 transform applied twice. Ref BeSl94. EIS § 2.7. [0,3; A7470]

M2980 3, 14, 115, 2086
Bicolored graphs in which colors are interchangeable. Ref ENVP B5 41 78. [2,1; A7140]

M2981 1, 1, 3, 14, 147, 3462, 294392
Egyptian fractions: partitions of 1 into parts $1/n$. Ref SI72. UPNT D11. rgw. [1,3; A2966]

M2982 1, 3, 14, 240, 63488, 4227858432, 18302628885633695744,
33895313892515354759047080037148786680
Self-complementary Boolean functions of n variables. Ref PGEC 12 561 63. [1,2; A1320, N1204]

M2983 1, 3, 15, 21, 15, 33, 1365, 3, 255, 399, 165, 69, 1365, 3, 435, 7161
Denominators of cosecant numbers. Cf. M4403. Ref NO24 458. ANN 36 640 35. DA63 2 187. [0,2; A1897, N1205]

M2984 1, 3, 15, 26, 39, 45, 74, 104, 111, 117, 122, 146, 175, 183, 195, 219, 296, 314, 333, 357, 386, 471, 488, 549, 554, 555, 579, 584, 585, 608, 626, 646, 657, 794, 831, 842, 914
$\phi(n) = \phi(\sigma(n))$. Ref AAMS 14 415 93. [1,2; A6872]

M2985 3, 15, 27, 51, 147, 243, 267, 347, 471, 747, 2163, 3087, 5355, 6539, 7311
$17.2^n + 1$ is prime. Ref PAMS 9 674 58. Rie85 381. [1,1; A2259, N1206]

M2986 1, 3, 15, 35, 315, 693, 3003, 6435, 109395, 230945, 969969, 2028117, 16900975, 35102025, 145422675, 300540195, 9917826435, 20419054425, 83945001525
Numerators in expansion of $(1-x)^{-3/2}$. Ref PR33 156. AS1 798. [0,2; A1803, N1207]

M2987 1, 3, 15, 45, 189, 588, 2352, 7560, 29700, 98010, 382239, 1288287, 5010005, 17177160, 66745536, 232092432, 901995588, 3173688180, 12342120700, 43861998180
Walks on square lattice. Ref GU90. [2,2; A5560]

M2988 1, 3, 15, 60, 260, 1092, 4641, 19635, 83215, 352440, 1493064, 6324552, 26791505, 113490195, 480752895, 2036500788, 8626757644, 36543528780
Fibonomial coefficients: $F(n)F(n+1)F(n+2)/2$. Ref FQ 6 82 68. BR72 74. [0,2; A1655, N1208]
$$\text{G.f.: } (1 + x - x^2)^{-1}(1 - 4x - x^2)^{-1}.$$

M2989 1, 3, 15, 73, 387, 2106
E-trees with at most 3 colors. Ref AcMaSc 2 109 82. [1,2; A7142]

M2990 1, 3, 15, 75, 363, 1767, 8463, 40695, 193983, 926943, 4404939, 20967075, 99421371, 471987255, 2234455839, 10587573027, 50060937987, 236865126051
n-step self-avoiding walks on cubic lattice. Equals ½M4202. Ref JCP 39 411 63. PPS 92 649 67. JPA 5 659 72. [1,2; A2902, N1210]

M2991 1, 1, 1, 3, 15, 75, 435, 3045, 24465, 220185, 2200905, 24209955, 290529855, 3776888115, 52876298475, 793144477125, 12690313661025, 215735332237425
Expansion of $\exp(-x^2/2) / (1-x)$. Ref R1 85. [0,4; A0266, N1211]

M2992 0, 0, 0, 0, 1, 3, 15, 79, 474, 3207, 24087, 198923, 1791902, 17484377
Asymptotic expansion of Hankel function. Ref CL45 XXXV. [1,6; A2514, N1212]

M2993 1, 3, 15, 81, 422, 2124, 10223, 47813, 218130, 977354, 4315130, 18833538
Dissections of a polygon. Ref AEQ 18 387 78. [5,2; A3448]

M2994 1, 3, 15, 82, 495, 3144, 20875, 142773, 1000131, 7136812, 51702231, 379234623,
2810874950, 21020047557, 158398829121
Directed rooted trees with n nodes. Ref LeMi91. [1,2; A6964]

M2995 1, 3, 15, 84, 495, 3003, 18564, 116280, 735471, 4686825, 30045015, 193536720,
1251677700, 8122425444, 52860229080, 344867425584, 2254848913647
Binomial coefficients $C(3n,n)$. See Fig M1645. Ref AS1 828. [0,2; A5809]

M2996 1, 3, 15, 86, 534, 3481, 23502, 162913, 1152870, 8294738, 60494549, 446205905,
3322769321, 24946773111, 188625900446, 1435074454755
Fixed 3-dimensional polyominoes with n cells. Ref RE72 108. dhr. [1,2; A1931, N1213]

M2997 1, 3, 15, 91, 612, 4389, 32890, 254475, 2017356, 16301164, 133767543,
1111731933, 9338434700, 79155435870, 676196049060, 5815796869995
From generalized Catalan numbers. Ref LNM 952 280 82. [0,2; A6632]

M2998 1, 3, 15, 93, 639, 4653, 35169, 272835, 2157759, 17319837, 140668065,
1153462995, 9533639025, 79326566595, 663835030335, 5582724468093
$\Sigma C(n,k)^2 . C(2k,k)$, $k = 0 .. n$. Ref AIP 9 345 60. SIAR 17 168 75. [0,2; A2893, N1214]

M2999 1, 3, 15, 104, 164, 194, 255, 495, 584, 975, 2204, 2625, 2834, 3255, 3705, 5186,
5187, 10604, 11715, 13365, 18315, 22935, 25545, 32864, 38804, 39524, 46215, 48704
$\phi(n) = \phi(n+1)$. Ref AMM 56 22 49. MI72. [1,2; A1274, N1215]

M3000 1, 3, 15, 104, 495, 975, 22935, 32864, 57584, 131144, 491535
Residues mod n are isomorphic to residues mod $n+1$. Ref MOC 27 448 73. [1,2; A3276]

M3001 3, 15, 105, 315, 6930, 18018, 90090, 218790, 2078505, 4849845, 22309287,
50702925, 1825305300, 4071834900, 18032411700, 39671305740, 347123925225
Coefficients of Legendre polynomials. Ref PR33 156. AS1 798. [0,1; A1801, N1216]

M3002 1, 1, 3, 15, 105, 945, 10395, 135135, 2027025, 34459425, 654729075,
13749310575, 316234143225, 7905853580625, 213458046676875, 6190283353629375
Double factorials: $(2n+1)!! = 1.3.5....(2n+1)$. Ref AMM 55 425 48. MOC 24 231 70.
[0,3; A1147, N1217]

M3003 1, 3, 15, 105, 947, 10472, 137337, 2085605, 36017472, 697407850, 14969626900,
352877606716, 9064191508018, 252024567201300, 7542036496650006
Expansion of $\ln(1+\ln(1+\ln(1+x)))$. Ref SMA 11 353 45. [0,2; A0268, N1218]

M3004 0, 3, 15, 120, 528, 4095, 17955, 139128, 609960, 4726275, 20720703, 160554240,
703893960, 5454117903, 23911673955, 185279454480, 812293020528, 6294047334435
Solution to a diophantine equation. Ref TR July 1973 p. 74. jos. [0,2; A6454]

M3005 1, 3, 15, 133, 2025, 37851, 1030367, 36362925, 1606008513
Toroidal semi-queens on a $(2n+1) \times (2n+1)$ board. Ref VA91 118. [0,2; A6717]

M3006 1, 1, 3, 15, 138, 2021, 43581, 1295493, 50752145, 2533755933, 157055247261,
11836611005031, 1066129321651668, 113117849882149725, 13965580274228976213
4-valent labeled graphs with n nodes. Ref SIAA 4 192 83. [0,3; A5816]

M3007 1, 3, 15, 159, 3903, 214143, 25098495
Certain subgraphs of a directed graph. Ref DM 14 119 76. [1,2; A5016]

M3008 1, 1, 1, 0, 3, 15, 203, 3785
Simple tournaments with n nodes. Ref DM 11 65 75. [1,5; A3505]

M3009 1, 1, 3, 15, 219, 7839, 777069, 208836207, 156458382975, 328208016021561,
1946879656265710431, 32834193098697741359313
Labeled Eulerian oriented graphs with n nodes. Ref CN 40 216 83. [1,3; A7081]

M3010 1, 3, 15, 3814279
Benford numbers: $a(n) = e \uparrow e \uparrow \cdots \uparrow e$ (n times). Ref NAMS 38 300 91. [0,2; A4002]

M3011 3, 16, 51, 126, 266, 504, 882, 1452, 2277, 3432, 5005, 7098, 9828, 13328, 17748,
23256, 30039, 38304, 48279, 60214, 74382, 91080, 110630, 133380, 159705, 190008
From expansion of $(1+x+x^2)^n$. Ref JCT 1 372 66. C1 78. [3,1; A0574, N1219]

M3012 3, 16, 57, 184, 601, 2036, 7072, 25088, 90503, 330836, 1222783, 4561058,
17145990
n-step walks on hexagonal lattice. Ref JPA 6 352 73. [3,1; A5550]

M3013 0, 0, 3, 16, 65, 238, 866, 3138
E-trees with exactly 2 colors. Ref AcMaSc 2 109 82. [1,3; A7143]

M3014 3, 16, 67, 251, 888, 3023, 10038, 32722
Partially labeled trees with n nodes. Ref R1 138. [3,1; A0269, N1220]

M3015 1, 1, 1, 3, 16, 75, 309, 1183, 4360, 15783, 56750, 203929, 734722, 2658071,
9662093, 35292151, 129513736, 477376575, 1766738922, 6563071865, 24464169890
$C(2n-2a\ n-1)/n - 2^{n-1} + n$. Ref JCT B21 75 76. [2,4; A4303]

M3016 0, 3, 16, 75, 356, 1770, 9306
Tumbling distance for n-input mappings. Ref PRV A32 2343 85. [0,2; A5947]

M3017 1, 3, 16, 75, 361, 1728, 8281
Area of nth triple of squares around a triangle. Ref PYTH 14 81 75. [1,2; A5386]

M3018 0, 1, 3, 16, 95, 666, 5327, 47944, 479439, 5273830, 63285959, 822717468, 11518044551, 172770668266, 2764330692255, 46993621768336, 845885191830047
$a(n) = (n+1)a(n-1) + (-1)^n$. [1,3; A6347]

M3019 1, 1, 0, 3, 16, 95, 672, 5397, 48704
Discordant permutations. Ref SMA 19 118 53. [0,4; A0270, N1221]

M3020 0, 0, 1, 3, 16, 96, 675, 5413, 48800, 488592, 5379333, 64595975, 840192288, 11767626752, 176574062535, 2825965531593, 48052401132800, 865108807357216
Sums of ménage numbers. Ref AH21 2 79. CJM 10 478 58. R1 198. [3,4; A0271, N1222]

$$a(n) = (n-1) \, a(n-2) + (n-1) \, a(n-1) + a(n-3).$$

M3021 1, 3, 16, 101, 756, 6607, 65794, 733833, 9046648
Forests with n nodes and height at most 2. Ref JCT 5 102 68. [1,2; A0949, N1223]

M3022 3, 16, 111, 2548, 14385, 672360, 10351845, 270594968, 2631486186, 310710613080
Coefficients for step-by-step integration. Ref JACM 11 231 64. [0,1; A2404, N1224]

M3023 1, 1, 3, 16, 112, 1020, 10222, 109947, 1230840, 14218671, 168256840, 2031152928
Dissecting a polygon into n 7-gons. Ref DM 11 388 75. [1,3; A5419]

M3024 1, 1, 3, 16, 124, 1256, 15576, 226248, 3729216, 68179968, 1361836800, 29501349120, 693638208000, 17815908096000, 502048890201600
Infinitesimal generator of $x(x+1)$. Ref EJC 1 132 80. [1,3; A5119]

M3025 1, 3, 16, 125, 1176, 12847, 160504, 2261289, 35464816
Forests with n nodes and height at most 3. Ref JCT 5 102 68. [1,2; A0950, N1225]

M3026 1, 3, 16, 125, 1296, 16087, 229384, 3687609, 66025360
Forests with n nodes and height at most 4. Ref JCT 5 102 68. [1,2; A0951, N1226]

M3027 1, 3, 16, 125, 1296, 16807, 262144, 4782969, 100000000, 2357947691, 61917364224, 1792160394037, 56693912375296, 1946195068359375
n^{n-2}. See Fig M0791. Ref BA9. R1 128. [2,2; A0272, N1227]

M3028 1, 3, 16, 137, 1826, 37777, 1214256, 60075185, 4484316358
Labeled topologies with n points. Ref MSM 11 243 74. [0,2; A6057]

M3029 1, 3, 16, 139, 1750, 29388, 623909
Bicoverings of an n-set. Ref SMH 3 147 68. [1,2; A2719, N1228]

M3030 1, 1, 3, 16, 145, 2111, 47624, 1626003, 82564031, 6146805142
Connected labeled topologies with n points. Ref MSM 11 243 74. [0,3; A6058]

M3031 0, 1, 3, 16, 185, 10886, 10552451
Edges in graph of maximal intersecting families of sets. Ref Loeb94a. Meye94. [1,3; A7006]

M3032 1, 1, 3, 16, 218, 9608, 1540944, 882033440, 1793359192848,
13027956824399552, 341260431952972580352, 3252290938505588611197440
Directed graphs with n nodes. See Fig M3032. Ref MIT 17 20 55. MAN 174 70 67. HA69 225. [0,3; A0273, N1229]

||||||||||||||||||||||||||||||||||||||||||||||||||||||||||||||||||||||||||||||||||||||||||||||||||||||||||||||||||||||||

Figure M3032. RELATIONS.

A **relation** R on a set S is any subset of $S \times S$, and xRy means $(x,y) \in R$ or "x is related to y." A relation is **reflexive** if xRx for all x in S, **symmetric** if $xRy \Rightarrow yRx$, **antisymmetric** if xRy and $yRx \Rightarrow x = y$, and **transitive** if xRy and $yRz \Rightarrow xRz$.

The most important types of relations are: (1) unrestricted, or digraphs with loops of length 1 allowed (M1980); (2) symmetric, or graphs with loops of length 1 allowed (M1650, M2868); (3) reflexive, or digraphs (M3032, illustrated below); (4) reflexive symmetric, or graphs (M1253, Fig. M1253); (5) reflexive transitive, or topologies (M2817, Fig. M2817). For the connection between digraphs and topologies, see [Bl1 117]); (6) reflexive symmetric transitive, or partitions (M0663, Fig. M0663); (7) reflexive antisymmetric transitive, or partially ordered sets (M1495, Fig. M1495). Generating functions are known for cases (1)–(4) and (6), but not (5) or (7) (see [PAMS 4 486 53], [MAN 174 53 67], [HP73]).

||||||||||||||||||||||||||||||||||||||||||||||||||||||||||||||||||||||||||||||||||||||||||||||||||||||||||||||||||||||||

M3033 1, 3, 16, 272, 11456
n-dimensional space groups in largest crystal class. Ref SC80 34. [1,2; A5031]

M3034 1, 1, 3, 16, 547, 538811, 620245817465, 692770666469127829226736,
1025344764595988314871439243086711931108916434521
Numerators of convergents to Lehmer's constant. Cf. M1545. Ref DUMJ 4 334 38. jww. [0,3; A2794, N1230]

M3035 3, 17, 29, 31, 43, 61, 67, 71, 83, 97, 107, 109, 113, 149, 151, 163, 181, 191, 193, 199, 227, 229, 233, 257, 269, 283, 307, 311, 313, 337, 347, 359, 389, 431, 433, 439, 443
Primes with -10 as primitive root. Ref AS1 846. [1,1; A7348]

M3036 3, 17, 29, 43, 73, 127, 179, 197, 251, 277, 281, 307, 349, 359, 397, 433, 521, 547, 557, 577, 593, 701, 757, 811, 853, 857, 863, 881, 919, 953, 1009, 1051, 1091, 1217, 1249
Primes of form $x^3 + y^3 + z^3$. Ref SI64 108. [1,1; A7490]

M3037 1, 3, 17, 99, 577, 3363, 19601, 114243, 665857, 3880899, 22619537, 131836323, 768398401, 4478554083, 26102926097, 152139002499, 886731088897
$a(n) = 6a(n-1) - a(n-2)$. Bisection of M2665. Ref NCM 4 166 1878. QJMA 45 14 14. ANN 36 644 35. AMM 75 683 68. [0,2; A1541, N1231]

M3038 0, 0, 0, 3, 17, 131, 915, 6553, 47026, 341888, 2505499, 18534827
One-sided 3-dimensional polyominoes with n cells. Ref CJN 18 366 75. [1,4; A6759]

M3039   1, 3, 17, 136, 2388, 80890, 5114079, 573273505, 113095167034,
39582550575765, 24908445793058442, 28560405143495819079
3-connected graphs with n nodes. Ref JCT B32 29 82. JCT B57 306 93. [4,2; A6290]

M3040   1, 3, 17, 142, 1569, 21576, 355081, 6805296, 148869153, 3660215680,
99920609601, 2998836525312, 98139640241473, 3478081490967552
$\Sigma n! \, n^{n-k-1} / (n-k)!$, $k = 1 .. n$. Ref AMS 26 515 55. KN1 1 112. [1,2; A1865, N1232]

M3041   1, 1, 3, 17, 155, 2073, 38227, 929569, 28820619, 1109652905, 51943281731,
2905151042481, 191329672483963, 14655626154768697, 1291885088448017715
Genocchi numbers: expansion of $\tan(x/2)$. See Fig M4019. Ref MOC 1 386 45. FMR 1 73.
C1 49. GKP 528. [1,3; A1469, N1233]

M3042   3, 17, 577, 665857, 886731088897, 1572584048032918633353217,
4946041176255201878775086487573351061418968498177
$a(n) = 2a(n-1)^2 - 1$. Ref AMM 61 424 54. TCS 65 219 89. [0,1; A1601, N1234]

M3043   1, 3, 18, 7, 1, 25, 7, 539, 25, 7, 22, 442, 225, 192, 13, 15, 26914, 244, 50, 5552, 30,
553, 7, 4493, 83342, 83, 65, 899, 3807, 64, 556, 20, 106, 132, 2277, 15, 1788, 5063, 27
Sum of n squares starting here is a square. Ref AMM 101 439 94. [1,2; A7475]

M3044   3, 18, 60, 150, 315, 588, 1008, 1620, 2475, 3630, 5148, 7098, 9555, 12600, 16320,
20808, 26163, 32490, 39900, 48510, 58443, 69828, 82800, 97500, 114075, 132678
Paraffins. Ref BER 30 1923 1897. [1,1; A6011]

$$\text{G.f.:} \quad 3 \, (1 + x) \, / \, (1 - x)^5.$$

M3045   3, 18, 61, 225, 716, 2272
Alkyls with n carbon atoms. Ref ZFK 93 437 36. [1,1; A0648, N1235]

M3046   1, 3, 18, 90, 270, 1134, 5670, 2430, 7290, 133650, 112266, 1990170, 9950850,
2296350, 984150
Denominators of generalized Bernoulli numbers. Cf. M3731. Ref DUMJ 34 614 67. [0,2;
A6568]

M3047   3, 18, 105, 636, 3807, 23094, 140469, 857736, 5251163, 32230218
Expansion of susceptibility series related to Potts model. Ref JPA 12 L230 79. [1,1;
A7277]

M3048   1, 3, 18, 110, 795, 6489, 59332, 600732, 6674805, 80765135, 1057289046,
14890154058, 224497707343, 3607998868005, 61576514013960, 1112225784377144
Permutations of length n by rises. Ref DKB 263. R1 210 (divided by 2). [3,2; A0274,
N1236]

$$a(n) = (1 + n) \, a(n-1) + (3 + n) \, a(n-2) + (3 - n) \, a(n-3) + (2 - n) \, a(n-4).$$

M3049   1, 3, 18, 136, 1170, 10962, 109158, 1138032, 12298392, 136803060, 1558392462
Hamiltonian rooted triangulations with n internal nodes. Ref DM 6 167 73. [0,2; A3122]

M3050 1, 3, 18, 153, 1638, 20898, 307908, 5134293, 95518278, 1967333838
Feynman diagrams of order $2n$. Ref PRV D18 1949 78. [1,2; A5412]

M3051 1, 3, 18, 172, 2433
Finite difference measurements. Ref SIAD 1 342 88. [2,2; A5192]

M3052 1, 1, 3, 18, 180, 2700, 56700, 1587600, 57153600, 2571912000, 141455160000,
9336040560000, 728211163680000, 66267215894880000, 6958057668962400000
$n!(n-1)!/2^{n-1}$. Ref SCS 12 122 81. [1,3; A6472]

M3053 1, 3, 18, 190, 3285, 88851, 3640644, 220674924
Precomplete Post functions. Ref SMD 10 619 69. JCT A14 6 73. [2,2; A2824, N1237]

M3054 3, 18, 1200, 33601536
Switching networks. Ref JFI 276 323 63. [1,1; A0853, N1238]

M3055 3, 18, 5778, 192900153618, 7177905237579946589743592924684178
$a(n+1)=a(n)(a(n)^2-3)$. Ref AMM 44 645 37. FQ 11 436 73. [0,1; A1999, N1239]

M3056 1, 3, 19, 117, 721, 4443, 27379, 168717, 1039681, 6406803, 39480499,
243289797, 1499219281, 9238605483, 56930852179, 350823718557, 2161873163521
$a(n)=6a(n-1)+a(n-2)$. Ref rkg. [0,2; A5667]

M3057 1, 3, 19, 147, 1251, 11253, 104959, 1004307, 9793891, 96918753, 970336269,
9807518757, 99912156111, 1024622952993, 10567623342519, 109527728400147
Apéry numbers: $\Sigma\, C(n,k)^2 . C(n+k,k)$, $k=0 \ldots n$. Ref AST 61 12 79. JNT 25 201 87.
[0,2; A5258]

M3058 1, 3, 19, 149, 2581, 84151, 5201856, 577050233, 113372069299,
39618015318982, 24916462761069296, 28563626972509456884
Series-reduced 2-connected graphs with n nodes. Ref JCT B32 31 82. JCT B57 299 93.
[4,2; A6289]

M3059 0, 0, 0, 1, 3, 19, 150, 2589, 84242, 5203110, 577076528, 113373005661,
39618075274687, 24916469690421103, 28563628406172313565
Connected unlabeled graphs with n nodes and degree ≥ 3. Ref rwr. [1,5; A7112]

M3060 0, 0, 0, 0, 1, 3, 19, 150, 2590, 84245, 5203135, 577076735, 113373008891,
39618075369549, 24916469695937480, 28563628406766988588
Unlabeled graphs with n nodes and degree ≥ 3. Ref rwr. [0,6; A7111]

M3061 1, 3, 19, 183, 2371, 38703, 763099
Semiorders on n elements. Ref MSH 62 79 78. [1,2; A6531]

M3062 1, 3, 19, 193, 2721, 49171, 1084483, 28245729, 848456353, 28875761731,
1098127402131, 46150226651233, 2124008553358849, 106246577894593683
Denominators of convergents to e. Cf. M4444. Ref BAT 17 1871. MOC 2 69 46. [0,2;
A1517, N1240]

$$a(n) = (4n-6)\,a(n-1) + a(n-2).$$

M3063 1, 1, 3, 19, 195, 3031, 67263, 2086099, 89224635, 5254054111, 426609529863, 47982981969979, 7507894696005795, 1641072554263066471
Labeled connected bipartite graphs. Ref JCT 6 17 69. CJM 31 65 79. NR82. [0,3; A1832, N1241]

M3064 1, 3, 19, 198, 2906, 55018, 1275030, 34947664, 1105740320, 39661089864
Planted evolutionary trees of magnitude n. Ref CN 44 85 85. [1,2; A7151]

M3065 1, 1, 3, 19, 211, 3651, 90921, 3081513, 136407699, 7642177651, 528579161353, 44237263696473, 4405990782649369, 515018848029036937, 69818743428262376523
Coefficients of a Bessel function. Ref AMM 71 493 64. BAMS 80 881 74. [0,3; A0275, N1242]

M3066 1, 1, 3, 19, 217, 3961, 105963, 3908059, 190065457, 11785687921, 907546301523, 84965187064099, 9504085749177097, 1251854782837499881
Salié numbers (expansion of cosh x /cos x). Ref C1 87. [0,3; A5647]

M3067 1, 3, 19, 219, 3991, 106623
Graded partially ordered sets with n elements. Ref JCT 6 17 69. [1,2; A1833, N1243]

M3068 1, 1, 3, 19, 219, 4231, 130023, 6129859, 431723379, 44511042511, 6611065248783, 1396281677105899, 414864951055853499, 171850728381587059351
Labeled partially ordered sets with n elements. Ref C1 60. CN 8 180 73. DM 53 148 85. ErSt89. [0,3; A1035, N1244]

M3069 1, 3, 19, 225, 3441, 79259, 2424195
Special permutations. Ref JNT 5 48 73. [3,2; A3111]

M3070 1, 1, 3, 19, 233, 4851, 158175, 7724333, 550898367, 56536880923
Connected labeled topologies on n points. Ref CN 8 180 73. MSM 11 243 74. [0,3; A1929, N1245]

M3071 1, 3, 19, 271, 7365, 326011, 21295783, 1924223799
Permutations of objects alike in pairs. Ref R1 17. [0,2; A3011]

M3072 1, 3, 20, 35, 126, 231, 3432, 6435, 24310, 46189, 352716, 676039, 2600150, 5014575, 155117520, 300540195, 1166803110
Coefficients of Legendre polynomials. Ref PR33 157. FMR 1 362. [2,2; A2461, N1246]

M3073 3, 20, 75, 210, 490, 1008, 1890, 3300, 5445, 8580, 13013
Nonseparable planar tree-rooted maps. Ref JCT B18 243 75. [1,1; A6411]

M3074 0, 3, 20, 119, 696, 4059, 23660, 137903, 803760, 4684659, 27304196, 159140519, 927538920, 5406093003, 31509019100, 183648021599, 1070379110496
Pythagorean triangles with consecutive legs (lesser given): $a(n)=6a(n-1)-a(n-2)+2$. Cf. M3955. Ref MLG 2 322 10. FQ 6(3) 104 68. [0,2; A1652, N1247]

M3075 3, 20, 130, 924, 7308, 64224, 623376, 6636960, 76998240, 967524480,
13096736640, 190060335360, 2944310342400, 48503818137600, 846795372595200
Associated Stirling numbers. Ref R1 75. C1 256. [4,1; A0276, N1248]

$$\text{E.g.f.:} \quad (3 + 2\,x - 6\ln(1-x)) \,/\, (1-x)^{-4}.$$

M3076 1, 1, 3, 20, 210, 3024, 55440, 1235520, 32432400, 980179200, 33522128640,
1279935820800, 53970627110400, 2490952020480000, 1249034513126400000
Planar embedded labeled trees with n nodes: $(2n-3)!/(n-1)!$. Ref LeMi91. [1,3;
A6963]

M3077 1, 1, 3, 20, 364, 17017, 2097018, 674740506, 568965009030, 1255571292290712,
7254987185250544104, 109744478168199574282739
Fibonomial Catalan numbers. Ref FQ 10 363 72. [0,3; A3150]

M3078 1, 3, 20, 996, 9333312
Post functions of n variables. Ref ZML 7 198 61. [1,2; A2857, N1249]

M3079 1, 3, 21, 23, 842, 1683
$n.16^n + 1$ is prime. Ref JRM 21 191 89. [1,2; A7648]

M3080 3, 21, 32, 79, 144, 155, 173, 202, 220, 231
Related to representations as sums of Fibonacci numbers. Ref FQ 11 357 73. [1,1; A6133]

M3081 1, 3, 21, 151, 1257, 12651, 151933, 2127231, 34035921, 612646867,
12252937701, 269564629863, 6469551117241, 168208329048891, 4709833213369677
$a(n+1) = (2n+3)\,a(n) - 2n\,a(n-1) + 8n$. Ref AMM 101 Problem 10403 94. [0,2;
A7566]

M3082 1, 3, 21, 185, 2010, 25914, 386407, 6539679, 123823305, 2593076255,
59505341676, 1484818160748, 40025880386401, 1159156815431055
Partitions into pairs. Ref PLIS 23 65 78. [1,2; A6199]

M3083 1, 3, 21, 231, 3495, 67455, 1584765, 43897455, 1400923755, 50619052575,
2042745514425, 91066568444775, 4444738893770175, 235731740255186175
A class of rooted trees with n nodes. Ref SZ 27 32 78. [1,2; A5373]

M3084 1, 1, 3, 21, 282, 6210, 202410, 9135630, 545007960, 41514583320,
3930730108200, 452785322266200, 62347376347779600, 10112899541133589200
Stochastic matrices of integers. Ref PSAM 15 101 63. SS70. [1,3; A0681, N1250]

$$a(n) = (n-1)^2 a(n-1) - \tfrac{1}{2}(n-1)(n-2)^2 a(n-2).$$

M3085 1, 1, 3, 21, 315, 9765, 615195
Certain subgraphs of a directed graph. Ref DM 14 118 76. [1,3; A5329]

M3086 1, 3, 21, 545, 30368
Trivalent graphs of girth exactly 7 and $2n$ nodes. Ref gr. [12,2; A6927]

M3087   1, 1, 3, 21, 651, 457653, 210065930571, 44127887745696109598901,
   1947270476915296449559659317606103024276803403
Binary trees of height n. Ref RSE 59(2) 159 39. FQ 11 437 73. [0,3; A1699, N1251]

M3088   1, 3, 21, 6615, 64595475
Stable feedback shift registers with n stages. Ref RO67 238. [2,2; A1139, N1252]

M3089   1, 3, 22, 66, 70, 81, 94, 115, 119, 170, 210, 214, 217, 265, 282, 310, 322, 343, 345,
   357, 364, 382, 385, 400, 472, 497, 510, 517, 527, 642, 651, 679, 710, 742, 745, 782, 795
Sum of divisors is a square. Ref B1 8. [1,2; A6532]

M3090   3, 22, 71, 169, 343, 628, 1068, 1717, 2640, 3914, 5629, 7889, 10813, 14536,
   19210, 25005, 32110, 40734, 51107, 63481, 78131, 95356, 115480, 138853, 165852
$C(n,5)+C(n,4)-C(n,3)+1$, $n \geq 7$. Ref NET 96. DKB 241. MMAG 61 28 88. [6,1;
A5288]

M3091   1, 3, 22, 85, 254, 644, 1448, 2967, 5645, 10109, 17214, 28093
Simple triangulations of a disk. Ref JCT B16 137 74. [0,2; A4305]

M3092   3, 22, 118, 383, 571, 635, 70529, 375687, 399380, 575584, 699357, 1561065,
   1795712, 194445473, 253745996, 3199003690, 3727084011, 6607433185, 16248462801
Pierce expansion for $1 / \pi$. Ref FQ 22 332 84. jos. [0,1; A6283]

M3093   1, 0, 3, 22, 192, 2046, 24853, 329406
Partition function for cubic lattice. Ref JMP 3 185 62. [0,3; A1393, N1253]

M3094   3, 22, 201, 2160, 24680, 285384, 3278484, 37154172
Maximally extended polygons of length $2n$ on cubic lattice. Ref JPA 22 2642 89. [1,1;
A6783]

M3095   1, 0, 3, 22, 207, 2412, 31754, 452640, 6840774, 108088232
$2n$-step polygons on cubic lattice. Ref JMP 3 188 62. [0,3; A1409, N1254]

M3096   1, 3, 22, 262, 4336, 91984, 2381408, 72800928, 2566606784, 102515201984,
   4575271116032, 225649908491264, 12187240730230208, 715392567595384832
Greg trees with n nodes. Ref SZ 27 31 78. LNM 829 122 80. MSS 34 127 90. [1,2; A5264]

M3097   3, 22, 333, 355, 103993, 104348, 208341, 312689, 833719, 1146408, 4272943,
   5419351, 80143857, 165707065, 245850922, 411557987, 1068966896, 2549491779
Numerators of convergents to π. See Fig M3097. Cf. M4456. Ref ELM 2 7 47. Beck71
171. [0,1; A2485, N1255]

Figure M3097. CONTINUED FRACTIONS.

Unlike the decimal expansion of a number, which depends on the arbitrary choice of 10 for the base, the **continued fraction** expansion of a number is "natural" or "intrinsic" [B1 257], [NZ66 151], [Gold94]. Here is the beginning of the continued fraction expansion of π (cf. Fig. M2218):

$$\pi = 3.14159\ldots$$
$$= 3 + \cfrac{1}{7 + \cfrac{1}{15 + \cfrac{1}{1 + \cfrac{1}{292 + \cdots}}}}$$

The terms in this expansion: 3, 7, 15, 1, 292, ... form sequence M2646. No pattern is known. By truncating a continued fraction at the n-th step we obtain the n-th **convergent** to the number. The successive convergents to π are

$$3, \frac{22}{7}, \frac{333}{106}, \frac{355}{113}, \frac{103993}{33102}, \ldots$$

whose numerators and denominators form M3097, M4456, respectively.

M3098 1, 3, 23, 36, 39, 56, 75, 83, 119, 120, 176, 183, 228, 633, 1520
$2^{2n+1} - 2^{n+1} + 1$ is prime. Cf. M2703. Ref CUNN. [1,2; A6598]

M3099 1, 3, 23, 117, 1609, 9747, 184607, 1257728
Minimal discriminant of number field of degree n. Ref STNB 2 133 90. [1,2; A6557]

M3100 1, 3, 23, 153, 1077, 8490, 75234, 742710, 8084990
Cycles in the complement of a path. Ref DM 55 277 85. [4,2; A6184]

M3101 1, 3, 23, 165, 3802, 21385, 993605, 15198435, 394722916, 3814933122,
447827009070
Coefficients for step-by-step integration. Ref JACM 11 231 64. [0,2; A2398, N1256]

M3102 0, 0, 1, 3, 23, 177, 1553, 14963, 157931
Polygons formed from n points on circle, no 2 adjacent. Ref IDM 26 118 19. [3,4; A2816, N1257]

M3103 3, 23, 275, 4511, 92779, 2306599
Minimal discriminant of number field of degree n. Ref Hass80 617. STNB 2 133 90. [2,1; A6555]

M3104 1, 3, 24, 150, 825, 4205, 20384, 95472, 436050, 1954150, 8629528, 37665030
Dissections of a polygon. Ref AEQ 18 386 78. [3,2; A3443]

M3105 1, 3, 24, 159, 2044, 36181
Pseudo-bricks with n nodes. Ref JCT B32 29 82. [4,2; A6292]

M3106   3, 24, 216, 1824, 15150
Card matching. Ref R1 193. [1,1; A0279, N1258]

M3107   1, 3, 24, 320, 6122, 153762, 4794664, 178788528, 7762727196, 384733667780,
21434922419504, 1326212860090560, 90227121642144424, 6694736236093168200
$\Sigma\,(-1)^{n-k}C(n,k)\,C((k+1)^2, n)$, $k = 0 \,.\,.\, n$. Ref hwg. [0,2; A3236]

M3108   3, 24, 1676, 22920064
Switching networks. Ref JFI 276 324 63. [1,1; A0856, N1259]

M3109   3, 25, 69, 135, 223, 333, 465, 619, 795, 993, 1213, 1455, 1719, 2005, 2313, 2643,
2995, 3369, 3765, 4183, 4623, 5085, 5569, 6075, 6603, 7153, 7725, 8319, 8935, 9573
$11n^2 + 11n + 3$. Ref LNM 751 68 79. [0,1; A6222]

M3110   3, 25, 155, 1005, 7488, 64164, 619986, 6646750, 78161249, 999473835,
13801761213, 204631472475, 3241541125110
Permutations of length n by rises. Ref DKB 264. [4,1; A0544, N1260]

M3111   1, 3, 25, 253, 3121, 46651, 823537, 16777209, 387420481, 9999999991,
285311670601, 8916100448245, 302875106592241, 11112006825558003
$n^n - n + 1$. Ref EUR 41 7 81. [1,2; A6091]

M3112   1, 3, 25, 299, 4785, 95699, 2296777, 64309755, 2057912161, 74084837795,
2963393511801, 130389314519243, 6258687096923665, 325451729040030579
Expansion of $e^{-x}/(1-4x)$. Ref R1 83. [0,2; A1907, N1261]

M3113   1, 1, 3, 25, 543, 29281, 3781503, 1138779265, 783702329343,
1213442454842881, 4175098976430598143
n-node acyclic digraphs. Ref HA73 254. [0,3; A3024]

M3114   1, 3, 25, 765, 3121, 233275, 823537, 117440505, 387420481, 89999999991,
285311670601, 98077104930805, 302875106592241, 144456088732254195
Pile of coconuts problem: $(n-1)(n^n-1)$, n even; $n^n - n + 1$, n odd. Ref AMM 35 48 28.
[1,2; A2021, N1262]

M3115   1, 1, 3, 26, 646, 45885, 9304650
Alternating sign matrices. Ref LNM 1234 292 86. [1,3; A5156]

M3116   1, 3, 27, 143, 3315, 20349, 260015, 1710855, 92116035, 631165425, 8775943605,
61750730457
Coefficients of Legendre polynomials. Ref MOC 3 17 48. [0,2; A1796, N1263]

M3117   0, 3, 28, 210, 1506, 10871, 80592, 618939, 4942070, 41076508, 355372524,
3198027157, 29905143464, 290243182755, 2920041395248, 30414515081650
Minimal covers of an n-set. Ref DM 5 249 73. [2,2; A3466]

M3118   1, 3, 28, 510, 18631, 1351413
Certain subgraphs of a directed graph. Ref DM 14 118 76. [2,2; A5328]

M3119   1, 3, 29, 289, 1627, 27769, 18044381, 145511171, 1514611753, 142324922009
Related to numerical integration formulas. Ref MOC 11 198 57. [1,2; A2669, N1264]

M3120   3, 29, 322, 3571, 39603, 439204, 4870847, 54018521, 599074578, 6643838879,
73681302247, 817138163596, 9062201101803, 100501350283429
Related to Bernoulli numbers. Ref RCI 141. [0,1; A1947, N1265]

$$\text{G.f.:}\ (3 - 4x)\ /\ (1 - 11x + x^2).$$

M3121   1, 0, 1, 3, 29, 2101, 7011349, 1788775603133, 53304526022885278659
Connected 2-plexes. Ref DM 6 384 73. [1,4; A3190]

M3122   3, 29, 15786, 513429610, 339840390654894740,
3835158804626209465840185663503380249
Egyptian fraction for $1/e$. [0,1; A6526]

M3123   1, 3, 30, 70, 315, 693, 12012, 25740, 109395, 230945, 1939938, 4056234,
16900975, 35102025, 1163381400, 2404321560, 9917826435, 20419054425
Coefficients of Legendre polynomials. Ref PR33 156. AS1 798. [0,2; A1800, N1266]

M3124   1, 3, 30, 175, 4410, 29106, 396396, 2760615, 156434850
Coefficients of Legendre polynomials. Ref PR33 157. FMR 1 362. [0,2; A2463, N1267]

M3125   1, 3, 30, 420, 6930, 126126, 2450448, 49884120, 1051723530, 22787343150,
504636071940, 11377249621920, 260363981732400, 6034149862347600
$(3n)!\ /\ (n+1)(n!)^3$. [0,2; A7004]

M3126   1, 3, 30, 630, 22680, 1247400, 97297200, 10216206000, 1389404016000,
237588086736000, 49893498214560000, 12623055048283680000
$(2n+1)!\ /\ 2^n$. [0,2; A7019]

M3127   3, 31, 171, 1575, 8403, 77206, 411771, 3828187, 20176803, 185374380,
988663371, 9083344725, 48444505203, 445083891551, 2373780754971
Free subsets of multiplicative group of $GF(7^n)$. Ref SFCA92 2 15. [1,1; A7233]

M3128   1, 1, 3, 31, 8401, 100130704103
Ternary trees with n nodes. Ref CMB 11 90 68. [0,3; A2707, N1268]

M3129   3, 31, 314159, 31415926535897932384626433832795028841
Primes in decimal expansion of π. Ref mg. [1,1; A5042]

M3130   3, 32, 225, 1320, 7007, 34944, 167076, 775200, 3517470, 15690048
Partitions of a polygon by number of parts. Ref CAY 13 95. [5,1; A2059, N1269]

M3131 0, 3, 33, 270, 2025, 14868, 109851, 827508, 6397665
Transfer impedances of an n-terminal network. Ref BSTJ 18 301 39. [2,2; A3129]

M3132 0, 0, 0, 3, 33, 338, 3580, 39525, 452865, 5354832, 65022840, 807560625, 10224817515, 131631305614
Simple quadrangulations. Ref JCT 4 275 68. [1,4; A1507, N1270]

M3133 3, 33, 564, 8976, 155124, 2791300, 51395172
Specific heat for cubic lattice. Ref PRV 129 102 63. [0,1; A2916, N1271]

M3134 1, 3, 33, 731, 25857, 1311379, 89060065, 7778778091, 849264442881, 113234181108643, 18073465545032353, 3395124358886313595
Expansion of sin(sin(sin x)). [0,2; A3715]

M3135 3, 33, 903, 46113, 3784503, 455538993, 75603118503, 16546026500673, 4616979073434903, 1599868423237443153, 674014138103352845703
Glaisher's H numbers. Ref PLMS 31 229 1899. FMR 1 76. [1,1; A2112, N1272]

M3136 1, 1, 3, 33, 13699, 19738610121
Nonantipodal balanced colorings of n-cube. Ref JALC 1 263 92. [1,3; A6854]

M3137 0, 0, 0, 0, 3, 35, 412, 4888, 57122, 667959, 7799183
4-dimensional polyominoes with n cells. Ref CJN 18 367 75. [1,5; A6767]

M3138 1, 3, 35, 1395, 200787, 109221651, 230674393235, 1919209135381395, 63379954960524853651, 8339787869494479328087443
Gaussian binomial coefficient $[2n,n]$ for $q=2$. Ref GJ83 99. ARS A17 328 84. [0,2; A6098]

M3139 0, 3, 36, 135, 360, 798, 1568, 2826, 4770, 7645, 11748, 17433
Tree-rooted planar maps. Ref JCT B18 256 75. [1,2; A6428]

M3140 1, 3, 36, 270, 4320, 17010, 5443200, 204120, 2351462400, 1515591000, 2172751257600, 354648294000, 10168475885568000, 7447614174000
Related to expansion of gamma function. Cf. M5399. Ref AMM 97 827 90. [1,2; A5446]

M3141 3, 36, 46764, 102266868132036, 1069559300034650646049671039050649693658764
A continued cotangent. Ref NBS B80 288 76. [0,1; A6268]

M3142 1, 1, 1, 3, 37, 1, 13, 638
Queens problem. Ref SL26 49. [1,4; A2563, N1273]

M3143 1, 3, 37, 959, 41641, 2693691, 241586893, 28607094455, 4315903789009, 807258131578995, 183184249105857781, 49548882107764546223
Expansion of sin(tanh x). [0,2; A3717]

M3144 1, 3, 37, 1015, 47881, 3459819, 354711853, 48961863007, 8754050024209, 1967989239505875, 543326939019354421, 180718022989699819207
Expansion of tan(sinh x). [0,2; A3716]

M3145 1, 1, 1, 3, 38, 135, 4315, 48125, 950684, 7217406, 682590930
Coefficients for step-by-step integration. Ref JACM 11 231 64. [0,4; A2405, N1274]

M3146 1, 0, 1, 3, 38, 680
Labeled Eulerian graphs with n nodes. Ref BW78 392. [1,4; A5780]

M3147 3, 40, 336, 2304, 14080, 79872, 430080, 2228224, 11206656, 55050240, 265289728, 1258291200, 5888802816, 27246198784, 124822487040, 566935683072
Coefficients of Chebyshev polynomials: $n(2n+1)2^{2n-2}$. Ref LA56 518. [1,1; A2700, N1275]

M3148 3, 40, 546, 7728, 112035, 1650792, 24608948, 370084832, 5603730876, 85316186400, 1304770191802, 20029132137840, 308437355259930
Quadrinomial coefficients. Ref C1 78. [2,1; A5724]

M3149 1, 1, 3, 41, 1035, 40721, 2291331, 174783865, 17394878523, 2192620580129, 341767803858867, 64587124941406473, 14555427555355014123
Reversion of g.f. for Euler numbers. Cf. M4019. [1,3; A7313]

M3150 3, 43, 73, 487, 2579, 8741
$(15^n - 1)/14$ is prime. Ref MOC 61 928 93. [1,1; A6033]

M3151 1, 3, 43, 95, 12139, 25333, 81227, 498233, 121563469, 246183839, 32808117961
Numerators of coefficients for numerical differentiation. Cf. M2133. Ref PHM 33 13 42. [1,2; A2551, N1276]

M3152 3, 45, 252, 28350, 1496880, 3405402000, 17513496000, 7815397590000, 5543722023840000, 235212205868640000, 206559082608278400000
Denominators of coefficients for repeated integration. Cf. M5136. Ref SAM 28 56 49. [0,1; A2682, N1277]

M3153 3, 45, 3411, 1809459, 7071729867, 208517974495911, 47481903377454219975, 85161307642554753639601848
Point-self-dual nets with $2n$ nodes. Ref CCC 2 32 77. rwr. JGT 1 295 77. [1,1; A4105]

M3154 1, 3, 48, 675, 9408, 131043, 1825200, 25421763, 354079482, 4931690986, 68689594335, 956722629712, 13325427221632
Standard deviation of 1,...,n is an integer. Cf. M4948. Ref dab. [1,2; A7654]

M3155 0, 3, 48, 765, 12192, 194307, 3096720, 49353213, 786554688, 12535521795, 199781794032, 3183973182717, 50743789129440, 808716652888323
$a(n) = 16a(n-1) - a(n-2)$. Ref NCM 4 167 1878. TH52 281. [0,2; A1080, N1278]

M3156 1, 3, 48, 3400, 955860, 1034141596, 4338541672792, 71839019692720536
Unilaterally connected digraphs with n nodes. Ref HA73 270. [1,2; A3029]

M3157 1, 3, 51, 3614, 991930, 1051469032, 4364841320040, 71943752944978224
n-node digraphs with a source. Ref HA73 270. [1,2; A3028]

M3158 3, 52, 575, 5470, 49303, 436446, 3850752, 3406392, 303790797
n-step walks on f.c.c. lattice. Ref JPA 6 351 73. [3,1; A5547]

M3159 3, 53, 680, 8064, 96370, 1200070, 15778800, 220047400, 3257228485,
51125192475, 849388162448
Permutations of length n by rises. Ref DKB 264. [6,1; A1279, N1279]

M3160 1, 3, 54, 3750, 1009680
Labeled mating digraphs with n nodes. Ref RE89. [1,2; A6025]

M3161 1, 3, 54, 3834, 1027080, 1067245748, 4390480560744, 72022346390883864
Weakly connected digraphs with n nodes. Ref HA73 270. [1,2; A3027]

M3162 1, 3, 55, 8103, 8886111, 72004899337, 4311231547115195,
1907346572495099690525, 6235149080811616882909238709
Nearest integer to exp n^2. Ref MNAS 14(5) 14 25. FW39. FMR 1 230. [0,2; A2818,
N1280]

M3163 1, 3, 57, 2763, 250737, 36581523, 7828053417, 2309644635483,
898621108880097, 445777636063460643, 274613643571568682777
Generalized Euler numbers. Ref QJMA 45 201 14. MOC 21 689 67. [0,2; A0281, N1281]

M3164 3, 59, 131, 251, 419, 659, 1019, 971, 1091, 2099, 1931, 1811, 3851, 3299, 2939,
3251, 4091, 4259, 8147, 5099, 9467, 6299, 6971, 8291, 8819, 14771, 22619, 9539, 13331
Smallest prime $\equiv 3 \bmod 8$ where $Q(\sqrt{-p})$ has class number $2n+1$. Cf. M5407. Ref MOC
24 492 70. BU89 224. [0,1; A2148, N1282]

M3165 3, 60, 630, 5040, 34650, 216216
Coefficients for extrapolation. Ref SE33 93. [0,1; A2738, N1283]

M3166 0, 3, 60, 650, 5352, 37681, 239752, 1421226, 7996160, 43219990, 226309800,
1154900708
Tree-rooted planar maps. Ref JCT B18 257 75. [1,2; A6432]

M3167 0, 3, 60, 1197, 23880, 476403, 9504180, 189607197, 3782639760, 75463188003,
1505481120300, 30034159217997, 599177703239640, 11953519905574803
$a(n) = 20a(n-1) - a(n-2)$. Ref NCM 4 167 1878. MTS 65(4, Supplement) 8 56. [0,2;
A1084, N1284]

M3168 1, 3, 60, 7848
Connected regular graphs of degree 5 with $2n$ nodes. Ref OR76 135. [3,2; A6821]

M3169   3, 70, 3783
Finite automata. Ref CJM 17 112 65. [1,1; A0282, N1285]

M3170   1, 3, 72, 439128, 84722519069640072,
  608130213374088941214747405817720857404971722895128
Denominators of a continued fraction. Ref NBS B80 288 76. [0,2; A6270]

M3171   3, 73, 8599, 400091364, 371853741549033970,
  253461181173408820488703379557217678
Continued cotangent for π. Ref DUMJ 4 339 38. jos. [0,1; A2667, N1286]

M3172   1, 1, 3, 107, 1095, 41897, 3027637, 34528445, 11832720271, 1190157296815,
  22592230600813, 23107531656941541, 2633888933338158633
Expansion of tan(sin x). [0,3; A3705]

M3173   0, 0, 3, 131, 1830, 16990, 127953, 851361, 5231460, 30459980, 170761503,
  931484191, 4979773890, 26223530970, 136522672653, 704553794621, 3611494269120
Trees of subsets of an n-set. Ref MBIO 54 9 81. [1,3; A5175]

M3174   3, 147, 1383123, 489735485064147, 245597025618959718190041238775763,
  247062114274836300381127305147102564467751924522387062291401805739987
$a(n+1) = a(n) + 4^{n-1} a(n)^2$. Ref JLMS 28 286 53. [3,1; A3009]

M3175   1, 3, 196, 3406687200
Invertible Boolean functions of n variables. Ref JACM 10 27 63. [1,2; A0724, N1287]

M3176   1, 1, 3, 275, 15015, 968167, 77000363, 7433044411, 843598411471,
  107426835190735, 14072980460605907, 1424712499632406371
Expansion of sin(tan x). [0,3; A3706]

M3177   1, 3, 340, 246295, 796058676, 9736032295374, 432386386904461704,
  70004505120317453723895, 41988978212639552393332333300
Bipartite blocks. Ref CJM 31 67 79. [1,2; A5335]

M3178   1, 3, 355, 297619, 1120452771, 15350524923547, 738416821509929731,
  126430202628042630866787, 78847417416749666369637926851
Bipartite blocks. Ref CJM 31 67 79. [1,2; A5336]

M3179   1, 3, 567, 43659, 392931, 1724574159, 2498907956391, 1671769422825579,
  88417613265912513891, 21857510418232875496803
Expansion of Weierstrass P-function. Ref MOC 16 477 62. [1,2; A2306, N1288]

M3180   1, 3, 840, 54486432000
Invertible Boolean functions of n variables. Ref JACM 10 27 63. [1,2; A0723, N1289]

M3181   1, 3, 4523, 11991, 18197, 141683, 1092489, 3168099, 6435309, 12489657,
  17906499, 68301841, 295742437, 390117873, 542959199
Square in base 2 is a palindrome. Ref JRM 5 13 72. rhh. [1,2; A3166]

M3182 3, 26861, 26879, 616841, 617039, 617269, 617471, 617521, 617587, 617689, 617723, 622813, 623387, 623401, 623851, 623933, 624031, 624097, 624191, 624241
Where prime race $4n - 1$ vs. $4n + 1$ changes leader. Ref rgw. [1,1; A7350]

SEQUENCES BEGINNING . . ., 4, 0, . . . TO . . ., 4, 12, . . .

M3183 0, 0, 0, 0, 0, 0, 0, 0, 0, 4, 0, 0, 0, 0, 0, 0, 0, 0, 0, 0, 0, 0, 0, 0, 0, 1, 0, 0, 0, 0, 0, 0, 0, 9, 0, 0, 0, 0, 0, 0, 0, 0, 0, 0, 0, 0, 0, 0, 0, 6, 0, 0, 0, 0, 0, 0, 0, 9, 0, 0, 0, 0, 0, 0, 0, 0, 0, 0, 0
Theta series of h.c.p. w.r.t. tetrahedral hole. Ref JCP 83 6530 85. [0,10; A5873]

M3184 1, 0, 0, 4, 0, 0, 0, 0, 12, 0, 0, 12, 0, 0, 0, 0, 6, 0, 0, 12, 0, 0, 0, 0, 24, 0, 0, 16, 0, 0, 0, 0, 12, 0, 0, 24, 0, 0, 0, 0, 24, 0, 0, 12, 0, 0, 0, 0, 8, 0, 0, 24, 0, 0, 0, 0, 48, 0, 0, 36, 0, 0, 0, 0
Theta series of diamond lattice. Ref JMP 28 1653 87. [0,4; A5925]

M3185 0, 0, 0, 4, 0, 0, 0, 8, 0, 112, 256, 156, 896, 3536, 5472, 5400, 49088, 115008, 47776, 555492, 1976736, 2563424, 4446272, 29452776, 61952896, 4795392, 374024448
Susceptibility for b.c.c. lattice. Ref JMP 6 298 65. JPA 6 1511 73. [1,4; A3194]

M3186 4, 0, 0, 4, 28, 80, 120, 120, 264, 1080, 3120
n-step polygons on Kagomé lattice. Ref PRV 114 53 59. [3,1; A5397]

M3187 4, 0, 0, 8, 60, 144, 416, 1248, 4200, 13248, 42936, 138072, 452840, 1480944, 4883688, 16114784, 53457696, 177637248
Susceptibility for cubic lattice. Ref JMP 6 298 65. JPA 6 1511 73. [3,1; A2915, N1290]

M3188 0, 0, 1, 0, 0, 4, 0, 1, 0, 28, 1, 118, 294, 184, 441, 1486, 273, 6464, 8969
Percolation series for directed b.c.c. lattice. Ref JPA 16 3146 83. [2,6; A6838]

M3189 4, 0, 12, 8, 48, 96, 320, 888, 2748, 8384, 26340, 83568, 268864, 873648
Susceptibility for honeycomb. Ref JPA 5 636 72. [3,1; A2978]

M3190 1, 0, 4, 0, 14, 8, 40, 32, 105, 112, 284, 320, 702, 840
Expansion of a modular function. Ref PLMS 9 385 59. [−3,3; A6710]

M3191 0, 4, 0, 16, 0, 64, 96, 488, 1392, 5064, 17856, 65576, 231728, 863664, 3313392
Susceptibility for diamond lattice. Ref JMP 6 298 65. JPA 6 1511 73. [1,2; A3195]

M3192 1, 0, 4, 0, 16, 24, 122, 348, 1266, 4464, 16394, 57932, 215916, 828348
Low temperature antiferromagnetic susceptibility for diamond lattice. Ref DG74 422. [0,3; A7216]

M3193 0, 4, 0, 16, 32, 156, 608, 2688, 12064, 55956, 266656
Susceptibility for square lattice. Ref JPA 5 636 72. [1,2; A2979]

M3194 0, 1, 4, 1, 2, 5, 4, 5, 8, 1, 2, 5, 2, 3, 6, 5, 6, 9, 4, 5, 8, 5, 6, 9, 8, 9, 12, 1, 2, 5, 2, 3, 6, 5, 6, 9, 2, 3, 6, 3, 4, 7, 6, 7, 10, 5, 6, 9, 6, 7, 10, 9, 10, 13, 4, 5, 8, 5, 6, 9, 8, 9, 12, 5, 6, 9, 6
Sum of squares of digits of ternary representation of n. Ref jos. [0,3; A6287]

M3195 1, 4, 1, 4, 2, 1, 3, 5, 6, 2, 3, 7, 3, 0, 9, 5, 0, 4, 8, 8, 0, 1, 6, 8, 8, 7, 2, 4, 2, 0, 9, 6, 9,
8, 0, 7, 8, 5, 6, 9, 6, 7, 1, 8, 7, 5, 3, 7, 6, 9, 4, 8, 0, 7, 3, 1, 7, 6, 6, 7, 9, 7, 3, 7, 9, 9, 9, 0, 7, 3
Decimal expansion of square root of 2. Ref PNAS 37 65 51. MOC 22 899 68. [1,2; A2193,
N1291]

M3196 0, 1, 1, 4, 1, 5, 1, 12, 6, 7, 1, 16, 1, 9, 8, 32, 1, 21, 1, 24, 10, 13, 1, 44, 10, 15, 27,
32, 1, 31, 1, 80, 14, 19, 12, 60, 1, 21, 16, 68, 1, 41, 1, 48, 39, 25, 1, 112, 14, 45, 20, 56, 1
$a(1) = 0$, $a(\text{ prime }) = 1$, $a(mn) = m.a(n) + n.a(m)$. Ref CMB 4 117 61. CMCN 5(8) 6 73.
[1,4; A3415]

M3197 1, 1, 4, 1, 6, 4, 8, 1, 13, 6, 12, 4, 14, 8, 24, 1, 18, 13, 20, 6, 32, 12, 24, 4, 31, 14, 40,
8, 30, 24, 32, 1, 48, 18, 48, 13, 38, 20, 56, 6, 42, 32, 44, 12, 78, 24, 48, 4, 57, 31, 72, 14
Sum of odd divisors of n. Ref RCI 187. [1,3; A0593, N1292]

M3198 4, 1, 10, 2, 16, 3, 22, 4, 28, 5, 34, 6, 40, 7, 46, 8, 52, 9, 58, 10, 64, 11, 70, 12, 76,
13, 82, 14, 88, 15, 94, 16, 100, 17, 106, 18, 112, 19, 118, 20, 124, 21, 130, 22, 136, 23
Image of n under the '$3x + 1$' map. See Fig M2629. Ref UPNT 16. [1,1; A6370]

$$\text{G.f.: } (4x + x^2 + 2x^3) \ / \ (1 - x^2)^2.$$

M3199 1, 4, 1, 12, 186, 4, 86, 4860
Queens problem. Ref SL26 49. [1,2; A2564, N1293]

M3200 1, 4, 1, 16, 16, 120, 8, 728
Queens problem. Ref SL26 49. [1,2; A2568, N1294]

M3201 1, 4, 2, 4, 8, 64, 64, 256
Numerator of nth power of Hermite constant for dimension n. See Fig M2209. Cf. M2209.
Ref Cass71 332. GrLe87 410. SPLAG 20. [1,2; A7361]

M3202 1, 1, 4, 2, 7, 3, 508, 1, 5, 5, 1, 1, 1, 2, 1, 1, 24, 1, 1, 1, 3, 3, 30, 4, 10, 158, 6, 1, 1, 2,
12, 1, 10, 1, 1, 3, 2, 1, 1, 89, 1, 1, 2, 1, 1, 1, 3, 1, 2, 1, 7, 1, 2, 18, 1, 17, 2, 2, 10, 14, 3, 1, 2
Continued fraction for cube root of 6. [1,3; A2949]

M3203 1, 1, 4, 2, 7, 5, 15, 6, 37, 13, 36, 32, 37, 34, 73, 58, 183, 150, 262, 186, 1009, 420,
707, 703, 760, 1180, 4639
Polygonal graphs. Ref SL26 21. [4,3; A2560, N1295]

M3204 1, 4, 2, 8, 5, 4, 10, 8, 9, 0, 14, 16, 10, 4, 0, 8, 14, 20, 2, 0, 11, 20, 32, 16, 0, 4, 14, 8,
9, 20, 26, 0, 2, 28, 0, 16, 16, 28, 22, 0, 14, 16, 0, 40, 0, 28, 26, 32, 17, 0, 32, 16, 22, 0, 10
Expansion of $\Pi(1 - x^k)^4$. Ref KNAW 59 207 56. [0,2; A0727, N1296]

M3205 0, 1, 4, 2, 8, 13, 28, 26, 56, 69, 48, 134, 80, 182, 84, 312, 280, 204, 332, 142, 816,
91, 196, 780, 224, 526
Related to representation as sums of squares. Ref QJMA 38 56 07. [0,3; A2291, N1297]

M3206 1, 4, 3, 1, 3, 5, 1, 3, 5, 3, 3, 1, 5, 3, 1, 3, 3, 5, 3, 1, 3, 5, 1, 3, 3, 5, 3, 1, 5, 3, 1, 3, 5,
  3, 3, 1, 3, 5, 1, 3, 5, 3, 3, 1, 5, 3, 1, 3, 5, 3, 3, 1, 3, 5, 1, 3, 3, 5, 3, 1, 5, 3, 1, 3, 3, 5, 3, 1, 3
A continued fraction. Ref JNT 11 217 79. [0,2; A6467]

M3207 0, 4, 3, 2, 3, 1, 2, 2, 1, 2, 3, 1, 3, 2, 3, 1, 2, 1, 2, 2, 2, 2, 2, 1, 3, 3, 2, 2, 3, 1, 2, 2, 3,
  2, 2, 1, 3, 2, 3, 2, 3, 2, 3, 2, 1, 2, 3, 1, 3, 2, 2, 3, 3, 3, 2, 3, 2, 2, 3, 4, 1, 2, 2, 2, 3, 3, 1, 3, 2, 2
Fibonacci frequency of n. Ref HM68. MOC 23 460 69. ACA 16 109 69. [1,2; A1178, N1298]

M3208 4, 3, 3, 6, 4, 5, 4, 6, 6, 6, 10, 11, 10, 12, 10, 11, 10, 12, 12, 8, 12, 11, 11, 14, 12, 13,
  12, 14, 14, 8, 12, 11, 11, 14, 12, 13, 12, 14, 14, 5, 9, 8, 8, 11, 9, 10, 9, 11, 11, 9, 13, 12, 12
Number of letters in n (in Russian). [1,1; A6994]

M3209 4, 3, 3, 11, 35, 395, 13859, 5474699, 75873867299, 415386585433442699,
  315169866609617578165202672699
Knopfmacher expansion of 2/3: $a(n+1)=a(n-1)(a(n)+1)-1$. Ref Knop. [0,1; A7568]

M3210 4, 3, 4, 2, 9, 4, 4, 8, 1, 9, 0, 3, 2, 5, 1, 8, 2, 7, 6, 5, 1, 1, 2, 8, 9, 1, 8, 9, 1, 6, 6, 0, 5,
  0, 8, 2, 2, 9, 4, 3, 9, 7, 0, 0, 5, 8, 0, 3, 6, 6, 6, 5, 5, 6, 6, 1, 1, 4, 4, 5, 3, 7, 8, 3, 1, 6, 5, 8, 6, 4
Decimal expansion of common logarithm of e. Ref PNAS 26 211 40. [0,1; A2285, N1299]

M3211 1, 1, 4, 3, 4, 4, 8, 11, 4, 4, 12, 48, 12, 8, 16, 25, 16, 4, 20, 0, 32, 12, 24, 248, 4, 12,
  4, 208, 28, 16, 32, 41, 48, 16, 32, 400, 36, 20, 48, 88, 40, 32, 44, 544, 16, 24, 48, 732, 8, 4
Coefficients of a Dirichlet series. Ref EM 6 38 60. [1,3; A2558, N1300]

M3212 1, 4, 3, 8, 5, 24, 5, 13, 9, 20, 7, 48, 13, 16, 13, 26, 17, 52, 19, 37, 21, 44, 13, 96, 25,
  34, 27, 32, 13, 124, 17, 52, 33, 41, 19, 104, 35, 52, 37, 65, 25, 123, 17, 73, 39, 92, 41, 183
Smallest k such that $\phi(n+k) = \phi(k)$. Ref AS1 840. [1,2; A7015]

M3213 0, 0, 1, 1, 4, 3, 8, 8, 12, 13, 22, 17, 28, 31, 36, 36, 51, 47, 64, 61, 70, 77, 98, 85,
  103, 112, 125, 124, 151, 138, 167, 167, 184, 197, 218, 198, 233, 248, 269, 258, 297, 284
Problem E2817. Ref AMM 87 137 80. jos. [1,5; A4125]

M3214 0, 1, 1, 4, 3, 11, 8, 29, 21, 76, 55, 199, 144, 521, 377, 1364, 987, 3571, 2584, 9349,
  6765, 24476, 17711, 64079, 46368, 167761, 121393, 439204, 317811, 1149851, 832040
$a(n)=3a(n-2)-a(n-4)$. Ref LNM 748 57 79. [0,4; A5013]

M3215 1, 4, 3, 11, 15, 13, 17, 24, 23, 73, 3000, 11000, 15000, 101, 104, 103, 111, 115,
  113, 117, 124, 123, 173, 473, 373, 1104, 1103, 1111, 1115, 1113, 1117, 1124, 1123, 1173
Smallest number requiring n letters in English. [3,2; A1166, N1301]

M3216 1, 4, 3, 32, 75, 216, 3577, 5888, 15741, 106300, 13486539, 9903168,
  42194238652, 710986864, 796661595, 127626606592, 450185515446285
Cotesian numbers. Ref QJMA 46 63 14. [1,2; A2178, N1302]

M3217 1, 1, 4, 3, 192, 20, 11520, 315, 573440, 36288, 928972800, 1663200,
  54499737600, 74131200, 1322526965760, 68108040000
Values of an integral. Ref PHM 36 295 45. MOC 19 114 65. [1,3; A2298, N1303]

M3218 1, 4, 4, 0, 4, 8, 0, 0, 4, 4, 8, 0, 0, 8, 0, 0, 4, 8, 4, 0, 8, 0, 0, 0, 0, 12, 8, 0, 0, 8, 0, 0, 4,
     0, 8, 0, 4, 8, 0, 0, 8, 8, 0, 0, 0, 8, 0, 0, 0, 4, 12, 0, 8, 8, 0, 0, 0, 0, 8, 0, 0, 8, 0, 0, 4, 16, 0, 0
Theta series of square lattice. See Fig M3218. Ref SPLAG 106. [0,2; A4018]

||||||||||||||||||||||||||||||||||||||||||||||||||||||||||||||||||||||||||||||||||||||||||||||||||||||||||||||||||||||||||||||||||||

Figure M3218. THETA SERIES.

The **theta series** of a lattice L is the generating function $\sum_{u \in L} q^{u \cdot u} = \sum A_n q^n$ in which
the exponents n are the squared lengths, or **norms**, of the lattice vectors, and the coefficients
A_n give the number of lattice vectors of norm n [SPLAG 45]. For the two-dimensional square
lattice, visible on any piece of squared paper, the sequence begins

$$1 + 4q + 4q^2 + 4q^4 + 8q^5 + 4q^8 + \cdots,$$

and is equal to $\vartheta_3(q)^2$, where ϑ_3 is a Jacobi theta function [SPLAG 106], [WhWa Chap. XXI].
In this case the exponents (M0968) are those numbers that can be written as the sum of two
squares, and the coefficients form M3218. More generally we may form the theta series with
respect to (w.r.t.) any point w of the space:

$$\sum_{u \in \Lambda} q^{(u-w) \cdot (u-w)}.$$

M0931 and M3319 give the coefficients of the theta series of this lattice w.r.t. respectively the
midpoint of an edge joining two lattice points, and a point of the space maximally distant from
the lattice (a **deep hole** in the lattice).

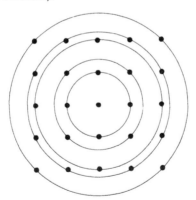

||||||||||||||||||||||||||||||||||||||||||||||||||||||||||||||||||||||||||||||||||||||||||||||||||||||||||||||||||||||||||||||||||||

M3219 4, 4, 0, 22, 44, 32, 100, 352, 492, 166, 2268, 4914, 3212, 11083
Percolation series for b.c.c. lattice. Ref SSP 10 921 77. [0,1; A6805]

M3220 1, 4, 4, 2, 2, 4, 9, 5, 7, 0, 3, 0, 7, 4, 0, 8, 3, 8, 2, 3, 2, 1, 6, 3, 8, 3, 1, 0, 7, 8, 0, 1, 0,
     9, 5, 8, 8, 3, 9, 1, 8, 6, 9, 2, 5, 3, 4, 9, 9, 3, 5, 0, 5, 7, 7, 5, 4, 6, 4, 1, 6, 1, 9, 4, 5, 4, 1, 6, 8
Decimal expansion of cube root of 3. Ref SMA 18 175 52. [1,2; A2581, N1304]

M3221 1, 4, 4, 2, 6, 9, 5, 0, 4, 0, 8, 8, 8, 9, 6, 3, 4, 0, 7, 3, 5, 9, 9, 2, 4, 6, 8, 1, 0, 0, 1, 8, 9,
     2, 1, 3, 7, 4, 2, 6, 6, 4, 5, 9, 5, 4, 1, 5, 2, 9, 8, 5, 9, 3, 4, 1, 3, 5, 4, 4, 9, 4, 0, 6, 9, 3, 1, 1, 0
Decimal expansion of $\log_2 e$. [1,2; A7525]

M3222 4, 4, 4, 4, 4, 5, 6, 4, 4, 4, 3, 4, 8, 8, 8, 8, 8, 8, 8, 7, 13, 14, 14, 14, 14, 15, 16, 14, 14,
8, 14, 15, 15, 15, 15, 16, 17, 15, 15, 7, 13, 14, 14, 14, 14, 15, 16, 14, 14, 7, 13, 14, 14, 14
Number of letters in n (in German). [1,1; A7208]

M3223 0, 0, 1, 1, 4, 4, 4, 4, 5, 5, 5, 5, 5, 5, 5, 5, 5, 8, 8, 8, 8, 8, 8, 8, 8, 8, 8, 8, 8, 8, 8, 8, 8, 9,
9, 9, 9, 9, 9, 9, 9, 9, 9, 9, 9, 9, 9, 9, 9, 9, 9, 9, 9, 9, 9, 9, 9, 9, 9, 9, 9, 9, 9, 9, 9, 10, 10, 10
Diagonal length function. Ref SIAC 20 161 91. [0,5; A6264]

M3224 1, 4, 4, 4, 5, 8, 10, 12, 14, 16, 21, 24
Minimal number of knights to cover $n \times n$ board. See Fig M3224. Ref GA78 194. [1,2;
A6075]

‖‖

Figure M3224. ATTACKING KNIGHTS.

M3224 gives the minimal number of knights needed to attack or occupy every square of an
$n \times n$ board. M0884 gives the number of distinct solutions. The unique solution for an 8×8
board is:

‖‖

M3225 1, 0, 4, 4, 5, 0, 12, 16, 21, 16, 24, 20, 17, 0, 32, 48, 65, 64, 84, 84, 85, 64, 92, 96,
101, 80, 88, 68, 49, 0, 80, 128, 177, 192, 244, 260, 277, 256, 316, 336, 357, 336, 360, 340
Σ k AND $n - k$, $k = 1 .. n - 1$. Ref mlb. [2,3; A6581]

M3226 4, 4, 6, 7, 8, 9, 11, 12, 13, 14
Ramsey numbers. Ref ADM 41 80 89. [1,1; A6672]

M3227 4, 4, 8, 12, 4, 12, 12, 12, 16, 16, 8, 8, 28, 12, 20, 24, 8, 16, 28, 12, 16, 28, 20, 32,
20, 16, 16, 32, 20, 24, 28, 8, 36, 44, 12, 32, 36, 16, 24, 20, 28, 20, 56, 28, 16, 40, 20, 40
Theta series of b.c.c. lattice w.r.t. deep hole. Ref JCP 83 6532 85. [0,1; A4024]

M3228 4, 4, 8, 12, 20, 24, 28, 16, 40, 20, 56, 20, 12, 60, 80, 28, 84, 56, 52, 16, 28, 112, 84,
132, 112, 140, 156, 96, 144, 176, 160, 136, 140, 44, 76, 88, 204, 152, 220, 24, 252, 120
y such that $p^2 = x^2 + 3y^2$. Cf. M4773. Ref CU27 79. L1 60. [7,1; A2368, N1337]

M3229 1, 1, 4, 4, 8, 24, 32, 40, 120, 296
Graceful permutations of length n. Ref WiYo87. [1,3; A6967]

M3230 1, 1, 4, 4, 9, 8, 55, 21, 105, 62, 429, 196
Isonemal fabrics of period exactly n. Ref HW84 88. [2,3; A5441]

M3231 1, 4, 4, 10, 4, 16, 4, 20, 10, 16, 4, 40, 4, 16, 16, 35, 4, 40, 4, 40, 16, 16, 4, 80, 10,
   16, 20, 40, 4, 64, 4, 56, 16, 16, 16, 100, 4, 16, 16, 80, 4, 64, 4, 40, 40, 16, 4, 140, 10, 40
Inverse Moebius transform applied thrice to all 1's sequence. Ref EIS § 2.7. [1,2; A7426]

M3232 1, 1, 4, 4, 10, 11, 22, 25, 44, 51, 83, 98, 149, 177, 259, 309, 436, 521, 716, 857,
   1151, 1376, 1816, 2170, 2818, 3361, 4309, 5132
Partitions with at least 1 odd and 1 even part. Ref AMM 79 508 72. [3,3; A6477]

M3233 1, 1, 4, 4, 13, 19, 39, 59, 112, 169, 294, 448, 735, 1110, 1757
Representation degeneracies for Neveu-Schwarz strings. Ref NUPH B274 547 86. [1,3; A5301]

M3234 1, 0, 1, 0, 4, 4, 18, 26, 86, 158, 462, 976, 2665, 6082, 16040, 38338, 99536,
   244880, 631923, 1583796, 4081939
Dyck paths of knight moves. Ref DAM 24 218 89. [0,5; A5222]

M3235 0, 1, 4, 4, 32, 16, 56, 80, 192, 98, 740, 704, 96, 224, 2440, 3520, 2624, 351, 780,
   10632, 2688, 2960, 9496, 18176, 14208, 3934, 12552, 9856, 24608, 9760, 2720, 25344
Related to representation as sums of squares. Ref QJMA 38 320 07. [1,3; A2611, N1305]

M3236 1, 1, 4, 5, 6, 4, 8, 13, 13, 6, 12, 20, 14, 8, 24, 29
Generalized divisor function. Ref PLMS 19 111 19. [1,3; A2129, N1307]

M3237 1, 4, 5, 6, 7, 8, 10, 16, 18, 19, 21, 31, 32, 33, 42, 46, 56, 57, 66, 70, 79, 82, 91, 96,
   104, 105, 107, 116, 129, 130, 131, 141, 158, 165, 168, 179, 180, 182, 191, 204, 205, 206
$a(n)$ is smallest number which is uniquely $a(j)+a(k)$, $j<k$. Ref GU94. [1,2; A3666]

M3238 0, 4, 5, 6, 8, 16, 27, 49, 92, 168, 320, 613, 1177, 2262, 4432, 8696, 17072, 33531,
   65885, 130593, 258924
A generalized Conway-Guy sequence. Ref MOC 50 312 88. [0,2; A6756]

M3239 1, 4, 5, 6, 9, 12, 15, 16, 17, 20, 21, 22, 25, 26, 27, 30, 33, 36, 37, 38, 41, 44, 47, 48,
   49, 52, 55, 58, 59, 60, 63, 64, 65, 68, 69, 70, 73, 76, 79, 80, 81, 84, 85, 86, 89, 90, 91, 94
A self-generating sequence. Ref FQ 10 500 72. [1,2; A3156]

M3240 1, 1, 4, 5, 6, 10, 15, 21, 31, 46, 67, 98, 144, 211, 309, 453, 664, 973, 1426, 2090,
   3063, 4489, 6579, 9642, 14131, 20710, 30352, 44483, 65193, 95545, 140028, 205221
$a(n)=a(n-1)+a(n-3)$. Ref JA66 91. FQ 6(3) 68 68. [0,3; A1609, N1308]

M3241 1, 4, 5, 6, 11, 12, 13, 14, 15, 22, 23, 24, 25, 26, 27, 28, 37, 38, 39, 40, 41, 42, 43,
   44, 45, 56, 57, 58, 59, 60, 61, 62, 63, 64, 65, 66, 79, 80, 81, 82, 83, 84, 85, 86, 87, 88, 89
Take 1, skip 2, take 3, etc. Cf. M0821. Ref HO85a 177. [1,2; A7606]

M3242 0, 1, 0, 0, 0, 4, 5, 6, 11, 31, 72, 157, 312, 700, 1472, 3446, 7855
Self-avoiding walks on square lattice. Ref JCT A13 181 72. [4,6; A6144]

M3243 0, 4, 5, 7, 11, 12, 16, 23, 26, 31, 33, 37, 38, 44, 49, 56, 73, 78, 80, 85, 95, 99, 106,
124, 128, 131, 136, 143, 169, 188, 197, 203, 220, 221, 226, 227, 238, 247, 259, 269, 276
No 3-term arithmetic progression. Ref UPNT E10. [0,2; A5487]

M3244 4, 5, 9, 10, 11, 14, 19, 20, 23, 24, 25, 26, 32, 33, 37, 38, 39, 41, 42, 48, 50, 53, 54,
55, 59, 63, 64, 65, 69, 70, 76, 77, 80, 83, 85, 86, 89, 99, 102, 104, 108, 110, 113, 114, 115
Not the sum of 3 hexagonal numbers (probably finite). Ref AMM 101 170 94. [1,1; A7536]

M3245 4, 5, 9, 13, 14, 17, 19, 21, 24, 25, 27, 35, 37, 43, 45, 47, 57, 67, 69, 73, 77, 83, 93,
101, 105, 109, 113, 115, 123, 125, 133, 149, 153, 163, 173, 197, 201, 205, 209, 211, 213
$a(n)$ is smallest number which is uniquely $a(j)+a(k)$, $j<k$ (periodic mod 192). Ref JCT
A12 33 72. EXPM 1 58 92. GU94. [1,1; A6844]

M3246 1, 4, 5, 9, 14, 23, 37, 60, 97, 157, 254, 411, 665, 1076, 1741, 2817, 4558, 7375,
11933, 19308, 31241, 50549, 81790, 132339, 214129, 346468, 560597, 907065, 1467662
$a(n)=a(n-1)+a(n-2)$. Ref FQ 3 129 65. BR72 53. Robe92 224. [0,2; A0285, N1309]

M3247 1, 4, 5, 10, 14, 41, 94, 154, 158, 500
13.4^n+1 is prime. Ref PAMS 9 674 58. Rie85 381. [1,2; A2257, N1310]

M3248 1, 4, 5, 10, 16, 19, 20, 26, 29, 31, 35, 41, 43, 49, 50, 55, 56, 59, 70, 71, 80, 85, 94,
95, 100, 101, 106, 109, 110, 121, 149, 154, 160, 166, 175, 179, 184, 190, 191, 200, 205
$4n^2+9$ is prime. Ref KK71 1. [1,2; A2970]

M3249 1, 4, 5, 11, 7, 20, 9, 26, 18, 28, 13, 55, 15, 36, 35, 57, 19, 72, 21, 77, 45, 52, 25,
130, 38, 60, 58, 99, 31, 140, 33, 120, 65, 76, 63, 198, 39, 84, 75, 182, 43, 180, 45, 143
Inverse Moebius transform applied twice to natural numbers. Ref EIS § 2.7. [1,2; A7429]

M3250 1, 1, 4, 5, 11, 16, 29, 45, 76, 121, 199, 320, 521, 841, 1364, 2205, 3571, 5776,
9349, 15125, 24476, 39601, 64079, 103680, 167761, 271441, 439204, 710645, 1149851
Expansion of $(1+x^2)/(1-x^2)(1-x-x^2)$. Ref EUR 11 22 49. [0,3; A1350, N1311]

M3251 0, 1, 1, 4, 5, 11, 20, 36, 65, 119, 218, 412, 770, 1466, 2784, 5322, 10226, 19691,
38048, 73665, 142927, 277822, 540851, 1054502, 2058507, 4023164
Integers $\leq 2^n$ of form $2x^2+5y^2$. Ref MOC 20 563 66. [0,4; A0286, N1312]

M3252 1, 1, 4, 5, 11, 22, 57, 51, 156, 158, 566, 499, 1366
No-3-in-line problem on $n \times n$ grid. Ref GK68. Wels71 124. LNM 403 7 74. GA89 69.
JCT A60 307 92. [2,3; A0769, N1313]

M3253 0, 0, 0, 1, 1, 4, 5, 13, 18, 39, 57, 112, 169, 313, 482, 859, 1341, 2328, 3669, 6253,
9922, 16687, 26609, 44320, 70929, 117297, 188226, 309619, 497845, 815656, 1313501
$F(n)-2^{[n/2]}$. Ref rkg. [0,6; A5672]

M3254 1, 1, 4, 5, 14, 23, 52, 97, 202, 395, 800, 1589, 3190, 6367, 12748, 25481, 50978,
101939, 203896, 407773, 815566, 1631111, 3262244, 6524465, 13048954, 26097883
$a(n) = a(n-1) + 2.a(n-2) + (-1)^n$. Ref GKP 327. [4,3; A6904]

$$\text{G.f.: } (1+x+x^2) / (1-2x)(1+x)^2.$$

M3255 0, 0, 0, 0, 1, 1, 1, 4, 5, 14, 28, 86, 211, 648, 1878, 5941, 18326, 58746
Trivalent 3-connected bipartite planar graphs with $4n$ nodes. Ref JCT B38 295 85. [2,8;
A7084]

M3256 1, 1, 4, 5, 15, 19, 45, 52, 118, 137, 281, 316, 625, 695, 1331
Expansion of a modular function. Ref PLMS 9 385 59. [-4,3; A2509, N1314]

M3257 0, 0, 0, 4, 5, 15, 21, 44, 66, 120, 187, 319, 507, 840, 1348, 2204, 3553, 5776, 9329,
15124, 24454, 39600, 64055, 103679, 167735, 271440, 439176, 710644, 1149821
Strict (-1)st-order maximal independent sets in cycle graph. Ref YaBa94. [1,4; A7390]

M3258 1, 0, 4, 5, 15, 28, 60, 117, 230, 440, 834, 1560, 2891, 5310, 9680, 17527, 31545,
56468, 100590, 178395, 315106, 554530, 972564, 1700400, 2964325, 5153868, 8938300
Generalized Lucas numbers. Ref FQ 15 252 77. [2,3; A6491]

M3259 0, 1, 4, 5, 16, 17, 20, 21, 64, 65, 68, 69, 80, 81, 84, 85, 256, 257, 260, 261, 272,
273, 276, 277, 320, 321, 324, 325, 336, 337, 340, 341, 1024, 1025, 1028, 1029, 1040
Moser-de Bruijn sequence: sums of distinct powers of 4. Ref MMAG 35 37 62. MOC 18
537 64. TCS 98 188 92. [1,3; A0695, N1315]

M3260 1, 4, 5, 29, 34, 63, 286, 349, 635, 5429, 6064, 90325, 96389, 1054215, 2204819,
3259034, 15240955, 186150494, 387541943, 1348776323, 3085094589, 4433870912
Convergents to cube root of 2. Ref AMP 46 105 1866. L1 67. [1,2; A2352, N1316]

M3261 1, 1, 0, 4, 5, 96, 427, 6448, 56961, 892720, 11905091, 211153944, 3692964145,
75701219608, 1613086090995
Series-reduced labeled trees with n nodes. Ref MAT 15 188 68. LeMi91. rcr. [1,4; A5512]

M3262 0, 0, 0, 4, 6, 0, 0, 0, 0, 0, 0, 12, 8, 0, 0, 0, 0, 0, 0, 12, 24, 0, 0, 0, 0, 0, 0, 16, 0, 0, 0,
0, 0, 0, 0, 24, 30, 0, 0, 0, 0, 0, 0, 12, 24, 0, 0, 0, 0, 0, 0, 24, 24, 0, 0, 0, 0, 0, 0, 36, 0, 0, 0, 0
Theta series of diamond lattice w.r.t. deep hole. Ref JMP 28 1653 87. [0,4; A5927]

M3263 1, 4, 6, 4, 3, 12, 16, 16, 6, 8, 18, 28, 26, 20, 2, 12, 23, 32, 36, 28, 6, 4, 22, 20, 39,
32, 32, 12, 2, 16, 12, 24, 40, 28, 34, 0, 6, 16, 0, 40, 6, 36, 26, 32, 5, 0, 20
Coefficients of powers of η function. Ref JLMS 39 435 64. [4,2; A1482, N1317]

M3264 4, 6, 6, 9, 2, 0, 1, 6, 0, 9, 1, 0, 2, 9, 9, 0, 6, 7, 1, 8, 5, 3, 2, 0, 3, 8, 2, 0, 4, 6, 6, 2, 0,
1, 6, 1, 7, 2, 5, 8, 1, 8, 5, 5, 7, 7, 4, 7, 5, 7, 6, 8, 6, 3, 2, 7, 4, 4, 5, 6, 5, 1, 3, 4, 3, 0, 0, 4, 1, 3
Decimal expansion of Feigenbaum bifurcation velocity. Ref JPA 12 275 79. MOC 57 438
91. [1,1; A6890]

M3265 4, 6, 7, 7, 8, 9, 9, 10, 10, 10, 11, 11, 12, 12, 12, 13, 13, 13, 13, 14, 14, 14, 15, 15,
  15, 15, 16, 16, 16, 16, 16, 17, 17, 17, 17, 18, 18, 18, 18, 18, 19, 19, 19, 19, 19, 19, 20
Chromatic number of surface of connectivity n. Ref CJM 4 480 52. PNAS 60 438 68. IJM
21 429 77. [1,1; A0703, N1318]

M3266 1, 4, 6, 7, 7, 12, 12, 19, 21, 26
Pair-coverings with largest block size 3. Ref ARS 11 90 81. [3,2; A6185]

M3267 1, 4, 6, 7, 8, 9, 9, 10, 10, 11, 11, 11, 11, 12, 12, 12, 12, 13, 13, 13
Smallest coprime dissection of $n \times n$ quilt. Ref PCPS 60 367 64. UPG C3. [1,2; A5670]

M3268 4, 6, 7, 9, 10, 11, 12, 14, 15, 16, 17, 18, 19, 20, 22, 23, 24, 25, 26, 27, 28, 29, 30,
  31, 32, 33, 35, 36, 37, 38, 39, 40, 41, 42, 43, 44, 45, 46, 47, 48, 49, 50, 51, 52, 53, 54, 56
Non-Fibonacci numbers. Ref FQ 3 183 65. [1,1; A1690, N1319]

M3269 0, 1, 4, 6, 7, 9, 10, 15, 16, 22, 24, 25, 28, 31, 33, 36, 40, 42, 49
Of the form $x^2 + 6y^2$. Ref EUL (1) 1 425 11. [1,3; A2481, N1320]

M3270 1, 4, 6, 7, 9, 13, 21, 46, 71, 109, 168, 265, 417
Spiral sieve using Fibonacci numbers. Ref FQ 12 395 74. [1,2; A5621]

M3271 1, 4, 6, 7, 13, 14, 16, 20, 21, 23, 25, 27, 29, 32, 34, 42, 45, 49, 51, 53, 59, 60, 70,
  75, 78, 81, 84, 85, 86, 87, 88, 90, 93, 95, 96, 104, 109, 114, 115, 116, 124, 125, 135, 137
Elliptic curves. Ref JRAM 212 23 63. [1,2; A2151, N1321]

M3272 4, 6, 8, 9, 10, 12, 14, 15, 16, 18, 20, 21, 22, 24, 25, 26, 27, 28, 30, 32, 33, 34, 35,
  36, 38, 39, 40, 42, 44, 45, 46, 48, 49, 50, 51, 52, 54, 55, 56, 57, 58, 60, 62, 63, 64, 65
Composite numbers. Ref HW1 2. [1,1; A2808, N1322]

M3273 1, 4, 6, 8, 11, 13, 16, 18, 23, 25, 28, 30, 35, 37, 40, 42, 47, 49, 52, 54, 59, 61, 64,
  66, 71, 73, 76, 78, 83, 85, 88, 90, 95, 97, 100, 102, 107, 109, 112, 114, 119, 121, 124, 126
$a(n)$ is smallest number $\neq a(j) + a(k)$, $j < k$. Ref GU94. [1,2; A3662]

M3274 4, 6, 9, 10, 14, 15, 21, 22, 25, 26, 33, 34, 35, 38, 39, 46, 49, 51, 55, 57, 58, 62, 65,
  69, 74, 77, 82, 85, 86, 87, 91, 93, 94, 95, 106, 111, 115, 118, 119, 121, 122, 123, 129
Product of two primes (sometimes called semi-primes). Ref EUR 17 8 54. MMAG 47 167
74. [1,1; A1358, N1323]

M3275 4, 6, 9, 10, 15, 16, 18, 24, 27, 28, 30, 34, 42, 45, 46, 51, 52, 54, 58, 66, 69, 78, 81,
  82, 87, 88, 90, 99, 100, 102, 106, 114, 123, 130, 132, 135, 136, 150, 153, 154, 159, 160
If n appears so do $2n - 2$ and $3n - 3$. [1,1; A5659]

M3276 1, 4, 6, 9, 11, 14, 17, 19, 22, 25, 27, 30, 32, 35, 38, 40, 43, 45, 48, 51, 53, 56, 59,
  61, 64, 66, 69, 72, 74, 77, 79, 82, 85, 87, 90, 93, 95, 98, 100, 103, 106, 108, 111, 114, 116
Related to Fibonacci representations. Ref FQ 11 386 73. [1,2; A3259]

M3277 1, 4, 6, 9, 12, 14, 17, 19, 22
First column of Wythoff array. Ref Morr80. Kimb91. [1,2; A7065]

M3278 1, 4, 6, 9, 12, 14, 17, 19, 22, 25, 27, 30, 33, 35, 38, 40, 43, 46, 48, 51, 53, 56, 59, 61, 64, 67, 69, 72, 74, 77, 80, 82, 85, 88, 90, 93, 95, 98, 101, 103, 106, 108, 111, 114, 116
From a 3-way splitting of positive integers: $[[n\tau]\tau]$. Cf. M2715. Ref BR72 62. Robe92 10. [1,2; A3622]

M3279 1, 4, 6, 9, 12, 14, 17, 20, 22, 25, 27, 30, 33, 35
First column of array associated with lexicographically justified array. Ref Kimb93. [1,2; A7073]

M3280 1, 4, 6, 9, 12, 15, 17, 20, 22, 25, 28, 30, 33
First column of array associated with reverse lexicographically justified array. Ref Kimb93. [1,2; A7074]

M3281 4, 6, 11, 14, 21, 24, 26, 29, 31, 39, 44, 46, 51, 54, 76, 79, 89, 94, 99, 101, 111, 119, 124, 129, 131, 136, 146, 149, 154, 156, 164, 176, 179, 181, 194, 201, 211, 214, 229, 231
$(4n^2 + 1)/5$ is prime. Ref EUL (1) 3 24 17. [1,1; A2732, N1324]

M3282 0, 1, 0, 1, 1, 4, 6, 14, 28, 60
Fixed points in planted trees. Ref PCPS 85 413 79. [1,6; A5202]

M3283 0, 1, 1, 4, 6, 15, 30, 74, 160, 379, 867, 2057, 4817, 11465, 27214, 65102, 155753, 374208, 900073, 2170500, 5240723, 12676162, 30697119, 74435204, 180679171
Symmetry sites in all planted 1,3-trees with $2n$ nodes. Ref DAM 5 157 83. CN 41 149 84. rwr. [1,4; A7135]

M3284 1, 1, 4, 6, 16, 28, 64, 120, 256, 496
Dual pairs of integrals arising from reflection coefficients. Ref JPA 14 365 81. [1,3; A7179]

M3285 0, 1, 1, 4, 6, 18, 35, 93, 214, 549, 1362, 3534, 9102, 23951, 63192, 168561, 451764, 1219290, 3305783, 9008027, 24643538, 67681372, 186504925, 515566016
Average forests of planted trees. Ref JCT B27 118 79. [1,4; A5199]

M3286 1, 1, 4, 6, 18, 42, 118, 314, 895, 2521, 7307, 21238, 62566, 185310, 553288, 1660490, 5011299, 15190665, 46244031, 141296042, 433204573, 1332261200
Ethylene derivatives with n carbon atoms. Ref BA76 44. [2,3; A5959]

M3287 1, 1, 1, 4, 6, 19, 43, 121
One-sided triangular polyominoes with n cells. Ref jm. [1,4; A6534]

M3288 1, 1, 1, 1, 4, 6, 19, 49, 150, 442, 1424, 4522, 14924, 49536, 167367, 570285, 1965058, 6823410, 23884366, 84155478, 298377508, 1063750740, 3811803164
One-sided triangulations of the disk. Ref AMM 64 153 57. PLMS 14 759 64. DM 11 387 75. [1,5; A1683, N1325]

M3289 1, 1, 4, 6, 23
n-dimensional crystal families. Ref BB78 52. [0,3; A4032]

M3290   1, 0, 4, 6, 24, 66, 214, 676, 2209, 7296, 24460, 82926, 284068, 981882, 3421318, 12007554, 42416488, 150718770, 538421590, 1932856590, 6969847486
Rooted polyhedral graphs with n edges. Ref CJM 15 265 63. [6,3; A0287, N1326]

M3291   1, 1, 1, 1, 1, 1, 1, 4, 7, 4, 4, 4, 7, 4, 13, 7, 19, 7, 7, 7, 19, 19, 19, 16, 31, 19, 28, 19, 49, 31, 28, 31, 64, 43, 37, 127, 61, 52, 52, 52, 49, 100, 37, 112, 64, 67, 61, 76, 61, 76, 61
Class numbers of cubic fields. Ref MOC 28 1140 74. [1,8; A5472]

M3292   4, 7, 8, 9, 10, 11, 12, 12, 13, 13, 14, 15, 15, 16, 16, 16, 17, 17, 18, 18, 19, 19, 19, 20, 20, 20, 21, 21, 21, 22, 22, 22, 23, 23, 23, 24, 24, 24, 24, 25, 25, 25, 25, 26, 26, 26
Chromatic number of surface of genus n: $[(7+\sqrt{(1+48n)})/2]$. Ref PNAS 60 438 68. IJM 21 429 77. [0,1; A0934, N1327]

M3293   4, 7, 8, 10, 26, 32, 70, 74, 122, 146, 308, 314, 386, 512, 554, 572, 626, 635, 728, 794, 842, 910, 914, 1015, 1082, 1226, 1322, 1330, 1346, 1466, 1514, 1608, 1754, 1994
$\phi(n) = \phi(n+2)$. Ref AMM 56 22 49. [1,1; A1494, N1328]

M3294   1, 4, 7, 8, 11, 17, 20, 20, 23, 29, 35, 38, 39, 45, 51, 51, 54, 63, 69, 72, 78, 84, 87, 87, 90, 99, 111, 115, 115, 127, 133, 133, 136, 142, 151, 157, 163, 169, 178, 178, 184, 199
Nonnegative solutions of $x^2 + y^2 + z^2 \leq n$. Ref PNISI 13 39 47. [0,2; A0606, N1329]

M3295   1, 4, 7, 8, 17, 21, 29
Diameter of integral set of n points in plane. Ref DCG 9 430 93. [3,2; A7285]

M3296   1, 4, 7, 9, 11, 12, 14, 16, 19, 20, 23, 24, 27, 28, 31, 32, 35, 40, 39, 40, 45, 48, 51, 52, 55, 56, 59, 59, 64, 65, 70, 73, 75, 79, 80, 84, 85, 88, 89, 91, 93, 96, 98, 101, 103, 106
Atomic weights of the elements. [1,2; A7656]

M3297   1, 4, 7, 9, 12, 14, 17, 19, 22, 25, 27, 30, 33, 35
First column of array associated with monotonic justified array. Ref Kimb93. [1,2; A7072]

M3298   1, 4, 7, 9, 12, 14, 17, 20, 22, 25
First column of Stolarsky array. Ref FQ 15 224 77. PAMS 117 317 93. [1,2; A7064]

M3299   1, 4, 7, 9, 12, 15, 17, 20, 22, 25, 28, 30, 33, 36
First column of dual Wythoff array. Ref Morr80. Kimb91. [1,2; A7066]

M3300   1, 4, 7, 10, 13, 17, 22, 25, 30, 35, 40, 46, 53, 57, 61
Zarankiewicz's problem. Ref TI68 132. LNM 110 141 69. C1 291. [1,2; A1197, N1330]

M3301   4, 7, 10, 13, 19, 28, 31, 34, 40, 43, 52, 70, 73, 76, 82, 85, 91, 97, 103, 112, 115, 124, 127, 136, 145, 148, 157, 166, 175, 187, 190, 199, 202, 223, 241, 244, 259, 265, 271
$(n^2 + n + 1)/3$ is prime. Ref CU23 1 248. [1,1; A2640, N1331]

M3302   4, 7, 10, 16, 28, 52, 100, 196, 388, 772, 1540, 3076, 6148, 12292, 24580, 49156, 98308, 196612, 393220, 786436, 1572868, 3145732, 6291460, 12582916, 25165828
Bode numbers: $4 + 3 \cdot 2^{n-1}$. Ref SKY 43 281 72. McL1. [0,1; A3461]

M3303 4, 7, 11, 15, 18, 19, 23, 25, 27, 31, 32, 33, 35, 41, 47, 49, 55, 57, 61, 63, 75, 87, 89, 91, 105, 119, 121, 125, 129, 133, 139, 147, 153, 161, 185, 189, 203, 206, 213, 225, 233
$a(n)$ is smallest number which is uniquely $a(j)+a(k)$, $j<k$ (periodic mod 11301098). Ref GU94. [1,1; A3670]

M3304 1, 1, 4, 7, 11, 20, 35, 59, 99, 165, 270, 443
Restricted solid partitions. Ref JCT A13 144 72. [1,3; A2974]

M3305 1, 4, 7, 12, 16, 23, 28, 35, 40, 47
Davenport-Schinzel numbers. Ref ARS 1 47 76. UPNT E20. [1,2; A5005]

M3306 0, 0, 0, 0, 0, 0, 0, 0, 4, 7, 12, 18, 37, 53, 75, 100, 152
Biplanar crossing number of complete graph on n nodes. Ref PGCT 18 280 71. [1,9; A7333]

M3307 1, 1, 1, 1, 4, 7, 13, 25, 49, 94, 181, 349, 673, 1297, 2500, 4819, 9289, 17905, 34513, 66526, 128233, 247177, 476449, 918385, 1770244, 3412255, 6577333, 12678217
Tetranacci numbers: $a(n)=a(n-1)+a(n-2)+a(n-3)+a(n-4)$. Ref FQ 2 260 64. [0,5; A0288, N1332]

M3308 1, 4, 7, 13, 25, 49, 97, 193, 385, 769, 1537, 3073, 6145, 12289, 24577, 49153, 98305, 196609, 393217, 786433, 1572865, 3145729, 6291457, 12582913, 25165825
$3 \cdot 2^n + 1$. Ref MOC 30 660 76. [0,2; A4119]

M3309 1, 4, 7, 14, 16, 31, 29, 50, 52, 74, 67, 119, 92, 137, 142, 186, 154, 247, 191, 294, 266, 323, 277, 455, 341, 446, 430, 553, 436, 686, 497, 714, 634, 752, 674, 1001, 704, 935
Inverse Moebius transform of triangular numbers. Ref BeSl94. EIS § 2.7. [1,2; A7437]

M3310 1, 1, 4, 7, 14, 23, 41, 63, 104, 152, 230, 327, 470, 647, 897, 1202, 1616, 2117, 2775, 3566, 4580, 5787, 7301, 9092, 11298, 13885, 17028, 20688, 25076, 30154, 36172
Certain partially ordered sets of integers. Ref P4BC 123. [0,3; A3404]

M3311 1, 1, 4, 7, 16, 26, 50, 76, 126, 185, 280, 392, 561, 756, 1032, 1353, 1782, 2277, 2920, 3652, 4576, 5626, 6916, 8372, 10133
n-bead necklaces with 6 red beads. Ref JAuMS 33 12 82. [6,3; A5513]

M3312 1, 4, 7, 18, 26, 68
Triangulations. Ref WB79 337. [0,2; A5509]

M3313 1, 1, 4, 7, 19, 32, 68, 114, 210, 336, 562
$3 \times n$ binary matrices. Ref PGEC 22 1050 73. [0,3; A6381]

M3314 1, 1, 4, 7, 19, 40, 97, 217, 508, 1159, 2683, 6160, 14209, 32689, 75316, 173383, 399331, 919480, 2117473, 4875913, 11228332, 25856071, 59541067, 137109280
$a(n)=a(n-1)+3a(n-2)$. Ref FQ 15 24 77. [0,3; A6130]

M3315 1, 4, 7, 29, 199, 5778, 1149851, 6643838879, 7639424778862807,
5075510735900469455482320́4
$L(L(n))$, where L is a Lucas number. [1,2; A5371]

M3316 1, 4, 7, 31, 871, 756031, 571580604871, 326704387862983487112031,
106735757048926752040856495274871386126283608871
A nonlinear recurrence. Ref AMM 70 403 63. FQ 11 431 73. [0,2; A0289, N1333]

M3317 1, 1, 4, 7, 33
n-dimensional crystal systems. Ref BB78 52. Enge93 1021. [0,3; A4031]

M3318 4, 8, 1, 2, 1, 1, 8, 2, 5, 0, 5, 9, 6, 0, 3, 4, 4, 7, 4, 9, 7, 7, 5, 8, 9, 1, 3, 4, 2, 4, 3, 6, 8,
4, 2, 3, 1, 3, 5, 1, 8, 4, 3, 3, 4, 3, 8, 5, 6, 6, 0, 5, 1, 9, 6, 6, 1, 0, 1, 8, 1, 6, 8, 8, 4, 0, 1, 6, 3
Natural logarithm of golden ratio. Cf. M4046. Ref RS8 XVIII. [0,1; A2390, N1334]

M3319 4, 8, 4, 8, 8, 0, 12, 8, 0, 8, 8, 8, 4, 8, 0, 8, 16, 0, 8, 0, 4, 16, 8, 0, 8, 8, 0, 8, 8, 8, 4,
16, 0, 0, 0, 16, 8, 8, 8, 0, 0, 12, 8, 0, 8, 16, 0, 8, 8, 0, 16, 0, 0, 0, 16, 12, 8, 8, 0, 8, 8, 0, 0, 8
Theta series of square lattice w.r.t. deep hole. See Fig M3218. Ref SPLAG 106. [0,1;
A5883]

M3320 1, 1, 4, 8, 5, 22, 42, 40, 120, 265, 286, 764, 1729, 2198, 5168, 12144, 17034,
37702, 88958, 136584, 288270, 682572, 1118996, 2306464, 5428800, 9409517
Subgroups of index n in modular group. Ref MOC 30 845 76. [1,3; A5133]

M3321 1, 1, 4, 8, 6, 9, 8, 3, 5, 4, 9, 9, 7, 0, 3, 5, 0, 0, 6, 7, 9, 8, 6, 2, 6, 9, 4, 6, 7, 7, 7, 9, 2,
7, 5, 8, 9, 4, 4, 4, 3, 8, 5, 0, 8, 8, 9, 0, 9, 7, 7, 9, 7, 5, 0, 5, 5, 1, 3, 7, 1, 1, 1, 1, 8, 4, 9, 3, 6, 0
Decimal expansion of fifth root of 2. [1,3; A5531]

M3322 4, 8, 8, 16, 12, 8, 24, 16, 16, 24, 16, 16, 28, 32, 8, 32, 32, 16, 40, 16, 16, 40, 40, 32,
36, 16, 24, 48, 32, 24, 40, 48, 16, 56, 32, 16, 64, 40, 32, 32, 36, 40, 48, 48, 32, 48, 48, 16
Theta series of cubic lattice w.r.t. square. Ref SPLAG 107. [0,1; A5877]

M3323 4, 8, 9, 16, 19, 20, 21, 26, 30, 31, 33, 38, 42, 43, 50, 54, 55, 60, 65, 67, 77, 81, 84,
88, 89, 90, 96, 99, 100, 101, 111, 112, 113, 120, 125, 131, 135, 138, 142, 154, 159, 160
Not the sum of 3 pentagonal numbers. Ref AMM 101 171 94. [1,1; A3679]

M3324 4, 8, 9, 16, 21, 25, 27, 32, 35, 36, 39, 49, 50, 55, 57, 63, 64, 65, 75, 77, 81, 85, 93,
98, 100, 111, 115, 119, 121, 125, 128, 129, 133, 143, 144, 155, 161, 169, 171, 175, 183
Duffinian numbers: n composite and relatively prime to $\sigma(n)$. Ref JRM 12 112 79. Robe92
64. [1,1; A3624]

M3325 1, 4, 8, 9, 16, 25, 27, 32, 36, 49, 64, 72, 81, 100, 108, 121, 125, 128, 144, 169, 196,
200, 216, 225, 243, 256, 288, 289, 324, 343, 361, 392, 400, 432, 441, 484, 500, 512, 529
Powerful numbers (1): if $p \mid n$ then $p^2 \mid n$. Ref AMM 77 848 70. [1,2; A1694, N1335]

M3326 1, 4, 8, 9, 16, 25, 27, 32, 36, 49, 64, 81, 100, 121, 125, 128, 144, 169, 196, 216,
225, 243, 256, 289, 324, 343, 361, 400, 441, 484, 512, 529, 576, 625, 676, 729, 784, 841
Perfect powers. Ref FQ 8 268 70. GKP 66. [1,2; A1597, N1336]

M3327   4, 8, 12, 17, 21, 25, 30, 34, 38, 43, 47, 51, 55, 60, 64, 68, 73, 77, 81, 86, 90, 94, 98, 103, 107, 111, 116, 120, 124, 129, 133, 137, 141, 146, 150, 154, 159, 163, 167, 172, 176
A Beatty sequence. Cf. M0615. Ref CMB 2 188 59. [1,1; A1956, N1338]

M3328   1, 4, 8, 12, 17, 22, 27, 32, 37, 42, 47, 53, 58, 64, 69, 75, 81, 86, 92, 98, 104
Davenport-Schinzel numbers. Ref PLC 1 250 70. ARS 1 47 76. UPNT E20. [1,2; A2004, N1339]

M3329   1, 4, 8, 14, 21, 30, 40, 52, 65, 80, 96, 114, 133, 154, 176, 200, 225, 252, 280, 310, 341, 374, 408, 444, 481, 520, 560, 602, 645, 690, 736, 784, 833, 884, 936, 990, 1045
Expansion of $(1+2x) / (1-x)^2 (1-x^2)$. Ref mlb. [0,2; A6578]

$$M3329 + M0998 = M1581 = n(n+1).$$

M3330   1, 4, 8, 16, 25, 40, 56, 80, 105, 140, 176, 224
Dissections of a polygon. Ref AEQ 18 387 78. [5,2; A3451]

$$\text{G.f.:}\ (1 + 2x - x^2) / (1 - x)^4 (1 + x)^2.$$

M3331   1, 4, 8, 16, 32, 54, 100, 182, 328, 494, 984, 1572, 2656, 4212, 8162
Cluster series for honeycomb. Ref PRV 133 A315 64. DG72 225. [0,2; A3199]

M3332   1, 4, 8, 16, 32, 56, 96, 160, 256, 404, 624, 944, 1408, 2072, 3008, 4320, 6144, 8648, 12072, 16720, 22976, 31360, 42528, 57312, 76800, 102364, 135728, 179104
$f(x^2)^2 = \frac{1}{2}(f(x) + 1/f(x))$. [0,2; A7096]

M3333   4, 8, 16, 32, 64, 128, 144, 216, 288, 432, 864, 1296, 1728, 2592, 3456, 5184, 7776, 10368, 15552, 20736, 31104, 41472, 62208, 86400, 108000, 129600, 216000, 259200
Highly powerful numbers. Ref CN 37 300 83. PAMS 91 181 84. [1,1; A5934]

M3334   4, 8, 17, 33, 34, 35, 66, 67, 69, 133, 134, 135, 137, 138, 139, 265, 266, 267, 270, 275, 277, 531, 533, 537, 539, 549, 551, 555, 1061, 1063, 1067, 1075, 1076, 1077, 1078
Positions of remoteness 5 in Beans-Don't-Talk. Ref MMAG 59 267 86. [1,1; A5697]

M3335   1, 1, 4, 8, 18, 32, 58, 94, 151, 227, 338, 480, 676, 920, 1242, 1636
Restricted partitions. Ref CAY 2 279. [0,3; A1977, N1342]

M3336   1, 4, 8, 20, 21, 56, 60, 96, 105, 220, 152, 364, 301, 360, 464, 816, 549, 1140, 760, 1036, 1221, 2024, 1196, 2200, 2041, 2484, 2184, 4060, 2205, 4960, 3664, 4224, 4641
Σ l.c.m. $\{ k, n-k \}$, $k = 1 .. n-1$. Ref mlb. [2,2; A6580]

M3337   4, 8, 20, 92, 2744, 950998216
Boolean functions of n variables. Ref JACM 13 153 66. [1,1; A0585, N1343]

M3338   1, 4, 8, 21, 39, 92, 170, 331, 593, 1176, 2118, 3699
Nonzeros in character table of S_n. Ref jmckay. [1,2; A6908]

M3339 0, 4, 8, 21, 52, 65, 96, 1, 5, 9, 31, 53, 75, 97, 101, 501, 505, 905, 909, 319, 323,
723, 727, 137, 141, 541, 545, 945, 949, 359, 363, 763, 767, 177, 181, 581, 585, 985, 989
Add 4, then reverse digits! Ref Robe92 15. [0,2; A3608]

M3340 1, 4, 8, 22, 42, 103, 199, 441, 859, 1784, 3435, 6882, 13067, 25366, 47623, 90312,
167344, 311603, 570496, 1045896, 1893886, 3426466, 6140824, 10984249, 19499214
Conjugacy classes in $GL(n,q)$. Ref TAMS 80 408 55. [1,2; A3606]

M3341 1, 1, 4, 8, 22, 51, 136, 335, 871, 2217, 5749, 14837, 38636, 100622, 263381,
690709, 1817544, 4793449, 12675741, 33592349, 89223734, 237455566, 633176939
Alkyl derivatives of benzene with n carbon atoms. Ref ZFK 93 422 36. BA76 22. [6,3;
A0639, N1344]

M3342 4, 8, 24, 40, 60, 88
Restricted postage stamp problem. Ref LNM 751 326 82. [1,1; A6640]

M3343 0, 0, 1, 1, 4, 8, 25, 53, 164, 348, 1077, 2285, 7072, 40051, 46437, 98521, 304920,
646920, 2002201, 4247881, 13147084, 27892928, 86327905
A ternary continued fraction. Ref TOH 37 441 33. [0,5; A0964, N1345]

M3344 1, 0, 1, 1, 4, 8, 37, 184, 1782, 31026, 1148626, 86539128, 12798435868,
3620169692289, 1940367005824561, 1965937435288738165
Connected Eulerian graphs with n nodes. Ref PTGT 151. MR 44 #6557. HP73 117. rwr.
[1,5; A3049]

M3345 1, 4, 8, 38, 209, 1400, 10849, 95516
Hit polynomials. Ref RI63. [2,2; A1889, N1346]

M3346 1, 0, 4, 8, 39, 152, 672, 3016, 13989, 66664
Low temperature antiferromagnetic susceptibility for square lattice. Ref DG74 422. [0,3;
A7215]

M3347 1, 4, 8, 48, 10, 224, 80, 448, 231, 40, 248, 1408, 1466, 2240, 80, 1280, 4766, 924,
1944, 480, 9600, 6944, 2704, 8704, 15525, 5864, 3984, 14080, 25498, 2240, 10816
Related to representation as sums of squares. Ref QJMA 38 190 07. [1,2; A2470, N1347]

M3348 1, 4, 9, 1, 2, 3, 4, 6, 8, 1, 1, 1, 1, 1, 2, 2, 2, 3, 3, 4, 4, 5, 5, 6, 6, 7, 7, 8, 9, 9, 1, 1,
1, 1, 1, 1, 1, 1, 1, 1, 1, 1, 2, 2, 2, 2, 2, 2, 2, 2, 2, 2, 3, 3, 3, 3, 3, 3, 3, 3, 3, 4, 4, 4, 4, 4, 4
Initial digits of squares. [1,2; A2993]

M3349 1, 1, 1, 1, 4, 9, 2, 0, 1, 45, 4, 3, 2, 7, 0, 1, 36792, 11320425, 2, 24, 267972, 352849,
10, 0, 17, 135531, 12, 189709, 21012, 371631, 1008, 32, 8, 34255, 0, 71847
From fundamental unit of $Z[(-n)^{1/4}]$. Ref MOC 48 49 87. [1,5; A6830]

M3350 1, 4, 9, 11, 14, 16, 21, 25, 30, 36, 41, 44, 49, 52, 54, 64, 69, 71, 81, 84, 86, 92, 100,
105, 120, 121, 126, 136, 141, 144, 149, 164, 169, 174, 189, 196, 201, 208, 216, 225, 230
Epstein's Put or Take a Square game. Ref UPNT E26. [1,2; A5241]

M3351 1, 1, 4, 9, 11, 16, 29, 49, 76, 121, 199, 324, 521, 841, 1364, 2209, 3571, 5776,
9349, 15129, 24476, 39601, 64079, 103684, 167761, 271441, 439204, 710649, 1149851
A Fielder sequence. Ref FQ 6(3) 68 68. [1,3; A1638, N1348]

M3352 4, 9, 11, 23, 32, 39, 44, 51, 53, 60, 65, 72, 86, 93, 95, 114, 123, 156, 170, 179, 186,
200, 207, 212, 219, 228, 233, 240, 249, 261, 270, 303, 317, 333, 338, 345, 375, 389, 401
$(n^2+n+1)/7$ is prime. Ref CU23 1 250. [1,1; A2641, N0446]

M3353 1, 1, 4, 9, 16, 22, 36, 65, 112, 186, 309, 522, 885, 1492, 2509, 4225, 7124, 12010,
20236, 34094, 57453, 96823, 163163, 274946, 463316, 780755, 1315687, 2217112
A Fielder sequence. Ref FQ 6(3) 68 68. [1,3; A1639, N1349]

M3354 1, 4, 9, 16, 25, 35, 46, 58, 71, 85, 100, 116, 133, 151, 170, 190, 211, 233, 256, 280,
305, 331, 358, 386, 415, 445, 476, 508, 541, 575, 610, 646, 683, 721, 760, 800, 841, 883
Expansion of $(1+x-x^5)/(1-x)^3$. Ref SIAR 12 296 70. [0,2; A4120]

M3355 1, 4, 9, 16, 25, 36, 49, 64, 81, 1, 2, 5, 10, 17, 26, 37, 50, 65, 82, 4, 5, 8, 13, 20, 29,
40, 53, 68, 85, 9, 10, 13, 18, 25, 34, 45, 58, 73, 90, 16, 17, 20, 25, 32, 41, 52, 65, 80, 97
Sum of squares of digits of n. Ref CJM 12 374 60. [1,2; A3132]

M3356 1, 4, 9, 16, 25, 36, 49, 64, 81, 100, 121, 144, 169, 196, 225, 256, 289, 324, 361,
400, 441, 484, 529, 576, 625, 676, 729, 784, 841, 900, 961, 1024, 1089, 1156, 1225, 1296
The squares. See Fig M2535. Ref BA9. [1,2; A0290, N1350]

M3357 0, 1, 4, 9, 16, 26, 39, 56, 78, 106, 141, 184, 236, 299, 374, 465, 570, 696, 843,
1014, 1212, 1441, 1708, 2014, 2365, 2769, 3226, 3749, 4343, 5016, 5774, 6630, 7596
Generated by a sieve. Ref PC 4 41-15 76. [0,3; A6508]

M3358 1, 1, 4, 9, 16, 28, 43, 73, 130, 226, 386, 660, 1132, 1947, 3349, 5753, 9878, 16966,
29147, 50074, 86020, 147764, 253829, 436036, 749041, 1286728, 2210377, 3797047
A Fielder sequence. Ref FQ 6(3) 68 68. [1,3; A1640, N1352]

M3359 1, 1, 4, 9, 16, 31, 64, 129, 256, 511, 1024, 2049, 4096, 8191, 16384, 32769, 65536,
131071, 262144, 524289, 1048576, 2097151, 4194304, 8388609, 16777216, 33554431
$\Sigma\ 2^k C(n-k,2k).n/(n-k)$, $k = 0..[n/3]$. Ref AMM 102 Problem 10424 95. [1,3;
A7679]

M3360 1, 4, 9, 17, 28, 43, 62, 86
n-covers of a 2-set. Ref DM 81 151 90. [1,2; A5744]

M3361 4, 9, 19, 37, 73, 143, 279, 548, 1079, 2132, 4223, 8384, 16673, 33203, 66190,
132055, 263619, 526502, 1051899, 2102137, 4201783, 8399828, 16794048, 33579681
Related to Waring's problem: $2^n+[1.5^n]-2$. Ref HAR 1 668. BPNR 239. [2,1; A2804,
N1353]

M3362 4, 9, 21, 40, 74, 125, 209, 330, 515, 778, 1160, 1690, 2439, 3457, 4857, 6735,
9264, 12607, 17040, 22826, 30391, 40165, 52788, 68938, 89589, 115778, 148957
Bipartite partitions. Ref ChGu56 11. [0,1; A2762, N1354]

M3363 1, 1, 4, 9, 22, 46, 102, 206, 427, 841, 1658, 3173, 6038, 11251, 20807, 37907, 68493, 122338, 216819, 380637, 663417, 1147033, 1969961, 3359677, 5694592
Solid partitions of n. Ref MOC 24 956 70. [0,3; A2835, N1355]

M3364 1, 1, 4, 9, 25, 64, 169, 441, 1156, 3025, 7921, 20736, 54289, 142129, 372100, 974169, 2550409, 6677056, 17480761, 45765225, 119814916, 313679521, 821223649
$F(n)^2$. Ref HO85a 130. [0,3; A7598]

M3365 1, 1, 4, 9, 28, 71, 202
Connected graphs with one cycle. Ref R1 150. [4,3; A0368, N1356]

M3366 1, 4, 9, 32, 65, 192, 385, 1024, 2049, 5120, 10241, 24576, 49153, 114688, 229377, 524288, 1048577, 2359296, 4718593, 10485760, 20971521, 46137344, 92274689
Longest walk on edges of n-cube. Ref clm. [1,2; A5985]

M3367 1, 4, 9, 32, 132, 597
Planar maps without faces of degree 1 or 2. Ref SIAA 4 174 83. [2,2; A6393]

M3368 1, 4, 9, 34, 161, 830
Planar maps without faces of degree 1 or 2. Ref SIAA 4 174 83. [2,2; A6392]

M3369 1, 4, 9, 49, 144, 441, 1444, 11449, 44944, 991494144, 4914991449, 149991994944, 9141411449911441, 199499144494999441, 9914419419914449449
Squares with digits 1, 4, 9 (probably finite). Ref VA91 234. [1,2; A6716]

M3370 1, 4, 9, 61, 52, 63, 94, 46, 18, 1, 121, 441, 961, 691, 522, 652, 982, 423, 163, 4, 144, 484, 925, 675, 526, 676, 927, 487, 148, 9, 169, 4201, 9801, 6511, 5221, 6921, 9631
Squares written backwards. [1,2; A2942, N1357]

M3371 1, 4, 9, 121, 484, 676, 10201, 12321, 14641, 40804, 44944, 69696, 94249, 698896, 1002001, 1234321, 4008004, 5221225, 6948496, 100020001, 102030201, 104060401
Palindromic squares. Ref JRM 3 94 70. [1,2; A2779, N1358]

M3372 0, 1, 1, 1, 4, 9, 196, 16641, 639988804, 177227652025317609, 7258990646358542780528129597781 6196
Partial quotients in c.f. expansion of Cahen's constant. Ref MFM 111 122 91. [0,5; A6280]

M3373 1, 4, 10, 12, 22, 26, 30, 46, 54, 62, 66, 78, 94, 110, 126, 134, 138, 158, 162, 186, 190, 222, 254, 270, 278, 282, 318, 326, 330, 374, 378, 382, 402, 446, 474, 510, 542, 558
A grasshopper sequence: closed under $n \rightarrow 2n+2$ and $6n+6$. Ref Pick91 353. [0,2; A7319]

M3374 4, 10, 14, 20, 24, 30, 36, 40, 46, 50, 56, 60, 66, 72, 76, 82, 86, 92, 96, 102, 108, 112, 118, 122, 128, 132, 138, 150, 160, 169, 176, 186, 192, 196, 202, 206, 212, 218, 222
Winning moves in Fibonacci nim. Ref FQ 3 62 65. [1,1; A1581, N1359]

M3375 4, 10, 15, 22, 32, 33, 46, 48, 66, 68, 69, 94, 98, 99, 102, 134, 138, 140, 141, 147, 190, 198, 200, 201, 206, 207, 210, 270, 278, 282, 284, 285, 296, 297, 300, 309, 382, 398
If n appears then so do $2n+2$ and $3n+3$. [1,1; A5662]

M3376 4, 10, 17, 18, 30, 34, 69, 109, 111, 189, 192, 193, 194, 195, 311, 763, 898, 900, 2215, 2810, 2811, 2812, 2813, 3417, 4260, 6000, 6002, 6003, 6004, 23331, 31569, 31601
Related to gaps between primes. Ref MOC 13 122 59. [1,1; A1549, N1360]

M3377 1, 4, 10, 17, 27, 40, 54, 71, 100, 121, 144, 170, 207, 237, 270, 314, 351, 400, 441, 484, 540, 587, 647, 710, 764, 831, 1000, 1061, 1134, 1210, 1277, 1357, 1440, 1524
Squares written in base 9. Ref TH52 98. [1,2; A2442, N1361]

M3378 1, 4, 10, 19, 31, 46, 64, 85, 109, 136, 166, 199, 235, 274, 316, 361, 409, 460, 514, 571, 631, 694, 760, 829, 901, 976, 1054, 1135, 1219, 1306, 1396, 1489, 1585, 1684, 1786
Centered triangular numbers: $(3n^2 + 3n + 2)/2$. See Fig M3826. Ref INOC 24 4550 85. [1,2; A5448]

M3379 1, 4, 10, 19, 31, 47, 68, 92, 120, 153, 190, 232, 279, 332, 392, 454, 521, 593, 670, 753
Optimal cost of search tree. Ref SIAC 17 1213 88. [1,2; A7077]

M3380 1, 4, 10, 20, 34, 52, 74, 100, 130, 164, 202, 244, 290, 340, 394, 452, 514, 580, 650, 724, 802, 884, 970, 1060, 1154, 1252, 1354, 1460, 1570, 1684, 1802, 1924, 2050, 2180
Points on surface of tetrahedron: $2n^2 + 2$. Ref MF73 46. Coxe74. INOC 24 4550 85. [0,2; A5893]

M3381 1, 4, 10, 20, 34, 56, 80, 120, 154, 220
Compositions into 4 relatively prime parts. Ref FQ 2 250 64. [3,2; A0742, N1362]

M3382 1, 4, 10, 20, 35, 56, 84, 120, 165, 220, 286, 364, 455, 560, 680, 816, 969, 1140, 1330, 1540, 1771, 2024, 2300, 2600, 2925, 3276, 3654, 4060, 4495, 4960, 5456, 5984
Tetrahedral numbers: $C(n+3,3)$. See Fig M3382. Ref D1 2 4. RS3. B1 194. AS1 828. [0,2; A0292, N1363]

$$\text{G.f.: } 1 / (1 - x)^4.$$

Figure M3382. PYRAMIDAL NUMBERS.

The number of balls in a pyramid of height r and a p-sided base is $r(r + 1)\{(r - 1)(p - 2) + 3\}/6$. These are the **pyramidal** numbers [B1 194]. When $p = 3$ we obtain the tetrahedral numbers, M3382, shown here. M3844, M4116, M4374, M4498 are also of this type. Similarly, M3853, M4135, M4385, M4506, M4617, M4699 are 4-dimensional pyramidal numbers, and M4387 is a 5-dimensional version.

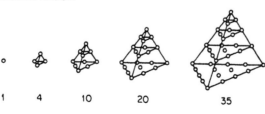

1   4   10   20   35

M3383 1, 4, 10, 20, 36, 64, 120, 240, 496, 952
Expansion of bracket function. Ref FQ 2 254 64. [4,2; A0749, N1364]

M3384 0, 1, 4, 10, 21, 40, 72, 125, 212, 354, 585, 960, 1568, 2553, 4148, 6730, 10909,
17672, 28616, 46325, 74980, 121346, 196369, 317760, 514176, 831985, 1346212
Hit polynomials. Ref RI63. [0,3; A1891, N1365]

$$\text{G.f.: } x\,(1 + x) \;/\; (1 - x - x^2)\,(1 - x)^2.$$

M3385 1, 4, 10, 22, 43, 76, 124, 190, 277, 388, 526, 694, 895, 1132, 1408, 1726, 2089,
2500, 2962, 3478, 4051, 4684, 5380, 6142, 6973, 7876, 8854, 9910, 11047, 12268, 13576
Paraffins. Ref BER 30 1922 1897. [0,2; A6001]

$$\text{G.f.: } (1 + 2\,x^3) \;/\; (1 - x)^4.$$

M3386 1, 4, 10, 23, 40, 68, 108, 167, 241, 345, 482, 653, 869
From Størmer's problem. Ref IJM 8 66 64. [1,2; A2071, N1366]

M3387 4, 10, 23, 45, 83, 142, 237, 377, 588, 892, 1330, 1943, 2804, 3982, 5595, 7768,
10686, 14555, 19674, 26371, 35112, 46424, 61015, 79705, 103579, 133883, 172243
Bipartite partitions. Ref ChGu56 19. [0,1; A2766, N1367]

M3388 1, 4, 10, 23, 48, 94, 166, 285, 464, 734, 1109, 1646
Restricted partitions. Ref CAY 2 280. [0,2; A1980, N1368]

M3389 1, 1, 4, 10, 24, 49, 94, 169, 289, 468, 734, 1117, 1656
Restricted partitions. Ref CAY 2 280. [0,3; A1979, N1369]

M3390 1, 4, 10, 24, 70, 208, 700, 2344, 8230, 29144, 104968, 381304, 1398500, 5162224,
19175140, 71582944, 268439590, 1010580544, 3817763740, 14467258264
n-bead necklaces with 4 colors. See Fig M3860. Ref R1 162. IJM 5 658 61. [0,2; A1868,
N1370]

M3391 4, 10, 26, 44, 70, 108, 162, 220
Postage stamp problem. Ref CJN 12 379 69. AMM 87 208 80. [1,1; A1214, N1559]

M3392 1, 1, 4, 10, 26, 59, 140, 307, 684, 1464, 3122, 6500, 13426, 27248, 54804, 108802,
214071, 416849, 805124, 1541637, 2930329, 5528733, 10362312, 19295226, 35713454
Solid partitions of n. Ref MOC 24 956 70. [0,3; A0293, N1371]

M3393 1, 1, 4, 10, 26, 59, 141, 310, 692, 1483, 3162, 6583, 13602, 27613, 55579, 110445,
217554, 424148, 820294, 1572647, 2992892, 5652954, 10605608, 19765082
Related to solid partitions. Ref PNISI 26 135 60. PCPS 63 1100 67. [1,3; A0294, N1372]

$$\text{G.f.: } \Pi\,(1 - x^{k(k-1)/2})^{-1}.$$

M3394 0, 0, 1, 4, 10, 29, 55, 153, 307, 588, 1018, 2230
Zeros in character table of S_n. Ref jmckay. [1,4; A6907]

M3395 4, 10, 30, 65, 173, 343, 778, 1518, 3088, 5609
Restricted partitions. Ref JCT 9 373 70. [2,1; A2220, N1374]

M3396 1, 4, 10, 30, 85, 246, 707, 2037, 5864, 16886, 48620
Distributive lattices. Ref MSH 53 19 76. MSG 121 121 76. [0,2; A6357]

M3397 4, 10, 30, 100, 354, 1300, 4890, 18700, 72354, 282340, 1108650, 4373500,
17312754, 68711380, 273234810, 1088123500, 4338079554, 17309140420
$1^n + 2^n + 3^n + 4^n$. Ref AS1 813. [0,1; A1551, N1375]

M3398 1, 1, 4, 10, 31, 91, 274, 820, 2461, 7381, 22144, 66430, 199291, 597871, 1793614,
5380840, 16142521, 48427561, 145282684, 435848050, 1307544151, 3922632451
Coloring a circuit with 4 colors. Ref TAMS 60 355 46. BE74. [0,3; A6342]

$$\text{G.f.:} \quad (1 - 2x) / (1 - x^2)(1 - 3x).$$

M3399 0, 0, 0, 0, 0, 0, 4, 10, 34, 96, 284, 782, 4226, 6198
Chiral trees with n nodes. Ref TET 32 356 76. [1,7; A5630]

M3400 1, 4, 10, 34, 112, 398, 1443, 5387, 20482, 79177, 310102, 1228187, 4910413,
19792582, 80343445, 328159601, 1347699906, 5561774999, 23052871229
Elementary maps with n nodes. Ref TAMS 60 355 46. BE74. CONT 98 185 89. frb. [2,2; A6343]

$$\Sigma (n - k - 1)^{-1} C(n, k) C(2n - 3k - 4, n - 2k - 2); \ k = 0..[(n-2)/2].$$

M3401 1, 1, 1, 4, 10, 40, 171, 831, 4147, 21822, 117062, 642600, 3582322, 20256886,
115888201, 669911568, 3907720521, 22979343010, 136107859377, 811430160282
Simplicial 3-clusters with n cells. Ref DM 40 216 82. [1,4; A7173]

M3402 1, 0, 1, 1, 4, 10, 53, 292, 2224, 18493, 167504, 1571020
4-connected polyhedral graphs with n nodes. Ref Dil92. [4,5; A7027]

M3403 1, 4, 10, 56, 29, 332, 30, 1064, 302, 1940, 288, 1960, 1071, 1192, 1938, 736, 2000,
1488, 5014, 7288, 4170, 10644, 8482, 11184, 12647, 15544
Related to representation as sums of squares. Ref QJMA 38 56 07. [0,2; A2290, N1376]

M3404 1, 1, 4, 10, 136, 720, 44224, 703760
Self-complementary digraphs with n nodes. Ref HP73 140. [1,3; A3086]

M3405 4, 11, 15, 22, 26, 29, 33, 40, 44, 51, 58, 62, 69, 73, 76, 80, 87, 91, 98, 102, 105,
109, 116, 120, 127, 134, 138, 145, 149, 152, 156, 163, 167, 174, 178, 181, 185, 192, 196
Related to Fibonacci representations. Ref FQ 11 385 73. [1,1; A3250]

M3406   1, 4, 11, 16, 24, 29, 33, 35, 39, 45, 47, 51, 56, 58, 62, 64, 69, 73, 78, 80, 84, 89, 94,
99, 105, 112, 117, 123, 127, 132, 137, 142, 147, 158, 164, 169, 174, 181, 183, 193, 198
T is the first, fourth, eleventh, ... letter in this sentence (Aronson's sequence). See Fig
M2629. Ref HO85 44. [1,2; A5224]

M3407   4, 11, 17, 24, 28, 35, 41, 48, 55, 61, 68, 72, 79, 85, 92, 98, 105, 109, 116, 122, 129,
136, 142, 149, 153, 160, 166, 173, 177, 184, 190, 197, 204, 210, 217, 221, 228, 234
A self-generating sequence. Ref FQ 10 49 72. [1,1; A3146]

M3408   1, 4, 11, 20, 31, 44, 61, 100, 121, 144, 171, 220, 251, 304, 341, 400, 441, 504, 551,
620, 671, 744, 1021, 1100, 1161, 1244, 1331, 1420, 1511, 1604, 1701, 2000, 2101, 2204
Squares written in base 8. Ref TH52 95. [1,2; A2441, N1378]

M3409   1, 1, 4, 11, 23, 79, 148, 533, 977, 3553, 6484, 23627, 43079, 157039, 286276,
1043669, 1902497, 6936001, 12643492, 46094987, 84025463, 306335887, 558412276
$a(2n) = a(2n-1) + 3a(2n-2)$, $a(2n+1) = 2a(2n) + 3a(2n-1)$. Ref MQET 1 12 16.
[0,3; A2537, N1379]

M3410   0, 1, 4, 11, 24, 45, 76, 119, 176, 249, 340, 451, 584, 741, 924, 1135, 1376, 1649,
1956, 2299, 2680, 3101, 3564, 4071, 4624, 5225, 5876, 6579, 7336, 8149, 9020, 9951
$(n^3 + 2n)/3$. Ref GA66 246. [0,3; A6527]

M3411   1, 4, 11, 24, 50, 80, 154, 220, 375, 444, 781, 952, 1456, 1696, 2500, 2466, 4029,
4500, 6175, 6820, 9086, 9024, 12926, 13988, 17875, 19180, 24129, 21480, 31900, 33856
Regions in regular n-gon with all diagonals drawn. Ref WP 10 62 72. PoRu94. [3,2;
A7678]

M3412   1, 4, 11, 25, 49, 86, 139, 211
Paraffins. Ref BER 30 1922 1897. [1,2; A6004]

M3413   1, 4, 11, 25, 50, 91, 154, 246, 375, 550, 781, 1079, 1456, 1925, 2500, 3196, 4029,
5016, 6175, 7525, 9086, 10879, 12926, 15250, 17875, 20826, 24129, 27811, 31900
$C(n,4) + C(n-1,2)$. Ref HO73 102. [3,2; A6522]

M3414   4, 11, 26, 52, 98, 171, 289, 467, 737, 1131, 1704, 2515, 3661, 5246, 7430, 10396,
14405, 19760, 26884, 36269, 48583, 64614, 85399, 112170, 146526, 190362
Bipartite partitions. Ref ChGu56 11. [0,1; A2763, N1380]

M3415   1, 4, 11, 26, 56, 114, 223, 424, 789, 1444
Arrays of dumbbells. Ref JMP 11 3098 70; 15 214 74. [1,2; A2940, N1381]

M3416   0, 1, 4, 11, 26, 57, 120, 247, 502, 1013, 2036, 4083, 8178, 16369, 32752, 65519,
131054, 262125, 524268, 1048555, 2097130, 4194281, 8388584, 16777191, 33554406
Eulerian numbers $2^n - n - 1$. See Fig M3416. Ref R1 215. DB1 151. [1,3; A0295, N1382]

||||||||||||||||||||||||||||||||||||||||||||||||||||||||||||||||||||||||||||||||||||||||||||||||||||||||||||||||||||||||||||||||||||||||||

Figure M3416.   EULER'S TRIANGLE.

```
1
1    1
1    4     1
1   11    11     1
1   26    66    26     1
1   57   302   302    57    1
1  120  1191  2416  1191  120   1
```

The k-th entry in the n-th row is the **Eulerian number** $A(n, k)$ [R1 215], [C1 243], [GKP 253]. $A(n, k)$ is the number of permutations of n objects with $k - 1$ rises (i.e. permutations π_1, $\pi_2, \ldots, \pi_n$ with $k - 1$ places where $\pi_i < \pi_{i+1}$). The columns of this table give M3416, M4795, M5188, M5317, M5379, M5422, M5457. Also

$$A(n, k) = (n - k + 1)A(n - 1, k - 1) + kA(n - 1, k),$$

$$A(n, k) = \sum_{j=0}^{k} (-1)^j \binom{n + 1}{j} (k - j)^n,$$

$$x^n = \sum_{k=1}^{n} A(n, k) \binom{x + k - 1}{n}.$$

||||||||||||||||||||||||||||||||||||||||||||||||||||||||||||||||||||||||||||||||||||||||||||||||||||||||||||||||||||||||||||||||||||||||||

M3417   1, 4, 11, 28, 67, 152, 335, 724, 1539, 3232, 6727, 13900, 28555, 58392, 118959, 241604, 489459, 989520, 1997015, 4024508, 8100699, 16289032, 32726655, 65705268
Expansion of $1/(1-x)(1-2x)(1-x-2x^3)$. Ref DT76. [0,2; A3230]

M3418   1, 1, 4, 11, 28, 69, 168, 407, 984, 2377, 5740, 13859, 33460, 80781, 195024, 470831, 1136688, 2744209, 6625108, 15994427, 38613964, 93222357, 225058680
Polynomials of height n: $a(n) = 2a(n-1) + a(n-2) + 2$. Ref CR41 103. smd. [1,3; A5409]

M3419   1, 4, 11, 29, 54, 99, 163, 239, 344, 486, 648, 847, 1069, 1355, 1680, 2046, 2446, 2911, 3443, 4022, 4662, 5395, 6145, 6998, 7913, 8913, 10006, 11194, 12437, 13751
Nonnegative solutions to $x^2 + y^2 + z^2 \leq n$. Ref PNISI 13 37 47. [0,2; A0604, N1383]

M3420   1, 4, 11, 29, 76, 199, 521, 1364, 3571, 9349, 24476, 64079, 167761, 439204, 1149851, 3010349, 7881196, 20633239, 54018521, 141422324, 370248451, 969323029
Bisection of Lucas sequence. Cf. M2341. Ref FQ 9 284 71. [0,2; A2878, N1384]

M3421   0, 0, 1, 4, 11, 34, 107, 368, 1284, 4654, 17072, 63599, 238590, 901970, 3426575, 13079254, 50107908
Polyominoes with n cells. Ref CJN 18 367 75. [1,4; A6765]

M3422 0, 0, 0, 0, 0, 0, 4, 11, 35, 101, 290, 804, 2256, 6296, 17689, 49952, 142016,
406330, 1169356, 3390052
Paraffins with n carbon atoms. Ref JACS 54 1544 32. [1,7; A0626, N1386]

M3423 1, 0, 1, 1, 4, 11, 41, 162, 715, 3425, 17722, 98253, 580317, 3633280, 24011157,
166888165, 1216070380, 9264071767, 73600798037, 608476008122, 5224266196935
Expansion of exp $(e^x - 1 - x)$. Ref PoSz72 1 228. FQ 14 69 76. ANY 319 464 79. [0,5;
A0296, N1387]

M3424 1, 4, 11, 60, 362, 2987
3-edge-colored connected trivalent graphs with $2n$ nodes. Ref RE58. [1,2; A2831, N1388]

M3425 4, 11, 64, 5276
Switching networks. Ref JFI 276 324 63. [1,1; A0880, N1389]

M3426 1, 1, 4, 11, 66, 302, 2416, 15619, 156190, 1310354, 15724248, 162512286,
2275172004, 27971176092, 447538817472, 6382798925475, 114890380658550
Maximal Eulerian numbers. Ref C1 243. EJC 13 399 92. [1,3; A6551]

M3427 4, 11, 79, 7621
Switching networks. Ref JFI 276 322 63. [1,1; A0850, N1390]

M3428 1, 1, 1, 1, 1, 4, 11, 135, 4382, 312356
Primitive sorting networks on n elements. Ref KN91. jb. [1,6; A6248]

M3429 4, 12, 12, 16, 24, 12, 24, 36, 12, 28, 36, 24, 36, 36, 24, 24, 60, 36, 28, 48, 12, 60,
60, 24, 48, 48, 36, 48, 60, 24, 52, 84, 48, 24, 60, 36, 48, 96, 36, 72, 48, 36, 72, 60, 48, 52
Theta series of f.c.c. lattice w.r.t. tetrahedral hole. Ref JCP 83 6526 85. [0,1; A5886]

M3430 4, 12, 15, 21, 35, 40, 45, 60, 55, 80, 72, 99, 91, 112, 105, 140, 132, 165, 180, 168,
195, 221, 208, 209, 255, 260, 252, 231, 285, 312, 308, 288, 299, 272, 275, 340, 325
y such that $p^2 = x^2 + y^2$, $x \le y$. Cf. M2442. Ref CU27 77. L1 60. [5,1; A2365, N1391]

M3431 1, 4, 12, 22, 34, 51, 100, 121, 144, 202, 232, 264, 331, 400, 441, 514, 562, 642,
1024, 1111, 1200, 1261, 1354, 1452, 1552, 1654, 2061, 2200, 2311, 2424, 2542, 2662
Squares written in base 7. Ref TH52 93. [1,2; A2440, N1392]

M3432 4, 12, 24, 44, 71, 114, 165, 234, 326, 427, 547, 708, 873, 1094
Postage stamp problem. Ref CJN 12 379 69. [1,1; A1209, N1568]

M3433 1, 4, 12, 24, 52, 108, 224, 412, 844, 1528, 3152
Cluster series for square lattice. Ref PRV 133 A315 64. DG72 225. [0,2; A3203]

M3434 4, 12, 25, 44, 70, 104, 147, 200, 264, 340, 429, 532, 650, 784, 935, 1104, 1292,
1500, 1729, 1980, 2254, 2552, 2875, 3224, 3600, 4004, 4437, 4900, 5394, 5920, 6479
Expansion of $(2-x)^2 / (1-x)^4$. Ref R1 150. FQ 15 194 77. [0,1; A0297, N1393]

M3435 1, 4, 12, 27, 54, 96, 160, 250, 375, 540, 756, 1029, 1372, 1792, 2304, 2916, 3645, 4500, 5500, 6655, 7986, 9504, 11232, 13182, 15379, 17836, 20580, 23625, 27000, 30720
3-voter voting schemes with n linearly ranked choices. Ref Loeb94b. [1,2; A7009]

$$\text{G.f.:} \quad (1 - x^3) / (1 - x)^4 (1 - x^2)^2.$$

M3436 1, 4, 12, 28, 55, 96, 154, 232, 333, 460, 616, 804, 1027, 1288, 1590, 1936, 2329, 2772, 3268, 3820, 4431, 5104, 5842, 6648, 7525, 8476, 9504, 10612, 11803, 13080
Paraffins. Ref BER 30 1922 1897. [0,2; A6000]

$$\text{G.f.:} \quad (1 + 2 x^2) / (1 - x)^4.$$

M3437 1, 4, 12, 28, 68, 164, 396, 940, 2244, 5324, 12668, 29940, 71012, 167468, 396204
n-step walks on square lattice. Ref JCP 34 1261 61. [0,2; A2932, N1394]

M3438 1, 1, 4, 12, 30, 66, 132
The coding-theoretic function $A(n,4,6)$. See Fig M0240. Ref PGIT 36 1335 90. [6,3; A4036]

M3439 1, 4, 12, 30, 70, 159, 339, 706, 1436, 2853
Partitions into non-integral powers. Ref PCPS 47 215 51. [1,2; A0298, N1395]

M3440 0, 0, 1, 4, 12, 31, 67, 132, 239, 407, 657, 1019, 1523, 2211, 3126, 4323, 5859, 7806, 10236, 13239, 16906, 21346, 26670, 33010, 40498, 49290, 59543, 71438, 85158
Graphs on n nodes with 3 cliques. Ref AMM 80 1124 73; 82 997 75. JLMS 8 97 74. rkg. [1,4; A5289]

M3441 1, 4, 12, 31, 71, 147, 285, 519, 902, 1502
Restricted partitions. Ref CAY 2 281. [0,2; A1982, N1396]

M3442 1, 4, 12, 32, 64, 128, 192, 256, 256
Minimal determinant of n-dimensional norm 4 lattice. Ref SPLAG 180. [0,2; A5104]

M3443 1, 4, 12, 32, 76, 168, 352, 704, 1356, 2532, 4600, 8160, 14176, 24168, 40512, 66880, 108876, 174984, 277932, 436640, 679032, 1046016, 1597088, 2418240, 3632992
Coefficients of an elliptic function. Ref CAY 9 128. [0,2; A1934, N1397]

$$\text{G.f.:} \quad \Pi \, (1 - x^k)^{-c(k)}, \; c(k)=4,2,4,2,4,2,....$$

M3444 1, 4, 12, 32, 80, 192, 448, 1024, 2304, 5120, 11264, 24576, 53248, 114688, 245760, 524288, 1114112, 2359296, 4980736, 10485760, 22020096, 46137344
$n.2^{n-1}$. Ref RSE 62 190 46. BIO 46 422 59. AS1 796. [1,2; A1787, N1398]

M3445 1, 4, 12, 32, 88, 240, 652, 1744, 4616, 12208, 32328, 85408, 224608, 588832
n-step walks on Kagomé lattice. Ref PRV 114 53 59. [0,2; A1665, N1399]

M3446 4, 12, 36, 96, 264, 648, 1584, 3576, 7872, 15360, 29184, 51120, 90384, 158448, 286296, 509808, 904296, 1556304
Strongly asymmetric sequences of length n. Ref MOC 25 159 71. [1,1; A2842, N1400]

M3447 1, 4, 12, 36, 100, 276, 740, 1972, 5172, 13492, 34876, 89764, 229628, 585508, 1486308, 3763460, 9497380, 23918708, 60080156, 150660388, 377009300, 942105604
Susceptibility for square lattice. Ref JPA 5 629 72. DG74 380. [0,2; A2906, N1401]

M3448 1, 4, 12, 36, 100, 284, 780, 2172, 5916, 16268, 44100, 120292, 324932, 881500, 2374444, 6416596, 17245332, 46466676, 124658732, 335116620, 897697164
n-step self-avoiding walks on square lattice. Twice M1621. Ref JPA 20 1847 87. JPA 26 1519 93. MINT 16 29 94. [0,2; A1411, N1402]

M3449 1, 4, 12, 36, 108, 264, 708, 1668, 4536, 10926, 28416
Cluster series for diamond lattice. Ref PRV 133 A315 64. DG72 225. [0,2; A3212]

M3450 1, 4, 12, 36, 108, 316, 916, 2628, 7500, 21268, 60092, 169092, 474924, 1329188, 3715244, 10359636, 28856252, 80220244, 222847804, 618083972, 1713283628
Trails of length n on square lattice. Ref JPA 18 576 85. [0,2; A6817]

M3451 1, 4, 12, 36, 108, 324, 948, 2772, 8076, 23508, 67980, 196548, 566820, 1633956, 4697412, 13501492, 38742652, 111146820, 318390684, 911904996, 2608952940
Susceptibility for diamond lattice. Ref JPA 6 1520 73. DG74 381. [0,2; A3119]

M3452 1, 4, 12, 36, 108, 324, 948, 2796, 8196, 24060, 70188, 205284, 597996, 1744548, 5073900, 14774652, 42922452, 124814484, 362267652, 1052271732, 3051900516
n-step self-avoiding walks on diamond lattice. Ref PHA 29 381 63. JPA 22 2809 89. [0,2; A1394, N1403]

M3453 1, 4, 12, 38, 125, 414, 1369, 4522, 14934, 49322, 162899, 538020, 1776961, 5868904, 19383672, 64019918, 211443425, 698350194, 2306494009, 7617832222
Nonintersecting rook paths joining opposite corners of $3xn$ board. Ref ARS 6 168 78. [1,2; A6192]

M3454 1, 1, 4, 12, 41, 126, 428, 1416, 4857, 16753, 58785, 207868, 742899, 2674010, 9694799, 35356240, 129644789, 477633711, 1767263189, 6564103612, 24466266587
$\Sigma \mu(k).C(n/k)$, $k \mid n$ (μ = Moebius, C = Catalan). Ref MAB 11(6) 13. CRB 109. [1,3; A2996]

M3455 1, 4, 12, 43, 143, 504, 1768, 6310, 22610, 81752, 297160, 1086601
Dissections of a polygon. Ref AEQ 18 387 78. [4,2; A3444]

M3456 1, 1, 4, 12, 44, 155, 580, 2128, 8092, 30276, 116304, 440484, 1703636, 6506786, 25288120, 97181760, 379061020, 1463609356, 5724954544, 22187304112
Quadrinomial coefficients. Ref FQ 7 347 69. C1 78. [0,3; A5190]

M3457 4, 12, 44, 172, 772, 3308, 14924, 64956, 294252, 1301044
n-step walks on cubic lattice. Ref PCPS 58 99 62. [1,1; A0759, N1404]

M3458   4, 12, 80, 3984, 37333248, 25626412338274304
Boolean functions of n variables. Ref HA65 147. [1,1; A0369, N1405]

M3459   1, 4, 12, 132, 3156, 136980, 10015092, 1199364852, 234207001236,
75018740661780
Colored graphs. Ref CJM 22 596 70. rcr. [1,2; A2029, N1406]

SEQUENCES BEGINNING . . ., 4, 13, . . ., . . ., 4, 14, . . .

M3460   1, 4, 13, 36, 87, 190, 386, 734, 1324
$3 \times n$ binary matrices. Ref CPM 89 217 64. PGEC 22 1050 73. SLC 19 79 88. [0,2; A2727, N1407]

M3461   1, 4, 13, 36, 93, 225, 528, 1198, 2666, 5815, 12517, 26587, 55933, 116564,
241151, 495417, 1011950, 2055892, 4157514, 8371318, 16792066, 33564256, 66875221
n-node trees of height 4. Ref IBMJ 4 475 60. KU64. [5,2; A0299, N1408]

M3462   1, 1, 1, 4, 13, 36, 181, 848, 3865, 23824, 140521, 871872, 6324517, 44942912,
344747677, 2860930816, 23853473329, 213856723200, 1996865965009
Expansion of $\exp(x \cosh x)$. [0,4; A3727]

M3463   1, 4, 13, 40, 121, 364, 1093, 3280, 9841, 29524, 88573, 265720, 797161, 2391484,
7174453, 21523360, 64570081, 193710244, 581130733, 1743392200, 5230176601
$(3^n - 1)/2$. Ref BPNR 60. Ribe91 53. [1,2; A3462]

M3464   1, 4, 13, 41, 131, 428, 1429, 4861, 16795, 58785, 208011, 742899, 2674439,
9694844, 35357669, 129644789, 477638699, 1767263189, 6564120419, 24466267019
Catalan numbers − 1. Ref MOC 22 390 68. [2,2; A1453, N1409]

M3465   1, 4, 13, 41, 134, 471, 1819, 7778
Rhyme schemes. Ref ANY 319 463 79. [1,2; A5002]

M3466   1, 4, 13, 42, 131, 402
Paraffins with n carbon atoms. Ref ZFK 93 437 36. [1,2; A0640, N1410]

M3467   1, 4, 13, 44, 163, 666, 2985, 14550, 76497, 430746, 2582447, 16403028,
109918745, 774289168, 5715471605, 44087879136, 354521950931, 2965359744446
From expansion of falling factorials. Ref JCT A24 316 78. [1,2; A5490]

M3468   1, 4, 13, 50, 203, 1154, 6627, 49356, 403293, 3858376, 33929377, 460614670,
5168544119, 64518640406, 946910125319, 16124114481720, 221243980745433
Sums of logarithmic numbers. Ref TMS 31 78 63. jos. [1,2; A2746, N1411]

M3469   1, 1, 4, 13, 53, 228, 1037, 4885, 23640, 116793, 586633, 2986616, 15377097,
79927913, 418852716, 2210503285, 11738292397, 62673984492, 336260313765
Generalized Fibonacci numbers. Ref LNM 622 186 77. [0,3; A6604]

M3470 1, 1, 4, 13, 58, 279, 1406, 7525
A subclass of $2n$-node trivalent planar graphs without triangles. Ref JCT B45 309 88. [8,3; A6798]

M3471 1, 1, 4, 13, 64, 315, 1727, 9658, 55657, 325390, 1929160, 11555172, 69840032, 425318971, 2607388905, 16077392564, 99646239355, 620439153165, 3879069845640
Dissections of a polygon. Ref DM 11 387 75. [1,3; A5035]

M3472 1, 1, 4, 13, 130, 1210, 33880, 925771, 75913222, 6174066262, 1506472167928, 366573514642546, 267598665689058580, 195168545232713290660
Gaussian binomial coefficient $[n, n/2]$ for $q = 3$. Ref GJ83 99. ARS A17 328 84. [0,3; A6104]

M3473 1, 4, 14, 2, 1, 1, 3, 2, 29, 2, 1, 7, 1, 5, 2, 1, 1, 19, 12, 77, 2, 16, 2, 1, 1, 15, 1, 1, 3, 14, 5, 1, 3, 2, 1, 1, 1, 1, 1, 1, 5, 1, 463, 1, 379, 3, 5, 3, 11, 1, 7, 7, 1, 1, 2, 1, 1, 1, 2, 1, 1, 1
Continued fraction for fifth root of 3. [1,2; A3117]

M3474 1, 4, 14, 38, 76, 136, 218, 330, 472, 652, 870, 1134
Maximal length rook tour on $n \times n$ board. Ref GA86 76. [1,2; A6071]

M3475 1, 4, 14, 40, 101, 236, 518, 1080, 2162, 4180, 7840, 14328, 25591, 44776, 76918, 129952, 216240, 354864, 574958, 920600, 1457946, 2285452, 3548550, 5460592
Coefficients of an elliptic function. Ref CAY 9 128. [0,2; A1938, N1412]

$$\text{G.f.: } \Pi (1 - x^k)^{-c(k)}, \; c(k) = 4, 4, 4, 0, 4, 4, 4, 0, \dots .$$

M3476 1, 4, 14, 40, 105, 256, 594, 1324, 2860, 6020, 12402, 25088, 49963, 98160, 190570, 366108, 696787, 1315072, 2463300, 4582600, 8472280, 15574520, 28481220
Convolved Fibonacci numbers. Ref RCI 101. FQ 15 118 77. [0,2; A1872, N1413]

$$\text{G.f.: } (1 - x - x^2)^{-4} .$$

M3477 1, 4, 14, 42, 123, 351, 988, 2761, 7682, 21313, 59029, 163314, 451529, 1247842, 3447574, 9523375, 26303825, 72646588, 200627795, 554056162
Irreducible positions of size n in Montreal solitaire. Ref JCT A60 56 92. [6,2; A7076]

M3478 1, 4, 14, 44, 128, 352, 928, 2368, 5888, 14336, 34304, 80896, 188416, 434176, 991232, 2244608, 5046272, 11272192, 25034752, 55312384, 121634816, 266338304
Exponential-convolution of natural numbers with themselves. Ref BeSl94. [0,2; A7466]

M3479 1, 4, 14, 44, 133, 388, 1116, 3168, 8938, 25100, 70334, 196824, 550656, 1540832, 4314190, 12089368, 33911543, 95228760, 267727154, 753579420, 2123637318
Powers of rooted tree enumerator. Ref R1 150. [1,2; A0300, N1414]

M3480 1, 4, 14, 44, 133, 392, 1140, 3288, 9438, 27016, 77220, 220584, 630084, 1800384, 5147328, 14727168, 42171849, 120870324, 346757334, 995742748, 2862099185
Column of Motzkin triangle. Ref JCT A23 293 77. [3,2; A5323]

M3481 1, 4, 14, 45, 140, 427, 1288, 3858, 11505, 34210
Directed animals of size n. Ref AAM 9 340 88. [3,2; A5775]

M3482 1, 4, 14, 48, 164, 560, 1912, 6528, 22288, 76096, 259808, 887040, 3028544,
10340096, 35303296, 120532992, 411525376, 1405035520, 4797091328, 16378294272
First row of 2-shuffle of spectral array $W(\sqrt{2})$. Ref FrKi94. [1,2; A7070]

$$\text{G.f.: } 1 \ / \ (1 \ - \ 4x \ + \ 2x^2).$$

M3483 1, 4, 14, 48, 165, 572, 2002, 7072, 25194, 90440, 326876, 1188640, 4345965,
15967980, 58929450, 218349120, 811985790, 3029594040, 11338026180, 42550029600
$4C(2n+1,n-1)/(n+3)$. Ref CAY 13 95. FQ 14 397 76. DM 14 84 76. [1,2; A2057,
N1415]

M3484 1, 4, 14, 49, 174, 628, 2298, 8504, 31758, 119483, 452284, 1720774, 6574987,
25214332, 96997223, 374153699, 1446677555
Permutations by inversions. Ref NET 96. DKB 241. MMAG 61 28 88. rkg. [4,2; A1894,
N1416]

M3485 0, 1, 4, 14, 56, 256, 1324, 7664, 49136, 345856, 2652244, 22014464, 196658216,
1881389056, 19192151164, 207961585664, 2385488163296, 28879019769856
Entringer numbers. Ref NAW 14 241 66. DM 38 268 82. [0,3; A6212]

M3486 4, 14, 56, 331, 1324, 12284, 49136
Related to Euler numbers. Ref JIMS 14 146 22. [1,1; A2735, N1417]

M3487 0, 0, 1, 1, 4, 14, 67, 428, 3515, 31763, 307543, 3064701
Polyhedral graphs with n nodes and minimal degree 4. Ref Dil92. [4,5; A7025]

M3488 1, 4, 14, 69, 396, 2503
Triangulations. Ref WB79 336. [0,2; A5501]

M3489 1, 1, 1, 4, 14, 74, 434, 2876, 19848
Regular flexagons with $3n$ triangles. Ref frb. [1,4; A7282]

M3490 1, 1, 4, 14, 80, 496, 3904, 34544, 354560, 4055296, 51733504, 724212224,
11070525440, 183218384896, 3266330312704, 62380415842304, 1270842139934720
Expansion of $\ln(1+\tan x)$. [0,3; A3707]

M3491 4, 14, 104, 1498, 32876, 950054, 33304122
Fanout-free functions of n variables. Ref PGEC 27 315 78. [1,1; A5743]

M3492 1, 0, 1, 1, 4, 14, 114, 2335, 172958
Self-dual threshold functions of n variables. Ref MU71 38. [1,5; A3184]

M3493 1, 1, 4, 14, 129, 1980
Connected regular bipartite graphs of degree 4 with $2n$ nodes. Ref OR76 135. [4,3; A6824]

M3494   4, 14, 194, 37634, 1416317954, 2005956546822746114,
40238616677410360228256356556102100994
$a(n) = a(n-1)^2 - 2$. Ref D1 1 399. JLMS 28 285 53. FQ 11 432 73. [0,1; A3010]

M3495   4, 15, 52, 151, 372, 799, 1540, 2727, 4516, 7087, 10644, 15415, 21652, 29631,
39652, 52039, 67140, 85327, 106996, 132567, 162484, 197215, 237252, 283111, 335332
From expansion of falling factorials. Ref JCT A24 316 78. [4,1; A5492]

$$a(n) = 5\,a(n-1) - 10\,a(n-2) + 10\,a(n-3) - 5\,a(n-4) + a(n-5).$$

M3496   1, 4, 15, 54, 189, 648, 2187, 7290, 24057, 78732, 255879, 826686, 2657205,
8503056, 27103491, 86093442, 272629233, 860934420, 2711943423, 8523250758
$n.3^{n-4}$. Ref JCT B24 208 78. [3,2; A6234]

M3497   1, 4, 15, 54, 193, 690, 2476, 8928, 32358, 117866, 431381, 1585842, 5853849,
21690378, 80650536, 300845232, 1125555054, 4222603968, 15881652606
A simple recurrence. Ref IFC 16 351 70. [0,2; A1559, N1418]

M3498   4, 15, 55, 58, 74, 109, 110, 119, 140, 175, 245, 294, 418, 435, 452, 474, 492, 528,
535, 550, 562, 588, 644, 688, 702, 714, 740, 747, 753, 818, 868, 908, 918, 1098
Tetrahedral numbers which are sum of 2 tetrahedrals. Ref MOC 16 484 62. AB71 112.
[1,1; A2311, N1419]

M3499   1, 4, 15, 56, 209, 780, 2911, 10864, 40545, 151316, 564719, 2107560, 7865521,
29354524, 109552575, 408855776, 1525870529, 5694626340, 21252634831
$a(n) = 4a(n-1) - a(n-2)$. Ref MMAG 40 78 67. MOC 24 180 70; 25 799 71. [0,2;
A1353, N1420]

M3500   1, 4, 15, 56, 210, 792, 3003, 11440, 43758, 167960, 646646, 2496144, 9657700,
37442160, 145422675, 565722720, 2203961430, 8597496600, 33578000610
Binomial coefficients $C(2n, n-1)$. See Fig M1645. Ref LA56 517. AS1 828. PLC 1 292
70. [1,2; A1791, N1421]

M3501   1, 4, 15, 58, 226, 882, 3457, 13606, 53683
Value of an urn. Ref DM 5 307 73. [1,2; A3126]

M3502   1, 4, 15, 59, 209, 780
Complexity of a $2 \times n$ grid. Ref JCT B24 210 78. [1,2; A7342]

M3503   1, 1, 4, 15, 62, 262, 1148, 5123, 23316, 155684
Blobs with vertical symmetry. Ref AEQ 31 54 86. [0,3; A7161]

M3504   1, 1, 4, 15, 62, 271, 1247, 5938, 29113, 145815
Skeins with $2n+1$ edges. Ref AEQ 31 56 86. [0,3; A7167]

M3505   0, 1, 4, 15, 64, 325, 1956, 13699, 109600, 986409, 9864100, 108505111,
1302061344, 16926797485, 236975164804, 3554627472075, 56874039553216
$a(n) = n(a(n-1) + 1)$. Ref jkh. [0,3; A7526]

M3506 1, 4, 15, 76, 373, 2676, 17539, 152860, 1383561, 14658148, 143131351,
2070738924, 24754959805, 341745565396, 5260157782923, 92358395065276
Sums of logarithmic numbers. Ref TMS 31 79 63. jos. [0,2; A2750, N1422]

M3507 1, 1, 4, 15, 76, 455, 3186, 25487, 229384, 2293839, 25232230, 302786759,
3936227868, 55107190151, 826607852266, 13225725636255, 224837335816336
The game of Mousetrap with n cards: $a(n) = (n-1)(a(n-1) + a(n-2))$. Ref QJMA 15
241 1878. jos. GN93. [1,3; A2467, N1423]

M3508 4, 15, 276, 5534533
Switching networks. Ref JFI 276 324 63. [1,1; A0881, N1425]

M3509 4, 15, 609, 845029, 1010073215739, 1300459886313272270974271,
1939680952094609786557359582286462958434022504402
Egyptian fraction for $1/\pi$. Ref hpr. rgw. [0,1; A6524]

M3510 1, 4, 16, 40, 136, 304, 880, 1768, 4936, 9112, 25216
Words of length n in a certain language. Ref DM 40 231 82. [0,2; A7057]

M3511 1, 1, 1, 1, 4, 16, 46, 106, 316, 1324, 5356, 18316, 63856, 272416, 1264264,
5409496, 22302736, 101343376, 507711376, 2495918224, 11798364736, 58074029056
Degree n even permutations of order dividing 2. Ref CJM 7 168 55. [0,5; A0704, N1427]

$$\text{E.g.f.: } e^x \cosh(x^2/2).$$

M3512 1, 1, 1, 1, 0, 4, 16, 46, 111, 228, 379, 389, 393, 3810, 14169, 39735, 91861,
172623, 225378, 10246, 1347935, 5843671, 17779693, 43942706, 89033228, 133666868
Reversion of g.f. for number of trees with n nodes. Cf. M0791. [1,6; A7315]

M3513 4, 16, 48, 108, 216, 384, 640, 1000
Paraffins. Ref BER 30 1923 1897. [1,1; A6009]

M3514 4, 16, 52, 144, 420
5-colorings of cyclic group of order n. Ref MMAG 63 212 90. [1,1; A7688]

M3515 1, 4, 16, 56, 197, 680
Projective plane trees with n nodes. Ref LNM 406 348 74. [5,2; A6079]

M3516 4, 16, 64, 246, 944, 3532, 13252, 48825
Percolation series for b.c.c. lattice. Ref SSP 10 921 77. [1,1; A6811]

M3517 0, 0, 0, 1, 4, 16, 64, 252, 1018, 4182, 17510, 74510
Identity matched trees with n nodes. Ref DM 88 97 91. [1,5; A5755]

M3518 1, 4, 16, 64, 256, 1024, 4096, 16384, 65536, 262144, 1048576, 4194304,
16777216, 67108864, 268435456, 1073741824, 4294967296, 17179869184
Powers of 4. Ref BA9. [0,2; A0302, N1428]

M3519 4, 16, 64, 416, 4544, 23488, 207616, 4205056, 198295552, 2574439424
Susceptibility for square lattice. Ref PHL A25 208 67. [1,1; A5401]

M3520 4, 16, 64, 736, 11584, 43072, 607232, 50435584, 1204185088
Susceptibility for diamond lattice. Ref PPS 86 13 65. [1,1; A2923, N1429]

M3521 1, 4, 16, 68, 304, 1412, 6752, 33028, 164512, 831620, 4255728, 22004292,
114781008, 603308292, 3192216000, 16989553668, 90890869312, 488500827908
Royal paths in a lattice (convolution of M1659). Ref CRO 20 18 73. [1,2; A6319]

M3522 1, 4, 16, 69, 348, 2016, 13357, 99376, 822040, 7477161, 74207208, 797771520,
9236662345, 114579019468, 1516103040832, 21314681315997
Permutations by length of runs. Ref DKB 261. [1,2; A0303, N1430]

M3523 0, 1, 4, 16, 72, 522, 3642, 30753
$(x \to x^2)$-free subsets of symmetric group. Ref SFCA92 2 17. [1,3; A7234]

M3524 1, 1, 1, 4, 16, 78, 457, 2938, 20118
Triangulations of the disk. Ref PLMS 14 765 64. [0,4; A2713, N1431]

M3525 1, 0, 0, 0, 4, 16, 80, 672, 4752, 48768, 440192, 5377280, 59245120, 839996160,
10930514688, 176547098112, 2649865335040, 48047352500224, 817154768973824
Permutations with no hits on 2 main diagonals. Ref R1 187. Sim92. [0,5; A3471]

M3526 1, 0, 0, 0, 4, 16, 80, 672, 4896, 49920, 460032, 5598720, 62584320, 885381120,
11644323840, 187811205120, 2841958748160, 51481298534400, 881192033648640
Restricted permutations. Ref MU06 3 468. Sim92. [0,5; A2777, N1432]

M3527 1, 4, 16, 85, 646, 6664, 86731, 1354630, 24607816
Binary phylogenetic trees with n labels. Ref LNM 884 198 81. [1,2; A6681]

M3528 4, 16, 88, 520, 3112, 18664, 111976, 671848, 4031080, 24186472, 145118824,
870712936, 5224277608, 31345665640, 188073993832, 1128443962984
$a(n)=6a(n-1)-8$. Ref PGEC 11 140 62. [0,1; A5618]

M3529 4, 16, 88, 538, 3568, 24596
Triangulations of the disk. Ref PLMS 14 759 64. [0,1; A5495]

M3530 4, 16, 152, 2368, 47688, 1156000, 32699080, 1057082752, 38444581640,
1553526946144, 69054999618888, 3348574955346496, 175908582307762312
Cascade-realizable functions of n variables. Ref PGEC 24 688 75. [1,1; A5749]

M3531 4, 16, 152, 2680, 68968, 2311640, 95193064, 4645069336, 261938616104,
16756882325464, 1198897678224232, 94851206834082200, 8221740727881348520
Disjunctively-realizable functions of n variables. Ref PGEC 24 689 75. [1,1; A5739]

M3532   4, 16, 160, 3112, 89488, 3358600, 154925968
Fanout-free functions of n variables. Ref PGEC 27 315 78. [1,1; A5741]

M3533   4, 16, 392, 1966074
Switching networks. Ref JFI 276 324 63. [1,1; A0874, N1433]

M3534   0, 0, 4, 17, 61, 214, 758, 2723, 9908, 36444, 135266, 505859, 1903888, 7204872,
27394664, 104592935, 400795842
Fixed polyominoes with n cells. Ref CJN 18 367 75. [1,3; A6762]

M3535   4, 17, 65, 230, 736, 2197, 6093
4-covers of an n-set. Ref DM 81 151 90. [1,1; A5784]

M3536   4, 17, 69, 290, 1174, 4762, 20011, 84101, 340461, 1378277, 5590589, 22676645,
95292383, 400440122, 1682945112, 7072978202, 28633110562, 115913692522
$a(n) = 1 + a([n/2]) a(\lceil n/2 \rceil)$. Ref clm. [1,1; A5511]

M3537   0, 1, 4, 17, 70, 282, 1136, 4583, 18457, 74131
Value of an urn. Ref DM 5 307 73. [1,3; A3127]

M3538   1, 4, 17, 72, 305, 1292, 5473, 23184, 98209, 416020, 1762289, 7465176,
31622993, 133957148, 567451585, 2403763488, 10182505537, 43133785636
$a(n) = 4a(n-1) + a(n-2)$. Ref TH52 282. [0,2; A1076, N1434]

M3539   1, 4, 17, 76, 354, 1704, 8421, 42508, 218318, 1137400, 5996938, 31940792,
171605956, 928931280, 5061593709, 27739833228, 152809506582, 845646470616
Walks on cubic lattice (binomial transform of M2850). Ref GU90. EIS § 2.7. [0,2; A5572]

M3540   1, 4, 17, 77, 372, 1915, 10481, 60814, 372939, 2409837, 16360786, 116393205,
865549453, 6713065156, 54190360453, 454442481041, 3952241526188
From expansion of falling factorials (binomial transform of M2851). Ref JCT A24 316 78.
EIS § 2.7. [0,2; A5494]

M3541   1, 4, 18, 32, 160, 324, 1456, 2048, 13122, 25600, 117128, 209952, 913952,
2119936, 9447840, 13107200, 86093440
Generalized Euler Φ function. Ref MOC 28 1168 74. [1,2; A3474]

M3542   1, 4, 18, 88, 455, 2448, 13566, 76912, 444015, 2601300, 15426840, 92431584,
558685348, 3402497504, 20858916870, 128618832864, 797168807855, 4963511449260
From generalized Catalan numbers. Ref LNM 952 279 82. [0,2; A6629]

$$\text{G.f.:}\quad {}_3F_2([2,5/3,4/3];\ [3,5/2];\ 27x/4).$$

M3543   1, 4, 18, 89, 466, 2537
Rooted planar maps. Ref CJM 15 542 63. [1,2; A0305, N1435]

M3544 1, 4, 18, 96, 265, 672, 1617, 3776, 8577, 19080, 41745
Rook polynomials. Ref JAuMS A28 375 79. [1,2; A5777]

M3545 1, 4, 18, 96, 600, 4320, 35280, 322560, 3265920, 36288000, 439084800,
    5748019200, 80951270400, 1220496076800, 19615115520000, 334764638208000
n.n!. Ref JRAM 198 61 57. [1,2; A1563, N1436]

M3546 1, 1, 4, 18, 105, 636, 4710, 38508, 352902
Unreformed permutations. Ref GN93. [1,3; A7711]

M3547 1, 1, 4, 18, 110, 810, 7040, 70280, 792200, 9945000, 137550400, 2077719600,
    34026132400, 600433397200, 11356783360000, 229193571984000, 4915556301968000
Expansion of tan(ln(1+x)). [0,3; A3708]

M3548 1, 0, 1, 4, 18, 112, 820, 6912, 66178, 708256, 8372754, 108306280, 1521077404,
    23041655136, 374385141832, 6493515450688, 119724090206940
Stochastic matrices of integers. Ref DUMJ 35 659 68. [0,4; A0986, N1437]

M3549 1, 1, 4, 18, 120, 960, 9360, 105840, 1370880, 19958400
From a Fibonacci-like differential equation. Ref FQ 27 306 89. [0,3; A5442]

M3550 1, 4, 18, 126, 1160, 15973, 836021
Semigroups of order *n*. Ref PL65. MAL 2 2 67. SGF 14 71 77. [1,2; A1423, N1438]

M3551 0, 1, 4, 18, 166, 7579, 7828352, 2414682040996
Spectrum of a certain 3-element algebra. Ref Berm83. [0,3; A7153]

M3552 1, 4, 19, 66, 219, 645, 1813, 4802, 12265, 30198, 72396, 169231, 387707, 871989,
    1930868, 4215615, 9091410, 19389327, 40944999, 85691893, 177898521
Trees of diameter 8. Ref IBMJ 4 476 60. KU64. [9,2; A0306, N1440]

M3553 1, 4, 19, 91, 436, 2089, 10009, 47956, 229771, 1100899, 5274724, 25272721,
    121088881, 580171684, 2779769539, 13318676011, 63813610516, 305749376569
Pythagoras' theorem generalized. Ref BU71 75. [1,2; A4253]

$$\text{G.f.:}\quad (1 - x) \,/\, (1 - 5x + x^2).$$

M3554 0, 0, 0, 0, 1, 4, 19, 93, 539, 3474, 24856, 192972, 1613219, 14410374, 136920388,
    1378542639, 14663082556, 164340455701, 1936286904952, 23932267735948
Dependable interval graphs with *n* nodes. Ref TAMS 272 422 82. pjh. [1,6; A5978]

M3555 1, 4, 19, 98, 531, 2971, 16997, 98830, 581788, 3458249, 20718292, 124929233,
    757421601, 4613459330, 28213402944, 173141766742, 1065820341078
n-node animals on f.c.c. lattice (invert M2925). Ref PE90. DU92 42. [1,2; A6194]

M3556 1, 1, 4, 19, 100, 562, 3304, 20071, 124996, 793774, 5120632, 33463102,
    221060008, 1473830308, 9904186192, 67015401391, 456192667396, 3122028222934
Shifts left when INVERT transform applied thrice. Ref BeS194. EIS § 2.7. [0,3; A7564]

M3557 1, 1, 4, 19, 109, 742, 5815, 51193, 498118, 5296321, 60987817, 754940848,
9983845261, 140329768789, 2087182244308, 32725315072135, 539118388883449
Coincides with its 3rd order binomial transform. Ref DM 21 320 78. EIS § 2.7. [0,3;
A4212]

$$Lgd. e.g.f.: \quad e^{3x}.$$

M3558 1, 4, 19, 179, 16142
Matrices with n columns whose rows do not cover each other. Ref hofri. [2,2; A7411]

M3559 4, 19, 556, 2945786
Switching networks. Ref JFI 276 322 63. [1,1; A0844, N1441]

M3560 4, 19, 632, 19245637
Switching networks. Ref JFI 276 324 63. [1,1; A0863, N1442]

M3561 4, 19, 5779, 192900153619, 7177905237579946589743592924684179
$a(n) = a(n-1)^3 - 3a(n-1)^2 + 3$. Ref CRP 83 1287 1876. D1 1 397. [0,1; A2813, N1443]

M3562 4, 20, 56, 120, 220, 364, 560, 816, 1140, 1540, 2024, 2600, 3276, 4060, 4960,
5984, 7140, 8436, 9880, 11480, 13244, 15180, 17296, 19600, 22100, 24804, 27720
$2n(n+1)(2n+1)/3$. Ref MOC 4 23 50. [1,1; A2492, N1444]

M3563 0, 4, 20, 68, 196, 512, 1256, 2936, 6628, 14528, 31140, 65414, 135276, 275656,
555216, 1105726, 2182380, 4268906, 8290740, 15984420, 30638312, 58369924
First moment of site percolation series for hexagonal lattice. Ref JPA 21 3822 88. [0,2;
A6740]

M3564 4, 20, 84, 292, 980, 3052, 9316, 27396, 79412
Susceptibility for square lattice. Ref DG72 136. [1,1; A3489]

M3565 1, 4, 20, 110, 638, 3832, 23592, 147941, 940982, 6053180, 39299408, 257105146,
1692931066, 11208974860, 74570549714, 498174818986
Fixed n-celled polyominoes which need only touch at corners. Ref dhr. [1,2; A6770]

M3566 1, 4, 20, 120, 840, 6720, 60480, 604800, 6652800, 79833600, 1037836800,
14529715200, 217945728000, 3487131648000, 59281238016000, 1067062284288000
$n!/6$. Ref PEF 77 44 62. [3,2; A1715, N1445]

M3567 0, 1, 4, 20, 124, 920, 7940, 78040, 859580, 10477880, 139931620, 2030707640,
31805257340, 534514790680, 9591325648580, 182974870484120, 3697147584561340
$a(n) = 2n \cdot a(n-1) - (n-1)^2 a(n-2)$. Ref SE33 78. [0,3; A2793, N1446]

M3568 1, 4, 20, 127, 967, 8549, 85829, 962308, 11895252, 160475855, 2343491207,
36795832297, 617662302441, 11031160457672, 208736299803440, 4169680371133507
Natural numbers exponentiated twice. Ref BeLs94. [1,2; A7550]

M3569 1, 1, 0, 4, 20, 144, 630, 5696, 39366, 366400
Permutations of length n with spread 0. Ref JAuMS A21 489 76. [1,4; A4204]

M3570 1, 1, 4, 20, 148, 1348, 15104, 198144, 2998656
Expansion of $E(\operatorname{tr}(X'X)^n)$, X rectangular and Gaussian. Ref clm. CONT 158 151 92. [1,3; A1171, N1447]

M3571 1, 4, 20, 155, 1716, 24654, 434155, 9043990, 217457456
Binary phylogenetic trees with n labels. Ref LNM 884 198 81. [1,2; A6682]

M3572 4, 20, 264, 80104
Switching networks. Ref JFI 276 322 63. [1,1; A0847, N1448]

M3573 1, 1, 1, 4, 21, 122, 849, 6719, 59873
Hit polynomials. Ref RI63. [1,4; A1888, N1449]

M3574 1, 1, 4, 21, 126, 818, 5594, 39693, 289510, 2157150, 16348960, 125642146, 976789620, 7668465964, 60708178054, 484093913917, 3884724864390
$\Sigma\, C(n,k).C(2n+k,k-1)/n$, $k = 1 .. n$. Ref FQ 11 123 73. AEQ 31 52 86. [1,3; A3168]

M3575 4, 21, 127, 831, 5722, 40879, 300440, 2258455, 17291704, 134417955, 1058279251, 8422155293
Havender tableaux of height 2 with n columns. Ref GoBe89. [1,1; A7345]

M3576 0, 1, 4, 21, 134, 1001, 8544, 81901, 870274, 10146321, 128718044, 1764651461, 25992300894, 409295679481, 6860638482424, 121951698034461, 2291179503374234
$a(n) = n.a(n-1) + (n-4).a(n-2)$. Ref R1 188. [2,3; A1909, N1450]

M3577 4, 21, 143, 1061, 8363, 68906, 586081, 5096876, 45086079, 404204977, 3733002302, 33419857205, 308457624821, 2858876213963, 26639628671867
Number of primes with n digits. Cf. M3608. Ref Shan78 15. BPNR 179. Long87 77. [1,1; A6879]

M3578 1, 1, 4, 21, 148, 1305, 13806, 170401, 2403640, 38143377, 672552730, 13044463641, 276003553860, 6326524990825, 156171026562838, 4130464801497105
Expansion of $1/(1 - x\, e^x)$. Ref ARS 10 136 80. [0,3; A6153]

M3579 1, 1, 4, 21, 266, 7849
Connected regular graphs of degree 6 with n nodes. Ref OR76 135. [7,3; A6822]

M3580 4, 21, 1531, 44782251
Switching networks. Ref JFI 276 324 63. [1,1; A0868, N1451]

M3581 4, 21, 2914, 4379140552
Switching networks. Ref JFI 276 324 63. [1,1; A0875, N1452]

M3582 4, 22, 27, 58, 85, 94, 121, 166, 202, 265, 274, 319, 346, 355, 378, 382, 391, 438, 454, 483, 517, 526, 535, 562, 576, 588, 627, 634, 636, 645, 648, 654, 663, 666, 690, 706
Smith numbers: sum of digits = Σ sum of digits of prime factors. Ref TYCM 13 21 87. GA89 300. [1,1; A6753]

M3583 1, 4, 22, 107, 486, 2075, 8548, 33851, 130365, 489387, 1799700, 6499706, 23118465, 81134475, 281454170
Connected graphs with n nodes, $n+3$ edges. Ref SS67. [4,2; A1436, N1453]

M3584 1, 4, 22, 110, 515, 2272, 9777, 40752
Graphs with no isolated vertices. Ref LNM 952 101 82. [4,2; A6651]

M3585 1, 4, 22, 130, 807, 5163, 33742, 224002, 1505146, 10211027, 69814781, 480435484, 3324233772, 23108532996, 161288459289
Strict n-node animals on b.c.c. lattice. Ref DU92 41. [1,2; A7195]

M3586 1, 4, 22, 136, 897, 6168, 43670, 315956, 2324479, 17329828, 130605478, 993182984, 7610051579, 58689316888, 455159096044
Primitive n-node animals on b.c.c. lattice. Ref DU92 41. [1,2; A7196]

M3587 1, 1, 4, 22, 140, 969, 7084, 53820, 420732, 3362260, 27343888, 225568798, 1882933364, 15875338990, 134993766600, 1156393243320, 9969937491420
Dissections of a polygon: $C(4n,n)/(3n+1)$. Ref DM 11 388 75. [0,3; A2293, N1454]

M3588 1, 4, 22, 140, 970, 7196, 56092, 452064, 3735700, 31484244, 269613896, 2339571468, 20529434520
n-step walks on f.c.c. lattice. Ref JPA 6 351 73. [1,2; A3287]

M3589 1, 1, 4, 22, 147, 1074, 8216, 64798, 521900, 4272967, 35447724, 297308810, 2516830890, 21476307960, 184530904560, 1595190209002, 13863857007924
Dissections of a polygon. Ref DM 11 388 75. [1,3; A5039]

M3590 1, 1, 4, 22, 154, 1304, 12915, 146115, 1855570, 26097835, 402215465, 6734414075, 121629173423, 2355470737637, 48664218965021, 1067895971109199
Coefficients of iterated exponentials. Ref SMA 11 353 45. PRV A32 2342 85. [0,3; A0307, N1455]

M3591 1, 4, 22, 166, 1726, 24814, 494902
Related to partially ordered sets. Ref JCT 6 17 69. [0,2; A1827, N1456]

M3592 0, 1, 1, 4, 22, 178, 2278, 46380, 1578060, 92765486, 9676866173, 1821391854302, 625710416245358, 395761853562201960, 464128290507379386872
Rooted nonseparable graphs with n nodes. Ref rwr. [1,4; A4115]

M3593 4, 22, 190, 3250, 136758, 17256831
Incidence matrices. Ref CPM 89 217 64. SLC 19 79 88. [1,1; A2728, N1457]

M3594 1, 4, 23, 156, 1162, 9192, 75819, 644908, 5616182, 49826712, 448771622, 4092553752, 37714212564, 350658882768, 3285490743987, 30989950019532
Reversion of g.f. for squares. Ref DM 9 341 74. [1,2; A7297]

M3595 4, 24, 36, 48, 48, 144, 32, 60, 192, 108, 144, 72, 240, 288, 192
Frequency of nth largest distance in $N{\times}N{\times}N$ grid, $N > n$. Ref ReSk94. [1,1; A7544]

M3596   1, 4, 24, 84, 392, 1344, 5760, 19800, 81675, 283140, 1145144, 4008004,
16032016, 56632576, 225059328, 801773856, 3173688180, 11392726800, 44986664800
Walks on square lattice. Ref GU90. [3,2; A5561]

M3597   0, 4, 24, 104, 384, 1284, 4012, 11924, 34100, 94584, 255852, 677850, 1764482,
4523924, 11447870, 28636218, 70907326, 173991368, 423469988, 1023162920
First moment of bond percolation series for hexagonal lattice. Ref JPA 21 3822 88. [0,2;
A6736]

M3598   4, 24, 120, 560, 2520, 11088, 48048, 205920, 875160, 3695120, 15519504,
64899744, 270415600, 1123264800, 4653525600, 19234572480, 79342611480
Expansion of $4(1-4x)^{-3/2}$. Ref PLC 1 292 70. [0,1; A2011, N1458]

M3599   0, 4, 24, 140, 816, 4756, 27720, 161564, 941664, 5488420, 31988856, 186444716,
1086679440, 6333631924, 36915112104, 215157040700, 1254027132096
$a(n)=6a(n-1)-a(n-2)$. [0,2; A5319]

M3600   4, 24, 152, 1080, 8152, 63976, 518232, 4299728, 36360872, 312284536
n-step walks on f.c.c. lattice. Ref JPA 6 351 73. [2,1; A3288]

M3601   1, 4, 24, 176, 1456, 13056, 124032, 1230592, 12629760, 133186560, 1436098560,
15774990336, 176028860416, 1990947110912, 22783499599872, 263411369705472
Rooted maps with $2n$ nodes. Ref CJM 14 416 62. [1,2; A0309, N1460]

$$a(n) = 4 \, a(n-1) \, C(3n, 3) \, / \, C(2n+2, 3).$$

M3602   4, 24, 188, 1368, 10572
$2n$-step walks on diamond lattice. Ref PCPS 58 100 62. [1,1; A1397, N1461]

M3603   0, 0, 1, 4, 24, 188, 1705, 16980, 180670, 2020120, 23478426, 281481880,
3461873536, 43494961412, 556461655783
Simple quadrangulations. Ref JCT 4 275 68. [1,4; A1506, N1462]

M3604   1, 4, 24, 192, 1920, 23040, 322560, 5160960, 92897280, 1857945600,
40874803200, 980995276800, 25505877196800, 714164561510400
$2^{n-1} n!$. Ref PSPM 19 172 71. [1,2; A2866, N1463]

M3605   1, 0, 1, 4, 24, 193, 2420, 47912, 1600524, 93253226, 9694177479,
1822463625183, 625829508087155, 395785845695978077, 464137111800208818956
Rooted bridgeless graphs with n nodes. Ref JCT B33 302 82. [1,4; A7145]

M3606   1, 4, 24, 240, 4320, 146880, 9694080, 1260230400, 325139443200,
167121673804800, 171466837323724800, 351507016513635840000
$a(n)=(2^n+2)a(n-1)$. Ref CJM 16 665 64. SPLAG 151. [0,2; A6088]

M3607   4, 24, 304, 5440, 125824, 3566080, 119614464
Boolean functions of n variables by AND rank. Ref CACM 23 704 76. [2,1; A5756]

M3608   1, 4, 25, 168, 1229, 9592, 78498, 664579, 5761455, 50847534, 455052511,
4188054813, 37607912018, 346065536839, 3204941750802, 29844570422669
Primes with at most n digits. Cf. M3577. Ref Shan78 15. BPNR 179. Long87 77. [0,2;
A6880]

M3609   0, 1, 4, 25, 174, 1393, 12536, 125361, 1378970, 16547641, 215119332,
3011670649, 45175059734, 722800955745, 12287616247664, 221177092457953
$a(n) = (n+2)a(n-1) + (-1)^n$. [1,3; A6348]

M3610   1, 4, 25, 208, 2146, 26368, 375733, 6092032, 110769550, 2232792064
Feynman diagrams of order $2n$. Ref PRV D18 1949 78. [1,2; A5411]

M3611   1, 4, 25, 676, 458329, 210066388900, 44127887745906175987801,
19472704769152964495597034454938489304452791204
$a(n) = (a(n-1)+1)^2$. Ref FQ 11 437 73. [0,2; A4019]

M3612   1, 1, 4, 26, 234, 2696, 37919, 630521, 12111114, 264051201, 6445170229,
174183891471, 5164718385337, 166737090160871, 5822980248613990
Coefficients of iterated exponentials. Ref SMA 11 353 45. [0,3; A0310, N1464]

M3613   1, 1, 1, 4, 26, 236, 2752, 39208, 660032, 12818912, 282137824, 6939897856,
188666182784, 5617349020544, 181790703209728, 6353726042486272
Schroeder's fourth problem. Ref RCI 197. C1 224. [0,4; A0311, N1465]

G.f. $A(x)$ satisfies $e^{A(x)} = 2 A(x) - x + 1$.

M3614   0, 1, 4, 26, 236, 2760, 39572, 672592, 13227804, 295579520, 7398318500,
205075286784, 6236796259916, 206489747516416, 7393749269685300
Normalized total height of rooted trees with n nodes. Ref JAuMS 10 281 69. [1,3; A1863,
N1466]

M3615   1, 1, 4, 26, 255, 3642, 75606, 2316169, 106289210, 7321773414
Labeled topologies with n points. Ref MSM 11 243 74. [0,3; A6056]

M3616   1, 4, 26, 260, 3368, 53744, 1022320, 22522960
Bishops on an $n \times n$ board. Ref AH21 1 271. [1,2; A2465, N1467]

M3617   1, 0, 0, 0, 0, 4, 27, 172, 1141, 8017, 60319, 486372, 4196384, 38621356,
377949874, 3920335179, 42975606304, 496545261764, 6031989895262
Connected interval graphs with n nodes. Ref TAMS 272 422 82. pjh. [1,6; A5974]

M3618   1, 1, 4, 27, 248, 2830, 38232, 593859, 10401712, 202601898, 4342263000,
101551822350, 2573779506192, 70282204726396, 2057490936366320
Irreducible diagrams with $2n$ nodes. Ref CJM 4 25 52. JCT A24 361 78. [1,3; A0699,
N1468]

M3619   1, 4, 27, 256, 3125, 46656, 823543, 16777216, 387420489, 10000000000,
285311670611, 8916100448256, 302875106592253, 11112006825558016
n^n. Ref BA9. [1,2; A0312, N1469]

M3620   4, 27, 14056, 104751025086
Switching networks. Ref JFI 276 324 63. [1,1; A0869, N1470]

M3621   0, 0, 1, 4, 28, 85, 630, 3096, 23220, 123952, 1036080, 7230828, 66349440,
503745252, 5080269600
Bishops on an $n \times n$ board. Ref LNM 560 212 76. [3,4; A5634]

M3622   4, 28, 148, 704, 3176
Related to enumeration of rooted maps. Ref JCT A13 124 72. [2,1; A6302]

M3623   4, 28, 188, 1428, 10708
$2n$-step walks on diamond lattice. Ref PCPS 58 100 62. [1,1; A1396, N1471]

M3624   1, 4, 28, 196, 1324, 8980, 60028, 402412, 2675860, 17826340, 118145548,
784024780, 5184334996, 34313323804, 226516271020, 1496391824212
n-step self-avoiding walks on b.c.c. lattice. Ref PPS 92 649 67. [1,2; A2903, N1472]

M3625   1, 4, 28, 220, 1820, 15504, 134596, 1184040, 10518300, 94143280, 847660528,
7669339132, 69668534468, 635013559600, 5804731963800, 53194089192720
Binomial coefficients $C(4n,n)$. See Fig M1645. Ref AS1 828. dek. [0,2; A5810]

M3626   1, 4, 28, 256, 2716, 31504, 387136
$2n$-step polygons on diamond lattice. Ref AIP 9 345 60. [0,2; A2895, N1473]

M3627   1, 1, 4, 28, 280, 3640, 58240, 1106560, 24344320, 608608000, 17041024000,
528271744000, 17961239296000, 664565853952000, 26582634158080000
Expansion of $(1-3x)^{-1/3}$. [0,3; A7559]

M3628   1, 1, 4, 28, 301, 4466, 84974, 1974904, 54233540, 1718280152
Evolutionary trees of magnitude n. Ref CN 44 85 85. [1,3; A7152]

M3629   4, 28, 2272, 67170304
Switching networks. Ref JFI 276 321 and 588 63. [1,1; A0838, N1474]

M3630   1, 0, 0, 0, 1, 4, 29, 206, 1708, 15702
Hit polynomials. Ref RI63. [0,6; A1883, N1475]

M3631   1, 1, 4, 29, 355, 6942, 209527, 9535241, 642779354, 63260289423,
8977053873043, 1816846038736192, 519355571065774021, 207881393656668953041
Labeled topologies or transitive digraphs on n points. Ref JAuMS 8 194 68. Cl 229.
ErSt89. [0,3; A0798, N1476]

M3632   4, 30, 126, 393, 1016, 2304, 4740, 9042, 16236, 27742, 45474, 71955, 110448,
165104, 241128, 344964, 484500, 669294, 910822, 1222749, 1621224, 2125200
From expansion of $(1+x+x^2)^n$. Ref Cl 78. [4,1; A5715]

M3633 1, 4, 30, 220, 1855, 17304, 177996, 2002440, 24474285, 323060540, 4581585866, 69487385604, 1122488536715
Permutations of length n by rises. Ref DKB 263. [4,2; A0313, N1477]

M3634 1, 1, 4, 30, 330, 4719, 81796, 1643356, 37119160, 922268360, 24801924512, 713055329720, 21706243125300, 694280570551875, 23188541161342500
Dyck paths. Ref SC83. [0,3; A6149]

$$\text{G.f.:} \quad {}_4F_3([1,1/2,5/2,3/2]\,;\,[4,5,6]\,;\,64x).$$

M3635 1, 1, 4, 30, 336, 5040, 95040, 2162160, 57657600
Dissections of a disk. Ref CMA 2 25 70. MAN 191 98 71. [2,3; A1761, N1478]

M3636 1, 0, 0, 0, 0, 4, 31, 199, 1313, 9158, 68336, 546697, 4682870, 42818887, 416581477, 4298371842, 46896673051, 539527125454, 6528590200432
Identity interval graphs with n nodes. Ref TAMS 272 422 82. pjh. [1,6; A5216]

M3637 1, 4, 31, 244, 1921, 15124, 119071, 937444, 7380481, 58106404, 457470751, 3601659604, 28355806081, 223244789044, 1757602506271, 13837575261124
$a(n) = 8a(n-1) - a(n-2)$. Ref NCM 4 167 1878. [0,2; A1091, N1479]

M3638 1, 4, 31, 293, 3326, 44189, 673471, 11588884, 222304897, 4704612119, 108897613826, 2737023412199, 74236203425281, 2161288643251828
Hamiltonian circuits on n-octahedron. Ref JCT B19 2 75. [2,2; A3436]

$$a(n) = (-2n+4)\,a(n-2) - a(n-3) + (2n+2)\,a(n-1).$$

M3639 1, 1, 4, 31, 362, 5676
Mixed Husimi trees with n nodes. Ref PNAS 42 532 56. [1,3; A0314, N1480]

M3640 1, 4, 31, 379, 6556, 150349, 4373461, 156297964, 6698486371, 337789490599, 19738202807236, 1319703681935929, 99896787342523081, 8484301665702298804
Expansion of exp $(\cos x - 1)$. Ref JO61 150. [0,2; A5046]

M3641 1, 4, 31, 1294
4-dimensional polytopes with n vertices. Ref UPG B15. [5,2; A5841]

M3642 4, 31, 1921, 7380481, 108942999582721, 23737154316161495960243527681
$a(n) = 2a(n-1)^2 - 1$. Ref jos. [0,1; A5828]

M3643 4, 32, 200, 1120, 5880, 29568, 144144
Almost trivalent maps. Ref PLC 1 292 70. [0,1; A2012, N1481]

M3644 4, 32, 252, 2032, 16292, 132000, 1070716, 8729216, 71230324, 584550656
Expansion of susceptibility series related to Potts model. Ref JPA 12 L230 79. [1,1; A7278]

M3645 4, 32, 292, 2672, 24780, 232512, 2201948
n-step walks on f.c.c. lattice. Ref PCPS 58 100 62. [1,1; A0766, N1482]

M3646 1, 4, 32, 336, 4096, 54912, 786432, 11824384
Almost trivalent maps. Ref PLC 1 292 70. [0,2; A2005, N1483]

M3647 1, 1, 4, 32, 396, 6692, 143816, 3756104, 115553024, 4093236352, 164098040448,
7345463787136
Greg trees. Ref MSS 34 127 90. [1,3; A5263]

M3648 1, 4, 32, 416, 7552, 176128, 5018624, 168968192, 6563282944, 288909131776,
14212910809088, 772776684683264, 46017323176296448, 2978458881388183550
Trees of subsets of an n-set. Ref CACM 23 704 76. LNM 829 122 80. MBIO 54 8 81. [1,2;
A5172]

$$\text{E.g.f.: } -1/2 - W(- e^{-1/2 + x} / 2), \text{ where } W(z) = \Sigma\, n^{n-1} z^n.$$

M3649 1, 1, 4, 32, 436, 9012, 262760, 10270696, 518277560, 32795928016,
2542945605432, 237106822506952, 26173354092593696, 3375693096567983232
Ultradissimilarity relations on an n-set. Ref MET 27 130 80. EJC 5 313 84. ANAL 12 109
92. [1,3; A5121]

M3650 1, 1, 4, 32, 588, 21476, 1551368, 218218610
Labeled mating graphs with n nodes. Ref RE89. [1,3; A6024]

M3651 1, 4, 33, 456, 9460, 274800, 10643745, 530052880, 32995478376,
2510382661920, 229195817258100, 24730000147369440, 3113066087894608560
Related to Bessel functions. Ref PAMS 14 2 63. [2,2; A2190, N1484]

M3652 1, 1, 4, 33, 480, 11010, 367560, 16854390, 1016930880
From a distribution problem. Ref DUMJ 33 761 66. [0,3; A2018, N1485]

M3653 0, 4, 34, 113, 268, 524, 905, 1437, 2145, 3054, 4189, 5575, 7238, 9203, 11494,
14137, 17157, 20580, 24429, 28731, 33510, 38792, 44602, 50965, 57906, 65450, 73622
Nearest integer to $4\pi . n^3 / 3$. Ref PNISI 13 37 47. [0,2; A2101, N1486]

M3654 4, 34, 308, 3024, 31680, 349206, 4008004, 47530912, 579058896, 7215393640,
91644262864, 1183274479040, 15497363512800, 205519758825150
Walks on square lattice. Ref GU90. [1,1; A5569]

$$\text{G.f.: } {}_4F_3([2,17/5,5/2,3/2]; [4,5,12/5]; 16x).$$

M3655 1, 1, 4, 34, 496, 11056, 349504, 14873104, 819786496, 56814228736,
4835447317504, 495812444583424, 60283564499562496, 8575634961418940416
$|2^n(2^{2n}-1)B_{2n}/n|$, B_n = Bernoulli. Ref JFI 239 67 45. MOC 1 385 45. [1,3; A2105,
N1487]

M3656 4, 34, 8900, 15320103918
Switching networks. Ref JFI 276 324 63. [1,1; A0860, N1488]

M3657 4, 35, 166, 633, 2276, 8107, 29086, 105460, 386320, 1428664, 5327738, 20014741
n-step walks on hexagonal lattice. Ref JPA 6 352 73. [4,1; A5552]

M3658 1, 1, 4, 35, 541, 13062, 444767, 19912657, 1121041222, 77048430033,
6329916102841, 611728117464928, 68657066350744197, 8854866422322096893
Connected labeled interval graphs with n nodes. Ref TAMS 272 422 82. pjh. [1,3; A5973]

M3659 1, 1, 4, 35, 541, 13302, 489287, 25864897, 1910753782, 193328835393,
26404671468121, 4818917841228328, 1167442027829857677
Connected chordal graphs with n nodes. Ref GC 1 199 85. [1,3; A7134]

M3660 1, 4, 35, 1246
Combinatorial 3-manifolds. Ref DM 16 93 76. [6,2; A5026]

M3661 1, 4, 36, 144, 3600, 3600, 176400, 705600, 6350400, 1270080, 153679680,
153679680, 25971865920, 25971865920, 129859329600, 519437318400
Denominators of Σk^{-2}; $k = 1..n$. Cf. M4004. Ref KaWa 89. [1,2; A7407]

M3662 4, 36, 232, 1308, 6808, 33560, 159108
Susceptibility for hexagonal lattice. Ref DG72 136. [1,1; A3488]

M3663 4, 36, 308, 2764, 25404, 237164, 2237948
n-step walks on f.c.c. lattice. Ref PCPS 58 100 62. [1,1; A0765, N1489]

M3664 1, 4, 36, 400, 4900, 63504, 853776, 11778624, 165636900, 2363904400,
34134779536, 497634306624, 7312459672336, 108172480360000, 1609341595560000
$C(2n,n)^2$. Ref AIP 9 345 60. [0,2; A2894, N1490]

M3665 1, 4, 36, 480, 8400, 181440, 4656960, 138378240, 4670265600, 176432256000,
7374868300800, 337903056691200, 16838835658444800, 906706535454720000
Coefficients of orthogonal polynomials. Ref MOC 9 174 55. [1,2; A2690, N1491]

$$\text{E.g.f.:} \quad (1 - 2x) / (1 - 4x)^{3/2}.$$

M3666 1, 4, 36, 576, 14400, 518400, 25401600, 1625702400, 131681894400,
13168189440000, 1593350922240000, 229442532802560000, 38775788043632640000
$(n!)^2$. Ref RCI 217. [1,2; A1044, N1492]

M3667 1, 4, 36, 624, 18256, 814144, 51475776, 4381112064, 482962852096,
66942218896384, 11394877025289216, 2336793875186479104
Expansion of $\sinh x / \cos x$. Ref CMB 13 306 70. [0,2; A2084, N1493]

M3668 4, 36, 3178, 298908192
Switching networks. Ref JFI 276 324 63. [1,1; A1152, N1494]

M3669   1, 4, 37, 559, 11776, 318511, 10522639, 410701432, 18492087079,
943507142461, 53798399207356, 3390242657205889, 233980541746413697
Bessel polynomial $y_n(3)$. Ref RCI 77. [0,2; A1518, N1495]

M3670   1, 1, 1, 4, 38, 78, 5246, 11680, 2066056, 22308440, 1898577048, 48769559680,
3518093351728, 174500124820560, 11809059761527536, 1021558531563834368
$2^{1-n}a(n)$ fixed up to signs by Stirling-2 transform. Ref BeS194. EIS § 2.7. [1,4; A3633]

M3671   1, 1, 1, 4, 38, 728, 26704, 1866256, 251548592, 66296291072, 34496488594816,
35641657548953344, 7335459620676622208, 30127220264966408951808
Connected labeled graphs with n nodes. Ref CJM 8 407 56. dgc. [0,4; A1187, N1496]

M3672   4, 39, 190, 651, 1792, 4242, 8988, 17490, 31812
Rooted nonseparable maps on the torus. Ref JCT B18 241 75. [2,1; A6408]

M3673   0, 4, 40, 468, 5828
Sets with a congruence property. Ref MFC 15 315 65. [0,2; A2705, N1497]

M3674   1, 1, 4, 40, 672, 16128, 506880, 19768320, 922521600, 50185175040,
3120605429760, 218442380083200, 17004899126476800, 1457562782269440000
Embeddings of bouquet in surface of genus n. Ref JCT B47 301 89. [0,3; A5431]

$$n\,a(n) = 4\,(2\,n - 3)\,(n - 2)\,a(n - 1).$$

M3675   4, 40, 3264, 45826304
Switching networks. Ref JFI 276 322 63. [1,1; A0841, N1498]

M3676   1, 4, 40, 12096, 604800, 760320, 217945728000, 697426329600,
16937496576000, 30964207376793600, 187333454629601280000
Coefficients for central differences. Ref SAM 42 162 63. [1,2; A2677, N1499]

M3677   1, 4, 41, 614, 12281, 307024, 9210721, 322375234, 12895009361, 580275421244,
29013771062201, 1595757408421054, 95745444505263241
Expansion of $e^{-x}/(1-5x)$. Ref R1 83. [0,2; A1908, N1500]

M3678   1, 0, 1, 4, 41, 768, 27449, 1887284, 252522481, 66376424160, 34509011894545,
35645504882731588, 73356937912127722841, 301275024444053951967648
$\Sigma a(k)\,C(n,k) = 2\!\uparrow\! C(k,2)$. Ref clm. [0,4; A6129]

M3679   1, 1, 4, 41, 1981
Connected regular bipartite graphs of degree 5 with $2n$ nodes. Ref OR76 135. [5,3; A6825]

M3680   4, 44, 408, 3688, 33212, 298932
Coefficients of elliptic function cn. Ref Cay95 56. TM93 4 92. JCT A29 123 80. [2,1;
A2754, N1501]

M3681 1, 4, 44, 580, 8092, 116304, 1703636, 25288120, 379061020, 5724954544, 86981744944, 1327977811076, 20356299454276, 313095240079600
Quadrinomial coefficients. Ref C1 78. [0,2; A5721]

M3682 1, 4, 46, 1064, 35792, 1673792, 103443808, 8154999232, 798030483328
Related to Latin rectangles. Ref BCMS 33 125 41. [2,2; A1623, N1502]

M3683 1, 0, 1, 4, 46, 1322, 112519, 32267168, 34153652752
Self-dual threshold functions of n variables. Ref PGEC 17 806 68. MU71 38. [1,4; A2077, N1503]

M3684 4, 47, 240, 831, 2282, 5362, 11256, 21690, 39072, 66649
Rooted toroidal maps. Ref JCT B18 250 75. [1,1; A6422]

M3685 4, 48, 224, 448, 40, 1408, 2240, 1280, 924, 480, 6944, 8704, 5864, 14080, 2240, 33772, 19064, 11088, 54432, 4480, 38400, 43648, 75712, 124928, 62100, 70368
Bisection of M3347. Ref QJMA 38 191 07. [1,1; A2287, N1504]

M3686 4, 49, 273, 1023, 3003, 7462, 16422, 32946, 61446, 108031, 180895, 290745, 451269, 679644, 997084, 1429428, 2007768, 2769117, 3757117, 5022787, 6625311
Central factorial numbers. Ref RCI 217. [3,1; A0596, N1505]

M3687 4, 51, 46218, 366543984720
Switching networks. Ref JFI 276 323 63. [1,1; A0854, N1506]

M3688 0, 0, 0, 0, 4, 52, 709, 8946, 108761, 1296258, 15308897
One-sided 4-dimensional polyominoes with n cells. Ref CJN 18 366 75. [1,5; A6760]

M3689 1, 1, 4, 55, 2008, 153040, 20933840, 4662857360, 1579060246400, 772200774683520, 523853880779443200, 477360556805016931200
Stochastic matrices of integers. Ref SS70. C1 125. SIAA 4 193 83. [0,3; A1500, N1507]

M3690 1, 1, 1, 4, 56, 9408, 16942080, 535281401856, 377597570964258816, 7580721483160132811489280
Reduced Latin squares of order n. See Fig M2051. Ref R1 210. RY63 53. FY63 22. RMM 193. DM 11 94 75. C1 183. bdm. [1,4; A0315, N1508]

M3691 4, 60, 550, 4004, 25480, 148512, 813960, 4263600, 18573816
Partitions of a polygon by number of parts. Ref CAY 13 95. [6,1; A2060, N1509]

M3692 4, 60, 588, 4636, 31932, 200364, 1174492, 6538492, 34965772, 181084796, 913687100, 4511834156, 21880671292, 104497300828, 492527133804
Almost-convex polygons of perimeter $2n$ on square lattice. Ref EG92. [6,1; A7220]

M3693 1, 4, 64, 2304, 147456, 14745600, 2123366400, 416179814400,
106542032486400, 34519618525593600, 13807847410237440000
Central factorial numbers: $4^n(n!)^2$. Ref OP80 7. FMR 1 110. RCI 217. [0,2; A2454, N1510]

M3694 1, 4, 72, 2896, 203904, 22112000, 3412366336, 709998153728,
191483931951104, 64956739430973440, 27065724289967718400
Expansion of tan(tan x). [0,2; A3718]

M3695 0, 4, 74, 43682, 160297810086
Essentially n-ary operations in a certain 3-element algebra. Ref Berm83. [0,2; A7157]

M3696 4, 74, 63440, 244728561176
Switching networks. Ref JFI 276 324 and 588 63. [1,1; A0857, N1511]

M3697 4, 75, 604, 3150, 12480, 40788, 115500, 292578, 677820, 1459315
Nonseparable planar tree-rooted maps. Ref JCT B18 243 75. [1,1; A6412]

M3698 1, 4, 75, 3456, 300125, 42467328, 8931928887, 2621440000000,
1025271882697689, 515978035200000000, 325063112540091870659
$n^{n-2}(n+2)^{n-1}$. Ref JCT B24 208 78. [1,2; A6236]

M3699 1, 4, 76, 439204, 84722519070079276,
60813021337408894121474740581772094212749079297440404
A continued cotangent. Ref NBS B80 288 76. [0,2; A6267]

M3700 4, 79, 900, 7885, 59080, 398846, 2499096, 14805705, 83969600, 459868530
Rooted toroidal maps. Ref JCT B18 251 75. [1,1; A6425]

M3701 1, 4, 80, 3904, 354560, 51733504, 11070525440, 3266330312704,
1270842139934720, 6304242777638805504, 388362339077351014400
Multiples of Euler numbers. Ref QJMA 44 110 13. FMR 1 75. [1,2; A2436, N1512]

M3702 0, 4, 80, 4752, 440192, 59245120, 10930514688, 2649865335040,
817154768973824, 312426715251262464, 145060238642780180480
Permutations with no hits on 2 main diagonals. Ref R1 187. [1,2; A0316, N1513]

M3703 0, 4, 82, 43916, 160297985274
Spectrum of a certain 3-element algebra. Ref Berm83. [0,2; A7154]

M3704 1, 4, 96, 5888, 686080, 130179072, 36590059520, 7405376630685696,
4071967909087792857088, 4980673081258443273955966976
Labeled odd degree trees with $2n$ nodes. Ref rwr. [1,2; A7106]

M3705 4, 104, 1020, 6092, 26670, 94128, 283338, 754380, 1821534, 4061200
Nonseparable toroidal tree-rooted maps. Ref JCT B18 243 75. [0,1; A6415]

M3706   1, 4, 108, 27648, 86400000, 4031078400000, 3319766398771200000,
    55696437941726556979200000, 2157794122294185620916802682880000
Hyperfactorials: $\Pi\ k^k$, $k = 1 \ . \ . \ n$. Ref FMR 1 50. GKP 477. [1,2; A2109, N1514]

M3707   4, 112, 8432, 909288, 121106960, 18167084064, 2956370702688,
    510696155882492, 9234303960644064, 17311893232788414400
Golygons of length $8n$. Ref VA91 92. [1,1; A6718]

M3708   4, 120, 1230, 7424, 32424, 113584, 338742, 893220, 2136618, 4721728, 9770904
Tree-rooted toroidal maps. Ref JCT B18 258 75. [1,1; A6434]

M3709   1, 4, 120, 3024, 151200, 79200, 1513512000, 1513512000, 51459408000,
    74662922880, 18068427336960, 133196739984000, 1215553449093984000
Coefficients for numerical differentiation. Ref OP80 21. SAM 22 120 43. [2,2; A2702,
N1515]

M3710   1, 4, 120, 12096, 3024000, 1576143360, 1525620096000, 2522591034163200,
    6686974460694528000, 27033456071346536448000
Special determinants. Ref BMG 6 105 65. [1,2; A1332, N1516]

M3711   0, 4, 120, 33600, 18446400, 18361728000, 30199104936000,
    76326119565696000, 280889824362219072000, 14434284290455783335360000
Labeled connected rooted trivalent graphs with $2n$ nodes. Ref LNM 686 342 78. [1,2;
A6607]

M3712   4, 124, 217, 561, 781, 1541, 1729, 1891, 2821, 4123, 5461, 5611, 5662, 5731,
    6601, 7449, 7813, 8029, 8911, 9881, 11041, 11476, 12801, 13021, 13333, 13981, 14981
Pseudoprimes to base 5. Ref UPNT A12. [1,1; A5936]

M3713   4, 128, 16384, 4456448
Generalized tangent numbers. Ref MOC 21 690 67. [1,1; A0318, N1517]

M3714   1, 4, 129, 43968
Commutative groupoids with n elements. Ref LE70 246. [1,2; A1425, N1518]

M3715   1, 4, 130, 33880, 75913222, 1506472167928, 267598665689058580,
    4270287769691766796799640080, 6129263888495201102915629695046
Gaussian binomial coefficient $[2n,n]$ for $q = 3$. Ref GJ83 99. ARS A17 328 84. [0,2;
A6103]

M3716   0, 4, 135, 1368, 7350, 28400, 89073, 241220, 585057, 1301420, 2699125
Tree-rooted planar maps. Ref JCT B18 256 75. [1,2; A6429]

M3717   4, 136, 44224, 179228736, 9383939974144
Relational systems on n nodes. Ref OB66. [1,1; A1374, N1519]

M3718 4, 140, 4056, 129360, 4381848
Specific heat for cubic lattice. Ref PRV 129 102 63. [0,1; A2917, N1520]

M3719 4, 152, 2630, 31500, 303534, 2530976, 19030428, 132386340, 866782510,
 5405853200
Tree-rooted toroidal maps. Ref JCT B18 258 75. [1,1; A6439]

M3720 0, 4, 175, 3324, 42469, 429120, 3711027, 28723640, 204598130, 1366223880,
 8664086470
Tree-rooted planar maps. Ref JCT B18 257 75. [1,2; A6433]

M3721 1, 4, 192, 100352, 557568000, 32565539635200
Complexity of an $n \times n$ grid. Ref JCT B24 210 78. [1,2; A7341]

M3722 4, 272, 55744, 23750912, 17328937984, 19313964388352, 30527905292468224,
 64955605537174126592, 179013508069217017790464
Generalized tangent numbers. Ref MOC 21 690 67. [1,1; A0320, N1521]

M3723 1, 1, 4, 302, 2569966041123963092
Invertible Boolean functions. Ref PGEC 13 530 64. [1,3; A1537, N1522]

M3724 1, 4, 324, 21233664, 3240000000000000000,
 257860619962263388654298726400000000000000000
$\Pi \ k \uparrow (2 \uparrow (k-1))$, $k = 1 .. n$. Ref JCMCC 1 146 87. [1,2; A5832]

M3725 1, 4, 384, 42467328, 20776019874734407680,
 16575091270477789938706015460369010524160000000
Complexity of tensor sum of n graphs. Ref JCT B24 209 78. [1,2; A6237]

M3726 4, 32896, 3002399885885440, 14178431955039103827204744901417762816
Relational systems on n nodes. Ref OB66. [1,1; A1376, N1523]

SEQUENCES BEGINNING . . ., 5, . . .

M3727 0, 0, 0, 0, 0, 0, 0, 0, 0, 5, 0, 12, 0, 21, 5, 32, 17, 45, 38, 65, 70, 99, 115, 156, 180,
 247, 279, 385, 435, 590, 682, 896, 1067, 1360, 1657, 2073, 2553, 3173, 3913, 4865, 5986
Strict 3rd-order maximal independent sets in cycle graph. Ref YaBa94. [1,10; A7392]

M3728 1, 1, 0, 5, 1, 0, 5, 2, 8, 18, 19, 7, 16, 13, 6, 34, 27, 56, 12, 69, 11, 73, 20, 70, 70, 72,
 57, 1, 30, 95, 71, 119, 56, 67, 94, 86, 151, 108, 21, 106, 48, 72, 159, 35, 147, 118, 173
Wilson remainders $((p-1)! + 1)/p \bmod p$. Ref JLMS 28 253 53. AFM 4 481 61. Robe92
244. [2,4; A2068, N1524]

M3729 0, 5, 1, 6, 11, 61, 66, 17, 22, 72, 77, 28, 33, 83, 88, 39, 44, 94, 99, 401, 604, 906,
 119, 421, 624, 926, 139, 441, 644, 946, 159, 461, 664, 966, 179, 481, 684, 986, 199, 402
Add 5, then reverse digits! Ref Robe92 15. [0,2; A7397]

M3730 0, 1, 1, 1, 5, 1, 7, 8, 5, 19, 11, 23, 35, 27, 64, 61, 85, 137, 133, 229, 275, 344, 529,
599, 875, 1151, 1431, 2071, 2560, 3481, 4697, 5953, 8245, 10649, 14111, 19048, 24605
$a(n+6) = -a(n+5)+a(n+4)+3a(n+3)+a(n+2)-a(n+1)-a(n)$. Ref JLMS 8 166
33. [0,5; A1945, N1525]

M3731 1, 1, 1, 1, 1, 1, 5, 1, 7, 13, 307, 479, 1837, 100921, 15587, 23737
Numerators of generalized Bernoulli numbers. Cf. M3046. Ref DUMJ 34 614 67. [0,6;
A6569]

M3732 1, 1, 1, 5, 1, 41, 31, 461, 895, 6481, 22591, 107029, 604031, 1964665, 17669471,
37341149, 567425279, 627491489, 19919950975, 2669742629, 759627879679
Expansion of $e^{x(1-x)}$. Ref JMSJ 1(4) 240 50. R1 209. [0,4; A0321, N1526]

M3733 0, 1, 5, 2, 4, 6, 11, 3, 13, 5, 10, 7, 7, 12, 12, 4, 9, 14, 14, 6, 6, 11, 11, 8, 16, 8, 70,
13, 13, 13, 67, 5, 18, 10, 10, 15, 15, 15, 23, 7, 69, 7, 20, 12, 12, 12, 66, 9, 17, 17, 17, 9, 9
Number of halving steps to reach 1 in '$3x+1$' problem. See Fig M2629. Ref UPNT E16.
rwg. [1,3; A6666]

M3734 0, 0, 0, 0, 0, 0, 0, 0, 0, 0, 5, 2, 9, 1, 7, 7, 2, 4, 9
Decimal expansion of Bohr radius (meters). Ref FiFi87. Lang91. [0,11; A3671]

M3735 5, 2, 26, 86, 362, 1430, 5738, 22934, 91754, 366998, 1468010, 5872022,
23488106, 93952406, 375809642, 1503238550, 6012954218, 24051816854
Generalization of the golden ratio (expansion of $(5-13x)/(1+x)(1-4x)$). Ref JRM 8 207
76. [0,1; A7572]

M3736 1, 1, 1, 5, 3, 60, 487
Hadamard matrices of order $4n$. See Fig M3736. Ref Kimu94a. Kimu94b. [1,4; A7299]

||||||||||||||||||||||||||||||||||||||||||||||||||||||||||||||||||||||||||||||||||||||||||||||||||||||||||||||||||||||||||

Figure M3736. HADAMARD MATRICES.

   A **Hadamard matrix** is a matrix of +1's and −1's whose rows have scalar product 0 with
each other, for example

$$
\begin{array}{cccc}
+1 & +1 & +1 & +1 \\
+1 & -1 & +1 & -1 \\
+1 & +1 & -1 & -1 \\
+1 & -1 & -1 & +1
\end{array}
$$

   M3736 gives the number of inequivalent matrices of order $4n$, for $n \le 8$. The next term is
not known. The unique matrix of order 12 can be obtained from the first 12 rows of the code
shown in Fig. M0240 by replacing 0's by −1's. See also Fig. M2051.

||||||||||||||||||||||||||||||||||||||||||||||||||||||||||||||||||||||||||||||||||||||||||||||||||||||||||||||||||||||||||

M3737   1, 5, 3, 251, 95, 19087, 5257, 1070017, 25713, 26842253, 4777223,
703604254357, 106364763817, 1166309819657, 25221445, 8092989203533249
Numerators of coefficients for numerical integration. Cf. M2015. Ref SAM 22 49 43. [1,2;
A2208, N1527]

M3738   0, 0, 0, 5, 4, 8, 5, 7, 9, 9, 0, 3
Decimal expansion of electron mass (mass units). Ref FiFi87. BoBo90 v. [0,4; A3672]

M3739   5, 4, 8, 7, 6, 11, 8, 9, 14, 18, 13, 11, 17, 16, 12, 13, 14, 28, 19, 14, 18, 16, 27, 22,
31, 16, 17, 26, 19, 24, 24, 23, 22, 28, 37, 41, 27, 32, 21, 26, 22, 23, 31, 22, 44, 48, 23
x such that $p = x^2 - 5y^2$. Cf. M0136. Ref CU04 1. L1 55. [5,1; A2340, N1528]

M3740   5, 4, 9, 5, 3, 9, 3, 1, 2, 9, 8, 1, 6, 4, 4, 8, 2, 2, 3, 3, 7, 6, 6, 1, 7, 6, 8, 8, 0, 2, 9, 0, 7,
7, 8, 8, 3, 3, 0, 6, 9, 8, 9, 8, 1, 2, 6, 3, 0, 0, 6, 4, 7, 9, 1, 0, 9, 0, 1, 5, 1, 3, 0, 4, 5, 7, 6, 6, 3, 1
Decimal expansion of $|\ln(\gamma)|$. Cf. M3755. Ref RS8 XVIII. [0,1; A2389]

M3741   1, 1, 5, 4, 18, 19, 73, 73, 278, 283, 1076, 1090, 4125, 4183, 15939, 16105, 61628,
62170, 239388, 240907, 932230, 936447
Rotationally symmetric polyominoes with n cells. Ref DM 36 203 81. [4,3; A6747]

M3742   1, 5, 5, 10, 15, 6, 5, 25, 15, 20, 9, 45, 5, 25, 20, 10, 15, 20, 50, 35, 30, 55, 50, 15,
80, 1, 50, 35, 45, 15, 5, 50, 25, 55, 85, 51, 50, 10, 40, 65, 10, 10, 115, 50, 115, 100, 85, 80
Expansion of $\Pi(1-x^n)^5$. Ref KNAW 59 207 56. [0,2; A0728, N1529]

M3743   1, 0, 1, 1, 5, 5, 21, 40, 176, 500, 2053, 7532, 31206, 124552, 521332
Bipartite polyhedral graphs with n nodes. Ref Dil92. [8,5; A7028]

M3744   5, 6, 5, 6, 5, 5, 5, 7, 6, 5, 5, 8, 7, 10, 10, 9, 9, 11, 10, 10, 9, 11, 12, 11, 12, 11, 11, 13,
12, 11, 9, 11, 12, 11, 12, 11, 11, 13, 12, 11, 8, 10, 11, 10, 11, 10, 10, 12, 11, 10, 8, 10, 11
Letters in ordinal numbers. [1,1; A6944]

M3745   1, 1, 5, 6, 7, 7, 9, 53, 60, 66, 83, 83, 136, 136, 185, 185, 185, 312, 312, 312, 3064,
3718, 3718, 3718, 8096, 9826, 12384, 16602, 16602, 16602, 16760, 16760, 182424
Class numbers of quadratic fields. Ref MOC 24 445 70. [3,3; A2141, N1530]

M3746   1, 5, 6, 7, 8, 9, 10, 12, 20, 22, 23, 24, 26, 38, 39, 40, 41, 52, 57, 69, 70, 71, 82, 87,
98, 102, 113, 119, 129, 130, 133, 144, 160, 161, 162, 163, 175, 196, 205, 208, 209, 222
$a(n)$ is smallest number which is uniquely $a(j) + a(k)$, $j < k$. Ref GU94. [1,2; A3667]

M3747   5, 6, 7, 13, 14, 15, 20, 21, 22, 23, 24, 28, 29, 30, 31, 34, 37, 38, 39, 41, 45, 46, 47,
52, 53, 54, 55, 56, 60, 61, 62, 63, 65, 69, 70, 71, 77, 78, 79, 80, 84, 85, 86, 87, 88, 92, 93
Congruent numbers. Ref MOC 28 304 74. UPNT D27. [1,1; A3273]

M3748   5, 6, 7, 13, 14, 15, 21, 22, 23, 29, 30, 31, 34, 37, 38, 39, 41, 46, 47, 53, 55, 61, 62,
65, 69, 70, 71, 77, 78, 79, 85, 86, 87, 93, 94, 95, 101, 102, 103, 109, 110, 111, 118, 119
Primitive congruent numbers. Ref MOC 28 304 74. UPNT D27. [1,1; A6991]

M3749 5, 6, 10, 13, 15, 22, 35, 37, 51, 58, 91, 115, 123, 187, 235, 267, 403, 427.
Imaginary quadratic fields with class number 2 (a finite sequence). Ref LNM 751 226 79.
BPNR 142. [1,1; A5847]

M3750 1, 5, 6, 16, 8, 30, 10, 42, 24, 40, 14, 96, 16, 50, 48, 99, 20, 120, 22, 128, 60, 70, 26,
252, 46, 80, 82, 160, 32, 240, 34, 219, 84, 100, 80, 384, 40, 110, 96, 336, 44, 300, 46, 224
Inverse Moebius transform applied thrice to natural numbers. Ref BeSl94. EIS § 2.7. [1,2;
A7430]

M3751 1, 0, 5, 6, 21, 40, 93, 190, 396, 796, 1586, 3108, 6025, 11552, 21947, 41346,
77311, 143580, 265013, 486398, 888122, 1613944, 2920100, 5261880, 9445905
Generalized Lucas numbers. Ref FQ 15 252 77. [3,3; A6492]

M3752 1, 5, 6, 25, 76, 376, 625, 9376, 90625, 109376, 890625, 2890625, 7109376,
12890625, 87109376, 212890625, 787109376, 1787109376, 8212890625, 18212890625
Automorphic numbers: n^2 ends with n. See Fig M5405. Ref JRM 1 178 68. [1,2; A3226]

M3753 1, 5, 6, 353, 72, 1141
Smallest number such that $a(n)^n$ is sum of n n-th powers. Ref Well86 164. [1,2; A7666]

M3754 0, 1, 5, 7, 4, 11, 8, 1, 5, 7, 17, 19, 13, 2, 20, 23, 19, 14, 25, 7, 23, 11, 13, 28, 22, 17,
29, 26, 32, 16, 35, 1, 5, 37, 35, 13, 29, 34, 31, 19, 2, 28, 10, 23, 25, 32, 43, 29, 1, 31, 11
x such that $p = (x^2 + 27y^2)/4$. Cf. M0058. Ref CU04 1. L1 55. [3,3; A2338, N1531]

M3755 5, 7, 7, 2, 1, 5, 6, 6, 4, 9, 0, 1, 5, 3, 2, 8, 6, 0, 6, 0, 6, 5, 1, 2, 0, 9, 0, 0, 8, 2, 4, 0, 2,
4, 3, 1, 0, 4, 2, 1, 5, 9, 3, 3, 5, 9, 3, 9, 9, 2, 3, 5, 9, 8, 8, 0, 5, 7, 6, 7, 2, 3, 4, 8, 8, 4, 8, 6, 7
Decimal expansion of Euler's constant γ. Ref MOC 17 175 63. [0,1; A1620, N1532]

M3756 1, 1, 5, 7, 7, 7, 9, 53, 73, 83, 83, 83, 157, 157, 185, 185, 185, 1927, 2295, 2273,
5313, 5313, 7173, 9529, 18545, 18545, 18545, 18545, 22635, 22635, 66011, 121725
Class numbers of quadratic fields. Ref MOC 24 445 70. [3,3; A1989, N1535]

M3757 5, 7, 9, 10, 13, 14, 16, 18, 21, 25, 26, 28, 33, 36, 38, 40, 43, 44, 50, 54, 57, 61, 64,
68, 75, 77, 81, 84, 88, 91, 97, 100, 102, 108, 117, 122, 124, 128, 130, 135, 144, 148, 150
Rational points on elliptic curves over $GF(q)$. Ref HW84 51. [2,1; A5523]

M3758 5, 7, 9, 11, 12, 13, 16, 17, 17, 19, 19, 22, 21, 23, 24, 26, 27, 29, 27, 28, 29, 32, 31,
31, 33, 32, 34, 33, 37, 37, 37, 37, 39, 41, 39, 41, 43, 41, 41, 42, 43, 44, 46, 43, 44, 47, 49
x such that $p = (x^2 - 5y^2)/4$. Cf. M0109. Ref CU04 1. L1 55. [5,1; A2342, N1534]

M3759 5, 7, 9, 12, 16, 22, 29, 39, 52, 69, 92, 123, 164, 218, 291, 388, 517, 690, 920, 1226,
1635, 2180, 2907, 3876, 5168, 6890, 9187, 12249, 16332, 21776, 29035, 38713, 51618
Josephus problem. Ref JNT 26 208 87. [1,1; A5427]

M3760 5, 7, 11, 13, 17, 37, 41, 67, 97, 101, 103, 107, 191, 193, 223, 227, 277, 307, 311,
347, 457, 461, 613, 641, 821, 823, 853, 857, 877, 881, 1087, 1091, 1277, 1297, 1301
Prime triplets: n; $n + 2$ or $n + 4$; $n + 6$ all prime. Ref Rie85 65. rgw. [1,1; A7529]

M3761   5, 7, 11, 23, 47, 59, 83, 107, 167, 179, 227, 263, 347, 359, 383, 467, 479, 503, 563, 587, 719, 839, 863, 887, 983, 1019, 1187, 1283, 1307, 1319, 1367, 1439, 1487, 1523
n and $(n-1)/2$ are prime. Ref AS1 870. [1,1; A5385]

M3762   1, 5, 7, 13, 11, 23, 15, 29, 25, 35, 23, 55, 27, 47, 47, 61, 35, 77, 39, 83, 63, 71, 47, 119, 61, 83, 79, 111, 59, 143, 63, 125, 95, 107, 95, 181, 75, 119, 111, 179, 83, 191, 87
Related to planar partitions. Ref MES 52 115 24. [1,2; A2659, N1536]

G.f. of Moebius transf.: $(1 + 2x - x^2) / (1 - x)^2$.

M3763   5, 7, 13, 19, 31, 43, 61, 73, 103, 109, 139, 151, 181, 193, 199, 229, 241, 271, 283, 313, 349, 421, 433, 463, 523, 571, 601, 619, 643, 661, 811, 823, 829, 859, 883, 1021
Greater of twin primes. Cf. M2476. Ref AS1 870. [1,1; A6512]

M3764   5, 7, 13, 23, 29, 31, 37, 47, 53, 61, 71, 79, 101, 103, 109, 127, 149, 151, 157, 167, 173, 181, 191, 197, 199, 223, 229, 239, 263, 269, 271, 277, 293, 311, 317, 349, 359, 367
Inert rational primes in $Q(\sqrt{-2})$. Ref Hass80 498. [1,1; A3628]

M3765   5, 7, 15, 27, 57, 114, 243, 506, 1102, 2381, 5269, 11686, 26277, 59348, 135317, 310064, 715475, 1659321, 3870414, 9071915, 21372782, 50591199, 120332237
Trees with stability index n. Ref LNM 403 51 74. [1,1; A3429]

M3766   5, 7, 17, 19, 29, 31, 41, 43, 53, 67, 79, 89, 101, 103, 113, 127, 137, 139, 149, 151, 163, 173, 197, 199, 211, 223, 233, 257, 269, 271, 281, 283, 293, 307, 317, 331, 353, 367
Inert rational primes in $Q(\sqrt{3})$. Ref Hass80 498. [1,1; A3630]

M3767   5, 7, 19, 31, 53, 67, 293, 641
$(10^p + 1)/11$ is prime. Ref CUNN. [1,1; A1562, N1537]

M3768   1, 1, 1, 5, 7, 21, 33, 429, 715, 2431, 4199, 29393, 52003, 185725, 334305, 9694845, 17678835, 64822395, 119409675, 883631595, 1641030105, 6116566755
Numerators in expansion of $(1-x)^{1/2}$. [1,4; A2596, N1538]

M3769   1, 1, 1, 5, 7, 37, 104, 782
Quartering an $n \times n$ chessboard. See Fig M3987. Cf. M3987. Ref PC 1 7-1 73. GA69 189. trp. [1,4; A6067]

M3770   0, 1, 1, 1, 1, 1, 5, 8, 7, 1, 19, 43, 55, 27, 64, 211, 343, 307, 85, 911, 1919, 2344, 989, 3151, 9625, 15049, 12609, 5671, 39296, 85609, 100225, 33977, 154007, 437009
$a(n+6) = -3a(n+5)-5a(n+4)-5a(n+3)-5a(n+2)-3a(n+1)-a(n)$. Ref EJC 4 213 83. [0,7; A5120]

M3771   1, 5, 8, 7, 4, 0, 1, 0, 5, 1, 9, 6, 8, 1, 9, 9, 4, 7, 4, 7, 5, 1, 7, 0, 5, 6, 3, 9, 2, 7, 2, 3, 0, 8, 2, 6, 0, 3, 9, 1, 4, 9, 3, 3, 2, 7, 8, 9, 9, 8, 5, 3, 0, 0, 9, 8, 0, 8, 2, 8, 5, 7, 6, 1, 8, 2, 5, 2, 1
Decimal expansion of cube root of 4. [1,2; A5480]

M3772   0, 5, 8, 8, 2, 3, 5, 2, 9, 4, 1, 1, 7, 6, 4, 7, 0, 5, 8, 8, 2, 3, 5, 2, 9, 4, 1, 1, 7, 6, 4, 7, 0,
  5, 8, 8, 2, 3, 5, 2, 9, 4, 1, 1, 7, 6, 4, 7, 0, 5, 8, 8, 2, 3, 5, 2, 9, 4, 1, 1, 7, 6, 4, 7, 0, 5, 8, 8, 2
Decimal expansion of 1/17. [0,2; A7450]

M3773   1, 5, 8, 10, 11, 12, 12, 13, 13
Pair-coverings with largest block size 4. Ref ARS 11 90 81. [4,2; A6186]

M3774   1, 5, 8, 11, 15, 18, 22, 25, 29, 32, 35, 39, 42, 46, 49, 52, 56, 59, 63, 66, 69, 73, 76,
  80, 83, 87, 90, 93, 97, 100, 104, 107, 110, 114, 117, 121, 124, 128, 131, 134, 138, 141
Wythoff game. Ref CMB 2 188 59. [0,2; A1954, N1539]

M3775   5, 8, 11, 15, 19, 23, 27, 32, 37, 43, 49, 54, 59, 64
Zarankiewicz's problem. Ref LNM 110 141 69. [2,1; A6620]

M3776   5, 8, 12, 13, 17, 21, 24, 28, 29, 33, 37, 40, 41, 44, 53, 56, 57, 60, 61, 65, 69, 73, 76,
  77, 85, 88, 89, 92, 93, 97, 101, 104, 105, 109, 113, 120, 124, 129, 133, 136, 137, 140, 141
Discriminants of real quadratic fields. Ref Ribe72 97. [1,1; A3658]

M3777   5, 8, 12, 13, 17, 21, 24, 28, 29, 33, 37, 41, 44, 53, 56, 57, 61, 69, 73, 76, 77, 88, 89,
  92, 93, 97, 101, 109, 113, 124, 129, 133, 137, 141, 149, 152, 157, 161, 172, 173, 177, 181
Discriminants of real quadratic fields with unique factorization. Ref BU89 236. [1,1;
A3656]

M3778   5, 8, 12, 13, 17, 21, 24, 28, 29, 33, 37, 41, 44, 57, 73, 76.
Discriminants of real quadratic Euclidean fields (a finite sequence). Ref LE56 2 57. AMM
75 948 68. ST70 294. [1,1; A3246]

M3779   5, 8, 12, 14, 21, 22, 26, 33, 39, 40, 42, 50, 60, 63, 64, 75, 76, 78, 82, 96, 98, 114,
  117, 118, 123, 124, 126, 147, 148, 150, 154, 162, 177, 186, 189, 190, 194, 222, 225, 226
$n \in S$ implies $2n - 2$, $3n - 3 \in S$. [1,1; A5661]

M3780   5, 8, 12, 18, 24, 30, 36, 42, 52, 60, 68, 78, 84, 90, 100, 112, 120, 128, 138, 144,
  152, 162, 172, 186, 198, 204, 210, 216, 222, 240, 258, 268, 276, 288, 300, 308, 320, 330
Sum of 2 successive primes. Ref EUR 26 12 63. [1,1; A1043, N0968]

M3781   5, 8, 13, 17, 29, 37, 40, 41, 53, 61, 65, 73, 85, 89, 97, 101, 104, 109, 113, 137, 145,
  149, 157, 173, 181, 185, 193, 197, 229, 232, 233, 241, 257, 265, 269, 277, 281, 293, 296
Discriminants of quadratic fields whose fundamental unit has norm -1. Ref BU89 236.
[1,1; A3653]

M3782   5, 8, 13, 17, 29, 37, 41, 53, 61, 73, 89, 97, 101, 109, 113, 137, 149, 157, 173, 181,
  193, 197, 233, 241, 269, 277, 281, 293, 313, 317, 337, 349, 353, 373, 389, 397, 409, 421
Discriminants of positive definite quadratic forms with class number 1. Ref BU89 236.
[1,1; A3655]

M3783 0, 1, 1, 5, 8, 31, 55, 203, 368, 1345, 2449, 8933, 16280, 59359, 108199
$a(2n) = a(2n-1) + 3a(2n-2)$, $a(2n+1) = 2a(2n) + 3a(2n-1)$. Ref MQET 1 12 16.
[0,4; A2536, N1540]

M3784 1, 5, 9, 17, 21, 29, 45, 177
$7.2^n - 1$ is prime. Ref MOC 22 421 68. Rie85 384. [1,2; A1771, N1541]

M3785 1, 5, 9, 17, 22, 34, 41, 53, 61, 73
Davenport-Schinzel numbers. Ref ARS 1 47 76. UPNT E20. [1,2; A5006]

M3786 1, 1, 1, 1, 1, 5, 9, 17, 33, 65, 129, 253, 497, 977, 1921, 3777, 7425, 14597, 28697,
    56417, 110913, 218049, 428673, 842749, 1656801, 3257185, 6403457, 12588865
Pentanacci numbers. Ref FQ 2 260 64. [0,6; A0322, N1542]

M3787 5, 9, 21, 37, 69, 69, 89, 137, 177, 421, 481, 657, 749, 885, 1085, 1305, 1353, 1489,
    1861, 2617, 2693, 3125, 5249, 5761, 7129, 8109, 9465, 9465, 10717, 12401, 12401
Lattice points in circles. Ref MOC 20 306 66. [1,1; A0323, N1543]

M3788 1, 1, 5, 9, 29, 65, 181, 441, 1165, 2929, 7589, 19305, 49661, 126881, 325525,
    833049, 2135149, 5467345, 14007941, 35877321, 91909085, 235418369, 603054709
$a(n) = a(n-1) + 4a(n-2)$. Ref FQ 15 24 77. [0,3; A6131]

M3789 1, 5, 9, 49, 2209, 4870849, 23725150497409, 562882766124611619513723649,
    316837008400094222150776738483768236006420971486980609
A nonlinear recurrence. Ref AMM 70 403 63. FQ 11 431 73. [0,2; A0324, N1544]

M3790 1, 1, 5, 9, 251, 475, 19087, 36799, 1070017, 2082753, 134211265, 262747265,
    703604254357, 1382741929621, 8164168737599, 5362709743125, 8092989203533249
Numerators of Cauchy numbers. Cf. M1559. Ref MT33 136. C1 294. [0,3; A2657, N1545]

M3791 1, 5, 10, 10, 0, 19, 35, 40, 25, 10, 45, 75, 80, 60, 15, 45, 85, 115, 115, 90, 21, 35,
    95, 130, 135, 135, 70, 35, 65, 105, 146, 120, 150, 90, 65, 25, 90, 115, 150, 125, 130, 45
Coefficients of powers of η function. Ref JLMS 39 435 64. [5,2; A1483, N1546]

M3792 5, 10, 11, 13, 17, 25, 32, 37, 47, 58, 71, 79, 95, 109, 119, 130, 134, 142, 149, 163,
    173, 184, 194, 214, 221, 226, 236, 247, 260, 268, 284, 298, 317, 328, 341, 349, 365, 379
$a(n) = a(n-1) +$ sum of digits of $a(n-1)$. Ref Robe92 65. [1,1; A7618]

M3793 5, 10, 11, 20, 25, 26, 38, 39, 54, 65, 70, 114, 130
Not the sum of 4 hexagonal numbers (probably 130 is last term). Ref AMM 101 170 94.
[1,1; A7527]

M3794 1, 5, 10, 14, 18, 22, 27, 31, 35, 40, 44, 48, 53, 57, 61, 65, 70, 74, 78, 83, 87, 91, 96,
    100, 104, 109, 113, 117, 121, 126, 130, 134, 139, 143, 147, 152, 156, 160, 164, 169, 173
Wythoff game. Ref CMB 2 188 59. [0,2; A1958, N1547]

M3795 5, 10, 15, 20, 26, 31, 36, 41, 47, 52, 57, 62, 68, 73, 78, 83, 89, 94, 99, 104, 109,
115, 120, 125, 130, 136, 141, 146, 151, 157, 162, 167, 172, 178, 183, 188, 193, 198, 204
A Beatty sequence: $[n(\sqrt{5} + 3)]$. Cf. M0540. Ref CMB 2 189 59. [1,1; A1962, N1548]

M3796 1, 5, 10, 16, 22, 29
Davenport-Schinzel numbers. Ref PLC 1 250 70. UPNT E20. [1,2; A5280]

M3797 1, 5, 10, 17, 16, 32, 22, 41, 37, 50
Related to planar partitions. Ref MES 52 115 24. [1,2; A2660, N1549]

M3798 1, 5, 10, 21, 21, 38, 29, 53, 46, 65
Related to planar partitions. Ref MES 52 115 24. [1,2; A2791, N1550]

M3799 1, 5, 10, 21, 26, 50, 50, 85, 91, 130, 122, 210, 170, 250, 260, 341, 290, 455, 362,
546, 500, 610, 530, 850, 651, 850, 820, 1050, 842, 1300, 962, 1365, 1220, 1450, 1300
Sum of squares of divisors of n. Ref AS1 827. [1,2; A1157, N1551]

M3800 1, 5, 10, 21, 26, 53, 50, 85, 91, 130
Related to planar partitions. Ref MES 52 115 24. [1,2; A2800, N1552]

M3801 1, 1, 5, 10, 29, 57, 126, 232, 440, 750, 1282, 2052, 3260, 4950, 7440, 10824,
15581, 21879, 30415, 41470, 56021, 74503, 98254
n-bead necklaces with 8 red beads. Ref JAuMS 33 12 82. [8,3; A5514]

M3802 1, 5, 10, 30, 74, 199, 515, 1355, 3540, 9276, 24276, 63565, 166405, 435665,
1140574, 2986074, 7817630, 20466835, 53582855, 140281751
From a definite integral. Ref PEMS 10 184 57. [1,2; A2571, N1553]

M3803 1, 0, 1, 1, 5, 10, 31, 72, 201, 509
Fixed points in rooted trees. Ref PCPS 85 410 79. [1,5; A5201]

M3804 5, 10, 40, 150, 624, 2580, 11160, 48750, 217000, 976248, 4438920, 20343700,
93900240, 435959820, 2034504992, 9536718750, 44878791360, 211927516500
Irreducible polynomials of degree n over $GF(5)$. Ref AMM 77 744 70. [1,1; A1692,
N1554]

M3805 1, 1, 5, 10, 56, 178
Column of Kempner tableau. There is a simple 2-D recurrence. Ref STNB 11 41 81. [1,3;
A5438]

M3806 5, 11, 13, 17, 23, 41, 43, 61, 67, 71, 73, 79, 89, 97, 101, 107, 127, 151, 157, 163,
173, 179, 181, 191, 211, 229, 239, 241, 257, 263, 269, 293, 313, 331, 347, 349, 353, 359
Inert rational primes in $Q(\sqrt{7})$. Ref Hass80 498. [1,1; A3632]

M3807 5, 11, 13, 19, 23, 29, 37, 47, 53, 59, 61, 67, 71, 83, 97, 101, 107, 131, 139, 149,
163, 167, 173, 179, 181, 191, 193, 197, 211, 227, 239, 263, 269, 293, 307, 311, 313, 317
Solution of a congruence. Ref Krai24 1 63. [1,1; A1915, N1555]

M3808   5, 11, 17, 23, 29, 30, 36, 42, 48, 54, 60, 61, 67, 73, 79, 85, 91, 92, 98, 104, 110, 116, 122, 123, 129, 135, 141, 147, 153, 154, 155, 161, 167, 173, 179, 185, 186, 192, 198
$n!$ never ends in this many 0's. Ref MMAG 27 55 53. rgw. [1,1; A0966, N1557]

M3809   5, 11, 17, 23, 29, 41, 47, 53, 59, 71, 83, 89, 101, 107, 113, 131, 137, 149, 167, 173, 179, 191, 197, 227, 233, 239, 251, 257, 263, 269, 281, 293, 311, 317, 347, 353, 359, 383
Primes of form $6n - 1$. Ref AS1 870. [1,1; A7528]

M3810   5, 11, 19, 29, 41, 71, 89, 109, 131, 181, 239, 271, 379, 419, 461, 599, 701, 811, 929, 991, 1259, 1481, 1559, 1721, 1979, 2069, 2161, 2351, 2549, 2861, 2969, 3079
Primes of form $n^2 - n - 1$. Ref PO20 249. L1 46. [1,1; A2327, N1558]

M3811   5, 11, 19, 31, 41, 59, 61, 71, 79, 109, 131, 149, 179, 191
Primes with a Fibonacci primitive root. Ref FQ 10 164 72. [1,1; A3147]

M3812   5, 11, 29, 97, 149, 211, 127, 1847, 541, 907, 1151, 1693, 2503, 2999, 4327, 5623, 1361, 9587, 30631, 19373, 16183, 15727, 81509, 28277, 31957, 19661, 35671, 82129
Upper prime of gap of $2n$ between primes. Cf. M2685. Ref MOC 21 485 67. [1,1; A1632, N1560]

M3813   1, 5, 11, 36, 95, 281, 781, 2245, 6336, 18061, 51205, 145601, 413351, 1174500, 3335651, 9475901, 26915305, 76455961, 217172736, 616891945, 1752296281
$a(n) = a(n-1) + 5a(n-2) + a(n-3) - a(n-4)$. [0,2; A5178]

M3814   1, 1, 5, 11, 41, 101, 301, 757, 1981, 4714, 11133
$4 \times n$ binary matrices. Ref PGEC 22 1050 73. [0,3; A6382]

M3815   1, 1, 5, 11, 82, 257, 130638, 130895, 785113, 4056460, 4841573, 8898033, 13739606, 36377245, 50116851, 86494096, 2125975155, 2212469251, 4338444406
Convergents to cube root of 6. Ref AMP 46 107 1866. L1 67. hpr. [1,3; A2359, N1561]

M3816   5, 11, 101, 191, 821, 1481, 1871, 2081, 3251, 3461, 5651, 9431, 13001, 15641, 15731, 16061, 18041, 18911, 19421, 21011, 22271, 25301, 31721, 34841, 43781, 51341
Prime quadruplets: p, $p + 2$, $p + 6$, $p + 8$ all prime. Ref Rade64 4. Rie85 65. rgw. [1,1; A7530]

M3817   0, 0, 1, 5, 11, 203, 17207, 3607, 1408301, 8181503, 137483257, 24971924401
Numerators of an asymptotic expansion. Cf. M1828. Ref SIAD 3 575 90. [0,4; A6572]

M3818   1, 5, 12, 22, 35, 51, 70, 92, 117, 145, 176, 210, 247, 287, 330, 376, 425, 477, 532, 590, 651, 715, 782, 852, 925, 1001, 1080, 1162, 1247, 1335, 1426, 1520, 1617, 1717
Pentagonal numbers $n(3n - 1)/2$. See Fig M2535. Ref D1 2 1. B1 189. HW1 284. FQ 8 84 70. [1,2; A0326, N1562]

M3819   1, 5, 12, 23, 39, 62, 91, 127
Partitions into non-integral powers. Ref PCPS 47 214 51. [3,2; A0327, N1563]

M3820   5, 12, 28, 54, 100, 170, 284, 450, 702, 1062, 1583, 2308, 3329, 4720, 6628, 9190,
12634, 17189, 23219, 31092, 41371, 54651, 71782, 93695, 121684, 157169
Bipartite partitions. Ref ChGu56 26. [0,1; A2767, N1564]

M3821   5, 12, 29, 57, 109, 189, 323, 522, 831, 1279, 1941, 2876, 4215, 6066, 8644, 12151,
16933, 23336, 31921, 43264, 58250, 77825, 103362, 136371, 178975, 233532
Bipartite partitions. Ref PCPS 49 72 53. ChGu56 1. [0,1; A0465, N1565]

M3822   1, 1, 5, 12, 45, 143, 511, 1768
Triangulations of an n-gon. Ref LNM 406 346 74. [4,3; A6078]

M3823   5, 13, 17, 29, 37, 41, 53, 61, 73, 89, 97, 101, 109, 113, 137, 149, 157, 173, 181,
193, 197, 229, 233, 241, 257, 269, 277, 281, 293, 313, 317, 337, 349, 353, 373, 389, 397
Primes of form $4n + 1$. Ref AS1 870. [1,1; A2144, N1566]

M3824   5, 13, 21, 29, 33, 37, 41, 57, 65, 69, 77, 85, 93, 101, 105, 109, 113, 129, 137, 141,
157, 165, 177, 181, 185, 193, 201, 209, 213, 217, 221, 229, 237, 253, 257, 265, 281, 285
n, $n + 1$, $n + 2$ are square-free. Ref Halm91 28. [1,1; A7675]

M3825   1, 5, 13, 23, 29, 30, 31, 40, 61, 77, 78, 60, 47, 70, 104, 138, 125, 90, 85, 100, 174,
184, 156
Expansion of a modular form. Ref JNT 25 205 87. [0,2; A6353]

M3826   1, 5, 13, 25, 41, 61, 85, 113, 145, 181, 221, 265, 313, 365, 421, 481, 545, 613, 685,
761, 841, 925, 1013, 1105, 1201, 1301, 1405, 1513, 1625, 1741, 1861, 1985, 2113, 2245
Centered square numbers: $n^2 + (n - 1)^2$. See Fig M3826. Ref MMAG 35 162 62. SIAR 12
277 70. INOC 24 4550 85. [1,2; A1844, N1567]

Figure M3826.   CENTERED POLYGONAL NUMBERS.

The figures show definitions for the **centered triangular** (M3378) and **centered square**
numbers (M3826). See also the **centered pentagonal** numbers (M4112).

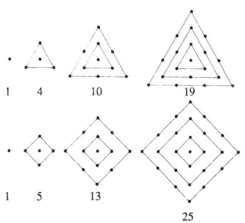

M3827 0, 1, 5, 13, 27, 48, 78, 118, 170, 235, 315, 411, 525, 658, 812, 988, 1188, 1413, 1665, 1945, 2255, 2596, 2970, 3378, 3822, 4303, 4823, 5383, 5985, 6630, 7320, 8056 $[n(n+2)(2n+1)/8]$. Ref MAG 46 55 62; 55 440 71. MMAG 47 290 74. [0,3; A2717, N1569]

M3828 5, 13, 29, 37, 53, 61, 101, 109, 149, 157, 173, 181, 197, 229, 269, 277, 293, 317, 349, 373, 389, 397, 421, 461, 509, 541, 557, 613, 653, 661, 677, 701, 709, 733, 757, 773 Primes of form $8n+5$. Ref AS1 870. [1,1; A7521]

M3829 1, 5, 13, 29, 49, 81, 113, 149, 197, 253, 317, 377, 441, 529, 613, 709, 797, 901, 1009, 1129, 1257, 1373, 1517, 1653, 1793, 1961, 2121, 2289, 2453, 2629, 2821, 3001 Points of norm $\leq n$ in square lattice. Ref PNISI 13 37 47. MOC 16 287 62. SPLAG 106. [0,2; A0328, N1570]

M3830 5, 13, 29, 53, 173, 229, 293, 733, 1093, 1229, 1373, 2029, 2213, 3253, 4229, 4493, 5333, 7229, 7573, 9029, 9413, 10613, 13229, 13693, 15629, 18229, 18773, 21613, 24029 Primes of form n^2+4. Ref MOC 28 1143 74. [1,1; A5473]

M3831 5, 13, 30, 59, 109, 187, 312, 497, 775, 1176, 1753, 2561, 3694, 5245, 7366, 10223, 14056, 19137, 25853, 34637, 46092, 60910, 80009, 104462, 135674, 175274 Bipartite partitions. Ref ChGu56 32. [0,1; A2768, N1571]

M3832 1, 1, 5, 13, 33, 73, 151, 289, 526, 910, 1514 Restricted partitions. Ref CAY 2 281. [0,3; A1981, N1572]

M3833 1, 5, 13, 35, 49, 126, 161, 330, 301, 715, 757, 1365, 1377, 2380, 1837, 3876, 3841, 5985, 5941, 8855, 7297, 12650, 12481, 17550, 17249, 23751, 16801, 31465, 30913 Number of intersections of diagonals of regular n-gon. Cf. M0724. Ref PoRu94. [4,2; A6561]

M3834 1, 5, 13, 45, 121, 413, 1261, 4221, 13801, 46365, 155497, 527613, 1792805, 6126293, 20986153, 72121853, 248396793, 857416949, 2964896877 Related to series-parallel networks. Ref AAP 4 123 72. [1,2; A6349]

M3835 1, 5, 13, 52, 121, 455, 1093, 4305, 9841, 36905, 88573, 348728, 797161, 2989355, 7174453, 28585712, 64570081, 242137805, 581130733, 2288202262 Free subsets of multiplicative group of $GF(3^n)$. Ref SFCA92 2 15. [1,2; A7231]

M3836 5, 13, 131, 149, 1699 $(7^n-1)/6$ is prime. Ref CUNN. MOC 61 928 93. [1,1; A4063]

M3837 1, 5, 13, 132, 233, 305, 1404, 910, 1533 Coefficients of a modular function. Ref GMJ 8 29 67. [-6,2; A5764]

M3838 5, 13, 563 Wilson primes: $(p-1)! \equiv -1 \bmod p^2$. Ref B1 52. Well86 163. VA91 73. [1,1; A7540]

M3839   5, 14, 20, 29, 35, 39, 45, 54, 60, 69, 78, 84, 93, 99, 103, 109, 118, 124, 133, 139,
143, 149, 158, 164, 173, 182, 188, 197, 203, 207, 213, 222, 228, 237, 243, 247, 253, 262
Related to Fibonacci representations. Ref FQ 11 385 73. [1,1; A3248]

M3840   5, 14, 27, 41, 44, 65, 76, 90, 109, 125, 139, 152, 155, 169, 186, 189, 203, 208, 209,
219, 227, 230, 237, 265, 275, 298, 307, 311, 314, 321, 324, 329, 344, 377, 413, 419, 428
$C(2n,n)$ is divisible by $(n+1)^2$. Ref JIMS 18 97 29. [0,1; A2503]

M3841   0, 5, 14, 28, 48, 75, 110, 154, 208, 273, 350, 440, 544, 663, 798, 950, 1120, 1309,
1518, 1748, 2000, 2275, 2574, 2898, 3248, 3625, 4030, 4464, 4928, 5423, 5950, 6510
$n(n+4)(n+5)/6$. Ref AS1 796. [0,2; A5586]

M3842   5, 14, 28, 48, 75, 110, 154, 208, 273, 350, 440, 544, 663, 798, 950, 1120, 1309,
1518, 1748, 2000, 2275, 2574, 2898, 3248, 3625, 4030, 4464, 4928, 5423, 5950, 6510
Walks on square lattice. Ref GU90. [0,1; A5555]

$$\text{G.f.:}\ \ (5 - 6x + 2x^2)\,(1 - x)^{-4}.$$

M3843   1, 5, 14, 29, 50, 77, 110, 149, 194, 245, 302, 365, 434, 509, 590, 677, 770, 869,
974, 1085, 1202, 1325, 1454, 1589, 1730, 1877, 2030, 2189, 2354, 2525, 2702, 2885
Points on surface of square pyramid: $3n^2 + 2$. Ref Coxe74. INOC 24 4552 85. [0,2; A5918]

M3844   1, 5, 14, 30, 55, 91, 140, 204, 285, 385, 506, 650, 819, 1015, 1240, 1496, 1785,
2109, 2470, 2870, 3311, 3795, 4324, 4900, 5525, 6201, 6930, 7714, 8555, 9455, 10416
Square pyramidal numbers: $n(n+1)(2n+1)/6$. See Fig M3382. Ref D1 2 2. B1 194. AS1
813. [1,2; A0330, N1574]

M3845   5, 14, 35, 71, 126, 211
Postage stamp problem. Ref CJN 12 379 69. AMM 87 208 80. [1,1; A1215, N1706]

M3846   1, 5, 14, 53, 178, 685
Triangulations. Ref WB79 336. [0,2; A5504]

M3847   1, 1, 5, 14, 64, 189, 826
n-dimensional Bravais lattices. Ref BB78 52. MOC 43 574 84. Enge93 1022. [0,3; A4030]

M3848   5, 14, 1026, 4324, 311387, 6425694, 579783114, 4028104212, 7315072725560
Related to zeros of Bessel function. Ref MOC 1 406 45. [4,1; A0331, N1575]

M3849   1, 5, 15, 34, 65, 111, 175, 260, 369, 505, 671, 870, 1105, 1379, 1695, 2056, 2465,
2925, 3439, 4010, 4641, 5335, 6095, 6924, 7825, 8801, 9855, 10990, 12209, 13515
Paraffins. Ref BER 30 1922 1897. [0,2; A6003]

$$\text{G.f.:}\ \ (1 - x^3)\,/\,(1 - x)^5.$$

M3850   1, 5, 15, 35, 69, 121, 195, 295, 425, 589, 791, 1035, 1325, 1665, 2059, 2511, 3025,
3605, 4255, 4979, 5781, 6665, 7635, 8695, 9849, 11101, 12455, 13915, 15485, 17169
Centered tetrahedral numbers. Ref INOC 24 4550 85. [0,2; A5894]

M3851 1, 5, 15, 35, 70, 125, 200, 255, 275
Expansion of bracket function. Ref FQ 2 254 64. [5,2; A0750, N1576]

M3852 1, 5, 15, 35, 70, 125, 210, 325, 495
Compositions into 5 relatively prime parts. Ref FQ 2 250 64. [3,2; A0743, N1577]

M3853 1, 5, 15, 35, 70, 126, 210, 330, 495, 715, 1001, 1365, 1820, 2380, 3060, 3876,
4845, 5985, 7315, 8855, 10626, 12650, 14950, 17550, 20475, 23751, 27405, 31465
Binomial coefficients $C(n,4)$. See Fig M1645. Ref D1 2 7. RS3. B1 196. AS1 828. [4,2;
A0332, N1578]

M3854 0, 1, 5, 15, 36, 75, 141, 245, 400, 621, 925, 1331, 1860, 2535, 3381, 4425, 5696,
7225, 9045, 11191, 13700, 16611, 19965, 23805, 28176, 33125, 38701, 44955, 51940
$(n^4 + 11n^2)/12$. Ref BER 30 1923 1897. GA66 246. [0,3; A6008]

M3855 1, 5, 15, 37, 77, 141, 235, 365, 537, 757, 1031, 1365, 1765, 2237, 2787, 3421,
4145, 4965, 5887, 6917, 8061, 9325, 10715, 12237, 13897, 15701, 17655, 19765, 22037
$n^3 + 3n + 1$. Ref JCT A24 316 78. [0,2; A5491]

M3856 1, 5, 15, 40, 98, 237, 534, 1185, 2554, 5391
Partitions into non-integral powers. Ref PCPS 47 215 51. [1,2; A0333, N1579]

M3857 0, 1, 5, 15, 43, 119, 327, 895, 2447, 6687, 18271, 49919, 136383, 372607,
1017983, 2781183, 7598335, 20759039, 56714751, 154947583, 423324671, 1156544511
Tower of Hanoi with cyclic moves only. Ref IPL 13 118 81. GKP 18. [0,3; A5665]

M3858 1, 5, 15, 45, 120, 326, 835, 2145, 5345, 13220, 32068, 76965, 181975, 425490,
982615, 2245444, 5077090, 11371250
4-dimensional partitions of n. Ref PCPS 63 1099 67. [1,2; A0334, N1580]

M3859 1, 5, 15, 45, 120, 331, 855, 2214, 5545, 13741, 33362, 80091, 189339, 442799,
1023192, 2340904, 5302061, 11902618, 26488454, 58479965, 128120214, 278680698
Euler transform of M3382. Ref PCPS 63 1100 67. EIS § 2.7. [1,2; A0335, N1581]

M3860 1, 5, 15, 45, 165, 629, 2635, 11165, 48915, 217045, 976887, 4438925, 20346485,
93900245, 435970995, 2034505661, 9536767665, 44878791365, 211927736135
n-bead necklaces with 5 colors. See Fig M3860. Ref R1 162. IJM 5 658 61. [0,2; A1869,
N1582]

M3861 5, 15, 55, 140, 448, 1022, 2710, 6048, 14114, 28831
Restricted partitions. Ref JCT 9 373 70. [2,1; A2221, N1583]

M3862 1, 5, 15, 55, 190, 671, 2353, 8272, 29056, 102091, 358671
Distributive lattices. Ref MSH 53 19 76. MSG 121 121 76. [0,2; A6358]

‖‖

Figure M3860. NECKLACES, POLYNOMIALS.

The figure illustrates M0564, the number of different necklaces that can be made from beads of two colors, when the necklaces can be rotated but not turned over. This number is $a_n = \sum \phi(d) 2^{n/d}$, where $\phi(d)$ is the Euler totient function (M0299) and the sum is over all divisors of n. It is also the number of binary irreducible polynomials whose degree divides n, an important sequence in digital circuitry ([BE2 70], [CMA 1 358 69], [MS78 115]), and each necklace is labeled with the corresponding polynomial. (Let α be a primitive element of $GF(2^n)$. The polynomial with α^i as a root corresponds to the necklace found by reading i in binary!) The number of irreducible polynomials of degree exactly n (or the number of necklaces of length n that are not repetitions of a shorter necklace) is given by M0116. If turning over is allowed, the number of different necklaces is given by M0563. In M0564 there are two different colors of beads. If instead there are 3, 4 or 5 colors we obtain M2548, M3390, M3860.

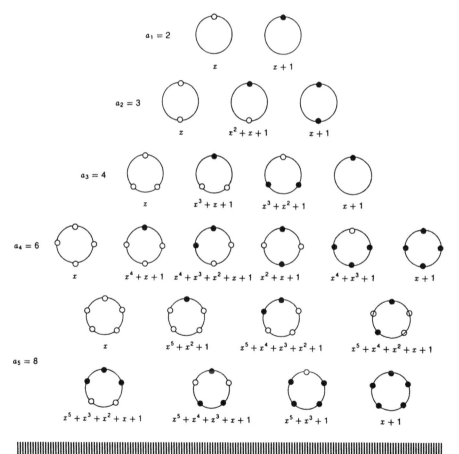

‖‖

M3863 5, 15, 55, 225, 979, 4425, 20515, 96825, 462979, 2235465, 10874275, 53201625, 261453379, 1289414505, 6376750435, 31605701625, 156925970179
$1^n + 2^n + \cdots + 5^n$. Ref AS1 813. [0,1; A1552, N1584]

M3864 5, 16, 36, 70, 126, 216, 345, 512, 797
Postage stamp problem. Ref CJN 12 379 69. AMM 87 208 80. [1,1; A1210, N1707]

M3865 1, 5, 16, 40, 85, 161, 280, 456, 705, 1045, 1496, 2080, 2821, 3745, 4880, 6256,
   7905, 9861, 12160, 14840, 17941, 21505, 25576, 30200, 35425, 41301, 47880, 55216
Paraffins. Ref BER 30 1923 1897. [0,2; A6007]

$$\text{G.f.:} \quad (1 - x^4) \; / \; (1 - x)^5 \; (1 - x^2).$$

M3866 0, 0, 0, 1, 5, 16, 42, 99, 219, 466, 968, 1981, 4017, 8100, 16278, 32647, 65399,
   130918, 261972, 524097, 1048365, 2096920, 4194050, 8388331, 16776915, 33554106
$2^n - 1 - n(n+1)/2$. Ref MFM 73 18 69. [0,5; A2662, N1585]

M3867 1, 5, 16, 45, 121, 320, 841, 2205, 5776, 15125, 39601, 103680, 271441, 710645,
   1860496, 4870845, 12752041, 33385280, 87403801, 228826125, 599074576
Alternate Lucas numbers − 2. Ref PGEC 18 281 71. FQ 13 51 75. [1,2; A4146]

M3868 1, 5, 16, 51, 127, 340, 798, 1830, 3916, 8569
Homogeneous primitive partition identities with largest part n. Ref DGS94. [3,2; A7343]

M3869 5, 16, 56, 224, 1024, 5296, 30656, 196544, 1383424, 10608976, 88057856,
   786632864, 7525556224, 76768604656, 831846342656, 9541952653184
Entringer numbers. Ref NAW 14 241 66. DM 38 268 82. [0,1; A6217]

M3870 1, 0, 1, 1, 5, 16, 65, 260, 1085, 4600, 19845, 86725, 383251, 1709566, 7687615,
   34812519, 158614405, 726612216, 3344696501, 15462729645, 71763732545
Noncommutative $SL(2,C)$-invariants of degree n in 5 variables. Ref JALG 93 189 85. [0,5;
A7043]

M3871 1, 5, 16, 86, 448, 3580
3-edge-colored trivalent graphs with $2n$ nodes. Ref RE58. [1,2; A2830, N1586]

M3872 1, 5, 17, 46, 116, 252, 533, 1034, 1961
Coefficients of a modular function. Ref GMJ 8 29 67. [−1,2; A3295]

M3873 1, 5, 17, 49, 125, 297, 669, 1457, 3093, 6457, 13309, 27201, 55237, 111689,
   225101, 452689, 908885, 1822809, 3652701, 7315553, 14645349, 29311081, 58650733
Number of elements in $Z[i]$ whose 'smallest algorithm' is $\leq n$. Ref JALG 19 290 71. hwl.
[0,2; A6457]

M3874 1, 5, 17, 49, 129, 321, 769, 1793, 4097, 9217, 20481, 45057, 98305, 212993,
   458753, 983041, 2097153, 4456449, 9437185, 19922945, 41943041, 88080385
Expansion of $1 \; / \; (1-x)(1-2x)^2$. Ref HB67 16. [0,2; A0337, N1587]

M3875 1, 5, 17, 61, 217, 773, 2753, 9805, 34921, 124373, 442961, 1577629, 5618809,
   20011685, 71272673, 253841389, 904069513, 3219891317, 11467812977, 40843221565
Subsequences of $[1,...,2n+1]$ in which each odd number has an even neighbor. Ref
GuMo94. [0,2; A7483]
$$a(n) = 3\,a(n-1) + 2\,a(n-2).$$

M3876 1, 1, 1, 5, 17, 83, 593, 2893, 36101, 172195, 3421285, 15520165, 467129785, 1954015955, 86971636825, 323371713725, 21196564551725, 66760541581475
$a(n+1) = a(n) - n(n-1)a(n-1)$. Ref DUMJ 26 580 59. [1,4; A2020, N1588]

M3877 5, 18, 42, 75, 117, 168, 228, 297, 375, 462, 558, 663, 777, 900, 1032, 1173, 1323, 1482, 1650, 1827, 2013, 2208, 2412, 2625, 2847, 3078, 3318, 3567, 3825, 4092, 4368
Expansion of $(5-2x)(1-x^3) / (1-x)^4$. Ref SMA 20 23 54. [3,1; A0338, N1589]

M3878 5, 18, 45, 45, 52, 139, 80, 89, 184, 145, 103, 312, 96, 225, 379
Number of triangles with integer sides and area $= n$ times perimeter. Ref AMM 99 176 92. [1,1; A7237]

M3879 1, 5, 18, 45, 100, 185, 323, 522, 804
Partitions into non-integral powers. Ref PCPS 47 214 51. [2,2; A0339, N1590]

M3880 0, 0, 0, 1, 5, 18, 52, 134, 318, 713, 1531, 3180, 6432, 12732, 24756, 47417, 89665, 167694, 310628, 570562, 1040226, 1883953, 3391799, 6073848, 10824096, 19204536
From variance of Fibonacci search. Ref BIT 13 93 73. [0,5; A6479]

M3881 1, 5, 18, 56, 160, 432, 1120, 2816, 6912, 16640, 39424, 92160, 212992, 487424, 1105920, 2490368, 5570560, 12386304, 27394048, 60293120, 132120576, 288358400
Coefficients of Chebyshev polynomials: $n(n+3)2^{n-3}$. Ref RSE 62 190 46. AS1 795. [1,2; A1793, N1591]

M3882 1, 5, 18, 58, 179, 543, 1636, 4916, 14757, 44281, 132854, 398574, 1195735, 3587219, 10761672, 32285032, 96855113, 290565357, 871696090, 2615088290
Expansion of $1/(1-x)^2(1-3x)$. Ref DKB 260. [0,2; A0340, N1592]

M3883 1, 5, 18, 82, 643, 15182, 7848984
Precomplete Post functions. Ref SMD 10 619 69. JCT A14 6 73. [1,2; A2826, N1593]

M3884 1, 5, 19, 61, 180, 498, 1323, 3405, 8557, 21103, 51248, 122898, 291579, 685562, 1599209, 3705122, 8532309, 19543867, 44552066, 101124867, 228640542
n-node trees of height 5. Ref IBMJ 4 475 60. KU64. [6,2; A0342, N1594]

M3885 1, 5, 19, 63, 185, 502, 1270, 3046, 6968, 15335
Coefficients of a modular function. Ref GMJ 8 29 67. [1,2; A3296]

M3886 1, 5, 19, 65, 210, 654, 1985, 5911, 17345, 50305, 144516, 411900, 1166209, 3283145, 9197455, 25655489, 71293590, 197452746, 545222465, 1501460635
Expansion of $(1-x)/(1-3x+x^2)^2$. [0,2; A1870, N1595]

M3887 1, 5, 19, 65, 211, 665, 2059, 6305, 19171, 58025, 175099, 527345, 1586131, 4766585, 14316139, 42981185, 129009091, 387158345, 1161737179, 3485735825
$3^n - 2^n$. Ref EUR 24 20 61. CRP 268 579 69. [1,2; A1047, N1596]

M3888 1, 5, 19, 66, 221, 728, 2380, 7753, 25213, 81927, 266110, 864201, 2806272, 9112264, 29587889, 96072133, 311945595, 1012883066, 3288813893, 10678716664
Random walks (binomial transform of M1396). Ref DM 17 44 77. EIS § 2.7. [0,2; A5021]

M3889 1, 5, 19, 67, 236, 797, 2678, 8833, 28908, 93569, 300748, 959374, 3042808,
9597679, 30134509
Connected graphs with n nodes, $n+2$ edges. Ref SS67. [4,2; A1435, N1597]

M3890 1, 5, 19, 71, 265, 989, 3691, 13775, 51409, 191861, 716035, 2672279, 9973081,
37220045, 138907099, 518408351, 1934726305, 7220496869, 26947261171
$a(n) = 4a(n-1) - a(n-2)$. Ref EUL (1) 1 375 11. MMAG 40 78 67. [0,2; A1834, N1598]

M3891 1, 1, 5, 19, 85, 381, 1751, 8135, 38165, 180325, 856945, 4091495, 19611175,
94309099, 454805755, 2198649549, 10651488789, 51698642405, 251345549849
Expansion of $(1 + x + \cdots + x^4)^n$. Ref FQ 7 347 69. [0,3; A5191]

M3892 1, 1, 5, 19, 101, 619, 4421, 35899, 326981, 3301819, 36614981, 442386619,
5784634181, 81393657019, 1226280710981, 19696509177019, 335990918918981
$n! - (n-1)! + (n-2)! - \cdots$. Ref AMM 95 699 88. [1,3; A5165]

M3893 0, 5, 20, 29, 45, 80, 101, 116, 135, 145, 165, 173, 236, 257, 397, 404, 445, 477,
540, 565, 580, 585, 629, 666, 836, 845, 885, 909, 944, 949, 954, 975, 1125, 1177
Epstein's Put or Take a Square game. Ref UPNT E26. [1,2; A5240]

M3894 5, 20, 49, 98, 174, 285, 440, 649, 923, 1274, 1715, 2260, 2924, 3723, 4674, 5795,
7105, 8624, 10373, 12374, 14650, 17225, 20124, 23373, 26999, 31030, 35495, 40424
Permutations by inversions: $C(n,4) + C(n,3) - C(n,2)$. Ref NET 96. DKB 241. MMAG
61 28 88. rkg. [5,1; A5287]

M3895 5, 20, 51, 105, 190, 315, 490, 726, 1035, 1430, 1925, 2535, 3276, 4165, 5220,
6460, 7905, 9576, 11495, 13685, 16170, 18975, 22126, 25650, 29575, 33930, 38745
Coefficient of x^4 in $(1 - x - x^2)^{-n}$. Ref FQ 14 43 76. [1,1; A6504]

M3896 0, 1, 5, 20, 51, 112, 221, 411, 720, 1221, 2003, 3206, 5021, 7728, 11698, 17472,
25766, 37580, 54254, 77617, 110087, 154942, 216488, 300456, 414365, 568113, 774571
From a partition triangle. Ref AMM 100 288 93. [2,3; A7045]

M3897 1, 5, 20, 52, 117, 225, 400, 656
Paraffins. Ref BER 30 1923 1897. [1,2; A6010]

M3898 1, 5, 20, 65, 185, 481, 1165, 2665, 5820, 12220, 24802, 48880, 93865, 176125,
323685, 583798, 1035060, 1806600, 3108085, 5276305, 8846884, 14663645, 24044285
Coefficients of an elliptic function. Ref CAY 9 128. [0,2; A1939, N1599]

$$\text{G.f.: } \Pi \, (1 - x^k)^{-c(k)}, \; c(k) = 5, 5, 5, 0, 5, 5, 5, 0, \ldots.$$

M3899 1, 5, 20, 65, 190, 511, 1295, 3130, 7285, 16435, 36122, 77645, 163730, 339535,
693835, 1399478, 2790100, 5504650, 10758050, 20845300, 40075630, 76495450
Convolved Fibonacci numbers. Ref RCI 101. FQ 15 118 77. DM 26 267 79. [0,2; A1873,
N1600]

$$\text{G.f.: } (1 - x - x^2)^{-5}.$$

M3900 0, 0, 0, 0, 0, 5, 20, 69, 200, 521
Unexplained difference between two partition g.f.s. Ref PCPS 63 1100 67. [1,6; A7327]

M3901 1, 5, 20, 70, 230, 721, 2200, 6575, 19385, 56575, 163952, 472645, 1357550, 3888820, 11119325, 31753269, 90603650, 258401245, 736796675, 2100818555
Powers of rooted tree enumerator. Ref R1 150. [1,2; A0343, N1601]

M3902 1, 5, 20, 70, 230, 726, 2235, 6765, 20240, 60060, 177177, 520455, 1524120, 4453320, 12991230, 37854954, 110218905, 320751445, 933149470, 2714401580
Column of Motzkin triangle. Ref JCT A23 293 77. [4,2; A5324]

M3903 1, 5, 20, 73, 271, 974, 3507, 12487
Graphs with no isolated vertices. Ref LNM 952 101 82. [4,2; A6650]

M3904 1, 5, 20, 75, 275, 1001, 3640, 13260, 48450, 177650, 653752, 2414425, 8947575, 33266625, 124062000, 463991880, 1739969550, 6541168950, 24647883000
$5C(2n, n-2)/(n+3)$. Ref QAM 14 407 56. MOC 29 216 75. FQ 14 397 76. [2,2; A0344, N1602]

M3905 1, 5, 20, 76, 285, 1068, 4015, 15159, 57486, 218895, 836604, 3208036, 12337630, 47572239, 183856635, 712033264
Permutations by inversions. Ref NET 96. DKB 241. MMAG 61 28 88. rkg. [5,2; A5283]

M3906 1, 5, 20, 84, 316, 1196, 4461, 16750, 62878, 237394, 899265, 342211, 13069026, 50091095, 192583152, 742560511, 2870523142, 11122817672, 43191285751
Asymmetric polyominoes with n cells. Ref DM 36 203 81. [4,2; A6749]

M3907 1, 1, 5, 20, 84, 354, 1540, 6704, 29610, 131745, 591049, 2669346
Restricted hexagonal polyominoes with n cells. Ref PEMS 17 11 70. [1,3; A2213, N1603]

M3908 0, 1, 5, 20, 84, 409, 2365, 16064, 125664, 1112073, 10976173, 119481284, 1421542628, 18348340113, 255323504917, 3809950976992, 60683990530208
$\Sigma \, n(n-1) \cdots (n-k+1)/k$, $k = 2 \ldots n$. Ref rkg. [1,3; A6231]

M3909 1, 5, 20, 96, 469, 3145, 20684, 173544, 1557105, 16215253, 159346604, 2230085528, 26985045333, 368730610729, 5628888393652, 97987283458928
Sums of logarithmic numbers. Ref TMS 31 78 63. jos. [1,2; A2745, N1604]

M3910 5, 20, 206, 54155
Switching networks. Ref JFI 276 324 63. [1,1; A0877, N1605]

M3911 1, 5, 20, 300, 9980, 616260, 65814020, 11878194300, 3621432947180, 1880516646144660
Colored graphs. Ref CJM 22 596 70. rcr. [1,2; A2030, N1606]

M3912 1, 5, 21, 33, 65, 85, 133
Coefficients of period polynomials. Ref LNM 899 292 81. [3,2; A6309]

M3913 1, 5, 21, 84, 330, 1287, 5005, 19448, 75582, 293930, 1144066, 4457400, 17383860, 67863915, 265182525, 1037158320, 4059928950, 15905368710
Binomial coefficients $C(2n+1, n-1)$. See Fig M1645. Ref CAY 13 95. AS1 828. [1,2; A2054, N1607]

M3914 1, 5, 21, 85, 341, 1365, 5461, 21845, 87381, 349525, 1398101, 5592405, 22369621, 89478485, 357913941, 1431655765, 5726623061, 22906492245
$(4^n - 1)/3$. Ref TH09 35. FMR 1 112. RCI 217. [1,2; A2450, N1608]

M3915 1, 5, 21, 93, 409, 1837, 8209, 36969, 166041, 748889, 3373941, 15248153, 68840633
Expansion of layer susceptibility series for cubic lattice. Ref JPA 12 2451 79. [0,2; A7287]

M3916 1, 5, 21, 119, 735, 4830, 33253
Triangulations of the disk. Ref PLMS 14 759 64. [0,2; A2711, N1609]

M3917 1, 1, 5, 21, 357, 5797, 376805, 24208613, 6221613541, 1594283908581, 1634141006295525, 1673768626404966885, 6857430062381149327845
Gaussian binomial coefficient $[n, n/2]$ for $q = 4$. Ref GJ83 99. ARS A17 328 84. [0,3; A6109]

M3918 1, 5, 22, 71, 186, 427, 888, 1704
Partitions into non-integral powers. Ref PCPS 47 214 51. [3,2; A0345, N1610]

M3919 1, 5, 22, 87, 317, 1053, 3250, 9343, 25207
$4 \times n$ binary matrices. Ref CPM 89 217 64. PGEC 22 1050 73. SLC 19 79 88. [0,2; A6148]

M3920 1, 5, 22, 93, 386, 1586, 6476, 26333, 106762, 431910, 1744436, 7036530, 28354132, 114159428, 459312152, 1846943453, 7423131482, 29822170718
$2^{2n+1} - C(2n+1, n+1)$. Ref BAMS 74 74 68. JCT A13 215 72. [0,2; A0346, N1611]

M3921 1, 1, 5, 22, 116, 612, 3399, 19228, 111041, 650325, 3856892, 23107896, 139672312, 850624376, 5214734547, 32154708216, 199292232035, 1240877862315
Dissections of a polygon. Ref DM 11 387 75. [1,3; A5033]

M3922 1, 1, 5, 22, 164, 1030, 8885, 65954, 614404, 5030004, 49145460, 429166584, 4331674512, 39599553708, 409230997461
Rooted planar maps with n edges. Ref BAMS 74 74 68. WA71. JCT A13 215 72. [0,3; A6294]

M3923 0, 0, 5, 22, 258, 1628, 18052
Bishops on an $n \times n$ board. Ref LNM 560 212 76. [0,3; A5632]

M3924 1, 5, 22, 1001, 2882, 15251, 720027, 7081807, 7451547, 26811862, 54177145, 1050660501, 1085885801, 1528888251, 2911771192
Pentagonal palindromes. Ref AMM 48 211 41. [1,2; A2069, N1612]

M3925 1, 5, 23, 17, 719, 5039, 1753, 2999, 125131, 7853, 479001599, 3593203,
87178291199, 1510259, 6880233439, 256443711677, 478749547, 78143369
Largest factor of $n!$ $-$ 1. Ref SMA 14 25 48. MOC 26 570 72. [2,2; A2582, N1613]

M3926 5, 23, 527, 277727, 77132286527, 5949389624883225721727,
35395236908668169265765137996816180039862527
$a(n) = a(n-1)^2 - 2$. [0,1; A3487]

M3927 1, 5, 23, 1681, 257543, 67637281, 27138236663, 15442193173681
Glaisher's T numbers. Ref FMR 1 76. jcpm. [1,2; A2811, N1615]

M3928 5, 24, 79, 223, 579, 1432
Total preorders. Ref MSH 53 20 76. [3,1; A6328]

M3929 1, 5, 24, 84, 251, 653, 1543
Partitions into non-integral powers. Ref PCPS 47 214 51. [4,2; A0347, N1616]

M3930 1, 5, 24, 115, 551, 2640, 12649, 60605, 290376, 1391275, 6665999, 31938720,
153027601, 733199285, 3512968824, 16831644835, 80645255351, 386394631920
Pythagoras' theorem generalized: $a(n+1) = 5.a(n) - a(n-1)$. Ref BU71 75. [1,2; A4254]

M3931 1, 5, 24, 122, 680, 4155, 27776
Total preorders. Ref MSH 53 20 76. [3,2; A6326]

M3932 1, 5, 24, 128, 835, 6423, 56410, 554306, 6016077, 71426225, 920484892,
12793635300, 190730117959, 3035659077083
Permutations of length n by rises. Ref DKB 263. [3,2; A0349, N1617]

M3933 1, 1, 1, 5, 24, 133, 846, 5661, 39556
Triangulations of the disk. Ref PLMS 14 759 64. [0,4; A2709, N1618]

M3934 1, 5, 24, 391, 9549, 401691
Superpositions of cycles. Ref AMA 131 143 73. [3,2; A3224]

M3935 0, 1, 5, 25, 29, 41, 49, 61, 65, 85, 89, 101, 125, 145, 149, 245, 265, 365, 385, 485,
505, 601, 605, 625, 649, 701, 725, 745, 749, 845, 865, 965, 985, 1105, 1205, 1249, 1345
$F(n)$ ends with n. Ref FQ 4 156 66. [0,3; A0350, N1619]

M3936 1, 5, 25, 65, 265, 605, 2125, 4345, 14665, 27965, 93025
Words of length n in a certain language. Ref DM 40 231 82. [0,2; A7058]

M3937 1, 5, 25, 125, 625, 3125, 15625, 78125, 390625, 1953125, 9765625, 48828125,
244140625, 1220703125, 6103515625, 30517578125, 152587890625
Powers of 5. Ref BA9. [0,2; A0351, N1620]

M3938 1, 5, 25, 129, 681, 3653, 19825, 108545, 598417, 3317445, 18474633, 103274625,
579168825, 3256957317, 18359266785, 103706427393, 586889743905, 3326741166725
$\Sigma\, C(n, k+1).C(n+k, k)$, $k = 0 \, . \, . \, n-1$. Ref AMM 43 29 36. [1,2; A2002, N1621]

M3939 1, 5, 25, 149, 1081, 9365, 94585, 1091669, 14174521, 204495125, 3245265145,
  56183135189, 1053716696761, 21282685940885, 460566381955705
Simplices in barycentric subdivisions of n-simplex. Ref SKA 11 95 28. MMAG 37 132 64.
[0,2; A2050, N1622]

$$\text{E.g.f.:} \quad (1 - e^x) / (1 - 2\, e^{-x}).$$

M3940 5, 25, 625, 625, 90625, 890625, 2890625, 12890625, 212890625, 8212890625
Idempotents: $a(n)^2 \equiv a(n) \bmod 10^n$. Ref Schut91. [1,1; A7185]

M3941 1, 5, 26, 97, 265, 362, 1351, 13775, 70226, 262087, 716035, 978122
Related to Bernoulli numbers. Ref ANN 36 645 35. [0,2; A2316, N1624]

M3942 5, 26, 119, 538, 2310, 9882
Triangulations of the disk. Ref PLMS 14 759 64. [0,1; A5499]

M3943 1, 5, 26, 139, 758, 4194, 23460, 132339, 751526, 4290838, 24607628, 141648830,
  817952188, 4736107172, 27487711752, 159864676803
Walks on cubic lattice. Ref GU90. [0,2; A5573]

M3944 1, 5, 26, 154, 1044, 8028, 69264, 663696, 6999840, 80627040, 1007441280,
  13575738240, 196287356160, 3031488633600, 49811492505600
Generalized Stirling numbers. Ref PEF 77 7 62. [0,2; A1705, N1625]

$$\text{E.g.f.:} \quad -\ln\,(1 - x) / (1 - x)^2.$$

M3945 1, 1, 5, 26, 205, 1936, 22265, 297296, 4544185, 78098176, 1491632525,
  31336418816, 718181418565, 17831101321216, 476768795646785
Alternating 3-signed permutations. Ref EhRe94. [0,3; A7286]

$$\text{G.f.:} \quad (\sin x + \cos 2x) / \cos 3x.$$

M3946 1, 5, 26, 272, 722, 5270, 5260, 37358
2×2 matrices with entries mod n. Ref MMAG 41 59 68. aw. [1,2; A6045]

M3947 1, 1, 5, 27, 502, 2375, 95435, 1287965, 29960476, 262426878, 28184365650
Coefficients for step-by-step integration. Ref JACM 11 231 64. [0,3; A2401, N1626]

M3948 5, 27, 1204, 85617952
Switching networks. Ref JFI 276 324 63. [1,1; A0878, N1627]

M3949 5, 28, 156, 863, 4571, 22952, 108182
5-covers of an n-set. Ref DM 81 151 90. [1,1; A5785]

M3950 1, 5, 28, 180, 1320, 10920, 100800, 1028160, 11491200, 139708800, 1836172800,
  25945920000, 392302310400, 6320426112000, 108101081088000, 1956280854528000
$a(n+1) = (n-1)\,a(n) + n.n!$. Ref RAIRO 12 58 78. [2,2; A6157]

M3951 5, 28, 190, 1340, 9065, 57512, 344316, 1966440, 10813935, 57672340,
  299893594, 1526727748, 7633634645, 37580965520, 182536112120, 876173330832
Minimal covers of an n-set. Ref DM 5 249 73. [3,1; A3467]

$$\text{G.f.: } 1 + (1 - 4x)^{-4} + 3(1 - x)^{-4}.$$

M3952 1, 5, 29, 19, 2309, 30029, 8369, 929, 46027, 81894851, 876817, 38669,
  304250263527209, 92608862041, 59799107, 1143707681, 69664915493
Largest factor of $2.3.5.7 \cdots - 1$. Ref SMA 14 26 48. Krai52 2. MOC 26 568 72. MMAG
48 93 75. jls. [1,2; A2584, N1628]

M3953 5, 29, 29, 29, 29, 29, 29, 29, 23669, 23669, 23669, 23669, 23669, 23669, 1508789,
  5025869, 9636461, 9636461, 9636461, 37989701, 37989701, 37989701, 37989701
Sequence of prescribed quadratic character. Ref MOC 24 446 70. [3,1; A1990, N1632]

M3954 5, 29, 118, 418, 1383, 4407, 13736, 42236, 128761, 390385, 1179354, 3554454
Permutations of length n by number of runs. Ref DKB 260. [4,1; A0352, N1629]

M3955 1, 1, 5, 29, 169, 985, 5741, 33461, 195025, 1136689, 6625109, 38613965,
  225058681, 1311738121, 7645370045, 44560482149, 259717522849, 1513744654945
Pythagorean triangles with consecutive legs (hypotenuse): $a(n) = 6a(n-1) - a(n-2)$. Cf.
M3074. Ref AMM 4 25 1897. MLG 2 322 10. FQ 6(3) 104 68. [0,3; A1653, N1630]

M3956 1, 1, 5, 29, 201, 1657, 15821, 170389, 2032785, 26546673, 376085653,
  5736591885, 93614616409, 1625661357673, 29905322979421, 580513190237573
Coincides with its 4th order binomial transform. Ref DM 21 320 78. BeSl94. EIS § 2.7.
[0,3; A4213]

$$\text{Lgd.e.g.f.: } e^{4x}.$$

M3957 1, 1, 5, 29, 233, 2329, 27949, 391285, 6260561, 112690097, 2253801941,
  49583642701, 1190007424825, 30940193045449, 866325405272573
Expansion of $e^{-x} / (1 - 2x)$. Ref LU91 1 223. R1 83. [0,3; A0354, N1631]

M3958 5, 30, 115, 425, 1396, 4440
Alkyls with n carbon atoms. Ref ZFK 93 437 36. [1,1; A0649, N1633]

M3959 1, 5, 30, 186, 1164, 7344, 46732, 299604, 1932900, 12537542, 81705782,
  534663812, 3511466838, 23136724382, 152888000934, 1012925595468
Sum of squared spans of n-step polygons on square lattice. Ref JPA 21 L167 88. [1,2;
A6773]

M3960 1, 5, 30, 210, 1680, 15120, 151200, 1663200, 19958400, 259459200, 3632428800,
  54486432000, 871782912000, 14820309504000, 266765571072000, 5068545850368000
$n!/24$. Ref PEF 77 61 62. [4,2; A1720, N1634]

M3961 1, 5, 31, 141, 659, 3005, 13739, 62669, 285931, 1304285
Worst case of a Jacobi symbol algorithm. Ref JSC 10 605 90. [0,2; A5826]

M3962   0, 0, 1, 5, 31, 203, 1501, 12449, 114955, 1171799, 13082617, 158860349,
2085208951, 29427878435, 444413828821, 7151855533913
The game of Mousetrap with n cards. Ref QJMA 15 241 1878. GN93. [2,4; A2469,
N1635]

M3963   1, 5, 31, 209, 1476, 10739, 79780, 601905, 4595485, 35419710, 275109858,
2150537435, 16901814190, 133452123796, 1057920031536
n-node animals on b.c.c. lattice. Ref PE90. DU92 41. [1,2; A7197]

M3964   5, 31, 211, 1031, 2801, 4651, 5261, 6841, 8431, 14251, 17891, 20101, 21121,
22621, 22861, 26321, 30941, 33751, 36061, 41141, 46021, 48871, 51001, 58411, 61051
Quintan primes: $p = (x^5 - y^5)/(x - y)$. Ref CU23 2 200. [1,1; A2649, N1636]

M3965   0, 1, 5, 31, 227, 1909, 18089, 190435, 2203319, 27772873, 378673901,
5551390471, 87057596075, 1453986832381, 25762467303377, 482626240281739
$a(n) = n.a(n-1) + (n-5)a(n-2)$. Ref R1 188. [3,3; A1910, N1637]

M3966   1, 1, 5, 31, 257, 2671, 33305, 484471, 8054177, 150635551, 3130287705
From Fibonacci sums. Ref FQ 5 48 67. [0,3; A0556, N1638]

M3967   0, 5, 32, 178, 1024, 6320, 42272, 306448, 2401024, 20253440, 183194912,
1769901568, 18198049024, 198465167360, 2288729963552, 27831596812288
Entringer numbers. Ref NAW 14 241 66. DM 38 268 82. [0,2; A6214]

M3968   1, 5, 32, 288, 3413, 50069, 873612, 17650828, 405071317, 10405071317,
295716741928, 9211817190184, 312086923782437, 11424093749340453
Σn^n. Ref AMM 53 471 46. [1,2; A1923, N1639]

M3969   1, 1, 5, 32, 385, 7573, 181224
Trivalent graphs of girth exactly 6 and $2n$ nodes. Ref gr. [7,3; A6926]

M3970   1, 0, 0, 0, 1, 5, 33, 236, 1918, 17440
Hit polynomials. Ref RI63. [0,6; A1887, N1640]

M3971   1, 5, 33, 287, 3309, 50975, 1058493
Related to partially ordered sets. Ref JCT 6 17 69. [0,2; A1828, N1641]

M3972   1, 5, 34, 258, 2136, 19320, 190800, 2051280
Terms in certain determinants. Ref PLMS 10 122 1879. [1,2; A2776, N1642]

M3973   5, 35, 140, 420, 1050, 2310, 4620, 8580
Related to binomial moments. Ref JO39 449. [3,1; A0910, N1643]

M3974   1, 5, 35, 140, 720, 2700, 12375, 45375, 196625, 715715, 3006003, 10930920,
45048640, 164105760, 668144880, 2441298600, 9859090500, 36149998500
Walks on square lattice. Ref GU90. [4,2; A5562]

M3975 0, 5, 35, 189, 924, 4290, 19305
Coefficients for extrapolation. Ref SE33 93. [0,2; A2737, N1644]

M3976 1, 5, 35, 225, 67375, 66693, 955040625, 1861234375
Denominators of coefficients in an asymptotic expansion. Cf. M2268. Ref JACM 3 14 56.
[0,2; A2074, N1645]

M3977 1, 1, 5, 35, 285, 2530, 23751, 231880, 2330445, 23950355, 250543370,
2658968130, 28558343775, 309831575760, 3390416787880, 37377257159280
$C(5n,n)/(4n+1)$. Ref AMP 1 198 1841. DM 11 388 75. [0,3; A2294, N1646]

M3978 1, 5, 35, 294, 2772, 28314, 306735, 3476330, 40831076, 493684828, 6114096716,
77266057400, 993420738000, 12964140630900, 171393565105575, 2291968851019650
Hamiltonian rooted maps with $2n$ nodes: $(2n)!(2n+1)!/n!^2(n+1)!(n+2)!$. Ref CJM
14 416 62. [1,2; A0356, N1647]

M3979 1, 1, 5, 35, 315, 3455, 44590, 660665, 11035095, 204904830, 4183174520,
93055783320, 2238954627848, 57903797748386, 1601122732128779
Coefficients of iterated exponentials. Ref SMA 11 353 45. PRV A32 2342 85. [0,3; A0357,
N1648]

M3980 5, 35, 1260, 4620, 30030, 90090, 1021020, 2771340, 14549535, 37182145,
1487285800, 3650610600, 17644617900, 42075627300, 396713057400
Coefficients of Legendre polynomials. Ref PR33 156. AS1 798. [0,1; A1802, N1649]

M3981 5, 35, 2266, 30564722
Switching networks. Ref JFI 276 324 63. [1,1; A0871, N1650]

M3982 0, 1, 5, 36, 329, 3655, 47844, 721315, 12310199, 234615096, 4939227215,
113836841041, 2850860253240, 77087063678521, 2238375706930349
$a(n)=(2n+1)a(n-1)+a(n-2)$. Ref CJM 8 308 56. [0,3; A0806, N1651]

M3983 1, 5, 36, 3406, 14694817, 727050997716715,
2074744506784679417243551677046
Continued cotangent for square root of 2. Ref DUMJ 4 339 38. jos. [0,2; A2666, N1652]

M3984 5, 37, 150, 449, 1113, 2422, 4788, 8790, 15213, 25091, 39754, 60879
Rooted planar maps. Ref JCT B18 249 75. [1,1; A6468]

M3985 1, 5, 37, 353, 4081, 55205, 854197, 14876033, 288018721, 6138913925,
142882295557, 3606682364513, 98158402127761, 2865624738913445
$a(n) = n(2n-1)!! - \Sigma a(k)(2n-2-k)!!$. [1,2; A4208]

M3986 1, 1, 5, 37, 457, 8169, 188685, 5497741, 197920145, 8541537105, 432381471509,
25340238127989, 1699894200469849, 129076687233903673, 10989863562589199389
Expansion of $\cos(\sin x)$. [0,3; A3709]

M3987 1, 5, 37, 782, 44240
Quartering a $2n \times 2n$ chessboard. See Fig M3987. Bisection of M3769. Ref PC 1 7-1 73. GA69 189. trp. [1,2; A3213]

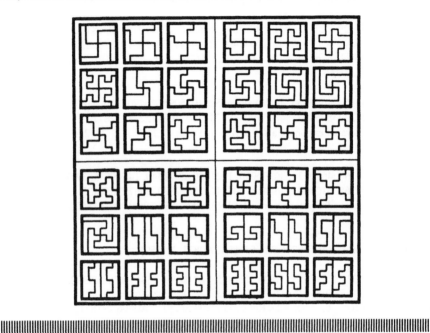

Figure M3987. QUARTERING A CHESSBOARD.

How many ways are there to dissect an $n \times n$ board into four congruent pieces? (If n is odd, omit the center square.) This is M3769, and the even-numbered terms form M3987. Not many terms are known. Here is the 6th term of M3769 and the third term of M3987:

M3988 0, 5, 39, 272, 1869, 12815, 87840, 602069, 4126647, 28284464, 193864605, 1328767775, 9107509824, 62423800997, 427859097159, 2932589879120
$a(n) = 7a(n-1) - a(n-2) + 4$. Ref DM 9 89 74. [0,2; A3482]

M3989 1, 5, 40, 260, 1820, 12376, 85085, 582505, 3994320, 27372840, 187628376, 1285992240, 8814405145, 60414613805, 41408893560, 2838203264876
Fibonomial coefficients. Ref FQ 6 82 68. BR72 74. [0,2; A1656, N1653]

M3990 0, 0, 0, 0, 0, 0, 0, 0, 0, 1, 5, 40, 343, 2979
1-supertough but non-1-Hamiltonian simplicial polyhedra with n nodes. Ref Dil92. [1,11; A7036]

M3991 1, 1, 5, 40, 440, 6170, 105315, 2120610, 49242470, 1296133195, 38152216495, 1242274374380, 44345089721923, 1722416374173854, 72330102999829054
Coefficients of iterated exponentials. Ref SMA 11 353 45. [0,3; A0359, N1654]

M3992   1, 5, 40, 644, 21496, 1471460
Certain subgraphs of a directed graph. Ref DM 14 119 76. [2,2; A5330]

M3993   1, 5, 40, 801, 46821, 9185102, 6163297995, 14339791693249
Related to number of digraphs. Ref HP73 124. [1,2; A3084]

M3994   1, 5, 41, 73, 193, 1181, 6481, 16493, 21523361, 530713, 42521761, 570461, 769,
    4795973261, 647753, 47763361, 926510094425921, 1743831169, 282429005041
Largest factor of $9^n + 1$. Ref Krai24 2 89. CUNN. [0,2; A2592, N1655]

M3995   1, 5, 41, 545, 11681, 402305, 22207361, 1961396225, 276825510401,
    62368881977345
3-colored labeled graphs on n nodes. Ref CJM 12 413 60. rcr. [1,2; A0685, N1656]

M3996   1, 5, 41, 685, 19921, 887765, 56126201, 4776869245, 526589630881,
    72989204937125, 12424192360405961, 2547879762929443405
Expansion of $\tan x \cdot \cosh x$. [0,2; A3719]

M3997   1, 5, 42, 462, 6006, 87516, 1385670, 23371634, 414315330, 7646001090
3-dimensional Catalan numbers. Ref CN 75 124 90. [1,2; A5789]

M3998   1, 5, 43, 557, 10075
Values of Gandhi polynomials. Ref DUMJ 41 308 74. [0,2; A5989]

M3999   1, 5, 45, 385, 3710, 38934, 444990, 5506710, 73422855, 1049946755,
    16035550531, 260577696015
Permutations of length n by rises. Ref DKB 263. [5,2; A1260, N1657]

$$(n - 1)\, a(n) = (n + 3)\, (a(n - 1)\, n + a(n - 2)\, n - a(n - 1) + 2\, a(n - 2)).$$

M4000   5, 45, 420, 4130, 42480, 453350, 4986860, 56251230, 648055650, 7601584050,
    90556803600
Hamiltonian rooted triangulations with n internal nodes. Ref DM 6 167 73. [0,1; A5979]

M4001   1, 1, 5, 45, 585, 9945, 208845, 5221125, 151412625, 4996616625, 184874815125,
    7579867420125, 341094033905625, 16713607661375625, 885821206052908125
Expansion of $(1 - 4x)^{-1/4}$. [0,3; A7696]

M4002   5, 46, 19930, 69945183326
Switching networks. Ref JFI 276 324 63. [1,1; A0872, N1658]

M4003   1, 5, 47, 641, 11389, 248749, 6439075, 192621953, 6536413529, 248040482741,
    10407123510871, 478360626529345, 23903857657114837, 1290205338991689821
$(2n)!\ \Sigma(-1)^k C(n,k)\ /\ (n+k)!$. Ref CN 33 80 81. [1,2; A6902]

M4004   1, 5, 49, 205, 5269, 5369, 266681, 1077749, 9778141, 1968329, 239437889,
    240505109, 40799043101, 40931552621, 205234915681, 822968714749
Numerators of Σk^{-2}; $k = 1..n$. Cf. M3661. Ref KaWa 89. [1,2; A7406]

M4005   1, 5, 49, 485, 4801, 47525, 470449, 4656965, 46099201, 456335045, 4517251249,
   44716177445, 442644523201, 4381729054565, 43374646022449, 429364731169925
$a(n) = 10a(n-1) - a(n-2)$. Ref EUL (1) 1 374 11. TH52 281. [0,2; A1079, N1659]

M4006   1, 5, 49, 725, 14641, 300125, 20134393, 282300416
Minimal discriminant of totally real number field of degree n. Ref Hass80 617. STNB 2
133 90. [1,2; A6554]

M4007   1, 5, 49, 809, 20317, 722813, 34607305, 2145998417, 167317266613,
   16020403322021, 1848020950359841, 252778977216700025, 40453941942593304589
Glaisher's G numbers. Ref PLMS 31 224 1899. FMR 1 76. [1,2; A2111, N1660]

M4008   1, 5, 49, 820, 21076, 773136, 38402064, 2483133696, 202759531776,
   20407635072000, 2482492033152000, 359072203696128000, 60912644957448192000
Central factorial numbers. Ref RCI 217. [0,2; A1819, N1661]

M4009   0, 1, 1, 0, 5, 51, 3634, 374119, 73161880, 26545249985, 17904840957826,
   22602069719494379, 53938847227326533032, 24610794547947272758874483
Series-reduced connected labeled graphs with n nodes. Ref JCT B19 282 75. [0,5; A3515]

M4010   1, 5, 52, 1522, 145984, 48464496, 56141454464, 229148550030864,
   3333310786076963968, 174695272746749919580928
Relations on n nodes. Cf. M1980. Ref PAMS 4 494 53. MIT 17 19 55. MAN 174 66 67
(divided by 2). [1,2; A1173, N1662]

M4011   5, 53, 157, 173, 211, 257, 263, 373, 563, 593, 607, 653, 733, 947, 977, 1103, 1123,
   1187, 1223, 1367, 1511, 1747, 1753, 1907, 2287, 2417, 2677, 2903, 2963, 3307, 3313
Primes which are average of their neighbors. Ref AS1 870. [1,1; A6562]

M4012   5, 53, 173, 173, 293, 2477, 9173, 9173, 61613, 74093, 74093, 74093, 170957,
   360293, 679733, 2004917, 2004917, 69009533, 138473837, 237536213, 32426677
Sequence of prescribed quadratic character. Ref MOC 24 449 70. [3,1; A1992, N1663]

M4013   1, 1, 5, 55, 1001, 26026, 884884, 37119160, 1844536720, 105408179176,
   6774025632340
Dyck paths. Ref SC83. [0,3; A6150]

M4014   5, 56, 580, 5894, 60312
Closed meanders with 3 components. See Fig M4587. Ref SFCA91 292. [3,1; A6658]

M4015   1, 5, 56, 945, 28569, 1421360
Pseudo-bricks with n nodes. Ref JCT B32 29 82. [4,2; A6293]

M4016   5, 57, 352, 1280, 3522, 7970, 15872, 29184, 49410, 79042
Generalized class numbers. Ref MOC 21 689 67. [1,1; A0362, N1664]

M4017   1, 5, 58, 1274, 41728, 1912112, 116346400, 9059742176, 877746364288
Related to Latin rectangles. Ref BCMS 33 125 41. [2,2; A1624, N1665]

M4018 5, 61, 479, 3111, 18270, 101166, 540242, 2819266, 14494859, 73802835, 373398489, 1881341265
Permutations of length n by number of runs. Ref DKB 260. [4,1; A0363, N1666]

M4019 1, 1, 5, 61, 1385, 50521, 2702765, 199360981, 19391512145, 2404879675441, 370371188237525, 69348874393137901, 15514534163557086905
Euler numbers: expansion of sec x. See Fig M4019. Ref AS1 810. MOC 21 675 67. [0,3; A0364, N1667]

||||||||||||||||||||||||||||||||||||||||||||||||||||||||||||||||||||||||||||||||||||||||||||||||||||||||||||||||||||||||||||||||||||||||

Figure M4019.   EULER, TANGENT, GENOCCHI NUMBERS.

The **Euler** numbers E_n, M1492, give the number of permutations of n objects which first rise and then alternately fall and rise. Only the second rows of the permutations are shown.

$E_1 = 1$ 1

$E_2 = 1$ 1 2

$E_3 = 2$ 1 3 2   2 3 1

$E_4 = 5$ 1 3 2 4   1 4 2 3   2 3 1 4   2 4 1 3   3 4 1 2

$E_5 = 16$ 1 3 2 5 4   1 4 2 5 3   1 4 3 5 2   1 5 2 4 3
             1 5 3 4 2   2 3 1 5 4   2 4 1 5 3   2 4 3 5 1
             2 5 1 4 3   2 5 3 4 1   3 4 1 5 2   3 4 2 5 1
             3 5 1 4 2   3 5 2 4 1   4 5 1 3 2   4 5 2 3 1

The even-numbered Euler numbers E_{2n} form M4019, and have generating function

$$\sec x = 1 + 1\frac{x^2}{2!} + 5\frac{x^4}{4!} + 61\frac{x^6}{6!} + \cdots$$

(Sometimes these are called Euler numbers instead of M1492.)

The odd-numbered Euler numbers are the **tangent** numbers, $T_n = E_{2n-1}$, M2096 and have generating function

$$\tan x = x + 2\frac{x^3}{3!} + \frac{16x^5}{5!} + 272\frac{x^7}{7!} + \cdots .$$

The Bernoulli numbers of Fig. M4189 are related to the Euler numbers by

$$B_n = \frac{2nE_{2n-1}}{2^{2n}(2^{2n} - 1)} .$$

Also related are the **Genocchi** numbers $G_n = 2^{2-2n}nE_{2n-1}$, M3041, with generating function

$$\tan\frac{x}{2} = 1\frac{x}{2!} + 1\frac{x^3}{4!} + 3\frac{x^5}{6!} + 17\frac{x^7}{8!} + \cdots .$$

References: [Jo39], [DB1], [C1], [GKP].

||||||||||||||||||||||||||||||||||||||||||||||||||||||||||||||||||||||||||||||||||||||||||||||||||||||||||||||||||||||||||||||||||||||||

M4020   1, 5, 73, 1445, 33001, 819005, 21460825, 584307365, 16367912425,
    468690849005, 13657436403073, 403676083788125, 12073365010564729
Apéry numbers: $\Sigma\,(C(n,k)\,C(n+k,k))^2$, $k = 0 \ldots n$. Ref AST 61 13 79. Ape81. JNT 25
201 87. [0,2; A5259]

M4021   5, 84, 650, 3324, 13020, 42240, 118998, 300300, 693693, 1490060, 3011580
Rooted planar maps. Ref JCT B18 257 75. [1,1; A6471]

M4022   5, 93, 1030, 8885, 65954, 442610, 2762412, 16322085, 92400330, 505403910,
    2687477780, 13957496098
Rooted planar maps with n edges. Ref BAMS 74 74 68. WA71. JCT A13 215 72. [3,1;
A0365, N1669]

M4023   1, 1, 5, 103, 329891, 36846277, 1230752346353, 336967037143579,
    48869596859895986087, 105133911935073745000051862069
Wilson quotients: $((p-1)! + 1)/p$. Ref BPNR 277. [2,3; A7619]

M4024   1, 5, 109, 32297, 2147321017, 9223372023970362989,
    170141183460469231667123699502996689125
Covers of an n-set. Ref C1 165. CN 8 515 73. DM 5 247 73. MMAG 67 143 94. [1,2;
A3465]

M4025   5, 111, 5232, 49910, 3527745, 76435695, 2673350008, 33507517680,
    4954123399050
Coefficients for step-by-step integration. Ref JACM 11 231 64. [2,1; A2400, N1670]

M4026   0, 5, 117, 535, 1463, 3105, 5665, 9347, 14355, 20893, 29165, 39375, 51727,
    66425, 83673, 103675, 126635, 152757, 182245, 215303, 252135, 292945, 337937
From continued fraction for $\zeta(3)$. Ref LNM 751 68 79. [0,2; A6221]

M4027   5, 120, 1840, 27552, 421248, 6613504, 106441472, 1750927872
Almost trivalent maps. Ref PLC 1 292 70. [0,1; A2008, N1671]

M4028   5, 140, 2744420, 20670535451567121260,
    883192109405810771118595679733598486261240651506783773980
A continued cotangent. Ref NBS B80 288 76. [0,1; A6269]

M4029   1, 5, 205, 22265, 4544185, 1491632525, 718181418565, 476768795646785,
    417370516232719345, 465849831125196593045, 645702241048404020542525
Multiples of Euler numbers. Ref MES 28 51 1898. FMR 1 75. hpr. [1,2; A2438, N1672]

M4030   1, 5, 205, 36317, 23679901, 56294206205, 502757743028605,
    17309316971673776957, 233350840061464687473462l
2-colored graphs. Ref CJM 31 66 79. [1,2; A5333]

M4031   5, 210, 3150, 27556, 170793, 829920, 3359356, 11786190, 36845718
Nonseparable planar tree-rooted maps. Ref JCT B18 243 75. [1,1; A6413]

M4032 5, 229, 401, 577, 1129, 1297, 7057, 8761, 14401, 32401, 41617, 57601, 90001
Primes $p \equiv 1$ (mod 4) where class number of $Q(\sqrt{p})$ increases. Ref MOC 23 214 69. [1,1; A2142, N1673]

M4033 1, 5, 253, 39299, 13265939
Coefficients of lemniscate function. Ref HUR 2 372. [2,2; A2770, N1675]

M4034 1, 5, 259, 3229, 117469, 7156487, 2430898831, 60967921, 141433003757, 25587296781661
Numerators of coefficients for numerical differentiation. Cf. M5177. Ref OP80 23. PHM 33 14 42. [1,2; A2554, N1676]

M4035 1, 5, 357, 376805, 6221613541, 1634141006295525, 6857430062381149327845, 4602505140835762067965487 72325, 4942053077477465038530751310018 23990245
Gaussian binomial coefficient $[2n,n]$ for $q=4$. Ref GJ83 99. ARS A17 328 84. [0,2; A6108]

M4036 0, 5, 360, 7350, 73700, 474588, 2292790, 9046807, 30676440, 92393015
Tree-rooted planar maps. Ref JCT B18 256 75. [1,2; A6430]

M4037 5, 365, 35645, 3492725, 342251285, 33537133085, 3286296790925, 322023548377445, 31555021444198565, 3092070077983031805
Both the sum of 2 and of 3 consecutive squares. Ref GA88 22. [0,1; A7667]

M4038 1, 5, 393, 131473, 117316993, 219639324573
An occupancy problem. Ref JACM 24 593 77. [0,2; A6700]

M4039 1, 1, 1, 1, 1, 5, 691, 7, 3617, 43867, 174611, 854513, 236364091, 8553103, 23749461029, 8615841276005, 7709321041217, 2577687858367
Numerators of Bernoulli numbers B_{2n}. See Fig M4189. Cf. M4189. Ref DA63 2 230. AS1 810. [0,6; A0367, N1677]

SEQUENCES BEGINNING . . ., 6, . . .

M4040 0, 0, 0, 6, 0, 0, 0, 0, 0, 6, 0, 6, 0, 0, 0, 12, 0, 6, 0, 0, 0, 0, 0, 12, 0, 0, 0, 18, 0, 0, 0, 0, 0, 12, 0, 12, 0, 0, 0, 24, 0, 6, 0, 0, 0, 0, 0, 12, 0, 0, 0, 24, 0, 0, 0, 0, 0, 24, 0, 6, 0, 0, 0, 36, 0
Theta series of h.c.p. w.r.t. octahedral hole. Ref JCP 83 6531 85. [0,4; A5872]

M4041 6, 0, 0, 21, 60, 90, 182, 378, 861, 1737, 3458, 6717, 13377, 25877, 49949, 95085, 180254, 338003, 631124, 1168226, 2151409, 3934674, 7159108, 12948649, 23307439
Solid partitions. Ref PNISI 26 135 60. [3,1; A2044, N2214]

M4042 1, 6, 0, 6, 6, 0, 0, 12, 0, 6, 0, 0, 6, 12, 0, 0, 6, 0, 0, 12, 0, 12, 0, 0, 0, 6, 0, 6, 12, 0, 0, 12, 0, 0, 0, 0, 6, 12, 0, 12, 0, 0, 0, 12, 0, 0, 0, 0, 6, 18, 0, 0, 12, 0, 0, 0, 0, 12, 0, 0, 0, 12, 0
Theta series of planar hexagonal lattice. See Fig M2336. Ref SPLAG 111. [0,2; A4016]

M4043 0, 6, 0, 30, 24, 168, 288, 1170, 2760
$2n$-step polygons on honeycomb. Ref PRV 114 53 59. [2,2; A5396]

M4044 1, 6, 1, 2, 1, 1, 1, 3, 25, 1, 4, 3, 3, 7, 52, 1, 2, 3, 2, 15, 2, 2, 4, 16, 2, 7, 1, 1, 1, 10,
21, 1, 1, 1, 141, 2, 4, 1, 4, 2, 1, 1, 17, 1, 3, 3, 4, 1, 3, 1, 3, 2, 1, 1, 2, 33, 1, 6, 6, 1, 2, 4, 1
Continued fraction for fifth root of 2. [1,2; A2950]

M4045 1, 6, 1, 2, 6, 16, 18, 6, 22, 3, 28, 15, 2, 3, 6, 5, 21, 46, 42, 16, 13, 18, 58, 60, 6, 33,
22, 35, 8, 6, 13, 9, 41, 28, 44, 6, 15, 96, 2, 4, 34, 53, 108, 3, 112, 6, 48, 22, 5, 42, 21, 130
Periods of reciprocals of integers prime to 10. Ref PCPS 3 204 1878. L1 12. [3,2; A2329,
N1678]

M4046 1, 6, 1, 8, 0, 3, 3, 9, 8, 8, 7, 4, 9, 8, 9, 4, 8, 4, 8, 2, 0, 4, 5, 8, 6, 8, 3, 4, 3, 6, 5, 6, 3,
8, 1, 1, 7, 7, 2, 0, 3, 0, 9, 1, 7, 9, 8, 0, 5, 7, 6, 2, 8, 6, 2, 1, 3, 5, 4, 4, 8, 6, 2, 2, 7, 0, 5, 2, 6
Decimal expansion of golden ratio $\tau = (1 + \sqrt{5})/2$. Ref FQ 4 161 66. [1,2; A1622, N1679]

M4047 1, 6, 2, 5, 2, 4, 2, 7, 7, 4, 7, 4, 7, 6, 3, 4, 3, 9, 3, 9, 3, 9, 3, 11, 6, 6, 6, 9, 6, 6, 6, 8, 6,
8, 3, 18, 3, 14, 3, 5, 3, 6, 3, 6, 3, 6, 3, 11, 5, 11, 5, 11, 5, 11, 5, 5, 5, 11, 5, 11, 5, 5, 3, 5, 3
Time for juggler sequence starting at n to converge. Ref Pick91 232. [2,2; A7320]

M4048 1, 6, 2, 5, 2, 13, 7, 10, 7, 4, 7, 6, 3, 9, 3, 9, 3, 12, 3, 9, 6, 9, 6, 19, 6, 9, 6, 9, 6, 16, 3,
5, 3, 8, 3, 16, 3, 5, 3, 14, 3, 11, 14, 11, 14, 5, 14, 14, 14, 14, 14, 5, 14, 5, 14, 11, 8, 11, 8, 8
Time for juggler sequence starting at n to converge. Ref Pick91 233. [2,2; A7321]

M4049 6, 2, 5, 5, 4, 5, 6, 3, 7, 6, 8, 4, 7, 7, 6, 7, 8, 5, 9, 8, 11, 7, 10, 10, 9, 10, 11, 8, 12, 11,
11, 7, 10, 10, 9, 10, 11, 8, 12, 11, 10, 6, 9, 9, 8, 9, 10, 7, 11, 10, 11, 7, 10, 10, 9, 10, 11, 8
Segments to represent n on calculator display. [0,1; A6942]

M4050 1, 1, 1, 6, 2, 6, 16, 18, 22, 28, 15, 3, 5, 21, 46, 13, 58, 60, 33, 35, 8, 13, 41, 44, 96,
4, 34, 53, 108, 112, 42, 130, 8, 46, 148, 75, 78, 81, 166, 43, 178, 180, 95, 192, 98, 99, 30
Periods of reciprocals of primes. Ref PRS A22 203 1874. L1 15. [2,4; A2371, N1680]

M4051 0, 1, 6, 3, 82, 84, 444, 769, 1110, 2643, 860, 2901, 1176, 6277, 1170, 21315, 2308,
14244, 29442, 15540, 58194, 13338, 31886, 4080, 176682, 70715, 51240
Related to representation as sums of squares. Ref QJMA 38 312 07. [0,3; A2610, N1681]

M4052 6, 4, 3, 12, 8, 12, 30, 20, 30, 72, 46, 60, 156, 96, 117, 300, 188, 228, 552, 344, 420,
1008, 603, 732, 1770, 1048, 1245, 2976, 1776, 2088, 4908, 2900, 3420, 7992, 4658, 5460
McKay-Thompson series of class 6E for Monster. Ref PLMS 9 384 59. CALG 18 257 90.
FMN94. [1,1; A7258]

M4053 0, 6, 4, 12, 20, 42, 32, 40, 48, 78, 84, 116, 148, 210, 176, 176, 176, 214, 212, 252,
292, 378, 368, 408, 448, 542, 580, 676, 772, 930, 832, 800, 768, 806, 772, 812, 852, 970
$\Sigma\ k$ XOR $n - k$, $k = 1 . . n - 1$. Ref mlb. [2,2; A6582]

M4054 6, 6, 0, 0, 8, 27, 78, 45, 21, 62, 0, 584, 903, 155, 66, 998, 2357, 9803, 8273
Percolation series for f.c.c. lattice. Ref SSP 10 921 77. [0,1; A6806]

M4055 1, 0, 0, 0, 0, 0, 1, 0, 0, 0, 0, 6, 6, 0, 0, 8, 42, 114, 66, 24, 123, 134
Partition function for f.c.c. lattice. Ref AIP 9 279 60. [0,12; A2892, N1682]

M4056 6, 6, 0, 1, 6, 1, 8, 1, 5, 8, 4, 6, 8, 6, 9, 5, 7, 3, 9, 2, 7, 8, 1, 2, 1, 1, 0, 0, 1, 4, 5, 5, 5, 7, 7, 8, 4, 3, 2, 6, 2, 3
Decimal expansion of twin prime constant. Ref MOC 15 398 61. [0,1; A5597]

M4057 6, 6, 2, 6, 0, 7, 5, 5
Decimal expansion of Planck constant (joule sec). Ref FiFi87. Lang91. [-33,1; A3676]

M4058 1, 6, 6, 4, 6, 12, 28, 72, 198, 572, 1716, 5304, 16796, 54264, 178296, 594320, 2005830, 6843420, 23571780, 81880920, 286583220, 1009864680, 3580429320
Expansion of $(1 - 4x)^{3/2}$. Ref TH09 164. FMR 1 55. [0,2; A2421, N1683]

M4059 1, 6, 7, 4, 5, 26, 27, 24, 25, 30, 31, 28, 29, 18, 19, 16, 17, 22, 23, 20, 21, 106, 107, 104, 105, 110, 111, 108, 109, 98, 99, 96, 97, 102, 103, 100, 101, 122, 123, 120, 121, 126
Base -2 representation for n read as binary number. Ref GA86 101. [0,2; A5351]

M4060 0, 1, 6, 7, 10, 11, 12, 13, 18, 19, 20, 21, 24
Even number of 1's in binary, ignoring last bit. Ref PSAM 43 44 91. [1,3; A6364]

M4061 1, 1, 6, 7, 18, 29, 59, 92, 171, 267, 457, 709, 1155, 1763
Representation degeneracies for Neveu-Schwarz strings. Ref NUPH B274 547 86. [2,3; A5302]

M4062 1, 6, 7, 20, 27, 47, 74, 269, 6799, 7068, 35071, 112281, 371914, 2715679, 141587222, 144302901, 430193024, 1434881973, 3299956970, 50934236523
Convergents to fifth root of 2. Ref AMP 46 115 1866. L1 67. hpr. [1,2; A2361, N1684]

M4063 1, 0, 6, 7, 28, 54, 135, 286, 627, 1313, 2730, 5565, 11212, 22304, 43911, 85614, 165490, 317373, 604296, 1143054, 2149074, 4017950, 7473180, 13832910, 25490115
Generalized Lucas numbers. Ref FQ 15 252 77. [3,3; A6493]

M4064 1, 6, 8, 4, 10, 12, 14, 15, 9, 18, 22, 20, 26, 21, 24, 16, 34, 27, 38, 30, 28, 33, 46, 32, 25, 39, 35, 40, 58, 42, 62, 45, 44, 51, 48, 36, 74, 57, 52, 50, 82, 56, 86, 55, 60, 69, 94, 54
Ron's sequence. Ref MMAG 60 180 87. GKP 147. [1,2; A6255]

M4065 1, 6, 8, 10, 12, 14, 15, 18, 20, 21, 22, 26, 27, 28, 32, 33, 34, 35, 36, 38, 39, 44, 45, 46, 48, 50, 51, 52, 55, 57, 58, 62, 63, 64, 65, 68, 69, 74, 75, 77, 80, 82, 85, 86, 87, 91, 92
A 2-way classification of integers. Cf. M0520. Ref CMB 2 89 59. Robe92 22. [1,2; A0379, N1685]

M4066 1, 6, 8, 10, 12, 15, 17, 19, 24, 26, 28, 33, 35, 37, 42, 44, 46, 51, 53, 55, 60, 62, 64, 69, 71, 73, 78, 80, 82, 87, 89, 91, 96, 98, 100, 105, 107, 109, 114, 116, 118, 123, 125, 127
$a(n)$ is smallest number $\neq a(j) + a(k)$, $j < k$. Ref AMM 99 671 92. GU94. [1,2; A3663]

M4067 6, 8, 10, 12, 16, 18, 20, 24, 26, 33, 32, 36, 42, 46, 48, 50, 52, 53, 60, 66, 68, 74, 78, 82, 90, 92, 97, 100, 104, 106, 114, 118, 120, 126, 136, 140, 144, 148, 150, 156, 166, 170
Rational points on curves of genus 2 over $GF(q)$. Ref CRP 296 398 83. HW84 51. [2,1; A5525]

M4068 1, 6, 8, 10, 14, 15, 21, 22, 26, 27, 33, 34, 35, 38, 39, 46, 51, 55, 57, 58, 62, 65, 69,
74, 77, 82, 85, 87, 91, 93, 95, 111, 115, 119, 123, 125, 133, 143, 145, 155, 161, 185, 187
Multiplicatively perfect numbers: product of divisors is n^2. Ref IrRo82 19. [1,2; A7422]

M4069 1, 6, 8, 16, 25, 42, 44, 56, 69, 94, 108, 136, 165, 210, 208, 224, 241, 278, 296, 336,
377, 442, 460, 504, 549, 622, 668, 744, 821, 930, 912, 928, 945, 998, 1016, 1072, 1129
$\Sigma\, k$ OR $n-k$, $k = 1 \ldots n-1$. Ref mlb. [2,2; A6583]

M4070 6, 8, 24, 0, 30, 24, 24, 0, 48, 24, 48, 0, 30, 32, 72, 0, 48, 48, 24, 0, 96, 24, 72, 0, 54,
48, 72, 0, 48, 72, 72, 0, 96, 24, 96, 0, 48, 56, 96, 0, 102, 72, 48, 0, 144, 48, 48, 0, 48, 72
Theta series of f.c.c. lattice w.r.t. octahedral hole. Ref JCP 83 6527 85. [0,1; A5887]

M4071 6, 8, 40, 176, 1421, 10352, 93114, 912920, 9929997, 117970704, 1521176826,
21150414880, 315400444070, 5020920314016, 84979755347122
Discordant permutations. Ref SMA 20 23 54. [3,1; A0380, N1686]

M4072 6, 8, 180, 32, 10080, 3456, 453600, 115200, 47900160, 71680, 217945728000,
36578304000, 2241727488000, 45984153600, 2000741783040000
Denominators of coefficients for repeated integration. Cf. M5066. Ref PHM 38 336 47.
[1,1; A2689, N1687]

M4073 1, 6, 8, 262, 2448, 17997702, 44082372248, 5829766629386380698502,
256989942683351711945337288361248
A simple recurrence. Ref MMAG 37 167 64. [1,2; A0955, N1688]

M4074 6, 9, 3, 1, 4, 7, 1, 8, 0, 5, 5, 9, 9, 4, 5, 3, 0, 9, 4, 1, 7, 2, 3, 2, 1, 2, 1, 4, 5, 8, 1, 7, 6,
5, 6, 8, 0, 7, 5, 5, 0, 0, 1, 3, 4, 3, 6, 0, 2, 5, 5, 2, 5, 4, 1, 2, 0, 6, 8, 0, 0, 0, 9, 4, 9, 3, 3, 9
Decimal expansion of natural logarithm of 2. Ref MOC 17 177 63. [0,1; A2162, N1689]

M4075 1, 6, 9, 4, 6, 54, 40, 168, 81, 36, 564, 36, 638, 240, 54, 1136, 882, 486, 556, 24,
360, 3384, 840, 1512, 3089, 3828, 729, 160, 4638, 324, 4400, 1440, 5076, 5292, 240, 324
Expansion of 6-dimensional cusp form. Ref SPLAG 204. [1,2; A7332]

$$\text{G.f.:} \quad x \prod (1-x^k)^6 (1-x^{3k})^6.$$

M4076 1, 6, 9, 10, 30, 0, 11, 42, 0, 70, 18, 54, 49, 90, 0, 22, 60, 0, 110, 0, 81, 180, 78, 0,
130, 198, 0, 182, 30, 90, 121, 84, 0, 0, 210, 0, 252, 102, 270, 170, 0, 0, 69, 330, 0, 38, 420
Expansion of $\prod(1-x^k)^6$. Ref KNAW 59 207 56. [0,2; A0729, N1691]

M4077 1, 6, 9, 13, 19, 37, 58, 97, 143, 227, 328, 492, 688, 992, 1364, 1903, 2551, 3473,
4586, 6097, 7911, 10333, 13226, 16988, 21454, 27172, 33938, 42437, 52423, 64833
A generalized partition function. Ref PNISI 17 237 51. [1,2; A2598, N1693]

M4078 1, 6, 9, 16, 66, 54, 98, 300, 243, 364, 1128, 828, 1221
McKay-Thompson series of class 6c for Monster. Ref FMN94. [0,2; A7262]

M4079 1, 6, 9, 22, 40, 43, 48, 56, 61, 64
Related to representations as sums of Fibonacci numbers. Ref FQ 11 357 73. [1,2; A6132]

M4080 6, 9, 27, 45, 45, 57, 75, 81, 87, 105, 123, 135, 135, 165, 169, 189, 195, 209, 231,
237, 267, 267, 267, 315, 315, 333, 345, 363, 369, 405, 411, 429, 441, 465, 483, 485, 525
Largest number not a sum of distinct primes $\geq p$, p prime. Ref NMT 21 138 73. Robe92 73.
[2,1; A7414]

M4081 1, 6, 10, 13, 15, 16, 16, 18
Pair-coverings with largest block size 5. Ref ARS 11 90 81. [5,2; A6187]

M4082 6, 10, 14, 15, 21, 22, 26, 33, 34, 35, 38, 39, 46, 51, 55, 57, 58, 62, 65, 69, 74, 77,
82, 85, 86, 87, 91, 93, 94, 95, 106, 111, 115, 118, 119, 122, 123, 129, 133, 134, 141, 142
Products of two distinct primes. [1,1; A6881]

M4083 6, 10, 14, 19, 25, 30, 36, 43, 51, 57
Zarankiewicz's problem. Ref LNM 110 142 69. [2,1; A6617]

M4084 6, 10, 15, 20, 21, 28, 35, 36, 45, 55, 56, 66, 70, 78, 84, 91, 105, 120, 126, 136, 153,
165, 171, 190, 210, 220, 231, 252, 253, 276, 286, 300, 325, 330, 351, 364, 378, 406, 435
Nontrivial binomial coefficients $C(n,k)$, $1 < k < n - 1$. See Fig M1645. [1,1; A6987]

M4085 6, 10, 22, 34, 48, 60, 78, 84, 90, 114, 144, 120, 168, 180, 234, 246, 288, 240, 210,
324, 300, 360, 474, 330, 528, 576, 390, 462, 480, 420, 570, 510, 672, 792, 756, 876
Smallest even number which is sum of two odd primes in n ways. Ref ew. [1,1; A1172,
N1694]

M4086 1, 6, 11, 17, 22, 27, 32, 37, 43, 48, 53, 58, 64, 69, 74, 79, 85, 90, 95, 100, 106, 111,
116, 121, 126, 132, 137, 142, 147, 153, 158, 163, 168, 174, 179, 184, 189, 195, 200, 205
Wythoff game. Ref CMB 2 189 59. [0,2; A1964, N1695]

M4087 6, 11, 20, 36, 65, 119, 218, 400, 735, 1351, 2484, 4568, 8401, 15451, 28418,
52268, 96135, 176819, 325220, 598172, 1100209, 2023599, 3721978, 6845784
Restricted permutations. Ref CMB 4 32 61 (divided by 4). [4,1; A0382, N1696]

M4088 1, 1, 1, 1, 1, 1, 6, 11, 21, 41, 81, 161, 321, 636, 1261, 2501, 4961, 9841, 19521,
38721, 76806, 152351, 302201, 599441, 1189041, 2358561, 4678401, 9279996
Hexanacci numbers. Ref FQ 2 302 64. [0,7; A0383, N1697]

M4089 1, 6, 11, 27, 27, 66, 51, 112, 102, 162, 123, 297, 171, 306, 297, 453, 291, 612, 363,
729, 561, 738, 531, 1232, 678, 1026, 922, 1377, 843, 1782, 963, 1818, 1353, 1746, 1377
Inverse Moebius transform applied twice to squares. Ref EIS § 2.7. [1,2; A7433]

M4090 1, 1, 6, 11, 36, 85, 235, 600, 1590, 4140, 10866, 28416, 74431, 194821, 510096,
1335395, 3496170, 9153025, 23963005, 62735880
From a definite integral. Ref PEMS 10 184 57. [1,3; A2570, N1698]

M4091   1, 6, 11, 71, 4691, 21982031, 483209576974811,
233491495280173380882643611671
A nonlinear recurrence. Ref AMM 70 403 63. FQ 11 431 73. [0,2; A1543, N1699]

M4092   1, 6, 12, 8, 6, 24, 24, 0, 12, 30, 24, 24, 8, 24, 48, 0, 6, 48, 36, 24, 24, 48, 24, 0, 24,
30, 72, 32, 0, 72, 48, 0, 12, 48, 48, 48, 30, 24, 72, 0, 24, 96, 48, 24, 24, 72, 48, 0, 8, 54, 84
Theta series of cubic lattice. Ref SPLAG 107. [0,2; A5875]

M4093   6, 12, 15, 18, 20, 24, 28, 30, 35, 36, 40, 42, 45, 48, 54, 56, 60, 63, 66, 70, 72, 75,
77, 78, 80, 84, 88, 90, 91, 96, 99, 100, 102, 104, 105, 108, 110, 112, 114, 117, 120, 126
Numbers having divisors d, e with $d < e < 2d$. Ref UPNT E3. [1,1; A5279]

M4094   6, 12, 18, 20, 24, 28, 30, 36, 40, 42, 48, 54, 56, 60, 66, 72, 78, 80, 84, 88, 90, 96,
100, 102, 104, 108, 112, 114, 120, 126, 132, 138, 140, 144, 150, 156, 160, 162, 168, 174
Pseudoperfect numbers: $n \mid \sigma(n)$. Ref UPNT B2. [1,1; A5835]

M4095   1, 6, 12, 18, 20, 24, 28, 30, 36, 40, 42, 48, 54, 56, 60, 66, 72, 78, 80, 84, 88, 90, 96,
100, 104, 108, 112, 120, 126, 132, 140, 144, 150, 156, 160, 162, 168, 176, 180, 192, 196
Practical numbers (second definition): all $k \leq n$ are sums of proper divisors of n. Ref HO73
113. [1,2; A7620]

M4096   6, 12, 20, 30, 42, 56, 60, 72, 90, 105, 110, 132, 140, 156, 168, 182, 210, 240, 252,
272, 280, 306, 342, 360, 380, 420, 462, 495, 504, 506, 552, 600, 630, 650, 660, 702, 756
Denominators in Leibniz triangle. Ref Well86 35. [1,1; A7622]

M4097   6, 12, 23, 45, 46, 89, 91, 92, 93, 177, 179, 183, 185, 354, 355, 359, 367, 707, 708,
709, 711, 717, 718, 719, 733, 739, 1415, 1417, 1433, 1435, 1437, 1438, 1465, 1469, 1479
Positions of remoteness 6 in Beans-Don't-Talk. Ref MMAG 59 267 86. [1,1; A5694]

M4098   6, 12, 24, 60, 72, 168, 192, 324, 360, 660, 576, 1092, 1008, 1440, 1536, 2448,
1944, 3420, 2880, 4032, 3960, 6072, 4608, 7500, 6552, 8748, 8064, 12180, 8640, 14880
Index of modular group Γ_n. Ref GU62 15. [2,1; A1766, N1700]

M4099   1, 0, 0, 6, 12, 40, 180, 1512, 11760, 38880, 20160, 2106720, 22381920,
173197440, 703999296, 1737489600, 86030380800, 1149696737280, 11455162974720
Expansion of $((1+x)^x)^x$. [0,4; A7121]

M4100   0, 6, 12, 84, 1200, 3120, 249312, 920928, 86274816, 1232035584
Specific heat for square lattice. Ref PHL A25 208 67. [1,2; A5402]

M4101   1, 0, 6, 12, 90, 360, 2040, 10080, 54810, 290640
$2n$-step polygons on hexagonal lattice. Ref AIP 9 345 60. [0,3; A2898, N1701]

M4102   1, 6, 12, 96, 1560, 4848, 28848, 248352, 1446240, 12905664, 99071040,
649236480, 4924099200, 49007023872, 304778309376, 2301818168832
Symmetries in planted (1,4) trees on $3n - 1$ vertices. Ref GTA91 849. [1,2; A3613]

M4103 0, 6, 12, 156, 1680, 21264, 592032, 5712096, 390388992
Specific heat for diamond lattice. Ref PPS 86 10 65. [1,2; A2922, N1702]

M4104 1, 6, 14, 23, 34
Davenport-Schinzel numbers. Ref PLC 1 250 70. UPNT E20. [1,2; A5281]

M4105 1, 1, 6, 14, 47, 111, 280, 600, 1282, 2494, 4752, 8524, 14938, 25102, 41272,
65772, 102817, 156871, 235378, 346346, 502303
n-bead necklaces with 10 red beads. Ref JAuMS 33 12 82. [6,3; A5515]

M4106 1, 6, 15, 19, 24, 42, 73, 127, 208, 337, 528, 827, 1263, 1902, 2819, 4133, 5986,
8578, 12146, 17057, 23711, 32708, 44726, 60713, 81800, 109468, 145526, 192288
A generalized partition function. Ref PNISI 17 236 51. [1,2; A2599, N1703]

M4107 1, 6, 15, 20, 9, 24, 65, 90, 75, 6, 90, 180, 220, 180, 66, 110, 264, 360, 365, 264, 66,
178, 375, 510, 496, 414, 180, 60, 330, 570, 622, 582, 390, 220, 96, 300, 621, 630, 705
Coefficients of powers of η function. Ref JLMS 39 435 64. [6,2; A1484, N1704]

M4108 1, 6, 15, 28, 45, 66, 91, 120, 153, 190, 231, 276, 325, 378, 435, 496, 561, 630, 703,
780, 861, 946, 1035, 1128, 1225, 1326, 1431, 1540, 1653, 1770, 1891, 2016, 2145, 2278
Hexagonal numbers: $n(2n - 1)$. See Fig M2535. Ref D1 2 2. B1 189. [1,2; A0384, N1705]

M4109 1, 6, 15, 29, 49, 76, 111, 155, 209, 274, 351, 441, 545, 664, 799, 951, 1121, 1310,
1519, 1749, 2001, 2276, 2575, 2899, 3249, 3626, 4031, 4465, 4929, 5424, 5951, 6511
$(n + 3)(n^2 + 6n + 2)/6$. Ref NET 96. MMAG 61 28 88. rkg. [0,2; A5286]

M4110 6, 15, 35, 77, 143, 221, 323, 437, 667, 899, 1147, 1517, 1763, 2021, 2491, 3127,
3599, 4087, 4757, 5183, 5767, 6557, 7387, 8633, 9797, 10403, 11021, 11663, 12317
Product of successive primes. Ref EUR 45 24 85. [1,1; A6094]

M4111 1, 1, 1, 6, 15, 255, 1897, 92263, 1972653, 213207210
Halving an $n \times n$ chessboard. Ref GA69 189. [1,4; A3155]

M4112 1, 6, 16, 31, 51, 76, 106, 141, 181, 226, 276, 331, 391, 456, 526, 601, 681, 766,
856, 951, 1051, 1156, 1266, 1381, 1501, 1626, 1756, 1891, 2031, 2176, 2326, 2481, 2641
Centered pentagonal numbers: $(5n^2 + 5n + 2)/2$. See Fig M3826. Ref INOC 24 4550 85.
[0,2; A5891]

M4113 1, 6, 17, 38, 70, 116, 185, 258, 384, 490, 686, 826, 1124, 1292, 1705, 1896, 2491,
2670, 3416, 3680, 4602, 4796, 6110, 6178, 7700, 7980, 9684, 9730, 12156, 11920, 14601
Convolution of M2329. Ref SMA 19 39 53. [1,2; A0385, N1708]

M4114 0, 0, 1, 6, 17, 59, 195, 703, 2499, 9188, 33890, 126758, 476269, 1802311,
6849776, 26152417, 100203193
One-sided polyominoes with n cells. Ref CJN 18 366 75. [1,4; A6758]

M4115 1, 6, 18, 38, 66, 102, 146, 198, 258, 326, 402, 486, 578, 678, 786, 902, 1026, 1158, 1298, 1446, 1602, 1766, 1938, 2118, 2306, 2502, 2706, 2918, 3138, 3366, 3602, 3846
Points on surface of octahedron: $4n^2 + 2$. Ref MF73 46. Coxe74. INOC 24 4550 85. [0,2; A5899]

M4116 1, 6, 18, 40, 75, 126, 196, 288, 405, 550, 726, 936, 1183, 1470, 1800, 2176, 2601, 3078, 3610, 4200, 4851, 5566, 6348, 7200, 8125, 9126, 10206, 11368, 12615, 13950
Pentagonal pyramidal numbers: $n^2(n+1)/2$. See Fig M3382. Ref D1 2 2. B1 194. [1,2; A2411, N1709]

M4117 1, 6, 18, 48, 126, 300, 750, 1686, 4074, 8868
Cluster series for hexagonal lattice. Ref PRV 133 A315 64. DG72 225. [0,2; A3202]

M4118 1, 6, 18, 48, 126, 300, 762, 1668, 4216, 8668, 21988, 43058
Cluster series for square lattice. Ref PRV 133 A315 64. DG72 225. [0,2; A3198]

M4119 1, 6, 18, 50, 156, 508, 1724, 6018, 21440, 77632, 284706, 1055162, 3944956, 14858934
n-step walks on hexagonal lattice. Ref JPA 6 352 73. [2,2; A3290]

M4120 6, 18, 52, 114, 216, 388
Postage stamp problem. Ref CJN 12 379 69. AMM 87 208 80. [1,1; A1216, N1831]

M4121 1, 6, 18, 54, 150, 426, 1158, 3204, 8682, 23724, 64194, 174378, 470856, 1274430, 3434826, 9272346, 24953004, 67230288, 180705126, 486152604, 1305430884
n-step spirals on cubic lattice. Ref JPA 20 492 87. [0,2; A6779]

M4122 1, 6, 18, 54, 162, 456, 1302, 3630, 10158, 27648, 77022, 206508
Cluster series for diamond lattice. Ref PRV 133 A315 64. DG72 225. [0,2; A3208]

M4123 1, 6, 18, 54, 162, 474, 1398, 4074, 11898, 34554, 100302, 290334, 839466
n-step walks on hexagonal lattice. Ref JCP 34 1261 61. [0,2; A2933, N1711]

M4124 6, 18, 54, 168, 534, 1732, 5706, 19038, 64176, 218190, 747180, 2574488, 8918070, 31036560, 108457488, 380390574, 1338495492
Magnetization for honeycomb lattice. Ref DG74 420. [0,1; A7206]

M4125 0, 0, 0, 0, 0, 0, 6, 18, 66, 208, 646, 1962, 5962, 18014, 54578, 165650, 504220, 1539330, 4712742, 14475936
Paraffins with n carbon atoms. Ref JACS 54 1105 32. [1,7; A0623, N1712]

M4126 1, 0, 1, 1, 6, 18, 111, 839, 11076, 260327, 11698115, 1005829079, 163985322983, 50324128516939, 29000032348355991, 31395491269119883535
Connected rooted strength 1 Eulerian graphs with n nodes. Ref rwr. [1,5; A7126]

M4127 1, 6, 18, 132, 810, 5724, 42156, 323352, 2550042, 20559660, 168680196
Internal energy series for cubic lattice. Ref DG72 425. [0,2; A3496]

M4128 1, 6, 19, 44, 85, 146, 231, 344, 489, 670, 891, 1156, 1469, 1834, 2255, 2736, 3281, 3894, 4579, 5340, 6181, 7106, 8119, 9224, 10425, 11726, 13131, 14644, 16269, 18010
Octahedral numbers: $(2n^3 + n)/3$. Ref Coxe74. INOC 24 4550 85. [1,2; A5900]

M4129 1, 6, 19, 45, 90, 161, 266, 414, 615, 880, 1221, 1651, 2184, 2835, 3620, 4556, 5661, 6954, 8455, 10185, 12166, 14421, 16974, 19850, 23075, 26676, 30681, 35119
From expansion of $(1 + x + x^2)^n$. Ref C1 78. [2,2; A5712]

M4130 0, 6, 20, 12, 70, 900, 22, 352
Queens problem. Ref SL26 49. [1,2; A2566, N1713]

M4131 1, 0, 6, 20, 15, 36, 0, 84, 195, 100, 240, 0, 461, 1020, 540, 1144, 0, 1980, 4170, 2040, 4275, 0, 6984, 14340, 6940
McKay-Thompson series of class 5a for Monster. Ref FMN94. [-1,3; A7253]

M4132 6, 20, 28, 70, 88, 104, 272, 304, 368, 464, 496, 550, 572, 650, 748, 836, 945, 1184, 1312, 1376, 1430, 1504, 1575, 1696, 1870, 1888, 1952, 2002, 2090, 2205, 2210, 2470
Primitive non-deficient numbers. Ref AJM 35 426 13. [1,1; A6039]

M4133 6, 20, 28, 88, 104, 272, 304, 350, 368, 464, 490, 496, 550, 572, 650, 748, 770, 910, 945, 1184, 1190, 1312, 1330, 1376, 1430, 1504, 1575, 1610, 1696, 1870, 1888, 1952
Primitive pseudoperfect numbers. Ref UPNT B2. [1,1; A6036]

M4134 6, 20, 45, 84, 140, 216, 315, 440, 594, 780, 1001, 1260, 1560, 1904, 2295, 2736, 3230, 3780, 4389, 5060, 5796, 6600, 7475, 8424, 9450, 10556, 11745, 13020, 14384
Walks on square lattice. Ref GU90. [0,1; A5564]

$$\text{G.f.: } (6 - 4x + x^2)(1 - x)^{-4}.$$

M4135 1, 6, 20, 50, 105, 196, 336, 540, 825, 1210, 1716, 2366, 3185, 4200, 5440, 6936, 8721, 10830, 13300, 16170, 19481, 23276, 27600, 32500, 38025, 44226, 51156, 58870
4-dimensional pyramidal numbers: $n^2(n^2 - 1)/12$. See Fig M3382. Ref B1 195. [2,2; A2415, N1714]

M4136 6, 20, 52, 108, 211, 388
Postage stamp problem. Ref CJN 12 379 69. AMM 87 208 80. [1,1; A1211, N1836]

M4137 1, 6, 20, 134, 915, 7324, 65784, 657180, 7223637, 86637650, 1125842556, 15757002706, 236298742375, 3780061394232, 64251145312880, 1156374220457784
From ménage numbers. Ref R1 198. [3,2; A0386, N1715]

M4138 1, 0, 6, 20, 135, 924, 7420, 66744, 667485, 7342280, 88107426, 1145396460, 16035550531, 240533257860, 3848532125880, 65425046139824
Rencontres numbers. Ref R1 65. [2,3; A0387, N1716]

M4139 6, 20, 180, 1106, 9292, 82980, 831545, 9139482, 109595496, 1423490744, 19911182207, 298408841160, 4770598226296, 81037124739588
Discordant permutations. Ref SMA 20 23 54. [4,1; A0388, N1717]

M4140   1, 0, 6, 21, 40, 5, 504, 4697, 39808, 362151, 3627800, 39915469, 478999872,
6227018603, 87178288456, 1307674364625, 20922789883904, 355687428091087
$n! - n^3$. [0,3; A7339]

M4141   1, 6, 21, 55, 120, 231, 406, 666, 1035, 1540, 2211, 3081, 4186, 5565, 7260, 9316,
11781, 14706, 18145, 22155, 26796, 32131, 38226, 45150, 52975, 61776, 71631, 82621
Doubly triangular numbers: $C(n+2,2)+3C(n+3,4)$. Ref TCPS 9 477 1856. SIAC 4 477
75. ANS 4 1178 76. [0,2; A2817, N1718]

M4142   1, 6, 21, 56, 126, 252, 462, 792, 1287, 2002, 3003, 4368, 6188, 8568, 11628,
15504, 20349, 26334, 33649, 42504, 53130, 65780, 80730, 98280, 118755, 142506
Binomial coefficients $C(n,5)$. See Fig M1645. Ref D1 2 7. RS3. B1 196. AS1 828. [5,2;
A0389, N1719]

M4143   1, 6, 21, 71, 216, 657, 1907, 5507, 15522, 43352, 119140, 323946, 869476,
2308071, 6056581
5-dimensional partitions of n. Ref PCPS 63 1099 67. [1,2; A0390, N1720]

M4144   1, 6, 21, 71, 216, 672, 1982, 5817, 16582, 46633, 128704, 350665, 941715,
2499640, 6557378, 17024095, 43756166, 111433472, 281303882, 704320180
Euler transform of M3853. Ref PCPS 63 1100 67. EIS § 2.7. [1,2; A0391, N1721]

M4145   1, 6, 21, 76, 249, 814, 2521, 7824, 23473, 70590, 207345, 610356, 1765959,
511006, 14643993, 41958852, 118976633, 337823486
Related to self-avoiding walks on square lattice. Ref JPA 22 3624 89. [1,2; A6814]

M4146   1, 6, 21, 88, 330, 1302, 5005
Triangulations of the disk. Ref PLMS 14 759 64. [0,2; A5498]

M4147   6, 21, 91, 266, 994, 2562, 7764, 19482, 51212, 116028
Restricted partitions. Ref JCT 9 373 70. [2,1; A2222, N1722]

M4148   1, 6, 21, 91, 371, 1547, 6405, 26585, 110254, 457379, 1897214
Distributive lattices. Ref MSH 53 19 76. MSG 121 121 76. [0,2; A6359]

M4149   6, 21, 91, 441, 2275, 12201, 67171, 376761, 2142595, 12313161, 71340451,
415998681, 2438235715, 14350108521, 84740914531, 501790686201, 2978035877635
$1^n + 2^n + \cdots + 6^n$. Ref AS1 813. [0,1; A1553, N1723]

M4150   6, 21, 325, 1950625
Smallest n-hyperperfect number: m such that $m = n(\sigma(m) - m - 1) + 1$. Ref MOC 34 639
80. Robe92 177. [1,1; A7594]

M4151   1, 6, 22, 64, 162, 374, 809, 1668, 3316, 6408, 12108, 22468, 41081, 74202,
132666, 235160, 413790, 723530, 1258225, 2177640, 3753096, 6444336, 11028792
From rook polynomials. Ref SMA 20 18 54. [0,2; A1925, N1724]

M4152   1, 6, 22, 64, 163, 382, 848, 1816, 3797, 7814, 15914, 32192, 64839, 130238,
261156, 523128, 1047225, 2095590, 4192510, 8386560, 16774891, 33551806, 67105912
$2^n - C(n,0) - \cdots - C(n,3)$. Ref MFM 73 18 69. [4,2; A2663, N1725]

M4153   1, 6, 22, 65, 171, 420, 988, 2259, 5065, 11198, 24498, 53157, 114583, 245640,
524152, 1113959, 2359125, 4980546, 10485550, 22019865, 46137091, 96468716
Minimal covers of an n-set. Ref DM 5 249 73. [2,2; A3469]

$$\text{G.f.:}\quad (1 - x - x^2) \,/\, (1 - x)^3 (1 - 2x)^2.$$

M4154   0, 1, 1, 6, 22, 130, 822, 6202, 52552, 499194, 5238370, 60222844, 752587764,
10157945044, 147267180508, 2282355168060, 37655004171808, 658906772228668
$a(n) = (n-1)(a(n-1) + a(n-2)) - (n-1)(n-2)a(n-3)/2$. Ref PLMS 17 29 17.
EDMN 34 3 44. [1,4; A2137, N1726]

M4155   6, 22, 159, 1044, 9121, 78132, 748719, 7161484, 70800861, 699869892,
6978353179, 69580078524, 695292156201, 6947835288052, 69465637212039
Number of n-tuples that are final digits of squares. Ref AMM 67 1002 60. [1,1; A0993,
N1727]

M4156   1, 6, 23, 65, 156, 336, 664, 1229
n-covers of a 3-set. Ref DM 81 151 90. [1,2; A5745]

M4157   1, 6, 23, 84, 283, 930, 2921, 9096, 27507, 82930, 244819, 722116, 2096603,
6087290, 17458887, 50090544, 142317089, 404543142
Related to self-avoiding walks on square lattice. Ref JPA 22 3624 89. [1,2; A6815]

M4158   6, 24, 45, 480, 10080, 24192, 907200, 1036800, 239500800, 106444800,
9906624000, 475517952000, 15692092416000, 4828336128000, 8002967132160000
Denominators of coefficients for repeated integration. Cf. M4457. Ref PHM 38 336 47.
[1,1; A2688, N1728]

M4159   6, 24, 60, 120, 210, 336, 504, 720, 990, 1320, 1716, 2184, 2730, 3360, 4080, 4896,
5814, 6840, 7980, 9240, 10626, 12144, 13800, 15600, 17550, 19656, 21924, 24360
$n(n+1)(n+2)$. [0,1; A7531]

M4160   0, 1, 6, 24, 70, 165, 336, 616, 1044, 1665, 2530, 3696, 5226, 7189, 9660, 12720,
16456, 20961, 26334, 32680, 40110, 48741, 58696, 70104, 83100, 97825, 114426
$(n^4 + n^2 + 2n)/4$. Ref GA66 246. [0,3; A6528]

M4161   1, 6, 24, 80, 240, 672, 1792, 4608, 11520, 28160, 67584, 159744, 372736, 860160,
1966080, 4456448, 10027008, 22413312, 49807360, 110100480, 242221056, 530579456
$C(n,2).2^{n-2}$. Ref RSE 62 190 46. AS1 796. MFM 74 62 70. [2,2; A1788, N1729]

M4162   1, 6, 24, 90, 318, 1098, 3696, 12270, 40224, 130650, 421176, 1348998, 4299018,
13635630, 43092888, 135698970, 426144654
Susceptibility for hexagonal lattice. Ref JPA 5 632 72. [0,2; A2919, N1730]

M4163   1, 6, 24, 90, 324, 1166, 4138, 14730, 51992, 183898, 646980, 2279702, 8002976, 28127418, 98585096, 345848306, 1210704274, 4241348770, 14833284544
n-step spirals on cubic lattice. Ref JPA 20 492 87. [0,2; A6780]

M4164   1, 6, 24, 90, 336, 1254, 4680, 17466, 65184, 243270, 907896, 3388314, 12645360, 47193126, 176127144, 657315450, 2453134656, 9155223174, 34167758040
$a(n) = 4a(n-1) - a(n-2)$. Ref MOC 24 180 70. [0,2; A1352, N1731]

M4165   1, 6, 25, 60, 203, 3710, 21347
Related to Weber functions. Ref KNAW 66 751 63. [2,2; A1664, N1732]

M4166   1, 6, 25, 90, 300, 954, 2939, 8850, 26195, 76500, 221016, 632916, 1799125, 5082270, 14279725, 39935214, 111228804, 308681550, 853904015, 2355364650
Expansion of $1/(1-3x+x^2)^2$. [0,2; A1871, N1733]

M4167   1, 6, 25, 90, 301, 966, 3025, 9330, 28501, 86526, 261625, 788970, 2375101, 7141686, 21457825, 64439010, 193448101, 580606446, 1742343625, 5228079450
Stirling numbers of second kind. See Fig M4981. Ref AS1 835. DKB 223. [3,2; A0392, N1734]

M4168   6, 25, 325, 561, 703, 817, 1105, 1825, 2101, 2353, 2465, 3277, 4525, 4825, 6697, 8321, 10225, 10585, 10621, 11041, 11521, 12025, 13665, 14089, 16725, 16806, 18721
Pseudoprimes to base 7. Ref UPNT A12. [1,1; A5938]

M4169   1, 6, 26, 71, 155, 295, 511, 826, 1266, 1860, 2640, 3641, 4901, 6461, 8365, 10660, 13396, 16626, 20406, 24795, 29855, 35651, 42251, 49726, 58150, 67600, 78156
Generalized Stirling numbers. Ref PEF 77 7 62. [1,2; A1701, N1735]

M4170   1, 6, 26, 94, 308, 941, 2744, 7722, 21166, 56809, 149971, 390517, 1005491, 2564164, 6485901, 16289602, 40659669, 100934017, 249343899, 613286048
n-node trees of height 6. Ref IBMJ 4 475 60. KU64. [7,2; A0393, N1736]

M4171   1, 6, 26, 100, 364, 1288, 4488, 15504, 53296, 182688, 625184, 2137408, 7303360, 24946816, 85196928, 290926848, 993379072, 3391793664, 11580678656, 39539651584
Random walks. Ref DM 17 44 77. TCS 9 105 79. [3,2; A5022]

$$\text{G.f.:} \quad x^3 \ / \ (1-2x)(1-4x+2x^2).$$

M4172   6, 26, 192, 3014
Bicolored graphs in which colors are interchangeable. Ref ENVP B5 41 78. [2,1; A7139]

M4173   1, 6, 27, 98, 309, 882, 2330, 5784, 13644, 30826, 67107, 141444, 289746, 578646, 1129527, 2159774, 4052721, 7474806, 13569463, 24274716, 42838245, 74644794
Coefficients of an elliptic function. Ref CAY 9 128. [0,2; A1940, N1737]

$$\text{G.f.:} \quad \Pi \, (1 - x^k)^{-c(k)}, \quad c(k) = 6,6,6,0,6,6,6,0,....$$

M4174 1, 6, 27, 98, 315, 924, 2534, 6588, 16407, 39430, 91959, 209034, 464723,
1013292, 2171850, 4584620, 9546570, 19635840, 39940460, 80421600, 160437690
Convolved Fibonacci numbers. Ref RCI 101. [0,2; A1874, N1738]

$$\text{G.f.:}\quad (1 - x - x^2)^{-6}.$$

M4175 1, 6, 27, 104, 369, 1236, 3989, 12522, 38535, 116808, 350064, 1039896, 3068145,
9004182, 26314773, 76652582, 222705603, 645731148, 1869303857, 5404655358
Powers of rooted tree enumerator. Ref R1 150. [1,2; A0395, N1739]

M4176 1, 6, 27, 104, 369, 1242, 4037, 12804, 39897, 122694, 373581, 1128816, 3390582,
10136556, 30192102, 89662216, 265640691, 785509362, 2319218869, 6839057544
Column of Motzkin triangle. Ref JCT A23 293 77. [5,2; A5325]

M4177 1, 6, 27, 110, 429, 1638, 6188, 23256, 87210, 326876, 1225785, 4601610,
17298645, 65132550, 245642760, 927983760, 3511574910, 13309856820, 50528160150
$6C(2n+1, n-2)/(n+4)$. Ref FQ 14 397 76. DM 14 84 76. [2,2; A3517]

M4178 1, 6, 27, 111, 440, 1717, 6655, 25728, 99412, 384320, 1487262, 5762643,
22357907, 86859412, 337879565
Permutations by inversions. Ref NET 96. MMAG 61 28 88. rkg. [6,2; A5284]

M4179 1, 6, 27, 122, 516, 2148, 8792, 35622, 143079, 570830, 2264649, 8942436,
35169616, 137839308
Susceptibility for honeycomb. Ref PHA 28 934 62. DG74 421. [3,2; A2912, N1740]

M4180 6, 27, 488, 7974, 149796, 2725447, 56970432, 1151053821, 25279412332,
543871341927, 12411512060544, 278163517356594, 6498314231705568
Witt vector *3!. Ref SLC 16 106 88. [1,1; A6174]

M4181 1, 6, 28, 120, 495, 2002, 8008, 31824, 125970, 497420, 1961256, 7726160,
30421755, 119759850, 471435600, 1855967520, 7307872110, 28781143380
Binomial coefficients $C(2n, n-2)$. See Fig M1645. Ref LA56 517. AS1 828. [2,2; A2694,
N1741]

M4182 1, 6, 28, 120, 496, 672, 8128, 30240, 32760, 523776, 2178540, 23569920,
33550336, 45532800
Multiply-perfect numbers: n divides $\sigma(n)$. Ref B1 22. Robe92 176. rgw. [1,2; A7691]

M4183 1, 6, 28, 120, 496, 2016, 8128, 32640, 130816, 523776, 2096128, 8386560,
33550336, 134209536, 536854528, 2147450880, 8589869056, 34359607296
$2^{n-1}(2^n - 1)$. Ref HO73 113. [1,2; A6516]

M4184 0, 1, 6, 28, 125, 527, 2168, 8781, 35155, 139531, 550068
Spheroidal harmonics. Ref MES 52 75 24. [0,3; A2693, N1742]

M4185 6, 28, 140, 270, 496, 672, 1638, 2970, 6200, 8128, 8190, 18600, 18620, 27846, 30240, 32760, 55860, 105664, 117800, 167400, 173600, 237510, 242060, 332640
Harmonic or Ore numbers: harmonic mean of divisors is integral. See Fig M4299. Ref AMM 61 95 54. UPNT B2. [1,1; A1599, N1743]

M4186 6, 28, 496, 8128, 33550336, 8589869056, 137438691328, 2305843008139952128, 2658455991569831744654692615953842176
Perfect numbers. See Fig M0062. Ref SMA 19 128 53. B1 19. NAMS 18 608 71. CUNN. [1,1; A0396, N1744]

M4187 1, 6, 29, 108, 393, 1298, 4271, 13312, 41469, 125042, 376747, 1111144, 3274475, 9505054, 27573041, 79086964, 226727667, 644301026
Related to self-avoiding walks on square lattice. Ref JPA 22 3624 89. [1,2; A6816]

M4188 1, 1, 0, 1, 6, 29, 150, 841, 5166, 34649, 252750, 1995181, 16962726, 154624469
Quasi-alternating permutations of length n. Equals ½ M2027. Ref NET 113. C1 261. [0,5; A0708, N1745]

M4189 1, 6, 30, 42, 30, 66, 2730, 6, 510, 798, 330, 138, 2730, 6, 870, 14322, 510, 6, 1919190, 6, 13530, 1806, 690, 282, 46410, 66, 1590, 798, 870, 354, 56786730, 6, 510
Denominators of Bernoulli numbers B_{2n}. See Fig M4189. Cf. M4039. Ref DA63 2 230. AS1 810. [0,2; A2445, N1746]

‖‖‖

Figure M4189. BERNOULLI NUMBERS.

The **Bernoulli** numbers arise in numerical analysis, number theory and combinatorics, and are defined by the generating function

$$\frac{x}{e^x - 1} = \sum_{n=0}^{\infty} B_n \frac{x^n}{n!} \, ,$$

so that $B_0 = 1$, $B_1 = -\frac{1}{2}$, $B_{2m+1} = 0$ for $m \geq 1$, and

$$B_2 = \frac{1}{6}, B_4 = -\frac{1}{30}, B_6 = \frac{1}{42}, B_8 = -\frac{1}{30} \, , \ldots \, .$$

They satisfy the recurrence

$$\sum_{k=0}^{n} \binom{n+1}{k} B_k = 0 \, , \quad (n \geq 1)$$

and appear in Bernoulli's formula

$$\sum_{k=0}^{m-1} k^n = \frac{1}{n+1} \sum_{k=0}^{n} \binom{n+1}{k} B_k m^{n+1-k} \, ,$$

and more generally in the Euler-Maclaurin summation formula. References: [C1 48], [AS1 804], [KN1 1 108], [GKP 269]. M4039 and M4189 give respectively the numerators and denominators of $|B_{2n}|$, and M4435 gives the nearest integer to $|B_{2n}|$.

‖‖‖

M4190   6, 30, 42, 54, 60, 66, 78, 90, 100, 102, 114, 126, 140, 148, 194, 196, 208, 220, 238, 244, 252, 274, 288, 292, 300, 336, 348, 350, 364, 374, 380, 382, 386, 388, 400, 420
Beginnings of periodic unitary aliquot sequences. Ref RI72 14. [1,1; A3062]

M4191   1, 6, 30, 84, 90, 132, 5460, 360, 1530, 7980, 13860, 8280, 81900, 1512, 3480, 114576
Denominators of Bernoulli numbers. Ref DA63 2 208. [0,2; A2444, N1747]

M4192   1, 6, 30, 114, 438, 1542, 5754, 19574, 71958
Cluster series for cubic lattice. Ref PRV 133 A315 64. DG72 225. [0,2; A3211]

M4193   0, 6, 30, 126, 510, 2046, 8190, 32766, 131070, 524286, 2097150, 8388606, 33554430, 134217726, 536870910, 2147483646, 8589934590, 34359738366
$2^{2n+1} - 2$. Ref QJMA 47 110 16. FMR 1 112. DA63 2 283. [0,2; A2446, N1748]

M4194   1, 6, 30, 126, 534, 2214, 9246, 38142, 157974, 649086, 2674926
n-step walks on cubic lattice. Ref JCP 34 1261 61. [0,2; A2934, N1749]

M4195   1, 6, 30, 128, 486, 1692, 5512, 17040, 50496, 144512, 401664, 1089024, 2890240, 7529472, 19298304, 48754688, 121602048, 299827200, 731643904, 1768685568
Exponential-convolution of triangular numbers with themselves. Ref BeSl94. [0,2; A7465]

M4196   1, 6, 30, 138, 606, 2586, 10818, 44574, 181542, 732678, 2935218, 11687202, 46296210, 182588850, 717395262, 2809372302, 10969820358
Susceptibility for hexagonal lattice. Ref JPA 5 627 72. DG74 380. [0,2; A2920, N1750]

M4197   1, 6, 30, 138, 618, 2730, 11946, 51882, 224130, 964134, 4133166, 17668938, 75355206, 320734686, 1362791250, 5781765582, 24497330322, 103673967882
n-step self-avoiding walks on hexagonal lattice. Ref JPA 18 L201 85. DG89 56. GW91. [0,2; A1334, N1751]

M4198   1, 6, 30, 140, 630, 2772, 12012, 51480, 218790, 923780, 3879876, 16224936, 67603900, 280816200, 1163381400, 4808643120, 19835652870, 81676217700
$(2n+2)!/(2.n!(n+1)!)$. Ref OP80 21. SE33 92. JO39 449. SAM 22 120 43. LA56 514. [0,2; A2457, N1752]

M4199   1, 6, 30, 144, 666, 3024, 13476, 59328, 258354, 1115856, 4784508, 20393856, 86473548, 365034816, 1534827960, 6431000832, 26862228450
Susceptibility for cubic lattice. Ref PTRS 273 607 73. DG72 404. [0,2; A3279]

M4200   1, 6, 30, 146, 714, 3534, 17718, 89898, 461010, 2386390, 12455118, 65478978, 346448538, 1843520670, 9859734630, 52974158938, 285791932578, 1547585781414
Royal paths in a lattice. Ref CRO 20 18 73. [0,2; A6320]

M4201   1, 6, 30, 150, 726, 3510, 16710, 79494, 375174, 1769686, 8306862, 38975286, 182265822, 852063558, 3973784886, 18527532310, 86228667894, 401225391222
Susceptibility for cubic lattice. Ref JPA 5 651 72. DG74 381. [0,2; A2913, N1753]

M4202 1, 6, 30, 150, 726, 3534, 16926, 81390, 387966, 1853886, 8809878, 41934150, 198842742, 943974510, 4468911678, 21175146054, 100121875974, 473730252102
n-step self-avoiding walks on cubic lattice. Ref JPA 20 1847 87. [0,2; A1412, N1754]

M4203 1, 6, 30, 150, 738, 3570, 17118, 81498, 385710, 1817046, 8528478, 39903462, 186198642, 866861394, 4027766490, 18681900270, 86518735722
Trails of length n on hexagonal lattice. Ref JPA 18 576 85. [0,2; A6818]

M4204 1, 6, 30, 150, 750, 3726, 18438, 90966, 447918, 2201622, 10809006, 52999446, 259668942, 1271054982, 6218232414, 30399142614
Trails of length n on cubic lattice. Ref JPA 18 576 85. [0,2; A6819]

M4205 6, 30, 174, 1158, 8742, 74046, 696750, 7219974, 81762438, 1005151902, 13336264686, 189992451270, 2893180308774, 46904155833918, 806663460996462
$\Sigma(n+3)!C(n,k)$, $k = 0 .. n$. Ref CJM 22 26 70. [0,1; A1341, N1755]

M4206 0, 0, 0, 6, 30, 180, 840, 5460, 30996, 209160, 1290960, 9753480, 69618120, 571627056, 4443697440, 40027718640, 346953934320, 3369416698080
Degree n permutations of order exactly 4. Ref CJM 7 159 55. [1,4; A1473, N1756]

M4207 1, 6, 30, 192, 1560, 15120, 171360, 2217600, 32296320
From solution to a difference equation. Ref FQ 25 363 87. [1,2; A5922]

M4208 1, 1, 1, 6, 31, 120, 337, 784, 24705, 288000, 2451679, 14032128, 17936543, 2173889536, 42895630065, 583266662400, 5396647099903, 5119183650816
Expansion of $\ln(1+\cos(x).x)$. [0,4; A3728]

M4209 1, 6, 31, 156, 781, 3906, 19531, 97656, 488281, 2441406, 12207031, 61035156, 305175781, 1525878906, 7629394531, 38146972656, 190734863281, 953674316406
$(5^n - 1)/4$. [1,2; A3463]

M4210 1, 6, 31, 160, 856, 4802, 28337, 175896, 1146931
Driving-point impedances of an n-terminal network. Ref BSTJ 18 301 39. [2,2; A3128]

M4211 1, 1, 6, 31, 806, 20306, 2558556, 320327931, 200525284806, 125368356709806, 391901483074853556, 1224770494838892134806, 191382637523525284984778556
Gaussian binomial coefficient $[n,n/2]$ for $q = 5$. Ref GJ83 99. ARS A17 329 84. [0,3; A6115]

M4212 6, 32, 109, 288, 654, 1337
Partitions into non-integral powers. Ref PCPS 47 215 51. [5,1; A0397, N1757]

M4213 1, 6, 32, 175, 1012, 6230, 40819
Generalized Stirling numbers of second kind. Ref FQ 5 366 67. [2,2; A0558, N1758]

M4214   1, 6, 33, 182, 1020, 5814, 33649, 197340, 1170585, 7012200, 42364476,
257854776, 1579730984, 9734161206, 60290077905, 375138262520, 2343880406595
From generalized Catalan numbers. Ref LNM 952 279 82. [0,2; A6630]

$$\text{G.f.:} \quad {}_3F_2([2,8/3,7/3];[4,7/2];27x/4).$$

M4215   6, 34, 1154, 1331714, 1773462177794, 3145168096065837266706434,
98920823525104037575501729751467021228379369963354
$a(n) = a(n-1)^2 - 2$. Ref AJM 1 313 1878. D1 1 376. MMAG 48 210 75. [0,1; A3423]

M4216   1, 6, 35, 180, 921, 4626, 23215, 116160, 581141, 2906046, 14531595, 72659340,
363302161, 1816516266, 9082603175, 45413037720, 227065275981, 1135326467286
Expansion of $1/(1-x)(1-4x^2)(1-5x)$. Ref AMM 3 244 1896. [1,2; A2041, N1759]

M4217   1, 6, 35, 204, 1189, 6930, 40391, 235416, 1372105, 7997214, 46611179,
271669860, 1583407981, 9228778026, 53789260175, 313506783024, 1827251437969
$a(n) = 6a(n-1) - a(n-2)$. Ref D1 2 10. MAG 47 237 63. B1 193. FQ 9 95 71. [0,2;
A1109, N1760]

M4218   1, 6, 35, 225, 1624, 13132, 118124, 1172700, 12753576, 150917976, 1931559552,
26596717056, 392156797824, 6165817614720, 102992244837120
Stirling numbers of first kind. See Fig M4730. Ref AS1 833. DKB 226. [3,2; A0399,
N1762]

M4219   6, 36, 150, 540, 1806, 5796, 18150, 55980, 171006, 519156, 1569750, 4733820,
14250606, 42850116, 128746950, 386634060, 1160688606, 3483638676
Differences of 0. Ref VO11 31. DA63 2 212. R1 33. [3,1; A1117, N1763]

M4220   6, 36, 200, 1170, 7392, 50568, 372528, 2936070
Labeled trees of height 2 with n nodes. Ref IBMJ 4 478 60. [3,1; A0551, N1764]

M4221   6, 36, 208, 1171, 6474, 35324, 190853
Percolation series for f.c.c. lattice. Ref SSP 10 921 77. [1,1; A6812]

M4222   1, 6, 36, 216, 1260, 7206, 40650, 227256, 1262832, 6983730, 38470220,
211220800, 1156490000, 6317095284, 34435495872
Expansion for generalized walks on hexagonal lattice. Ref JPA 17 L458 84. [0,2; A7274]

M4223   1, 6, 36, 216, 1296, 7776, 46440, 276054, 1633848, 9633366, 56616140,
331847200, 1940717000, 11327957196, 66010769382
Expansion for generalized walks on hexagonal lattice. Ref JPA 17 L458 84. [0,2; A7275]

M4224   1, 6, 36, 216, 1296, 7776, 46656, 279936, 1679616, 10077696, 60466176,
362797056, 2176782336, 13060694016, 78364164096, 470184984576, 2821109907456
Powers of 6. Ref BA9. [0,2; A0400, N1765]

M4225 1, 6, 36, 240, 1800, 15120, 141120, 1451520, 16329600, 199584000, 2634508800,
37362124800, 566658892800, 9153720576000, 156920924160000
Lah numbers: $\frac{1}{2}(n-1)\, n!$. Ref R1 44. C1 156. [2,2; A1286, N1766]

M4226 0, 0, 0, 0, 0, 0, 1, 6, 37, 195, 979, 4663, 21474, 96496, 425365
n-celled polyominoes with holes. Ref PA67. JRM 2 182 69. [1,8; A1419, N1767]

M4227 0, 1, 6, 37, 228, 1405, 8658, 53353, 328776, 2026009, 12484830, 76934989,
474094764, 2921503573, 18003116202, 110940200785, 683644320912, 4212806126257
Convergents to square root of 10. Ref rkg. [0,3; A5668]

$$\text{G.f.:}\quad x \,/\, (1 - 6\,x - x^2).$$

M4228 1, 6, 37, 236, 1517, 9770, 62953, 405688, 2614457, 16849006, 108584525,
699780452, 4509783909, 29063617746, 187302518353, 1207084188912
Hamiltonian circuits on $2n \times 4$ rectangle. Ref JPA 17 445 84. [1,2; A5389]

M4229 1, 6, 39, 258, 1719, 11496, 77052, 517194, 3475071, 23366598, 157206519,
1058119992, 7124428836, 47983020624, 323240752272, 2177956129818
$\Sigma\, C(3k,k).C(3n-3k,n-k)$, $k = 0 \,.\,.\, n$. Ref dek. [0,2; A6256]

M4230 1, 6, 39, 272, 1995, 15180, 118755, 949344, 7721604, 63698830, 531697881,
4482448656, 38111876530, 326439471960, 2814095259675, 24397023508416
From generalized Catalan numbers. Ref LNM 952 280 82. [0,2; A6633]

M4231 1, 6, 39, 320, 3281, 40558, 586751, 9719616, 181353777, 3762893750,
85934344775, 2141853777856, 57852105131809, 1683237633305502
Functors of degree n from free abelian groups to abelian groups. Ref JPAA 91 49 94. dz.
[1,2; A7322]

M4232 0, 1, 1, 6, 39, 390, 4815, 73080, 1304415, 26847450, 625528575, 16279193700,
468022452975, 14731683916950, 503880434632575, 18609309606888000
Planted binary phylogenetic trees with n labels. Ref LNM 884 196 81. [0,4; A6678]

M4233 6, 40, 112, 1152, 2816, 13312, 10240, 557056, 1245184, 5505024, 12058624,
104857600, 226492416, 973078528, 2080374784, 23622320128, 30064771072
Denominator of $(2n-1)!! \,/\, (2n+1).(2n)!!$. Ref PHM 33 13 42. MOC 3 17 48. [1,1;
A2595, N1768]

M4234 6, 40, 155, 456, 1128, 2472, 4950, 9240, 16302, 27456, 44473, 69680, 106080,
157488, 228684, 325584, 455430, 627000, 850839, 1139512, 1507880, 1973400
Quadrinomial coefficients. Ref JCT 1 372 66. C1 78. [3,1; A1919, N1769]

M4235 6, 40, 174, 644, 2268, 8020, 28666, 103696, 379450, 1402276, 5227366, 19633732
n-step walks on hexagonal lattice. Ref JPA 6 352 73. [4,1; A5553]

M4236   1, 1, 6, 40, 285, 2126, 16380, 129456, 1043460, 8544965, 70893054, 594610536,
5033644070, 42952562100, 369061673400, 3190379997272, 27727712947836
Dissecting a polygon into n pentagons. Ref DM 11 388 75. [1,3; A5037]

M4237   1, 6, 40, 320, 2946
Genus 1 maps with n edges. Ref SIAA 4 169 83. [2,2; A6387]

M4238   0, 1, 6, 40, 360, 4576, 82656, 2122240, 77366400, 4002843136, 293717546496,
30558458490880
2-colored labeled graphs on n nodes. Ref CJM 12 412 60. rcr. [1,3; A0683, N1770]

M4239   1, 6, 41, 293, 2309, 19975, 189524, 1960041, 21993884, 266361634, 3465832370,
48245601976, 715756932697, 11277786883706, 188135296650845
Permutations of length n by length of runs. Ref DKB 261. [3,2; A0402, N1771]

M4240   1, 6, 41, 331, 3176, 35451, 447981, 6282416, 96546231, 1611270851,
28985293526, 558413253581, 11458179765541, 249255304141006, 5725640423174901
Coincides with its 5th order binomial transform. Ref DM 21 320 78. BeSl94. EIS § 2.7.
[0,2; A5011]

$$\text{Lgd.e.g.f.:}\ e^{5x}.$$

M4241   1, 1, 6, 41, 365, 3984, 51499, 769159, 13031514, 246925295, 5173842311,
118776068256, 2964697094281, 79937923931761, 2315462770608870
Partitions into pairs. Ref PLIS 23 65 78. [1,3; A6198]

M4242   1, 1, 6, 41, 816, 54121, 14274660, 14153665099, 51048862475458,
667165739670566962, 31770009199858957846460
Rooted strength 3 Eulerian graphs with n nodes. Ref rwr. [1,3; A7130]

M4243   1, 6, 42, 336, 3024, 30240, 332640, 3991680, 51891840, 726485760,
10897286400, 174356582400, 2964061900800, 53353114214400, 1013709170073600
$n!/5!$. Ref PEF 107 5 63. [5,2; A1725, N1772]

M4244   6, 43, 336, 2864, 25326, 223034, 1890123
6-covers of an n-set. Ref DM 81 151 90. [1,1; A5786]

M4245   6, 44, 145, 336, 644, 1096, 1719, 2540, 3586, 4884, 6461, 8344, 10560, 13136,
16099, 19476, 23294, 27580, 32361, 37664, 43516, 49944, 56975, 64636, 72954, 81956
Discordant permutations. Ref SMA 20 23 54. [3,1; A0561, N1773]

M4246   6, 44, 351, 3093, 33445
Semigroups of order n with 3 idempotents. Ref MAL 2 2 67. SGF 14 71 77. [3,1; A5591]

M4247   0, 1, 6, 44, 430, 5322, 79184, 1381144
Total diameter of labeled trees with n nodes. Ref IBMJ 4 478 60. [1,3; A1852, N1774]

M4248 1, 6, 45, 60, 90, 420, 630, 1512, 3780, 5460, 7560, 8190, 9100, 15925, 16632, 27300, 31500, 40950, 46494, 51408, 55125, 64260, 66528, 81900, 87360, 95550, 143640
Unitary harmonic numbers. Ref PAMS 51 7 75. [1,2; A6086]

M4249 1, 6, 45, 365, 3101, 27144, 242636, 2202873, 20241055, 187766940, 1755409652, 16517284570, 156265005369, 1485269469971, 14174126304850
Strict n-node animals on cubic lattice. Ref DU92 40. [1,2; A7193]

M4250 0, 1, 6, 45, 374, 3300, 30282, 285682, 2751258, 26921589, 266797836, 2671518873, 26981681412, 274493963898, 2809920769440, 28919412629031
Primitive n-node animals on cubic lattice. Ref DU92 40. [0,3; A7194]

M4251 1, 6, 45, 420, 4725, 62370, 945945, 16216200, 310134825, 6547290750, 151242416325, 3794809718700, 102776096548125, 2988412653476250
Expansion of $(1+x)/(1-2x^{5/2})$. Ref RCI 77. [0,2; A1879, N1775]

M4252 1, 6, 46, 450, 5650, 91866, 1957066
Related to partially ordered sets. Ref JCT 6 17 69. [0,2; A1829, N1776]

M4253 1, 6, 46, 452, 4852
Genus 1 maps with n edges. Ref SIAA 4 169 83. [2,2; A6386]

M4254 0, 6, 48, 168, 480, 966, 2016, 3360, 5616, 8550, 13200, 17832, 26208, 34566, 45840, 59520, 78336, 95526, 123120, 147240, 181776, 219846, 267168, 307488, 372000
2×2 matrices with entries mod n. Ref tb. [1,2; A5353]

M4255 6, 48, 390, 3216, 26844, 229584, 2006736, 17809008
Specific heat for cubic lattice. Ref PRV 129 102 63. [0,1; A2918, N1777]

M4256 6, 48, 408, 3600, 42336, 781728, 13646016, 90893568, 1798204416
Susceptibility for hexagonal lattice. Ref PHL A25 208 67. [1,1; A5399]

M4257 6, 48, 528, 7920, 149856, 3169248, 77046528, 2231209728, 71938507776, 2446325534208
Susceptibility for cubic lattice. Ref PRV 164 801 67. [1,1; A2170, N1778]

M4258 6, 50, 225, 735, 1960, 4536, 9450, 18150, 32670, 55770, 91091, 143325, 218400, 323680, 468180, 662796, 920550, 1256850, 1689765, 2240315, 2932776, 3795000
Stirling numbers of first kind. See Fig M4730. Ref AS1 833. DKB 226. [1,1; A1303, N1779]

M4259 1, 6, 50, 518, 6354, 89782, 1429480
Vertex diagrams of order $2n$. Ref NUPH B127 181 77. [1,2; A5416]

M4260 1, 1, 6, 51, 506, 5481, 62832, 749398, 9203634, 115607310, 1478314266, 19180049928, 251857119696, 3340843549855, 44700485049720, 602574657427116
Dissections of a polygon: $C(6n,n)/(5n+1)$. Ref AMP 1 198 1841. DM 11 388 75. [0,3; A2295, N1780]

M4261 1, 1, 6, 51, 561, 7556, 120196, 2201856, 45592666, 1051951026, 26740775306,
742069051906, 22310563733864, 722108667742546, 25024187820786357
Coefficients of iterated exponentials. Ref SMA 11 353 45. PRV A32 2342 85. [0,3; A0405,
N1781]

M4262 0, 6, 52, 411, 3392, 30070, 287802
Tumbling distance for n-input mappings. Ref PRV A32 2343 85. [0,2; A5948]

M4263 1, 6, 55, 610, 7980, 120274, 2052309, 39110490, 823324755, 18974858540,
475182478056, 12848667150956, 373081590628565, 11578264139795430
Partitions into pairs. Ref PLIS 23 65 78. [1,2; A6200]

M4264 1, 1, 6, 57, 741, 12244, 245755, 5809875, 158198200, 4877852505,
168055077875, 6400217406500, 267058149580823, 12118701719205803
Coefficients of iterated exponentials. Ref SMA 11 353 45. [0,3; A0406, N1782]

M4265 0, 6, 58, 328, 1452, 5610, 19950, 67260, 218848, 695038, 2170626, 6699696,
20507988, 62407890, 189123286, 571432036, 1722945672, 5187185766, 15600353130
Second-order Eulerian numbers. Ref JCT A24 28 78. GKP 256. [2,2; A4301]

M4266 6, 60, 90, 120, 36720, 73440, 12646368, 22276800, 44553600
Infinitary multi-perfect numbers. Ref MOC 54 405 90. glc. [1,1; A7358]

M4267 6, 60, 90, 36720, 12646368, 22276800
Infinitary perfect numbers. Ref MOC 54 405 90. glc. [1,1; A7357]

M4268 6, 60, 90, 87360, 146361946186458562560000
Unitary perfect numbers. Ref NAMS 18 630 71. CMB 18 115 75. UPNT B3. [1,1; A2827,
N1783]

M4269 0, 6, 60, 314, 1240, 4166, 12600, 35324, 93576, 236944, 578764, 1371478,
3169380, 7165478, 15901324, 34705018, 74661832, 158529158, 332756408, 691084378
Second moment of site percolation series for hexagonal lattice. Ref JPA 21 3822 88. [0,2;
A6741]

M4270 1, 6, 60, 840, 15120, 332640, 8648640, 259459200, 8821612800, 335221286400,
14079294028800, 647647525324800, 32382376266240000, 1748648318376960000
$(2n+1)!/n!$. Ref MOC 3 168 48; 9 174 55. CMA 2 25 70. MAN 191 98 71. [0,2; A0407,
N1784]

M4271 1, 6, 60, 1368, 15552, 201240, 2016432, 21582624
Folding a $3 \times n$ strip of stamps. See Fig M4587. Ref CJN 14 77 71. [0,2; A1416, N1785]

M4272 1, 6, 60, 1820, 136136, 27261234, 14169550626, 19344810307020,
69056421075989160, 64569385948729842556, 158032048562207386967144416
Central Fibonomial coefficients. Ref FQ 6 82 68. BR72 74. [0,2; A3267]

M4273 1, 6, 63, 616, 6678, 77868, 978978, 13216104, 190899423, 2939850914, 48106651593
Permutations of length n by rises. Ref DKB 263. [6,2; A1261, N1786]

M4274 1, 6, 65, 1092, 25272, 749034, 2710844
M-trees on n nodes. Ref LNM 1234 207 85. [1,2; A6959]

M4275 0, 6, 68, 442, 2218, 9528, 36834, 131856, 445000, 1433294, 4444006, 13349510, 39041224, 111583236, 312618368, 860662498, 2333112020, 6238124024, 16474149036
Second moment of bond percolation series for hexagonal lattice. Ref JPA 21 3822 88. [0,2; A6737]

M4276 1, 6, 71, 1456, 45541, 2020656, 120686411, 9336345856, 908138776681, 108480272749056, 15611712012050351, 2664103110372192256
2 up, 2 down, 2 up, ... permutations of length $2n + 1$. Ref prs. [1,2; A5981]

M4277 6, 72, 690, 6192, 53946, 466800, 4053816, 35450940, 312411672
n-step walks on f.c.c. lattice. Ref JPA 6 351 73. [3,1; A5548]

M4278 0, 0, 0, 0, 0, 0, 0, 0, 0, 0, 0, 0, 1, 6, 72, 847, 9801
Non-Hamiltonian 1-tough simplicial polyhedra with n nodes. Ref Dil92. [1,14; A7031]

M4279 1, 6, 72, 1320, 32760, 1028160, 39070080, 1744364160, 89513424000, 5191778592000, 335885501952000, 23982224839372800, 1873278229119897600
Dissections of a ball: $(3n+3)!/(2n+3)!$. Ref CMA 2 25 70. MAN 191 98 71. [0,2; A1763, N1788]

M4280 1, 1, 6, 72, 1322, 32550, 1003632, 37162384, 1605962556, 79330914540, 4409098539560, 272297452742304, 18499002436677336, 1371050716542451672
$\Sigma (-1)^{n-k} C(n,k).C(k^2,n)$, $k = 0 .. n$. Ref hwg. [0,3; A3235]

M4281 1, 0, 1, 6, 72, 2320, 245765
Partitions of 1 into unique parts $1/n$. Ref mlb. [1,4; A6585]

M4282 1, 6, 80, 30240, 1814400, 2661120, 871782912000, 3138418483200, 84687482880000, 170303140572364800, 1124000727777607680000
Denominators of coefficients for central differences. Cf. M5035. Ref SAM 42 162 63. [2,2; A2676, N1789]

M4283 6, 90, 945, 9450, 93555, 638512875, 18243225, 325641566250, 38979295480125, 1531329465290625, 13447856940643125, 201919571963756521875
Denominator of Σk^{-2n}, $k \geq 1$. Ref FMR 1 84. JCAM 21 253 88. [1,1; A2432, N1790]

M4284 1, 6, 90, 1680, 34650, 756756, 17153136, 399072960, 9465511770, 227873431500, 5550996791340, 136526995463040, 3384731762521200
$(3n)!/n!^3$. [0,2; A6480]

M4285   1, 6, 90, 1860, 44730, 1172556, 32496156, 936369720, 27770358330,
842090474940, 25989269017140, 813689707488840, 25780447171287900
$2n$-step polygons on cubic lattice. Ref AIP 9 345 60. PTRS 273 586 73. [0,2; A2896,
N1791]

M4286   0, 1, 6, 90, 2040, 67950, 3110940, 187530840, 14398171200, 1371785398200,
158815387962000, 21959547410077200, 3574340599104475200
Stochastic matrices of integers. Ref SS70. DUMJ 33 763 66. [1,3; A1499, N1792]

M4287   1, 6, 90, 2520, 113400, 7484400, 681080400, 81729648000, 12504636144000,
2375880867360000, 548828480360160000, 151476660579404160000
$(2n)!/2^n$. Ref QJMA 47 110 16. FMR 1 112. DA63 2 283. PSAM 15 101 63. [1,2; A0680,
N1793]

M4288   1, 1, 6, 91, 2548, 111384, 6852768, 553361016, 55804330152, 6774025632340
Dyck paths. Ref SC83. [0,3; A6151]

M4289   1, 0, 1, 6, 91, 2820, 177661, 22562946, 5753551231, 2940064679040,
3007686166657921, 6156733583148764286, 25211824022994189751171
Certain subgraphs of a directed graph (inverse binomial transform of M1986). Ref DM 14
118 76. EIS § 2.7. [1,4; A5327]

M4290   6, 96, 960, 7680
A traffic light problem. Ref BIO 46 422 59. [4,1; A6044]

M4291   6, 96, 1200, 14400, 176400, 2257920, 30481920, 435456000, 6586272000,
105380352000
Coefficients of Laguerre polynomials. Ref AS1 799. [3,1; A1805, N1794]

M4292   6, 104, 1345, 16344, 200452
Paths through an array. Ref EJC 5 52 84. [3,1; A6676]

M4293   0, 0, 0, 0, 0, 6, 104, 2009, 36585
5-dimensional polyominoes with n cells. Ref CJN 18 367 75. [1,6; A6768]

M4294   1, 1, 6, 114, 5256, 507720, 93616560, 30894489360, 17407086641280,
16152167106391680, 23990233574783750400, 5573509644870749203200
$n!$ times number of posets with n elements. Cf. M3068. Ref LNM 403 21 74. [0,3; A3425]

M4295   0, 0, 6, 120, 1980, 32970, 584430, 11204676, 233098740, 5254404210,
127921380840, 3350718545460, 94062457204716, 2819367702529560
Bessel polynomial $y_n''(1)$. Ref RCI 77. [0,3; A1516, N1795]

M4296   1, 1, 6, 120, 5250, 395010, 45197460, 7299452160, 1580682203100,
441926274289500, 154940341854097800, 66565404923242024800
$a(n) = (n-1)^2 a(n-2) - 3C(n-1,3)a(n-4)$. Ref MU06 3 282. EDMN 34 4 44. [0,3;
A2370, N1796]

M4297 1, 6, 120, 30240, 14182439040, 154345556085770649600
First n-fold perfect number. Ref B1 22. UPNT B2. BR73 138. [1,2; A7539]

M4298 6, 130, 2380, 44100, 866250, 18288270, 416215800, 10199989800,
   268438920750, 7562120816250, 227266937597700, 7262844156067500
Associated Stirling numbers. Ref TOH 37 259 33. JO39 152. C1 256. [0,1; A0907, N1797]

M4299 1, 6, 140, 270, 672, 1638, 2970, 6200, 8190, 18600, 18620, 27846, 30240, 32760,
   55860, 105664, 117800, 167400, 173600, 237510, 242060, 332640, 360360, 539400
Both harmonic and arithmetic means of divisors are integral. See Fig M4299. Ref AMM 55
615 48. glc. [1,2; A7340]

||||||||||||||||||||||||||||||||||||||||||||||||||||||||||||||||||||||||||||||||||||||||||||||||||||||||||||||

Figure M4299. HARMONIC NUMBERS.

   There are two completely different sequences that are called **harmonic** numbers. The first
is the sequence of partial sums $\sum_{m=1}^{n} \frac{1}{m}$:

$$1, \frac{3}{2}, \frac{11}{6}, \frac{25}{12}, \frac{137}{60}, \frac{49}{20}, \ldots$$

[KN1 1 615]. The numerators and denominators form M2885, M1589 respectively. M1249
shows quite dramatically how slowly the sum diverges. The n-th harmonic number (with the
above definition) is the maximal distance that a stack of n cards can project beyond the edge
of a table without toppling [GKP 259]:

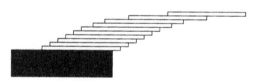

   The second definition of harmonic numbers (also called **Ore** numbers, after [AMM 55 615 48])
is that these are the numbers for which the harmonic mean of the divisors of n is an integer:
M4185. The harmonic means themselves are M0609. M4299 gives the numbers for which both
the harmonic and arithmetic means of the divisors are integral.

||||||||||||||||||||||||||||||||||||||||||||||||||||||||||||||||||||||||||||||||||||||||||||||||||||||||||||||

M4300 1, 6, 154, 66344, 15471166144, 663447306235471066144,
   15471166144731621543116634473062154710661 44
Octal formula for dragon curve of order n. Ref ScAm 216(4) 118 67. GA78 216. [1,2;
A3460]

M4301 1, 6, 168, 10672, 1198080, 208521728, 51874413568, 17449541107712,
   7622674735988736, 4193561606973095936, 2836052065377836597248
Expansion of tan(tan(tan x)). [0,2; A3720]

M4302 1, 6, 168, 20160, 9999360, 20158709760, 163849992929280,
   5348063769211699200, 6996123100331976642547200
Nonsingular binary matrices. Ref JSIAM 20 377 71. [1,2; A2884, N1798]

M4303 1, 6, 210, 223092870, 3217644767340672907899084554130
A highly composite sequence. Ref AMM 74 874 67. [0,2; A2037, N1799]

M4304 6, 216, 3706, 44060, 417486, 3386912, 24554562, 163587572, 1020918342,
   6050527496, 34395992698, 188971062108, 1009130882494, 5261414979024
Almost-convex polygons of perimeter $2n$ on square lattice. Ref EG92. [8,1; A7221]

M4305 0, 6, 240, 1020, 78120, 279930, 40353600, 134217720, 31381059600
Pile of coconuts problem. Ref AMM 35 48 28. [2,2; A2022, N1800]

M4306 0, 6, 350, 43260, 14591171
Singular $n \times n$ (0,1)-matrices. Ref JCT 3 198 67. [2,2; A0409, N1801]

M4307 6, 360, 10080, 259200, 239500800, 145297152000, 15692092416000,
   16005934264320000, 8515157028618240000, 3372002183332823040000
Denominators of coefficients for repeated integration. Cf. M4421. Ref SAM 28 56 49. [0,1;
A2684, N1802]

M4308 0, 6, 425, 65625, 27894671
Singular $n \times n$ (0,1)-matrices. Ref JCT 3 198 67. [2,2; A0410, N1803]

M4309 6, 438, 3962646
Post functions. Ref JCT 4 298 68. [2,1; A1328, N1804]

M4310 6, 480, 196560, 153498240, 214951968000
3-edge-colored connected trivalent graphs with $2n$ nodes. Ref RE58. [2,1; A6713]

M4311 6, 480, 197820, 154103040, 215643443400
3-edge-colored labeled trivalent graphs with $2n$ nodes. Ref RE58. [2,1; A6712]

M4312 6, 522, 152166, 93241002, 97949265606, 157201459863882,
   357802951084619046, 1096291279711115037162, 4350684698032741048452486
Generalized tangent numbers. Ref MOC 21 690 67. [1,1; A0411, N1805]

M4313 1, 6, 720, 1512000, 53343360000, 31052236723200000,
   29541557827511092800000, 456696058907168107347640320000000
An ill-conditioned determinant. Ref MOC 9 155 55. hpr. [1,2; A2204, N1806]

M4314 1, 6, 806, 2558556, 200525284806, 391901483074853556,
   1913826375235252849847856, 23362736428829686448189697999416056
Gaussian binomial coefficient $[2n,n]$ for $q=5$. Ref GJ83 99. ARS A17 329 84. [0,2;
A6114]

M4315 1, 6, 924, 81738720000, 256963707943061374889193111552000
Invertible Boolean functions of n variables. Ref PGEC 13 530 64. [1,2; A0652, N1807]

M4316 6, 1230, 134355076
Post functions. Ref JCT 4 296 68. [2,1; A1324, N1808]

M4317 6, 2862, 537157696
Post functions. Ref JCT 4 297 68. [2,1; A1326, N1809]

M4318 1, 1, 6, 5972, 1225533120
Symmetric Latin squares of order $2n$ with constant diagonal. Ref JRM 5 202 72. [1,3; A3191]

SEQUENCES BEGINNING . . ., 7, . . .

M4319 1, 7, 0, 9, 9, 7, 5, 9, 4, 6, 6, 7, 6, 6, 9, 6, 9, 8, 9, 3, 5, 3, 1, 0, 8, 8, 7, 2, 5, 4, 3, 8, 6, 0, 1, 0, 9, 8, 6, 8, 0, 5, 5, 1, 1, 0, 5, 4, 3, 0, 5, 4, 9, 2, 4, 3, 8, 2, 8, 6, 1, 7, 0, 7, 4, 4, 4, 2, 9
Decimal expansion of cube root of 5. [1,2; A5481]

M4320 0, 0, 0, 0, 0, 0, 0, 0, 0, 0, 0, 0, 0, 0, 7, 0, 16, 0, 27, 0, 40, 7, 55, 23, 72, 50, 91, 90, 119, 145, 165, 217, 240, 308, 357, 427, 531, 592, 779, 832, 1120, 1189, 1582, 1720, 2211
Strict 5th-order maximal independent sets in cycle graph. Ref YaBa94. [1,14; A7393]

M4321 1, 1, 7, 1, 31, 1, 127, 17, 73, 31, 2047, 1, 8191, 5461, 4681, 257, 131071, 73, 524287
From generalized Bernoulli numbers. Ref SAM 23 211 44. [2,3; A2678, N1810]

M4322 7, 2, 1, 1, 3, 18, 5, 1, 1, 6, 30, 8, 1, 1, 9, 42, 11, 1, 1, 12, 54, 14, 1, 1, 15, 66, 17, 1, 1, 18, 78, 20, 1, 1, 21, 90, 23, 1, 1, 24, 102, 26, 1, 1, 27, 114, 29, 1, 1, 30, 126, 32, 1, 1, 33
Continued fraction for e^2. Ref PE29 138. [1,1; A1204, N1811]

M4323 0, 1, 7, 2, 5, 8, 16, 3, 19, 6, 14, 9, 9, 17, 17, 4, 12, 20, 20, 7, 7, 15, 15, 10, 23, 10, 111, 18, 18, 18, 106, 5, 26, 13, 13, 21, 21, 21, 34, 8, 109, 8, 29, 16, 16, 16, 104, 11, 24, 24
Number of halving and tripling steps to reach 1 in '$3x+1$' problem. See Fig M2629. Ref UPNT E16. rwg. [1,3; A6577]

M4324 1, 0, 0, 7, 2, 7, 6, 4, 7, 0
Decimal expansion of proton mass (mass units). Ref FiFi87. BoBo90 v. [1,4; A3677]

M4325 0, 0, 7, 2, 9, 7, 3, 5, 3, 0, 8
Decimal expansion of fine-structure constant. Ref FiFi87. Lang91. [0,3; A3673]

M4326 1, 7, 3, 2, 0, 5, 0, 8, 0, 7, 5, 6, 8, 8, 7, 7, 2, 9, 3, 5, 2, 7, 4, 4, 6, 3, 4, 1, 5, 0, 5, 8, 7, 2, 3, 6, 6, 9, 4, 2, 8, 0, 5, 2, 5, 3, 8, 1, 0, 3, 8, 0, 6, 2, 8, 0, 5, 5, 8, 0, 6, 9, 7, 9, 4, 5, 1, 9, 3
Decimal expansion of square root of 3. Ref PNAS 37 444 51. MOC 22 234 68. [1,2; A2194, N1812]

M4327 1, 1, 1, 1, 7, 3, 97, 275, 2063, 15015, 53409, 968167, 752343, 77000363,
166831871, 7433044411, 43685848289, 843598411471, 9398558916159
Expansion of exp(tanh x). [0,5; A3723]

M4328 0, 7, 5, 65, 11, 63, 9, 17, 61, 69, 7, 15, 59, 23, 67, 93, 31, 13, 57, 57, 21, 65, 21, 91,
29, 73, 11, 55, 11, 55, 19, 63, 63, 19, 45, 89, 27, 27, 71, 27, 53, 53, 9, 35, 53, 79, 17, 61
Shortest wins at Beanstalk. Ref MMAG 59 262 86. [0,2; A5692]

M4329 1, 1, 1, 1, 7, 5, 85, 335, 1135, 15245, 13475, 717575, 4256825, 29782325,
525045275, 243258625, 56809006625, 415670267875, 5068080417875
Logarithmic transform of Fibonacci numbers. Ref BeSl94. EIS § 2.7. [1,5; A7553]

M4330 1, 1, 1, 7, 5, 145, 5, 6095, 5815, 433025, 956375, 46676375, 172917875,
7108596625, 38579649875, 1454225641375, 10713341611375, 384836032842625
$a(n) = a(n-1) - (n-1)(n-2)a(n-2)$. Ref DUMJ 26 573 59. hpr. [1,4; A2019, N1813]

M4331 0, 1, 7, 5, 3635, 557485, 7596391, 19681954039, 32139541115
Coefficients of Green function for cubic lattice. Ref PTRS 273 593 73. [0,3; A3299]

M4332 1, 7, 7, 2, 4, 5, 3, 8, 5, 0, 9, 0, 5, 5, 1, 6, 0, 2, 7, 2, 9, 8, 1, 6, 7, 4, 8, 3, 3, 4, 1, 1, 4,
5, 1, 8, 2, 7, 9, 7, 5, 4, 9, 4, 5, 6, 1, 2, 2, 3, 8, 7, 1, 1, 2, 8, 2, 1, 3, 8, 0, 7, 7, 8, 9, 8, 5, 2, 9, 1
Decimal expansion of square root of π. Ref RS8 XVIII. [1,2; A2161, N1814]

M4333 7, 7, 127, 463, 463, 487, 1423, 33247, 73327, 118903, 118903, 118903, 454183,
773767, 773767, 773767, 773767, 86976583, 125325127, 132690343, 788667223
Sequence of prescribed quadratic character. Ref MOC 24 444 70. [3,1; A1988, N1888]

M4334 7, 8, 9, 9, 9, 7, 8, 8, 8, 7, 7, 8, 9, 11, 9, 8, 12, 12, 12, 9, 15, 14, 15, 15, 15, 13, 14,
14, 14, 9, 16, 15, 16, 16, 16, 14, 15, 15, 15, 11, 18, 17, 18, 18, 18, 16, 17, 17, 17, 9, 19, 18
Letters in ordinal numbers (in French). [1,1; A6969]

M4335 0, 1, 7, 8, 16, 19, 20, 23, 111, 112, 115, 118, 121, 124, 127, 130, 143, 144, 170,
178, 181, 182, 208, 216, 237, 261, 267, 275, 278, 281, 307, 310, 323, 339, 350, 353, 374
'3x+1' records (iterations). See Fig M2629. Ref GEB 400. ScAm 250(1) 12 84. CMWA 24
96 92. [1,3; A6878]

M4336 1, 7, 8, 23, 31, 54, 85, 309, 7810, 8119, 40286, 128977, 427217, 3119496,
162641009, 165760505, 494162019, 1648246562, 3790655143, 58508073707
Convergents to fifth root of 2. Ref AMP 46 115 1866. L1 67. hpr. [1,2; A2362, N1815]

M4337 7, 9, 40, 74, 1526, 5436, 2323240, 29548570, 5397414549030, 873117986721660,
2913208381320760028721924, 762335018736884842676898606570
$a(n+1) = a(n-1)^2 - a(n)$, $a(n+2) = a(n)^2 - a(n-1)$. Ref ChLi 49(6) 14 94.
[0,1; A7449]

M4338 7, 10, 14, 16, 20, 24, 28, 28, 32
Rational points on curves of genus 3 over $GF(q)$. Ref STNB 22 1 83. HW84 51. [2,1;
A5526]

M4339 7, 11, 19, 23, 31, 43, 47, 59, 67, 71, 83, 103, 107, 127, 131, 139, 151, 163, 167, 179, 191, 199, 211, 227, 239, 251, 263, 271, 283, 307, 311, 331, 347, 367, 379, 383, 419
Prime determinants of forms with class number 2. Ref IAS 2 178 35. [1,1; A2052, N1816]

M4340 7, 11, 26, 45, 83, 125, 140, 182, 197, 201, 216, 239, 258, 311, 330, 353, 444, 467, 482, 486, 524, 539, 558, 600, 752, 771, 843, 881, 885, 923, 980, 999, 1071, 1113
$(n^2 + n + 1)/19$ is prime. Ref CU23 1 252. [1,1; A2643, N1817]

M4341 1, 7, 11, 27, 77, 107, 111, 127, 177, 777, 1127, 1177, 1777, 7777, 11777, 27777, 77777, 107777, 111777, 127777, 177777, 777777, 1127777, 1177777, 1777777, 7777777
Smallest number requiring n syllables in English. [1,2; A2810, N1818]

M4342 7, 13, 17, 23, 27, 33, 37, 53, 63, 67, 77, 87, 97, 103, 113, 127, 137, 147, 153, 163, 167, 197, 223, 227, 247, 263, 267, 277, 283, 287, 297, 303, 323, 347, 363, 367, 373, 383
$(n^2 + 1)/10$ is prime. Ref EUL (1) 3 25 17. [1,1; A2733, N1047]

M4343 1, 7, 13, 17, 84, 57, 93, 81, 63
Coefficients of a modular function. Ref GMJ 8 29 67. [-4,2; A5762]

M4344 7, 13, 19, 31, 37, 43, 61, 67, 73, 79, 97, 103, 109, 127, 139, 151, 157, 163, 181, 193, 199, 211, 223, 229, 241, 271, 277, 283, 307, 313, 331, 337, 349, 367, 373, 379, 397
Primes of form $6n + 1$. Ref RE75 1. AS1 870. [1,1; A2476, N1819]

M4345 7, 13, 19, 37, 79, 97, 139, 163, 313, 349, 607, 709, 877, 937, 1063, 1129, 1489, 1567, 1987, 2557, 2659, 3313, 3547, 4297, 5119, 5557, 7489, 8017, 8563, 9127, 9319
Primes of form $n^2 + 3n + 9$. Ref MOC 28 1140 74. [1,1; A5471]

M4346 1, 7, 13, 97, 8833, 77968897, 6079148431583233, 36956045653220845240164417232897
A nonlinear recurrence. Ref AMM 70 403 63. [0,2; A1544, N1820]

M4347 1, 7, 14, 7, 49, 21, 35, 41, 49, 133, 98, 21, 126, 112, 176, 105, 126, 140, 35, 147, 259, 98, 420, 224, 238, 455, 273, 14, 322, 406, 35, 7, 637, 196, 245, 181, 574, 462, 147
Expansion of $\Pi(1 - x^n)^7$. Ref KNAW 59 207 56. [0,2; A0730, N1821]

M4348 7, 14, 19, 29, 40, 44, 52, 59, 73, 83, 94, 107, 115, 122, 127, 137, 148, 161, 169, 185, 199, 218, 229, 242, 250, 257, 271, 281, 292, 305, 313, 320, 325, 335, 346, 359, 376
$a(n + 1) = a(n) +$ sum of digits of $a(n)$. Ref PC 4 37-12 76. jos. [1,1; A6507]

M4349 7, 15, 23, 28, 31, 39, 47, 55, 60, 63, 71, 79, 87, 92, 95, 103, 111, 112, 119, 124, 127, 135, 143, 151, 156, 159, 167
The sum of 4 (but no fewer) squares. Ref D1 2 261. [1,1; A4215]

M4350 7, 15, 29, 57, 109, 213
Maximal cycles in trivalent graphs. Ref ARS 11 292 81. [3,1; A6188]

M4351 1, 7, 15, 292, 436, 20766, 78629, 179136, 12996958, 878783625
Increasing partial quotients of π. Ref MOC 31 1044 77. rwg. [1,2; A7541]

M4352 7, 16, 49, 212, 1158, 7584, 57720, 499680, 4843440, 51932160, 610001280,
   7787404800, 107336275200, 1588369305600, 25113574886400, 422465999155200
$\Sigma(n+k)!C(3,k)$, $k = 0 \ .. \ 3$. Ref CJM 22 26 70. [-1,1; A1345, N1822]

M4353 7, 17, 19, 23, 29, 47, 59, 61, 97, 109, 113, 131, 149, 167, 179, 181, 193, 223, 229,
   233, 257, 263, 269, 313, 337, 367, 379, 383, 389, 419, 433, 461, 487, 491, 499, 503, 509
Primes with 10 as a primitive root. Ref Krai24 1 61. HW1 115. AS1 864. [1,1; A1913,
N1823]

M4354 7, 17, 23, 31, 41, 47, 71, 73, 79, 89, 97, 103, 113, 127, 137, 151, 167, 191, 193,
   199, 223, 233, 239, 241, 257, 263, 271, 281, 311, 313, 337, 353, 359, 367, 383, 401, 409
Primes $\equiv \pm 1$ mod 8. Ref AS1 870. [1,1; A1132, N1824]

M4355 0, 0, 0, 0, 0, 1, 0, 1, 7, 17, 30, 49, 124, 321, 761, 1721, 3815
Self-avoiding walks on square lattice. Ref JCT A13 181 72. [4,9; A6142]

M4356 7, 17, 31, 43, 79, 89, 113, 127, 137, 199, 223, 233, 257, 281, 283, 331, 353, 401,
   449, 463, 487, 521, 569, 571, 593, 607, 617, 631, 641, 691, 739, 751, 809, 811, 823, 857
Primes with 3 as smallest primitive root. Ref Krai24 1 57. AS1 864. [1,1; A1123, N1825]

M4357 1, 7, 18, 24, 24, 6, 66, 258, 1014, 3906, 14760, 54696, 198510, 704010, 2431110,
   8130096, 26103624, 79292226, 221534442, 532863372, 870102906
Compressibility of hard-hexagon lattice gas model. Ref JPA 21 L986 88. [0,2; A7236]

M4358 1, 7, 18, 34, 55, 81, 112, 148, 189, 235, 286, 342, 403, 469, 540, 616, 697, 783,
   874, 970, 1071, 1177, 1288, 1404, 1525, 1651, 1782, 1918, 2059, 2205, 2356, 2512, 2673
Heptagonal numbers $n(5n-3)/2$. See Fig M2535. Ref D1 2 2. B1 189. [1,2; A0566,
N1826]

M4359 7, 18, 44, 88, 169, 296, 507, 824, 1314, 2029, 3083, 4578, 6714, 9676, 13795,
   19408, 27053, 37302, 51029, 69180, 93139, 124447, 165259, 218021, 286068, 373207
Bipartite partitions. Ref ChGu56 11. [0,1; A2764, N1827]

M4360 1, 1, 7, 19, 25, 67, 205, 3389, 24469
Numerators of coefficients of Green's function for cubic lattice. Cf. M2116. Ref PTRS 273
590 73. [0,3; A3282]

M4361 7, 19, 26, 37, 44, 56, 63
A card-arranging problem. Ref GA88 81. [1,1; A6063]

M4362 1, 7, 19, 37, 61, 91, 127, 169, 217, 271, 331, 397, 469, 547, 631, 721, 817, 919,
   1027, 1141, 1261, 1387, 1519, 1657, 1801, 1951, 2107, 2269, 2437, 2611, 2791, 2977
Hex numbers: $3n(n+1)+1$. See Fig M2535. Ref INOC 24 4550 85. AMM 95 701 88.
GA88 18. [0,2; A3215]

M4363 1, 7, 19, 37, 61, 127, 271, 331, 397, 547, 631, 919, 1657, 1801, 1951, 2269, 2437, 2791, 3169, 3571, 4219, 4447, 5167, 5419, 6211, 7057, 7351, 8269, 9241, 10267, 11719
Cuban primes: $p = (x^3 - y^3)/(x - y)$, $x = y + 1$. Ref MES 41 144 12. CU23 1 259. [1,2; A2407, N1828]

M4364 1, 1, 1, 7, 19, 41, 751, 989, 2857, 16067, 2171465, 1364651, 6137698213, 90241897, 105930069, 15043611773, 55294720874657, 203732352169
Cotesian numbers. Ref QJMA 46 63 14. [1,4; A2177, N1829]

M4365 7, 19, 47, 97, 189, 339, 589, 975, 1576, 2472, 3804, 5727, 8498, 12400, 17874, 25433, 35818, 49908, 68939, 94378, 128234, 172917, 231630, 308240, 407804, 536412
Bipartite partitions. Ref PCPS 49 72 53. ChGu56 1. [0,1; A0491, N1830]

M4366 7, 19, 53, 149, 421, 1193, 3387, 9627, 27383, 77923
Keys. Ref MAG 53 11 69. [1,1; A2714, N1832]

M4367 1, 7, 19, 57, 81, 251, 437, 691, 739, 1743, 3695, 6619, 8217, 9771, 14771, 15155, 16831, 18805, 26745, 30551, 41755, 46297, 54339, 72359, 86407, 96969, 131059
Lattice points in spheres. Ref MOC 20 306 66. [0,2; A0413, N1833]

M4368 1, 1, 7, 19, 72, 196, 561, 1368, 3260, 7105, 14938, 29624, 56822, 104468, 186616, 322786, 544802, 896259, 1444147
n-bead necklaces with 12 red beads. Ref JAuMS 33 12 82. [6,3; A5516]

M4369 1, 1, 7, 19, 73, 241, 847, 2899, 10033, 34561, 119287, 411379, 1419193, 4895281, 16886527, 58249459, 200931553, 693110401, 2390878567, 8247309139, 28449011113
$a(n) = 2a(n-1) + 5a(n-2)$. Ref MQET 1 11 16. [0,3; A2533, N1834]

M4370 0, 0, 1, 7, 20, 48, 100, 194, 352, 615, 1034, 1693, 2705, 4239, 6522, 9889, 14786, 21844, 31913, 46165, 66162, 94035, 132600, 185637, 258128, 356674, 489906, 669173
From a partition triangle. Ref AMM 100 288 93. [1,4; A7044]

M4371 1, 7, 21, 35, 28, 21, 105, 181, 189, 77, 140, 385, 546, 511, 252, 203, 693, 1029, 1092, 798, 203, 581, 1281, 1708, 1687, 1232, 413, 602, 1485, 2233, 2366, 2009, 1099
Coefficients of powers of η function. Ref JLMS 39 435 64. [7,2; A1485, N1835]

M4372 0, 1, 1, 7, 21, 112, 456, 2603, 13203
Nontrivial Baxter permutations of length $2n - 1$. Ref MAL 2 25 67. [1,4; A1185, N1837]

M4373 7, 21, 112, 588, 3360, 19544, 117648, 720300, 4483696, 28245840, 179756976, 1153430600, 7453000800, 48444446376, 316504099520
Irreducible polynomials of degree n over $GF(7)$. Ref AMM 77 744 70. [1,1; A1693, N1838]

M4374 1, 7, 22, 50, 95, 161, 252, 372, 525, 715, 946, 1222, 1547, 1925, 2360, 2856, 3417, 4047, 4750, 5530, 6391, 7337, 8372, 9500, 10725, 12051, 13482, 15022, 16675, 18445
Hexagonal pyramidal numbers: $n(n+1)(4n-1)/6$. See Fig M3382. Bisection of M2640. Ref D1 2 2. B1 194. [1,2; A2412, N1839]

M4375   7, 22, 153, 15209
Switching networks. Ref JFI 276 321 63. [1,1; A0835, N1840]

M4376   7, 23, 31, 47, 71, 79, 103, 127, 151, 167, 191, 199, 223, 239, 263, 271, 311, 359,
367, 383, 431, 439, 463, 479, 487, 503, 599, 607, 631, 647, 719, 727, 743, 751, 823, 839
Primes of form $8n + 7$. Ref AS1 870. [1,1; A7522]

M4377   7, 23, 47, 71, 199, 167, 191, 239, 383, 311, 431, 647, 479, 983, 887, 719, 839,
1031, 1487, 1439, 1151, 1847, 1319, 3023, 1511, 1559, 2711, 4463, 2591, 2399, 3863
Smallest prime $\equiv 7$ mod 8 where $Q(\sqrt{-p})$ has class number $2n + 1$. Cf. M4402. Ref MOC
24 492 70. BU89 224. [0,1; A2146, N1841]

M4378   7, 23, 61, 127, 199, 337, 479, 677, 937, 1193, 1511, 1871, 2267, 2707, 3251, 3769,
4349, 5009, 5711, 6451, 7321, 8231, 9173, 10151, 11197, 12343, 13487, 14779
A special sequence of primes. Ref ACA 6 372 61. [2,1; A1275, N1842]

M4379   1, 7, 23, 63, 159, 383, 895, 2047, 4607, 10239, 22527, 49151, 106495, 229375,
491519, 1048575, 2228223, 4718591, 9961471, 20971519, 44040191, 92274687
Woodall numbers $n.2^n - 1$. Ref BR73 159. [1,2; A3261]

M4380   7, 23, 69, 165, 345
Postage stamp problem. Ref CJN 12 379 69. AMM 87 208 80. [1,1; A5342]

M4381   7, 23, 71, 311, 479, 1559, 5711, 10559, 18191, 31391, 307271, 366791, 366791,
2155919, 2155919, 2155919, 6077111, 6077111, 98538359, 120293879, 131486759
Negative pseudo-squares mod p. Ref MOC 24 436 70. [2,1; A1984, N2226]

M4382   7, 23, 71, 311, 479, 1559, 5711, 10559, 18191, 31391, 366791, 366791, 366791,
4080359, 12537719, 30706079, 36415991, 82636319, 120293879, 120293879
Smallest prime such that first n primes are residues. Ref RS9 XV. MOC 24 436 70. [1,1;
A2223, N1843]

M4383   1, 7, 24, 50, 58, 3, 120, 200, 39, 402, 728, 246, 1200
Expansion of a modular function. Ref PLMS 9 384 59. [-2,2; A6707]

M4384   1, 7, 25, 63, 129, 231, 377, 575, 833, 1159, 1561, 2047, 2625, 3303, 4089, 4991,
6017, 7175, 8473, 9919, 11521, 13287, 15225, 17343, 19649, 22151, 24857, 27775
Expansion of $(1+x)^3/(1-x)^4$. Ref SIAR 12 277 70. C1 81. [0,2; A1845, N1844]

M4385   1, 7, 25, 65, 140, 266, 462, 750, 1155, 1705, 2431, 3367, 4550, 6020, 7820, 9996,
12597, 15675, 19285, 23485, 28336, 33902, 40250, 47450, 55575, 64701, 74907, 86275
4-dimensional pyramidal numbers: $(3n + 1).C(n + 2, 3)/4$. See Fig M3382. Ref AS1 835.
DKB 223. B1 195. [1,2; A1296, N1845]

M4386   1, 7, 25, 66, 143, 273, 476, 775, 1197, 1771, 2530, 3510, 4750, 6293, 8184, 10472,
13209, 16450, 20254, 24682, 29799, 35673, 42375, 49980, 58565, 68211, 79002
Fermat coefficients. Ref MMAG 27 141 54. [5,2; A0970, N1846]

M4387 1, 7, 27, 77, 182, 378, 714, 1254, 2079, 3289, 5005, 7371, 10556, 14756, 20196,
27132, 35853, 46683, 59983, 76153, 95634, 118910, 146510, 179010, 217035, 261261
5-dimensional pyramidal numbers: $n(n+1) \cdots (n+3)(2n+3)/5!$. See Fig M3382. Ref
AS1 797. [1,2; A5585]

M4388 0, 1, 7, 27, 101, 337, 1151, 3483, 12965, 43773, 148529, 505605, 1727771,
5920823, 20345445, 70073901, 241849929, 836230109, 2896104951, 10044664507
Related to series-parallel networks. Ref AAP 4 123 72. [1,3; A6350]

M4389 1, 1, 7, 27, 321, 2265, 37575, 390915, 8281665, 114610545, 2946939975,
51083368875, 1542234996225, 32192256321225, 1114841223671175
$a(n+1) = a(n)+2n(2n+1)a(n-1)$. Ref FQ 10 171 72. [0,3; A3148]

M4390 1, 7, 28, 84, 210, 462, 924, 1716, 3003, 5005, 8008, 12376, 18564, 27132, 38760,
54264, 74613, 100947, 134596, 177100, 230230, 296010, 376740, 475020, 593775
Figurate numbers or binomial coefficients $C(n,6)$. Ref D1 2 7. RS3. B1 196. AS1 828.
[6,2; A0579, N1847]

M4391 1, 7, 28, 105, 357, 1197, 3857, 12300, 38430, 118874, 362670, 1095430, 3271751,
9673993
6-dimensional partitions of n. Ref PCPS 63 1099 67. [1,2; A0416, N1848]

M4392 1, 7, 28, 105, 357, 1232, 4067, 13301, 42357, 132845, 409262, 1243767, 3727360,
11036649, 32300795, 93538278, 268164868, 761656685, 2144259516, 5986658951
Euler transform of M4142. Ref PCPS 63 1100 67. EIS § 2.7. [1,2; A0417, N1849]

M4393 1, 7, 28, 140, 784, 4676, 29008, 184820, 1200304, 7907396, 52666768,
353815700, 2393325424, 16279522916, 111239118928, 762963987380, 5249352196144
$1^n + 2^n + \cdots + 7^n$. Ref AS1 813. [0,2; A1554, N1850]

M4394 7, 29, 57, 227, 455, 1821, 3641, 14563
Positions of remoteness 2 in Beans-Don't-Talk. Ref MMAG 59 267 86. [1,1; A5698]

M4395 1, 7, 29, 93, 256, 638, 1486, 3302, 7099, 14913, 30827, 63019, 127858, 258096,
519252, 1042380, 2089605, 4185195, 8377705, 16764265, 33539156, 67090962
$2^n - C(n,0) - \cdots - C(n,4)$. Ref MFM 73 18 69. [4,2; A2664, N1851]

M4396 1, 7, 29, 94, 263, 667, 1577, 3538, 7622
Arrays of dumbbells. Ref JMP 11 3098 70; 15 214 74. [1,2; A2941, N1852]

M4397 1, 7, 31, 37, 109, 121, 127, 133, 151, 157, 403, 421, 511, 529, 631, 637, 661, 679,
1579, 1621, 1633, 1969, 1981, 2017, 2041, 2047, 2053, 2071, 2077, 2143, 2149, 2167
Good numbers. Ref MOC 18 541 64. [1,2; A0696, N1853]

M4398 7, 31, 43, 67, 73, 79, 103, 127, 163, 181, 223, 229, 271, 277, 307, 313, 337, 349,
409, 421, 439, 457, 463, 499
Related to Kummer's conjecture. Ref Hass64 482. [1,1; A0921, N1854]

M4399 1, 7, 31, 115, 391, 1267, 3979, 12271, 37423, 113371, 342091, 1029799, 3095671,
9298147, 27914179, 83777503, 251394415, 754292827, 2263072411, 6789560412
Number of elements in $Z[\omega]$ whose 'smallest algorithm' is $\leq n$. Ref JALG 19 290 71. hwl.
[0,2; A6458]

M4400 1, 7, 31, 127, 73, 23, 8191, 151, 131071, 524287, 337, 47, 601, 262657, 233,
2147483647, 599479, 71, 223, 79, 13367, 431, 631, 2351, 4432676798593, 103
Smallest primitive factor of $2^{2n+1} - 1$. Ref Krai24 2 84. CUNN. [0,2; A2184, N1855]

M4401 1, 7, 31, 127, 73, 89, 8191, 151, 131071, 524287, 337, 178481, 1801, 262657,
2089, 2147483647, 599479, 122921, 616318177, 121369, 164511353, 2099863, 23311
Largest factor of $2^{2n+1} - 1$. Ref Krai24 2 84. CUNN. [0,2; A2588, N1856]

M4402 7, 31, 127, 487, 1423, 1303, 2143, 2647, 4447, 5527, 5647, 6703, 5503, 11383,
8863, 13687, 13183, 12007, 22807, 18127, 21487, 22303, 29863, 25303, 27127
Largest prime $\equiv 7 \bmod 8$ with class number $2n + 1$. Cf. M4377. Ref MOC 24 492 70. [0,1;
A2147, N1857]

M4403 1, 1, 7, 31, 127, 2555, 1414477, 57337, 118518239, 5749691557, 91546277357,
1792042792463, 1982765468311237, 286994504449393, 3187598676787461083
Numerators of cosecant numbers. Cf. M2983. Ref NO24 458. ANN 36 640 35. DA63 2
187. [0,3; A1896, N1858]

M4404 0, 1, 7, 31, 145, 659, 3013, 13739, 62685, 285931
Worst case of a Jacobi symbol algorithm. Ref JSC 10 605 90. [0,3; A5825]

M4405 1, 7, 32, 120, 400, 1232, 3584, 9984, 26880, 70400, 180224, 452608, 1118208,
2723840, 6553600, 15597568, 36765696, 85917696, 199229440, 458752000
Coefficients of Chebyshev polynomials. Ref RSE 62 190 46. AS1 795. [0,2; A1794,
N1859]

$$\text{G.f.:}\ (1 - x) / (1 - 2x)^4.$$

M4406 1, 7, 33, 123, 257, 515, 925, 1419, 2109, 3071, 4169, 5575, 7153, 9171, 11513,
14147, 17077, 20479, 24405, 28671, 33401, 38911, 44473, 50883, 57777, 65267, 73525
Points of norm $\leq n$ in cubic lattice. Ref PNISI 13 37 47. MOC 16 287 62. SPLAG 107.
[0,2; A0605, N1860]

M4407 1, 1, 7, 33, 715, 4199, 52003, 334305, 17678835, 119409675, 1641030105,
11435320455, 322476036831
Coefficients of Legendre polynomials. Ref MOC 3 17 48. [0,3; A1795, N1861]

M4408 1, 7, 34, 136, 487, 1615, 5079, 15349, 45009, 128899, 362266, 1002681, 2740448,
7411408, 19865445, 52840977, 139624510, 366803313, 958696860, 2494322662
n-node trees of height 7. Ref IBMJ 4 475 60. KU64. [8,2; A0418, N1862]

M4409   7, 34, 143, 560, 2108, 7752, 28101, 100947, 360526, 1282735, 4552624, 16131656, 57099056, 201962057, 714012495, 2523515514, 8916942687, 31504028992
Random walks. Ref DM 17 44 77. [1,1; A5023]

M4410   7, 34, 1056, 5884954
Switching networks. Ref JFI 276 320 63. [1,1; A0829, N1863]

M4411   1, 7, 35, 140, 483, 1498, 4277, 11425, 28889, 69734, 161735, 362271, 786877, 1662927, 3428770, 6913760, 13660346, 26492361, 50504755, 94766875, 175221109
Coefficients of an elliptic function. Ref CAY 9 128. [0,2; A1941, N1864]

$$\text{G.f.: } \Pi\,(1 - x^k)^{-c(k)},\ c(k) = 7,7,7,0,7,7,7,0,....$$

M4412   1, 7, 35, 140, 490, 1554, 4578, 12720, 33705, 85855, 211519, 506408, 1182650, 2702350, 6056850, 13343820, 28947240, 61926900, 130814600, 273163100, 564415390
Convolved Fibonacci numbers. Ref RCI 101. DM 26 267 79. [0,2; A1875, N1865]

$$\text{G.f.: } (1 - x - x^2)^{-7}.$$

M4413   1, 7, 35, 154, 637, 2548, 9996, 38760, 149226, 572033, 2187185, 8351070, 31865925, 121580760, 463991880, 1771605360, 6768687870, 25880277150
$7C(2n, n-3)/(n+4)$. Ref QAM 14 407 56. MOC 29 216 75. FQ 14 397 76. [3,2; A0588, N1866]

M4414   1, 7, 35, 155, 649, 2640, 10569, 41926, 165425, 650658, 2554607, 10020277, 39287173, 154022930
Permutations by inversions. Ref NET 96. MMAG 61 28 88. rkg. [7,2; A5285]

M4415   1, 7, 35, 155, 651, 2667, 10795, 43435, 174251, 698027, 2794155, 11180715, 44731051, 178940587, 715795115, 2863245995, 11453115051, 45812722347
Gaussian binomial coefficient $[n,2]$ for $q = 2$. Ref GJ83 99. ARS A17 328 84. [2,2; A6095]

M4416   1, 7, 35, 156, 670, 2886, 12797
Rhyme schemes. Ref ANY 319 463 79. [1,2; A5003]

M4417   1, 7, 36, 165, 715, 3003, 12376, 50388, 203490, 817190, 3268760, 13037895, 51895935, 206253075, 818809200, 3247943160, 12875774670, 51021117810
Binomial coefficients $C(2n+1, n-2)$. See Fig M1645. Ref AS1 828. [2,2; A3516]

M4418   7, 37, 58, 163, 4687, 30178, 30493, 47338
Extreme values of Dirichlet series. Ref PSPM 24 277 73. [1,1; A3521]

M4419   1, 7, 37, 176, 794, 3473, 14893, 63004, 263950, 1097790, 4540386, 18696432, 76717268
Rooted planar maps. Ref JCT B18 249 75. [2,2; A6419]

M4420 0, 0, 1, 7, 37, 197, 1172, 8018, 62814, 556014, 5488059, 59740609, 710771275,
9174170011, 127661752406, 1904975488436, 30341995265036, 513771331467372
$\Sigma (k-1)!.C(n,k)/2$, $k = 3 .. n$. Ref PIEE 115 763 68. DM 55 272 85. [1,4; A2807,
N1867]

M4421 1, 7, 37, 199, 40321, 5512813, 136601407, 32373535937, 4039314145093,
377880467185583, 123905113265594071
Numerators of coefficients for repeated integration. Cf. M4307. Ref SAM 28 56 49. [0,2;
A2683, N1868]

M4422 0, 7, 41, 84, 19, 62, 96, 301, 803, 18, 52, 95, 201, 802, 908, 519, 625, 236, 342,
943, 59, 66, 37, 44, 15, 22, 92, 99, 601, 806, 318, 523, 35, 24, 13, 2, 9, 61, 86, 39, 64, 17
Add 7, then reverse digits! Ref Robe92 15. [0,2; A7398]

M4423 1, 7, 41, 239, 1393, 8119, 47321, 275807, 1607521, 9369319, 54608393,
318281039, 1855077841, 10812186007, 63018038201, 367296043199, 2140758220993
NSW numbers: $a(n) = 6a(n-1) - a(n-2)$. Bisection of M2665. Ref AMM 4 25 1897.
IDM 10 236 03. ANN 36 644 35. BPNR 288. [0,2; A2315, N1869]

M4424 1, 1, 7, 41, 479, 59, 266681, 63397, 514639, 178939, 10410343, 18500393,
40799043101, 1411432849
Coefficients for numerical differentiation. Ref OP80 21. SAM 22 120 43. [2,3; A2701,
N1870]

M4425 1, 7, 43, 259, 1555, 9331, 55987, 335923, 2015539, 12093235, 72559411,
435356467, 2612138803, 15672832819, 94036996915, 564221981491, 3385331888947
$(6^n - 1)/5$. [1,2; A3464]

M4426 1, 7, 45, 323, 2621, 23811, 239653, 2648395, 31889517, 415641779, 5830753109,
87601592187, 1403439027805, 23883728565283, 430284458893701
Permutations of length n by rises. Sequence M2070 divided by 2. Ref DKB 263. [4,2;
A1266, N1871]

M4427 1, 1, 7, 45, 465, 5775, 88515, 1588545, 32852925
Binary phylogenetic trees with n labels. Ref LNM 884 198 81. [1,3; A6680]

M4428 7, 46, 4336, 134281216
Switching networks. Ref JFI 276 320 63. [1,1; A0823, N1872]

M4429 1, 7, 47, 342, 2754, 24552, 241128, 2592720, 30334320, 383970240, 5231113920,
76349105280, 1188825724800, 19675048780800, 344937224217600
Generalized Stirling numbers. Ref PEF 77 26 62. [0,2; A1711, N1873]

$$\text{E.g.f.: } -\ln (1 - x) / (1 - x)^3.$$

M4430 1, 1, 7, 47, 497, 6241, 95767, 1704527, 34741217, 796079041, 20273087527,
567864586607, 17352768515537, 574448847467041, 20479521468959287
Alternating 4-signed permutations. Ref EhRe94. [0,3; A6873]

M4431 1, 7, 49, 343, 2401, 16807, 117649, 823543, 5764801, 40353607, 282475249,
1977326743, 13841287201, 96889010407, 678223072849, 4747561509943
Powers of 7. Ref BA9. [0,2; A0420, N1874]

M4432 1, 7, 49, 415, 4321, 53887, 783889, 13031935
From solution to a difference equation. Ref FQ 25 83 87. [1,2; A5924]

M4433 1, 7, 50, 390, 3360, 31920, 332640, 3780000, 46569600, 618710400, 8821612800,
134399865600, 2179457280000, 37486665216000, 681734237184000
Expansion of $(1+2x) / (1-x)^5$. Ref rkg. [0,2; A5460]

M4434 1, 7, 55, 505, 5497, 69823, 1007407, 16157905, 284214097, 5432922775,
112034017735, 2476196276617, 58332035387017, 1457666574447247
Related to symmetric groups. Ref DM 21 320 78. [0,2; A5012]

$$\text{Lgd.e.g.f.: } e^{6x}.$$

M4435 1, 0, 0, 0, 0, 0, 0, 1, 7, 55, 529, 6192, 86580, 1425517, 27298231, 601580874,
15116315767, 429614643061, 13711655205088, 488332318973593
Nearest integer to Bernoulli number B_{2n}. See Fig M4189. Ref DA63 2 236. AS1 810. [0,9;
A2882, N1875]

M4436 1, 7, 56, 504, 5040, 55440, 665280, 8648640, 121080960, 1816214400,
29059430400, 494010316800, 8892185702400, 168951528345600, 3379030566912000
$n!/6!$. Ref PEF 107 19 63. [6,2; A1730, N1876]

M4437 0, 1, 7, 58, 519, 4856, 46780, 460027, 4593647, 46416730, 473464492,
4866762231, 50346419064, 523649493732, 5471647249551, 57402510799673
n-node animals on cubic lattice. Ref DU92 40. [0,3; A6193]

M4438 1, 7, 58, 838, 25171, 1610977
Certain subgraphs of a directed graph. Ref DM 14 119 76. [2,2; A5332]

M4439 1, 7, 61, 661, 8953, 152917, 3334921
Related to partially ordered sets. Ref JCT 6 17 69. [0,2; A1830, N1877]

M4440 7, 63, 254, 710, 1605, 3157, 5628, 9324, 14595, 21835, 31482, 44018, 59969,
79905, 104440, 134232, 169983, 212439, 262390, 320670, 388157, 465773, 554484
Series expansion for rectilinear polymers on square lattice. Ref JPA 12 2137 79. [2,1;
A7291]

$$\text{G.f.: } (7 + 28x + 9x^2) / (1 - x)^5.$$

M4441 1, 7, 66, 916, 16816
Semi-regular digraphs on *n* nodes. Ref KNAW 75 330 72. [2,2; A3286]

M4442 1, 1, 7, 70, 819, 10472, 141778, 1997688, 28989675, 430321633, 6503352856,
99726673130, 1547847846090, 24269405074740, 383846168712104
Dissections of a polygon: $C(7n,n)/(6n+1)$. Ref AMP 1 198 1841. DM 11 389 75. [0,3;
A2296, N1878]

M4443 1, 1, 7, 70, 910, 14532, 274778, 5995892, 148154860, 4085619622,
124304629050, 4133867297490, 149114120602860, 5796433459664946
Coefficients of iterated exponentials. Ref SMA 11 353 45. PRV A32 2342 85. [0,3; A1669,
N1879]

M4444 1, 1, 7, 71, 1001, 18089, 398959, 10391023, 312129649, 10622799089,
403978495031, 16977719590391, 781379079653017, 39085931702241241
Numerators of convergents to *e*. CF. M3062. Ref BAT 17 1871. MOC 2 69 46. [0,3;
A2119, N1880]

M4445 1, 7, 72, 891, 12672, 202770, 3602880, 70425747, 1503484416, 34845294582
Feynman diagrams of order $2n$. Ref PRV D18 1949 78. [1,2; A5413]

M4446 0, 0, 7, 74, 882, 11144, 159652, 2571960, 46406392, 928734944, 20436096048,
409489794464, 12752891909920, 357081983435904, 10712466529388608
Symmetric permutations. Ref LU91 1 222. JRM 7 181 74. LNM 560 201 76. [1,3; A0901,
N1881]

M4447 1, 1, 7, 77, 1155, 21973, 506989, 13761937, 429853851, 15192078027,
599551077881, 26140497946017, 1248134313062231, 64783855286002573
Coefficients of iterated exponentials. Ref SMA 11 353 45. PRV A32 2342 85. [0,3; A1765,
N1882]

M4448 7, 85, 1660, 48076, 1942416, 104587344, 7245893376, 628308907776,
66687811660800, 8506654697548800
Differences of reciprocals of unity. Ref DKB 228. [1,1; A0424, N1883]

M4449 1, 7, 85, 1777, 63601, 3882817, 403308865, 71139019777, 21276992674561
4-colored labeled graphs on *n* nodes. Ref CJM 12 413 60. rcr. [1,2; A0686, N1884]

M4450 1, 7, 88, 1731, 55094, 2806539
Pseudo-bricks with *n* nodes. Ref JCT B32 29 82. [4,2; A6291]

M4451 1, 7, 93, 1419, 25225, 472037, 9501537, 196190781, 4219610242, 92198459515,
2068590840349, 46897782768404, 1083052539395723
Witt vector *3!/3!. Ref SLC 16 107 88. [1,2; A6178]

M4452 7, 97, 997, 9973, 99991, 999983, 9999991, 99999989, 999999937, 9999999967, 99999999977, 999999999989, 9999999999971, 99999999999973, 999999999999989
Largest *n*-digit prime. Ref JRM 22 278 90. [1,1; A3618]

M4453 1, 1, 7, 97, 2063, 53409, 752343, 166831871, 43685848289, 9398558916159, 2116926930779225, 524586454143030495, 1446202903788768229905
Expansion of cos(tan *x*). [0,3; A3710]

M4454 1, 1, 7, 97, 2911, 180481, 22740607, 5776114177, 2945818230271, 3010626231336961, 6159741269315422207, 252179807565773388515457
Certain subgraphs of a directed graph (inverse binomial transform of M1986). Ref DM 14 118 76. EIS § 2.7. [1,3; A5014]

M4455 0, 7, 104, 1455, 20272, 282359, 3932760, 54776287, 762935264, 10626317415, 148005508552, 2061450802319, 28712305723920, 399910829332567
$a(n) = 14a(n-1) - a(n-2) + 6$. Ref AMM 53 465 46. [0,2; A1921, N1885]

M4456 1, 7, 106, 113, 33102, 33215, 66317, 99532, 265381, 364913, 1360120, 1725033, 25510582, 52746197, 78256779, 131002976, 340262731, 811528438, 1963319607
Denominators of convergents to π. See Fig M3097. Cf. M3097. Ref ELM 2 7 47. Beck71 171. [0,2; A2486, N1886]

M4457 1, 1, 1, 7, 107, 199, 6031, 5741, 1129981, 435569, 35661419, 1523489833, 45183033541, 12597680311, 19055094997949, 9331210633373, 104148936040729
Denominators of coefficients for repeated integration. Cf. M4158. Ref PHM 38 336 47. [1,4; A2687, N1887]

M4458 1, 1, 7, 127, 4369, 243649, 20036983, 2280356863, 343141433761, 65967241200001, 15773461423793767, 4591227123230945407
From inverse error function. Ref PJM 13 470 63. [0,3; A2067, N1889]

M4459 0, 7, 128, 975, 4608, 16340, 48384, 124303, 281600, 583746, 1146240, 2125108, 3691008, 6151880, 10055424, 15914895, 24136704, 35748899, 52583040
Related to representation as sums of squares. Ref QJMA 38 349 07. [1,2; A2614, N1890]

M4460 0, 0, 0, 0, 0, 7, 153, 3350, 65973
One-sided 5-dimensional polyominoes with *n* cells. Ref CJN 18 366 75. [1,6; A6761]

M4461 1, 7, 169, 14911, 4925281, 6195974527
(0,1)-matrices by 1-width. Ref DM 20 110 77. [1,2; A5019]

M4462 1, 7, 305, 33367, 6815585, 2237423527, 1077270776465, 715153093789687, 626055764653322945, 698774745485355051847, 9685533613874204366950 25
Multiples of Euler numbers. Ref MES 28 51 1898. FMR 1 75. hpr. [1,2; A2437, N1891]

M4463   7, 322, 33385282, 37210469265847998489922,
 5152232359967762949673799032952863895658354830437805361558104353682
$a(n+1)=a(n)(a(n)^2-3)$. Ref AMM 44 645 37. [0,1; A2000, N1892]

M4464   1, 7, 791, 3748629, 151648960887729, 13234975445675611138595307148089,
 41444465282455711991644958522615049159671653083333293470875123
Denominators of an expansion for π. Ref AMM 54 138 47. jww. [0,2; A1467, N1893]

M4465   1, 7, 1734, 89512864
Groupoids with n elements. Ref LE70 246. [1,2; A1424, N1894]

SEQUENCES BEGINNING . . ., 8, . . . AND . . ., 9, . . .

M4466   1, 8, 1, 7, 1, 2, 0, 5, 9, 2, 8, 3, 2, 1, 3, 9, 6, 5, 8, 8, 9, 1, 2, 1, 1, 7, 5, 6, 3, 2, 7, 2, 6,
 0, 5, 0, 2, 4, 2, 8, 2, 1, 0, 4, 6, 3, 1, 4, 1, 2, 1, 9, 6, 7, 1, 4, 8, 1, 3, 3, 4, 2, 9, 7, 9, 3, 1, 3, 0
Decimal expansion of cube root of 6. [1,2; A5486]

M4467   1, 1, 8, 1, 26, 8, 48, 1, 73, 26, 120, 8, 170, 48, 208, 1, 290, 73, 360, 26, 384, 120,
 528, 8, 651, 170, 656, 48, 842, 208, 960, 1, 960, 290, 1248, 73, 1370, 360, 1360, 26, 1682
Related to the divisors of n. Ref QJMA 20 164 1884. [1,3; A2173, N1895]

M4468   1, 8, 2, 6, 1, 2, 3, 5, 7, 1, 1, 1, 2, 2, 3, 4, 4, 5, 6, 8, 9, 1, 1, 1, 1, 1, 1, 1, 2, 2, 2, 2, 3, 3,
 3, 4, 4, 5, 5, 5, 6, 6, 7, 7, 8, 9, 9, 1, 1, 1, 1, 1, 1, 1, 1, 1, 1, 1, 1, 1, 2, 2, 2, 2, 2, 2, 2, 2, 3, 3, 3
Initial digits of cubes. [1,2; A2994]

M4469   1, 1, 8, 2, 26, 27, 24, 136, 135, 162, 568, 486, 624
Expansion of a modular function. Ref PLMS 9 384 59. [−2,3; A6708]

M4470   1, 8, 3, 6, 1, 5, 2, 7, 0, 1
Decimal expansion of proton-to-electron mass ratio. Ref RMP 59 1141 87. Lang91. [4,2;
A5601]

M4471   1, 8, 3, 8, 6, 8, 3, 6, 6, 2
Decimal expansion of neutron-to-electron mass ratio. Ref RMP 59 1142 87. Lang91. [4,2;
A6833]

M4472   0, 0, 8, 4, 8, 16, 24, 44, 80, 144, 264, 484, 888, 1632, 3000, 5516, 10144, 18656,
 34312, 63108, 116072, 213488, 392664, 722220, 1328368, 2443248, 4493832, 8265444
$a(n+3)=a(n+2)+a(n+1)+a(n)-4$. Ref CMB 7 262 64. JCT 7 315 69. [0,3; A0803,
N2232]

M4473 1, 0, 0, 8, 6, 0, 0, 0, 12, 0, 0, 24, 8, 0, 0, 0, 6, 0, 0, 24, 24, 0, 0, 0, 24, 0, 0, 32, 0, 0, 0, 0, 12, 0, 0, 48, 30, 0, 0, 0, 24, 0, 0, 24, 24, 0, 0, 0, 8, 0, 0, 48, 24, 0, 0, 0, 48, 0, 0, 72, 0
Theta series of body-centered cubic lattice. Ref SPLAG 116. [0,4; A4013]

M4474 1, 0, 0, 8, 6, 6, 4, 9, 0, 4
Decimal expansion of neutron mass (mass units). Ref FiFi87. BoBo90 v. [1,4; A3675]

M4475 8, 8, 17, 19, 300, 1991, 2492, 7236, 10586, 34588, 63403, 70637, 1236467, 5417668, 5515697, 5633167, 7458122, 9637848, 9805775, 41840855, 58408380
Engel expansion of $\pi - 3$. [1,1; A6784]

M4476 0, 8, 10, 11, 12, 16, 32, 54, 97, 183, 334, 636, 1218, 2339, 4495, 8807, 17280, 33924, 66630, 130921, 259503
A generalized Conway-Guy sequence. Ref MOC 50 312 88. [0,2; A6757]

M4477 1, 8, 10, 11, 18, 80, 81, 88, 100, 101, 108, 110, 111, 118, 180, 181, 188, 800, 801, 808, 810, 811, 818, 880, 881, 888, 1001, 1008, 1010, 1011, 1018, 1080, 1081, 1088, 1100
Horizontally symmetric numbers. [1,2; A7284]

M4478 1, 8, 10, 45, 297, 2322, 2728, 4445, 4544, 4949, 5049, 5455, 5554, 7172, 27100, 44443, 55556, 60434, 77778, 143857, 208494, 226071, 279720, 313390
Kaprekar triples. Ref Well86 151. rpm. [1,2; A6887]

M4479 1, 8, 10, 80, 231, 248, 1466, 80, 4766, 1944, 9600, 2704, 15525, 3984, 25498, 10816, 29760, 800, 1994, 11728, 29362, 5560, 2310, 1952, 21649, 38128, 192854, 2480
Bisection of M3347. Ref QJMA 38 190 07. [0,2; A2286, N1896]

M4480 1, 8, 11, 69, 88, 96, 101, 111, 181, 609, 619, 689, 808, 818, 888, 906, 916, 986, 1001, 1111, 1691, 1881, 1961, 6009, 6119, 6699, 6889, 6969, 8008, 8118, 8698, 8888
Strobogrammatic numbers: the same upside down. Ref MMAG 34 184 61. [1,2; A0787, N1897]

M4481 1, 8, 11, 88, 101, 111, 181, 808, 818, 888, 1001, 1111, 1881, 8008, 8118, 8888, 10001, 10101, 10801, 11011, 11111, 11811, 18081, 18181, 18881, 80008, 80108, 80808
Mirror symmetry about middle. [1,2; A6072]

M4482 0, 0, 0, 0, 0, 0, 0, 0, 0, 0, 0, 0, 0, 0, 0, 0, 0, 0, 0, 0, 1, 8, 12, 26, 160, 441
Simple perfect squared squares of order n. See Fig M4482. Ref BoDu92. [1,22; A6983]

||||||||||||||||||||||||||||||||||||||||||||||||||||||||||||||||||||||||||||||||||||||||||||||||||||||||||||||||||||||||||||||||||||

Figure M4482. SQUARING A SQUARE.

M4482 gives the number of ways that a square can be dissected into n different squares such that no proper subset of the squares forms a rectangle or square. The smallest example, containing 21 squares, was found by A. J. W. Duijvestijn, and is shown below. Not many terms are known. See also M2496, M4614.

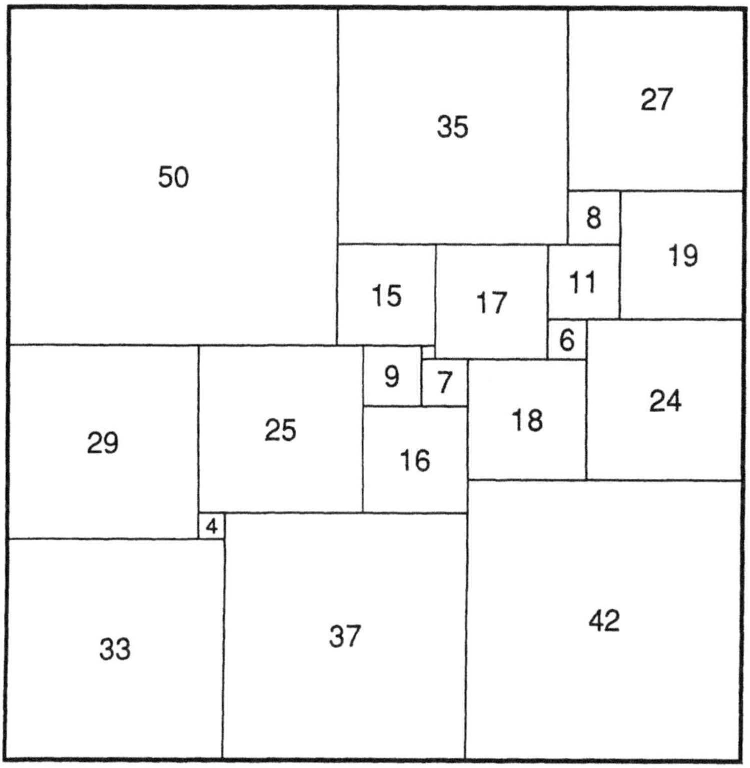

||||||||||||||||||||||||||||||||||||||||||||||||||||||||||||||||||||||||||||||||||||||||||||||||||||||||||||||||||||||||||||||||||||

M4483 1, 8, 12, 64, 210, 96, 1016, 512, 2043, 1680, 1092, 768, 1382, 8128, 2520, 4096, 14706, 16344, 39940, 13440, 12192, 8736, 68712, 6144, 34025, 11056
Related to representation as sums of squares. Ref QJMA 38 198 07. [1,2; A2288, N1898]

M4484 8, 13, 17, 22, 29, 34, 40, 47, 56
Zarankiewicz's problem. Ref LNM 110 142 69. [3,1; A6613]

M4485 8, 13, 18, 24, 31, 38, 46, 55
Zarankiewicz's problem. Ref LNM 110 143 69. [2,1; A6619]

M4486 1, 8, 15, 212, 865, 31560, 397285, 8760472, 73512810, 7619823960
Coefficients for step-by-step integration. Ref JACM 11 231 64. [0,2; A2406, N1899]

M4487 0, 1, 1, 8, 16, 224, 608, 13320, 41760, 1366152
From a Fibonacci-like differential equation. Ref FQ 27 309 89. [0,4; A5445]

M4488 1, 8, 20, 0, 70, 64, 56, 0, 125, 160, 308, 0, 110, 0, 520, 0, 57, 560, 0, 0, 182, 512,
880, 0, 1190, 448, 884, 0, 0, 0, 1400, 0, 1330, 1000, 1820, 0, 646, 1280, 0, 0, 1331, 2464
Expansion of $\Pi(1-x^k)^8$. Ref KNAW 59 207 56. [0,2; A0731, N1900]

M4489 0, 8, 20, 36, 64, 80, 112
Excess of a Hadamard matrix of order $4n$. Ref KNAW 80 361 77. [0,2; A4118]

M4490 1, 8, 20, 38, 63, 96, 138, 190, 253, 328, 416, 518, 635
Rooted planar maps. Ref JCT B18 248 75. [2,2; A6416]

M4491 0, 1, 0, 0, 8, 20, 96, 656, 5568, 48912, 494080, 5383552, 65097600
Permutations of length n with 1 fixed and 1 reflected point. Ref Sim92. [0,5; A7016]

M4492 8, 21, 29, 42, 50, 55, 63, 76, 84, 97, 110, 118, 131, 139, 144, 152, 164, 173, 186,
194, 199, 207, 220, 228, 241, 254, 262, 275, 283, 288, 296, 309, 317, 330, 338, 343, 351
Related to Fibonacci representations. Ref FQ 11 385 73. [1,1; A3249]

M4493 1, 8, 21, 40, 65, 96, 133, 176, 225, 280, 341, 408, 481, 560, 645, 736, 833, 936,
1045, 1160, 1281, 1408, 1541, 1680, 1825, 1976, 2133, 2296, 2465, 2640, 2821, 3008
Octagonal numbers: $n(3n-2)$. See Fig M2535. Ref D1 2 1. B1 189. [1,2; A0567, N1901]

M4494 1, 0, 1, 8, 22, 51, 342, 2609, 16896, 99114
A queen-placing problem on an $n \times n$ board. Ref SIAR 14 173 72. ACA 23 117 73. [1,4; A2968]

M4495 8, 23, 57, 119, 231, 415, 719, 1189, 1915, 2997, 4595, 6898, 10198, 14833, 21303,
30211, 42393, 58869, 81028, 110551, 149683, 201160, 268539, 356167, 469630
Bipartite partitions. Ref ChGu56 11. [0,1; A2765, N1902]

M4496 8, 24, 24, 32, 48, 24, 48, 72, 24, 56, 72, 48, 72, 72, 48, 48, 120, 72, 56, 96, 24, 120,
120, 48, 96, 96, 72, 96, 120, 48, 104, 168, 96, 48, 120, 72, 96, 192, 72, 144, 96, 72, 144
Theta series of cubic lattice w.r.t. deep hole. Ref SPLAG 107. [0,1; A5878]

M4497 1, 8, 26, 56, 98, 152, 218, 296, 386, 488, 602, 728, 866, 1016, 1178, 1352, 1538,
1736, 1946, 2168, 2402, 2648, 2906, 3176, 3458, 3752, 4058, 4376, 4706, 5048, 5402
Points on surface of cube: $6n^2+2$. Ref MF73 46. Coxe74. INOC 24 4550 85. [0,2; A5897]

M4498 1, 8, 26, 60, 115, 196, 308, 456, 645, 880, 1166, 1508, 1911, 2380, 2920, 3536,
4233, 5016, 5890, 6860, 7931, 9108, 10396, 11800, 13325, 14976, 16758, 18676, 20735
Heptagonal pyramidal numbers: $n(n+1)(5n-2)/6$. See Fig M3382. Ref D1 2 2. B1 194.
[1,2; A2413, N1904]

M4499 1, 8, 27, 64, 125, 216, 343, 512, 729, 1000, 1331, 1728, 2197, 2744, 3375, 4096,
4913, 5832, 6859, 8000, 9261, 10648, 12167, 13824, 15625, 17576, 19683, 21952, 24389
The cubes. Ref BA9. [1,2; A0578, N1905]

M4500 8, 27, 384, 12100, 736128, 70990416, 9939419136, 1896254551296,
472882821120000, 149328979405056000, 582558352265557644800
Structure constants for certain representations of S_n. Ref JCT A66 115 94. [1,1; A7235]

M4501 8, 28, 32, 56, 64, 68, 72, 92
Determinants of indecomposable indefinite ternary forms. Ref SPLAG 399. [1,1; A6377]

M4502 1, 8, 28, 56, 62, 0, 148, 328, 419, 280, 140, 728, 1232, 1336, 848, 224, 1582, 2688,
3072, 2408, 742, 1568, 3836, 5264, 5306, 3744, 924, 2576, 5686, 7792, 8092, 6272
Coefficients of powers of η function. Ref JLMS 39 435 64. [8,2; A1486, N1906]

M4503 1, 8, 28, 64, 126, 224, 344, 512, 757, 1008, 1332, 1792, 2198, 2752, 3528, 4096,
4914, 6056, 6860, 8064, 9632, 10656, 12168, 14336, 15751, 17584, 20440, 22016, 24390
Expansion of 8-dimensional cusp form. Ref SPLAG 187. [1,2; A7331]

$$\text{G.f.:}\quad x\,\Pi\,(1-x^{2k-1})^8(1-x^{4k})^8.$$

M4504 1, 8, 28, 64, 134, 288, 568, 1024, 1809
McKay-Thompson series of class 6F for Monster. Ref FMN94. [0,2; A7259]

M4505 8, 28, 89, 234, 512
Postage stamp problem. Ref CJN 12 379 69. AMM 87 208 80. [1,1; A5343]

M4506 1, 8, 30, 80, 175, 336, 588, 960, 1485, 2200, 3146, 4368, 5915, 7840, 10200,
13056, 16473, 20520, 25270, 30800, 37191, 44528, 52900, 62400, 73125, 85176, 98658
4-dimensional figurate numbers: $n.C(n+2,3)$. See Fig M3382. Bisection of M2723. Ref
B1 195. [1,2; A2417, N1907]

M4507 0, 0, 8, 30, 192, 1344, 10800, 97434, 976000, 10749024, 129103992, 1679495350,
23525384064, 353028802560, 5650370001120, 96082828074162, 1729886440780800
From ménage polynomials. Ref R1 197. [2,3; A0425, N1908]

M4508 1, 8, 31, 85, 190, 360, 610, 956, 1415, 2005, 2745, 3658, 4762
Putting balls into 5 boxes. Ref SIAR 12 296 70. [8,2; A5338]

M4509 8, 32, 48, 64, 104, 96, 112, 192, 144, 160, 256, 192, 248, 320, 240, 256, 384, 384,
304, 448, 336, 352, 624, 384, 456, 576, 432, 576, 640, 480, 496, 832, 672, 544, 768, 576
Theta series of D_4 lattice w.r.t. deep hole. Ref SPLAG 118. [0,1; A5879]

M4510 1, 8, 32, 108, 348, 1068, 3180, 9216
Cluster series for square lattice. Ref PRV 133 A315 64. [0,2; A3201]

M4511 1, 0, 0, 8, 33, 168, 962, 5928, 38907, 268056
Partition function for f.c.c. lattice. Ref PHM 2 745 57. [0,4; A1407, N1909]

M4512 8, 33, 168, 970, 6168, 42069, 301376
n-step polygons on f.c.c. lattice. Ref JMP 7 1567 66. [3,1; A5398]

M4513 1, 8, 35, 110, 287, 632
Triangles in complete graph on n nodes. Ref vm. [3,2; A6600]

M4514 1, 8, 35, 111, 287, 644, 1302, 2430, 4257, 7084, 11297, 17381, 25935, 37688,
  53516, 74460, 101745, 136800, 181279, 237083, 306383, 391644, 495650, 621530
$C(n+3,6)+C(n+1,5)+C(n,5)$. Ref LI68 20. MMAG 49 181 76. [3,2; A5732]

M4515 0, 1, 1, 1, 8, 35, 211, 1459, 11584, 103605, 1030805, 11291237, 135015896,
  1749915271, 24435107047, 365696282855, 5839492221440, 99096354764009
From ménage numbers. Ref MES 32 63 02. R1 198. [1,5; A0426, N1910]

M4516 8, 36, 102, 231, 456, 819, 13722, 2178, 4620
From generalized Catalan numbers. Ref LNM 952 288 82. [0,1; A6636]

M4517 1, 8, 36, 120, 330, 792, 1716, 3432, 6435, 11440, 19448, 31824, 50388, 77520,
  116280, 170544, 245157, 346104, 480700, 657800, 888030, 1184040, 1560780, 2035800
Binomial coefficients $C(n,7)$. See Fig M1645. Ref D1 2 7. RS3. B1 196. AS1 828. [7,2;
A0580, N1911]

M4518 1, 8, 36, 148, 554, 2024, 7134, 24796, 84625, 285784, 953430, 3151332, 10314257
7-dimensional partitions of n. Ref PCPS 63 1099 67. [1,2; A0427, N1912]

M4519 1, 8, 36, 148, 554, 2094, 7624, 27428, 96231, 332159, 1126792, 3769418,
  12437966, 40544836, 130643734, 416494314, 1314512589, 4110009734, 12737116845
Euler transform of M4390. Ref PCPS 63 1100 67. EIS § 2.7. [1,2; A0428, N1913]

M4520 8, 36, 204, 1296, 8772, 61776, 446964, 3297456, 24684612, 186884496,
  1427557524, 10983260016, 84998999652, 660994932816, 5161010498484
$1^n + 2^n + \cdots + 8^n$. Ref AS1 813. [0,1; A1555, N1914]

M4521 0, 0, 1, 1, 8, 36, 229, 1625, 13208, 120288, 1214673, 13469897, 162744944,
  2128047988, 29943053061, 451123462673, 7245940789072, 123604151490592
$a(n)=(n-3)a(n-1)+(n-2)(2a(n-2)+a(n-3))$. Ref PLMS 31 341 30. SPS 37-40-4
209 66. [1,5; A0757, N1915]

M4522 1, 8, 40, 160, 560, 1792, 5376, 15360, 42240, 112640, 292864, 745472, 1863680,
  4587520, 11141120, 26738688, 63504384, 149422080, 348651520, 807403520
$C(n+3,3).2^n$. Ref RSE 62 190 46. AS1 796. MFM 74 62 70. [0,2; A1789, N1916]

M4523   1, 8, 40, 168, 676, 2672, 10376, 39824, 151878, 576656, 2181496, 8229160, 30974700, 116385088, 436678520, 1636472360, 6126647748
$2n$-celled polygons with perimeter n on square lattice. Ref JSP 58 480 90. [1,2; A6726]

M4524   8, 40, 176, 748, 3248, 14280, 63768, 285296, 1285688
n-step walks on cubic lattice. Ref PCPS 58 99 62. [1,1; A0760, N1917]

M4525   1, 8, 43, 188, 728, 2593, 8706, 27961, 86802, 262348, 776126, 2256418, 6466614, 18311915, 51334232, 142673720, 393611872, 1078955836, 2941029334
n-node trees of height 8. Ref IBMJ 4 475 60. KU64. [9,2; A0429, N1918]

M4526   8, 43, 196, 820, 3264, 12597, 47652, 177859, 657800, 2417416, 8844448, 32256553, 117378336, 426440955, 1547491404, 5610955132, 20332248992
Random walks. Ref DM 17 44 77. [1,1; A5024]

M4527   0, 8, 44, 152, 372, 824, 1544, 2712, 4448
No-3-in-line problem on $n \times n$ grid. Ref GK68. Wels71 124. LNM 403 7 74. [2,2; A0938, N1919]

M4528   1, 8, 44, 192, 718, 2400, 7352, 20992, 56549, 145008, 356388, 844032, 1934534, 4306368, 9337704, 19771392, 40965362, 83207976, 165944732, 325393024, 628092832
Convolution of M3475. Ref AS1 591. [1,2; A5798]

M4529   1, 8, 44, 208, 910, 3808, 15504, 62016, 245157, 961400, 3749460, 14567280, 56448210, 218349120, 843621600, 3257112960, 12570420330, 48507033744
$8C(2n+1,n-3)/(n+5)$. Ref FQ 14 397 76. DM 14 84 76. [3,2; A3518]

M4530   1, 8, 44, 208, 984, 4584, 21314, 98292, 448850, 2038968, 9220346, 41545564, 186796388, 828623100
Susceptibility for diamond lattice. Ref JPA 6 1511 73. DG74 421. [0,2; A3220]

M4531   8, 44, 309, 2428, 21234, 205056, 2170680, 25022880, 312273360, 4196666880, 60451816320, 929459059200, 15196285843200, 263309095526400
5th differences of factorial numbers. Ref JRAM 198 61 57. [-1,1; A1689, N1920]

M4532   1, 8, 45, 220, 1001, 4368, 18564, 77520, 319770, 1307504, 5311735, 21474180, 86493225, 347373600, 1391975640, 5567902560, 22239974430, 88732378800
Binomial coefficients $C(2n,n-3)$. See Fig M1645. Ref LA56 517. AS1 828. [3,2; A2696, N1921]

M4533   1, 8, 45, 416, 1685, 31032, 1603182, 13856896, 132843888, 6551143600
Coefficients for numerical integration. Ref MOC 6 217 52. [1,2; A2686, N1922]

M4534   1, 8, 48, 256, 1280, 6144, 28672, 131072, 589824, 2621440, 11534336, 50331648, 218103808, 939524096, 4026531840, 17179869184, 73014444032, 309237645312
Coefficients of Chebyshev polynomials: $(n+1)4^n$. Ref LA56 516. [0,2; A2697, N1923]

M4535 1, 8, 48, 264, 1408, 7432, 39152, 206600, 1093760, 5813000, 31019568,
166188552, 893763840, 4823997960, 26124870640, 141926904328, 773293020928
Royal paths in a lattice. Ref CRO 20 18 73. [0,2; A6321]

M4536 1, 8, 49, 288, 1681, 9800, 57121, 332928, 1940449, 11309768, 65918161,
384199200, 2239277041, 13051463048, 76069501249, 443365544448, 2584123765441
$n(n+1)/2$ is square: $a(n+1) = 6.a(n)-a(n-1)+2$. Ref D1 2 10. MAG 47 237 63. B1
193. FQ 9 95 71. [1,2; A1108, N1924]

M4537 8, 50, 2908, 115125476
Switching networks. Ref JFI 276 322 63. [1,1; A0851, N1925]

M4538 8, 52, 288, 1424, 6648, 29700, 128800, 545600
Series-parallel numbers. Ref R1 142. [3,1; A0432, N1926]

M4539 1, 8, 52, 320, 1938, 11704, 70840, 430560, 2629575, 16138848, 99522896,
616480384, 3834669566, 23944995480, 150055305008, 943448717120, 5949850262895
From generalized Catalan numbers. Ref LNM 952 279 82. [0,2; A6631]

$$\text{G.f.:} \quad {}_3F_2([3,8/3,10/3]; [5,9/2]; 27x/4).$$

M4540 0, 1, 8, 54, 384, 3000, 25920, 246960, 2580480, 29393280, 362880000,
4829932800, 68976230400, 1052366515200, 17086945075200, 294226732800000
$n^2.n!$. Ref PLMS 10 122 1879. [0,3; A2775, N1927]

M4541 1, 0, 8, 56, 64, 12672, 309376, 2917888, 163782656, 12716052480,
495644917760, 4004259037184, 1359174582304768, 153146435763437568
Expansion of $\sin(\sinh x)$. [0,3; A3722]

M4542 1, 8, 56, 248, 1232, 5690, 26636, 113552
Cluster series for b.c.c. lattice. Ref PRV 133 A315 64. DG72 225. [0,2; A3210]

M4543 1, 8, 56, 384, 2536, 16512, 105664, 669696, 4201832, 26183808, 162073408,
998129664, 6117389760, 37346353152, 227164816896, 1377490599936
Susceptibility for b.c.c. lattice. Ref DG72 404. [0,2; A3494]

M4544 1, 8, 56, 392, 2648, 17864, 118760, 789032, 5201048, 34268104, 224679864,
1472595144, 9619740648, 62823141192, 409297617672, 2665987056200
Susceptibility for b.c.c. lattice. Ref JPA 5 651 72. DG74 381. [0,2; A2914, N1928]

M4545 1, 8, 56, 392, 2648, 17960, 120056, 804824, 5351720, 35652680, 236291096,
1568049560, 10368669992, 68626647608, 453032542040, 2992783648424
n-step self-avoiding walks on b.c.c. lattice. Ref JPA 5 659 72; 22 2809 89. [0,2; A1666,
N1929]

M4546 8, 56, 464, 3520, 27768
$(2n+1)$-step walks on diamond lattice. Ref PCPS 58 100 62. [1,1; A1398, N1930]

M4547 1, 8, 57, 1292, 7135, 325560, 4894715, 125078632, 1190664342, 137798986920
Coefficients for step-by-step integration. Ref JACM 11 231 64. [1,2; A2402, N1931]

M4548 1, 8, 58, 444, 3708, 33984, 341136, 3733920, 44339040, 568356480, 7827719040,
115336085760, 1810992556800, 30196376985600, 532953524275200
Permutations by descents. Ref SE33 83. NMT 7 16 59. JCT A24 28 78. [1,2; A2538, N1932]

M4549 1, 8, 60, 416, 2791, 18296, 118016, 752008, 4746341, 29727472
Susceptibility for square lattice. Ref JCP 38 811 63. DG74 421. [0,2; A2927, N1933]

M4550 1, 8, 60, 444, 3599, 32484, 325322, 3582600, 43029621, 559774736, 7841128936,
117668021988, 1883347579515
Permutations of length n by rises. Ref DKB 263. [4,2; A1267, N1934]

M4551 1, 8, 60, 480, 4200
Coefficients of Gandhi polynomials. Ref DUMJ 41 311 74. [2,2; A5990]

M4552 0, 8, 61, 96, 401, 904, 219, 722, 37, 54, 26, 43, 15, 32, 4, 21, 92, 1, 9, 71, 97, 501,
905, 319, 723, 137, 541, 945, 359, 763, 177, 581, 985, 399, 704, 217, 522, 35, 34, 24, 23
Add 8, then reverse digits! Ref Robe92 15. [0,2; A7399]

M4553 8, 61, 5020, 128541455, 162924332716605980,
28783052231699298507846309644849796
Denominator of Egyptian fraction for $\pi - 3$. Ref AMM 54 138 47. jww. [0,1; A1466, N1935]

M4554 1, 8, 63, 496, 3905, 30744, 242047, 1905632, 15003009, 118118440, 929944511,
7321437648, 57641556673, 453811015736, 3572846569215, 28128961537984
$a(n) = 8a(n-1) - a(n-2)$. Ref NCM 4 167 1878. [0,2; A1090, N1936]

M4555 1, 8, 64, 512, 4096, 32768, 262144, 2097152, 16777216, 134217728, 1073741824,
8589934592, 68719476736, 549755813888, 4398046511104, 35184372088832
Powers of 8. Ref BA9. [0,2; A1018, N1937]

M4556 1, 8, 67, 602, 5811, 60875, 690729, 8457285, 111323149, 1569068565,
23592426102, 377105857043, 6387313185590, 114303481217895, 2155348564851616
Permutations of length n by length of runs. Ref DKB 261. [4,2; A0434, N1938]

M4557 8, 72, 2160, 15504, 220248, 1564920, 89324640, 640807200, 9246847896,
67087213336, 1957095947664
Coefficients of Legendre polynomials. Ref MOC 3 17 48. [2,1; A1799, N1939]

M4558 0, 1, 8, 78, 944, 13800, 237432, 4708144, 105822432, 2660215680, 73983185000,
2255828154624, 74841555118992, 2684366717713408, 103512489775594200
Normalized total height of rooted trees with n nodes. Ref JAuMS 10 281 69. [1,3; A0435, N1940]

M4559 1, 8, 80, 1088, 19232, 424400
Bicoverings of an n-set. Ref SMH 3 145 68. [2,2; A2718, N1941]

M4560 1, 8, 80, 4374, 9800, 123200, 336140, 11859210, 11859210, 177182720,
1611308699, 3463199999, 63927525375
Every sequence of 2 numbers $> a(n)$ contains a prime $> p(n)$. Ref IJM 8 66 64. AMM 79
1087 72. [2,2; A2072, N1942]

M4561 1, 8, 84, 992, 12514, 164688, 2232200, 30920128, 435506703, 6215660600,
89668182220, 1305109502496, 19138260194422, 282441672732656
Expansion of Jacobi nome (reversion of M4528). Ref AS1 591. clm. [1,2; A5797]

M4562 8, 88, 840, 6888, 54824, 412712, 3065096, 22134152
Susceptibility for b.c.c. lattice. Ref DG72 136. [1,1; A3492]

M4563 1, 8, 88, 1216, 19160, 327232, 5896896, 110393856, 2126213592, 41861519680,
838733719616
Internal energy series for b.c.c. lattice. Ref DG72 425. [0,2; A3497]

M4564 8, 88, 2992, 23408, 354200, 2641320, 156641760, 1159149024, 17161845272,
127234370120
Coefficients of Legendre polynomials. Ref MOC 3 17 48. [2,1; A6750]

M4565 1, 1, 8, 92, 1240, 18278, 285384, 4638348, 77652024, 1329890705, 23190029720,
410333440536, 7349042994488, 132969010888280, 2426870706415800
Shifts left when convolved thrice. Ref BeSl94. [0,3; A7556]

M4566 8, 96, 1664, 36800, 1008768, 32626560, 1221399040, 51734584320,
2459086364672, 129082499311616
Susceptibility for b.c.c. lattice. Ref PRV 164 801 67. [1,1; A2168, N1943]

M4567 0, 0, 8, 102, 948, 7900, 62928, 491832
Colored series-parallel networks. Ref R1 159. [1,3; A1575, N1944]

M4568 1, 8, 104, 1092, 12376, 136136, 1514513, 16776144, 186135312, 2063912136,
22890661872, 253854868176, 2815321003313, 31222272414424, 34620798314872
Fibonomial coefficients. Ref FQ 6 82 68. BR72 74. [0,2; A1657, N1945]

M4569 1, 8, 105, 1456, 20273, 282360, 3932761, 54776288, 762935265, 10626317416,
148005508553, 2061450802320, 28712305723921, 399910829332568
$a(n) = 15a(n-1) - 15a(n-2) + a(n-3)$. Ref AMM 53 465 46. [0,2; A1922, N1946]

M4570 8, 106, 49008, 91901007752
Switching networks. Ref JFI 276 322 63. [1,1; A0845, N1947]

M4571 1, 0, 8, 112, 128, 109824, 8141824, 353878016, 9666461696, 5151942574080,
825073851170816, 76429076694827008, 2051308253366714368
Expansion of tan(tanh x). [0,3; A3721]

M4572 8, 120, 16880, 1791651440
Switching networks. Ref JFI 276 322 63. [1,1; A0848, N1948]

M4573 1, 8, 127, 2024, 32257, 514088, 8193151, 130576328, 2081028097, 33165873224,
528572943487, 8424001222568, 134255446617601, 2139663144659048
$a(n) = 16a(n-1) - a(n-2)$. Ref NCM 4 167 1878. TH52 281. [0,2; A1081, N1949]

M4574 0, 0, 0, 0, 0, 0, 0, 0, 1, 8, 135, 2557
Non-Hamiltonian polyhedra with n faces. Ref Dil92. [1,10; A7032]

M4575 1, 0, 0, 0, 0, 8, 152
Asymmetric graphs with n nodes. Ref ST90. [1,6; A3400]

M4576 8, 152, 2200, 28520, 347416, 4068024, 46360392
Susceptibility for f.c.c. lattice. Ref DG72 136. [1,1; A3491]

M4577 8, 176, 265728, 2199038984192
Switching networks. Ref JFI 276 321 63. [1,1; A0839, N1951]

M4578 8, 192, 11904, 1125120, 153262080, 28507207680, 6951513784320,
2153151603671040, 826060810479206400, 38460018899291 9961600
Hamiltonian circuits on n-octahedron. Ref JCT B19 2 75. [2,1; A3435]

M4579 1, 8, 216, 1728, 216000, 24000, 8232000, 65856000, 16003008000, 16003008000,
21300003648000, 21300003648000, 46796108014656000, 46796108014656000
Denominators of Σk^{-3}; $k = 1..n$. Ref KaWa 89. [1,2; A7409]

M4580 1, 8, 216, 8000, 343000, 16003008, 788889024, 40424237568, 2131746903000,
114933031928000, 6306605327953216, 351047164190381568, 19774031697705428416
$C(2n,n)^3$. Ref AIP 9 345 60. [0,2; A2897, N1952]

M4581 8, 222, 2337, 31941, 33371313, 311123771, 7149317941, 22931219729,
112084656339, 3347911118189, 11613496501723, 97130517917327, 531832651281459
Write down all the prime divisors in previous term! Ref hj. [1,1; A6919]

M4582 8, 288, 366080, 1468180471808
Switching networks. Ref JFI 276 322 63. [1,1; A0842, N1953]

M4583 1, 8, 343, 1331, 1030301, 1367631, 1003003001, 10662526601, 1000300030001,
1030607060301, 1334996994331, 1000030000300001, 10333949994933301
Palindromic cubes. Cf. M1736. Ref JRM 3 97 70. [1,2; A2781, N1954]

M4584 1, 8, 352, 38528, 7869952, 2583554048, 1243925143552, 825787662368768,
722906928498737152, 806875574817679474688, 111838908784308346 1066752
Generalized Euler numbers. Ref MOC 21 689 67. [0,2; A0436, N1955]

M4585 1, 8, 432, 131072, 204800000, 1565515579392, 56593444029595648,
9444732965739290427392, 714664660949440653104 1460224
Discriminant of Chebyshev polynomial $T_n(x)$. Ref AS1 795. [1,2; A7701]

M4586   0, 0, 1, 8, 1024, 5, 1071, 116503103764643, 1209889024954, 1184, 11131, 39, 7, 82731770
$a(n)2^{n+2} + 1$ divides nth Fermat number. Ref BPNR 71. Rie85 377. [0,4; A7117]

M4587   1, 8, 1368, 300608, 186086600
Folding an $n \times n$ sheet of stamps. See Fig M4587. Ref CJN 14 77 71. [0,2; A1418, N1956]

||||||||||||||||||||||||||||||||||||||||||||||||||||||||||||||||||||||||||||||||||||||||||||||||||||||

Figure M4587.   FOLDING STAMPS, MEANDERS.

M1455 gives the number of ways of folding a strip of n (unmarked, ungummed) stamps. Only the first 16 terms are known:

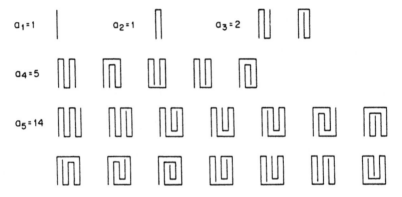

M1614, M1420, M1205 (really all the same sequence) consider marked stamps, and M0323, M0879, M1206, M1211, M1891, M4271, M4587, M1901 study related problems. A similar problem asks for the number m_n of ways a river flowing from the South-West to the East can cross a West-East road n times: V. I. Arnol'd calls these diagrams **meanders**. This is M0874 (only 21 terms are known), illustrated here. If the ends of the river are joined we get a **closed meander**. Let M_n be the number of closed meanders with $2n$ crossings (in fact $M_n = m_{2n-1}$). This is M1862. Seventeen terms of this sequence are known. M2037, M4014, M2025, M0840, M2286, M4921, M0374, M1871 are related to these.

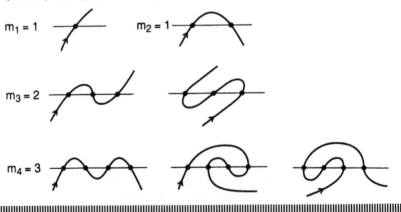

||||||||||||||||||||||||||||||||||||||||||||||||||||||||||||||||||||||||||||||||||||||||||||||||||||||

M4588   1, 8, 1920, 193536, 154828800, 1167851520, 892705701888000,
1428329123020800, 768472460034048000, 4058540589291090739200
Denominators of coefficients for central differences. Cf. M4894. Ref SAM 42 162 63. [1,2; A2672, N1957]

M4589   8, 2080, 22386176, 11728394650624, 31482461991446167552
Relational systems on n nodes. Ref OB66. [1,1; A1375, N1958]

M4590   0, 0, 0, 0, 0, 0, 0, 0, 0, 0, 0, 0, 0, 0, 0, 0, 0, 9, 0, 20, 0, 33, 0, 48, 0, 65, 9, 84, 29,
105, 62, 128, 110, 153, 175, 189, 259, 247, 364, 340, 492, 483, 645, 693, 834, 989, 1081
Strict 7th-order maximal independent sets in cycle graph. Ref YaBa94. [1,18; A7394]

M4591   0, 1, 9, 1, 1, 1, 5, 1, 1, 1, 1, 26, 1, 1, 3, 5, 1, 3, 1, 1, 44, 1, 1, 5, 5, 1, 5, 1, 1, 62, 1, 1,
7, 5, 1, 7, 1, 1, 80, 1, 1, 9, 5, 1, 9, 1, 1, 98, 1, 1, 11, 5, 1, 11, 1, 1, 116, 1, 1, 13, 5, 1, 13, 1
Continued fraction for $e/3$. Ref KN1 2 601. [1,3; A6084]

M4592   1, 9, 1, 2, 9, 3, 1, 1, 8, 2, 7, 7, 2, 3, 8, 9, 1, 0, 1, 1, 9, 9, 1, 1, 6, 8, 3, 9, 5, 4, 8, 7, 6,
0, 2, 8, 2, 8, 6, 2, 4, 3, 9, 0, 5, 0, 3, 4, 5, 8, 7, 5, 7, 6, 6, 2, 1, 0, 6, 4, 7, 6, 4, 0, 4, 4, 7, 2, 3
Decimal expansion of cube root of 7. [1,2; A5482]

M4593   9, 1, 5, 9, 6, 5, 5, 9, 4, 1, 7, 7, 2, 1, 9, 0, 1, 5, 0, 5, 4, 6, 0, 3, 5, 1, 4, 9, 3, 2, 3, 8, 4,
1, 1, 0, 7, 7, 4, 1, 4, 9, 3, 7, 4, 2, 8, 1, 6, 7, 2, 1, 3, 4, 2, 6, 6, 4, 9, 8, 1, 1, 9, 6, 2, 1, 7, 6, 3
Decimal expansion of Catalan's constant. Ref FE90. [0,1; A6752]

M4594   1, 0, 0, 9, 7, 3, 2, 5, 3, 3, 7, 6, 5, 2, 0, 1, 3, 5, 8, 6, 3, 4, 6, 7, 3, 5, 4, 8, 7, 6, 8, 0, 9,
5, 9, 0, 9, 1, 1, 7, 3, 9, 2, 9, 2, 7, 4, 9, 4, 5, 3, 7, 5, 4, 2, 0, 4, 8, 0, 5, 6, 4, 8, 9, 4, 7, 4, 2, 9
A random sequence. Ref RA55. [1,4; A2205, N1959]

M4595   1, 0, 9, 8, 6, 1, 2, 2, 8, 8, 6, 6, 8, 1, 0, 9, 6, 9, 1, 3, 9, 5, 2, 4, 5, 2, 3, 6, 9, 2, 2, 5, 2,
5, 7, 0, 4, 6, 4, 7, 4, 9, 0, 5, 5, 7, 8, 2, 2, 7, 4, 9, 4, 5, 1, 7, 3, 4, 6, 9, 4, 3, 3, 3, 6, 3, 7, 4, 9
Decimal expansion of natural logarithm of 3. Ref RS8 2. [1,3; A2391, N1960]

M4596   9, 8, 6, 9, 6, 0, 4, 4, 0, 1, 0, 8, 9, 3, 5, 8, 6, 1, 8, 8, 3, 4, 4, 9, 0, 9, 9, 9, 8, 7, 6, 1, 5,
1, 1, 3, 5, 3, 1, 3, 6, 9, 9, 4, 0, 7, 2, 4, 0, 7, 9, 0, 6, 2, 6, 4, 1, 3, 3, 4, 9, 3, 7, 6, 2, 2, 0, 0
Decimal expansion of π^2. Ref RS8 XVIII. [1,1; A2388, N1961]

M4597   1, 9, 9, 3, 9, 9, 3, 9, 9, 1, 18, 9, 9, 9, 9, 9, 9, 9, 6, 9, 18, 6, 9, 9, 6, 9, 9, 4, 9, 9, 12,
18, 18, 3, 9, 9, 3, 9, 9, 3, 18, 18, 12, 18, 9, 5, 9, 9, 9, 9, 18, 6, 18, 18, 2, 9, 9, 9, 9, 9, 12, 5
Sum of digits of $na(n)$ is n ($= $ M5054/n). Ref jhc. [1,2; A7471]

M4598   9, 10, 15, 16, 21, 22, 25, 26, 27, 28, 33, 34, 35, 36, 39, 40, 45, 46, 49, 50, 51, 52,
55, 56, 57, 58, 63, 64, 65, 66, 69, 70, 75, 76, 77, 78, 81, 82, 85, 86, 87, 88, 91, 92, 93, 94
n and $n-1$ are composite. Ref AS1 844. [1,1; A5381]

M4599   9, 10, 30, 6, 25, 96, 60, 250, 45, 150, 544, 360, 1230, 184, 675, 2310, 1410, 4830,
750, 2450, 8196, 4920, 16180, 2376, 7875, 25644, 15000, 48720, 7126, 22800, 73221
McKay-Thompson series of class 5B for Monster. Ref CALG 18 257 90. FMN94. [1,1; A7252]

M4600 9, 14, 20, 27, 33, 41, 49, 57
Zarankiewicz's problem. Ref LNM 110 143 69. [2,1; A6624]

M4601 9, 14, 21, 27, 34, 43, 50, 61
Zarankiewicz's problem. Ref TI68 146. LNM 110 143 69. C1 291. [3,1; A1198, N1962]

M4602 0, 9, 18, 162, 2520, 33192, 1019088, 7804944, 723961728, 2596523904
Specific heat for cubic lattice. Ref PRV 164 801 67. [1,2; A2169, N1963]

M4603 0, 9, 18, 306, 3240, 49176, 1466640, 13626000, 1172668032, 75256704
Specific heat for hexagonal lattice. Ref PHL A25 208 67. [1,2; A5400]

M4604 1, 9, 24, 46, 75, 111, 154, 204, 261, 325, 396, 474, 559, 651, 750, 856, 969, 1089,
   1216, 1350, 1491, 1639, 1794, 1956, 2125, 2301, 2484, 2674, 2871, 3075, 3286, 3504
Enneagonal numbers: $n(7n-5)/2$. See Fig M2535. Ref B1 189. [1,2; A1106]

M4605 1, 9, 28, 73, 126, 252, 344, 585, 757, 1134, 1332, 2044, 2198, 3096, 3528, 4681,
   4914, 6813, 6860, 9198, 9632, 11988, 12168, 16380, 15751, 19782, 20440, 25112, 24390
Sum of cubes of divisors of n. Ref AS1 827. [1,2; A1158, N1964]

M4606 9, 29, 35, 42, 48, 113, 120, 126, 152, 185, 204, 224, 237, 243, 276, 302, 308, 321,
   341, 386, 399, 419, 432, 477, 503, 510, 516, 542, 549, 588, 633, 659, 666, 705, 731
$(n^2+n+1)/13$ is prime. Ref CU23 1 251. [1,1; A2642, N1131]

M4607 1, 9, 30, 65, 114, 177, 254, 345, 450, 569, 702, 849, 1010, 1185, 1374, 1577, 1794,
   2025, 2270, 2529, 2802, 3089, 3390, 3705, 4034, 4377, 4734, 5105, 5490, 5889, 6302
Points on surface of tricapped prism: $7n^2+2$. Ref INOC 24 4552 85. [0,2; A5919]

M4608 9, 30, 69, 133, 230, 369, 560, 814, 1143, 1560, 2079, 2715, 3484, 4403, 5490,
   6764, 8245, 9954, 11913, 14145, 16674, 19525, 22724, 26298, 30275, 34684, 39555
Powers of rooted tree enumerator. Ref R1 150. [1,1; A0439, N1965]

M4609 1, 9, 30, 70, 135, 231, 364, 540, 765, 1045, 1386, 1794, 2275, 2835, 3480, 4216,
   5049, 5985, 7030, 8190, 9471, 10879, 12420, 14100, 15925, 17901, 20034, 22330, 24795
Octagonal pyramidal numbers: $n(n+1)(2n-1)/2$. Ref D1 2 2. B1 194. [1,2; A2414, N1966]

M4610 9, 30, 180, 980, 8326, 70272, 695690, 7518720, 89193276, 1148241458,
   15947668065, 237613988040, 3780133322620, 63945806121448
Discordant permutations. Ref SMA 20 23 54. [4,1; A0440, N1967]

M4611 1, 9, 33, 82, 165, 291, 469, 708, 1017, 1405, 1881, 2454, 3133, 3927, 4845, 5896,
   7089, 8433, 9937, 11610, 13461, 15499, 17733, 20172, 22825, 25701, 28809, 32158
Tricapped prism numbers. Ref INOC 24 4552 85. [0,2; A5920]

M4612 9, 33, 91, 99, 259, 451, 481, 561, 657, 703, 909, 1233, 1729, 2409, 2821, 2981,
   3333, 3367, 4141, 4187, 4521, 5461, 6533, 6541, 6601, 7107, 7471, 7777, 8149, 8401
Pseudoprimes to base 10. Ref UPNT A12. [1,1; A5939]

M4613   1, 9, 34, 95, 210, 406, 740, 1161, 1920, 2695, 4116, 5369, 7868, 9690, 13640,
16116, 22419, 25365, 34160, 38640, 50622, 55154, 73320, 77225, 100100, 107730
Related to the divisor function. Ref SMA 19 39 53. [1,2; A0441, N1968]

M4614   0, 0, 0, 0, 0, 0, 0, 0, 1, 0, 0, 9, 34, 104, 283, 957, 3033, 9519
Imperfect squared rectangles of order n. See Fig M4482. Ref GA61 207. cjb. [1,12; A2881,
N1969]

M4615   9, 34, 112, 326, 797
Postage stamp problem. Ref CJN 12 379 69. AMM 87 208 80. [1,1; A5344]

M4616   1, 9, 35, 91, 189, 341, 559, 855, 1241, 1729, 2331, 3059, 3925, 4941, 6119, 7471,
9009, 10745, 12691, 14859, 17261, 19909, 22815, 25991, 29449, 33201, 37259, 41635
Centered cube numbers: $n^3 + (n-1)^3$. Ref AMM 82 819 75. INOC 24 4550 85. [0,2;
A5898]

M4617   1, 9, 35, 95, 210, 406, 714, 1170, 1815, 2695, 3861, 5369, 7280, 9660, 12580,
16116, 20349, 25365, 31255, 38115, 46046, 55154, 65550, 77350, 90675, 105651
4-dimensional figurate numbers: $(5n-1).C(n+2,3)/4$. See Fig M3382. Ref B1 195. [1,2;
A2418, N1970]

M4618   1, 9, 36, 84, 117, 54, 177, 540, 837, 755, 54, 1197, 2535, 3204, 2520, 246, 3150,
6426, 8106, 7011, 2844, 3549, 10359, 15120, 15804, 11403, 2574, 8610, 18972, 25425
Coefficients of powers of η function. Ref JLMS 39 435 64. [9,2; A1487, N1971]

M4619   1, 9, 36, 100, 225, 441, 784, 1296, 2025, 3025, 4356, 6084, 8281, 11025, 14400,
18496, 23409, 29241, 36100, 44100, 53361, 64009, 76176, 90000, 105625, 123201
Sums of cubes. Ref AS1 813. [1,2; A0537, N1972]

M4620   1, 1, 9, 39, 1141, 12721, 804309, 17113719, 1886573641, 65373260641,
11127809595009, 570506317184199, 138730500808639741, 9867549661639871761
Expansion of sinh x / cos x. Ref CMB 13 309 70. [0,3; A2085, N1973]

M4621   1, 9, 40, 125, 315, 686, 1344, 2430, 4125, 6655, 10296
Nonseparable toroidal tree-rooted maps. Ref JCT B18 243 75. [0,2; A6414]

M4622   1, 9, 41, 129, 321, 681, 1289, 2241, 3649, 5641, 8361, 11969, 16641, 22569,
29961, 39041, 50049, 63241, 78889, 97281, 118721, 143529, 172041, 204609, 241601
Expansion of $(1+x)^4/(1-x)^5$. Ref SIAR 12 277 70. C1 81. [0,2; A1846, N1974]

M4623   1, 9, 42, 132, 334, 728, 1428, 2584, 4389, 7084, 10963, 16380, 23751, 33563,
46376, 62832, 83657, 109668, 141778, 181001, 228459, 285384, 353127, 433160
Fermat coefficients. Ref MMAG 27 141 54. [6,2; A0971, N1975]

M4624   9, 42, 236, 1287, 7314, 41990, 245256, 1448655, 8649823, 52106040, 316360752
Perforation patterns for punctured convolutional codes (3,1). Ref SFCA92 1 9. [2,1;
A7227]

M4625 1, 9, 45, 55, 99, 297, 703, 999, 2223, 2728, 4950, 5050, 7272, 7777, 9999, 17344,
22222, 77778, 82656, 95121, 99999, 142857, 148149, 181819, 187110, 208495, 318682
Kaprekar numbers. Ref Well86 #297. rpm. [1,2; A6886]

M4626 1, 9, 45, 165, 495, 1287, 3003, 6435, 12870, 24310, 43758, 75582, 125970,
203490, 319770, 490314, 735471, 1081575, 1562275, 2220075, 3108105, 4292145
Binomial coefficients $C(n,8)$. See Fig M1645. Ref D1 2 7. RS3. B1 196. AS1 828. [8,2;
A0581, N1976]

M4627 1, 9, 45, 285, 2025, 15333, 120825, 978405, 8080425, 67731333, 574304985,
4914341925, 42364319625, 367428536133, 3202860761145, 28037802953445
$1^n + 2^n + \cdots + 9^n$. Ref AS1 813. [0,2; A1556, N1977]

M4628 1, 9, 46, 177, 571, 1632, 4270, 10446, 24244, 53942, 115954, 242240, 494087,
987503, 1939634, 3753007, 7167461, 13532608, 25293964, 46856332, 86110792
From rook polynomials. Ref SMA 20 18 54. [0,2; A1926, N1978]

M4629 0, 0, 0, 0, 0, 0, 0, 1, 1, 9, 48, 343, 2466, 18905
Noninscribable simplicial polyhedra with n nodes. Ref Dil92. [1,10; A7037]

M4630 1, 9, 49, 214, 800, 2685, 8274, 23829, 64843
Coefficients of a modular function. Ref GMJ 8 29 67. [2,2; A3297]

M4631 1, 9, 50, 220, 840, 2912, 9408, 28800, 84480, 239360, 658944, 1770496, 4659200,
12042240, 30638080, 76873728, 190513152, 466944000, 1133117440, 2724986880
Coefficients of Chebyshev polynomials. Ref AS1 795. [0,2; A6974]

M4632 1, 9, 50, 294, 1944, 14520, 121680, 1134000, 11652480, 130999680, 1600300800,
21115987200, 299376000000, 4539498163200, 73316942899200, 1256675067648000
$(2n+1)^2 n!$. Ref UM 45 82 94. [0,2; A7681]

M4633 1, 9, 50, 1225, 7938, 106722, 736164, 41409225, 295488050
Coefficients of Legendre polynomials. Ref PR33 157. FMR 1 362. [0,2; A2462, N1979]

M4634 1, 9, 51, 230, 863, 2864, 8609, 23883
n-covers of a 4-set. Ref DM 81 151 90. [1,2; A5746]

M4635 9, 53, 260, 1156, 4845, 19551, 76912, 297275, 1134705, 4292145, 16128061,
60304951, 224660626, 834641671, 3094322026, 11453607152, 42344301686
Random walks. Ref DM 17 44 77. [1,1; A5025]

M4636 1, 9, 53, 362, 2790, 24024, 229080, 2399760, 27422640, 339696000, 4536362880,
64988179200, 994447238400, 16190733081600, 279499828608000
4th differences of factorial numbers. Ref JRAM 198 61 57. [-1,2; A1688, N1980]

M4637 1, 9, 54, 273, 1260, 5508, 23256, 95931, 389367, 1562275, 6216210, 24582285,
96768360, 379629720, 1485507600, 5801732460, 22626756594, 88152205554
$9C(2n,n-4)/(n+5)$. Ref QAM 14 407 56. MOC 29 216 75. [4,2; A1392, N1981]

M4638   1, 9, 55, 286, 1362, 6143, 26729, 113471, 473471, 1951612, 7974660, 32384127,
130926391, 527657073, 2121795391, 8518575466, 34162154550, 136893468863
Convex polygons of length $2n$ on square lattice. Ref TCS 34 179 84. [5,2; A5770]

M4639   1, 9, 56, 300, 1485, 7007, 32032, 143208, 629850, 2735810, 11767536, 50220040,
212952285
Dissections of a polygon by number of parts. Ref CAY 13 95. AEQ 18 385 78. [5,2;
A2055, N1982]

M4640   1, 9, 61, 381, 2332, 14337, 89497, 569794, 3704504, 24584693, 166335677,
1145533650, 8017098273, 56928364553, 409558170361, 2981386305018
Permutations of length n by subsequences. Ref MOC 22 390 68. [3,2; A1454, N1983]

M4641   9, 64, 326, 1433, 5799, 22224, 81987, 293987
Partially labeled rooted trees with n nodes. Ref R1 134. [3,1; A0444, N1984]

M4642   1, 9, 66, 450, 2955, 18963, 119812, 748548, 4637205, 28537245, 174683718,
1064611782, 6464582943, 39132819495, 236256182280, 1423046656008
Spheroidal harmonics. Ref MES 52 75 24. [1,2; A2695, N1985]

$$\text{G.f.: } (1 - 6x + x^2)^{-3/2}.$$

M4643   1, 9, 66, 455, 3060, 20349, 134596, 888030, 5852925, 38567100, 254186856,
1676056044, 11058116888, 73006209045, 482320623240, 3188675231420
$C(3n+6, n)$. Ref DM 9 355 74. [0,2; A3408]

M4644   1, 9, 67, 525, 4651, 47229, 545707, 7087005, 102247051, 1622631549,
28091565547, 526858344285, 10641342962251, 230283190961469, 5315654681948587
Differences of 0. Ref SKA 11 95 28. [3,2; A2051, N1986]

M4645   0, 0, 1, 9, 70, 571, 4820, 44676, 450824, 4980274, 59834748, 778230060,
10896609768, 163456629604, 2615335902176, 44460874280032, 800296440705472
Asymmetric permutations. Ref LU91 1 222. JRM 7 181 74. LNM 560 201 76. [2,4;
A0899, N1987]

M4646   1, 9, 71, 580, 5104, 48860, 509004, 5753736, 70290936, 924118272,
13020978816, 195869441664, 3134328981120, 53180752331520, 953884282141440
Generalized Stirling numbers. Ref PEF 77 7 62. [0,2; A1706, N1988]

$$\text{E.g.f.: } \ln (1 - x)^2 \, / \, 2 \, (1 - x)^2.$$

M4647   1, 9, 72, 320, 1185, 3892, 11776, 33480, 90745, 236808
Rook polynomials. Ref JAuMS A28 375 79. [1,2; A5778]

M4648   1, 9, 72, 570, 4554, 36855, 302064, 2504304, 20974005, 177232627, 1509395976,
12943656180, 111676661460, 968786892675, 8445123522144, 73940567860896
From generalized Catalan numbers. Ref LNM 952 280 82. [0,2; A6634]

M4649 1, 9, 72, 600, 5400, 52920, 564480, 6531840, 81648000, 1097712000,
15807052800, 242853811200, 3966612249600, 68652904320000, 1255367393280000
Coefficients of Laguerre polynomials. Ref LA56 519. AS1 799. [2,2; A1809, N1989]

$$\text{E.g.f.:} \quad x\,(1 + \tfrac{1}{2}x)\,/\,(1 - x)^4.$$

M4650 1, 9, 72, 626, 6084, 64974
Characteristic polynomial of Pascal matrix. Ref FQ 15 204 77. [1,2; A6135]

M4651 1, 9, 74, 638, 5944, 60216, 662640, 7893840, 101378880, 1397759040,
20606463360, 323626665600, 5395972377600, 95218662067200, 1773217155225600
Generalized Stirling numbers. Ref PEF 77 44 62. [0,2; A1716, N1990]

M4652 9, 77, 1224, 7888, 202124, 1649375
First occurrences of 2 consecutive nth power residues. Ref MOC 18 397 64. EG80 87. [2,1; A0445, N1991]

M4653 1, 9, 81, 729, 6561, 59049, 531441, 4782969, 43046721, 387420489, 3486784401,
31381059609, 282429536481, 2541865828329, 22876792454961
Powers of 9. Ref BA9. [0,2; A1019, N1992]

M4654 0, 1, 9, 81, 835, 9990, 137466, 2148139, 37662381, 733015845, 15693217705,
366695853876, 9289111077324, 253623142901401, 7425873460633005
Bessel polynomial $y_n{}'(1)$. Ref RCI 77. [0,3; A1514, N1993]

M4655 1, 9, 81, 8505, 229635, 413343, 531972441, 227988189, 3419822835
Coefficients of Green function for cubic lattice. Ref PTRS 273 593 73. [0,2; A3302]

M4656 1, 9, 92, 920, 8928, 84448, 782464, 7130880, 64117760, 570166784, 5023524864,
43915595776, 381350330368, 3292451880960, 28283033157632, 241884640182272
$\Sigma\,(\,\Sigma\,C(n,k),\,k=0..m)^3$, $m=0..n$. Ref Calk94. [0,2; A7403]

$$n.2^{3n-1} + 2^{3n} - 3n.2^{n-2}\,C(2n,n).$$

M4657 9, 95, 420, 1225, 2834, 5652, 10165, 16940, 26625, 39949, 57722, 80835, 110260,
147050, 192339, 247342, 313355, 391755, 484000, 591629, 716262, 859600
Discordant permutations. Ref SMA 20 23 54. [3,1; A0562, N1994]

M4658 9, 96, 835, 7020, 58857, 497360, 4251804, 36765592, 321262541
n-step walks on f.c.c. lattice. Ref JPA 6 351 73. [3,1; A5545]

M4659 1, 1, 9, 101, 3223, 301597, 98198291, 112780875113, 458970424333059,
6669800460126763729, 349443329644003900650627
Rooted connected strength 3 Eulerian graphs with n nodes. Ref rwr. [1,3; A7133]

M4660 1, 9, 108, 3420, 114480, 7786800
Generators for symmetric group. Ref JCT 9 111 70. [2,2; A1691, N1995]

M4661   1, 9, 120, 2100, 45360, 1164240, 34594560, 1167566400, 44108064000,
1843717075200, 84475764172800, 4209708914611200, 226676633863680000
Coefficients of orthogonal polynomials. Ref MOC 9 174 55. [2,2; A2691, N1996]

$$\text{E.g.f.:} \quad (1 - x) / (1 - 4\,x)^{5/2}.$$

M4662   9, 148, 3493, 106431, 3950832, 172325014, 8617033285, 485267003023,
30363691715629, 2088698040637242, 156612539215405732, 12709745319947141220
Connected *n*-state finite automata with 2 inputs. Ref GTA85 683. [1,1; A6691]

M4663   1, 9, 165, 24651, 29522961, 286646256675, 21717897090413481,
129805366893186260768

40, 62082697145168772833294318409
Self-dual nets with 2*n* nodes. Ref CCC 2 32 77. rwr. JGT 1 295 77. [1,2; A4107]

M4664   1, 9, 175, 2025, 102235, 1356047, 37160123, 6771931925, 772428184055
Coefficients of Green function for cubic lattice. Ref PTRS 273 590 73. [0,2; A3280]

M4665   1, 1, 9, 177, 6097, 325249, 24807321, 2558036145, 342232522657,
57569080467073, 11879658510739497, 2948163649552594737
Expansion of cos(tanh *x*). [0,3; A3711]

M4666   1, 9, 198, 10710, 1384335, 416990763, 286992935964, 444374705175516,
1528973599758889005
n-node acyclic digraphs with 2 out-points. Ref HA73 254. [2,2; A3026]

M4667   0, 0, 0, 1, 9, 216, 7560, 357120, 22025430, 1720751760, 166198926600,
19453788144000, 2714247736061400, 445133524289731200, 84788348720139464400
Connected labeled 2-regular digraphs with *n* nodes. Ref rwr. [0,5; A7108]

M4668   1, 0, 0, 1, 9, 216, 7570, 357435, 22040361, 1721632024, 166261966956,
19459238879565, 2714812050902545, 445202898702992496, 84798391618743138414
Labeled 2-regular digraphs with *n* nodes. Ref rwr. [0,5; A7107]

M4669   1, 9, 225, 11025, 893025, 108056025, 18261468225, 4108830350625,
1187451971330625, 428670161650355625, 189043541287806830625
$(1.3.5...(2n+1))^2$. Ref RCI 217. [0,2; A1818, N1997]

M4670   1, 9, 251, 2035, 256103, 28567, 9822481, 78708473, 19148110939, 19164113947,
25523438671457, 25535765062457, 56123375845866029, 56140429821090029
Numerators of Σk^{-3}; $k = 1..n$. Ref KaWa 89. [1,2; A7408]

M4671   9, 259, 1974, 8778, 28743, 77077, 179452, 375972, 725781, 1312311, 2249170,
3686670, 5818995, 8892009, 13211704, 19153288, 27170913, 37808043
Central factorial numbers. Ref RCI 217. [2,1; A1823, N1998]

M4672   1, 9, 297, 7587, 1086939, 51064263, 5995159677, 423959714955,
281014370213715, 26702465299878195, 5723872792950096855
Spin-wave coefficients for cubic lattice. Ref PTRS 273 605 73. [0,2; A3303]

M4673   9, 2047, 1373653, 25326001, 3215031751, 2152302898747, 3474749660383,
341550071728321
Smallest odd number for which Miller-Rabin primality test on bases $< p$ fails. Ref MOC 25
1003 80. MINT 8 58 86. MOC 61 915 93. [2,1; A6945]

SEQUENCES BEGINNING . . ., 10, . . . TO . . ., 13, . . .

M4674   10, 0, 0, 0, 55, 150, 210, 280, 580, 1275, 2905, 5350, 9985, 17965, 33665, 62895,
117287, 214610, 389805, 700720, 1259890, 2250405, 4008717, 7092366, 12497237
Solid partitions. Ref PNISI 26 135 60. [4,1; A2045, N2307]

M4675   1, 1, 10, 2, 16, 2, 1, 4, 2, 1, 21, 1, 3, 5, 1, 2, 1, 1, 2, 11, 5, 1, 3, 1, 2, 27, 4, 1, 282, 8,
1, 2, 1, 1, 3, 1, 3, 2, 6, 4, 1, 2, 1, 5, 1, 1, 2, 1, 1, 1, 3, 2, 8, 1, 2, 2, 4, 5, 1, 1, 36, 1, 1, 1, 1, 2
Continued fraction for cube root of 7. Ref JRAM 255 126 72. [1,3; A5483]

M4676   10, 5, 6, 10, 20, 45, 110, 286, 780, 2210, 6460, 19380, 59432, 185725, 589950,
1900950, 6203100, 20470230, 68234100, 229514700, 778354200, 2659376850
Super ballot numbers: $60(2n)! / n!(n+3)!$. Ref JSC 14 181 92. [0,1; A7272]

M4677   0, 1, 10, 6, 1, 6, 2, 14, 4, 124, 2, 1, 2, 2039, 1, 9, 1, 1, 1, 262111, 2, 8, 1, 1, 1, 3, 1,
536870655, 4, 16, 3, 1, 3, 7, 1, 140737488347135, 8, 128, 2, 1, 1, 1, 7, 2, 1, 9, 1
Continued fraction for $\Sigma 2^{-F(k)}$, $k \geq 2$. Ref CJM 45 1067 93. [0,3; A6518]

M4678   1, 10, 11, 20, 21, 100, 101, 110, 111, 120, 121, 200, 201, 210, 211, 220, 221, 300,
301, 310, 311, 320, 321, 1000, 1001, 1010, 1011, 1020, 1021, 1100, 1101, 1110, 1111
Integers written in factorial base. Ref KN1 2 192. [1,2; A7623]

M4679   1, 10, 11, 100, 101, 110, 111, 1000, 1001, 1010, 1011, 1100, 1101, 1110, 1111,
10000, 10001, 10010, 10011, 10100, 10101, 10110, 10111, 11000, 11001, 11010, 11011
Natural numbers in base 2. [1,2; A7088]

M4680   10, 12, 15, 16, 18, 20, 24, 26, 28, 30, 32, 34, 35, 36, 40, 42, 44, 45, 48, 50, 52, 54,
56, 57, 58, 60, 63, 64
Orders of vertex-transitive graphs which are not Cayley graphs. Ref JAuMS A56 53 94.
bdm. [1,1; A6793]

M4681   10, 15, 20, 25, 32, 37, 43, 51
Zarankiewicz's problem. Ref LNM 110 142 69. [4,1; A6623]

M4682   1, 0, 1, 10, 17, 406, 1437, 20476, 44907, 1068404, 5112483, 230851094,
1942311373, 31916614874, 27260241361, 3826126294680, 37957167335671
Sums of logarithmic numbers. Ref TMS 31 77 63. jos. [1,4; A2744, N2001]

M4683   0, 10, 19, 199, 19999999999999999999999
Smallest number of additive persistence n. Ref JRM 7 134 74. [0,2; A6050]

M4684 1, 10, 20, 120, 440, 3200, 20460, 116600, 612700, 3628800
$n!$ in base n. [1,2; A6993]

M4685 1, 10, 22, 28, 30, 46, 52, 54, 58, 66, 70, 78, 82, 102, 106, 110, 126, 130, 136, 138,
148, 150, 166, 172, 178, 190, 196, 198, 210, 222, 226, 228, 238, 250, 262, 268, 270, 282
$\phi(x) = n$ has exactly 2 solutions. Ref AS1 840. [1,2; A7366]

M4686 1, 10, 25, 37, 42, 48, 79, 145, 244, 415, 672, 1100, 1722, 2727, 4193, 6428, 9658,
14478, 21313, 31304, 45329, 65311, 93074, 132026, 185413, 259242, 359395, 495839
A generalized partition function. Ref PNISI 17 236 51. [1,2; A2600, N2002]

M4687 0, 10, 25, 39, 77, 679, 6788, 68889, 2677889, 26888999, 3778888999,
277777788888899
Smallest number of persistence n. Probably finite. Ref JRM 6 97 73. [0,2; A3001]

M4688 10, 26, 34, 50, 52, 58, 86, 100
Noncototients. Ref UPNT B36. [1,1; A5278]

M4689 1, 10, 26, 75, 196, 520, 1361, 3570, 9346, 24475, 64076, 167760, 439201,
1149850, 3010346, 7881195, 20633236, 54018520, 141422321, 370248450, 969323026
Sum of squares of Lucas numbers. Ref BR72 20. [1,2; A5970]

M4690 1, 10, 27, 52, 85, 126, 175, 232, 297, 370, 451, 540, 637, 742, 855, 976, 1105,
1242, 1387, 1540, 1701, 1870, 2047, 2232, 2425, 2626, 2835, 3052, 3277, 3510, 3751
Decagonal numbers: $4n^2 - 3n$. See Fig M2535. Ref B1 189. [1,2; A1107]

M4691 1, 0, 10, 28, 0, 88, 4524, 0, 140692, 820496, 0, 128850048
Nonattacking queens on a $2n+1 \times 2n+1$ toroidal board. Ref AMM 101 637 94. [0,3; A7705]

M4692 1, 10, 30, 20, 10, 12, 20, 40, 90, 220, 572, 1560, 4420, 12920, 38760, 118864,
371450, 1179900, 3801900, 12406200, 40940460, 136468200, 459029400, 1556708400
Expansion of $(1-4x)^{5/2}$. Ref TH09 164. FMR 1 55. [0,2; A2422, N2003]

M4693 0, 1, 10, 33, 81, 148
Packing 3-dimensional cubes of side 2 in torus of side n. Ref SIAMP 4 98 71. [0,3; A3012]

M4694 1, 10, 34, 58, 73, 79, 86, 152, 265, 457, 763, 1268, 2058, 3308, 5236, 8220, 12731,
19546, 29685, 44702, 66714, 98806, 145154, 211756, 306667, 441249, 630771
A generalized partition function. Ref PNISI 17 236 51. [1,2; A2601, N2004]

M4695 0, 1, 10, 34, 80, 155, 266, 420, 624, 885, 1210, 1606, 2080, 2639, 3290, 4040,
4896, 5865, 6954, 8170, 9520, 11011, 12650, 14444, 16400, 18525, 20826, 23310, 25984
Enneagonal pyramidal numbers: $n(n+1)(7n-4)/6$. Ref B1 194. [0,3; A7584]

M4696 1, 10, 34, 206, 1351, 10543, 92708
Hit polynomials. Ref RI63. [3,2; A1890, N2005]

M4697 0, 1, 10, 35, 84, 165, 286, 455, 680, 969, 1330, 1771, 2300, 2925, 3654, 4495,
5456, 6545, 7770, 9139, 10660, 12341, 14190, 16215, 18424, 20825, 23426, 26235
$n(4n^2-1)/3$. Ref CC55 742. RCI 217. JO61 7. [0,3; A0447, N2006]

M4698 10, 35, 271, 29821
Switching networks. Ref JFI 276 318 63. [1,1; A0820, N2007]

M4699 1, 10, 40, 110, 245, 476, 840, 1380, 2145, 3190, 4576, 6370, 8645, 11480, 14960,
19176, 24225, 30210, 37240, 45430, 54901, 65780, 78200, 92300, 108225, 126126
4-dimensional figurate numbers: $(6n-2)$ $C(n+2,3)/4$. See Fig M3382. Ref B1 195. [1,2;
A2419, N2008]

M4700 1, 0, 10, 40, 315, 2464, 22260, 222480, 2447445, 29369120, 381798846,
5345183480, 80177752655, 1282844041920, 21808348713320, 392550276838944
Rencontres numbers. Ref R1 65. [3,3; A0449, N2009]

M4701 10, 42, 198, 1001, 5304, 29070, 163438, 937365, 5462730, 32256120, 192565800
Perforation patterns for punctured convolutional codes (3,1). Ref SFCA92 1 9. [2,1;
A7226]

M4702 1, 10, 44, 135, 336, 728, 1428, 2598, 4455, 7282, 11440, 17381, 25662, 36960,
52088, 72012, 97869, 130986, 172900, 225379, 290444, 370392, 467820, 585650
Quadrinomial coefficients. Ref C1 78. [2,2; A5720]

M4703 1, 10, 45, 120, 200, 162, 160, 810, 1530, 1730, 749, 1630, 4755, 7070, 6700, 2450,
5295, 14070, 20010, 19350, 10157, 6290, 25515, 40660, 44940, 34268, 9180, 24510
Coefficients of powers of η function. Ref JLMS 39 435 64. [10,2; A1488, N2010]

M4704 1, 10, 45, 141, 357, 784, 1554, 2850, 4917, 8074, 12727, 19383, 28665, 41328,
58276, 80580, 109497, 146490, 193249, 251713, 324093, 412896, 520950, 651430
From expansion of $(1+x+x^2)^n$. Ref C1 78. [3,2; A5714]

M4705 1, 10, 46, 186, 706, 2568, 9004, 30894, 103832, 343006
Cluster series for hexagonal lattice. Ref PRV 133 A315 64. DG72 225. [0,2; A3197]

M4706 10, 46, 556, 160948
Switching networks. Ref JFI 276 320 63. [1,1; A0832, N2011]

M4707 1, 10, 50, 175, 490, 1176, 2520, 4950, 9075, 15730, 26026, 41405, 63700, 95200,
138720, 197676, 276165, 379050, 512050, 681835, 896126, 1163800, 1495000, 1901250
$C(n,3).C(n-1,3)/4$. Ref CRO 10 30 67. [4,2; A6542]

M4708 1, 10, 50, 238, 1114, 4998, 22562, 98174, 434894, 1855346
Cluster series for cubic lattice. Ref PRV 133 A315 64. DG72 225. [0,2; A3207]

M4709 1, 10, 50, 385, 3130, 28764, 291900, 3249210, 39367395, 515874470,
7270929806, 109691447395, 1763782644020, 30114243100760, 544123405603800
From ménage numbers. Ref R1 198. [4,2; A0450, N2012]

M4710 10, 53, 242, 377, 1491, 1492, 6801, 14007, 100823, 559940, 1148303
$2^{a(n)}$ contains n consecutive 0's. Ref pdm. [1,1; A6889]

M4711 0, 0, 0, 0, 0, 0, 0, 1, 1, 10, 53, 383, 2809, 21884
Non-1-Hamiltonian simplicial polyhedra with n nodes. Ref Dil92. [1,10; A7035]

M4712 1, 10, 55, 220, 715, 2002, 5005, 11440, 24310, 48620, 92378, 167960, 293930,
497420, 817190, 1307504, 2042975, 3124550, 4686825, 6906900, 10015005, 14307150
Binomial coefficients $C(n,9)$. See Fig M1645. Ref D1 2 7. RS3. B1 196. AS1 828. [9,2;
A0582, N2013]

M4713 1, 10, 55, 385, 3025, 25333, 220825, 1978405, 18080425, 167731333,
1574304985, 14914341925, 142364319625, 1367428536133, 13202860761145
$1^n + 2^n + \cdots + 10^n$. Ref AS1 813. [0,2; A1557, N2014]

M4714 10, 55, 1996, 11756666
Switching networks. Ref JFI 276 318 63. [1,1; A0814, N2015]

M4715 1, 10, 56, 234, 815, 2504, 7018, 18336
Arrays of dumbbells. Ref JMP 11 3098 70; 15 214 74. [1,2; A2889, N2016]

M4716 1, 10, 57, 234, 770, 2136, 5180, 11292, 22599, 42190, 74371, 124950, 201552,
313964, 474510, 698456, 1004445, 1414962, 1956829, 2661730, 3566766, 4715040
n-coloring a cube. Ref C1 254. [1,2; A6550]

M4717 0, 1, 10, 57, 272, 885, 2226, 4725, 8912, 15417, 24970, 38401, 56640, 80717,
111762, 151005, 199776, 259505, 331722, 418057, 520240, 640101, 779570, 940677
Cubes with sides of n colors. Ref GA66 246. [0,3; A6529]

M4718 1, 10, 60, 280, 1120, 4032, 13440, 42240, 126720, 366080, 1025024, 2795520,
7454720, 19496960, 50135040, 127008768, 317521920, 784465920, 1917583360
$2^{n-4} \cdot C(n,4)$. Ref RSE 62 190 46. AS1 796. MFM 74 62 70 (divided by 2). [4,2; A3472]

M4719 1, 0, 0, 0, 10, 60, 462, 3920, 36954, 382740
Kings on an $n \times n$ cylinder. Ref AMS 38 1253 67. [1,5; A2493, N2017]

M4720 0, 10, 60, 595, 4512, 44802, 457040, 5159517
Hit polynomials. Ref JAuMS A28 375 79. [4,2; A4309]

M4721 1, 10, 65, 350, 1700, 7752, 33915, 144210, 600875, 2466750, 10015005,
  40320150, 161280600, 641886000, 2544619500, 10056336264, 39645171810
$10C(2n+1, n-4)/(n+6)$. Ref FQ 14 397 76. [4,2; A3519]

M4722 1, 10, 65, 350, 1701, 7770, 34105, 145750, 611501, 2532530, 10391745,
  42355950, 171798901, 694337290, 2798806985, 11259666950, 45232115901
Stirling numbers of second kind. See Fig M4981. Ref AS1 835. DKB 223. [4,2; A0453, N2018]

M4723 10, 70, 308, 1092, 3414, 9834, 26752, 69784, 176306, 434382, 1048812, 2490636,
  5833006, 13500754, 30933368, 70255008, 158335434, 354419190, 788529700
Walks on square lattice. Ref GU90. [0,1; A5567]

M4724 1, 10, 70, 420, 2310, 12012, 60060, 291720, 1385670, 6466460, 29745716,
  135207800, 608435100, 2714556600, 12021607800, 52895074320, 231415950150
$(2n+3)!/(6.n!(n+1)!)$. Ref JO39 449. [0,2; A2802, N2019]

M4725 0, 0, 1, 10, 70, 431, 2534, 14820, 88267, 542912, 3475978
n-dimensional hypotheses allowing for conditional independence. Ref ANS 4 1171 76.
[0,4; A5465]

M4726 10, 76, 8416, 268496896
Switching networks. Ref JFI 276 317 63. [1,1; A0808, N2020]

M4727 10, 79, 340, 1071, 2772, 6258, 12768, 24090, 42702, 71929
Rooted planar maps. Ref JCT B18 251 75. [1,1; A6469]

M4728 1, 10, 79, 602, 4682, 38072
Total preorders. Ref MSH 53 20 76. [3,2; A6329]

M4729 10, 80, 365, 1246, 3535
Sequences by number of increases. Ref JCT 1 372 66. [2,1; A0575, N2021]

M4730 1, 10, 85, 735, 6769, 67284, 723680, 8409500, 105258076, 1414014888,
  20313753096, 310989260400, 5056995703824, 87077748875904, 1583313975727488
Stirling numbers of first kind. See Fig M4730. Ref AS1 833. DKB 226. [4,2; A0454, N2022]

Figure M4730. PERMUTATIONS, STIRLING NUMBERS OF 1ST KIND.

A **permutation** of n objects is any rearrangement of them, and is specified either by a table:

$$\begin{matrix} 1 & 2 & 3 & 4 & 5 \\ 3 & 5 & 4 & 1 & 2 \end{matrix}$$

or by a product of cycles: (134)(25), both of which mean replace 1 by 3, 3 by 4, 4 by 1, 2 by 5, and 5 by 2. The total number of permutations of n objects is the **factorial** number $n! = 1.2.3.4.\ldots.n$, M1675. The **Stirling** number of the **first kind**, $s(n,k) = \begin{bmatrix} n \\ k \end{bmatrix}$, is the number of permutations of n objects that contain exactly k cycles. The first few values are illustrated as follows:

n \ k	1	2		3	4	Total
1	(1)					1
2	(12)	(1)(2)				2
3	(123)	(1)(23)		(1)(2)(3)		6
	(132)	(2)(13)				
		(3)(12)				
4	(1234)	(1)(234)	(1)(243)	(1)(2)(34)	(1)(2)(3)(4)	24
	(1243)	(2)(134)	(2)(143)	(1)(3)(24)		
	(1324)	(3)(124)	(3)(142)	(1)(4)(23)		
	(1342)	(4)(123)	(4)(132)	(2)(3)(14)		
	(1423)	(12)(34)	(13)(24)	(2)(4)(13)		
	(1432)	(14)(23)		(3)(4)(12)		

The Stirling numbers of the first kind continue:

							row sums $n!$
1							1
1	1						2
2	3	1					6
6	11	6	1				24
24	50	35	10	1			120
120	274	225	85	15	1		720
720	1764	1624	735	175	21	1	5040

The columns of this table give M1675, M2902, M4218, M4730, M4983, M5114, M5202, while the diagonals give M2535, M1998, M4258, M5155. Also

$$s(n,k) = (n-1)s(n-1,k) + s(n-1,k-1),$$

$$x(x-1)\cdots(x-n+1) = \sum_{k=0}^{n}(-1)^{n-k}s(n,k)x^{k}.$$

There is a complicated exact formula for $s(n,k)$, see [C1 216]. References: [R1 148], [DKB 226], [C1 212], [As1 835], [GKP 245].

M4731 10, 88, 6616, 91666432
Switching networks. Ref JFI 276 320 63. [1,1; A0826, N2023]

M4732 1, 10, 91, 651, 4026, 22737
Coefficients for extrapolation. Ref SE33 93. [1,2; A2739, N2024]

M4733 1, 10, 91, 820, 7381, 66430, 597871, 5380840, 48427561, 435848050,
3922632451, 35303692060, 317733228541, 2859599056870, 25736391511831
$(9^n - 1)/8$. Ref TH09 36. FMR 1 112. RCI 217. [1,2; A2452, N2025]

M4734 1, 10, 98, 982, 10062, 105024, 1112757, 11934910, 129307100, 1412855500,
15548498902, 172168201088, 1916619748084, 21436209373224, 240741065193282
Rooted trees with n nodes on projective plane. Ref CMB 31 269 88. [1,2; A7137]

M4735 1, 10, 99, 1024, 11304, 133669, 1695429, 23023811, 333840443, 5153118154,
84426592621, 1463941342191, 26793750988542, 516319125748337
Permutations of length n by length of runs. Ref DKB 261. [5,2; A0456, N2027]

M4736 1, 10, 105, 1260, 17325, 270270, 4729725, 91891800, 1964187225, 45831035250,
1159525191825, 31623414322500, 924984868933125, 28887988983603750
Expansion of $(1+3x)/(1-2x)^{7/2}$. Equals ½M2124. Ref TOH 37 259 33. JO39 152. DB1
296. C1 256. [0,2; A0457, N2028]

M4737 0, 10, 120, 1335, 15708, 200610, 2790510
Tumbling distance for n-input mappings. Ref PRV A32 2343 85. [0,2; A5949]

M4738 0, 0, 10, 124, 890, 5060, 25410, 118524, 527530, 2276020, 9613010, 40001324,
164698170, 672961380, 2734531810, 11066546524, 44652164810, 179768037140
Trees of subsets of an n-set. Ref MBIO 54 9 81. [1,3; A5174]

M4739 10, 167, 1720, 14065, 100156, 649950, 3944928, 22764165, 126264820,
678405090, 3550829360
Rooted genus-1 maps with n edges. Ref BAMS 74 74 68. WA71. JCT A13 215 72. [3,1;
A6295]

M4740 1, 10, 167, 1720, 24164, 256116, 3392843, 36703824, 472592916, 5188948072,
65723863196, 729734918432, 9145847808784
Rooted genus-1 maps with n edges. Ref WA71. JCT A13 215 72. [2,2; A6297]

M4741 1, 1, 10, 180, 4620, 152880, 6168960, 293025600
Dissections of a ball. Ref CMA 2 25 70. MAN 191 98 71. [3,3; A1762, N2029]

M4742 10, 190, 1568, 8344, 33580, 111100, 317680, 811096
Rooted nonseparable maps on the torus. Ref JCT B18 241 75. [2,1; A6409]

M4743 10, 192, 1630, 8924, 36834, 124560, 362934, 941820, 2227368, 4881448,
889015725
Tree-rooted toroidal maps. Ref JCT B18 258 75. [1,1; A6436]

M4744   1, 10, 199, 3970, 79201, 1580050, 31521799, 628855930, 12545596801,
250283080090, 4993116004999, 99612037019890, 1987247624392801
$a(n) = 20a(n-1) - a(n-2)$. Ref NCM 4 167 1878. MTS 65(4, Supplement) 8 56. [0,2; A1085, N2030]

M4745   1, 1, 10, 215, 12231, 2025462
Representations of 1 as a sum of n unit fractions. Ref SI72. UPNT D11. [1,3; A2967]

M4746   10, 219, 4796, 105030, 2300104, 50371117, 1103102046, 24157378203,
529034393290, 11585586272312, 253718493496142, 5556306986017175
$a(n) = 22a(n-1) - 3a(n-2) + 18a(n-3) - 11a(n-4)$. Deviates from M4747 starting at 1403-th term. Ref jhc. jwr. [1,1; A7698]

M4747   10, 219, 4796, 105030, 2300104, 50371117, 1103102046, 24157378203,
529034393290, 11585586272312, 253718493496142, 5556306986017175

A Pisot sequence: $a(n) =$ nearest integer $a(n-1)^2/a(n-2)$. Deviates from M4746 starting at 1403-th term. Ref jhc. jwr. [1,1; A7699]

M4748   10, 240, 2246, 12656, 52164, 173776, 495820, 1256992, 2902702
Rooted toroidal maps. Ref JCT B18 250 75. [1,1; A6423]

M4749   1, 10, 259, 12916, 1057221, 128816766, 21878089479, 4940831601000,
1432009163039625, 518142759828635250, 228929627246078500875
Central factorial numbers. Ref RCI 217. [0,2; A1824, N2031]

M4750   0, 1, 10, 297, 13756, 925705, 85394646, 10351036465, 1596005408152,
305104214112561, 70830194649795010, 19629681235869138841
Permutations with no hits on 2 main diagonals. Ref R1 187. [1,3; A0459, N2032]

M4751   10, 340, 5846, 71372, 706068, 6052840, 46759630, 333746556, 2238411692
Rooted toroidal maps. Ref JCT B18 251 75. [1,1; A6426]

M4752   10, 378, 16576, 819420
Finite automata. Ref IFC 10 507 67. [1,1; A0591, N2033]

M4753   10, 438, 5893028544
Post functions. Ref JCT 4 298 68. [1,1; A1327, N2034]

M4754   10, 595, 11010, 111650, 773640, 4104225, 17838730, 66390610, 218140650
Nonseparable toroidal tree-rooted maps. Ref JCT B18 243 75. [0,1; A6441]

M4755   10, 705, 14478, 154420, 1092640, 5826492, 25240410, 93203561, 303143970,
889015725
Tree-rooted toroidal maps. Ref JCT B18 258 75. [1,1; A6435]

M4756   10, 805, 23730, 431319, 5862920, 65548890, 636520890, 5555779185,
44603489700
Tree-rooted toroidal maps. Ref JCT B18 258 75. [1,1; A6440]

M4757 10, 840, 257040, 137260200, 118257539400, 154678050727200
Trivalent bipartite labeled graphs with $2n$ nodes. Ref RE58. [3,1; A6714]

M4758 10, 970, 912670090, 76022378683214797814371873O
Extracting a square root. Ref AMM 44 645 37. jos. [0,1; A6242]

M4759 10, 1010, 1111110010, 1000010001110100011000101111010
Convert the last term from decimal to binary! Ref Pick92 352. [1,1; A6937]

M4760 1, 10, 3330, 178981952
Groupoids with n elements. Ref PAMS 17 736 66. LE70 246. [1,2; A1329, N2035]

M4761 1, 10, 3360, 1753920, 1812888000, 3158396841600, 8496995611104000,
33199738565849856000, 180116271096528678912000
Labeled trivalent cyclically 4-connected graphs with $2n$ nodes. Ref rwr. [2,2; A7101]

M4762 0, 0, 0, 10, 3360, 1829520, 2008360200, 3613037828400, 9777509703561600,
37906993895091921600, 2029135829760422261523200
Labeled trivalent graphs with $2n$ nodes and no triangles. Ref rwr. [0,4; A7103]

M4763 0, 10, 3360, 1829520, 2010023400, 3622767548400, 9820308795897600,
38117776055769009600, 204209112522362230483200
Triangle-free cubic graphs with n nodes. Ref CN 33 376 81. [4,2; A6903]

M4764 0, 1, 11, 0, 8, 13, 5, 9, 10, 1, 2, 9, 2, 10, 10, 7, 9, 9, 4, 1, 9, 10, 3, 2, 7, 9, 10, 3, 6,
11, 2, 3, 10, 8, 9, 12, 6, 7, 11, 8, 10, 2, 8, 4, 11, 8, 9, 10, 4, 8, 10, 7, 9, 11, 9, 8, 10, 5, 8, 13
Iterations until n reaches 1 or 4 under x goes to sum of squares of digits map. Ref Robe92
13. [1,3; A3621]

M4765 1, 1, 11, 5, 137, 7, 363, 761, 7129, 671, 83711, 6617, 1145993, 1171733, 1195757,
143327, 42142223, 751279, 275295799
Numerators of coefficients for numerical differentiation. Cf. M4822. Ref PHM 33 11 42.
BAMS 48 922 42. [2,3; A2547, N2036]

M4766 11, 7, 2, 2131, 1531, 385591, 16651, 15514861, 857095381, 205528443121,
1389122693971, 216857744866621, 758083947856951
Chains of length n of nearly doubled primes. Ref MOC 53 755 89. [1,1; A5603]

M4767 1, 1, 1, 11, 7, 389, 1031, 19039, 24431, 1023497, 4044079, 225738611,
1711460279, 29974303501, 4656373513, 3798866053319, 34131041040991
Sums of logarithmic numbers. Ref TMS 31 78 63. jos. [0,4; A2749, N2037]

M4768 1, 11, 12, 1121, 122111, 112213, 12221131, 1123123111, 12213111213113,
11221131132111311231, 12221231123121133112213111
Describe the previous term! [1,2; A7651]

M4769 11, 13, 17, 19, 31, 37, 53, 59, 71, 73, 79, 97, 113, 131, 137, 139, 151, 157, 173,
179, 191, 193, 197, 199, 211, 233, 239, 251, 257, 271, 277, 293, 311, 313, 317, 331, 337
Inert rational primes in $Q(\sqrt{-5})$. Ref Hass80 498. [1,1; A3626]

M4770 11, 13, 17, 23, 37, 47, 53, 67, 73, 103, 107, 157, 173, 233, 257, 263, 277, 353, 373,
563, 593, 607, 613, 647, 653, 733, 947, 977, 1097, 1103, 1123, 1187, 1223, 1283, 1297
$n-6$, n, $n+6$ are primes. Ref MMTC 62 471 69. [1,1; A6489]

M4771 11, 13, 19, 22, 23, 25, 26, 27, 29, 38, 39, 41, 43, 44, 46, 47, 50, 52, 53, 54, 55, 57,
58, 59, 61, 71, 76, 77, 78, 79, 82, 86, 87, 88, 91, 92, 94, 95, 99, 100, 103, 104, 106, 107
Flimsy numbers. Ref ACA 38 117 80. [1,1; A5360]

M4772 11, 13, 19, 29, 37, 41, 53, 59, 61, 67, 71, 79, 83, 101, 107, 131, 139, 149, 163, 173,
179, 181, 191, 197, 199, 211, 227, 239, 251, 269, 271, 293, 311, 317, 347, 349
Solution of a congruence. Ref Krai24 1 64. [1,1; A1916, N2038]

M4773 1, 11, 13, 23, 13, 11, 37, 61, 23, 71, 1, 97, 107, 73, 11, 143, 59, 131, 157, 191, 193,
83, 169, 13, 143, 121, 61, 229, 179, 71, 181, 241, 251, 359, 349, 347, 181, 313, 179, 431
x such that $p^2 = x^2 + 3y^2$. Cf. M3228. Ref CU27 79. L1 60. [7,2; A2367, N1377]

M4774 11, 14, 15, 17, 19, 20, 21, 24, 26, 27, 30, 32, 33, 34, 35, 36, 37, 38, 39, 40, 42, 43,
44, 45, 46, 48, 49, 50, 51, 52, 53, 54, 55, 56, 57, 58, 61, 62, 63, 64, 65, 66, 67, 69, 70, 72
Conductors of elliptic curves. Ref LNM 476 82 75. [1,1; A5788]

M4775 11, 17, 22, 28, 36, 43, 51, 61
Zarankiewicz's problem. Ref LNM 110 143 69. [3,1; A6618]

M4776 11, 17, 23, 30, 38, 46, 55
Zarankiewicz's problem. Ref LNM 110 143 69. [3,1; A6621]

M4777 1, 1, 11, 19, 7861, 301259, 451526509, 6427914623, 16794274237
Coefficients of Green function for cubic lattice. Ref PTRS 273 593 73. [0,3; A3284]

M4778 1, 11, 21, 1112, 1231, 11131211, 2112111331, 112331122112, 12212221231221,
11221113121132112211, 21222112132112111113312221
Describe the previous term, from the right! See Fig M2629. Ref jhc. JRM 25 189 93. [1,2;
A6711]

M4779 1, 11, 21, 1112, 3112, 211213, 312213, 212223, 114213, 31121314, 41122314,
31221324, 21322314, 21322314, 21322314, 21322314, 21322314, 21322314, 21322314
Summarize the previous term! See Fig M2629. Ref AMM 101 560 94. [1,2; A5151]

M4780 1, 11, 21, 1211, 111221, 312211, 13112221, 1113213211, 31131211131221,
13211311123113112211, 11131221133112132113212221
Describe the previous term! See Fig M2629. Ref CoGo87 176. [1,2; A5150]

M4781 1, 1, 11, 23, 379, 781, 1160, 5421, 12002, 17423, 377885, 395308, 1563809, 8214353, 9778162, 27770677, 37548839, 65319516, 168187871, 1915386097
Convergents to cube root of 7. Ref AMP 46 106 1866. L1 67. hpr. [1,3; A5485]

M4782 11, 26, 39, 47, 53, 61, 67, 76, 83, 89, 104, 106, 109, 116, 118, 121, 139, 147, 152, 155, 170, 186, 191, 200, 207, 211, 212, 214, 219, 222, 233, 236, 244, 249, 262, 277, 286
Elliptic curves. Ref JRAM 212 24 63. [1,1; A2154, N2040]

M4783 11, 29, 31, 41, 43, 53, 61, 71, 79, 101, 103, 113, 127, 131, 137, 149, 151, 157, 181, 191, 197, 211, 223, 229, 239, 241, 251, 271, 281, 293, 307, 313, 337, 379, 389, 401, 409
Class 2 – primes. Ref UPNT A18. [1,1; A5110]

M4784 11, 30, 77, 162, 323, 589, 1043, 1752, 2876, 4571, 7128, 10860, 16306, 24051, 35040, 50355, 71609, 100697, 140349, 193784, 265505, 360889, 487214, 653243
Bipartite partitions. Ref ChGu56 1. [0,1; A2755, N2041]

M4785 11, 31, 151, 911, 5951, 40051, 272611, 1863551, 12760031, 87424711, 599129311, 4106261531, 28144128251, 192901135711, 1322159893351
Related to factors of Fibonacci numbers. Ref JA66 20. [0,1; A1604, N2042]

M4786 1, 1, 11, 31, 161, 601, 2651, 10711, 45281, 186961, 781451, 3245551, 13524161, 56258281, 234234011, 974792551, 4057691201, 16888515361, 70296251531
$a(n) = 2a(n-1) + 9a(n-2)$. Ref MQET 1 11 16. [0,3; A2535, N2043]

M4787 11, 32, 87, 247, 716, 2155, 6694, 21461
From the graph reconstruction problem. Ref LNM 952 101 82. [4,1; A6655]

M4788 1, 11, 36, 92, 491, 2537
Rhyme schemes. Ref ANY 319 464 79. [1,2; A5000]

M4789 1, 0, 0, 1, 1, 1, 11, 36, 92, 491, 2557, 11353, 60105, 362506, 2169246, 13580815, 91927435, 650078097, 4762023647, 36508923530, 292117087090, 2424048335917
Expansion of $\exp(e^x - 1 - x - \frac{1}{2}x^2)$. Ref FQ 14 69 76. [0,7; A6505]

M4790 11, 37, 101, 271, 37, 4649, 137, 333667, 9091, 513239, 9901, 265371653, 909091, 2906161, 5882353, 5363222357, 333667, 111111111111111111111, 27961, 10838689
Largest factor of 11...1. Ref Krai52 40. CUNN. [0,1; A3020]

M4791 0, 1, 11, 38, 90, 175, 301, 476, 708, 1005, 1375, 1826, 2366, 3003, 3745, 4600, 5576, 6681, 7923, 9310, 10850, 12551, 14421, 16468, 18700, 21125, 23751, 26586
Decagonal pyramidal numbers: $n(n+1)(8n-5)/6$. Ref B1 194. [0,3; A7585]

M4792 11, 61, 181, 421, 461, 521, 991, 1621, 1871, 3001, 4441, 4621, 6871, 9091, 9931, 12391, 13421, 14821, 19141, 25951, 35281, 35401, 55201, 58321, 61681, 62071, 72931
Quintan primes: $p = (x^5 + y^5)/(x + y)$. Ref CU23 2 201. [1,1; A2650, N2044]

M4793 1, 11, 61, 231, 681, 1683, 3653, 7183, 13073, 22363, 36365, 56695, 85305, 124515, 177045, 246047, 335137, 448427, 590557, 766727, 982729, 1244979, 1560549
Expansion of $(1+x)^5/(1-x)^6$. Ref SIAR 12 277 70. C1 81. [0,2; A1847, N2045]

M4794 1, 11, 66, 286, 1001, 3003, 8008, 19448, 43758, 92378, 184756, 352716, 646646, 1144066, 1961256, 3268760, 5311735, 8436285, 13123110, 20030010, 30045015
Binomial coefficients $C(n,10)$. See Fig M1645. Ref D1 2 7. RS3. B1 196. AS1 828. [10,2; A1287, N2046]

M4795 0, 1, 11, 66, 302, 1191, 4293, 14608, 47840, 152637, 478271, 1479726, 4537314, 13824739, 41932745, 126781020, 382439924, 1151775897, 3464764515, 10414216090
Eulerian numbers. See Fig M3416. Ref R1 215. DB1 151. JCT 1 351 66. DKB 260. C1 243. [2,3; A0460, N2047]

M4796 1, 11, 72, 364, 1568, 6048, 21504, 71808, 228096, 695552, 2050048, 5870592, 16400384, 44843008, 120324096, 317521920, 825556992, 2118057984, 5369233408
Coefficients of Chebyshev polynomials. Ref AS1 795. [0,2; A6975]

M4797 1, 11, 77, 440, 2244, 10659, 48279, 211508, 904475, 3798795, 15737865, 64512240, 262256280, 1059111900, 4254603804, 17018415216, 67837293986
$11C(2n,n-5)/(n+6)$. Ref QAM 14 407 56. MOC 29 216 75. [5,2; A0589, N2048]

M4798 1, 11, 85, 575, 3661, 22631, 137845, 833375, 5019421, 30174551
Differences of reciprocals of unity. Ref DKB 228. [1,2; A1240, N2049]

M4799 1, 11, 87, 693, 5934, 55674, 572650, 6429470, 78366855, 1031378445, 14583751161, 220562730171, 3553474061452
Permutations of length n by rises. Ref DKB 264. [4,2; A1278, N2050]

M4800 11, 101, 181, 619, 16091, 18181, 19861, 61819, 116911, 119611, 160091, 169691, 191161, 196961, 686989, 688889, 1008001
Strobogrammatic primes. Ref JRM 15 281 83. [1,1; A7597]

M4801 1, 11, 101, 781, 5611, 39161, 270281, 1857451, 12744061, 87382901, 599019851, 4105974961, 28143378001, 192899171531, 1322154751061, 9062194370461
Related to factors of Fibonacci numbers. Ref JA66 20. [0,2; A1603, N2051]

M4802 1, 11, 101, 1111, 10001, 110011, 1010101, 11111111, 100000001, 1100000011,
10100000101, 111100001111, 1000100010001, 11001100110011, 101010101010101
Rows of Pascal's triangle mod 2. Ref Pick92 353. [0,2; A6943]

M4803 1, 11, 107, 1066, 11274, 127860, 1557660, 20355120, 284574960, 4243508640,
67285058400, 1131047366400, 20099588140800, 376612896038400
Generalized Stirling numbers. Ref PEF 77 61 62. [0,2; A1721, N2052]

M4804 1, 11, 111, 1111, 11111, 111111, 1111111, 11111111, 111111111, 1111111111,
11111111111, 111111111111, 1111111111111, 11111111111111, 111111111111111
Unary representation of natural numbers. [1,2; A0042]

M4805 1, 11, 113, 1099, 11060, 118484, 1366134, 16970322, 226574211, 3240161105,
49453685911, 802790789101
Permutations of length n by rises. Ref DKB 263. [5,2; A1268, N2053]

M4806 1, 11, 121, 1001, 11011, 121121, 1002001, 11022011, 121212121, 1000000001,
11000000011, 121000000121, 1001000001001, 11011000011011, 121121000121121
Rows of Pascal's triangle mod 3. Ref Pick92 353. [0,2; A6940]

M4807 1, 11, 121, 1331, 14641, 161051, 1771561, 19487171, 214358881, 2357947691,
25937424601, 285311670611, 3138428376721, 34522712143931
Powers of 11. Ref BA9. [0,2; A1020, N2054]

M4808 1, 11, 188, 2992, 51708, 930436, 17131724
Specific heat for cubic lattice. Ref JMP 3 187 62. [0,2; A1408, N2055]

M4809 1, 11, 191, 2497, 14797, 92427157, 36740617, 61430943169, 23133945892303,
16399688681447
Numerators of coefficients for numerical integration. Cf. M4880. Ref OP80 545. PHM 35
263 44. [0,2; A2195, N2056]

M4810 1, 11, 301, 15371, 1261501, 151846331, 25201039501, 5515342166891,
1538993024478301, 533289474412481051, 224671379367784281901
Glaisher's H' numbers. Ref PLMS 31 232 1899. FMR 1 76. [1,2; A2114, N2057]

M4811 11, 309, 5805, 95575, 1516785, 24206055, 396475975, 6733084365,
119143997490
Permutations of length n by rises. Ref DKB 264. [8,1; A1280, N2058]

M4812   1, 11, 361, 24611, 2873041, 512343611, 129570724921, 44110959165011,
19450718635716001, 10784052561125704811, 7342627959965776406281
Generalized tangent numbers. Ref QJMA 45 202 14. MOC 21 690 67. [1,2; A0464,
N2059]

M4813   1, 1, 11, 378, 27213, 3378680
An occupancy problem. Ref JACM 24 593 77. [0,3; A6698]

M4814   11, 1011, 1111110011, 100001000111010001000101111011
Convert the last term from decimal to binary! Ref Pick92 352. [1,1; A6938]

M4815   11, 1230, 47093135946
Post functions. Ref JCT 4 296 68. [1,1; A1323, N2060]

M4816   11, 111111111111111111, 1111111111111111111111111
Primes of form $(10^n - 1)/9$ (next terms are for $n = 317, 1031$). Cf. M2114. Ref CUNN.
[1,1; A4022]

M4817   1, 0, 0, 12, 0, 0, 6, 0, 2, 18, 0, 12, 6, 0, 0, 12, 0, 12, 6, 6, 12, 24, 6, 0, 0, 12, 0, 12, 0,
24, 12, 12, 2, 12, 6, 24, 6, 12, 0, 24, 0, 12, 0, 6, 24, 12, 12, 24, 6, 12, 0, 24, 0, 24, 18, 12
Theta series of hexagonal close-packing. Ref SPLAG 114. [0,4; A4012]

M4818   12, 2, 78, 24, 548, 228, 4050, 2030, 30960, 17670, 242402, 152520, 1932000,
1312844, 15612150, 11297052
Magnetization for hexagonal lattice. Ref DG74 420. [0,1; A7207]

M4819   1, 0, 12, 4, 129, 122, 1332, 960, 10919, 11372, 132900, 126396, 1299851, 1349784
Susceptibility for hexagonal lattice. Ref PHA 28 934 62. DG74 421. [0,3; A2911, N2061]

M4820   1, 1, 12, 4, 360, 40, 20160, 12096, 259200, 604800, 239500800, 760320,
43589145600, 217945728000, 1494484992000, 697426329600, 3201186852864000
From generalized Bernoulli numbers. Ref SAM 23 211 44. [2,3; A2679, N2062]

M4821   1, 12, 6, 24, 12, 24, 8, 48, 6, 36, 24, 24, 24, 72, 0, 48, 12, 48, 30, 72, 24, 48, 24, 48,
8, 84, 24, 96, 48, 24, 0, 96, 6, 96, 48, 48, 36, 120, 24, 48, 24, 48, 48, 120, 24, 120, 0, 96
Theta series of face-centered cubic lattice. Ref SPLAG 113. [0,2; A4015]

M4822   1, 1, 12, 6, 180, 10, 560, 1260, 12600, 1260, 166320, 13860, 2522520, 2702700,
2882880, 360360, 110270160, 2042040, 775975200
Denominators of coefficients for numerical differentiation. Cf. M4765. Ref PHM 33 11 42.
BAMS 48 922 42. [2,3; A2548, N2063]

‖‖‖

Figure M4822. DISALLOWED SEQUENCES.

A number of pleasant puzzle sequences are not in the table because they are finite or are not integers:

(1) ¼, ½, 1, 3, 6, 12, 24, 30, 120, 240, 1200, 2400, English money in 1950.

(2) 3, 8, 8, 4, 89, 75, 30, 28, ?, planetary diameters in thousands of miles.

(3) 8, 5, 4, 9, 1, 7, 6, 3, 2, 0; or 8, 8000000000, ..., 18, 18000000000, ..., 180000000, ..., 18000, ..., 80, ..., 88, ..., 85, ..., 84, ..., 11, ..., 15, ..., 5, ..., 4, ..., the numbers arranged in alphabetical order (in English).

(4) 12, 13, 14, 15, 20, 22, 30, 110, 1100, the number 12 written in bases 10, 9, 8, ..., 2.

(5) H, H, L, B, B, C, N, O, F, N, N, M, A, S, P, S, C, A, K, C, S, T, V, C, M, F, C, N, C, Z, G, G, A, S, B, K, R, S, Y, Z, N, M, T, R, R, P, A, C, I, S, S, T, I, X, C, B, L, C, P, N, P, S, E, G, T, D, H, E, T, Y, L, H, T, W, R, O, I, ... the initial letters of symbols for the chemical elements. (See however M3296!)

(6) 14, 18, 23, 28, 34, 42, 50, 59, 66, 72, 79, 86, 96, 103, 110, 116, 125, 137, 145, 157, 168, 181, 191, 207, 215, 225, 231, 238, 242, the local stops on the New York West Side subway.

(7) 1714, 1727, 1760, 1820, 1910, 1936, dates of the accessions of the Georges to the English throne.

(8) 1732, 1735, 1743, 1751, 1758, 1767, 1767, 1782, 1773, 1790, 1795, 1784, 1800, 1804, 1791, 1809, 1808, 1822, 1822, 1831, 1830, 1837, 1833, 1843, 1858, 1857, 1856, 1865, 1872, 1874, 1882, 1884, 1890, 1917, 1908, 1913, 1913, 1924, 1911, 1924, 1946, dates of birth of presidents of the U.S.A.

(9) W, A, J, M, M, A, J, B, H, T, P, T, F, P, B, L, J, G, H, G, A, C, H, C, M, R, T, W, H, C, H, R, T, E, K, J, N, F, C, R, B, C, presidents of the USA.

(10) 778, 846, 863, 921, 967, 1081, 1121, 1187, 1214, 1289, 1423, 1462, 1601, 1638, 1710, 1754, 1785, 1755, dates of birth of Kings Louis I, II, ... of France.

(11) The integers 1, 2, 3, ... drawn next to a mirror

(12) O, T, T, F, F, S, S, E, N, T, E, T, T, F, F, S, S, E, N, T, T, T, T, ..., the initial letters of the English names for the numbers.

‖‖‖

M4823   0, 0, 1, 0, 12, 14, 135, 276, 1520, 4056, 17778, 54392, 213522, 700362, 2601674, 8836812, 31925046, 110323056, 393008712, 1369533048
Susceptibility for cubic lattice. Ref JPA 6 1511 73. DG74 421. [1,5; A2926, N2064]

M4824   12, 18, 20, 24, 28, 30, 32, 40, 42, 44, 45, 48, 50, 52, 54, 56, 60, 63, 66, 68, 70, 72, 75, 76, 78, 80, 84, 88, 90, 92, 96, 98, 99, 102, 104, 105, 108, 110, 112, 114, 116, 117, 120
Product of proper divisors of $n = n^k$, $k > 1$. Ref B1 23. [1,1; A7624]

M4825   12, 18, 20, 24, 30, 36, 40, 42, 48, 54, 56, 60, 66, 70, 72, 78, 80, 84, 88, 90, 96, 100, 102, 104, 108, 112, 114, 120, 126, 132, 138, 140, 144, 150, 156, 160, 162, 168, 174, 176
Abundant numbers. See Fig M0062. Ref QJMA 44 274 13. UPNT B2. [1,1; A5101]

M4826   12, 18, 26, 33, 41
Zarankiewicz's problem. Ref LNM 110 144 69. [3,1; A6622]

M4827   12, 18, 31, 32, 54, 56, 80, 98, 104, 108, 114, 124, 126, 128, 132, 140, 152, 156, 182, 186, 210, 264, 272, 280, 308, 320, 342, 378, 390, 392, 399, 403, 408, 416, 440, 444
$\sigma(x) = n$ has exactly 2 solutions. Ref AS1 840. [1,1; A7371]

M4828   0, 0, 12, 24, 60, 180, 588, 1968, 6840, 24240, 87252, 318360, 1173744, 4366740, 16370700, 61780320, 234505140, 894692736, 3429028116, 13195862760, 50968206912
n-step polygons on hexagonal lattice. Ref JPA 5 665 72; 17 455 84. ajg. [1,3; A1335, N2065]

M4829   1, 12, 24, 96, 72, 168, 240, 336, 360, 504, 576, 1512, 1080, 1008, 720, 2304, 3600, 5376, 2520, 2160, 1440, 10416, 13392, 3360, 4032, 3024, 7056, 6720, 2880, 6480
Smallest k such that $\sigma(x) = k$ has exactly n solutions. Ref AS1 840. [1,2; A7368]

M4830   0, 12, 24, 168, 1440, 24480, 297024, 28017216, 533681664, 41156316672
Specific heat for b.c.c. lattice. Ref PRV 164 801 67. [1,2; A2167, N2066]

M4831   12, 30, 210, 371, 22737, 19733142, 48264275462, 9769214287853155785, 11308412892367501453788572548 5, 5271244267917980801966553649147604697542
Continued fraction for gamma function. Cf. M5308. Ref MOC 34 548 80. AS1 258. [0,1; A5147]

M4832   1, 12, 30, 427
Crystallographic orbits in n dimensions. Ref Enge93 1027. [0,2; A7308]

M4833   1, 12, 37, 76, 129, 196, 277, 372, 481, 604, 741, 892, 1057, 1236, 1429, 1636, 1857, 2092, 2341, 2604, 2881, 3172, 3477, 3796, 4129, 4476, 4837, 5212, 5601, 6004
Truncated square numbers: $7n^2 + 4n + 1$. Ref INOC 24 4550 85. [0,2; A5892]

M4834   1, 12, 42, 92, 162, 252, 362, 492, 642, 812, 1002, 1212, 1442, 1692, 1962, 2252, 2562, 2892, 3242, 3612, 4002, 4412, 4842, 5292, 5762, 6252, 6762, 7292, 7842, 8412
Points on surface of cuboctahedron (or icosahedron): $10n^2 + 2$. Ref RO69 109. MF73 46. Coxe74. INOC 24 4550 85. [0,2; A5901]

M4835   0, 1, 12, 42, 100, 195, 336, 532, 792, 1125, 1540, 2046, 2652, 3367, 4200, 5160, 6256, 7497, 8892, 10450, 12180, 14091, 16192, 18492, 21000, 23725, 26676, 29862
Hendecagonal pyramidal numbers: $n(n+1)(3n-2)/2$. Ref B1 194. [0,3; A7586]

M4836 1, 12, 48, 16, 414, 960, 672, 4800, 2721, 9064, 8880, 6912, 2398, 13440, 29280,
  30976, 10878, 57228, 9360, 252384, 53760, 177600, 113952, 107520, 436131, 16488
Related to representation as sums of squares. Ref QJMA 38 325 07. [1,2; A2612, N2067]

M4837 1, 12, 48, 124, 255, 456, 742, 1128, 1629, 2260, 3036, 3972, 5083, 6384, 7890,
  9616, 11577, 13788, 16264, 19020, 22071, 25432, 29118, 33144, 37525, 42276, 47412
Icosahedral numbers: $n(5n^2 - 5n + 2)/2$. [1,2; A6564]

M4838 12, 48, 180, 792, 3444, 15000, 64932, 280200, 1204572, 5159448, 22043292,
  93952428, 399711348
Self-avoiding walks on hexagonal lattice, with additional constraints. Ref JPA 13 3530 80.
[2,1; A7200]

M4839 1, 12, 48, 252, 1440, 8544, 52416, 330588, 2130240, 13961808, 92784384,
  623772288, 4234688640, 28990262016, 199908428544, 1387276513308
Internal energy series for f.c.c. lattice. Ref DG72 425. [0,2; A3498]

M4840 1, 0, 12, 48, 540, 4320, 42240, 403200, 4038300, 40958400
n-step polygons on f.c.c. lattice. Ref AIP 9 345 60. [0,3; A2899, N2068]

M4841 1, 12, 54, 88, 99, 540, 418, 648, 594, 836, 1056, 4104, 209, 4104, 594, 4256, 6480,
  4752, 298, 5016, 17226, 12100, 5346, 1296, 9063, 7128, 19494, 29160, 10032, 7668
Expansion of $\Pi(1 - x^k)^{12}$. Ref QJMA 38 56 07. KNAW 59 207 56. GMJ 8 29 67. [0,2;
A0735, N2069]

M4842 1, 12, 54, 188, 636, 2168, 7556, 26826, 96724, 353390, 1305126, 4864450,
  18272804
n-step walks on hexagonal lattice. Ref JPA 6 352 73. [3,2; A5549]

M4843 12, 60, 210, 630, 1736, 4536, 11430
Labeled trees of diameter 3 with n nodes. Ref IBMJ 4 478 60. [4,1; A0554, N2070]

M4844 0, 1, 12, 61, 240, 841, 2772, 8821, 27480, 84481, 257532, 780781, 2358720,
  7108921, 21392292, 64307941, 193185960, 580082161, 1741295052, 5225982301
Trees of subsets of an n-set. Ref MBIO 54 9 81. [1,3; A5173]

$$\text{G.f.: } x(1 + 6x) / (1 - x)(1 - 2x)(1 - 3x).$$

M4845 1, 12, 66, 220, 483, 660, 252, 1320, 4059, 6644, 6336, 240, 12255, 27192, 35850,
  27972, 2343, 50568, 99286, 122496, 96162, 11584, 115116, 242616, 315216, 283800
Coefficients of powers of η function. Ref JLMS 39 436 64. [12,2; A1490, N2071]

M4846 1, 12, 66, 232, 639, 1596, 3774, 8328, 17283, 34520, 66882, 125568, 229244,
  409236, 716412, 1231048, 2079237, 3459264, 5677832, 9200232, 14729592, 23325752
McKay-Thompson series of class 4D for Monster. Ref CALG 18 257 90. FMN94. [0,2;
A7249]

M4847 1, 12, 66, 245, 715, 1768, 3876, 7752, 14421, 25300, 42287, 67860, 105183,
158224, 231880, 332112, 466089, 642341, 870922, 1163580, 1533939, 1997688
Fermat coefficients. Ref MMAG 27 141 54. [7,2; A0972, N2072]

M4848 1, 12, 66, 312, 1368, 5685
Cluster series for honeycomb. Ref PRV 133 A315 64. [0,2; A3200]

M4849 1, 12, 75, 384, 1805, 8100, 35287, 150528, 632025, 2620860, 10759331,
43804800, 177105253, 711809364, 2846259375, 11330543616, 44929049777
Complexity of doubled cycle. Ref JCT B24 208 78. [1,2; A6235]

M4850 1, 12, 78, 364, 1365, 4368, 12376, 31824, 75582, 167960, 352716, 705432,
1352078, 2496144, 4457400, 7726160, 13037895, 21474180, 34597290, 54627300
Binomial coefficients $C(n,11)$. See Fig M1645. Ref D1 2 7. RS3. B1 196. AS1 828. [11,2;
A1288, N2073]

M4851 1, 12, 81, 372, 1332, 3984, 10420, 24540, 53145, 107436, 205065, 372792,
649936, 1092672, 1779408, 2817288, 4350105, 6567660, 9716905, 14114892, 20163924
4-voter voting schemes with n linearly ranked choices. Ref Loeb94b. [1,2; A7010]

M4852 12, 84, 468, 2332, 11068, 51472, 237832, 1095384
n-step walks on cubic lattice. Ref PCPS 58 99 62. [1,1; A0761, N2074]

M4853 1, 12, 84, 504, 3012, 17142
Cluster series for f.c.c. lattice. Ref PRV 133 A315 64. DG72 225. [0,2; A3209]

M4854 1, 12, 90, 520, 2535, 10908, 42614, 153960, 521235, 1669720, 5098938,
14931072, 42124380, 114945780, 304351020, 784087848, 1970043621, 4837060800
Coefficients of a modular function. Ref GMJ 8 29 67. [5,2; A5758]

$$\text{G.f.: } \Pi \, (1 - x^k)^{-12}.$$

M4855 1, 12, 90, 560, 3150, 16632, 84084, 411840, 1969110, 9237800, 42678636,
194699232, 878850700, 3931426800, 17450721000, 76938289920, 337206098790
Coefficients for numerical differentiation. Ref OP80 21. SE33 92. SAM 22 120 43. LA56
514. [0,2; A2544, N2075]

$$\text{G.f.: } (1 + 2\,x) \,/\, (1 - 4\,x)^{5/2}.$$

M4856 1, 12, 103, 736, 4571, 25326, 127415, 588687
n-covers of a 5-set. Ref DM 81 151 90. [1,2; A5771]

M4857 0, 1, 0, 0, 1, 12, 104, 956
Irreducible posets. Ref PAMS 45 298 74. [0,6; A3431]

M4858 1, 12, 110, 945, 8092, 70756
Generalized Stirling numbers of second kind. Ref FQ 5 366 67. [3,2; A0559, N2076]

M4859   1, 12, 114, 940, 7568, 61728, 512996, 4334884, 37164700, 322624804
n-step walks on f.c.c. lattice. Ref JPA 6 351 73. [2,2; A5543]

M4860   1, 12, 114, 1012, 8775, 75516, 649264, 5593068, 48336171, 419276660,
3650774820, 31907617560, 279871768995, 2463161027292, 21747225841440
From generalized Catalan numbers. Ref LNM 952 280 82. [0,2; A6635]

M4861   1, 12, 119, 1175, 12154, 133938, 1580508, 19978308, 270074016, 3894932448,
59760168192, 972751628160, 16752851775360, 304473528961920
Generalized Stirling numbers. Ref PEF 77 26 62. [0,2; A1712, N2077]

M4862   12, 120, 720, 3360, 13440, 48384, 161280, 506880, 1520640, 4392960, 12300288,
33546240, 89456640, 233963520, 601620480, 1524105216, 3810263040, 9413591040
Coefficients of Hermite polynomials. Ref AS1 801. [0,1; A1816, N2078]

$$\text{G.f.:}\quad 12(1 - 2\,x)^{-5}.$$

M4863   1, 12, 120, 1200, 12600, 141120, 1693440, 21772800, 299376000, 4390848000,
68497228800, 1133317785600, 19833061248000, 366148823040000
Lah numbers: $n! C(n-1,2)/6$. Ref R1 44. C1 156. [3,2; A1754, N2079]

M4864   1, 12, 132, 847, 3921, 14506, 45402, 124707, 308407, 699766
$3 \times 3 \times 3$ partitions of n. Ref CJN 13 283 70. [0,2; A2721, N2080]

M4865   1, 12, 132, 1392, 14292, 144000, 1430592, 14057280, 136914804, 1323843936,
12722294736, 121625850240, 1157512059936, 10972654675200, 103654156958208
Susceptibility for f.c.c. lattice. Ref DG72 404. [0,2; A3495]

M4866   1, 12, 132, 1404, 14652, 151116, 1546332, 15734460, 159425580, 1609987708,
16215457188, 162961837500, 1634743178420
Susceptibility for f.c.c. lattice. Ref SSP 3 268 70. JPA 5 651 72. DG74 381. [0,2; A2921,
N2081]

M4867   1, 12, 132, 1404, 14700, 152532, 1573716, 16172148, 165697044, 1693773924,
17281929564, 176064704412, 1791455071068, 18208650297396, 184907370618612
n-step self-avoiding walks on f.c.c. lattice. Ref JCP 46 3481 67. JPA 12 L267 79. [0,2;
A1336, N2082]

M4868   1, 12, 137, 1602, 19710, 257400, 3574957, 52785901, 827242933, 13730434111,
240806565782, 4452251786946, 86585391630673
Permutations of length n by length of runs. Ref DKB 261. [6,2; A0467, N2083]

M4869   1, 12, 144, 1728, 20736, 248832, 2985984, 35831808, 429981696, 5159780352,
61917364224, 743008370688, 8916100448256, 106993205379072
Powers of 12. Ref BA9. [0,2; A1021, N2084]

M4870 1, 12, 144, 1750, 23420, 303240, 3641100, 46113200, 575360400, 7346545000, 112402762000, 1351035564000, 16432451210000, 221411634520000

Powers of ten written in base 8. Ref AS1 1017. [0,2; A0468, N2085]

M4871 1, 0, 12, 148, 2568, 53944

Partition function for b.c.c. lattice. Ref PHM 2 745 57. [0,3; A1406, N2086]

M4872 1, 12, 155, 2128, 30276, 440484, 6506786, 97181760, 1463609356, 22187304112, 338118529539, 5175023913008, 79492847013100, 1224838471521240

Quadrinomial coefficients. Ref C1 78. [1,2; A5723]

M4873 1, 12, 157, 1750, 17446, 164108, 1505099, 13720902, 125782441

Impedances of an n-terminal network. Ref BSTJ 18 301 39. [2,2; A3130]

M4874 1, 12, 180, 2800, 44100, 698544, 11099088, 176679360, 2815827300, 44914183600, 716830370256, 11445589052352, 182811491808400, 2920656969720000

Related to remainder in Gaussian quadrature. Ref MOC 1 53 43. [0,2; A0515, N2087]

M4875 12, 180, 3360, 75600, 1995840, 60540480, 2075673600, 79394515200, 3352212864000, 154872234316800, 7771770303897600, 420970891461120000

Coefficients of Hermite polynomials. Ref MOC 3 168 48. AS1 801. [2,1; A1814, N2088]

$$\text{E.g.f.: } (1 + 2\,x) \,/\, (1 - 4\,x)^{5/2}.$$

M4876 12, 216, 5248, 160675, 5931540, 256182290, 12665445248, 705068085303, 43631250229700, 2970581345516818, 220642839342906336, 17753181687544516980

n-state finite automata with 2 inputs. Ref GTA85 676. [1,1; A6689]

M4877 12, 240, 6624, 234720, 10208832, 526810176, 31434585600, 2127785025024, 161064469168128

Susceptibility for f.c.c. lattice. Ref PRV 164 801 67. [1,1; A2166, N2090]

M4878 1, 12, 288, 51840, 2488320, 209018880, 75246796800, 902961561600, 86684309913600, 514904800886784000, 86504006548979712000

Denominators of asymptotic series for gamma function: Stirling's formula. Cf. M4878. Ref MOC 22 619 68. [0,2; A1164, N2091]

M4879 1, 12, 360, 20160, 1814400, 239500800, 43589145600, 10461394944000, 3201186852864000, 1216451004088320000, 562000363888803840000

½(2n)!. Ref SAM 42 162 63. [1,2; A2674, N2092]

M4880 12, 720, 60480, 3628800, 95800320, 2615348736000, 4483454976000, 32011868528640000, 51090942171709440000, 15257928431370240000

Denominators of coefficients for numerical integration. Cf. M4809. Ref OP80 545. PHM 35 263 44. [0,1; A2196, N2093]

M4881 0, 12, 1360, 350000, 255036992, 571462430224
Degenerate simplices in n-cube. Ref AMM 86 49 79. [2,2; A4145]

M4882 1, 12, 2160, 6048000, 266716800000, 186313420339200000,
206790904792577064960000, 3653568471257344858781122560000000
Determinant of inverse Hilbert matrix. Ref AMM 90 306 83. [1,2; A5249]

M4883 12, 2772, 21624369228, 10111847525912679844170507482772
Denominators of a continued fraction. Ref NBS B80 288 76. [0,1; A6272]

M4884 12, 10206, 2148007936
Post functions. Ref JCT 4 295 68. [2,1; A1322, N2094]

M4885 13, 3, 41, 509, 2, 89, 1122659, 19099919, 85864769, 26089808579,
665043081119, 554688278429
Chains of length n of nearly doubled primes. Ref MOC 53 755 89. [1,1; A5602]

M4886 13, 11, 1093, 757, 3851, 797161, 4561, 34511, 363889, 368089, 1001523179,
391151, 8209, 20381027, 4404047, 2413941289, 2644097031, 17189128703, 7333
Largest factor of $3^{2n+1} - 1$. Ref Krai24 2 28. CUNN. [1,1; A2591, N2095]

M4887 13, 17, 31, 37, 71, 73, 79, 97, 107, 113, 149, 157, 167, 179, 199, 311, 337, 347,
359, 389, 701, 709, 733, 739, 743, 751, 761, 769, 907, 937, 941, 953, 967, 971, 983, 991
Emirps (primes whose reversal is a different prime). Ref GA85 230. [1,1; A6567]

M4888 13, 19, 23, 29, 49, 59, 79, 89, 103, 109, 111, 133, 199, 203, 209, 211, 233, 299,
311, 409, 411, 433, 499, 509, 511, 533, 599, 611, 709, 711, 733, 799, 809, 811, 833, 899
Primitive modest numbers. Ref JRM 17 140 84. [1,1; A7627]

M4889 13, 19, 29, 41, 43, 59, 61, 67, 79, 83, 89, 97, 101, 109, 131, 137, 139, 149, 167,
179, 197, 199, 211, 223, 229, 239, 241, 251, 263, 269, 271, 281, 283, 293, 307, 317, 349
Class 2+ primes. Ref UPNT A18. [1,1; A5106]

M4890 13, 19, 37, 61, 109, 157, 193, 241, 283, 367, 373, 379, 397, 487
Related to Kummer's conjecture. Ref Hass64 482. [1,1; A0922, N2096]

M4891 13, 25, 50, 51, 99, 101, 103, 199, 202, 403, 404, 405, 413, 797, 807, 809, 825,
1593, 1618, 3229, 3235, 3236, 3237, 3299, 6371, 6457, 6471, 6473, 6599
Positions of remoteness 4 in Beans-Don't-Talk. Ref MMAG 59 267 86. [1,1; A5696]

M4892 13, 31, 113, 311, 1031, 1033, 1103, 1181, 1301, 1381, 1831, 3011, 3083, 3301,
3803, 10333, 11003, 11083, 11833, 18013, 18133, 18803, 30011, 30881, 31033, 31081
Reflectable emirps. Cf. M4887. Ref JRM 15 253 83. [1,1; A7628]

M4893 1, 13, 37, 73, 121, 181, 253, 337, 433, 541, 661, 793, 937, 1093, 1261, 1441, 1633,
1837, 2053, 2281, 2521, 2773, 3037, 3313, 3601, 3901, 4213, 4537, 4873, 5221, 5581
Star numbers: $6n(n+1)+1$. See Fig M2535. Ref GA88 20. [0,2; A3154]

M4894 1, 1, 13, 41, 671, 73, 597871, 7913, 28009, 792451, 170549237, 19397633, 317733228541, 9860686403
Numerators of coefficients for central differences. Cf. M4588. Ref SAM 42 162 63. [1,3; A2673, N2097]

M4895 0, 1, 13, 46, 110, 215, 371, 588, 876, 1245, 1705, 2266, 2938, 3731, 4655, 5720, 6936, 8313, 9861, 11590, 13510, 15631, 17963, 20516, 23300, 26325, 29601, 33138
Dodecagonal pyramidal numbers: $n(n+1)(10n-7)/6$. Ref B1 194. [0,3; A7587]

M4896 1, 1, 1, 1, 13, 47, 73, 2447, 16811, 15551, 1726511, 18994849, 10979677, 2983409137, 48421103257, 135002366063, 10125320047141, 232033147779359
Coefficients of Airey's converging factor. Ref KNAW 66 751 63. PNAS 69 440 72. [0,5; A1662, N2098]

M4897 1, 0, 1, 1, 13, 51, 601, 4806, 39173, 775351
Series-reduced labeled trees with n nodes. Ref jr. [0,5; A2792, N2099]

M4898 1, 13, 55, 147, 309, 561, 923, 1415, 2057, 2869, 3871, 5083, 6525, 8217, 10179, 12431, 14993, 17885, 21127, 24739, 28741, 33153, 37995, 43287, 49049, 55301, 62063
Centered icosahedral (or cuboctahedral) numbers. Ref INOC 24 4550 85. CoSl95. [0,2; A5902]

M4899 1, 13, 57, 153, 323, 587, 967, 1483, 2157, 3009, 4061, 5333, 6847, 8623, 10683, 13047, 15737, 18773, 22177, 25969, 30171, 34803, 39887, 45443, 51493, 58057
Crystal ball numbers for h.c.p. Ref CoSl95. [0,2; A7202]

M4900 13, 61, 73, 193, 241, 541, 601, 1021, 1801, 1873, 1933, 2221, 3121, 3361, 4993, 5521, 6481, 8461, 9181, 9901, 10993, 11113, 12241, 12541, 13633, 14173, 17761, 20593
Sextan primes: $p = (x^6 + y^6)/(x^2 + y^2)$. Ref CU23 1 256. [1,1; A2647, N2100]

M4901 13, 72, 595, 4096, 39078, 379760, 4181826, 49916448, 647070333, 9035216428, 135236990388, 2159812592384, 36658601139066, 658942295734944
Discordant permutations. Ref SMA 20 23 54. [5,1; A0470, N2101]

M4902 1, 13, 73, 301, 1081, 3613, 11593, 36301, 111961, 342013, 1038313, 3139501, 9467641, 28501213, 85700233, 257493901, 773268121, 2321377213, 6967277353
Bitriangular permutations. Ref DUMJ 13 267 46. [4,2; A6230]

M4903 1, 13, 76, 295, 889, 2188, 4652, 8891, 15686
Putting balls into 7 boxes. Ref SIAR 12 296 70. [12,2; A5340]

M4904 1, 13, 85, 377, 1289, 3653, 8989, 19825, 40081, 75517, 134245, 227305, 369305, 579125, 880685, 1303777, 1884961, 2668525, 3707509, 5064793, 6814249, 9041957
Expansion of $(1+x)^6/(1-x)^7$. Ref SIAR 12 277 70. C1 81. [0,2; A1848, N2102]

M4905 13, 87, 4148, 153668757
Switching networks. Ref JFI 276 321 63. [1,1; A0836, N2103]

M4906 1, 1, 13, 93, 1245, 18093, 308605, 5887453, 124221373, 2864305277
Feynman diagrams of order $2n$. Ref PRV D18 1949 78. [1,3; A5414]

M4907 1, 13, 98, 560, 2688, 11424, 44352, 160512, 549120, 1793792, 5637632,
17145856, 50692096, 146227200, 412778496, 1143078912, 3111714816, 8341487616
Coefficients of Chebyshev polynomials. Ref AS1 795. [0,2; A6976]

M4908 1, 13, 104, 663, 3705, 19019, 92092, 427570, 1924065, 8454225, 36463440,
154969620, 650872404, 2707475148, 11173706960, 45812198536, 186803188858
$13C(2n, n-6)/(n+7)$. Ref QAM 14 407 56. MOC 29 216 75. [6,2; A0590, N2104]

M4909 1, 13, 108, 793, 5611, 39312
Eulerian circuits on checkerboard. Ref JCT B24 211 78. [1,2; A6239]

M4910 13, 109, 193, 433, 769, 1201, 1453, 2029, 3469, 3889, 4801, 10093, 12289, 13873,
18253, 20173, 21169, 22189, 28813, 37633, 43201, 47629, 60493, 63949, 65713, 69313
Cuban primes: $p = (x^3 - y^3)/(x-y)$, $x = y + 2$. Ref CU23 1 259. [1,1; A2648, N2105]

M4911 1, 13, 110, 758, 4617, 25895, 136949, 693369, 3395324, 16197548, 75675657,
347624505, 1574756959, 7051383905, 31266981002, 137492793602, 600295660953
Convex polygons of length $2n$ on square lattice. Ref TCS 34 179 84. [6,2; A5769]

M4912 1, 13, 130, 1210, 11011, 99463, 896260, 8069620, 72636421, 653757313,
5883904390, 52955405230, 476599444231, 4289397389563, 38604583680520
Gaussian binomial coefficient $[n,2]$ for $q = 3$. Ref GJ83 99. ARS A17 328 84. [2,2; A6100]

M4913 13, 158, 66336, 122544034314
Switching networks. Ref JFI 276 320 63. [1,1; A0830, N2106]

M4914 1, 13, 169, 2197, 28561, 371293, 4826809, 62748517, 815730721, 10604499373,
137858491849, 1792160394037, 23298085122481, 302875106592253
Powers of 13. Ref BA9. [0,2; A1022, N2107]

M4915 1, 13, 181, 2521, 35113, 489061, 6811741, 94875313, 1321442641, 18405321661,
256353060613, 3570537526921, 49731172316281, 692665874901013
From the solution to a Pellian. Ref AMM 56 174 49. [0,2; A1570, N2108]

M4916 13, 192, 1085, 3880, 10656, 24626, 50380, 94128, 163943, 270004, 424839,
643568, 944146, 1347606, 1878302, 2564152, 3436881, 4532264, 5890369, 7555800
Discordant permutations. Ref SMA 20 23 54. [3,1; A0563, N2109]

M4917 13, 237, 356026, 2932175712336
Switching networks. Ref JFI 276 320 63. [1,1; A0824, N2110]

M4918 0, 0, 0, 0, 13, 252, 3740, 51300, 685419, 9095856, 120872850, 1614234960,
21697730849
Simple quadrangulations. Ref JCT 4 275 68. [1,5; A1508, N2111]

M4919   1, 13, 273, 4641, 85085, 1514513, 27261234, 488605194, 8771626578,
   157373300370, 2824135408458, 50675778059634, 909348684070099
Fibonomial coefficients. Ref FQ 6 82 68. BR72 74. [0,2; A1658, N2112]

M4920   1, 1, 13, 4683, 102247563, 230283190977853
Dissimilarity relations on an n-set. Ref MET 27 130 80. [1,3; A6541]

SEQUENCES BEGINNING . . ., 14, . . . TO . . ., 24, . . .

M4921   14, 14, 36, 57, 155, 316, 902, 2053, 6059, 14810, 44842, 115009
Meanders in which first bridge is 7. See Fig M4587. Ref SFCA91 293. [3,1; A6662]

M4922   14, 19, 28, 47, 61, 75, 197, 742, 1104, 1537, 2208, 2508, 3684, 4788, 7385, 7647,
   7909, 31331, 34285, 34348, 55604, 62662, 86935, 93993, 120284, 129106, 147640
Repfigit numbers. Ref Pick91 229. [1,1; A7629]

M4923   14, 19, 47, 61, 75, 197, 742, 1104, 1537, 2580, 3684, 4788, 7385, 7647, 7909,
   31331, 34285, 34348, 56604, 86935, 120284, 129106, 147640, 156146, 174680, 183186
Primitive repfigit numbers. See Fig M5405. Ref JRM 1942 87. [1,1; A6576]

M4924   14, 21, 25, 30, 33, 38, 41, 43, 48, 50, 53, 56, 59, 61, 65, 67, 70, 72, 76, 77, 79, 83,
   85, 87, 89, 92, 95, 96, 99, 101, 104, 105, 107, 111, 112, 114, 116, 119, 121, 123, 124, 128
Nearest integer to imaginary part of zeros of Riemann zeta function. See Fig M2051. Ref
RS6 58. Edwa74 96. [1,1; A2410, N2113]

M4925   14, 21, 26, 32, 41, 48, 56, 67
Zarankiewicz's problem. Ref LNM 110 143 69. [4,1; A6614]

M4926   14, 21, 28, 36, 45
Zarankiewicz's problem. Ref LNM 110 144 69. [3,1; A6625]

M4927   14, 26, 34, 38, 50, 62, 68, 74, 76, 86, 90, 94, 98, 114, 118, 122, 124, 134, 142, 146,
   152, 154, 158, 170, 174, 182, 186, 188, 194, 202, 206, 214, 218, 230, 234, 236, 242, 244
Nontotients. Ref UPNT B36. [1,1; A5277]

M4928   14, 33, 382, 51, 6, 20, 10, 15, 14, 21, 28, 35, 182, 24, 26, 30, 142, 40, 34, 42, 20,
   57, 135, 70, 30, 99, 42, 66, 406, 88, 56, 60, 54, 93, 24, 105, 248, 147, 44, 63, 30, 80, 435
Smallest k such that $\sigma(n+k) = \sigma(k)$. Ref AS1 840. [1,1; A7365]

M4929   0, 14, 42, 90, 165, 275, 429, 637, 910, 1260, 1700, 2244, 2907, 3705, 4655, 5775,
   7084, 8602, 10350, 12350, 14625, 17199, 20097, 23345, 26970, 31000, 35464, 40392
$n(n+5)(n+6)(n+7)/24$. Ref AS1 796. [0,2; A5587]

M4930   14, 42, 90, 165, 275, 429, 637, 910, 1260, 1700, 2244, 2907, 3705, 4655, 5775,
   7084, 8602, 10350, 12350, 14625, 17199, 20097, 23345, 26970, 31000, 35464, 40392
Walks on square lattice. Ref GU90. [0,1; A5556]

$$\text{G.f.:} \quad (14 - 28\,x + 20\,x^2 - 5\,x^3) \,/\, (1 - x)^5.$$

M4931   1, 14, 50, 110, 194, 302, 434, 590, 770, 974, 1202, 1454, 1730, 2030, 2354, 2702, 3074, 3470, 3890, 4334, 4802, 5294, 5810, 6350, 6914, 7502, 8114, 8750, 9410, 10094
Points on surface of hexagonal prism: $12n^2 + 2$. Ref INOC 24 4552 85. [0,2; A5914]

M4932   0, 1, 14, 51, 124, 245, 426, 679, 1016, 1449, 1990, 2651, 3444, 4381, 5474, 6735, 8176, 9809, 11646, 13699, 15980, 18501, 21274, 24311, 27624, 31225, 35126, 39339
Stella octangula numbers: $n(2n\,2-1)$. Ref rkg. [0,3; A7588]

M4933   1, 14, 57, 148, 305, 546, 889, 1352, 1953, 2710, 3641, 4764, 6097, 7658, 9465, 11536, 13889, 16542, 19513, 22820, 26481, 30514, 34937, 39768, 45025, 50726, 56889
Hexagonal prism numbers: $(n+1)(3n^2 + 3n + 1)$. Ref INOC 24 4552 85. [0,2; A5915]

M4934   1, 14, 70, 140, 70, 28, 28, 40, 70, 140, 308, 728, 1820, 4760, 12920, 36176, 104006, 305900, 917700, 2801400, 8684340, 27293640, 86843400, 279409200
Expansion of $(1 - 4x)^{7/2}$. Ref TH09 164. FMR 1 55. [0,2; A2423, N2114]

M4935   1, 14, 84, 330, 1001, 2548, 5712, 11628, 21945, 38962, 65780, 106470, 166257, 251720, 371008, 534072, 752913, 1041846, 1417780, 1900514, 2513049, 3281916
From paths in the plane. Ref EJC 2 58 81. [0,2; A6858]

$$\text{G.f.:} \quad (1 + x)\,(1 + 6x + x^2)\,/\,(1 - x)^7.$$

M4936   14, 86, 518, 3110, 18662, 111974, 671846, 4031078
Functions realized by cascades of n gates. Ref BU77. [1,1; A5610]

M4937   1, 14, 98, 650, 4202, 26162, 163154, 984104, 6015512
Cluster series for b.c.c. lattice. Ref PRV 133 A315 64. DG72 225. [0,2; A3206]

M4938   1, 14, 102, 561, 2563, 10285, 37349, 125290
Coefficients of a modular function. Ref GMJ 8 29 67. [3,2; A5757]

M4939   1, 14, 112, 672, 3360, 14784, 59136, 219648, 768768, 2562560, 8200192, 25346048, 76038144, 222265344, 635043840, 1778122752, 4889837568, 13231325184
Expansion of $1/(1 - 2x)^7$. Ref MFM 74 62 70 (divided by 5). [0,2; A2409, N1668]

M4940   1, 14, 118, 780, 4466, 23276, 113620, 528840, 2375100, 10378056, 44381832, 186574864, 773564328, 3171317360, 12880883408, 51915526432, 207893871472
Binary trees of height n requiring 3 registers. Ref TCS 9 105 79. [7,2; A6223]

M4941   1, 14, 120, 825, 5005, 28028, 148512, 755820, 3730650, 17978180, 84987760, 395482815
Dissections of a polygon by number of parts. Ref CAY 13 95. AEQ 18 385 78. [6,2; A2056, N2115]

M4942   1, 14, 130, 700, 2635, 7826, 19684, 43800, 88725, 166870, 295526, 498004, 804895, 1255450, 1899080, 2796976, 4023849, 5669790, 7842250, 10668140, 14296051
Colored hexagons: $(n^6 + n^3 + 2n^2 + 2n)/6$. [1,2; A6565]

M4943   0, 1, 14, 135, 1228, 11069, 99642, 896803, 8071256, 72641337, 653772070,
  5883948671, 52955538084, 476599842805, 4289398585298, 38604587267739
$(3^{2n+1} - 8n - 3)/16$. Ref JCT A29 122 80. MOC 37 479 81. [0,3; A4004]

M4944   1, 1, 14, 135, 5478, 165826, 13180268, 834687179
Coefficients of elliptic function sn. Ref Cay95 56. TM93 4 92. [0,3; A2753, N2117]

M4945   1, 14, 147, 1408, 13013, 118482, 1071799, 9668036, 87099705, 784246870,
  7059619931, 63542171784, 571901915677, 5147206719578, 46325218390143
Expansion of $1/(1-x)(1-4x)(1-9x)$. Ref TH09 35. FMR 1 112. RCI 217. [0,2; A2451,
N2118]

M4946   1, 14, 154, 1696, 18684, 205832, 2267544, 24980352, 275195536, 3031685984,
  33398506528, 367933962880, 4053336963648, 44653503613184, 491924407670784
Hamiltonian cycles on $P_5 \times P_{2n}$: $a(n) = 11a(n-1) + 2a(n-3)$. Ref ARS 33 87 92. [1,2;
A6865]

M4947   1, 14, 155, 1665, 18424, 214676, 2655764, 34967140, 489896616, 7292774280,
  115119818736, 1922666722704, 33896996544384, 629429693586048
Generalized Stirling numbers. Ref PEF 77 7 62. [0,2; A1707, N2119]

$$\text{E.g.f.:}\quad -\ln(1-x)^3 \,/\, 6\,(x-1)^2.$$

M4948   0, 1, 14, 195, 2716, 37829, 526890, 7338631, 102213944, 1423656585,
  19828978246, 276182038859, 3846719565780
Standard deviation of M3154. Ref dab. [1,3; A7655]

M4949   1, 14, 196, 2744, 38416, 537824, 7529536, 105413504, 1475789056,
  20661046784, 289254654976, 4049565169664, 56693912375296, 793714773254144
Powers of 14. Ref BA9. [0,2; A1023, N2120]

M4950   14, 206, 957, 1334, 1364, 1634, 2685, 2974, 4364, 14841, 18873, 19358, 20145,
  24957, 33998, 36566, 42818, 56564, 64665, 74918, 79826, 79833, 84134, 92685
n and $n+1$ have same sum of divisors. Ref SI64 110. AS1 840. [1,1; A2961]

M4951   14, 254, 65534, 77575934, 103901883134
Functions realized by cascades of n gates. Ref BU77. [1,1; A5611]

M4952   1, 14, 273, 7645, 296296, 15291640, 1017067024, 84865562640, 8689315795776,
  1071814846360896, 156823829909121024, 26862299458337581056
Central factorial numbers. Ref RCI 217. [0,2; A1820, N2121]

M4953   14, 386, 5868, 65954, 614404, 5030004, 37460376, 259477218, 1697186964,
  10596579708, 63663115880
Rooted planar maps with n edges. Ref BAMS 74 74 68. WA71. JCT A13 215 72. [4,1;
A0473, N2122]

M4954   1, 14, 462, 24024, 1662804, 140229804, 13672405890, 1489877926680,
177295473274920
4-dimensional Catalan numbers. Ref CN 75 124 90. [1,2; A5790]

$$\text{G.f.:}\quad {}_4F_3([1,3/2,5/4,7/4];[3,4,5];256x).$$

M4955   14, 560, 11200, 197568, 3378944, 57573888
Almost trivalent maps. Ref PLC 1 292 70. [0,1; A2010, N2123]

M4956   1, 1, 14, 818, 141, 13063, 16774564, 1057052, 4651811, 778001383, 1947352646,
1073136102266, 72379420806883
Numerators of double sums of reciprocals. Ref RO00 316. FMR 1 117. [0,3; A2429,
N2124]

M4957   15, 17, 24, 37, 43, 57, 63, 65, 73, 79, 89, 101, 106, 122, 129, 131, 142, 145, 148,
151, 161, 164, 168, 171, 186, 195, 197, 198, 204, 217, 222, 223, 225, 229, 232, 233, 248
Elliptic curves. Ref JRAM 212 24 63. [1,1; A2155, N2125]

M4958   15, 20, 20, 6, 6, 19, 19, 5, 14, 20, 5, 20, 20, 6, 6, 19, 19, 5, 14, 20, 20, 20, 20, 20,
20, 20, 20, 20, 20, 20, 20, 20, 20, 20, 20, 6, 6, 6, 6, 6, 6, 6, 6, 6, 6, 6, 6, 6, 6
Name of n begins with $a(n)$-th letter. [1,1; A5606]

M4959   1, 15, 21, 33, 35, 39, 51, 55, 57, 65, 69, 77, 85, 87, 91, 93, 95, 115, 119, 133, 143,
145, 155, 161, 187, 203, 209, 217, 221, 247, 253, 299, 319, 323, 341, 377, 391, 403, 437
Liouville function $\lambda(n)$ is positive. Ref JIMS 7 71 43. [1,2; A2557, N2126]

M4960   15, 22, 31, 38, 46
Zarankiewicz's problem. Ref LNM 110 144 69. [4,1; A6615]

M4961   1, 15, 29, 12, 26, 12, 26, 9, 23, 7, 21, 4, 18, 2, 16, 30, 13, 27, 10, 24, 8, 22, 5, 19, 3,
17, 31, 14, 28, 11, 25, 11, 25, 8, 22, 6, 20, 3, 17, 1, 15, 29, 12, 26, 9, 23, 7, 21, 4, 18, 2, 16
Dates at fortnightly intervals from Jan. 1. Ref EUR 13 11 50. [1,2; A1356, N2127]

M4962   15, 32, 87, 192, 343, 672, 1290, 2176, 3705, 6336, 10214, 16320, 25905, 39936,
61227, 92928, 138160, 204576, 300756, 435328, 626727, 897408, 1271205
McKay-Thompson series of class 6C for Monster. Ref CALG 18 257 90. FMN94. [1,1;
A7256]

M4963   15, 40, 76, 124, 185, 260, 350, 456, 579, 720, 880, 1060, 1261, 1484, 1730, 2000,
2295, 2616, 2964, 3340, 3745, 4180, 4646, 5144, 5675, 6240, 6840, 7476, 8149, 8860
Putting balls into 4 boxes. Ref SIAR 12 296 70. [8,1; A5337]

$$\text{G.f.:}\quad (15 - 20x + 6x^2) / (1 - x)^4.$$

M4964   15, 45, 118, 257, 522, 975, 1752, 2998, 4987, 8043, 12693, 19584, 29719, 44324,
65210, 94642, 135805, 192699, 270822, 377048, 520624, 713123, 969784
Bipartite partitions. Ref ChGu56 1. [0,1; A2756, N2129]

M4965   1, 15, 51, 97, 127, 145, 152, 160, 273, 481, 811, 1372, 2250, 3692, 5924, 9472,
14887, 23310, 36005, 55314, 84042, 126998, 190138, 283108, 418175, 614429, 896439
A generalized partition function. Ref PNISI 17 236 51. [1,2; A2602, N2130]

M4966   1, 15, 60, 154, 315, 561, 910
n-step mappings with 4 inputs. Ref PRV A32 2342 85. [1,2; A5945]

M4967   15, 60, 450, 4500, 55125, 793800, 13097700
Expansion of an integral. Ref C1 167. [3,1; A1756, N2131]

M4968   1, 15, 65, 175, 369, 671, 1105, 1695, 2465, 3439, 4641, 6095, 7825, 9855, 12209,
14911, 17985, 21455, 25345, 29679, 34481, 39775, 45585, 51935, 58849, 66351, 74465
Rhombic dodecahedral numbers: $n^4 - (n-1)^4$. Ref AMM 82 819 75. INOC 24 4552 85.
[0,2; A5917]

M4969   1, 0, 15, 70, 630, 5544, 55650, 611820, 7342335, 95449640, 1336295961,
20044438050, 320711010620, 5452087178160, 98137569209940, 1864613814984984
Rencontres numbers. Ref R1 65. [4,3; A0475, N2132]

M4970   15, 72, 609, 4960, 46188, 471660, 5275941, 64146768, 842803767, 11902900380,
179857257960, 2895705788736, 49491631601635, 895010868095256
Discordant permutations. Ref SMA 20 23 54. [5,1; A0476, N2133]

M4971   1, 15, 73, 143, 208, 244, 265, 273, 282, 490, 838, 1426, 2367, 3908, 6356, 10246,
16327, 25812, 40379, 62748, 96660, 147833, 224446, 338584, 507293, 755612
A generalized partition function. Ref PNISI 17 235 51. [1,2; A2603, N2134]

M4972   0, 0, 0, 0, 0, 15, 75, 310, 1060, 3281
Unexplained difference between two partition g.f.s. Ref PCPS 63 1100 67. [1,6; A7328]

M4973   1, 15, 76, 275, 720, 1666, 3440, 6129, 11250, 17545, 28896, 41405, 65072, 85950,
128960, 162996, 238545, 286995, 404600, 482160, 662112, 756470, 1042560
Related to the divisor function. Ref SMA 19 39 53. [1,2; A0477, N2135]

M4974   1, 15, 90, 350, 1050, 2646, 5880, 11880, 22275, 39325, 66066, 106470, 165620,
249900, 367200, 527136, 741285, 1023435, 1389850, 1859550, 2454606, 3200450
Stirling numbers of second kind. See Fig M4981. Ref AS1 835. DKB 223. [1,2; A1297,
N2136]

M4975   1, 15, 90, 357, 1107, 2907, 6765, 14355, 28314, 52624, 93093, 157950, 258570,
410346, 633726, 955434, 1409895, 2040885, 2903428, 4065963, 5612805, 7646925
From expansion of $(1 + x + x^2)^n$. Ref C1 78. [4,2; A5716]

M4976   1, 15, 99, 429, 1430, 3978, 9690, 21318, 43263, 82225, 148005, 254475, 420732,
672452, 1043460, 1577532, 2330445, 3372291, 4790071, 6690585, 9203634
Fermat coefficients. Ref MMAG 27 141 54. [8,2; A0973, N2137]

M4977 1, 15, 105, 490, 1764, 5292, 13860, 32670, 70785, 143143, 273273, 496860, 866320, 1456560, 2372112, 3755844, 5799465, 8756055, 12954865, 18818646
Related to the coin tossing problem. Ref CRO 10 30 67. [0,2; A6857]

$$(4+n)!(5+n)! \; / \; 2880.n!(n+1)!.$$

M4978 15, 105, 490, 1918, 6825, 22935, 74316, 235092, 731731, 2252341, 6879678, 20900922, 63259533
Associated Stirling numbers. Ref R1 76. DB1 296. C1 222. [6,1; A0478, N2138]

M4979 1, 15, 113, 575, 2241, 7183, 19825, 48639, 108545, 224143, 433905, 795455, 1392065, 2340495, 3800305, 5984767, 9173505, 13726991, 20103025, 28875327
Expansion of $(1+x)^7/(1-x)^8$. Ref SIAR 12 277 70. C1 81. [0,2; A1849, N2139]

M4980 1, 15, 140, 1050, 6930, 42042, 240240, 1312740, 6928350, 35565530, 178474296, 878850700, 4259045700, 20359174500, 96172862400, 449608131720, 2082743551350
$(2n+4)!/(4!n!(n+1)!)$. Ref JO39 449. JCT B18 258 75. [0,2; A2803, N2140]

M4981 1, 15, 140, 1050, 6951, 42525, 246730, 1379400, 7508501, 40075035, 210766920, 1096190550, 5652751651, 28958095545, 147589284710, 749206090500
Stirling numbers of second kind. See Fig M4981. Ref AS1 835. DKB 223. [5,2; A0481, N2141]

M4982 1, 15, 155, 1395, 11811, 97155, 788035, 6347715, 50955971, 408345795, 3269560515, 26167664835, 209386049731, 1675267338435, 13402854502595
Gaussian binomial coefficient $[n,3]$ for $q=2$. Ref GJ83 99. ARS A17 328 84. [3,2; A6096]

M4983 1, 15, 175, 1960, 22449, 269325, 3416930, 45995730, 657206836, 9957703756, 159721605680, 2706813345600, 48366009233424, 909299905844112
Stirling numbers of first kind. See Fig M4730. Ref AS1 833. DKB 226. [5,2; A0482, N2142]

M4984 1, 15, 179, 2070, 24574, 305956, 4028156, 56231712, 832391136, 13051234944, 216374987520, 3785626465920, 69751622298240, 1350747863435520
Generalized Stirling numbers. Ref PEF 77 44 62. [0,2; A1717, N2143]

M4985 1, 15, 180, 2100, 25200, 317520, 4233600, 59875200, 898128000, 14270256000, 239740300800, 4249941696000, 79332244992000, 1556132497920000
Simplices in barycentric subdivision of n-simplex. Ref rkg. [1,2; A5461]

$$a(n) = n\,(n+1)\,(n+3)! \; / \; 48.$$

Figure M4981. STIRLING NUMBERS OF 2ND KIND, BELL NUMBERS.

The **Stirling** number of the **second kind**, $S(n, k) = \left\{{n \atop k}\right\}$, sometimes read as "$n$ heap k" gives the number of ways of partitioning n labeled objects into k nonempty subsets. The first few values are illustrated as follows:

n \ k	1	2	3	4	Total
1	1				1
2	12	1, 2			2
3	123	1, 23 2, 13 3, 12	1, 2, 3		5
4	1234	1, 234 2, 134 3, 124 4, 123 12, 34 13, 24 14, 23	1, 2, 34 1, 3, 24 1, 4, 23 2, 3, 14 2, 4, 13 3, 4, 12	1, 2, 3, 4	15

The numbers continue:

row sums
$B(n)$

							$B(n)$
1							1
1	1						2
1	3	1					5
1	7	6	1				15
1	15	25	10	1			52
1	31	90	65	15	1		203
1	63	301	350	140	21	1	877

The columns of this triangle give M2655, M4167, M4722, M4981, M5112, M5201, while the diagonals give M2535, M4385, M4974, M5222. Also

$$S(n, k) \;=\; kS(n-1, k) + S(n-1, k-1),$$

$$S(n, k) \;=\; \frac{1}{k!} \sum_{j=1}^{k} (-1)^{k-j} \binom{k}{j} j^n,$$

$$x^n \;=\; \sum_{k=0}^{n} S(n, k) x(x-1) \cdots (x-k+1).$$

References: [R1 48], [DKB 223], [C1 204], [AS1 835], [GKP 244]. The row sums in this triangle are the Bell or exponential numbers $B(n)$, M1484. $B(n)$ is also the number of equivalence relations on a set of n objects and has generating function

$$1 + x + 2\frac{x^2}{2!} + 5\frac{x^3}{3!} + \cdots = e^{e^x - 1}.$$

M4986   1, 15, 192, 2415, 30305
Complexity of a $3 \times n$ grid. Ref JCT B24 210 78. [1,2; A6238]

M4987   15, 200, 2672, 37600, 554880, 8514560, 134864640
Almost trivalent maps. Ref PLC 1 292 70. [0,1; A2007, N2144]

M4988   15, 210, 2380, 26432, 303660, 3678840, 47324376, 647536032, 9418945536,
   145410580224, 2377609752960, 41082721413120, 748459539843840
Associated Stirling numbers. Ref R1 75. C1 256. [6,1; A0483, N2145]

M4989   1, 15, 210, 3150, 51975, 945945, 18918900, 413513100, 9820936125,
   252070693875, 6957151150950, 205552193096250, 6474894082531875
Coefficients of Bessel polynomials $y_n(x)$. Ref RCI 77. [4,2; A1880, N2146]

$$\text{E.g.f.: } \ x\,(1 + x/2) \ / \ (1 - 2\,x)^{7/2}.$$

M4990   1, 15, 225, 3375, 50625, 759375, 11390625, 170859375, 2562890625,
   38443359375, 576650390625, 8649755859375, 129746337890625, 1946195068359375
Powers of 15. Ref BA9. [0,2; A1024, N2147]

M4991   1, 0, 0, 0, 0, 1, 15, 465, 19355, 1024380, 66462606, 5188453830, 480413921130,
   52113376310985, 6551246596501035, 945313907253606891, 155243722248524067795
4-valent labeled graphs with n nodes. Ref SIAA 4 192 83. [0,7; A5815]

M4992   1, 15, 528, 3990, 232305, 4262895, 128928632, 1420184304, 186936865290
Coefficients for step-by-step integration. Ref JACM 11 231 64. [2,2; A2403, N2148]

M4993   15, 575, 46760, 6998824, 1744835904, 673781602752, 381495483224064,
   303443622431870976
Differences of reciprocals of unity. Ref DKB 228. [1,1; A1236, N2149]

M4994   1, 16, 0, 256, 1054, 0, 0, 4096, 6561, 16864, 0, 0, 478, 0, 0, 65536, 63358, 104976,
   0, 269824, 0, 0, 0, 0, 720291, 7648, 0, 0, 1407838, 0, 0, 1048576, 0, 1013728, 0
Related to representation as sums of squares. Ref QJMA 38 304 07. [1,2; A2607, N2150]

M4995   1, 0, 16, 8, 0, 128, 28, 0, 576, 64, 0, 2048, 134, 0, 6304, 288, 0, 17408, 568, 0,
   44416, 1024, 0, 106496, 1809
McKay-Thompson series of class 6d for Monster. Ref FMN94. [-1,3; A7263]

M4996   16, 17, 20, 25, 32, 33, 34, 36, 39, 41, 43, 48, 50, 51, 52, 54, 55, 58, 61, 65, 66, 67,
   68, 69, 71, 74, 77, 78, 80, 83, 84, 85, 88, 89, 90, 93, 94, 96, 97, 99, 100, 101, 102, 105
$n^2 + n + 17$ is composite. [1,1; A7636]

M4997   0, 0, 0, 1, 0, 0, 16, 18, 0, 252, 576, 519, 3264, 12468, 20568, 26662, 215568,
   528576, 164616, 3014889, 10894920, 13796840, 29909616, 190423962, 399739840
Susceptibility for b.c.c. lattice. Ref JPA 6 1511 73. DG74 421. [1,7; A2925, N2151]

M4998   16, 23, 32, 43, 52
Zarankiewicz's problem. Ref LNM 110 144 69. [4,1; A6616]

M4999   16, 25, 33, 49, 52, 64, 73, 100, 121, 148, 169, 177
n consecutive odd numbers whose sum of squares is a square. Ref MMAG 40 198 67. [1,1; A1033, N2152]

M5000   16, 25, 37, 46, 58, 88, 109, 130, 142, 151, 184, 193, 205, 247, 268, 298, 310, 319, 331, 340, 382, 394, 403, 415, 424, 457, 478, 487, 541, 550, 604, 613, 688, 697, 709, 730
$(n^2+n+1)/21$ is prime. Ref CU23 1 252. [1,1; A2644, N1426]

M5001   1, 16, 58, 128, 226, 352, 506, 688, 898, 1136, 1402, 1696, 2018, 2368, 2746, 3152, 3586, 4048, 4538, 5056, 5602, 6176, 6778, 7408, 8066, 8752, 9466, 10208, 10978, 11776
Points on surface of truncated tetrahedron: $14n^2+2$. Ref Coxe74. INOC 24 4552 85. [0,2; A5905]

M5002   1, 16, 68, 180, 375, 676, 1106, 1688, 2445, 3400, 4576, 5996, 7683, 9660, 11950, 14576, 17561, 20928, 24700, 28900, 33551, 38676, 44298, 50440, 57125, 64376, 72216
Truncated tetrahedral numbers. Ref Coxe74. INOC 24 4552 85. [0,2; A5906]

M5003   16, 80, 1056, 320416
Switching networks. Ref JFI 276 318 63. [1,1; A0817, N2153]

M5004   1, 16, 81, 256, 625, 1296, 2401, 4096, 6561, 10000, 14641, 20736, 28561, 38416, 50625, 65536, 83521, 104976, 130321, 160000, 194481, 234256, 279841, 331776
Fourth powers. Ref BA9. [1,2; A0583, N2154]

M5005   16, 96, 344, 952, 2241, 4712, 9608, 16488, 30930
From generalized Catalan numbers. Ref LNM 952 288 82. [0,1; A6637]

M5006   1, 16, 104, 320, 260, 1248, 3712, 1664, 6890, 7280, 5568, 4160, 33176, 4640, 74240, 29824, 14035, 54288, 27040, 142720, 1508, 110240, 289536, 222720, 380770
Expansion of $\Pi(1-x^k)^{16}$. Ref KNAW 59 207 56. [0,2; A0739, N2155]

M5007   0, 16, 122, 800, 5296, 36976, 275792, 2204480, 18870016, 172585936, 1681843712, 17411416160, 190939611136, 2211961358896, 26999750469632
Entringer numbers. Ref NAW 14 241 66. DM 38 268 82. [0,2; A6215]

M5008   16, 125, 680, 3135, 13155, 51873, 195821
Partially labeled trees with n nodes. Ref R1 138. [4,1; A0485, N2156]

M5009   16, 128, 448, 1024, 2016, 3584, 5504, 81982, 12112, 16128, 21312, 28672, 35168, 44032, 56448, 65536, 78624, 96896, 109760, 129024, 154112, 170496, 194688
Theta series of E_8 lattice w.r.t. deep hole. Ref SPLAG 122. [1,1; A4017]

M5010   16, 144, 984, 5756, 30760, 155912, 766424
n-step walks on cubic lattice. Ref PCPS 58 99 62. [1,1; A0762, N2157]

M5011 16, 150, 926, 4788, 22548, 100530, 433162, 1825296, 7577120, 31130190,
126969558
Permutations of length n by number of runs. Ref DKB 260. [5,1; A0486, N2158]

M5012 1, 16, 150, 1104, 7077, 41504, 228810, 1205520, 6135690, 30391520, 147277676,
700990752
Rooted planar maps. Ref JCT B18 249 75. [2,2; A6420]

M5013 16, 160, 13056, 183305216
Switching networks. Ref JFI 276 317 and 588 63. [1,1; A0811, N2159]

M5014 0, 0, 0, 16, 177, 1874
E-trees with exactly 3 colors. Ref AcMaSc 2 109 82. [1,4; A7144]

M5015 1, 16, 177, 5548, 39615, 2236440, 40325915, 1207505768, 13229393814,
1737076976040
Coefficients for step-by-step integration. Ref JACM 11 231 64. [1,2; A2399, N2160]

M5016 1, 16, 181, 1821, 17557, 167449, 1604098, 15555398, 153315999, 1538907304,
15743413076, 164161815768, 1744049683213, 18865209953045
Permutations of length n by subsequences. Ref MOC 22 390 68. [4,2; A1455, N2161]

M5017 16, 192, 2016, 20160, 197940, 1930944
n-step walks on f.c.c. lattice. Ref PCPS 58 100 62. [1,1; A0767, N2162]

M5018 1, 16, 196, 2197, 22952, 223034, 2004975, 16642936
n-covers of a 6-set. Ref DM 81 151 90. [1,2; A5747]

M5019 1, 16, 200, 2400, 29400, 376320, 5080320, 72576000, 1097712000, 17563392000,
296821324800, 5288816332800, 99165306240000, 1952793722880000
Coefficients of Laguerre polynomials. Ref LA56 519. AS1 799. [3,2; A1810, N2163]

M5020 16, 240, 6448, 187184, 5474096, 160196400, 4688357168, 137211717424,
4015706384176
Functions realized by n-input cascades. Ref PGEC 27 790 78. [2,1; A5619]

M5021 1, 16, 256, 4096, 65536, 1048576, 16777216, 268435456, 4294967296,
68719476736, 1099511627776, 17592186044416, 281474976710656
Powers of 16. Ref BA9. [0,2; A1025, N2164]

M5022 16, 272, 2880, 24576, 185856, 1304832, 8728576, 56520704, 357888000,
2230947840, 13754155008
Permutations of length n by number of peaks. Ref DKB 261. [5,1; A0487, N2165]

M5023 16, 272, 3968, 56320, 814080, 12207360
Generalized tangent numbers. Ref TOH 42 152 36. [4,1; A2303, N2166]

M5024   16, 361, 3362, 16384, 55744, 152166, 355688, 739328, 1415232, 2529614
Generalized tangent numbers. Ref MOC 21 690 67. [1,1; A0488, N2167]

M5025   1, 16, 435, 7136, 99350
Card matching. Ref R1 193. [1,2; A0489, N2168]

M5026   16, 912, 30768, 870640, 22945056
Coefficients of elliptic function cn. Ref Cay95 56. TM93 4 92. JCT A29 123 80. [2,1; A6089]

M5027   1, 16, 1280, 249856
Generalized Euler numbers. Ref MOC 21 689 67. [0,2; A0490, N2169]

M5028   1, 16, 1296, 20736, 12960000, 12960000, 31116960000, 497871360000,
    40327580160000, 40327580160000, 590436101122560000, 590436101122560000
Denominators of Σk^{-4}; $k = 1..n$. Ref KaWa 89. [1,2; A7480]

M5029   1, 16, 9882
n-element algebras with 1 binary operation and 1 constant. Ref PAMS 17 737 66. [1,2; A6448]

M5030   1, 16, 19683, 4294967296, 298023223876953125,
    10314424798490535546171949056, 256923577752105887808861147722423562 1321607
$n \uparrow n^2$. Ref ELM 3 20 48. [1,2; A2489, N2170]

M5031   1, 16, 7625597484987
$n \uparrow n^n$. Ref ELM 3 20 48. [1,2; A2488, N2171]

M5032   17, 19, 73, 139, 907, 1907, 2029, 4801, 5153, 10867
$(11^n - 1)/10$ is prime. Ref CUNN. MOC 61 928 93. [1,1; A5808]

M5033   17, 27, 33, 52, 73, 82, 83, 103, 107, 137, 153, 162, 217, 219, 227, 237, 247, 258,
    268, 271, 282, 283, 302, 303, 313, 358, 383, 432, 437, 443, 447, 502, 548, 557, 558, 647
Not the sum of 4 tetrahedrals (a finite sequence). Ref MOC 12 142 58. [1,1; A0797, N2172]

M5034   17, 29, 61, 97, 109, 113, 149, 181, 193, 229, 233, 257, 269, 313, 337, 389, 433,
    461, 509, 541, 577, 593, 701, 709, 821, 857, 937, 941, 953, 977, 1021, 1033, 1069, 1097
Primes with both 10 and −10 as primitive root. Ref AS1 864. [1,1; A7349]

M5035   1, 1, 1, 17, 31, 1, 5461, 257, 73, 1271, 60787, 241, 22369621, 617093, 49981
Numerators of coefficients for central differences. Cf. M4282. Equals M2100/M0124. Ref SAM 42 162 63. [2,4; A2675, N2173]

M5036   1, 1, 1, 17, 31, 691, 5461, 929569, 3202291, 221930581, 4722116521,
    968383680827, 14717667114151, 2093660879252671, 86125672563301143
Related to Genocchi numbers. Ref AMP 26 5 1856. QJMA 46 38 14. FMR 1 73. [1,4; A2425, N2174]

M5037   17, 41, 73, 89, 97, 113, 137, 193, 233, 241, 257, 281, 313, 337, 353, 401, 409, 433, 449, 457, 521, 569, 577, 593, 601, 617, 641, 673, 761, 769, 809, 857, 881, 929, 937, 953
Primes of form $8n + 1$. Ref AS1 870. [1,1; A7519]

M5038   17, 50, 99, 164, 245, 342, 455, 584, 729, 890, 1067, 1260, 1469, 1694, 1935, 2192, 2465, 2754, 3059, 3380, 3717, 4070, 4439, 4824, 5225, 5642, 6075, 6524, 6989, 7470
Walks on cubic lattice. Ref GU90. [0,1; A5570]

M5039   17, 73, 241, 1009, 2641, 8089, 18001, 53881, 87481, 117049, 515761, 1083289, 3206641, 3818929, 9257329, 22000801, 48473881, 48473881, 175244281, 427733329
Pseudo-squares. Ref MOC 8 241 54; 24 434 70. [1,1; A2189, N2175]

M5040   17, 73, 241, 1009, 2689, 8089, 33049, 53881, 87481, 483289, 515761, 1083289, 3818929, 3818929, 9257329, 22000801, 48473881, 48473881, 175244281, 427733329
Smallest prime such that first n primes are residues. Ref RS9 XV. MOC 8 241 54; 24 434 70. [1,1; A2224, N2176]

M5041   1, 17, 82, 273, 626, 1394, 2402, 4369, 6643, 10642, 14642, 22386, 28562, 40834, 51332, 69905, 83522, 112931, 130322, 170898, 196964, 248914, 279842, 358258
Sum of 4th powers of divisors of n. Ref AS1 827. [1,2; A1159, N2177]

M5042   17, 97, 257, 337, 641, 881, 1297, 2417, 2657, 3697, 4177, 4721, 6577, 10657, 12401, 14657, 14897, 15937, 16561, 28817, 38561, 39041, 49297, 54721, 65537, 65617
Quartan primes: $p = x^4 + y^4$. Ref CU23 1 253. [1,1; A2645, N2178]

M5043   1, 17, 98, 354, 979, 2275, 4676, 8772, 15333, 25333, 39974, 60710, 89271, 127687, 178312, 243848, 327369, 432345, 562666, 722666, 917147, 1151403, 1431244
Sums of fourth powers. Ref AS1 813. [1,2; A0538, N2179]

M5044   1, 1, 1, 1, 1, 17, 107, 415, 1231, 56671, 924365, 11322001, 97495687, 78466897, 31987213451, 1073614991039, 26754505127713, 558657850929473
$a(n) = -\Sigma\ (n+k)!a(k)/(2k)!$, $k = 0..n-1$. Ref UM 45 82 94. [0,6; A7682]

M5045   1, 1, 17, 117, 1413, 46389, 1211085
Special permutations. Ref JNT 5 48 73. [3,3; A3109]

M5046   1, 17, 257, 241, 65537, 61681, 673, 15790321, 6700417, 38737, 4278255361, 2931542417, 22253377, 308761441, 54410972897, 4562284561, 67280421310721
Largest factor of $16^n + 1$. Ref Krai24 2 88. CUNN. [0,2; A2590, N2180]

M5047   17, 259, 2770, 27978, 294602, 3331790, 40682144, 535206440, 7557750635, 114101726625, 1834757172082
Permutations of length n by rises. Ref DKB 264. [6,1; A1282, N2181]

M5048   1, 17, 289, 4913, 83521, 1419857, 24137569, 410338673, 6975757441, 118587876497, 2015993900449, 34271896307633, 582622237229761
Powers of 17. Ref BA9. [0,2; A1026, N2182]

M5049 1, 17, 367, 27859, 1295803, 5329242827, 25198857127, 11959712166949,
11153239773419941, 31326450596954510807
Numerators of coefficients for numerical integration. Cf. M5178. Ref OP80 545. PHM 35
217 45. [0,2; A2197, N2183]

M5050 1, 17, 1393, 22369, 14001361, 14011361, 33654237761, 538589354801,
43631884298881, 43635917056897, 638913789210188977, 638942263173398977
Numerators of Σk^{-4}; $k = 1..n$. Ref KaWa 89. [1,2; A7410]

M5051 1, 17, 1835, 195013, 3887409, 58621671097
From higher order Bernoulli numbers. Ref NO24 463. [1,2; A1905, N2184]

M5052 18, 23, 28, 32, 35, 39, 42, 46, 49, 52, 55, 58, 60, 63, 66, 68, 71, 74, 76, 79, 81, 84,
86, 88, 91, 93, 95, 98, 100, 102, 104, 107, 109, 111, 113, 115, 118, 120, 122, 124, 126
Nearest integers to the Gram points. Ref RS6 58. [1,1; A2505, N2185]

M5053 0, 1, 18, 24, 27216, 5878656, 105815808, 346652587008, 693305174016
Coefficients of Green function for cubic lattice. Ref PTRS 273 593 73. [0,3; A3300]

M5054 1, 18, 27, 12, 45, 54, 21, 72, 81, 10, 198, 108, 117, 126, 135, 144, 153, 162, 114,
180, 378, 132, 207, 216, 150, 234, 243, 112, 261, 270, 372, 576, 594, 102, 315, 324, 111
Smallest number that is n times sum of its digits. Ref jhc. [1,2; A3634]

M5055 18, 45, 69, 96, 120, 147, 171, 198, 222, 249, 273, 300, 324, 351, 375, 402, 426,
453, 477, 504, 528, 555, 579, 606, 630, 657, 681, 708, 732, 759, 783, 810, 834, 861, 885
$a(n) = a(n-2) + a(n-3) - a(n-5)$. Ref JRAM 227 49 67. [1,1; A2798, N2186]

M5056 18, 72, 336, 1728, 9981, 57624, 359412, 2271552
Expansion of free energy series related to Potts model. Ref JPA 12 L230 79. [4,1; A7276]

M5057 0, 18, 108, 180, 5040, 162000, 14565600, 563253408, 17544639744,
750651187968
Specific heat for cubic lattice. Ref PRV 164 801 67. [1,2; A2165, N2187]

M5058 1, 18, 126, 420, 630, 252, 84, 72, 90, 140, 252, 504, 1092, 2520, 6120, 15504,
40698, 110124, 305900, 869400, 2521260, 7443720, 22331160, 67964400, 209556900
Expansion of $(1 - 4x)^{9/2}$. Ref TH09 164. FMR 1 55. [0,2; A2424, N2188]

M5059 1, 18, 160, 1120, 6912, 39424, 212992, 1105920, 5570560, 27394048, 132120576,
627048448, 2936012800, 13589544960, 62277025792, 282930970624, 1275605286912
Coefficients of Chebyshev polynomials: $n(2n-3)2^{2n-5}$. Ref LA56 516. [2,2; A2698, N2189]

M5060 1, 18, 245, 3135, 40369, 537628, 7494416, 109911300, 1698920916,
27679825272, 474957547272, 8572072384512, 162478082312064, 3229079010579072
Generalized Stirling numbers. Ref PEF 77 26 62. [0,2; A1713, N2190]

M5061 1, 18, 251, 3325, 44524, 617624, 8969148, 136954044, 2201931576,
   37272482280, 663644774880, 12413008539360, 243533741849280, 5003753991174720
Generalized Stirling numbers. Ref PEF 77 61 62. [0,2; A1722, N2191]

M5062 1, 18, 324, 5832, 104976, 1889568, 34012224, 612220032, 11019960576,
   198359290368, 3570467226624, 64268410079232, 1156831381426176
Powers of 18. Ref BA9. [0,2; A1027, N2192]

M5063 1, 18, 648, 2160, 1399680, 75582720, 149653785600, 2693768140800,
   8620058050560
Coefficients of Green function for cubic lattice. Ref PTRS 273 593 73. [0,2; A3298]

M5064 1, 1, 18, 1606, 565080, 734774776, 3523091615568, 63519209389664176,
   4400410978376102609280, 1190433705317814685295399296
Strongly connected digraphs with n nodes. Ref HA73 270. rwr. [1,3; A3030]

M5065 18, 2862, 158942078604
Post functions. Ref JCT 4 297 68. [1,1; A1325, N2193]

M5066 1, 1, 1, 19, 3, 863, 275, 33953, 8183, 3250433, 4671, 13695779093, 2224234463,
   132282840127, 2639651053, 111956703448001, 50188465, 2334028946344463
Numerators of logarithmic numbers. Cf. M2017. Ref SAM 22 49 43. PHM 38 336 47.
MOC 20 465 66. [1,4; A2206, N2194]

M5067 1, 1, 1, 1, 19, 9, 863, 1375, 33953, 57281, 3250433, 1891755, 13695779093,
   24466579093, 132282840127, 240208245823, 111956703448001, 4573423873125
Numerators of Cauchy numbers. Ref C1 294. [0,5; A6232]

M5068 19, 20, 22, 25, 29, 34, 38, 39, 40, 45, 47, 48, 55, 56, 57, 58, 60, 61, 63, 64, 65, 68,
   71, 74, 76, 77, 78, 82, 83, 85, 90, 91, 93, 94, 95, 96, 97, 102, 104, 107, 110, 112, 113, 114
$2n^2 - 2n + 19$ is composite. [1,1; A7640]

M5069 19, 23, 29, 37, 47, 59, 73, 89, 107, 127, 149, 173, 199, 227, 257, 359, 397, 479,
   523, 569, 617, 719, 773, 829, 887, 947, 1009, 1277, 1423, 1499, 1657, 1823, 1997, 2087
Primes of form $n^2 + n + 17$. [1,1; A7635]

M5070 19, 23, 31, 43, 59, 79, 103, 131, 163, 199, 239, 283, 331, 383, 439, 499, 563, 631,
   859, 1031, 1123, 1319, 1423, 1531, 1759, 1879, 2003, 2131, 2399, 2539, 2683, 3299
Primes of form $2n^2 - 2n + 19$. [1,1; A7639]

M5071 19, 27, 37, 46, 56
Zarankiewicz's problem. Ref LNM 110 144 69. [4,1; A6626]

M5072 19, 31, 47, 59, 61, 107, 337, 1061
$(19^n - 1)/18$ is prime. Ref MOC 61 928 93. [1,1; A6035]

M5073 19, 43, 43, 67, 67, 163, 163, 163, 163, 163, 163, 222643, 1333963, 1333963,
2404147, 2404147, 20950603, 51599563, 51599563, 96295483, 96295483, 146161723
Sequence of prescribed quadratic character. Ref MOC 24 440 70. [1,1; A1986, N2195]

M5074 19, 69, 280, 707, 2363, 3876, 8068, 11319, 19201, 36866, 45551, 75224, 101112,
117831, 152025, 215384, 293375, 327020, 428553
Steps to compute nth prime in PRIMEGAME (fast version). Cf. M2084. Ref MMAG 56 28
83. CoGo87 4. Oliv93 21. [1,1; A7546]

M5075 19, 69, 281, 710, 2375, 3893, 8102, 11361, 19268, 36981, 45680, 75417, 101354,
118093, 152344, 215797, 293897, 327571, 429229
Steps to compute nth prime in PRIMEGAME (slow version). Cf. M2084. Ref MMAG 56
28 83. CoGo87 4. Oliv93 21. [1,1; A7547]

M5076 19, 145, 100, 2191, 8592, 14516, 29080, 114575, 320417, 615566, 1125492,
2139700, 3664750, 5997448, 10103304, 15992719, 23857290, 36059435, 53341900
Related to representation as sums of squares. Ref QJMA 38 349 07. [1,1; A2615, N2196]

M5077 1, 19, 191, 1400, 8373, 43277, 199982, 844734
Coefficients of a modular function. Ref GMJ 8 29 67. [4,2; A5759]

M5078 1, 19, 205, 1795, 14221, 106819
Connected relations. Ref CRP 268 579 69. [1,2; A2501, N2197]

M5079 1, 19, 361, 6859, 130321, 2476099, 47045881, 893871739, 16983563041,
322687697779, 6131066257801, 116490258898219, 2213314919066161
Powers of 19. Ref BA9. [0,2; A1029, N2198]

M5080 1, 1, 19, 475, 1753, 1109769, 70784325, 2711086547, 1376283649103,
148592152807663, 21812320857733789, 12754009647903010101
Expansion of $\tan(x/\cosh x)$. [0,3; A3700]

M5081 1, 19, 916, 91212
Semi-regular digraphs on n nodes. Ref KNAW 75 330 72. [3,2; A5535]

M5082 1, 1, 19, 1513, 315523, 136085041, 105261234643, 132705221399353,
254604707462013571, 70592767752064467681, 2716778010767155313771539
Generalized Euler numbers. Ref ANN 36 649 35. [0,3; A2115, N2199]

M5083 0, 1, 20, 50, 92, 170, 284, 434, 620, 842, 1100, 1394, 1724, 2090, 2492, 2930,
3404, 3914, 4460, 5042, 5660, 6314, 7004, 7730, 8492, 9290, 10124, 10994, 11900
Dodecahedral surface numbers: $2((3n-7)^2 + 21)$. Ref rkg. [0,3; A7589]

M5084 1, 20, 62, 216, 641, 1636, 3778, 8248, 17277, 34664, 66878, 125312, 229252,
409676, 716420, 1230328, 2079227, 3460416, 5677816, 9198424, 14729608, 23328520
McKay-Thompson series of class 4C for Monster. Ref CALG 18 257 90. FMN94. [0,2;
A7248]

M5085 1, 20, 74, 24, 157, 124, 478, 1480, 1198, 3044, 480, 184, 2351, 1720, 3282, 5728,
2480, 1776, 10326, 9560, 8886, 9188, 11618, 23664, 16231, 23960
Related to representation as sums of squares. Ref QJMA 38 56 07. [0,2; A2292, N2201]

M5086 20, 74, 186, 388, 721, 1236, 1995, 3072, 4554, 6542, 9152, 12516, 16783, 22120,
28713, 36768, 46512, 58194, 72086, 88484, 107709, 130108, 156055, 185952
Powers of rooted tree enumerator. Ref R1 150. [1,1; A0529, N2202]

M5087 20, 75, 189, 392, 720, 1215, 1925, 2904, 4212, 5915, 8085, 10800, 14144, 18207,
23085, 28880, 35700, 43659, 52877, 63480, 75600, 89375, 104949, 122472, 142100
Walks on square lattice. Ref GU90. [0,1; A5565]

M5088 1, 20, 80, 144, 610, 448, 1120, 2240, 3423, 12200, 14800, 29440, 5470, 6272,
48800, 81664, 73090, 68460, 15600, 87840, 139776, 82880, 189920, 474112, 18525
Related to representation as sums of squares. Ref QJMA 38 311 07. [1,2; A2609, N2203]

M5089 1, 20, 84, 220, 455, 816, 1330, 2024, 2925, 4060, 5456, 7140, 9139, 11480, 14190,
17296, 20825, 24804, 29260, 34220, 39711, 45760, 52394, 59640, 67525, 76076, 85320
Dodecahedral numbers: $n(3n-1)(3n-2)/2$. [1,2; A6566]

M5090 1, 20, 130, 576, 2218, 8170, 29830, 109192, 402258, 1492746, 5578742, 20986424
n-step walks on hexagonal lattice. Ref JPA 6 352 73. [4,2; A5551]

M5091 1, 20, 131, 469, 1262, 2862, 5780, 10725, 18647, 30784, 48713, 74405
Rooted planar maps. Ref JCT B18 248 75. [2,2; A6417]

M5092 20, 154, 1676, 14292, 155690, 1731708, 21264624, 280260864, 3970116255,
60113625680, 969368687752, 16588175089420, 300272980075896
Discordant permutations. Ref SMA 20 23 54. [6,1; A0492, N2204]

M5093 20, 220, 23932, 2390065448
Switching networks. Ref JFI 276 320 63. [1,1; A0833, N2205]

M5094 1, 20, 225, 1925, 14014, 91728, 556920, 3197700, 17587350, 93486536,
483367885, 2442687975, 12109051500, 59053512000, 283963030560, 1348824395160
Dissections of a polygon by number of parts. Ref AEQ 18 385 78. [1,2; A7160]

$$(n + 5)(n - 1)n\, a(n) = 2(2n + 3)(n + 3)(n + 2)\, a(n - 1).$$

M5095 1, 20, 295, 4025, 54649, 761166, 11028590, 167310220, 2664929476,
44601786944, 784146622896, 14469012689040, 279870212258064, 5667093514231200
Generalized Stirling numbers. Ref PEF 77 7 62. [0,2; A1708, N2206]

$$\text{E.g.f.: } (\ln(1 - x))^4 / 24(1 - x)^2.$$

M5096 1, 20, 300, 4200, 58800, 846720, 12700800, 199584000, 3293136000,
57081024000, 1038874636800, 19833061248000, 396661224960000
Lah numbers: $n!\,C(n-1,3)/4!$. Ref R1 44. C1 156. [4,2; A1755, N2207]

M5097   1, 20, 307, 4280, 56914, 736568, 9370183, 117822512, 1469283166,
18210135416, 224636864830, 2760899996816, 33833099832484, 413610917006000
Rooted maps with n edges on torus. Ref WA71. JCT A13 215 72. CMB 31 269 88. [2,2;
A6300]

M5098   1, 20, 348, 6093, 108182, 1890123, 31500926, 490890304
n-covers of a 7-set. Ref DM 81 151 90. [1,2; A5748]

M5099   20, 371, 2588, 11097, 35645, 94457, 218124, 454220, 872648, 1571715, 2684936,
4388567, 6909867, 10536089, 15624200, 22611330, 32025950, 44499779
Discordant permutations. Ref SMA 20 23 54. [3,1; A0564, N2208]

M5100   1, 20, 400, 8902, 197742, 4897256, 120921506, 3284294545, 88867026005
Number of chess games with n moves. Ref ken. [0,2; A7545]

M5101   20, 484, 497760, 1957701217328
Switching networks. Ref JFI 276 320 63. [1,1; A0827, N2209]

M5102   20, 651, 8344, 64667, 361884, 1607125, 5997992
Rooted nonseparable maps on the torus. Ref JCT B18 241 75. [2,1; A6410]

M5103   1, 20, 784, 52480, 5395456, 791691264, 157294854144, 40683662475264,
13288048674471936
Central factorial numbers. Ref OP80 7. FMR 1 110. RCI 217. [2,2; A2455, N2210]

M5104   20, 831, 12656, 109075, 648792, 2978245, 11293436, 36973989
Rooted toroidal maps. Ref JCT B18 250 75. [1,1; A6424]

M5105   20, 1071, 26320, 431739, 5494896, 58677420, 550712668, 4681144391
Rooted toroidal maps. Ref JCT B18 251 75. [1,1; A6427]

M5106   1, 20, 1120, 3200, 3942400, 66560000, 10035200000
Denominators of an asymptotic expansion of an integral. Cf. 2305. Ref MOC 19 114 65.
[0,2; A2305, N2211]

M5107   1, 20, 1301, 202840, 61889101, 32676403052, 27418828825961,
34361404413755056, 61335081309931829401, 15022174068827565795794 0
3 up, 3 down, 3 up, ... permutations of length $2n + 1$. Ref prs. [1,2; A5982]

M5108   1, 1, 21, 31, 6257, 10293, 279025, 483127, 435506703, 776957575, 22417045555,
40784671953
Coefficients of Jacobi nome. Ref HER 477. MOC 3 234 48. [1,3; A2639, N2212]

M5109   21, 42, 65, 86, 109, 130, 151, 174, 195, 218, 239, 262, 283, 304, 327, 348, 371,
392, 415, 436, 457, 480, 501, 524, 545, 568, 589, 610, 633, 654, 677, 698, 721, 742, 763
Related to powers of 3. Ref AMM 64 367 57. [1,1; A1682, N2213]

M5110 1, 1, 21, 141, 10441, 183481, 29429661, 987318021, 276117553681,
  15085947275761, 6514632269358501, 526614587249608701, 324871912636292700121
Expansion of tan x /cosh x . [0,3; A3702]

M5111 1, 21, 171, 745, 2418, 7587, 20510, 51351, 122715, 277384, 598812, 1255761,
  2543973
McKay-Thompson series of class 6b for Monster. Ref FMN94. [0,2; A7261]

M5112 1, 21, 266, 2646, 22827, 179487, 1323652, 9321312, 63436373, 420693273,
  2734926558, 17505749898, 110687251039, 693081601779, 4306078895384
Stirling numbers of second kind. See Fig M4981. Ref AS1 835. DKB 223. [6,2; A0770,
N2215]

M5113 21, 301, 325, 697, 1333, 1909, 2041, 2133, 3901, 10693, 16513, 19521, 24601,
  26977, 51301, 96361, 130153, 159841, 163201, 176661, 214273, 250321, 275833
Hyperperfect numbers: $n = m(\sigma(n) - n - 1) + 1$ for some $m > 1$. Ref MOC 34 639 80.
Robe92 177. [1,1; A7592]

M5114 1, 21, 322, 4536, 63273, 902055, 13339535, 206070150, 3336118786,
  56663366760, 1009672107080, 18861567058880, 369012649234384
Stirling numbers of first kind. See Fig M4730. Ref AS1 833. DKB 226. [6,2; A1233,
N2216]

M5115 1, 21, 357, 5797, 93093, 1490853, 23859109, 381767589, 6108368805,
  97734250405, 1563749404581, 25019996065701, 400319959420837
Gaussian binomial coefficient $[n,2]$ for $q = 4$. Ref GJ83 99. ARS A17 328 84. [2,2; A6105]

M5116 1, 21, 378, 6930, 135135, 2837835, 64324260, 1571349780, 41247931725,
  1159525191825, 34785755754750, 1109981842719750, 37554385678684875
Coefficients of Bessel polynomials $y_n(x)$. Ref RCI 77. [5,2; A1881, N2217]

M5117 21, 483, 6468, 66066, 570570, 4390386, 31039008, 205633428, 1293938646,
  7808250450, 45510945480
Rooted genus-2 maps with n edges. Ref WA71. JCT A13 215 72. [4,1; A6298]

M5118 21, 483, 15018, 258972, 5554188, 85421118, 1558792200, 22555934280,
  375708427812, 5235847653036, 82234427131416
Rooted genus-2 maps with n edges. Ref WA71. JCT A13 215 72. [4,1; A6299]

M5119 1, 1, 21, 671, 180323, 20898423, 7426362705, 1874409465055
A series for π. Ref Luk69 36. [0,3; A6934]

M5120 0, 0, 0, 0, 21, 966, 27954, 650076, 13271982, 248371380, 4366441128,
  73231116024, 1183803697278, 18579191525700, 284601154513452
Rooted genus-2 maps with n edges. Ref WA71. JCT A13 215 72. JCT B53 297 91. [0,5;
A6301]

M5121   21, 2133, 19521, 176661
2-hyperperfect numbers: $n = 2(\sigma(n) - n - 1) + 1$. Ref MOC 34 639 80. Robe92 177. [1,1; A7593]

M5122   1, 21, 21000, 101, 121, 1101, 1121, 21121, 101101, 101121, 121121, 1101121, 1121121, 21121121, 101101121, 101121121, 121121121, 1101121121, 1121121121
Smallest number requiring n words in English. [1,2; A1167, N2218]

M5123   22, 67, 181, 401, 831, 1576, 2876, 4987, 8406, 13715, 21893, 34134, 52327, 78785, 116982, 171259, 247826, 354482, 502090, 704265, 979528, 1351109, 1849932
Bipartite partitions. Ref ChGu56 1. [0,1; A2757, N2219]

M5124   1, 22, 234, 2348, 22726, 214642, 1993002
Cluster series for f.c.c. lattice. Ref PRV 133 A315 64. DG72 225. [0,2; A3205]

M5125   1, 22, 305, 3410, 33621, 305382, 2619625, 21554170, 171870941, 1337764142, 10216988145, 76862115330, 571247591461, 4203844925302, 30687029023865
Minimal covers of an n-set. Ref DM 5 249 73. [3,2; A3468]

$$\text{G.f.:}\quad 1 / (1 - 4x)(1 - 5x)(1 - 6x)(1 - 7x).$$

M5126   1, 22, 328, 4400, 58140, 785304, 11026296, 162186912, 2507481216, 40788301824, 697929436800, 12550904017920, 236908271543040, 4687098165573120
Permutations by descents. Ref NMT 7 16 59. JCT A24 28 78. [1,2; A2539, N2221]

M5127   1, 22, 355, 5265, 77224, 1155420, 17893196, 288843260, 4876196776, 86194186584, 1595481972864, 30908820004608, 626110382381184
Generalized Stirling numbers. Ref PEF 77 44 62. [0,2; A1718, N2222]

M5128   1, 1, 23, 11, 563, 1627, 88069, 1423, 1593269, 7759469, 31730711, 46522243
Numerators of coefficients for numerical differentiation. Cf. M5139. Ref PHM 33 13 42. [1,3; A2549, N2223]

M5129   23, 24, 28, 31, 39, 44, 45, 46, 47, 50, 52, 56, 57, 60, 63, 67, 69, 70, 71, 79, 80, 85, 86, 88, 89, 90, 92, 93, 96, 97, 102, 107, 108, 112, 115, 116, 118, 119, 121, 122, 123, 126
$3n^2 - 3n + 23$ is composite. [1,1; A7638]

M5130   23, 29, 41, 59, 83, 113, 149, 191, 239, 293, 353, 419, 491, 569, 653, 743, 839, 941, 1049, 1163, 1283, 1409, 1823, 1973, 2129, 2459, 2633, 2999, 3191, 3389, 3593, 3803
Primes of the form $3n^2 - 3n + 23$. [1,1; A7637]

M5131   23, 31, 59, 83, 107, 139, 211, 283, 307, 331, 379, 499, 547, 643, 883, 907
Imaginary quadratic fields with class number 3 (a finite sequence). Ref LNM 751 226 79. [1,1; A6203]

M5132   23, 47, 73, 97, 103, 157, 167, 193, 263, 277, 307, 383, 397, 433, 503, 577, 647,
683, 727, 743, 863, 887, 937, 967, 983, 1033, 1093, 1103, 1153, 1163, 1223, 1367
Primes with 5 as smallest primitive root. Ref Krai24 1 57. AS1 864. [1,1; A1124, N2224]

M5133   23, 59, 67, 83, 89, 107, 173, 199, 227, 233, 263, 311, 317, 331, 349, 353, 367, 373,
383, 397, 419, 431, 463, 479, 503, 509, 523, 563, 569, 587, 607, 617, 619, 661, 683, 727
Class 3 – primes. Ref UPNT A18. [1,1; A5111]

M5134   23, 65, 261, 1370, 8742, 65304, 557400, 5343120, 56775600, 661933440,
8397406080, 115123680000, 1695705580800, 26701944192000, 447579574041600
$\Sigma\,(n+k)!\,C(4,k)$, $k = 0\,.\,.\,4$. Ref CJM 22 26 70. [-1,1; A1346]

M5135   23, 67, 89, 4567, 78901, 678901, 23456789, 45678901, 9012345678901,
789012345678901
Primes with consecutive digits. Ref JRM 5 254 72. [1,1; A6055]

M5136   1, 1, 1, 23, 263, 133787, 157009, 16215071, 2689453969, 26893118531,
5600751928169
Numerators of coefficients for repeated integration. Cf. M3152. Ref SAM 28 56 49. [0,4;
A2681, N2227]

M5137   0, 1, 23, 1477, 555273, 38466649, 1711814393, 48275151899, 28127429172349
Coefficients of Green function for cubic lattice. Ref PTRS 273 590 73. [0,3; A3281]

M5138   1, 23, 1681, 257543, 67637281, 27138236663, 15442193173681
Glaisher's T numbers. Ref QJMA 29 76 1897. FMR 1 76. [1,2; A2439, N2228]

M5139   1, 1, 24, 12, 640, 1920, 107520, 1792, 2064384, 10321920, 43253760, 64880640
Denominators of coefficients for numerical differentiation. Cf. M5128. Ref PHM 33 13 42.
[1,3; A2550, N2229]

M5140   1, 24, 24, 96, 24, 144, 96, 192, 24, 312, 144, 288, 96, 336, 192, 576, 24, 432, 312,
480, 144, 768, 288, 576, 96, 744, 336, 960, 192, 720, 576, 768, 24, 1152, 432, 1152
Theta series of D_4 lattice. See Fig M5140. Ref SPLAG 119. [0,2; A4011]

M5141   1, 0, 0, 0, 0, 24, 26, 0, 0, 72, 378, 1080, 665, 384, 1968, 2016, 25698, 39552, 3872,
20880, 65727, 379072, 1277646, 986856, 176978, 2163504, 1818996, 27871080
Susceptibility for f.c.c. lattice. Ref JPA 6 1510 73. DG74 421. [0,6; A2924, N2230]

M5142   24, 42, 48, 60, 84, 90, 224, 228, 234, 248, 270, 294, 324, 450, 468, 528, 558, 620,
640, 660, 810, 882, 888, 896, 968, 972, 1020, 1050, 1104, 1116, 1140, 1216, 1232, 1240
$\sigma(x) = n$ has exactly 3 solutions. Ref AS1 840. [1,1; A7372]

Figure M5140. 24-CELL.

The theta series of the 4-dimensional lattice D_4 begins $1 + 24q^2 + 24q^4 + \cdots$, whose coefficients give M5140. The 24 shortest vectors in this lattice form the vertices of the regular four-dimensional polytope known as the 24-cell ([SPLAG 216], [Coxe73 149]) and shown here:

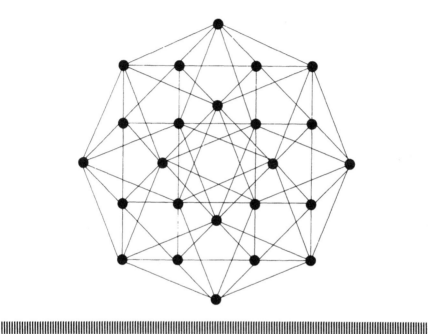

M5143 24, 44, 80, 144, 260, 476, 872, 1600, 2940, 5404, 9936, 18272, 33604, 61804, 113672, 209072, 384540, 707276, 1300880, 2392688, 4400836, 8094396, 14887912
Restricted permutations. Ref CMB 4 32 61. [4,1; A0496, N2231]

M5144 24, 54, 216, 648, 2376, 8100, 29232, 104544, 381672, 1397070, 5163480, 19170432, 71587080, 268423200, 1010595960, 3817704744, 14467313448
n-gons in cubic curve. Ref JEP 35 36 1884. [3,1; A5782]

M5145 1, 24, 72, 96, 168, 144, 288, 192, 360, 312, 432, 288, 672, 336, 576, 576, 744, 432, 936, 480, 1008, 768, 864
The modular form G_2. Ref JNT 25 205 87. [0,2; A6352]

M5146 24, 84, 264, 1128, 4728, 20304, 86496, 369732, 1573608, 6703068, 28474704, 120922272
Self-avoiding walks on hexagonal lattice. Ref JPA 13 3530 80. [3,1; A7201]

M5147 24, 144, 984, 7584, 65304, 622704, 6523224, 74542464, 923389464,
12331112784, 176656186584, 2703187857504, 44010975525144, 759759305162544
$\Sigma\,(n+4)!\,C(n,k)$, $k = 0 \ldots n$. Ref CJM 22 26 70. [0,1; A1342, N2233]

M5148 1, 24, 154, 580, 1665, 4025, 8624, 16884, 30810, 53130, 87450, 138424, 211939,
315315, 457520, 649400, 903924, 1236444, 1664970, 2210460, 2897125, 3752749
Generalized Stirling numbers. Ref PEF 77 7 62. [1,2; A1702, N2234]

M5149 0, 0, 24, 154, 1664, 15984, 173000, 2004486
Hit polynomials. Ref JAuMS A28 375 79. [4,3; A4308]

M5150 24, 240, 504, 480, 264, 65520, 24, 16320, 28728, 13200, 552, 131040, 24, 6960,
171864, 32640, 24, 138181680, 24, 1082400, 151704, 5520, 1128, 4455360, 264
Denominator of $B_{2k}\,/\,4k$. Ref LNCS 1326 127 86. [1,1; A6863]

M5151 24, 240, 1560, 8400, 40824, 186480, 818520, 3498000, 14676024, 60780720,
249401880, 1016542800, 4123173624, 16664094960, 67171367640, 270232006800
Differences of 0: $4!.S(n,4)$. Ref VO11 31. DA63 2 212. R1 33. [4,1; A0919, N2235]

M5152 24, 240, 2520, 26880, 304080, 3671136
3-line Latin rectangles. Ref R1 210. [4,1; A0536, N2236]

M5153 1, 24, 252, 1472, 4830, 6048, 16744, 84480, 113643, 115920, 534612, 370944,
577738, 401856, 1217160, 987136, 6905934, 2727432, 10661420, 7109760, 4219488
Ramanujan τ function. Ref PLMS 51 4 50. MOC 24 495 70. [1,2; A0594, N2237]

$$\text{G.f.:}\quad \Pi\,(1 - x^k)^{24}.$$

M5154 24, 264, 3312, 48240, 762096, 12673920, 218904768, 3891176352, 70742410800
$2n$-step polygons on cubic lattice. Ref JCP 34 1537 61. JPA 5 665 72. [2,1; A1413, N2238]

M5155 24, 274, 1624, 6769, 22449, 63273, 157773, 357423, 749463, 1474473, 2749747,
4899622, 8394022, 13896582, 22323822, 34916946, 53327946, 79721796, 116896626
Stirling numbers of first kind. See Fig M4730. Ref AS1 833. DKB 226. [1,1; A0915,
N2239]

M5156 1, 24, 276, 2024, 10602, 41952, 128500, 303048, 517155, 463496, 609684,
3757992, 9340852, 14912280, 12957624, 8669712, 59707149, 132295080, 183499244
Coefficients of powers of η function. Ref JLMS 39 439 64. [24,2; A6665]

M5157 1, 24, 276, 2048, 11202, 49152, 184024, 614400, 1881471, 5373952, 14478180,
37122048, 91231550, 216072192, 495248952, 1102430208, 2390434947, 5061476352
McKay-Thompson series of class 2B for Monster. Ref BLMS 11 334 79. LiSa92 100.
CALG 18 256 90. FMN94. [-1,2; A7191]

$$\text{G.f.:}\quad x^{-1}\,\Pi\,(1 + x^{2k+1})^{24}.$$

M5158   1, 24, 282, 2008, 10147, 40176, 132724, 381424, 981541, 2309384, 5045326,
10356424, 20158151, 37478624, 66952936, 115479776, 193077449, 313981688
4×4 stochastic matrices of integers. Ref SS70. CJN 13 283 70. SIAC 4 477 75. ANS 4
1179 76. [0,2; A1496, N2240]

M5159   24, 300, 3360, 38850, 475776, 6231960, 87530400
Labeled trees of height 3 with n nodes. Ref IBMJ 4 478 60. [4,1; A0552, N2241]

M5160   1, 24, 324, 3200, 25650, 176256, 1073720, 5930496, 30178575, 143184000,
639249300, 2705114880, 10914317934, 42189811200, 156883829400
Expansion of $\Pi \ (1 - x^k)^{-24}$. Cf. M5153. Ref RAM3 146. [0,2; A6922]

M5161   1, 24, 360, 4000, 39330, 367912, 3370604, 30630980, 277824572
n-step walks on f.c.c. lattice. Ref JPA 6 351 73. [3,2; A5546]

M5162   0, 24, 444, 4400, 32120, 195800, 1062500, 5326160, 25243904, 114876376,
507259276, 2189829808, 9292526920, 38917528600, 161343812980, 663661077072
Second-order Eulerian numbers. Ref JCT A24 28 78. GKP 256. [3,2; A6260]

M5163   0, 0, 0, 0, 0, 24, 570, 27900, 1827630, 152031600, 15453811800, 1884214710000,
271711218933000, 45788138466285600, 8922341314806519600
Connected labeled 2-regular oriented graphs with n nodes. Ref rwr. [0,6; A7110]

M5164   1, 0, 0, 0, 0, 24, 570, 27900, 1827630, 152031600, 15453884376, 1884221030160,
271711899360000, 45788222207669040, 8922353083744943700
Labeled 2-regular oriented graphs with n nodes. Ref rwr. [0,6; A7109]

M5165   24, 600, 10800, 176400, 2822400, 45722880, 762048000, 13172544000,
237105792000
Coefficients of Laguerre polynomials. Ref AS1 799. [4,1; A1806, N2242]

M5166   1, 24, 640, 7168, 294912, 2883584, 54525952, 167772160, 36507222016,
326417514496
Coefficients for numerical differentiation. Ref OP80 23. PHM 33 14 42. [0,2; A2553,
N2243]

M5167   0, 0, 1, 24, 640, 24000, 1367296, 122056704, 17282252800, 3897054412800
3-colored labeled graphs on n nodes. Ref CJM 12 412 60. rcr. [1,4; A6201]

M5168   1, 24, 852, 35744, 1645794, 80415216, 4094489992, 214888573248,
11542515402255, 631467591949480, 35063515239394764, 1971043639046131296
A sequence for π. Ref MOC 42 212 84. [1,2; A5149]

M5169   1, 24, 924, 26432, 705320, 18858840, 520059540, 14980405440, 453247114320,
14433720701400, 483908513388300, 17068210823664000, 632607429473019000
Associated Stirling numbers. Ref TOH 37 259 33. JO39 152. C1 256. [1,2; A1784, N2244]

M5170   0, 0, 1, 24, 936, 56640, 4968000, 598328640, 94916183040, 19200422062080,
   4826695329792000, 1476585999504000000, 540272647694971699200
Connected graphs with n nodes and n edges. Ref AMS 30 748 59. [0,4; A1866, N2245]

M5171   24, 972, 118592, 15210414, 2344956480, 377420590432, 67501965869568
Witt vector *4!. Ref SLC 16 106 88. [1,1; A6175]

M5172   24, 1344, 393120, 155185920, 143432634240
$4 \times n$ Latin rectangles. Ref FQ 11 246 73. [4,1; A3170]

M5173   24, 1920, 193536, 66355200, 13624934400, 243465191424000,
   4944216195072000, 9990141980442624000, 393917174842955880704000
Coefficients for numerical integration. Ref OP80 545. PHM 35 217 44. [0,1; A6685]

M5174   1, 24, 1920, 322560, 92897280, 40874803200, 25505877196800,
   21424936845312000, 23310331287699456000, 31888533201572855808000
$4^n(2n+1)!$. Ref SAM 42 162 63. [0,2; A2671, N2246]

M5175   1, 24, 2040, 297200, 68938800, 24046189440, 12025780892160,
   8302816499443200, 7673688777463632000, 9254768770160124288000
Stochastic matrices of integers. Ref RE58. SS70. [1,2; A1501, N2247]

M5176   1, 24, 4372, 96256, 1240002, 10698752, 74428120, 431529984, 2206741887,
   10117578752, 42616961892, 166564106240, 611800208702, 2125795885056
McKay-Thompson series of class 2A for Monster. Cf. M5369. Ref CALG 18 256 90.
FMN94. [-1,2; A7241]

M5177   1, 24, 5760, 322560, 51609600, 13624934400, 19837904486400, 2116043145216,
   20720294477955072, 15747423803245854720
Denominators of coefficients for numerical differentiation. Cf. M4034. Ref OP80 23. PHM
33 14 42. [1,2; A2555, N2249]

M5178   24, 5760, 967680, 464486400, 122624409600, 2678117105664000,
   64274810535936000, 149852129706639360000, 6696591972330299971968000
Denominators of coefficients for numerical integration. Cf. M5049. Ref OP80 545. PHM
35 217 45. [0,1; A2198, N2250]

M5179   1, 24, 196884, 21493760, 864299970, 20245856256, 333202640600,
   4252023300096, 44656994071935, 401490886656000, 3176440229784420
McKay-Thompson series of class 1A for Monster. Cf. M5477. Ref KNAW 56 398 53.
MOC 8 77 54. BLMS 11 326 79. CALG 18 256 90. [-1,2; A7240, N2372]

SEQUENCES BEGINNING . . ., 25, . . . ONWARDS

M5180   1, 24, 1058158080, 173008013097959424000
Group of $n \times n$ Rubik cube. Ref jhc. [0,2; A7458]

M5181 1, 25, 2, 4, 3, 22, 6, 8, 10, 5, 32, 83, 44, 14, 7, 66, 169, 11, 49595, 9, 69, 16, 24, 12, 43, 47, 7593, 15, 133, 109, 13, 198, 19, 33, 18, 23, 58, 65, 60, 93167, 68, 17, 1523, 39, 75
Step at which n is expelled in Kimberling's puzzle. Ref CRUX 17 (2) 44 91. [1,2; A6852]

M5182 1, 25, 169, 625, 1681, 3721, 7225, 12769, 21025, 32761, 48841, 70225, 97969, 133225, 177241, 231361, 297025, 375769, 469225, 579121, 707281, 855625, 1026169
Crystal ball numbers for D_4 lattice. Ref CoSl95. [0,2; A7204]

M5183 1, 25, 421, 6105, 83029, 1100902, 14516426, 192422979, 2579725656, 35098717902, 485534447114, 6835409506841, 97966603326993, 1429401763567226
Permutations of length n by subsequences. Ref MOC 22 390 68. [5,2; A1456, N2251]

M5184 1, 25, 445, 7140, 111769, 1767087, 28699460, 483004280, 8460980836, 154594537812, 2948470152264, 58696064973000, 1219007251826064
Generalized Stirling numbers. Ref PEF 77 26 62. [0,2; A1714, N2252]

M5185 1, 25, 450, 7350, 117600, 1905120, 31752000, 548856000, 9879408000, 185513328000, 3636061228800, 74373979680000, 1586644899840000
Coefficients of Laguerre polynomials. Ref LA56 519. AS1 799. [4,2; A1811, N2253]

M5186 1, 25, 490, 9450, 190575, 4099095, 94594500, 2343240900
Associated Stirling numbers. Ref AJM 56 92 34. DB1 296. [1,2; A0497, N2254]

M5187 1, 26, 165, 100, 570, 834, 1664, 856, 2151, 2460, 2316, 9588, 5270, 12160, 24930, 16912, 6498, 4086, 58732, 22440, 150144, 1464, 119112, 121416, 98825, 49604, 334179
Expansion of a modular form. Ref JNT 25 207 87. [1,2; A6354]

M5188 0, 1, 26, 302, 2416, 15619, 88234, 455192, 2203488, 10187685, 45533450, 198410786, 848090912, 3572085255, 14875399450, 61403313100, 251732291184
Eulerian numbers. See Fig M3416. Ref R1 215. DB1 151. JCT 1 351 66. DKB 260. C1 243. [3,3; A0498, N2255]

M5189 1, 26, 485, 8175, 134449, 2231012, 37972304, 668566300, 12230426076, 232959299496, 4623952866312, 95644160132976, 2060772784375824
Generalized Stirling numbers. Ref PEF 77 61 62. [0,2; A1723, N2256]

M5190 1, 26, 1768, 225096, 51725352, 21132802554, 15463799747936, 20604021770403328, 5092801940115851\5328, 2376444239489289941975\04
Labeled 3-connected graphs with n nodes. Ref JCT B32 8 82. [4,2; A5644]

M5191 1, 26, 1858, 236856, 53458832, 21494404400, 15580475076986, 20666605559464968, 5098732251586098023\6, 2377475649132323672026\56
Labeled irreducible 2-connected graphs with n edges. Ref JCT B32 8 82. [4,2; A5643]

M5192 27, 86, 243, 594, 1370, 2916, 5967, 11586, 21870, 39852, 71052, 123444, 210654, 352480, 581013, 942786, 1510254, 2388204, 3734964, 5777788, 8852004, 13434984
McKay-Thompson series of class 9A for Monster. Ref CALG 18 258 90. FMN94. [1,1; A7266]

M5193 1, 27, 184, 875, 2700, 7546, 17600, 35721, 72750, 126445, 223776, 353717, 595448, 843750, 1349120, 1827636, 2808837, 3600975, 5306000, 6667920, 9599172
Related to the divisor function. Ref SMA 19 39 53. [1,2; A0499, N2257]

M5194 0, 27, 378, 4536, 48600
Card matching. Ref R1 193. [1,2; A0535, N2258]

M5195 1, 27, 511, 8624, 140889, 2310945, 38759930, 671189310, 12061579816, 225525484184, 4392554369840, 89142436976320, 1884434077831824
Generalized Stirling numbers. Ref PEF 77 7 62. [0,2; A1709, N2259]

M5196 27, 10206, 1271126683458
Post functions. Ref JCT 4 295 68. [1,1; A1321, N2260]

M5197 1, 1, 1, 1, 1, 1, 28, 2, 8, 6, 992, 1, 3, 2, 16256, 2, 16, 16
Differential structures on n-sphere (assuming Poincaré conjecture). See Fig M2051. Ref ICM 50 62. ANN 77 504 63. [1,7; A1676, N2261]

M5198 1, 28, 92, 435, 1766, 7598, 31987, 135810, 574786, 2435653, 10316252, 43702500, 185123261, 784200368, 3321916912, 14071880655, 59609419066
Sum of cubes of Lucas numbers. Ref BR72 21. [1,2; A5971]

M5199 0, 1, 28, 153, 496, 1225, 2556, 4753, 8128, 13041, 19900, 29161, 41328, 56953, 76636, 101025, 130816, 166753, 209628, 260281, 319600, 388521, 468028, 559153
$n^2(2n^2-1)$. Ref CC55 742. JO61 7. [0,3; A2593, N2262]

M5200 28, 165, 1092, 7752, 57684, 444015, 3506100, 28242984, 231180144, 1917334783, 16077354108
Perforation patterns for punctured convolutional codes (4,1). Ref SFCA92 1 10. [2,1; A7228]

M5201 1, 28, 462, 5880, 63987, 627396, 5715424, 49329280, 408741333, 3281882604, 25708104786, 197462483400, 1492924634839, 11143554045652, 82310957214948
Stirling numbers of second kind. See Fig M4981. Ref AS1 835. DKB 223. [7,2; A0771, N2263]

M5202 1, 28, 546, 9450, 157773, 2637558, 44990231, 790943153, 14409322928, 272803210680, 5374523477960, 110228466184200, 2353125040549984
Stirling numbers of first kind. See Fig M4730. Ref AS1 834. DKB 226. [7,2; A1234, N2264]

M5203 28, 1468, 34680, 535452, 6302296, 61400920, 520788460, 3976323744, 27974148068, 184411212644, 1153389882896, 6908837566500, 39921952365008
Almost-convex polygons of perimeter $2n$ on square lattice. Ref EG92. [10,1; A7222]

M5204 1, 28, 2108, 227322, 30276740, 4541771016, 739092675672, 127674038970623, 23085759901610016, 4327973308197103600, 835531767841066680300
Serial isogons of order $8n$. Ref MMAG 64 324 91. [1,2; A7219]

M5205 29, 30, 32, 35, 39, 44, 50, 57, 58, 61, 63, 65, 72, 74, 76, 84, 87, 88, 89, 91, 92, 94, 95, 97, 99, 102, 107, 109, 113, 116, 118, 120, 122, 123, 125, 126, 127, 134, 138, 144, 145
$2n^2 + 29$ is composite. [1,1; A7642]

M5206 1, 29, 626, 13869, 347020
Characteristic polynomial of Pascal matrix. Ref FQ 15 204 77. [1,2; A6136]

M5207 30, 42, 66, 70, 78, 102, 105, 110, 114, 130, 138, 154, 165, 170, 174, 182, 186, 190, 195, 222, 230, 231, 238, 246, 255, 258, 266, 273, 282, 285, 286, 290, 310, 318, 322, 345
Product of 3 distinct primes. [1,1; A7304]

M5208 1, 30, 54, 96, 112, 114, 132, 156, 332, 342, 360, 376, 428, 430, 432, 448, 562, 588, 726, 738, 804, 850, 884, 1068, 1142, 1198, 1306, 1540, 1568, 1596, 1678, 1714, 1754
$n^{32} + 1$ is prime. [1,2; A6315]

M5209 30, 97, 267, 608, 1279, 2472, 4571, 8043, 13715, 22652, 36535, 57568, 89079, 135384, 202747, 299344, 436597, 629364, 897970, 1268634, 1776562, 2466961
Bipartite partitions. Ref ChGu56 2. [0,1; A2758, N2265]

M5210 1, 30, 330, 2145, 10010, 37128, 116280, 319770, 793155, 1808950, 3848130, 7719075, 14725620, 26898080, 47303520, 80454132, 132835365, 213578430
From paths in the plane. Ref EJC 2 58 81. [0,2; A6859]

$$G.f.: \quad (1+x)(1 + 19x + 56x^2 + 19x^3 + x^4) \; / \; (1-x)^{10}.$$

M5211 1, 30, 449, 4795, 41850, 319320, 2213665, 14283280, 87169790, 508887860, 2865204762
Rooted planar maps. Ref JCT B18 249 75. [2,2; A6421]

M5212 1, 30, 625, 11515, 203889, 3602088, 64720340, 1194928020, 22800117076, 450996059800, 9262414989464, 197632289814960, 4381123888865424
Generalized Stirling numbers. Ref PEF 77 44 62. [0,2; A1719, N2266]

M5213 1, 30, 630, 11760, 211680, 3810240, 69854400, 1317254400, 25686460800, 519437318400, 10908183686400, 237996734976000, 5394592659456000
Lah numbers: $n! C(n-1,4)/5!$. Ref R1 44. C1 156. [5,2; A1777, N2267]

M5214 1, 1, 30, 840, 1197504000, 60281712691200
Steiner triple systems on n elements. Ref C1 304. [1,3; A1201, N2268]

M5215 1, 30, 1023, 44473, 2475473, 173721912, 15088541896, 1593719752240, 201529405816816, 30092049283982400, 5242380158902146624
Central factorial numbers. Ref RCI 217. [0,2; A1821, N2269]

M5216 1, 30, 1260, 75600, 6237000, 681080400, 95351256000, 16672848192000, 3563821301040000, 914714133933600000, 277707211062240960000
Central differences of 0. Ref QJMA 47 110 16. FMR 1 112. DA63 2 283. [1,2; A2456, N2270]

M5217 1, 30, 30240, 1816214400, 10137091700736000, 7561714896123855667200000,
10251138855541810446097868390400000000
$C(n,2)!/n!$. Ref SCS 12 122 81. [3,2; A6473]

M5218 0, 30, 217800, 16294301520
Generalized tangent numbers of type 3^{2n+1}. Ref JCT A53 266 90. [0,2; A5801]

M5219 31, 37, 47, 61, 79, 101, 127, 157, 191, 229, 271, 317, 367, 421, 479, 541, 607, 677,
751, 829, 911, 997, 1087, 1181, 1279, 1381, 1487, 1597, 1951, 2207, 2341, 2621, 2767
Primes of form $2n^2 + 29$. [1,1; A7641]

M5220 31, 83, 293, 347, 671
Self-contained numbers. Ref UPNT E16. [1,1; A5184]

M5221 31, 223, 433, 439, 457, 727, 919, 1327, 1399, 1423, 1471, 1831, 1999, 2017, 2287,
2383, 2671, 2767, 2791, 2953, 3271, 3343, 3457, 3463, 3607, 3631, 3823, 3889
2 is a 6th power residue modulo p. Ref Krai24 1 59. [1,1; A1136, N2271]

M5222 1, 31, 301, 1701, 6951, 22827, 63987, 159027, 359502, 752752, 1479478,
2757118, 4910178, 8408778, 13916778, 22350954, 34952799, 53374629, 79781779
Stirling numbers of second kind. See Fig M4981. Ref AS1 835. DKB 223. [1,2; A1298,
N2272]

M5223 31, 304, 4230, 43880, 547338, 6924960, 94714620, 1375878816, 21273204330,
348919244768, 6056244249682, 110955673493568, 2140465858763844
Discordant permutations. Ref SMA 20 23 54. [7,1; A0500, N2273]

M5224 31, 307, 643, 5113, 21787, 39199, 360007, 4775569, 10318249, 65139031
Primes with large least nonresidues. Ref RS9 XVI. [1,1; A2225, N2274]

M5225 1, 31, 602, 10206, 166824, 2739240, 46070640, 801496080, 14495120640,
273158645760, 5368729766400, 110055327782400, 2351983118284800
Simplices in barycentric subdivision of n-simplex. Ref rkg. [3,2; A5462]

M5226 1, 31, 651, 11811, 200787, 3309747, 53743987, 866251507, 13910980083,
222984027123, 3571013994483, 57162391576563, 914807651274739
Gaussian binomial coefficient $[n,4]$ for $q = 2$. Ref GJ83 99. ARS A17 328 84. [4,2; A6097]

M5227 31, 696, 5823, 29380, 108933, 327840, 848380, 1958004, 4130895, 8107024,
14990889, 26372124, 44470165, 72305160, 113897310, 174496828, 260846703
Discordant permutations. Ref SMA 20 23 54. [3,1; A0565, N2275]

M5228 1, 31, 806, 20306, 508431, 12714681, 317886556, 7947261556, 198682027181,
4967053120931, 124176340230306, 3104408566792806, 77610214474995931
Gaussian binomial coefficient $[n,2]$ for $q = 5$. Ref GJ83 99. ARS A17 329 84. [2,2; A6111]

M5229 31, 3661, 1217776, 929081776, 1413470290176, 3878864920694016,
17810567950611972096
Differences of reciprocals of unity. Ref DKB 228. [1,1; A1237, N2276]

M5230 1, 32, 122, 272, 482, 752, 1082, 1472, 1922, 2432, 3002, 3632, 4322, 5072, 5882,
6752, 7682, 8672, 9722, 10832, 12002, 13232, 14522, 15872, 17282, 18752, 20282
Points on surface of dodecahedron: $30n^2 + 2$. Ref INOC 24 4550 85. [0,2; A5903]

M5231 1, 32, 243, 1024, 3125, 7776, 16807, 32768, 59049, 100000, 161051, 248832,
371293, 537824, 759375, 1048576, 1419857, 1889568, 2476099, 3200000, 4084101
5th powers. Ref BA9. [1,2; A0584, N2277]

M5232 0, 0, 0, 32, 348, 2836, 21225, 154741, 1123143, 8185403, 60088748, 444688325
Fixed 3-dimensional polyominoes with n cells. Ref CJN 18 367 75. [1,4; A6763]

M5233 32, 9784, 7571840
n-state Turing machines which halt. Ref AMM 81 736 74. [1,1; A4147]

M5234 33, 54, 284, 366, 834, 848, 918, 1240, 1504, 2910, 2913, 3304, 4148, 4187, 6110,
6902, 7169, 7912, 9359, 10250, 10540, 12565, 15085, 17272, 17814, 19004, 19688
$\sigma(n+2) = \sigma(n)$. Ref AS1 840. [1,1; A7373]

M5235 1, 1, 1, 1, 1, 1, 1, 1, 1, 1, 1, 33, 79, 79, 107, 107, 311, 487, 487, 665, 665, 857,
2293, 3523, 3523, 3523, 13909, 13909, 13909, 26713, 29351, 29351, 59801, 66287
Class numbers of quadratic fields. Ref MOC 24 441 70. [3,12; A1987, N2278]

M5236 33, 85, 93, 141, 201, 213, 217, 230, 242, 243, 301, 374, 393, 445, 603, 633, 663,
697, 902, 921, 1041, 1105, 1137, 1261, 1274, 1309, 1334, 1345, 1401, 1641, 1761, 1832
n, $n+1$, $n+2$ have same number of divisors. Ref AS1 840. UPNT B18. [1,1; A5238]

M5237 33, 128, 159, 267, 387, 713, 1152, 929, 994, 1240, 1770, 1943, 1950
Largest number not the sum of distinct nth order polygonal numbers. Ref Robe92 186.
[3,1; A7419]

M5238 1, 33, 153, 713, 2550, 7479, 20314, 51951, 122229, 276656, 601068, 1254105,
2541531
McKay-Thompson series of class 6a for Monster. Ref FMN94. [0,2; A7260]

M5239 1, 33, 155, 427, 909, 1661, 2743, 4215, 6137, 8569, 11571, 15203, 19525, 24597,
30479, 37231, 44913, 53585, 63307, 74139, 86141, 99373, 113895, 129767, 147049
Centered dodecahedral numbers. Ref INOC 24 4550 85. [0,2; A5904]

M5240 1, 33, 244, 1057, 3126, 8052, 16808, 33825, 59293, 103158, 161052, 257908,
371294, 554664, 762744, 1082401, 1419858, 1956669, 2476100, 3304182, 4101152
Sum of 5th powers of divisors of n. Ref AS1 827. [1,2; A1160, N2279]

M5241 1, 33, 276, 1300, 4425, 12201, 29008, 61776, 120825, 220825, 381876, 630708, 1002001, 1539825, 2299200, 3347776, 4767633, 6657201, 9133300, 12333300
Sums of 5th powers. Ref AS1 813. [1,2; A0539, N2280]

M5242 33, 524, 2322, 81912, 214181, 1182276, 3736614, 9972264, 24622002, 51265020, 106396576, 202547304, 357914103
Related to representation as sums of squares. Ref QJMA 38 305 07. [0,1; A2608, N2281]

M5243 33, 46728, 102266868085272, 1069559300034650646049671038948382825526728
Denominators of a continued fraction. Ref NBS B80 288 76. [0,1; A6274]

M5244 1, 1, 34, 7037, 6317926, 21073662977, 251973418941994, 10878710974408306717, 1727230695707098000548430
2-colored graphs. Ref CJM 31 66 79. [1,3; A5334]

M5245 35, 154, 424, 930, 1775, 3080, 4985, 7650, 11256
Putting balls into 6 boxes. Ref SIAR 12 296 70. [12,1; A5339]

M5246 35, 185, 217, 301, 481, 1105, 1111, 1261, 1333, 1729, 2465, 2701, 2821, 3421, 3565, 3589, 3913, 4123, 4495, 5713, 6533, 6601, 8029, 8365, 8911, 9331, 9881, 10585
Pseudoprimes to base 6. Ref UPNT A12. [1,1; A5937]

M5247 0, 0, 0, 0, 0, 35, 210, 1001, 3927, 13971
Unexplained difference between two partition g.f.s. Ref PCPS 63 1100 67. [1,6; A7329]

M5248 1, 35, 835, 17360, 342769, 6687009, 131590430, 2642422750, 54509190076, 1159615530788, 25497032420496, 580087776122400, 13662528306823824
Generalized Stirling numbers. Ref PEF 77 61 62. [0,2; A1724, N2282]

M5249 1, 35, 966, 24970, 631631, 15857205, 397027996, 9931080740, 248325446061, 6208571999575, 155218222621826, 3880490869237710, 97012589464171291
Central factorial numbers. Ref TH09 36. FMR 1 112. RCI 217. [0,2; A2453, N2283]

$$\text{G.f.:} \quad 1 \,/\, (1 - x)\,(1 - 9\,x)\,(1 - 25\,x).$$

M5250 1, 35, 1974, 172810, 21967231, 3841278805, 886165820604, 261042753755556, 95668443268795341, 42707926241367380631, 22821422608929422854674
Central factorial numbers. Ref RCI 217. [0,2; A1825, N2284]

M5251 1, 0, 0, 0, 35, 14700, 11832975, 15245900670, 29683109280825, 84114515340655800, 335974271076054435825, 18393165748441276904122750
Labeled disconnected trivalent graphs with $2n$ nodes. Ref rwr. [0,5; A7102]

M5252 36, 128, 386, 1024, 2488, 5632, 12031, 24576, 48308, 91904, 170110, 307200, 542872, 941056, 1602819, 2686976, 4439688, 7238272, 11657090, 18561024
McKay-Thompson series of class 8A for Monster. Ref CALG 18 258 90. FMN94. [1,1; A7265]

M5253 36, 330, 22060, 920737780
Switching networks. Ref JFI 276 318 63. [1,1; A0821, N2285]

M5254 36, 666, 384112, 735192450952
Switching networks. Ref JFI 276 318 63. [1,1; A0815, N2286]

M5255 36, 820, 7645, 44473, 191620, 669188, 1999370, 5293970, 12728936, 28285400,
58856655, 115842675, 217378200, 391367064, 679524340, 1142659012
Central factorial numbers. Ref RCI 217. [4,1; A0597, N2287]

M5256 1, 36, 841, 16465, 296326, 5122877, 87116283, 1477363967, 25191909848,
434119587475, 7583461369373, 134533482045389, 2426299018270338
Permutations of length n by subsequences. Ref MOC 22 390 68. [6,2; A1457, N2288]

M5257 1, 36, 882, 18816, 381024, 7620480, 153679680, 3161410560, 66784798080,
1454424491520, 32724551059200, 761589551923200, 18341615042150400
Coefficients of Laguerre polynomials. Ref LA56 519. AS1 799. [5,2; A1812, N2289]

M5258 36, 1072, 2100736, 17592201773056
Switching networks. Ref JFI 276 317 63. [1,1; A0809, N2290]

M5259 1, 36, 1225, 41616, 1413721, 48024900, 1631432881, 55420693056,
1882672131025, 63955431761796, 2172602007770041, 73804512832419600
Both triangular and square. Ref D1 2 10. MAG 47 237 63. B1 193. FQ 9 95 71. [1,2;
A1110, N2291]
$$\text{G.f.:} \quad (1 + x) \, / \, (1 - x) \, (1 - 34 \, x + x^2).$$

M5260 37, 59, 67, 101, 103, 131, 149, 157, 233, 257, 263, 271, 283, 293, 307, 311, 347,
353, 379, 389, 401, 409, 421, 433, 461, 463, 467, 491, 523, 541, 547, 557, 577, 587, 593
Irregular primes. Ref PNAS 40 31 54. BS66 430. [1,1; A0928, N2292]

M5261 37, 103, 113, 151, 157, 163, 173, 181, 193, 227, 233, 257, 277, 311, 331, 337, 347,
353, 379, 389, 397, 401, 409, 421, 457, 463, 467, 487, 491, 521, 523, 541, 547, 569, 571
Class 3+ primes. Ref UPNT A18. [1,1; A5107]

M5262 1, 1, 1, 1, 1, 1, 1, 1, 37, 111, 177, 177, 2753, 2753, 827, 827, 8386459, 8386459,
28033727, 28033727, 14529522883, 14529522883, 1799010587, 1799010587
Denominators of Van der Pol numbers. Cf. M1534. Ref JRAM 260 35 73. [0,9; A3164]

M5263 1, 37, 530, 5245, 42406
Related to enumeration of rooted maps. Ref JCT A13 124 72. [2,2; A6303]

M5264 1, 37, 1072, 32675, 1024028, 32463802, 1033917350, 32989068162,
1053349394128, 33643541208290
Hamiltonian circuits on $2n \times 6$ rectangle. Ref JPA 17 445 84. [1,2; A5390]

M5265 1, 37, 1261, 42841, 1455337, 49438621, 1679457781, 57052125937,
1938092824081, 65838103892821, 2236557439531837, 75977114840189641
Star-hex numbers. See Fig M2535. Ref GA88 22. JRM 16 192 83. [1,2; A6062]

M5266 1, 38, 201, 586, 1289, 2406, 4033, 6266, 9201, 12934, 17561, 23178, 29881,
37766, 46929, 57466, 69473, 83046, 98281, 115274, 134121, 154918, 177761, 202746
Truncated octahedral numbers. Ref Coxe74. INOC 24 4552 85. [0,2; A5910]

M5267 38, 264, 2016, 15504, 122661, 986700, 8064576, 66756144, 558689224,
4719593312, 40193414112
Perforation patterns for punctured convolutional codes (4,1). Ref SFCA92 1 10. [2,1;
A7229]

M5268 1, 38, 469, 3008, 12843, 42602, 119042, 293578, 658021, 1367170, 2670203
Rooted planar maps. Ref JCT B18 248 75. [2,2; A6418]

M5269 40, 41, 44, 49, 56, 65, 76, 81, 82, 84, 87, 89, 91, 96, 102, 104, 109, 117, 121, 122,
123, 126, 127, 130, 136, 138, 140, 143, 147, 155, 159, 161, 162, 163, 164, 170, 172, 173
$n^2 + n + 41$ is composite. Cf. M0473. [1,1; A7634]

M5270 1, 40, 90, 240, 200, 560, 400, 800, 730, 1240, 752, 1840, 1200, 2000, 1600, 2720,
1480, 3680, 2250, 3280, 2800, 4320, 2800, 5920, 2960, 5240, 3760, 6720, 4000, 7920
Theta series of D_5 lattice. Ref SPLAG 118. [0,2; A5930]

M5271 1, 40, 793, 12800, 193721
Eulerian circuits on checkerboard. Ref JCT B24 211 78. [1,2; A6240]

M5272 1, 40, 1210, 33880, 925771, 25095280, 678468820, 18326727760, 494894285941,
13362799477720, 360801469802830, 9741692640081640, 263026177881648511
Gaussian binomial coefficient $[n,3]$ for $q = 3$. Ref GJ83 99. ARS A17 328 84. [3,2; A6101]

M5273 41, 43, 47, 53, 61, 71, 83, 97, 113, 131, 151, 173, 197, 223, 251, 281, 313, 347,
383, 421, 461, 503, 547, 593, 641, 691, 743, 797, 853, 911, 971, 1033, 1097, 1163, 1231
Primes of form $n^2 + n + 41$. Ref BPNR 137. [0,1; A5846]

M5274 41, 43, 53, 59, 67, 71, 83, 89, 97, 103, 107, 109, 113, 131, 137, 149, 151, 157, 163,
167, 173, 181, 191, 193, 197, 199, 223, 233, 251, 257, 263, 271, 277, 281, 283, 293, 307
Primes not of form $\mid 3^x - 2^y \mid$ Ref MMAG 65 265 92. rgw. [1,1; A7643]

M5275 41, 109, 151, 229, 251, 271, 367, 733, 761, 971, 991, 1069, 1289, 1303, 1429,
1471, 1759, 1789, 1811, 1879, 2411, 2441, 2551, 2749, 2791, 3061, 3079, 3109, 3229
Primes with 6 as smallest primitive root. Ref Krai24 1 57. AS1 864. [1,1; A1125, N2293]

M5276 41, 313, 353, 1201, 3593, 4481, 7321, 8521, 10601, 14281, 14321, 14593, 21601,
26513, 32633, 41761, 41801, 42073, 42961, 49081, 56041, 66361, 67073, 72481, 90473
Half-quartan primes: $p = (x^4 + y^4)/2$. Ref CU23 1 254. [1,1; A2646, N2294]

M5277 42, 132, 297, 572, 1001, 1638, 2548, 3808, 5508, 7752, 10659, 14364, 19019,
24794, 31878, 40480, 50830, 63180, 77805, 95004, 115101, 138446, 165416, 196416
Walks on square lattice. Ref GU90. [0,1; A5557]

M5278 42, 139, 392, 907, 1941, 3804, 7128, 12693, 21893, 36535, 59521, 94664, 147794, 226524, 342006, 508866, 747753, 1085635, 1559725, 2218272, 3126541
Bipartite partitions. Ref ChGu56 2. [0,1; A2759, N2295]

M5279 1, 42, 1176, 28224, 635040, 13970880, 307359360, 6849722880, 155831195520, 3636061228800, 87265469491200, 2157837063782400, 55024845126451200
Lah numbers: $n! C(n-1,5)/6!$. Ref R1 44. C1 156. [6,2; A1778, N2297]

M5280 42, 1586, 31388, 442610, 5030004, 49145460, 429166584, 3435601554, 25658464260, 181055975100
Rooted planar maps with n edges. Ref BAMS 74 74 68. WA71. JCT A13 215 72. [5,1; A0502, N2298]

M5281 1, 42, 6006, 1662804, 701149020, 396499770810, 278607172289160, 231471904322784840
5-dimensional Catalan numbers. Ref CN 75 124 90. [1,2; A5791]

M5282 1, 1, 42, 9529, 6421892, 9652612995
An occupancy problem. Ref JACM 24 593 77. [0,3; A6699]

M5283 43, 109, 157, 229, 277, 283, 307, 499, 643, 691, 733, 739, 811, 997, 1021, 1051, 1069, 1093, 1459, 1579, 1597, 1627, 1699, 1723, 1789, 1933, 2179, 2203, 2251, 2341
2 is cubic residue modulo p. Ref Krai24 1 59. [1,1; A1133, N2299]

M5284 1, 44, 432, 1136, 610, 5568, 6048, 11456, 3423, 26840, 79920, 768, 5470, 77952, 263520, 61696, 73090, 150612, 84240, 692960, 139776, 1030080, 1025568
Related to representation as sums of squares. Ref QJMA 38 329 07. [1,2; A2613, N2300]

M5285 1, 0, 0, 0, 1, 44, 7570, 1975560, 749649145, 399035751464, 289021136349036, 277435664056527360, 345023964977303838105, 5450992365510258600229460
Labeled Eulerian 3-regular digraphs with n nodes. Ref rwr. [0,6; A7105]

M5286 1, 44, 4940800, 564083990621761115783168
Discriminants of Shapiro polynomials. Ref PAMS 25 115 70. [1,2; A1782, N2301]

M5287 0, 1, 45, 15913, 1073579193, 4611686005542975085, 85070591730234615801280047645054636261
Proper covers of an n-set. Ref MMAG 67 143 94. [1,3; A7537]

M5288 46, 92, 341, 1787, 9233, 45752, 285053, 1846955
A queens problem. Ref GA91 240. [4,1; A7630]

M5289 47, 139, 167, 179, 269, 277, 347, 461, 467, 499, 599, 643, 691, 709, 797, 827, 829, 839, 857, 863, 967, 997, 1013, 1019, 1039, 1063, 1069, 1151, 1163, 1181, 1289, 1367
Class 4− primes. Ref UPNT A18. [1,1; A5112]

M5290 1, 47, 2488, 138799, 7976456, 467232200, 27736348480, 1662803271215,
100442427373480, 6103747246289272, 372725876150863808, 22852464771010647496
A sequence for π. Ref MOC 42 211 84. [1,2; A5148]

M5291 48, 75, 140, 195, 1050, 1575, 1648, 1925, 2024, 2295, 5775, 6128, 8892, 9504,
16587, 20735, 62744, 75495, 186615, 196664, 199760, 206504, 219975, 266000, 309135
Betrothed (or quasi-amicable) numbers. Ref MOC 31 608 77. UPNT B5. [1,1; A5276]

M5292 1, 48, 186, 416, 738, 1152, 1658, 2256, 2946, 3728, 4602, 5568, 6626, 7776, 9018,
10352, 11778, 13296, 14906, 16608, 18402, 20288, 22266, 24336, 26498, 28752, 31098
Points on surface of truncated cube: $46n^2 + 2$. Ref INOC 24 4552 85. [0,2; A5911]

M5293 48, 264, 1680, 11640, 86352, 673104, 5424768, 44828400, 377810928,
3235366752, 28074857616, 246353214240
n-step polygons on f.c.c. lattice. Ref JCP 46 3481 67. [3,1; A1337, N2302]

M5294 48, 4096, 97152, 1130496, 8400704, 45785088
Theta series of Leech lattice w.r.t. deep hole of type A_1. Ref SPLAG 511. [0,1; A4034]

M5295 0, 0, 48, 48384, 58018928640
Hamiltonian paths on n-cube. Ref GA86 24. [1,3; A6070]

M5296 49, 81, 148, 169, 229, 257, 316, 321, 361, 404, 469, 473, 564, 568, 621, 697, 733,
756, 761, 785, 788, 837, 892, 940, 961, 985, 993, 1016, 1076, 1101, 1129, 1229, 1257
Discriminants of totally real cubic fields. Ref MOC 48 149 87. [1,1; A6832]

M5297 1, 49, 1513, 38281, 874886, 18943343, 399080475, 8312317976, 172912977525,
3615907795025, 76340522760097, 1631788075873114, 35378058306185002
Permutations of length n by subsequences. Ref MOC 22 390 68. [7,2; A1458, N2304]

M5298 49, 6877, 1854545, 807478656, 514798204147, 451182323794896,
519961864703259753, 762210147961330421167, 1384945048774500147047194
Connected n-state finite automata with 3 inputs. Ref GTA85 683. [1,1; A6692]

M5299 50, 65, 85, 125, 130, 145, 170, 185, 200, 205, 221, 250, 260, 265, 290, 305, 325,
338, 340, 365, 370, 377, 410, 425, 442, 445, 450, 481, 485, 493, 500, 505, 520, 530, 533
Sum of 2 nonzero squares in more than 1 way. Ref Well86 125. [1,1; A7692]

M5300 1, 50, 887, 8790, 59542, 307960, 1301610, 4701698, 14975675, 43025762,
113414717
Distributive lattices. Ref MSH 53 19 76. MSG 121 121 76. [0,2; A6360]

M5301 1, 50, 1660, 46760, 1217776, 30480800, 747497920, 18139003520, 437786795776
Differences of reciprocals of unity. Ref DKB 228. [1,2; A1241, N2305]

M5302   51, 204, 681, 1956, 5135, 12360, 28119, 60572, 125682, 251040, 487426, 920568,
1699611, 3070508, 5445510, 9490116, 16283793, 27537708, 45959775, 75760640
McKay-Thompson series of class 7A for Monster. Ref CALG 18 258 90. FMN94. [1,1;
A7264]

M5303   1, 52, 358, 304, 3435, 7556, 14532
n-step mappings with 5 inputs. Ref PRV A32 2342 85. [1,2; A5946]

M5304   52, 472, 3224, 18888, 101340, 511120, 2465904
Series-parallel numbers. Ref R1 142. [4,1; A0527, N2306]

M5305   1, 52, 834, 4760, 24703, 94980, 343998, 1077496, 3222915, 8844712, 23381058,
58359168, 141244796, 327974700, 742169724, 1627202744, 3490345477
McKay-Thompson series of class 4B for Monster. Ref CALG 18 257 90. FMN94. [0,2;
A7247]

M5306   1, 52, 5133, 655554, 97772875, 16019720210, 2812609211657, 518332479161091
Witt vector *4!/4!. Ref SLC 16 107 88. [1,2; A6179]

M5307   53, 71, 103, 107, 109, 149, 151, 157, 163, 167, 173, 181, 191, 193, 197, 199, 223,
233, 263, 271, 277, 281, 293, 311, 313, 317, 331, 347, 349, 353, 359, 367, 373, 379, 383
Primes not of form $\mid 3^x \pm 2^y \mid$ Ref MMAG 65 265 92. rgw. [1,1; A7644]

M5308   1, 1, 53, 195, 22999, 29944523, 109535241009, 29404527905795295658,
4553770304201134322102016914702, 2637081256939771900193199294564557877 9849
Continued fraction for gamma function. Cf. M4831. Ref MOC 34 548 80. AS1 258. [0,3;
A5146]

M5309   1, 0, 54, 72, 0, 432, 270, 0, 918, 720, 0, 2160, 936, 0, 2700, 2160, 0, 5184, 2214, 0,
5616, 3600, 0, 9504, 4590, 0, 9180, 6552, 0, 15120, 5184, 0, 14742, 10800, 0, 21600
Theta series of E_6* lattice. Ref SPLAG 127. [0,3; A5129]

M5310   54, 76, 243, 1188, 1384, 2916, 11934, 11580, 21870, 79704, 71022, 123444,
421308, 352544, 581013, 1885572, 1510236, 2388204, 7469928, 5777672, 8852004
McKay-Thompson series of class 3B for Monster. Ref CALG 18 256 90. FMN94. [1,1;
A7244]

M5311   1, 0, 0, 56, 14, 0, 0, 576, 84, 0, 0, 1512, 280, 0, 0, 4032, 574, 0, 0, 5544, 840, 0, 0,
12096, 1288, 0, 0, 13664, 2368, 0, 0, 24192, 3444, 0, 0, 27216, 3542, 0, 0, 44352, 4424, 0
Theta series of E_7* lattice. Ref SPLAG 125. [0,4; A5932]

M5312   1, 56, 311, 920, 2037, 3816, 6411, 9976, 14665, 20632, 28031, 37016, 47741,
60360, 75027, 91896, 111121, 132856, 157255, 184472, 214661, 247976, 284571
Truncated cube numbers. Ref INOC 24 4552 85. [0,2; A5912]

M5313 56, 576, 1512, 4032, 5544, 12096, 13664, 24192, 27216, 44352, 41832, 72576, 67536, 100800, 101304, 145728, 126504, 205632, 176456, 249984, 234360, 326592
Theta series of coset of E_7 lattice. Ref SPLAG 125. [0,1; A5931]

M5314 56, 1120, 18592, 300288, 4877824, 80349696, 1344154112
Almost trivalent maps. Ref PLC 1 292 70. [0,1; A2009, N2308]

M5315 1, 56, 1918, 56980, 1636635, 47507460, 1422280860
Associated Stirling numbers. Ref AJM 56 92 34. DB1 296. [1,2; A0504, N2309]

M5316 56, 7965, 2128064, 914929500, 576689214816, 500750172337212, 572879126392178688, 835007874759393878655, 1510492370204314777345000
n-state finite automata with 3 inputs. Ref GTA85 676. [1,1; A6690]

M5317 0, 1, 57, 1191, 15619, 156190, 1310354, 9738114, 66318474, 423281535, 2571742175, 15041229521, 85383238549, 473353301060, 2575022097600
Eulerian numbers. See Fig M3416. Ref R1 215. DB1 151. JCT 1 351 66. DKB 260. C1 243. [4,3; A0505, N2310]

M5318 60, 168, 360, 504, 660, 1092, 2448, 2520, 3420, 4080, 5616, 6048, 6072, 7800, 7920, 9828, 12180, 14880, 20160, 25308, 25920, 29120, 32736, 34440, 39732, 51888
Orders of non-cyclic simple groups. Ref DI58 309. LE70 137. ATLAS. [1,1; A1034, N2311]

M5319 60, 720, 6090, 47040, 363384, 2913120
Labeled trees of diameter 4 with n nodes. Ref IBMJ 4 478 60. [5,1; A0555, N2312]

M5320 1, 60, 10080, 3326400, 1816214400, 1482030950400, 1689515283456000, 2564684200286208000, 5001134190558105600000, 12182762888199545241600000
$(3n)!/(3!n!)$. [1,2; A1525]

M5321 61, 163, 487, 691, 1297, 1861, 4201, 4441, 4483, 5209, 5227, 9049, 9631, 12391, 14437, 16141, 16987, 61483, 63211, 65707, 65899, 67057, 69481, 92767, 94273, 96979
Primes whose reversal is a square. Ref JRM 17 173 85. [1,1; A7488]

M5322 61, 841, 7311, 51663, 325446, 1910706, 10715506, 58258210, 309958755, 1623847695
Permutations of length n by number of runs. Ref DKB 260. [6,1; A0506, N2313]

M5323 61, 1385, 19028, 206276, 1949762, 16889786, 137963364, 1081702420, 8236142455, 61386982075
Permutations of length n by number of runs. Ref DKB 260. [6,1; A0507, N2314]

M5324 61, 2763, 38528, 249856, 1066590, 3487246, 9493504, 22634496, 48649086, 96448478
Generalized class numbers. Ref MOC 21 689 67; 22 698 68. [1,1; A0508, N2315]

M5325 62, 63, 65, 75, 84, 95, 161, 173, 195, 216, 261, 266, 272, 276, 326, 371, 372, 377, 381, 383, 386, 387, 395, 411, 416, 422, 426, 431, 432, 438, 441, 443, 461, 466, 471, 476
No number is this multiple of the sum of its digits. Ref jhc. [1,1; A3635]

M5326 1, 63, 1932, 46620, 1020600, 21538440, 451725120, 9574044480, 207048441600, 4595022432000, 105006251750400, 2475732702643200
Simplices in barycentric subdivision of n-simplex. Ref rkg. [4,2; A5463]

M5327 1, 63, 2667, 97155, 3309747, 109221651, 3548836819, 114429029715, 3675639930963, 117843461817939, 3774561792168531, 120843139740969555
Gaussian binomial coefficient $[n,5]$ for $q = 2$. Ref GJ83 99. ARS A17 328 84. [5,2; A6110]

M5328 63, 22631, 30480800, 117550462624, 1083688832185344, 21006340945438768128
Differences of reciprocals of unity. Ref DKB 228. [1,1; A1238, N2316]

M5329 64, 625, 4016, 21256, 100407, 439646, 1823298
Partially labeled rooted trees with n nodes. Ref R1 134. [4,1; A0525, N2317]

M5330 1, 64, 729, 4096, 15625, 46656, 117649, 262144, 531441, 1000000, 1771561, 2985984, 4826809, 7529536, 11390625, 16777216, 24137569, 34012224, 47045881
6th powers. Ref BA9. [1,2; A1014, N2318]

M5331 64, 960, 135040, 14333211520
Switching networks. Ref JFI 276 318 63. [1,1; A0818, N2319]

M5332 64, 1024, 12480, 137472, 1443616
n-step walks on f.c.c. lattice. Ref PCPS 58 100 62. [1,1; A0768, N2320]

M5333 64, 1744, 48784, 1365904, 38245264, 1070867344, 29984285584, 839559996304, 23507679896464, 658215037100944, 18430021038826384, 516040589087138704
Functions realized by cascades of n gates. Ref BU77. [1,1; A5609]

$$\text{G.f.:} \quad (64 - 112\,x) \, / \, (1 - x)\,(1 - 28\,x).$$

M5334 64, 2304, 2928640, 11745443774464
Switching networks. Ref JFI 276 317 63. [1,1; A0812, N2321]

M5335 1, 65, 794, 4890, 20515, 67171, 184820, 446964, 978405, 1978405, 3749966, 6735950, 11562759, 19092295, 30482920, 47260136, 71397705, 105409929, 152455810
Sums of 6th powers. Ref AS1 813. [1,2; A0540, N2322]

M5336 1, 65, 1795, 36317, 636331
Connected relations. Ref CRP 268 579 69. [1,2; A2502, N2323]

M5337 70, 102, 114, 138, 174, 186, 222, 246, 258, 282, 318, 350, 354, 366, 372, 402, 426, 438, 444, 474, 490, 492, 498, 516, 534, 550, 564, 572, 582, 606, 618, 636, 642, 650, 654
Impractical numbers: even abundant numbers that are not practical(2). Ref rgw. [1,1; A7621]

M5338 0, 0, 0, 0, 0, 70, 490, 2632, 11606, 46375
Unexplained difference between two partition g.f.s. Ref PCPS 63 1100 67. [1,6; A7330]

M5339 70, 836, 4030, 5830, 7192, 7912, 9272, 10430, 10570, 10792, 10990, 11410, 11690, 12110, 12530, 12670, 13370, 13510, 13790, 13930, 14770, 15610, 15890, 16030
Weird numbers. Ref AMM 79 774 72. MOC 28 618 74. UPNT B2. [1,1; A6037]

M5340 70, 836, 4030, 5830, 7192, 7912, 9272, 10792, 17272, 45356, 73616, 83312, 91388, 113072, 243892, 254012, 338572, 343876, 388076, 519712, 539744, 555616
Primitive weird numbers. Ref AMM 79 774 72. MOC 28 618 74. HO73 115. UPNT B2. [1,1; A2975]

M5341 70, 1720, 24164, 256116, 2278660, 17970784, 129726760, 875029804, 5593305476, 34225196720, 201976335288
Rooted genus-1 maps with n edges. Ref WA71. JCT A13 215 72. [4,1; A6296]

M5342 70, 2330, 32130, 271285, 1655800, 7997850, 32332170, 113568455, 355905030
Tree-rooted toroidal maps. Ref JCT B18 258 75. [1,1; A6437]

M5343 1, 70, 16800, 9238320, 9520156800, 16305064776000, 42856575521760000, 163329351308323200000, 864876880105205071104000
Labeled trivalent 3-connected graphs with $2n$ nodes. Ref rwr. [2,2; A7100]

M5344 1, 70, 19320, 11052720, 11408720400, 19285018552800, 49792044478176000, 186348919238786304000, 970566620767088881536000
Labeled trivalent 2-connected graphs with $2n$ nodes. Ref rwr. [2,2; A7099]

M5345 0, 1, 70, 19320, 11166120, 11543439600, 19491385914000, 50233275604512000, 187663723374359232000, 975937986889287117696000
Connected trivalent labeled graphs with $2n$ nodes. Ref RE58. [1,3; A4109]

M5346 0, 1, 70, 19355, 11180820, 11555272575, 19506631814670, 50262958713792825, 187747837889699887800, 976273961160363172131825
Trivalent labeled graphs with $2n$ nodes. Ref RE58. SIAA 4 192 83. [1,3; A2829, N2324]

M5347 1, 70, 26599, 33757360, 107709888805, 726401013530416, 9197888739246870571, 200656681438694771057920
4 up, 4 down, 4 up, ... permutations of length $2n + 1$. Ref prs. [1,2; A5983]

M5348 71, 239, 241, 359, 431, 499, 599, 601, 919, 997, 1051, 1181, 1249, 1439, 1609, 1753, 2039, 2089, 2111, 2179, 2251, 2281, 2341, 2591, 2593, 2671, 2711, 2879, 3119
Primes with 7 as smallest primitive root. Ref Krai24 1 58. AS1 864. [1,1; A1126, N2325]

M5349 1, 72, 270, 720, 936, 2160, 2214, 3600, 4590, 6552, 5184, 10800, 9360, 12240, 13500, 17712, 14760, 25920, 19710, 26064, 28080, 36000, 25920, 47520, 37638, 43272
Theta series of E_6 lattice. Ref SPLAG 123. [0,2; A4007]

M5350 73, 313, 443, 617, 661, 673, 677, 691, 739, 757, 823, 887, 907, 941, 977, 1093, 1109, 1129, 1201, 1213, 1303, 1361, 1447, 1453, 1543, 1553, 1621, 1627, 1657, 1753
Class 4+ primes. Ref UPNT A18. [1,1; A5108]

M5351 0, 0, 0, 0, 0, 0, 0, 0, 0, 0, 0, 74, 1600, 43984, 1032208
Non-Hamiltonian polyhedra with n nodes. Ref Dil92. [1,11; A7033]

M5352 76, 288, 700, 1376, 2380, 3776, 5628, 8000, 10956, 14560, 18876, 23968, 29900, 36736, 44540, 53376, 63308, 74400, 86716
Walks on cubic lattice. Ref GU90. [0,1; A5571]

M5353 1, 76, 702, 5224, 23425, 98172, 336450, 1094152, 3188349, 8913752, 23247294, 58610304, 140786308
McKay-Thompson series of class 4a for Monster. Ref FMN94. [0,2; A7250]

M5354 78, 364, 1365, 4380, 12520, 32772, 80094, 185276, 409578, 871272, 1792754, 3582708, 6977100, 13277472, 24747867, 45267324, 81389908, 144048396
McKay-Thompson series of class 6B for Monster. Ref CALG 18 257 90. FMN94. [1,1; A7255]

M5355 79, 352, 1431, 4160, 13015, 31968, 81162, 183680, 412857, 864320, 1805030, 3564864, 7000753, 13243392, 24805035, 45168896, 81544240, 143832672
McKay-Thompson series of class 6A for Monster. Ref CALG 18 257 90. FMN94. [1,1; A7254]

M5356 0, 0, 0, 1, 80, 7040, 878080, 169967616, 53247344640
5-colored labeled graphs on n nodes. Ref CJM 12 412 60. rcr. [1,5; A6202]

M5357 81, 4375, 69295261, 37197234745112286741113, 433029035500163958643008531671880178799
Discriminants of period polynomials. Ref LNM 899 296 81. [3,1; A6312]

M5358 1, 82, 338, 2739, 17380, 122356, 829637, 5709318, 39071494, 267958135, 1836197336, 12586569192, 86266785673, 591288786874, 4052734152890
Sum of fourth powers of Lucas numbers. Ref BR72 21. [1,2; A5972]

M5359 1, 82, 707, 3108, 9669, 24310, 52871, 103496, 187017, 317338, 511819, 791660, 1182285, 1713726, 2421007, 3344528, 4530449, 6031074, 7905235, 10218676
Sums of fourth powers of odd numbers. Ref AMS 2 358 31 (divided by 2). CC55 742. [1,2; A2309, N2327]

M5360 1, 85, 5797, 376805, 24208613, 1550842085, 99277752549, 6354157930725, 406672215935205, 26027119554103525, 1665737215212030181
Gaussian binomial coefficient $[n,3]$ for $q = 4$. Ref GJ83 99. ARS A17 328 84. [3,2; A6106]

M5361 88, 326, 1631, 10112, 74046, 622704, 5900520, 62118720, 718709040,
9059339520, 123521086080, 1810829260800, 28397649772800, 474281518233600
$\Sigma(n+k)!\,C(5,k)$, $k = 0 \,.\,.\, 5$. Ref CJM 22 26 70. [-1,1; A1347, N2328]

M5362 91, 121, 286, 671, 703, 949, 1105, 1541, 1729, 1891, 2465, 2665, 2701, 2821,
3281, 3367, 3751, 4961, 5551, 6601, 7381, 8401, 8911, 10585, 11011, 12403, 14383
Pseudoprimes to base 3. Ref UPNT A12. [1,1; A5935]

M5363 1, 91, 8911, 873181, 85562821, 8384283271, 821574197731, 80505887094361,
7888755361049641, 773017519495770451, 75747828155224454551
Triangular hex numbers. Ref GA88 19. jos. [0,2; A6244]

M5364 96, 1776, 43776, 1237920, 37903776, 1223681760, 41040797376
$2n$-step polygons on b.c.c. lattice. Ref JPA 5 665 72. [2,1; A1667, N2330]

M5365 97, 139, 151, 199, 211, 331, 433
Related to Kummer's conjecture. Ref Hass64 482. [1,1; A0923, N2331]

M5366 1, 1, 97, 243, 12167, 577, 221874931
Coefficients of period polynomials. Ref LNM 899 292 81. [3,3; A6310]

M5367 101, 10000000000001, 100000000000000000000000102
Junction numbers. Ref GA88 116. [1,1; A6064]

M5368 1, 102, 162, 274, 300, 412, 562, 592, 728, 1084, 1094, 1108, 1120, 1200, 1558,
1566, 1630, 1804, 1876, 2094, 2162, 2164, 2238, 2336, 2388, 2420, 2494, 2524, 2614
$n^{64}+1$ is prime. Ref rgw. [1,2; A6316]

M5369 1, 104, 4372, 96256, 1240002, 10698752, 74428120, 431529984, 2206741887,
10117578752, 42616961892, 166564106240, 611800208702, 2125795885056
McKay-Thompson series of class 2A for Monster. Cf. M5176. Ref FMN94. [-1,2; A7267]

M5370 1, 105, 3490, 59542, 650644, 5157098, 32046856, 164489084, 723509159,
2801747767, 9748942554
Distributive lattices. Ref MSH 53 19 76. MSG 121 121 76. [0,2; A6361]

M5371 113, 281, 353, 577, 593, 617, 1033, 1049, 1097, 1153, 1193, 1201, 1481, 1601,
1889, 2129, 2273, 2393, 2473, 3049, 3089, 3137, 3217, 3313, 3529, 3673, 3833, 4001
2 is a quartic residue modulo p. Ref Krai24 1 59. [1,1; A1134, N2332]

M5372 114, 1140, 18018, 32130, 44772, 56430, 67158, 142310, 180180, 197340, 241110,
296010, 308220, 462330, 591030, 669900, 671580, 785148, 815100
Smaller of unitary amicable pair. Cf. M5389. Ref MOC 25 917 71. [1,1; A2952]

M5373 1, 115, 5390, 101275, 858650, 3309025, 4718075
5×5 stochastic matrices of integers. Ref SS70. ANS 4 1179 76. [0,2; A5466]

M5374 120, 210, 1540, 3003, 7140, 11628, 24310
Coincidences among binomial coefficients. Ref AMM 78 1119 71. [1,1; A3015]

M5375 120, 265, 579, 1265, 2783, 6208, 13909, 31337, 70985, 161545, 369024, 845825,
1944295, 4480285, 10345391, 23930320, 55435605, 128577253, 298529333
Permanent of a certain cyclic $n \times n$ (0,1) matrix. Ref CMB 7 262 64. JCT 7 315 69. [5,1;
A0804, N2333]

M5376 120, 672, 523776, 459818240, 1476304896, 51001180160
Triply perfect numbers. Ref BR73 138. UPNT B2. [1,1; A5820]

M5377 120, 1800, 16800, 126000, 834120, 5103000, 29607600, 165528000, 901020120,
4809004200, 25292030400, 131542866000, 678330198120, 3474971465400
Differences of 0. Ref VO11 31. DA63 2 212. R1 33. [5,1; A1118, N2334]

M5378 120, 2520, 43680, 757680, 13747104, 264181680
Labeled trees of height 4 with n nodes. Ref IBMJ 4 478 60. [5,1; A0553, N2335]

M5379 0, 1, 120, 4293, 88234, 1310354, 15724248, 162512286, 1505621508,
12843262863, 102776998928, 782115518299, 5717291972382, 40457344748072
Eulerian numbers. See Fig M3416. Ref R1 215. DB1 151. JCT 1 351 66. DKB 260. C1
243. [5,3; A0514, N2336]

M5380 120, 4320, 105840, 2257920, 45722880, 914457600, 18441561600, 379369267200
Coefficients of Laguerre polynomials. Ref AS1 799. [5,1; A1807, N2337]

M5381 1, 120, 6210, 153040, 2224955, 22069251, 164176640, 976395820, 4855258305,
20856798285, 79315936751
5×5 stochastic matrices of integers. Ref SIAC 4 477 75. [1,2; A3438]

M5382 1, 120, 7308, 303660, 11098780, 389449060, 13642629000, 486591585480,
17856935296200, 678103775949600, 26726282654771700
Associated Stirling numbers. Ref TOH 37 259 33. JO39 152. C1 256. [1,2; A1785, N2338]

M5383 120, 49500, 55480000, 75108093750, 124667171985024
Witt vector *5!. Ref SLC 16 106 88. [1,1; A6176]

M5384 1, 121, 11011, 925771, 75913222, 6174066262, 500777836042, 40581331447162,
3287582741506063, 266307564861468823, 21571273555248777493
Gaussian binomial coefficient $[n,4]$ for $q = 3$. Ref GJ83 99. ARS A17 328 84. [4,2; A6102]

M5385 1, 121, 11881, 1164241, 114083761, 11179044361, 1095432263641,
107341182792481, 10518340481399521, 1030690025994360601
Square star numbers. Ref GA88 22. [1,2; A6061]

M5386 1, 121, 12321, 1234321, 123454321, 12345654321, 1234567654321,
123456787654321, 12345678987654321, 1234567900987654321
Wonderful Demlo numbers. See Fig M5405. Ref TMS 6 68 38. [0,2; A2477, N2339]

$$\text{G.f.:} \quad (1+10x) \ / \ (1-x)(1-10x)(1-100x).$$

M5387 125, 1296, 8716, 47787, 232154, 1040014
Partially labeled trees with n nodes. Ref R1 138. [5,1; A0526, N2340]

M5388 1, 126, 756, 2072, 4158, 7560, 11592, 16704, 24948, 31878, 39816, 55944, 66584,
76104, 99792, 116928, 133182, 160272, 177660, 205128, 249480, 265104, 281736
Theta series of E_7 lattice. Ref SPLAG 123. [0,2; A4008]

M5389 126, 1260, 22302, 40446, 49308, 64530, 73962, 168730, 223020, 286500, 242730,
429750, 365700, 548550, 618570, 827700, 739620, 827652, 932100
Larger of unitary amicable pair. Cf. M5372. Ref MOC 25 917 71. [1,1; A2953]

M5390 1, 127, 149, 251, 331, 337, 373, 509, 599, 701, 757, 809, 877, 905, 907, 959, 977,
997, 1019, 1087, 1199, 1207, 1211, 1243, 1259, 1271, 1477, 1529, 1541, 1549, 1589
Odd numbers not of form $p + 2^x$. Ref Well86 136. [1,2; A6285]

M5391 1, 127, 6050, 204630, 5921520, 158838240, 4115105280, 105398092800,
2706620716800, 70309810771200, 1858166876966400
Simplices in barycentric subdivision of n-simplex. Ref rkg. [5,2; A5464]

M5392 1, 128, 2187, 16384, 78125, 279936, 823543, 2097152, 4782969, 10000000,
19487171, 35831808, 62748517, 105413504, 170859375, 268435456, 410338673
Seventh powers. Ref BA9. [1,2; A1015, N2341]

M5393 128, 12758, 5134240, 67898771, 11146309947
Largest number not the sum of distinct nth powers. Ref LE70 367. MOC 28 313 74. JRM
20 316 88. [2,1; A1661, N2342]

M5394 1, 129, 2316, 18700, 96825, 376761, 1200304, 3297456, 8080425, 18080425,
37567596, 73399404, 136147921, 241561425, 412420800, 680856256, 1091194929
Sums of 7th powers. Ref AS1 815. [1,2; A0541, N2343]

M5395 1, 130, 1270932917454
n-element algebras with 1 ternary operation. Ref PAMS 17 737 66. [1,2; A1331, N2344]

M5396 134, 760, 3345, 12256, 39350, 114096, 307060, 776000, 1867170, 4298600,
9540169, 20487360, 42756520, 86967184, 172859325, 336450560, 642489660
McKay-Thompson series of class 5A for Monster. Ref CALG 18 257 90. FMN94. [1,1;
A7251]

M5397 1, 135, 5478, 165826, 4494351, 116294673, 2949965020, 74197080276,
  1859539731885, 46535238000235, 1163848723925346, 29100851707716150
Coefficients of elliptic function sn. Ref Cay95 56. TM93 4 92. JCT A29 122 80. MOC 37
480 81. [2,2; A4005]

M5398 1, 136, 64573605
n-element algebras with 2 binary operations. Ref PAMS 17 736 66. [1,2; A1330, N2346]

M5399 1, 1, 1, 1, 1, 1, 139, 1, 571, 281, 163879, 5221, 5246819, 5459, 534703531,
  91207079, 4483131259, 2650986803, 432261921612371, 6171801683
Related to expansion of gamma function. Cf. M3140. Ref AMM 97 827 90. [1,7; A5447]

M5400 1, 1, 1, 139, 571, 163879, 5246819, 534703531, 4483131259, 432261921612371,
  6232523202521089, 25834629665134204969, 1579029138854919086429
Numerators of asymptotic series for gamma function. Cf. M4878. Ref MOC 22 619 68.
[0,4; A1163, N2347]

M5401 1, 141, 4713, 5795, 6611, 18496
Prime Cullen numbers: $n.2^n + 1$ is prime. Cf. 2064. Ref BPNR 283. [1,2; A5849]

M5402 151, 431, 6581, 67651, 241981, 2081921, 3395921
Primes with large least nonresidues. Ref RS9 XXIII. [1,1; A2226, N2348]

M5403 153, 1634, 4150, 548834, 1741725, 24678050, 146511208, 4679307774
Smallest number > 1 equal to sum of nth powers of its digits. Ref GA85 249. [3,1;
A3321]

M5404 1, 156, 20306, 2558556, 320327931, 40053706056, 5007031143556,
  625886840206056, 78236053707784181, 9779511680526143556
Gaussian binomial coefficient $[n,3]$ for $q = 5$. Ref GJ83 99. ARS A17 329 84. [3,2; A6112]

M5405 157, 192, 218, 220, 222, 224, 226, 243, 245, 247, 251, 278, 285, 286, 287, 312,
  355, 361, 366, 382, 384, 390, 394, 411, 434, 443, 478, 497, 499, 506, 508, 528, 529, 539
Apocalyptic powers: 2^n contains 666. See Fig M5405. Ref Pick92 337. [1,1; A7356]

|||||||||||||||||||||||||||||||||||||||||||||||||||||||||||||||||||||||||||||||||||||||||||||||||||||||||||||||||||

Figure M5405. SILLY SEQUENCES.

   M2239, M3752, M4923, M5386, M5405, etc. Some readers may feel that these sequences
should have been rejected. We are not proud of them. We certainly aren't going to repeat the
definitions here: you can look them up and decide for yourself!

|||||||||||||||||||||||||||||||||||||||||||||||||||||||||||||||||||||||||||||||||||||||||||||||||||||||||||||||||||

M5406 157, 262, 367, 412, 472, 487, 577, 682, 787, 877, 892, 907, 997, 1072, 1207, 1237,
  1312, 1402, 1522, 1567, 1627, 1657, 1732, 1852, 1942, 2047, 2062, 2152, 2194, 2257
$\phi(2n+1) < \phi(2n)$. Ref AMM 54 332 47. jos. [1,1; A1837, N2349]

M5407   163, 907, 2683, 5923, 10627, 15667, 20563, 34483, 37123, 38707, 61483, 90787,
  93307, 103387, 166147, 133387, 222643, 210907, 158923, 253507, 296587
Largest prime $\equiv 3 \bmod 8$ with class number $2n + 1$. Cf. M3164. Ref MOC 24 492 70. [0,1;
A2149, N2350]

M5408   1, 168, 7581, 160948, 2068224, 18561984, 127234008, 706987164, 3320153661,
  13583619496, 49530070161
Distributive lattices. Ref MSH 53 19 76. MSG 121 121 76. [0,2; A6363]

M5409   1, 169, 32761, 6355441, 1232922769, 239180661721, 46399815451081,
  9001325016847969, 1746210653453054881, 338755865444875798921
Square hex numbers. Ref GA88 19. [1,2; A6051]

M5410   196, 887, 1675, 7436, 13783, 52514, 94039, 187088, 1067869, 10755470,
  18211171, 35322452, 60744805, 111589511, 227574622, 454050344, 897100798
Reverse and add! 196 is conjectured to be first starter never leading to a palindrome. Ref
ScAm 250(4) 24 84. [1,1; A6960]

M5411   1, 196, 11196, 307960, 5157098, 60112692, 530962446, 3764727340,
  22326282261, 114158490576, 515063238810
Distributive lattices. Ref MSH 53 19 76. MSG 121 121 76. [0,2; A6362]

M5412   198, 7761798, 4676136464999866416198,
  10224946038730638447305617273857752108784394891639150859110579
Extracting a square root. Ref AMM 44 645 37. jos. [0,1; A6243]

M5413   211, 281, 421, 461, 521, 691, 881, 991, 1031, 1151, 1511, 1601, 1871, 1951, 2221,
  2591, 3001, 3251, 3571, 3851, 4021, 4391, 4441, 4481, 4621, 4651, 4691, 4751
Artiads. Ref PLMS 24 256 1893. JMA 15 118 66. [1,1; A1583, N2351]

M5414   220, 1184, 2620, 5020, 6232, 10744, 12285, 17296, 63020, 66928, 67095, 69615,
  79750, 100485, 122265, 122368, 141664, 142310, 171856, 176272, 185368, 196724
Smaller of amicable pair. See Fig M0062. Cf. M5435. Ref MOC 21 242 67; 47 S9 86. [1,1;
A2025, N2352]

M5415   1, 236, 32675, 4638576, 681728204, 102283239429, 15513067188008,
  2365714170297014
Hamiltonian circuits on $2n \times 8$ rectangle. Ref JPA 17 445 84. [1,2; A5391]

M5416   1, 240, 2160, 6720, 17520, 30240, 60480, 82560, 140400, 181680, 272160,
  319680, 490560, 527520, 743040, 846720, 1123440, 1179360, 1635120, 1646400
Theta series of E_8 lattice. Ref SPLAG 123. [0,2; A4009]

M5417   1, 0, 0, 240, 3060, 19584, 77760, 249120, 774180, 2110720, 4621824, 9294480,
  19873920, 40049280, 68181120, 110984160, 198425700, 342524160, 509271040
Theta series of Λ_{18} lattice. Ref SPLAG 157. [0,4; A5950]

M5418 241, 5521, 6481, 51361, 346561, 380881, 390001, 1678321, 4332721, 4654801,
5576881, 12707521, 39336721, 41432641, 42942001, 99990001, 167948881, 184970641
Duodecimal primes: $p = (x^{12} + y^{12})/(x^4 + y^4)$. Ref CU23 1 258. [1,1; A6687]

M5419 1, 241, 9361, 120161
Crystal ball numbers for E_8 lattice. Ref CoSl95. [0,2; A7205]

M5420 242, 3655, 4503, 5943, 6853, 7256, 8392, 9367, 10983, 11605, 11606, 12565,
12855, 12856, 12872, 13255, 13782, 13783, 14312, 16133, 17095, 18469, 19045, 19142
n, $n+1$, $n+2$, $n+3$ have same number of divisors. Ref AS1 840. UPNT B18. [1,1;
A6601]

M5421 1, 244, 3369, 20176, 79225, 240276, 611569, 1370944, 2790801, 5266900,
9351001, 15787344, 25552969, 39901876, 60413025, 89042176, 128177569, 180699444
Sums of 5th powers of odd numbers. Ref CC55 742. [1,2; A2594, N2354]

M5422 0, 1, 247, 14608, 455192, 9738114, 162512286, 2275172004, 27971176092,
311387598411, 3207483178157, 31055652948388, 285997074307300
Eulerian numbers. See Fig M3416. Ref R1 215. DB1 151. JCT 1 351 66. DKB 260. C1
243. [6,3; A1243, N2355]

M5423 1, 248, 4124, 34752, 213126, 1057504, 4530744, 17333248, 60655377,
197230000, 603096260, 1749556736, 4848776870, 12908659008, 33161242504
McKay-Thompson series of class 3C for Monster. Ref CALG 18 256 90. FMN94. [0,2;
A7245]

M5424 251, 571, 971, 1181, 1811, 2011, 2381, 2411, 3221, 3251, 3301, 3821, 4211, 4861,
4931, 5021, 5381, 5861, 6221, 6571, 6581, 8461, 8501, 9091, 9461
2 is a quintic residue modulo p. Ref Krai24 1 59. [1,1; A1135, N2356]

M5425 1, 253, 49141, 9533161, 1849384153, 358770992581, 69599723176621,
13501987525271953, 2619315980179582321, 508133798167313698381
Triangular star numbers. Ref GA88 20. [1,2; A6060]

M5426 1, 256, 6561, 65536, 390625, 1679616, 5764801, 16777216, 43046721,
100000000, 214358881, 429981696, 815730721, 1475789056, 2562890625, 4294967296
Eighth powers. Ref BA9. [1,2; A1016, N2357]

M5427 1, 257, 6818, 72354, 462979, 2142595, 7907396, 24684612, 67731333,
167731333, 382090214, 812071910, 1627802631, 3103591687, 5666482312
Sums of 8th powers. Ref AS1 815. [1,2; A0542, N2358]

M5428 257, 65537, 2070241, 100006561, 435746497, 815730977, 332507937,
1475795617, 2579667841, 4338014017, 5110698017, 6975822977, 16983628577
Octavan primes: $p = x^8 + y^8$. Ref CU23 1 258. [1,1; A6686]

M5429   263, 293, 368, 578, 683, 743, 788, 878, 893, 908, 998, 1073, 1103, 1208, 1238, 1268, 1403, 1418, 1502, 1523, 1658, 1733, 1838, 1943, 1964, 2048, 2063, 2153, 2228
$\phi(2n-1) < \phi(2n)$. Ref AMM 54 332 47. jos. [1,1; A1836, N2359]

M5430   1, 0, 272, 256, 3058, 2048, 11232, 7168, 32848, 16384, 67936, 32512, 139040, 59392, 217408, 95232, 385266, 147456, 528752, 226048, 819424, 315392, 1075040
Theta series of laminated lattice Λ_9. Ref SPLAG 157. [0,3; A5933]

M5431   272, 7936, 137216, 1841152, 21253376, 222398464, 2174832640, 20261765120, 182172651520
Permutations of length n by length of runs. Ref DKB 261. [7,1; A0517, N2360]

M5432   272, 24611, 515086, 4456448, 23750912, 93241002, 296327464, 806453248, 1951153920, 4300685074
Generalized tangent numbers. Ref MOC 21 690 67. [1,1; A0518, N2361]

M5433   1, 274, 48076, 6998824, 929081776, 117550462624, 14500866102976, 1765130436471424
Differences of reciprocals of unity. Ref DKB 228. [1,2; A1242, N2362]

M5434   276, 2048, 11202, 49152, 184024, 614400, 1881471, 5373952, 14478180, 37122048, 91231550, 216072192, 495248952, 1102430208, 2390434947, 5061476352
McKay-Thompson series of class 4A for Monster. Ref FMN94. [1,1; A7246]

M5435   284, 1210, 2924, 5564, 6368, 10856, 14595, 18416, 76084, 66992, 71145, 87633, 88730, 124155, 139815, 123152, 153176, 168730, 176336, 180848, 203432, 202444
Larger of amicable pair. See Fig M0062. Cf. M5414. Ref MOC 21 242 67; 47 S9 86. [1,1; A2046, N2363]

M5436   1, 0, 0, 0, 306, 0, 0, 0, 3024, 512, 0, 0, 13344, 0, 0, 0, 36594, 4608, 0, 0, 78048, 0, 0, 0, 149088, 18432, 0, 0, 256896, 0, 0, 0, 405072, 47616, 0, 0, 620306, 0, 0, 0, 900576
Theta series of P_{9a} packing. Ref SPLAG 140. [0,5; A5951]

M5437   324, 63, 1, 1023, 64, 1023, 1, 63, 1023, 1, 63, 1023, 1, 62, 1, 1023, 63, 1, 1023, 64, 1023, 1, 63, 1023, 1, 62, 1, 1023, 64, 1023, 1, 63, 1023, 1, 62, 1, 1023, 63, 1, 1023, 63, 1
A continued fraction. Ref JNT 11 216 79. [0,1; A6465]

M5438   331, 39139, 253243, 4397207, 21587171
Primes with large least nonresidues. Ref RS9 XXIV. [1,1; A2228, N2364]

M5439   1, 0, 336, 768, 4950, 6912, 22944, 27648, 75792, 72192, 181728, 158976, 393030, 317952, 682656, 557568, 1249686, 912384, 1881840, 1458432, 2979072, 2155776
Theta series of laminated lattice Λ_{10}. Ref SPLAG 157. ecp. [0,3; A6909]

M5440   341, 91, 15, 124, 35, 25, 9, 28, 33, 15, 65, 21, 15, 341, 51, 45, 25, 45, 21, 55, 69, 33, 25, 28, 27, 65, 87, 35, 49, 49, 33, 85, 35, 51, 91, 45, 39, 95, 91, 105, 205, 77, 45, 76
Smallest pseudoprime to base n. Ref B1 42. [2,1; A7535]

M5441 341, 561, 645, 1105, 1387, 1729, 1905, 2047, 2465, 2701, 2821, 3277, 4033, 4369,
4371, 4681, 5461, 6601, 7957, 8321, 8481, 8911, 10261, 10585, 11305, 12801, 13741
Sarrus numbers: pseudo-primes to base 2. Ref SPH 8 45 38. L1 48. SI64a 215. MOC 25
944 71. [1,1; A1567, N2365]

M5442 341, 561, 1105, 1729, 1905, 2047, 2465, 3277, 4033, 4681, 5461, 6601, 8321,
8481, 10241, 10585, 12801, 15709, 15841, 16705, 18705, 25761, 29341, 30121, 31621
Euler pseudoprimes: $2^{(n-1)/2} \equiv \pm 1 \bmod n$. Ref UPNT A12. rgw. [1,1; A6970]

M5443 341, 561, 1729, 1729, 399001, 399001, 1857241, 1857241, 6189121, 14469841,
14469841, 14469841
Least number for which Solovay-Strassen primality test on bases $< p$ fails. Ref bach. [2,1;
A7324]

M5444 341, 561, 11305, 825265, 45593065, 370851481, 38504389105, 7550611589521,
277960972890601, 32918038719446881, 1730865304568301265
Pseudoprimes with n prime factors. Ref rgep. [1,1; A7011]

M5445 1, 341, 93093, 24208613, 6221613541, 1594283908581, 408235958349285,
104514759495347685, 26756185103024942565, 6849609413493939400165
Gaussian binomial coefficient $[n,4]$ for $q=4$. Ref GJ83 99. ARS A17 328 84. [4,2; A6107]

M5446 353, 651, 2487, 2501, 2829, 3723, 3973, 4267, 4333, 4449, 4949, 5281, 5463,
5491, 5543, 5729, 6167, 6609, 6801, 7101, 7209, 7339, 7703, 8373, 8433, 8493, 8517
4th power of n = sum of four 4th powers. Ref MOC 27 492 73. [1,1; A3294]

M5447 1, 362, 130683, 47046242
Number of Go games with n moves. Ref herbt. [0,2; A7565]

M5448 1, 372, 768, 5684, 6144, 28608, 23040, 91956, 61440, 224680, 140544, 458688,
276480, 358240, 480768, 1467188, 798720, 2329012, 1251072, 3590952, 1843200
Theta series of packing P_{10c}. Ref SPLAG 140. [0,2; A4021]

M5449 389, 433, 563, 571, 643, 709, 997, 1061, 1171, 1483
From relations between Siegel theta series. Ref NMJ 121 87 91. [1,1; A6476]

M5450 399, 935, 2015, 2915, 3059, 4991, 5719, 7055, 8855
Lucas-Carmichael numbers: $p|n \rightarrow p+1|n+1$. Ref rgep. [1,1; A6972]

M5451 0, 0, 0, 0, 400, 8640, 129288, 1688424, 20762073, 248384816, 2937307716
Fixed 4-dimensional polyominoes with n cells. Ref CJN 18 367 75. [1,5; A6764]

M5452 1, 0, 432, 1632, 8700, 18048, 51072, 82880, 191926, 251648, 517568, 619104,
1204024, 1322368, 2326528, 2515904, 4396188, 4407552, 7238000, 7303456, 11911352
Theta series of laminated lattice $\Lambda_{11}^{\min}$. Ref SPLAG 157. ecp. [0,3; A6910]

M5453 1, 0, 438, 1536, 9372, 15360, 57896, 70656, 211638, 215040, 582648, 529920, 1316472, 1139712, 2619264, 2159616, 4815516, 3766272, 8165550, 6259200, 13070328
Theta series of laminated lattice Λ_{11}^{max}. Ref SPLAG 157. ecp. [0,3; A6911]

M5454 1, 472, 467133, 636430764, 1038934571875, 1903882757758426
Witt vector *5!/5!. Ref SLC 16 107 88. [1,2; A6180]

M5455 1, 492, 22590, 367400, 3764865, 28951452, 182474434, 990473160, 4780921725, 20974230680, 84963769662, 321583404672, 1147744866180
McKay-Thompson series of class 2a for Monster. Ref FMN94. [0,2; A7242]

M5456 1, 500, 7220, 36800, 118580, 288424, 589760, 1104000, 1884980, 2994740, 4618024
Theta series of P_{10b} packing. Ref SPLAG 140. [0,2; A5954]

M5457 0, 1, 502, 47840, 2203488, 66318474, 1505621508, 27971176092, 447538817472, 6382798925475, 83137223185370, 1006709967915228, 11485644635009424
Eulerian numbers. See Fig M3416. Ref R1 215. DB1 151. JCT 1 351 66. DKB 260. C1 243. [1,3; A1244, N2366]

M5458 1, 504, 270648, 144912096, 77599626552, 41553943041744, 22251789971649504, 11915647845248387520, 638072999141923648 8504
Expansion of a modular function. Ref RAM1 317. [0,2; A0706, N2367]

M5459 1, 512, 19683, 262144, 1953125, 10077696, 40353607, 134217728, 387420489, 1000000000, 2357947691, 5159780352, 10604499373, 20661046784, 38443359375
9th powers. Ref BA9. [1,2; A1017, N2368]

M5460 1, 513, 20196, 282340, 2235465, 12313161, 52666768, 186884496, 574304985, 1574304985, 3932252676, 9092033028, 19696532401, 40357579185, 78800938560
Sums of 9th powers. Ref AS1 815. [0,2; A7487]

M5461 561, 1105, 1729, 1905, 2047, 2465, 4033, 4681, 6601, 8321, 8481, 10585, 12801, 15841, 16705, 18705, 25761, 30121, 33153, 34945, 41041, 42799, 46657, 52633, 62745
Euler-Jacobi pseudoprimes: $2^{(n-1)/2} \equiv (2 / n)$ mod n. Ref Rie85. UPNT A12. rgep. rgw. [1,1; A6971]

M5462 561, 1105, 1729, 2465, 2821, 6601, 8911, 10585, 15841, 29341, 41041, 46657, 52633, 62745, 63973, 75361, 101101, 115921, 126217, 162401, 172081, 188461, 252601
Carmichael numbers: composite numbers n such that $a^{n-1} \equiv 1 \pmod{n}$ if a is prime to n. Ref SPH 8 45 38. SI64 51. MOC 25 944 71. UPNT A12. [1,1; A2997]

M5463 561, 41041, 825265, 321197185, 5394826801, 232250619601, 9746347772161, 1436697831295441, 60977817398996785, 7156857700403137441
Least Carmichael number with n factors. Ref MOC 61 381 93. [3,1; A6931]

M5464 1, 0, 566, 1280, 12188, 12800, 75304, 58880, 275126, 179200, 757240, 441600, 1711224, 949760, 3405696, 1799680, 6261404, 3138560, 10613550, 5216000, 16987640
Theta series of P_{11a} packing. Ref SPLAG 140. [0,3; A5953]

M5465 1, 0, 624, 3456, 17544, 47616, 130752, 252672, 560904, 887808, 1692576, 2412672, 4280736, 5564928, 9068928, 11460864, 17948424, 21310464, 32009904
Theta series of laminated lattice Λ_{12}^{min}. Ref SPLAG 157. ecp. [0,3; A6912]

M5466 631, 5531, 72661, 865957, 2375059, 32353609
Primes with large least nonresidues. Ref RS9 XXIII. [1,1; A2227]

M5467 1, 0, 632, 3328, 18440, 44032, 139872, 236032, 589576, 829440, 1803600, 2250496, 4499360, 5196800, 9676480, 10694144, 18865928, 19884032, 34147224
Theta series of laminated lattice Λ_{12}^{mid}. Ref SPLAG 157. ecp. [0,3; A6913]

M5468 1, 0, 648, 3072, 20232, 36864, 158112, 202752, 646920, 712704, 2025648, 1926144, 4936608, 4460544, 10891584, 9160704, 20700936, 17031168, 38421864
Theta series of laminated lattice Λ_{12}^{max}. Ref SPLAG 157. ecp. [0,3; A6914]

M5469 705, 2465, 2737, 3745, 4181, 5777, 6721, 10877, 13201, 15251, 24465, 29281, 34561, 35785, 51841, 54705, 64079, 64681, 67861, 68251, 75077, 80189, 90061, 96049
Lucas pseudoprimes. Ref BPNR 104. [1,1; A5845]

M5470 1, 714, 196677, 18941310, 809451144, 17914693608, 223688514048, 1633645276848, 6907466271384, 15642484909560, 1466561365176
6×6 stochastic matrices of integers. Ref SS70. ANS 4 1179 76. [0,2; A5467]

M5471 720, 1854, 4738, 12072, 30818, 79118, 204448, 528950, 1370674, 3557408, 9244418, 24043990, 62573616, 162925614, 424377730, 1105703640, 2881483458
Permanent of a certain cyclic $n \times n$ (0,1) matrix. Ref CMB 7 262 64. JCT 7 317 69. [5,1; A0805, N2369]

M5472 1, 0, 720, 13440, 97200, 455040, 1714320, 4821120, 12380400, 29043840, 58980960, 114076800, 219310320, 367338240, 621878400, 1037727360, 1583679600
Theta series of Eisenstein version of E_8 lattice. Ref JALG 52 248 78. [0,3; A4033]

M5473 720, 15120, 191520, 1905120, 16435440, 129230640, 953029440, 6711344640, 45674188560, 302899156560, 1969147121760, 12604139926560, 79694820748080
Differences of 0: $6!.S(n,6)$. Ref VO11 31. DA63 2 212. R1 33. [6,1; A0920, N2370]

M5474 1, 720, 202410, 20933840, 1047649905, 30767936616, 602351808741, 8575979362560, 94459713879600, 842286559093240, 6292583664553881
6×6 stochastic matrices of integers. Ref SIAC 4 477 75. [1,2; A3439]

M5475 725, 1125, 1600, 2000, 2048, 2225, 2304, 2525, 2624, 3600, 4205, 4225, 4352, 4400, 4525, 4752, 4913, 5125, 5225, 5725, 6125, 7056, 7168, 7225, 7232, 7488, 7600
Discriminants of totally real quartic fields. Ref JLMS 31 484 56. [1,1; A2769, N2371]

M5476 729, 47601, 3450897, 252034065, 18416334609
Functions realized by cascades of n gates. Ref BU77. [1,1; A5608]

M5477 1, 744, 196884, 21493760, 864299970, 20245856256, 333202640600, 4252023300096, 44656994071935, 401490886656000, 3176440229784420
Expansion of j-function. Cf. M5179. Ref KNAW 56 398 53. MOC 8 77 54. BLMS 11 326 79. CALG 18 256 90. [-1,2; A0521, N2372]

M5478 1, 0, 756, 4032, 20412, 60480, 139860, 326592, 652428, 1020096, 2000376, 3132864, 4445532, 7185024, 10747296, 13148352, 21003948, 27506304, 33724404
Theta series of Coxeter-Todd lattice K_{12}. Ref SPLAG 129. [0,3; A4010]

M5479 1, 781, 508431, 320327931, 200525284806, 125368356709806, 78360229974772306, 48975769621072897306, 30609934249224268600431
Gaussian binomial coefficient $[n,4]$ for $q = 5$. Ref GJ83 99. ARS A17 329 84. [4,2; A6113]

M5480 783, 8672, 65367, 371520, 1741655, 7161696, 26567946, 90521472, 288078201, 864924480, 2469235686, 6748494912, 17746495281, 45086909440, 111066966315
McKay-Thompson series of class 3A for Monster. Ref CALG 18 256 90. FMN94. [1,1; A7243]

M5481 1, 0, 840, 2560, 26376, 30720, 204960, 168960, 843528, 593920, 2625840, 1605120, 6435744, 3717120, 14118720, 7633920, 26992392, 14192640, 49806120
Theta series of P_{12a} packing. Ref SPLAG 139. [0,3; A5952]

M5482 1, 0, 1, 880, 275305224
Magic squares of order n. Ref ScAm 249(1) 118 76. GA88 216. [1,4; A6052]

M5483 1, 0, 888, 6432, 36392, 110720, 336992, 696512, 1656202, 2779392, 5603904
Theta series of laminated lattice Λ_{13}^{min}. Ref SPLAG 157. ecp. [0,3; A6915]

M5484 1, 0, 890, 6400, 36600, 110080, 337520, 698880, 1649610, 2780160, 5619792
Theta series of laminated lattice Λ_{13}^{mid}. Ref SPLAG 157. ecp. [0,3; A6916]

M5485 1, 0, 906, 6144, 38424, 102400, 359344, 651264, 1743434, 2596864, 5956560
Theta series of laminated lattice Λ_{13}^{max}. Ref SPLAG 157. ecp. [0,3; A6917]

M5486 945, 1575, 2205, 3465, 4095, 5355, 5775, 5985, 6435, 6825, 7245, 7425, 8085, 8415, 8925, 9135, 9555, 9765, 11655, 12705, 12915, 13545, 14805, 15015, 16695, 18585
Odd primitive abundant numbers. Ref AJM 35 422 13. [1,1; A6038]

M5487 999, 999999999999999999999999999999999999999999999999999999999999999999999999999999
The last sequence in the main table. Ref EIS §1.5. [1,1; A0015]

Bibliography

[AABB] J.-P. Allouche et al., "A sequence related to that of Thue-Morse," *Discrete Math.*, 1994, to appear.

[aam] A. A. Mcintosh, personal communication.

[AAM] *Advances in Applied Mathematics.*

[AAMS] *Abstracts of the American Mathematical Society.*

[AAP] *Advances in Applied Probability.*

[AB71] A. O. L. Atkin and B. J. Birch, editors, *Computers in Number Theory*, Academic Press, NY, 1971.

[ACA] *Acta Arithmetica.*

[ACC] *Acta Crystallographica.*

[AcMaSc] *Acta Mathematica Scientia.*

[Adam74] J. L. Adams, *Conceptual Blockbusting: A Guide to Better Ideas*, Freeman, San Francisco, 1974.

[ADM] *Annals of Discrete Mathematics.*

[ADV] *Advances in Mathematics.*

[AENS] *Annales Scientifiques de l'École Normale Supérieure*, Paris.

[AEQ] *Aequationes Mathematicae.*

[AFI] *American Federation of Information Processing Societies, Conference Proceedings.*

[AFM] *Arkiv för Matematik.*

[AH21] W. Ahrens, *Mathematische Unterhaltungen und Spiele*, Teubner, Leipzig, Vol. 1, 3rd ed., 1921; Vol. 2, 2nd ed., 1918.

[AIEE] *Transactions of the American Institute of Electrical Engineers.*

[Aign77] M. Aigner, editor, *Higher Combinatorics*, Reidel, Dordrecht, Holland, 1977.

[AIP] *Advances in Physics.*

[ajg] A. J. Guttmann, personal communication.

[AJG] *American Journal of Medical Genetics.*

[AJM] *American Journal of Mathematics.*

[ajs] A. J. Schwenk, personal communication.

[Algo93] *Algorithms Seminars 1992–1993*, Report #2130, INRIA, Rocquencourt, Dec. 1993.

[AMA] *Acta Mathematica* (Uppsala).

[AML] *Applied Mathematics Letters.*

[AMM] *American Mathematical Monthly.*

[AMN] *Archiv for Mathematik og Naturvidenskab.*

[AMP] *Archiv der Mathematik und Physik.*

[AMS] *Annals of Mathematical Statistics.*

[AN71] E. K. Annavaddar, *Determination of the Finite Groups Having Eight Conjugacy Classes*, Ph.D. Diss., Arizona State Univ., 1971.

[AN87] I. Anderson, *Combinatorics of Finite Sets*, Oxford Univ. Press, 1987.

[ANAL] *Analysis.*

[Andr76] G. E. Andrews, *The Theory of Partitions*, Addison-Wesley, Reading MA, 1976.

[Andr85] G. E. Andrews, *q-Series: Their Development and Application in Analysis, Number Theory, Combinatorics, Physics and Computer Algebra*, Amer. Math. Soc., Providence RI, 1985.

[ANN] *Annals of Mathematics.*

[ANP] *Annals of Probability.*

[ANS] *Annals of Statistics.*

[ANY] *Annals of the NY Academy of Sciences.*

[Ape81] R. Apéry, "Interpolation de fractions continues et irrationalité de certaines constantes," in *Mathématiques, Ministère universités (France), Comité travaux historiques et scientifiques, Bull. Section Sciences*, Vol. 3, pp. 243–246, 1981.

[ARC] *Archiv der Mathematik* (Basel).

[ARS] *Ars Combinatoria.*

[arw] A. R. Wilks, personal communication.

[AS1] M. Abramowitz and I. A. Stegun, *Handbook of Mathematical Functions*, National Bureau of Standards, Washington DC, 1964.

[ASB] *Annales de la Société Scientifique de Bruxelles.*

[AST] *Astérisque.*

[Atiy90] M. Atiyah, *The Geometry and Physics of Knots*, Cambr. Univ. Pr., 1990.

[ATLAS] J. H. Conway, R. T. Curtis, S. P. Norton, R. A. Parker and R. A. Wilson, *ATLAS of Finite Groups*, Oxford Univ. Press, 1985.

[aw] A. Wilansky, personal communication.

[B1] A. H. Beiler, *Recreations in the Theory of Numbers*, Dover, NY, 1964.

[BA4] E. L. Ince, *Cycles of Reduced Ideals in Quadratic Fields*, British Association Mathematical Tables, Vol. 4, London, 1934.

[BA6] British Association Mathematical Tables, Vol. 6, *Bessel Functions, Part 1, Functions of Order Zero and Unity*, Cambridge Univ. Press, 1937.

[BA62] D. St. P. Barnard, *50 Observer Brain Twisters*, Faber and Faber, London, 1962.

[BA76] A. T. Balaban, *Chemical Applications of Graph Theory*, Academic Press, NY, 1976.

[BA8] J. W. L. Glaisher, *Number-Divisor Tables*, British Association Mathematical Tables, Vol. 8, Cambridge Univ. Press, 1940.

[BA9] J. W. L. Glaisher et al., *Tables of Powers*, British Association Mathematical Tables, Vol. 9, Cambridge Univ. Press, 1940.

[bach] E. Bach, personal communication.

[BALK] *Mathematica Balkanica.*

[BAMS] *Bulletin of the American Mathematical Society.*

[BAR] *Reports of the British Association for the Advancement of Science.*

[BarN94] D. Bar-Natan, "On the Vassiliev knot invariants," preprint, 1994.

[BAT] *Reports of the British Association for the Advancement of Science, Transactions Section.*

[Batc27] P. M. Batchelder, *An Introduction to Finite Difference Equations*, Harvard Univ. Press, 1927.

[BB78] H. Brown, R. Bülow, J. Neubüser, H. Wondratschek and H. Zassenhaus, *Crystallographic Groups of Four-Dimensional Space*, Wiley, NY, 1978.

[BBD93] F. Bergeron, M. Bousquet-Mélou and S. Dulucq, "Standard paths in the composition poset," preprint, 1993.

[BCMS] *Bulletin of the Calcutta Mathematical Society.*

[BCW90] T. C. Bell, J. G. Cleary and I. H. Witten, *Text Compression*, Prentice Hall, Englewood Cliffs, NJ, 1990.

[bdm] B. D. McKay, personal communication.

[BE65] R. S. Berkowitz, editor, *Modern Radar*, Wiley, NY, 1965.

[BE68] E. R. Berlekamp, *Algebraic Coding Theory*, McGraw-Hill, NY, 1968.

[BE71] C. Berge, *Principles of Combinatorics*, Academic Press, NY, 1971.

[BE74] F. R. Bernhart, *Topics in Graph Theory Related to the Five Color Conjecture*, Ph.D. Dissertation, Kansas State Univ., 1974.

[Beck71] P. Beckmann, *A History of π*, Golem Press, Boulder, CO, 2nd ed., 1971.

[BER] *Chemische Berichte* (formerly *Berichte der Deutschen Chemischen Gesellschaft*).

[Berm83] J. Berman, "Free spectra of 3-element algebras," in R. S. Freese and O. C. Garcia, editors, *Universal Algebra and Lattice Theory (Puebla, 1982)*, LNM 1004, 1983.

[BeSl94] M. Bernstein and N. J. A. Sloane, "Some canonical sequences of integers," *Linear Algebra and Its Applications*, to appear.

[BF72] G. Berman and K. D. Fryer, *Introduction to Combinatorics*, Academic Press, NY, 1972.

[BF75] Y. M. M. Bishop, S. E. Fienberg and P. W. Holland, *Discrete Multivariate Analysis*, MIT Press, 1975.

[BFK94] F. Bergeron, L. Favreau, D. Krob, "Some conjectures on the enumeration of tableaux of bounded height," *Discrete Math.*, to appear.

[BhSk94] A. Bhansali and S. Skiena, "Analyzing integer sequences," in *Computational Support for Discrete Mathematics*, DIMACS Series in Discrete Mathematics and Computer Science, Vol. 15 (1994), pp. 1–16.

[BI67] G. Birkhoff, *Lattice Theory*, American Mathematical Society, Colloquium Publications, Vol. 25, 3rd ed., Providence, RI, 1967.

[BICA] *Bulletin of the Institute of Combinatorics and its Applications.*

[BIO] *Biometrika.*

[BIT] *Nordisk Tidskrift for Informationsbehandling.*

[BLL94] F. Bergeron, G. Labelle and P. Leroux, *Théorie des espèces et Combinatoire des Structures Arborescentes*, Publications du LACIM, Université du Québec à Montréal, 1994.

[BLMS] *Bulletin of the London Mathematical Society.*

[BMG] *Bulletin de la Société Mathématique de Grèce.*

[BO47] M. Boll, *Tables Numériques Universelles*, Dunod, Paris, 1947.

[BoBo90] J. Borwein and P. Borwein, *A Dictionary of Real Numbers*, Wadsworth, Pacific Grove, Calif., 1990.

[BoDu] C. J. Bouwkamp and A. J. W. Duijvestijn, *Catalogue of Simple Perfect Squared Squares of Orders 21 Through 25*, Eindhoven Univ. Technology, Dept. of Math., Report 92-WSK-03, Nov. 1992.

[BoHa92] M. Bousquet-Mélou and L. Habsieger, "Sur les matrices à signes alternants," preprint, Université Bordeaux 1, 1992.

[BPNR] P. Ribenboim, *The Book of Prime Number Records*, Springer-Verlag, NY, 2nd ed., 1989.

[br] B. Recamán, personal communication.

[BR72] A. Brousseau, *Fibonacci and Related Number Theoretic Tables*, Fibonacci Association, San Jose, CA, 1972.

[BR73] A. Brousseau, *Number Theory Tables*, Fibonacci Association, San Jose, CA, 1973.

[BR80] A. E. Brouwer, *The Enumeration of Locally Transitive Tournaments*, Math. Centr. Report ZW138, Amsterdam, April 1980.

[Bru65] V. Brun, "Un procédé qui ressemble au crible d'Eratostene," *An. Sti. Univ. Al. I. Cuza, Iasi Sect. Ia Mat.*, 1965, vol. 11B, pp. 47–53.

[BS65] R. G. Busacker and T. L. Saaty, *Finite Graphs and Networks*, McGraw-Hill, NY, 1965.

[BS66] Z. I. Borevich and I. R. Shafarevich, *Number Theory*, Academic Press, NY, 1966.

[BS71] W. A. Beyer, M. L. Stein and S. M. Ulam, *The Notion of Complexity*, Report LA-4822, Los Alamos Scientific Laboratory of the University of California, Los Alamos, NM, December 1971.

[BSMB] *Bulletin de la Société Mathématique de Belgique.*

[BSMF] *Bulletin de la Société Mathématique de France.*

[BSTJ] *The Bell System Technical Journal.*

[BSW94] M. Bernstein, N. J. A. Sloane and P. E. Wright, "Sublattices of the hexagonal lattice," *Discrete Math.*, submitted.

[BU71] J. C. Butcher, editor, *A Spectrum of Mathematics*, Auckland University Press, 1971.

[BU77] J. T. Butler, "Fanout-free networks of multivalued gates," in *Proc. Symposium Multiple-Valued Logic, Charlotte NC, 1977*, IEEE Press, NY, 1977, pp. 39–46.

[BU89] D. A. Buell, *Binary Quadratic Forms*, Springer-Verlag, NY, 1989.

[BW78] L. W. Beineke and R. J. Wilson, *Selected Topics in Graph Theory*, Academic Press, NY, 1978.

[C1] L. Comtet, *Advanced Combinatorics*, Reidel, Dordrecht, Holland, 1974.

[CACM] *Communications of the Association for Computing Machinery.*

[Cald94] C. K. Caldwell, *The Largest Known Primes*, available electronically from caldwell@utm.edu (or finger primes@math.utm.edu).

[CALG] *Communications in Algebra.*

[Calk94] N. J. Calkin, "A curious binomial identity," preprint, 1994.

[Capo90] R. M. Capocelli, editor, *Sequences* (Naples, 1988), Springer-Verlag, NY, 1990.

[CaRo91] S. C. Carter and R. W. Robinson, *Exponents of 2 in the Numbers of Unlabeled Graphs and Tournaments*, Report UGA-CS-TR-91-009, Comp. Sci. Dept., Univ. Georgia, Athens, GA, 1991.

[Cass71] J. W. S. Cassels, *An Introduction to the Geometry of Numbers*, Springer-Verlag, NY, 2nd ed., 1971.

[CAU] A. Cauchy, *Oeuvres Complètes*, Gauthier-Villars, Paris, 1882–1938.

[CAY] A. Cayley, *Collected Mathematical Papers*, Vols. 1–13, Cambridge Univ. Press, London, 1889–1897.

[Cay95] A. Cayley, *An Elementary Treatise on Elliptic Functions*, Bell, London, 1895.

[CBUL] *Computer Bulletin.*

[CC55] F. E. Croxton and D. J. Cowden, *Applied General Statistics*, 2nd ed., Prentice-Hall, Englewood Cliffs, NJ, 1955.

[CCC] *Proceedings Caribbean Conferences on Combinatorics and Computing*, Univ. West Indies, Barbados.

[CG] *Computers and Graphics.*

[ChGu56] M. S. Cheema and H. Gupta, *Tables of Partitions of Gaussian Integers*, National Institute of Sciences of India, Mathematical Tables, Vol. 1, New Delhi, 1956.

[CHIBA] *Journal of the College of Arts and Sciences, Chiba University* (Chiba, Japan).

[Chir93] Y. Chiricota, Équations différentielles combinatoires et calcul symbolique in *'FPSAC5': Formal Power Series and Algebraic Combinatorics, Fifth Conference*, Florence, June 21–25 1993, Vol. 2, p. 123.

[ChLi] *Chess Life.*

[cjb] C. J. Bouwkamp, personal communication.

[CJM] *Canadian Journal of Mathematics.*

[CJN] *The Computer Journal.*

[CJP] *Canadian Journal of Physics.*

[CK90] M. V. Connolly and W. J. Knight, "Search in an array in which probe costs grow exponentially or factorially," preprint, 1990.

[CL45] The Staff of the Computational Laboratory, *Tables of the Modified Hankel Functions of Order One-Third and of Their Derivatives*, Annals of the Computation Laboratory of Harvard University, Vol. 2, Harvard Univ. Press, Cambridge, Massachusetts, 1945.

[clm] C. L. Mallows, personal communication.

[CM84] H. S. M. Coxeter and W. O. J. Moser, *Generators and Relations for Discrete Groups*, 4th ed., Springer-Verlag, NY, reprinted 1984.

[CMA] *Combinatorial Mathematics and Its Applications*, Vol. 1, *Proceedings of a Conference held at University of North Carolina, Chapel Hill, April 1967* (R. C. Bose and T. A. Dowling, eds.). University of North Carolina Press, Chapel Hill, 1969. Vol. 2, *Proceedings of Second Chapel Hill Conference*, University of North Carolina, Chapel Hill, 1970.

[CMB] *Canadian Mathematical Bulletin.*

[CMCN] *Notes News and Comments, Canadian Mathematical Congress.*

[CMWA] *Computers & Mathematics with Applications.*

[CN] *Congressus Numerantium.*

[CO89] B. W. Conolly, "Meta-Fibonacci sequences," in S. Vajda, editor, *Fibonacci and Lucas Numbers, and the Golden Section*, Halstead Press, NY, 1989, pp. 127–138.

[CoGo87] T. M. Cover and B. Gopinath, editors, *Open Problems in Communication and Computation*, Springer-Verlag, NY, 1987.

[CoGu95] J. H. Conway and R. K. Guy, *The Book of Numbers*, in preparation.

[COMB] *Combinatorica*.

[CONT] *Contemporary Mathematics*.

[CoSl94] J. H. Conway and N. J. A. Sloane, "What are all the best sphere packings in low dimensions?," *Discrete Comput. Geom.*, to appear.

[CoSl95] J. H. Conway and N. J. A. Sloane, "Crystal balls," in preparation.

[Coxe73] H. S. M. Coxeter, *Regular Polytopes*, Dover, NY, 3rd edition, 1973.

[Coxe74] H. S. M. Coxeter, "Polyhedral numbers," in R. S. Cohen et al., editors, *For Dirk Struik*, Reidel, Dordrecht, 1974, pp. 25–35.

[CPM] *Časopis pro Pěstování Matematiky*.

[CR41] R. Courant and H. Robbins, *What is Mathematics?*, Oxford Univ. Press, 1941.

[CRB] *Comptes Rendus*, Congrès National des Sciences, Brussels.

[CRO] *Cahiers du Bureau Universitaire de Recherche Opérationnelle*, Institut de Statistique, Université de Paris.

[CRP] *Comptes Rendus Hebdomadaires des Séances de l'Académie des Sciences, Paris, Série A*.

[CRUX] *Crux Mathematicorum* (Ottawa).

[CRV] *Computing Reviews*.

[CSB] *Classification Society Bulletin*.

[CU04] A. J. C. Cunningham, *Quadratic Partitions*, Hodgson, London, 1904.

[CU23] A. J. C. Cunningham, *Binomial Factorisations*, Vols. 1–9, Hodgson, London, 1923–1929.

[CU27] A. J. C. Cunningham, *Quadratic and Linear Tables*, Hodgson, London, 1927.

[CU86] T. Curtright, "Counting symmetry patterns in the spectra of strings," in H. J. de Vega and N. Sánchez, editors, *String Theory, Quantum Cosmology and Quantum Gravity; Integrable and Conformal Invariant Theories*, World Scientific, Singapore, 1987, pp. 304–333.

[CUNN] J. Brillhart et al., *Factorizations of $b^n \pm 1$*, Contemporary Mathematics, Vol. 22, Amer. Math. Soc., Providence, RI, 2nd edition, 1985; and later supplements.

[DA63] H. T. Davis, *Tables of the Mathematical Functions*, Vols. 1 and 2, 2nd ed., 1963, Vol. 3 (with V. J. Fisher), 1962; Principia Press of Trinity Univ., San Antonio, TX.

[DA82] H. Davenport, *The Higher Arithmetic*, Cambridge Univ. Press, 5th edition, 1982.

[dab] D. A. Benaron, personal communication.

[DAM] *Discrete Applied Mathematics*.

[DB1] F. N. David and D. E. Barton, *Combinatorial Chance*, Hafner, NY, 1962.

[DCC] *Designs, Codes and Cryptography*.

[DCG] *Discrete and Computational Geometry*.

[DE17] C. F. Degen, *Canon Pellianus*, Hafniae, Copenhagen, 1817.

[DE87] M.-P. Delest, *Utilisation des Langages Algébriques et du Calcul Formel Pour le Codage et l'Enumeration des Polyominos*, Ph.D. Dissertation, Université Bordeaux I, May 1987.

[dek] D. E. Knuth, personal communication.

[DG72] C. Domb and M. S. Green, editors, *Phase Transitions and Critical Phenomena*, Vol. 2, Academic Press, NY, 1972.

[DG74] C. Domb and M. S. Green, editors, *Phase Transitions and Critical Phenomena*, Vol. 3, Academic Press, NY, 1974.

[DG89] C. Domb and J. L. Lebowitz, editors, *Phase Transitions and Critical Phenomena*, Vol. 13, Academic Press, NY, 1989.

[dgc] D. G. Cantor, personal communication.

[DGS94] P. Diaconis, R. L. Graham and B. Sturmfels, "Primitive partition identities," preprint, 1994.

[dhl] D. H. Lehmer, personal communication.

[dhr] D. H. Redelmeier, personal communication.

[dhw] D. H. Wiedemann, personal communication.

[DI57] L. E. Dickson, *Introduction to the Theory of Numbers*, Dover, NY, 1957.

[DI58] L. E. Dickson, *Linear Groups with an Exposition of the Galois Field Theory*, Dover, NY, 1958.

[DIA] *Diskretnyi Analiz* (Novosibirsk).

[Dil92] M. B. Dillencourt, *Polyhedra of small orders and their Hamiltonian properties*, Tech. Rep. 92–91, Info. and Comp. Sci. Dept., Univ. Calif. Irvine, 1992.

[DKB] F. N. David, M. G. Kendall and D. E. Barton, *Symmetric Function and Allied Tables*, Cambridge Univ. Press, London, 1966.

[DM] *Discrete Mathematics*.

[DO64] A. P. Domoryad, *Mathematical Games and Pastimes*, Macmillan, NY, 1964.

[DO86] L. Dorst, *Discrete Straight Line Segments: Parameters, Primitives and Properties*, Ph. D. Dissertation, Delft Univ. Technology, 1986.

[drh] D. R. Hofstadter, personal communication.

[dsk] D. S. Kluk, personal communication.

[DT76] D. E. Daykin and S. J. Tucker, *Introduction to Dragon Curves*, unpublished, 1976.

[DU92] I. Dutour, "Animaux dirigés et approximations de séries génératrices," Mémoire de DEA, June 1992, Université Bordeaux I.

[DUMJ] *Duke Mathematical Journal*.

[DVSS] *Det Kongelige Danske Videnskabernes Selskabs Skrifter*.

[DyWh94] W. M. Dymacek and T. Whaley, "Generating strings for bipartite Steinhaus graphs," *Discrete Math.*, to appear, 1994.

[dz] D. Zagier, personal communication.

[D1] L. E. Dickson, *History of the Theory of Numbers*, Carnegie Institute Public. 256, Washington, DC, Vol. 1, 1919; Vol. 2, 1920; Vol. 3, 1923.

[eco] E. Coven, personal communication.

[ecp] E. C. Pervin, personal communication.

[EDMN] *The Edinburgh Mathematical Notes*.

[Edwa74] H. M. Edwards, *Riemann's Zeta Function*, Academic Press, NY, 1974.

[EG80] P. Erdös and R. L. Graham, *Old and New Problems and Results in Combinatorial Number Theory*, L'Enseignement Math., Geneva, 1980.

[EG92] I.G. Enting, A.J. Guttmann, L.B. Richmond and N.C. Worwald, "Enumeration of almost-convex polygons on the square lattice," preprint 1992.

[EhRe94] R. Ehrenborg and M. A. Readdy, "Sheffer posets and R-signed permutations," LACIM, Univ. Québec Montréal, May 1994.

[EIS] N. J. A. Sloane and S. Plouffe, *The Encyclopedia of Integer Sequences*, Academic Press, San Diego, 1995 (this book).

[EJC] *European Journal of Combinatorics*.

[ELM] *Elemente der Mathematik*.

[EM] *L'Enseignement Mathématique*.

[Enge93] P. Engel, "Geometric crystallography," in P. M. Gruber and J. M. Wills, editors, *Handbook of Convex Geometry*, North-Holland, Amsterdam, Vol. B, pp. 989–1041.

[ENVP] *Environment and Planning*.

[ErSt89] M. Erné and K. Stege, "The number of partially ordered sets," preprint, 1989.

[EUL] L. Euler, *Opera Omnia*, Teubner, Leipzig, 1911.

[EUR] *Eureka, the Journal of the Archimedeans (Cambridge University Mathematical Society)*.

[ew] E. Wolman, personal communication.

[EXPM] *Experimental Mathematics*.

[FA90] D. C. Fielder and C. O. Alford, "An investigation of sequences derived from Hoggatt sums and Hoggatt triangles," in G. E. Bergum et al., editors, *Applications of Fibonacci Numbers: Proc. Third Internat. Conf. on Fibonacci Numbers and Their Applications, Pisa, July 25–29, 1988*, Kluwer, Dordrecht, Holland, Vol. 3, 1990, pp. 77–88.

[FE90] G. J. Fee, "Computation of Catalan's constant using Ramanujan's formula," in *Proc. Internat. Symposium on Symbolic and Algebraic Computation (ISSAC '90)*, 1990, pp. 157–160.

[Fell60] W. Feller, *An Introduction to Probability Theory and Its Applications*, Wiley, NY, Vol. 1, 2nd ed., 1960.

[Fel92] D. Feldman, "Counting plane trees," preprint, 1992.

[FI50] R. A. Fisher, *Contributions to Mathematical Statistics*, Wiley, NY, 1950.

[FI64] M. Fiedler, editor, *Theory of Graphs and Its Applications: Proceedings of the Symposium, Smolenice, Czechoslovakia, 1963*, Academic Press, NY, 1964.

[FiFi87] H. J. Fischbeck and K. Fischbeck, *Formulas, Facts and Constants*, Springer-Verlag, NY, 2nd ed., 1987.

[Fla91] P. Flajolet, "Pólya festoons," *J. Combin. Theory*, to appear.

[FMN94] D. Ford, J. McKay and S. P. Norton, "More on replicable functions," preprint, 1994.

[FMR] A. Fletcher, J. C. P. Miller, L. Rosenhead, and L. J. Comrie, *An Index of Mathematical Tables*, Vols. 1 and 2, 2nd ed., Blackwell, Oxford and Addison-Wesley, Reading, MA, 1962.

[FoMK91] D. Ford and J. McKay, personal communication, 1991.

[FP91] G. J. Fox and P. E. Parker, "The Lorentzian modular group and nonlinear lattices," preprint, 1991.

[FPSAC] A. Barlotti, M. Delest and R. Pinzani, editors, *Proceedings of the Fifth Conference on Formal Power Series and Algebraic Combinatorics*, Florence Italy, June 21–25 1993.

[FQ] *The Fibonacci Quarterly*.

[frb] F. R. Bernhart, personal communication.

[FrKi94] A. S. Fraenkel and C. Kimberling, "Generalized Wythoff arrays, shuffles, and interspersions," *Discrete Math.*, to appear.

[FVS] *Finska Vetenskaps-Societeten, Comment. Physico Math.*.

[FW39] Federal Works Agency, Work Projects Administration for the City of NY, *Tables of the Exponential Function*, National Bureau of Standards, Washington, DC, 1939.

[FY63] R. A. Fisher and F. Yates, *Statistical Tables for Biological, Agricultural and Medical Research*, 6th ed., Hafner, NY, 1963.

[GA01] C. F. Gauss, *Disquisitiones Arithmeticae*, 1801. English translation: Yale University Press, New Haven, CT, 1966.

[GA61] M. Gardner, *The 2nd Scientific American Book of Mathematical Puzzles and Diversions*, Simon and Schuster, NY, 1961.

[GA66] M. Gardner, *New Mathematical Diversions from Scientific American*, Simon and Schuster, NY, 1966.

[GA69] M. Gardner, *The Unexpected Hanging and Other Mathematical Diversions*, Simon and Schuster, NY, 1969.

[GA77] M. Gardner, *Mathematical Carnival*, Random House, NY, 1977.

[GA78] M. Gardner, *Mathematical Magic Show*, Random House, NY, 1978.

[GA83] M. Gardner, *Wheels, Life and Other Mathematical Amusements*, Freeman, NY, 1983.

[GA85] M. Gardner, *The Magic Numbers of Dr Matrix*, Prometheus, Buffalo, NY, 1985.

[GA86] M. Gardner, *Knotted Doughnuts and Other Mathematical Entertainments*, Freeman, NY, 1986.

[GA88] M. Gardner, *Time Travel and Other Mathematical Bewilderments*, Freeman, NY, 1988.

[GA89] M. Gardner, *Penrose Tiles to Trapdoor Ciphers*, Freeman, NY, 1989.

[GA89a] M. Gardner, *Whys and Wherefores*, Univ. Chicago Press, 1989.

[GA91] M. Gardner, *Fractal Music, Hypercards, and More*, Freeman, NY, 1991.

[gb] G. Butler, personal communication.

[GC] *Graphs and Combinatorics*.

[GEB] D. R. Hofstadter, *Gödel, Escher, Bach: An Eternal Golden Braid*, Vintage Books, NY, 1980.

[GeRo87] I. Gessel and G.-C. Rota, editors, *Classic Papers in Combinatorics*, Birkhäuser, Boston, 1987.

[GH55] W. H. Gottschalk and G. A. Hedlund, *Topological Dynamics*, American Mathematical Society, Colloquium Publications, Vol. 36, Providence, RI, 1955.

[GI89] D. Gusfield and R. W. Irving, *The Stable Marriage Problem: Structure and Algorithms*, MIT Press, 1989.

[GJ83] I. P. Goulden and D. M. Jackson, *Combinatorial Enumeration*, Wiley, NY, 1983.

[GK68] R. K. Guy and P. A. Kelly, *The No-Three-Line Problem*, Research Paper 33, Department of Mathematics, Univ. of Calgary, Calgary, Alberta, January 1968. Condensed version in *Canad. Math. Bull*, Vol. 11, pp. 527–531, 1968.

[GK90] D. H. Greene and D. E. Knuth, *Mathematics for the Analysis of Algorithms*, Birkhäuser, Boston, 3rd edition, 1990.

[GKP] R. L. Graham, D. E. Knuth and O. Patashnik, *Concrete Mathematics*, Addison-Wesley, Reading, MA, 1990.

[gla] G. Labelle, personal communication.

[glc] G. L. Cohen, personal communication.

[gm] G. Melançon, personal communication.

[GMD] *Geometriae Dedicata*.

[GMJ] *Glasgow Mathematical Journal* (formerly *Proceedings of the Glasgow Mathematical Association*).

[GN75] R. K. Guy and R. Nowakowski, "Discovering primes with Euclid," *Delta* (Waukesha), Vol. 5, pp. 49–63, 1975.

[GN93] R. K. Guy and R. J. Nowakowski, "Mousetrap," in D. Miklos, V.T. Sos and T. Szonyi, eds., *Combinatorics, Paul Erdos is Eighty*, Boylai Society Math. Studies, Vol. 1, pp. 193-206, 1993.

[GO61] H. W. Gould, *Exponential Binomial Coefficient Series*, Tech. Rep. 4, Math. Dept., West Virginia Univ., Morgantown, WV, Sept. 1961.

[GO65] S. W. Golomb, *Polyominoes*, Scribners, NY, 1965; second edition (*Polyominoes: Puzzles, Packings, Problems and Patterns*) to be published by Princeton Univ. Press, 1994.

[GO67] S. W. Golomb, *Shift Register Sequences*, Holden-Day, San Francisco, CA, 1967.

[GO71] H. W. Gould, "Research bibliography of two special number sequences," *Mathematica Monongaliae*, Vol. 12, 1971.

[GoBe89] D. Gouyou-Beauchamps, "Tableaux de Havender standards," in S. Brlek, editor, *Parallélisme: Modèles et Complexité*, LACIM, Université du Québec à Montréal, 1989.

[Gold94] J. R. Goldman, *Number Theory*, Peters, Wellesley, MA, 1994.

[gr] G. Royle, personal communication.

[GR38] D. A. Grave, *Traktat z Algebrichnogo Analizu (Monograph on Algebraic Analysis)*, Vol. 2, Vidavnitstvo Akademiia Nauk, Kiev, 1938.

[GR67] B. Grünbaum, *Convex Polytopes*, Wiley, NY, 1967.

[GR72] B. Grünbaum, *Arrangements and Spreads*, American Mathematical Society, Providence, RI, 1972.

[GR85] E. Grosswald, *Representations of Integers as Sums of Squares*, Springer-Verlag, NY, 1985.

[grauzy] G. Rauzy, personal communication.

[GrLe87] P. M. Gruber and C. G. Lekkerkerker, *Geometry of Numbers*, North-Holland, Amsterdam, 2nd ed., 1987.

[gs] G. Siebert, personal communication.

[GTA85] Y. Alavi et al., *Graph Theory with Applications to Algorithms and Computer Science*, Wiley, NY, 1985.

[GTA91] Y. Alavi et al., *Graph Theory, Combinatorics, and Applications*, Wiley, NY, 2 vols., 1991.

[GU58] R. K. Guy, "Dissecting a polygon into triangles," *Bull. Malayan Math. Soc.*, Vol. 5, pp. 57–60, 1958.

[GU60] R. K. Guy, "The crossing number of the complete graph," *Bull. Malayan Math. Soc.*, Vol. 7, pp. 68–72, 1960.

[GU62] R. C. Gunning, *Lectures on Modular Forms*, Princeton Univ. Press, Princeton, NJ, 1962.

[GU70] R. K. Guy et al., editors, *Combinatorial Structures and Their Applications* (*Proceedings Calgary Conference June 1969*), Gordon & Breach, NY, 1970.

[GU71] R. K. Guy, "Sedláček's Conjecture on Disjoint Solutions of $x + y = z$," in *Proc. Conf. Number Theory*, Pullman, WA, 1971, pp. 221–223. See also R. K. Guy, "Packing $[1, n]$ with solutions of $ax + by = cz$; the unity of combinatorics," in *Colloq. Internaz. Teorie Combinatorie*, Rome, 1973, *Atti Conv. Lincei*, Vol. 17, Part II, pp. 173–179, 1976.

[GU81] R. K. Guy, "Anyone for Twopins?," in D. A. Klarner, editor, *The Mathematical Gardner*, Prindle, Weber and Schmidt, Boston, 1981, pp. 2–15.

[GU90] R. K. Guy, "Catwalks, sandsteps and Pascal pyramids," preprint, 1990.

[GU94] R. K. Guy, "s-Additive sequences," preprint, 1994.

[GuMo94] R. K. Guy and W. O. J. Moser, "Numbers of subsequences without isolated odd members," *Fibonacci Quarterly*, submitted, 1994.

[GW91] A. J. Guttmann and J. Wang, "The extension of self-avoiding random walk series in 2 dimensions," preprint, 1991.

[HA26] K. Hayashi, *Sieben- und Mehrstellige Tafeln der Kreis- und Hyperbelfunktionen und deren Produkte Sowie der Gammafunktion*, Springer-Verlag, Berlin, 1926.

[HA65] M. A. Harrison, *Introduction to Switching and Automata Theory*, McGraw Hill, NY, 1965.

[HA67] F. Harary, editor, *Graph Theory and Theoretical Physics*, Academic Press, NY, 1967.

[HA69] F. Harary, *Graph Theory*, Addison-Wesley, Reading, MA, 1969.

[HA72] F. Harary, P. O'Neil, R. C. Read, and A. J. Schwenk, "The number of trees in a wheel," in D. J. A. Welsh and D. R. Woodall, editors, *Combinatorics*, Institute of Mathematics and Its Applications, Southend-on-Sea, England, 1972, pp. 155–163.

[HA73] F. Harary, editor, *New Directions in the Theory of Graphs*, Academic Press, NY, 1973.

[Hale94] T. C. Hales, "The status of the Kepler conjecture," *Math. Intelligencer*, Vol. 16, No. 3, pp. 47–58, 1994.

[Halm91] P. R. Halmos, *Problems for Mathematicians Young and Old*, Math. Assoc. America, 1991.

[HAR] G. H. Hardy, *Collected Papers*, Vols. 1–, Oxford Univ. Press, 1966–.

[Hass64] H. Hasse, *Vorlesungen über Zahlentheorie*, Springer-Verlag, NY, 1964.

[Hass80] H. Hasse, *Number Theory*, Springer-Verlag, NY, 1980.

[HB67] F. Harary and L. Beineke, editors, *A Seminar on Graph Theory*, Holt, NY, 1967.

[hd] H. Dubner, personal communication.

[Henr67] P. Henrici, "Quotient-difference algorithms," in A. Ralston and H. Wilf, eds., *Mathematical Methods for Digital Computers*, Wiley, NY, Vol. 2, pp. 35–62, 1967.

[Henr74] P. Henrici, *Applied and Computational Complex Analysis*, Wiley, NY, 3 vols., 1974–1986.

[HER] C. Hermite, *Oeuvres*, Vol. 4, Gauthier-Villars, Paris, 1917.

[herbt] H. Taylor, personal communication.

[Hirs76] M. W. Hirsch, *Differential Topolgy*, Springer-Verlag, NY, 1976.

[HIS] N. J. A. Sloane, *A Handbook of Integer Sequences*, Academic Press, NY, 1973.

[hj] H. Jaleebi, personal communication.

[HM68] B. H. Hannon and W. L. Morris, *Tables of Arithmetical Functions Related to the Fibonacci Numbers*, Report ORNL-4261, Oak Ridge National Laboratory, Oak Ridge, Tennessee, June 1968.

[HM85] F. Harary and J. S. Maybee, editors, *Graphs and Applications*, Wiley, NY, 1985.

[HM94] F. K. Hwang and C. L. Mallows, "Enumerating nested and consecutive partitions," *J. Comb. Theory* A (to appear).

[HO50] L. Hogben, *Choice and Chance by Cardpack and Chessboard*, Vol. 1, Chanticleer Press, NY, 1950.

[HO69] V. E. Hoggatt, Jr., *Fibonacci and Lucas Numbers*, Houghton, Boston, MA, 1969.

[HO70] R. Honsberger, *Ingenuity in Mathematics*, Random House, NY, 1970.

[HO73] R. Honsberger, *Mathematical Gems*, Math. Assoc. America, 1973.

[HO85] D. R. Hofstadter, *Metamagical Themas*, Basic Books, NY, 1985.

[HO85a] R. Honsberger, *Mathematical Gems III*, Math. Assoc. America, 1985.

[hofri] M. Hofri, personal communication.

[HP73] F. Harary and E. M. Palmer, *Graphical Enumeration*, Academic Press, NY, 1973.

[HP81] F. J. Hill and G. R. Peterson, *Introduction to Switching Theory and Logical Design*, Wiley, NY, 3rd ed., 1981.

[hpr] H. P. Robinson, personal communication.

[HS64] M. Hall, Jr. and J. K. Senior, *The Groups of Order 2^n ($n \leq 6$)*, Macmillan, NY, 1964.

[HuPi73] D. R. Hughes and F. C. Piper, *Projective Planes*, Springer-Verlag, NY, 1973.

[HUR] A. Hurwitz, *Mathematische Werke*, Vols. 1 and 2, Birkhaeuser, Basel, 1962–1963.

[HW1] G. H. Hardy and E. M. Wright, *An Introduction to the Theory of Numbers*, 3rd ed., Oxford Univ. Press, London and New York, 1954.

[HW84] F. C. Holroyd and R. J. Wilson, editors, *Geometrical Combinatorics*, Pitman, Boston, 1984.

[hwg] H. W. Gould, personal communication.

[hwl] H. W. Lenstra, Jr., personal communication.

[IAS] *Proceedings of the Indian Academy of Sciences, Section A.*

[IBMJ] *IBM Journal of Research and Development.*

[ICM] *Proceedings of the International Congress of Mathematicians.*

[IDM] *L'Intermédiaire des Mathématiciens.*

[IFC] *Information and Control.*

[IJCM] *International Journal of Computer Mathematics.*

[IJM] *Illinois Journal of Mathematics.*

[IJ1] *Indian Journal of Mathematics.*

[INOC] *Inorganic Chemistry.*

[INV] *Inventiones mathematicae.*

[IOWA] *Proceedings Iowa Academy of Science.*

[IPL] *Information Processing Letters.*

[IrRo82] K. Ireland and M. Rosen, *A Classical Introduction to Modern Number Theory*, Springer-Verlag, NY, 1982.

[ISJM] *Israel Journal of Mathematics.*

[IUMJ] *Indiana University Mathematics Journal.*

[JA62] N. Jacobson, *Lie Algebras*, Wiley, NY, 1962.

[JA66] D. Jarden, *Recurring Sequences*, Riveon Lematematika, Jerusalem, 1966.

[JA73] T. Janssen, *Crystallographic Groups*, North-Holland, Amsterdam, 1973.

[JACM] *Journal of the Association for Computing Machinery.*

[JACS] *Journal of the American Chemical Society.*

[JALC] *Journal of Algebraic Combinatorics.*

[JALG] *Journal of Algebra.*

[JAlgo] *Journal of Algorithms.*

[JAMS] *Journal of the American Mathematical Society.*

[jar] J. A. Reeds, personal communication.

[JAuMS] *The Journal of the Australian Mathematical Society.*

[jaw] J. A. Wright, personal communication.

[jb] J. Bokowski, personal communication.

[JCAM] *Journal of Computational and Applied Mathematics.*

[JCIS] *Journal of Combinatorics, Information ans System Science.*

[JCMCC] *Journal of Combinatorial Mathematics and Combinatorial Computing.*

[JCP] *The Journal of Chemical Physics.*

[jcpm] J. C. P. Miller, personal communication.

[JCT] *Journal of Combinatorial Theory.*

[JDG] *Journal of Differential Geometry.*

[JDM] *Journal de Mathématiques Pures et Appliquées.*

[JEP] *Journal de l'École Polytechnique*, Paris.

[JFI] *Journal of the Franklin Institute.*

[JGT] *Journal of Graph Theory.*

[jhc] J. H. Conway, personal communication.

[JIA] *Journal of the Institute of Actuaries.*

[JIMS] *The Journal of the Indian Mathematical Society.*

[JK85] I. M. James and E. H. Kronheimer, editors, *Aspects of Topology*, Cambridge Univ. Press, 1985.

[jkh] J. K. Horn, personal communication.

[JLMS] *Journal of the London Mathematical Society.*

[jls] J. L. Selfridge, personal communication.

[jm]   J. Meeus, personal communication.

[JMA]   *Journal of Mathematical Analysis and Applications.*

[jmckay]   J. McKay, personal communication.

[JMP]   *Journal of Mathematical Physics.*

[JMSJ]   *Journal of the Mathematical Society of Japan.*

[JNSM]   *The Journal of Natural Sciences and Mathematics.*

[JNT]   *Journal of Number Theory.*

[JO39]   C. Jordan, *Calculus of Finite Differences*, Budapest, 1939.

[JO61]   L. B. W. Jolley, *Summation of Series*, 2nd ed., Dover, NY, 1961.

[Jone91]   V. F. R. Jones, *Subfactors and Knots*, Amer. Math. Soc., 1991.

[jos]   J. O. Shallit, personal communication.

[JoTh80]   W. B. Jones and W. J. Thron, *Continued Fractions*, Addison-Wesley, Reading MA 1980.

[JPA]   *Journal of Physics A: Mathematical and General.*

[JPC]   *The Journal of Physical Chemistry.*

[jpropp]   J. Propp, personal communication.

[jql]   J. Q. Longyear, personal communication.

[jr]   John Riordan, personal communication.

[JRAM]   *Journal für die Reine und Angewandte Mathematik.*

[JRM]   *Journal of Recreational Mathematics.*

[jrs]   J. R. Stembridge, personal communication.

[JSC]   *Journal of Symbolic Computation.*

[JSIAM]   *SIAM Journal on Applied Mathematics* (formerly *Journal of the Society for Industrial and Applied Mathematics*).

[JSP]   *Journal of Statistical Physics.*

[jwr]   J. Wroblewski, personal communication.

[jww]   J. W. Wrench, Jr., personal communication.

[KA59]   D. R. Kaprekar, *Puzzles of the Self-Numbers*, 311 Devlali Camp, Devlali, India, 1959.

[KA67]   D. R. Kaprekar, *The Mathematics of the New Self Numbers (Part V)*, 311 Devlali Camp, Devlali, India, 1967.

[KAB]   *Sitzungsberichte Königlich Preussischen Akadamie Wissenschaften*, Berlin.

[Kauf87]   L. H. Kauffman, *On Knots*, Princeton Univ. Press, 1987.

[KaWa89]   E. Kaltofen and S. M. Watt, editors, *Computers and Mathematics*, Springer-Verlag, NY, 1989.

[kbrown]   K. Brown, personal communication.

[ken]   K. Thompson, personal communication.

[Kimb91]   C. Kimberling, "Stolarsky interspersions," *Ars Comb.*, to appear.

[Kimb92]   C. Kimberling, "Orderings of the set of all positive Fibonacci sequences," in G. E. Bergum et al., editors, *Applications of Fibonacci Numbers*, Vol. 5 (1993), pp. 405–416.

[Kimb94]   C. Kimberling, "Fractal sequences and interspersions," preprint, 1994.

[Kimb94a]   C. Kimberling, "Numeration systems and fractal sequences," preprint, 1994.

[Kimu94a]   H. Kimura, "Classification of Hadamard matrices of order 24 with Hall sets," *Discrete Math.*, 1994, to appear.

[Kimu94b] H. Kimura, "Hadamard matrices of order 24 without Hall sets are equivalent to the Paley matrix," *Discrete Math.*, 1994, to appear.

[KK71] E. Kogbetliantz and A. Krikorian, *Handbook of First Complex Prime Numbers*, Gordon and Breach, NY, 1971.

[klm] K. L. McAvaney, personal communication.

[KN1] D. E. Knuth, *The Art of Computer Programming*, Addison-Wesley, Reading, MA, Vol. 1, 2nd ed., 1973; Vol. 2, 2nd ed., 1981; Vol. 3, 1973; Vol. 4, In preparation.

[KN75] J. Knopfmacher, *Abstract Analytic Number Theory*, North-Holland, Amsterdam, 1975.

[KN91] D. E. Knuth, "Axioms and hulls," Lect. Notes Comp. Sci., Vol. 606.

[KNAW] *Proceedings of the Koninklijke Nederlandse Akademie van Wetenschappen, Series A.*

[Knop] A. Knopfmacher, "Rational numbers with predictable Engel product expansions," in G. E. Bergum et al., eds., *Applications of Fibonacci Numbers*, Vol. 5, pp. 421–427.

[Krai24] M. Kraitchik, *Recherches sur la Théorie des Nombres*, Gauthiers-Villars, Paris, Vol. 1, 1924, Vol. 2, 1929.

[Krai52] M. Kraitchik, *Introduction à la Théorie des Nombres*, Gauthier-Villars, Paris, 1952.

[Krai53] M. Kraitchik, *Mathematical Recreations*, Dover, NY, 2nd ed., 1953.

[KU64] H. Kumin, *The Enumeration of Trees by Height and Diameter*, M. A. Thesis, Department of Mathematics, Univ. of Texas, Austin, Texas, Jan. 1964. [See *Math. Comp.*, Vol. 25, p. 632, 1971.]

[kw] K. Wayland, personal communication.

[KYU] *Memoirs of the Faculty of Science, Kyusyu University, Series A.*

[L1] D. H. Lehmer, *Guide to Tables in the Theory of Numbers*, Bulletin No. 105, National Research Council, Washington, DC, 1941.

[LA56] C. Lanczos, *Applied Analysis*, Prentice-Hall, Englewood Cliffs, NJ, 1956.

[LA62] H. Langman, *Play Mathematics*, Hafner, NY, 1962.

[LA73] C. A. Landauer, *Simple Groups with 9, 10 and 11 Conjugate Classes*, Ph.D. Dissertation, California Inst. Tech., Pasadena, 1973.

[LA73a] T. Y. Lam, *The Algebraic Theory of Quadratic Forms*, Benjamin, Reading, MA, 1973.

[LA81] J. P. Lamoitier, *Fifty Basic Exercises*, SYBEX Inc., 1981.

[LA91] G. Labelle, "Sur la symétrie et l'asymétrie des structures combinatoires," preprint, 1991.

[LAA] *Linear Algebra and Its Applications.*

[Lang91] K. R. Lang, *Astrophysical Data: Planets and Stars*, Springer-Verlag, NY, 1991.

[LE56] W. J. LeVeque, *Topics in Number Theory*, Addison-Wesley, Reading, MA, 2 vols., 1956.

[LE59] R. S. Lehman, *A Study of Regular Continued Fractions*, Report 1066, Ballistic Research Laboratories, Aberdeen Proving Ground, Maryland, February 1959.

[LE70] J. Leech, editor, *Computational Problems in Abstract Algebra*, Pergamon, Oxford, 1970.

[LE77] W. J. LeVeque, *Fundamentals of Number Theory*, Addison-Wesley, Reading, MA, 1977.

[LeLe59] H. Levy and F. Lessman, *Finite Difference Equations*, Pitman, London, 1959.

[Lem00] E. Lemoine, "Note sur deux nouvelles décompositions des nombres entiers," *Assoc. française pour l'avancement des sciences*, Vol. 29, pp. 72–74, 1900.

[Lem82] E. Lemoine, "Décomposition d'un nombre entier N en ses puissances nièmes maxima," *C. R. Acad. Sci. Paris*, Vol. 95, pp. 719–722, 1882.

[LeMi91] P. Leroux and B. Miloudi, "Généralisations de la formule d'Otter," *Ann. Sci. Math. Québec*, Vol. 16, No. 1, pp. 53–80, 1992.

[LF60] A. V. Lebedev and R. M. Fedorova, *A Guide to Mathematical Tables*, Pergamon, Oxford, 1960.

[LI50] D. E. Littlewood, *The Theory of Group Characters and Matrix Representations of Groups*, 2nd ed., Oxford University Press, 1950.

[LI68] C. L. Liu, *Introduction to Combinatorial Analysis*, McGraw-Hill, NY, 1968.

[Lie92] S. Liedahl, *Q(i)-Division Rings and Admissibility*, Doctoral Dissertation, UCLA, 1992.

[LINM] *Atti della Reale Accademia Nazionale dei Lincei, Memorie della Classe di Scienze Fisiche, Matematiche e Naturali.*

[LINR] *Atti della Reale Accademia Nazionale dei Lincei, Rendiconti della Classe di Scienze Fisiche, Matematiche e Naturali.*

[LiSa92] M. Liebeck and J. Saxl, editors, *Groups, Combinatorics and Geometry (Durham, 1990)*, Cambridge Univ. Press, 1992.

[Liso93] P. Lisoněk, *Quasi-polynomials: a case study in experimental combinatorics*, Report RISC-Linz No. 93–18, Research Institute of Symbolic Computation, Johannes Kepler Univ., Linz, Austria, March 1993.

[LNCS] *Lecture Notes in Computer Science*, Springer-Verlag, NY.

[LNM] *Lecture Notes in Mathematics*, Springer-Verlag, NY.

[Loeb94] D. E. Loeb, "Challenges in playing multiplayer games," in Levy and Beal, editors, *Heuristic Programming in Artificial Intelligence* vol. 4, Ellis Horwood, to appear, 1994.

[Loeb94a] D. E. Loeb and A. Meyerowitz, "The maximal intersecting family of sets graph," in H. Barcelo and G. Kalai, editors, *Proceedings of the Conference on Jerusalem Combinatorics 1993*, to appear, AMS series Contemporary Mathematics, 1994.

[Loeb94b] D. E. Loeb, *On Games, Voting Schemes, and Distributive Lattices*, LaBRI Report 625–93, University of Bordeaux I, 1993.

[Long87] C. T. Long, *Elementary Introduction to Number Theory*, Prentice-Hall, Englewood Cliffs, NJ, 1987.

[Loth83] M. Lothaire, *Combinatorics on Words*, Addison-Wesley, Reading, MA, 1983.

[Loxt89] J. H. Loxton, "Spectral studies of automata," in G. Halász and V. T. Sós, editors, *Irregularities of Partitions*, Springer-Verlag, NY, 1989, pp. 115-128.

[LU79] H. Lüneburg, *Galoisfelder, Kreisteilungskorper und Schieberegisterfolgen*, B. I. Wissenschaftsverlag, Mannheim, 1979.

[LU91] E. Lucas, *Théorie des Nombres*, Gauthier-Villars, Paris, 1891.

[Luk69] Y. L. Luke, *The Special Functions and their Approximation*, Vol. 1, Academic Press, NY, 1969.

[Lunn74] W. F. Lunnon, "Q-D tables and zero-squares," unpublished manuscript, 1974.

[MA15] P. A. MacMahon, *Combinatory Analysis*, Cambridge Univ. Press, London and New York, Vol. 1, 1915 and Vol. 2, 1916.

[MA63] *Symposium on Mathematical Theory of Automata*, Polytechnic Institute of Brooklyn, 1963.

[MA71] *Proceedings Manitoba Conference on Numerical Mathematics*, University of Manitoba, Winnipeg, October 1971.

[MA77] D. Marcus, *Number Fields*, Springer-Verlag, 1977.

[MAB] *Memoires Acad. Bruxelles.*

[Mada66] J. S. Madachy, *Mathematics on Vacation*, Scribners, NY, 1966.

[MAG] *The Mathematical Gazette.*

[MAL] *Mathematical Algorithms.*

[MAN] *Mathematische Annalen.*

[Man82] B. B. Mandelbrot, *The Fractal Geometry of Nature*, Freeman, NY, 1982.

[MaRo94] R. Mathon and G. Royle, "Classification of the translation planes of order 49," *Designs, Codes and Cryptography*, in press, 1994.

[MASC] *Mathematical Scientist.*

[MAT] *Mathematika.*

[MAZ] *Mathematical Notes* (English translation of *Matematicheskie Zametki*).

[mb] M. Bernstein, personal communication.

[MBIO] *Mathematical Biosciences.*

[McEl77] R. J. McEliece, *The Theory of Information and Coding*, Addison-Wesley, Reading MA 1977.

[McL1] W. I. McLaughlin, "Note on a tetranacci alternative to Bode's law," preprint.

[md] M. B. Dillencourt, personal communication.

[MES] *The Messenger of Mathematics.*

[MET] *Metrika.*

[Meye94] A. Meyerowitz, "Maximal intersecting families," submitted to *Eur. J. Comb.*

[MF73] R. W. Marks and R. B. Fuller, *The Dymaxion World of Buckminster Fuller*, Anchor, NY, 1973.

[MFC] *Matematický Časopis* (formerly *Matematicko-Fyzikálny Časopis*).

[MFM] *Monatshefte für Mathematik.*

[mg] M. Gardner, personal communication.

[MI72] K. Miller, *Solutions of $\phi(n) = \phi(n + 1)$ for $1 \leq n \leq 500000$*, unpublished, 1972. [Cf. *Math. Comp.*, Vol. 27, p. 447, 1973].

[MINT] *Mathematical Intelligencer.*

[MIS] *Mathematics in School.*

[MIT] *Massachusetts Institute of Technology, Research Laboratory of Electronics, Quarterly Progress Reports.*

[mlb] M. Le Brun, personal communication.

[MLG] *The Mathematical Magazine.*

[MMAG] *Mathematics Magazine.*

[MMJ] *The Michigan Mathematical Journal.*

[MMTC] *Mathematics Teacher.*

[MNAS] *Memoirs of the National Academy of Sciences.*

[MNR] *Mathematische Nachrichten.*

[MO68] J. W. Moon, *Topics on Tournaments*, Holt, NY, 1968.

[MO78] E. J. Morgan, *Construction of Block Designs and Related Results*, Ph.D. Dissertation, Univ. Queensland, 1978.

[MOC] *Mathematics of Computation* (formerly *Mathematical Tables and Other Aids to Computation*).

[MOD] *Atti del Seminario Matematico e Fisico dell' Università di Modena*.

[Morr80] D. R. Morrison, "A Stolarsky array of Wythoff pairs," in *A Collection of Manuscripts Related to the Fibonacci Sequence*, Fibonacci Assoc., Santa Clara, CA, 1980, pp. 134–136.

[MQET] *Mathematical Questions and Solutions from the Educational Times*.

[MR] *Mathematical Reviews*.

[ms] M. Somos, personal communication.

[MS78] F. J. MacWilliams and N. J. A. Sloane, *The Theory of Error-Correcting Codes*, North-Holland, Amsterdam, 1978.

[MSC] *Mathematica Scandinavica*.

[MSG] *Mitteilungen aus dem Mathematischen Seminar Giessen*.

[MSH] *Mathématiques et Sciences Humaines* (Centre de Mathématique Sociale et de Statistique EPHE, Paris).

[MSM] *Manuscripta Mathematica*.

[MSS] *Manuscripta*.

[MST] *Mathematical Systems Theory*.

[mt] M. B. Thistlethwaite, personal communication.

[MT33] L. M. Milne-Thomson, *The Calculus of Finite Differences*, Macmillan, NY, 1933.

[MTS] *Mathesis*.

[MU06] T. Muir, *The Theory of Determinants in the Historical Order of Development*, 4 vols., Macmillan, NY, 1906–1923.

[MU60] T. Muir, *A Treatise on the Theory of Determinants*, Dover, NY, 1960.

[MU71] S. Muroga, *Threshold Logic and Its Applications*, Wiley, NY, 1971.

[MW63] L. F. Meyers and W. S.-Y. Wang, *Tree Representations in Linguistics*, Report 3, 1963, pp. 107–108, and L. F. Meyers, Corrections and additions to *Tree Representations in Linguistics*, Report 3, 1966, p. 138. Project on Linguistic Analysis, Ohio State University Research Foundation, Columbus, Ohio.

[MZT] *Mathematische Zeitschrift*.

[NA79] T. V. Narayana, *Lattice Path Combinatorics with Statistical Applications*, Univ. Toronto Press, 1979.

[NADM] *Nouvelles Annales de Mathématiques*.

[NAMS] *Notices of the American Mathematical Society*.

[NAT] *Nature*.

[NAW] *Nieuw Archief voor Wiskunde*.

[nbh] N. B. Hindman, personal communication.

[NBS] *Journal of Research of the National Bureau of Standards*.

[NCM] *Nouvelles Correspondance Mathématique*.

[NE72] M. Newman, *Integral Matrices*, Academic Press, NY, 1972.

[NET] E. Netto, *Lehrbuch der Combinatorik*, 2nd ed., Teubner, Leipzig, 1927.

[NMJ] *Nagoya Mathematical Journal*.

[NMT] *Nordisk Matematisk Tidskrift.*

[NO24] N. E. Nörlund, *Vorlesungen über Differenzenrechnung*, Springer-Verlag, Berlin, 1924.

[NR82] A. Nymeyer and R. W. Robinson, "Tabulation of the Numbers of Labeled Bipartite Blocks and Related Classes of Bicolored Graphs," unpublished manuscript, 1982.

[nsh] N. S. Hellerstein, personal communication.

[NUPH] *Nuclear Physics.*

[NZ66] I. Niven and H. S. Zuckerman, *An Introduction to the Theory of Numbers*, 2nd ed., Wiley, NY, 1966.

[OB66] W. Oberschelp, "Strukturzahlen in endlichen Relationssystemen," in *Contributions to Mathematical Logic (Proceedings 1966 Hanover Colloquium)*, pp. 199–213, North-Holland Publ., Amsterdam, 1968.

[Odly95] A. M. Odlyzko, *The 10^{20}-th Zero of the Riemann Zeta Function and 175 Million of its Neighbors*, manuscript in preparation.

[OG72] C. S. Ogilvy, *Tomorrow's Math*, 2nd ed., Oxford Univ. Press, 1972.

[OgAn66] C. S. Ogilvy and J. T. Anderson, *Excursions in Number Theory*, Oxford Univ. Press, 1966.

[Oliv93] D. Olivastro, *Ancient Puzzles*, Bantam Books, NY, 1993.

[ONAG] J. H. Conway, *On Numbers and Games*, Academic Press, NY, 1976.

[OP80] T. R. Van Oppolzer, *Lehrbuch zur Bahnbestimmung der Kometen und Planeten,*, Vol. 2, Engelmann, Leipzig, 1880.

[OR76] *Problèmes combinatoires et théorie des graphes (Orsay, 9–13 Juillet 1976)*, Colloq. Internat. du C.N.R.S., No. 260, Centre Nat. Recherche Scient., Paris, 1978.

[ORD] *Order.*

[P4BC] T. P. McDonough and V. C. Mavron, editors, *Combinatorics: Proceedings of the Fourth British Combinatorial Conference 1973*, London Mathematical Society, Lecture Note Series, Number 13, Cambridge University Press, NY, 1974.

[P5BC] C. J. Nash-Williams and J. Sheehan, editors, *Proceedings of the Fifth British Combinatorial Conference 1975*, Utilitas Math., Winnipeg, 1976.

[PA67] T. R. Parkin, L. J. Lander, and D. R. Parkin, "Polyomino enumeration results," *SIAM Fall Meeting, Santa Barbara, California, 1967.*

[pam] P. A. Morris, personal communication.

[PAMS] *Proceedings of the American Mathematical Society.*

[PC] *Popular Computing* (Calabasas, California).

[pcf] P. C. Fishburn, personal communication.

[PCPS] *Mathematical Proceedings of the Cambridge Philosophical Society* (formerly *Proceedings of the Cambridge Philosophical Society*).

[pdl] P. D. Lincoln, personal communication.

[pdm] P. D. Mitchelmore, personal communication.

[PE29] O. Perron, *Die Lehre von den Kettenbrüchen,*, 2nd ed., Teubner, Leipzig, 1929.

[PE57] J. T. Peters, *Ten-Place Logarithm Table*, Vols. 1 and 2, rev. ed. Ungar, NY, 1957.

[PE79] R. E. Peile, *Some Topics in Combinatorial Theory*, Ph. D. Dissertation, Oxford University, 1979.

[PE90] J.-G. Penaud, *Arbres et Animaux*, Ph.D. Dissertation, Université Bordeaux I, May 1990.

[PEF] *Publications Faculté d' Électrotechnique de l'Univ. Belgrade.*

[PEMS] *Proceedings of the Edinburgh Mathematical Society.*

[Pen91] R. Penrose, *The Emperor's New Mind*, Penguin Books, NY, 1991.

[PGAC] *IEEE Transactions on Automatic Control.*

[PGCT] *IEEE Transactions on Circuit Theory.*

[PGEC] *IEEE Transactions on Computers* (formerly *IEEE Transactions on Electronic Computers*).

[PGIT] *IEEE Transactions on Information Theory.*

[PH88] A. Phillips, "Simple Alternating Transit Mazes," preprint. Abridged version appeared as "La topologia dei labirinti," in M. Emmer, editor, *L'Occhio di Horus: Itinerari nell'Imaginario Matematico*, Istituto della Enciclopedia Italia, Rome, 1989, pp. 57–67.

[PHA] *Physica.*

[PHL] *Physics Letters.*

[PHM] *The Philosophical Magazine.*

[Pick91] C. Pickover, *Computers and the Imagination*, St. Martin's Press, NY, 1991.

[Pick92] C. Pickover, *Mazes for the Mind*, St. Martin's Press, NY, 1992.

[PIEE] *Proceedings of the Institution of Electrical Engineers.*

[pjc] P. J. Cameron, personal communication.

[pjh] P. J. Hanlon, personal communication.

[PJM] *Pacific Journal of Mathematics.*

[PL65] R. J. Plemmons, *Cayley Tables for All Semigroups of Order Less Than 7*, Department of Mathematics, Auburn Univ., 1965.

[PLC] Volume 1: *Proceedings of the Louisiana Conference on Combinatorics, Graph Theory and Computer Science*, edited R. C. Mullin et al., 1970. Volume 2: *Proceedings of the Second Louisiana Conference on Combinatorics, Graph Theory and Computing*, edited R. C. Mullin et al., 1971.

[PLIS] *Publications de l'Institut de Statistique de l'Université de Paris.*

[PLMS] *Proceedings of the London Mathematical Society.*

[Plou92] S. Plouffe, *Approximations de Séries Génératrices et Quelques Conjectures*, Dissertation, Université du Québec à Montréal, 1992.

[PME] *Pi Mu Epsilon Journal.*

[PNAS] *Proceedings of the National Academy of Sciences of the United States of America.*

[PNISI] *Proceedings of the Indian National Science Academy* (formerly *Proceedings of the National Institute of Sciences of India*), Part A.

[PO20] L. Poletti, *Tavole di Numeri Primi Entro Limiti Diversi e Tavole Affini*, Heopli, Milan, 1920.

[PO54] G. Pólya, *Induction and Analogy in Mathematics*, Princeton Univ. Press, 1954.

[PO74] K. A. Post, *Binary Sequences with Restricted Repetitions*, Report 74-WSK-02, Math. Dept., Tech. Univ. Eindhoven, May. 1974.

[PoRu94] B. Poonen and M. Rubinstein, "The number of intersection points made by the diagonals of a regular polygon," preprint, 1994.

[PoSz72] G. Pólya and G. Szegö, *Problems and Theorems in Analysis*, Springer-Verlag, NY, 2 vols., 1972.

[PPS] *Proceedings of the Physical Society.*

[PR33] G. Prévost, *Tables de Fonctions Sphériques*, Gauthier-Villars, Paris, 1933.

[prs] P. R. Stein, personal communication.

[PRS] *Proceedings of the Royal Society of London.*

[PRV] *Physical Review.*

[PSAM] *Proceedings of Symposia in Applied Mathematics*, American Mathematical Society, Providence, RI.

[PSPM] *Proceedings of Symposia in Pure Mathematics*, American Mathematical Society, Providence, RI.

[PTGT] F. Harary, editor, *Proof Techniques in Graph Theory*, Academic Press, NY, 1969.

[PTRS] *Philosophical Transactions of the Royal Society of London, Series A.*

[PURB] *Research Bulletin of the Panjab University (Science Section).*

[PW72] W. W. Peterson and E. J. Weldon, Jr., *Error-Correcting Codes*, MIT Press, Cambridge, MA, 2nd edition, 1972.

[PYTH] *Pythagoras* (Netherlands).

[QAM] *Quarterly of Applied Mathematics.*

[QJMA] *The Quarterly Journal of Pure and Applied Mathematics.*

[QJMO] *The Quarterly Journal of Mathematics, Oxford Series.*

[R1] J. Riordan, *An Introduction to Combinatorial Analysis*, Wiley, NY, 1958.

[RA55] The Rand Corporation, *A Million Random Digits with 100,000 Normal Deviates*, The Free Press, NY, 1955.

[Rade64] H. Rademacher, *Lectures on Elementary Number Theory*, Blaisdell, NY, 1964.

[RAIRO] *RAIRO Informatique Théorique et Applications.*

[RAM1] Srinivasa Ramanujan, *Coll. Papers*, Cambridge Univ. Pr., 1927.

[RAM2] G. H. Hardy, *Ramanujan*, Cambridge Univ. Press, 1940.

[RAM3] G. E. Andrews et al., editors, *Ramanujan Revisited*, Academic Press, NY, 1988.

[RCI] J. Riordan, *Combinatorial Identities*, Wiley, NY, 1968.

[rcr] R. C. Read, personal communication.

[RE58] R. C. Read, *Some Enumeration Problems in Graph Theory*, Ph.D. Dissertation, Department of Mathematics, Univ. London, 1958.

[RE72] R. C. Read, editor, *Graph Theory and Computing*, Academic Press, NY, 1972.

[RE75] K. G. Reuschle, *Tafeln Complexer Primzahlen*, Königl. Akademie der Wissenschaften, Berlin, 1875.

[RE89] R. C. Read, *The Enumeration of Mating-Type Graphs*, Report CORR 89–38, Dept. Combinatorics and Optimization, Univ. Waterloo, October, 1989.

[ReSk94] V. Reddy and S. Skiena, "Frequencies of large distances in integer lattices," in *Proc. Seventh Internat. Conf. Graph Theory, Combinatorics, Algorithms, and Applications*, Kalamazoo, MI, 1994, to appear.

[rgep] R. G. E. Pinch, personal communication.

[rgw] R. G. Wilson, personal communication.

[rhb] R. H. Buchholz, personal communication.

[rhh] R. H. Hardin, personal communication.

[RI63] J. Riordan, "The enumeration of permutations with three-ply staircase restric-tions," unpublished memorandum, Bell Telephone Laboratories, Murray Hill, NJ, October 1963.

[RI72] H. J. J. Te Riele, *Unitary Aliquot Sequences*, Report MR 139/72, Mathema-tisch Centrum, Amsterdam, September 1972.

[Ribe72] P. Ribenboim, *Algebraic Numbers*, Wiley, NY, 1972.

[Ribe91] P. Ribenboim, *The Little Book of Big Primes*, Springer-Verlag, NY, 1991.

[Rie85] H. Riesel, "Prime numbers and computer methods for factorization," *Progress in Mathematics*, Vol. 57, Birkhauser, Boston, 1985, Chap. 4.

[rkg] R. K. Guy, personal communication.

[rlg] R. L. Graham, personal communication.

[RLM] *Riveon Lematematika*.

[RMM] C. R. Rao, S. K. Mitra, and A. Matthai, editors, *Formulae and Tables for Statistical Work*, Statistical Publishing Society, Calcutta, India, 1966.

[RMP] *Reviews of Modern Physics*.

[RO00] H. A. Rothe, in C. F. Hindenburg, editor, *Sammlung Combinatorisch-Analytischer Abhandlungen*, Vol. 2, Chap. XI. Fleischer, Leipzig, 1800.

[RO67] P. Rosenthiehl, editor, *Theory of Graphs (International Symposium, Rome, 1966)*. Gordon & Breach, NY, 1967.

[RO69] S. Rosen, *Wizard of the Dome: R. Buckminster Fuller; Designer for the Future*, Little, Brown, Boston, 1969.

[Robe92] J. Roberts, *Lure of the Integers*, Math. Assoc. America, 1992.

[Rota75] G.-C. Rota et al., *Finite Operator Calculus*, Academic Press, NY, 1975.

[RoWa91] R. W. Robinson and T. R. Walsh, *Inversion of Cycle Index Sum Relations for 2- and 3-Connected Graphs*, Report UGA-CS-TR-91-008, Comp. Sci. Dept., Univ. Georgia, Athens, GA, 1991.

[rpm] R. P. Munafo, personal communication.

[rs] R. Schroeppel, personal communication.

[RS3] J. C. P. Miller, editor, *Table of Binomial Coefficients*, Royal Society Mathe-matical Tables, Vol. 3, Cambridge Univ. Press, 1954.

[RS4] H. Gupta et al., *Tables of Partitions*, Royal Society Mathematical Tables, Vol. 4, Cambridge Univ. Press, 1958.

[RS6] C. B. Haselgrove and J. C. P. Miller, *Tables of the Riemann Zeta Function*, Royal Society Mathematical Tables, Vol. 6, Cambridge Univ. Press, 1960.

[RS8] W. E. Mansell, *Tables of Natural and Common Logarithms*, Royal Society Mathematical Tables, Vol. 8, Cambridge Univ. Press, 1964.

[RS9] A. E. Western and J. C. P. Miller, *Tables of Indices and Primitive Roots*, Royal Society Mathematical Tables, Vol. 9, Cambridge Univ. Press, 1968.

[RSA] *Random Structures and Algorithms*.

[RSE] *Proceedings of the Royal Society of Edinburgh, Section A*.

[rwg] R. W. Gosper, personal communication.

[rwr] R. W. Robinson, personal communication.

[RY63] H. J. Ryser, *Combinatorial Mathematics*, Mathematical Association of Amer-ica, Carus Mathematical Monograph 14, 1963.

[SA81] A. Salomaa, *Jewels of Formal Language Theory*, Computer Science Press, Rockville, MD, 1981.

[sab] S. A. Burr, personal communication.

[SAM] *Studies in Applied Mathematics* (formerly *The Journal of Mathematics and Physics*).

[Satt94] U. Sattler, *Decidable Classes of Formal Power Series with Nice Closure Properties*, Diplomarbeit, Friedrich Alexander Universität, Erlangen Nürnberg, 1994.

[SaZi94] B. Salvy and P. Zimmermann, "Gfun: a Maple package for the manipulation of generating and holonomic functions in one variable," *ACM Trans. Mathematical Software*, 1994, in press.

[SC68] F. Schuh, *The Master Book of Mathematical Recreations*, Dover, NY, 1968.

[SC80] R. L. E. Schwarzenberger, *N-Dimensional Crystallography*, Pitman, London, 1980.

[SC83] M. de Sainte-Catherine, *Couplages et Pfaffiens en Combinatoire, Phys-ique et Informatique*, Ph.D Dissertation, Université Bordeaux I, 1983.

[ScAm] *Scientific American*.

[Scho74] B. Schoeneberg, *Elliptic Modular Functions* Springer-Verlag, NY, 1974.

[Schut91] C. P. Schut, *Idempotents*, Report AM-R9101, Centrum voor Wiskunde en Informatica, Amsterdam, 1991.

[SCS] *Journal of Statistical Computation and Simulation*.

[SE33] J. Ser, *Les Calculs Formels des Séries de Factorielles*, Gauthier-Villars, Paris, 1933.

[SeYa92] J. Seberry and M. Yamada, "Hadamard matrices, sequences, and block designs," in J. H. Dinitz and D. R. Stinson, eds., *Contemporary Design Theory*, Wiley NJ, 1992, pp. 431-560.

[SFCA91] *Séries Formelles et Combinatoire Algébrique*, Laboratoire Bordelais de Recherche Informatique, Université Bordeaux I, 1991.

[SFCA92] *Séries Formelles et Combinatoire Algébrique*, 4th colloquium, 15–19 Juin 1992, Montréal, Université du Québec à Montréal.

[SGF] *Semigroup Forum*.

[Shal94] J. O. Shallit, "On the dimension of the vector space of the automatic reals," preprint, 1994.

[Shan78] D. Shanks, *Solved and Unsolved Problems in Number Theory*, Chelsea, NY, 2nd edition, 1978.

[SI64] W. Sierpiński, *A Selection of Problems in the Theory of Numbers*, Macmillan, NY, 1964.

[SI64a] W. Sierpiński, *Elementary Theory of Numbers*, Państ. Wydaw. Nauk., Warsaw, 1964.

[SI72] D. Singmaster, "The number of representations of one as a sum of unit fractions," unpublished manuscript, 1972.

[SI74] W. Sierpiński, *Oeuvres Choisies*, Académie Polonaise des Sciences, Warsaw, Poland, 1974.

[SIAA] *SIAM Journal on Algebraic and Discrete Methods*.

[SIAC] *SIAM Journal on Computing*.

[SIAD] *SIAM Journal on Discrete Mathematics*.

[SIAJ] *SIAM Journal on Applied Mathematics.*

[SIAMP] SIAM-AMS Proceedings, published by American Mathematical Society, Providence, RI.

[SIAR] *SIAM Review.*

[Sim92] T. Simpson, "Permutations with unique fixed and reflected points," *Ars Combinatorica*, in press, 1994.

[SKA] *Skandinavisk Aktuarietidskrift.*

[skb] S. K. Bhaskar, personal communication.

[SKY] *Sky and Telescope.*

[SL26] M. A. Sainte-Laguë, *Les Réseaux (ou Graphes)*, Mémorial des Sciences Mathématiques, Fasc. 18., Gauthier-Villars, Paris, 1926.

[Slat66] L. J. Slater, *Generalized Hypergeometric Functions*, Cambridge Univ. Press, 1966.

[SLC] *Séminaire Lotharingien de Combinatoire*, Institut de Recherche Math. Avancée, Université Louis Pasteur, Strasbourg.

[Sloa94] N. J. A. Sloane, "An on-line version of the Encyclopedia of Integer Sequences," *Electronic J. Combinatorics*, Vol. 1, no. 1, 1994.

[sls] S. L. Snover, personal communication.

[SMA] *Scripta Mathematica.*

[smd] S. M. Diano, personal communication.

[SMD] *Soviet Mathematics–Doklady.*

[SMH] *Studia Scientiarum Mathematicarum Hungarica.*

[SMS] *Selecta Mathematica Sovietica.*

[SoGo94] H. Y. Song and S. W. Golomb, "Generalized Welch-Costas sequences and their application to Vatican arrays," in *Proc. 2nd International Conference on Finite Fields: Theory, Algorithms, and Applications (Las Vegas 1993)*, Contemp. Math., to appear, 1994.

[sp] S. Plouffe, personal communication.

[SPH] *Sphinx.*

[SPLAG] J. H. Conway and N. J. A. Sloane, *Sphere-Packings, Lattices and Groups*, Springer-Verlag, NY, 2nd ed., 1993.

[SpOl87] J. Spanier and K. B. Oldham, *An Atlas of Functions*, Hemisphere, NY, 1987.

[SPS] *Space Programs Summary*, Jet Propulsion Laboratory, California Institute of Technology, Pasadena, California.

[SS64] M. L. Stein and P. R. Stein, *Tables of the Number of Binary Decompositions of All Even Numbers Less Than 200,000 into Prime Numbers and Lucky Numbers*, Report LA-3106, Los Alamos Scientific Laboratory of the University of California, Los Alamos, NM, Sept. 1964.

[SS67] M. L. Stein and P. R. Stein, *Enumeration of Linear Graphs and Connected Linear Graphs up to $p = 18$ Points*, Report LA-3775, Los Alamos Scientific Laboratory of the University of California, Los Alamos, NM, October 1967.

[SS70] M. L. Stein and P. R. Stein, *Enumeration of Stochastic Matrices with Integer Elements*, Report LA-4434, Los Alamos Scientific Laboratory of the University of California, Los Alamos, NM, June 1970.

[SSP] *Solid State Physics (Journal of Physics C).*

[ssw] S. S. Wagstaff, Jr., personal communication.

[ST70]   H. M. Stark, *An Introduction to Number Theory*, Markham, Chicago, 1970.

[ST90]   P. Steinbach, *Field Guide to Simple Graphs*, Design Lab, Albuquerque NM, 1990.

[Stan86] R. P. Stanley, *Enumerative Combinatorics*, Wadsworth, Monterey, CA, Vol. 1, 1986.

[Stan89] R. P. Stanley, "A zonotope associated with graphical degree sequences," in *Applied Geometry and Discrete Combinatorics*, DIMACS Series in Discrete Math., Amer. Math. Soc., Vol. 4, pp. 555–570, 1991.

[STNB]   *Séminaire de Théorie des Nombres de Bordeaux.*

[SU70]   R. G. Schrandt and S. M. Ulam, "On recursively defined geometric objects and patterns of growth," in A. W. Burks, editor, *Essays on Cellular Automata*, Univ. Ill. Press, 1970, p. 238.

[Supp74] N. J. A. Sloane, *Supplement I to A Handbook of Integer Sequences*, Bell Labs, Murray Hill, NJ, March 1, 1974.

[SW91]   S. L. Snover, C. Wavereis and J. K. Williams, "Rep-tiling for triangles," *Discrete Math.*, to appear.

[SYL]   J. J. Sylvester, *Collected Mathematical Papers*, Vols. 1–4, Cambridge Univ. Press, 1904–1912.

[SZ]   *Systematic Zoology.*

[TAIT]   P. G. Tait, *Scientific Papers*, Cambridge Univ. Press, Vol. 1, 1898, Vol. 2, 1900.

[TAMS]   *Transactions of the American Mathematical Society.*

[tb]   T. Brenner, personal communication.

[TCPS]   *Transactions of the Cambridge Philosophical Society.*

[TCS]   *Theoretical Computer Science.*

[TET]   *Tetrahedron.*

[TH06]   A. Thue, *Über Unendliche Zeichenreihen*, Skrifter utg. av Videnskapsselskapet i Kristiania 1 Mat.–nat. Kl., 1906, No. 7.

[TH09]   T. N. Thiele, *Interpolationsrechnung*, Teubner, Leipzig, 1909.

[TH52]   V. Thébault, *Les Récréations Mathématiques*, Gauthier-Villars, Paris, 1952.

[Theo94] P. Théorêt, *Hyperbinomiales: Doubles Suites Satisfaisant Équations aux Différences Partielles de Dimension et D'ordre Deux de la Forme $H(n,k) = P(n,k)H(n-1,k)+Q(n,k)H(n-1,k-1)$*, Ph.D. Dissertation, Université du Québec à Montréal, May, 1994.

[Theo95] P. Théorêt, "Fonctions génératrices pour une classe d'équations aux différences partielles," *Ann. Sci. Math. Québec*, to appear.

[TI68]   P. Erdös and G. Katona, editors, *Theory of Graphs (Proceedings of the Colloquium, Tihany, Hungary)*, Academic Press, NY, 1968.

[tm]   T. Moreau, personal communication.

[TM93]   J. Tannery and J. Molk, *Eléments de la Théorie des Fonctions Elliptiques*, Vols. 1–4, Gauthier-Villars, Paris, 1893–1902. Reprinted by Chelsea, NY, 1972.

[TMJ]   *The Mathematica Journal.*

[TMS]   *The Mathematics Student.*

[TO51]   J. Todd, *Table of Arctangents*, National Bureau of Standards, Washington, DC, 1951.

[TO72] I. Tomescu, *Introducere in Combinatorica*, Editura Tehnica, Bucharest, 1972.

[TOH] *The Tôhoku Mathematical Journal.*

[TOMS] *ACM Transactions on Mathematical Software.*

[TR] *Technology Review.*

[trp] T. R. Parkin, personal communication.

[trsw] T. R. S. Walsh, personal communication.

[TU69] W. T. Tutte, editor, *Recent Progress in Combinatorics*, Academic Press, NY, 1969.

[TYCM] *The College Mathematics Journal* (formerly *Two-Year College Mathematics Journal*).

[Ulam60] S. M. Ulam, *Problems in Modern Mathematics*, Wiley, NY, 1960.

[UM] *Utilitas Mathematica.*

[UPG] H. T. Croft, K. J. Falconer and R. K. Guy, *Unsolved Problems in Geometry*, Springer-Verlag, NY, 1991. (References are to the sections.)

[UPNT] R. K. Guy, *Unsolved Problems in Number Theory*, Springer-Verlag, NY, 2nd ed., 1994. (References are to the sections.)

[VA91] I. Vardi, *Computational Recreations in Mathematica*, Addison-Wesley, Redwood City, CA, 1991.

[VDW] B. L. van der Waerden, *Modern Algebra*, Unger, NY, 2nd ed., Vols. 1–2, 1953.

[Vien83] G. Viennot, *Une Théorie Combinatoire des Polynômes Orthogonaux Généraux*, Publications du LACIM, Univ. Québec Montréal, 1983.

[vm] V. Meally, personal communication.

[VO11] A. H. Voigt, *Theorie der Zahlenreihen und der Reihengleichungen*, Goschen, Leipzig, 1911.

[vrp] V. R. Pratt, personal communication.

[WA71] T. R. S. Walsh, *Combinatorial Enumeration of Non-Planar Maps*, Ph.D. Dissertation, Univ. of Toronto, 1971.

[WAG91] S. Wagon, *Mathematica in Action*, Freeman, NY, 1991.

[Wate78] M. S. Waterman, "Secondary structure of single-stranded nucleic acids," *Studies in Foundations and Combinatorics*, Vol. 1, pp. 167–212, 1978.

[WB79] R. J. Wilson and L. W. Beineke, editors, *Applications of Graph Theory*, Academic Press, NY, 1979.

[wds] W. D. Smith, personal communication.

[Well71] M. B. Wells, *Elements of Combinatorial Computing*, Pergamon, Oxford, 1971.

[Well86] D. Wells, *The Penguin Dictionary of Curious and Interesting Numbers*, Penguin Books, NY, 1986.

[Wels71] D. J. A. Welsh, editor, *Combinatorial Mathematics and Its Applications*, Academic Press, NY, 1971.

[Wels76] D. J. A. Welsh, *Matroid Theory*, Academic Press, NY, 1976.

[wfl] W. F. Lunnon, personal communication.

[WhWa63] E. T. Whittaker and G. N. Watson, *A Course of Modern Analysis*, Cambridge Univ. Press, 4th ed., 1963.

[WI72] R. J. Wilson, *Introduction to Graph Theory*, Academic Press, NY, 1972.

[WI78] R. J. Wilson, editor, *Graph Theory and Combinatorics*, Pitman, London, 1978.

[WIEN] *Sitzungsbericht der Kaiserlichen Akademie der Wissenschaften Wien, Mathematisch-Naturwissenschaftlichen Klasse.*

[Wilf90] H. S. Wilf, *Generatingfunctionology*, Academic Press, NY, 1990.

[Wimp84] J. Wimp, *Computation with Recurrence Relations*, Pitman, Boston, 1984.

[WiYo87] H. S. Wilf and N. Yoshimura, "Ranking rooted trees and a graceful application," in *Discrete Algorithms and Complexity* (Proceedings of the Japan-US joint seminar June 4–6, 1986, Kyoto, Japan), edited by D. Johnson, T. Nishizeki, A. Nozaki and H. S. Wilf, Academic Press, NY, 1987, pp. 341–350.

[Woo68] P. E. Wood, Jr., *Switching Theory*, McGraw-Hill, NY, 1968.

[WP] *Wiskunde Post.*

[WW] E. R. Berlekamp, J. H. Conway and R. K. Guy, *Winning Ways*, Academic Press, NY, 2 vols., 1982.

[YaBa94] R. Yanco and A. Bagchi, "K-th order maximal independent sets in path and cycle graphs," *J. Graph Theory*, submitted, 1994.

[YAG] A. M. Yaglom and I. M. Yaglom, *Challenging Mathematical Problems with Elementary Solutions*, Vols. 1–2, Holden-Day, San Francisco, 1964–1967.

[ZA77] H. Zassenhaus, editor, *Number Theory and Algebra*, Academic Press, NY, 1977.

[ZAMM] *Zeitschrift für Angewandte Mathematik und Mechanik.*

[ZFK] *Zeitschrift für Kristallographie.*

[ZFW] *Zeitschrift für Wahrscheinlichkeitstheorie.*

[ZML] *Zeitschrift für Mathematische Logik und Grundlagen der Mathematik.*

Index

Not every sequence is listed. In particular, sequences whose description is a formula or recurrence have usually been omitted. The principal sequences are marked by asterisks (*).

Printed and bound by CPI Group (UK) Ltd, Croydon, CR0 4YY

03/10/2024

01040425-0019